Uranium – Past and Future Challenges

Broder J. Merkel · Alireza Arab
Editors

Uranium – Past and Future Challenges

Proceedings of the 7th International Conference on Uranium Mining and Hydrogeology

Springer

Editors
Broder J. Merkel
Alireza Arab
Institute for Geology
TU Bergakademie Freiberg
Freiberg
Germany

ISBN 978-3-319-11058-5 ISBN 978-3-319-11059-2 (eBook)
DOI 10.1007/978-3-319-11059-2

Library of Congress Control Number: 2014948732

Springer Cham Heidelberg New York Dordrecht London

Printed on acid-free paper

Springer is part of Springer Science+Business Media (www.springer.com)

Preface

In October 1995 the First International Conference on Uranium Mining and Hydrogeology took place with 350 participants from 30 countries (e.g. Canada, China, Czech Republic, Russia, and United States) at TU Bergakademie Freiberg in cooperation with the Saxonian State Ministry for Regional Development and the Environment and a significant financial support from this state authority. The aim of the conference had been to discuss the danger for surface- and groundwater by former uranium mines, uranium treatment plants, heaps and tailings and appropriate cleanup technologies as well as modeling tools. At this time shortly after the end of the cold war most uranium mines were or had already been closed down, the environmental risk of the uranium brownfields became obvious and public awareness rose.

Although the closure of uranium mining and milling facilities was a worldwide phenomenon at that time Freiberg was probably the most suitable city for the conference. This is because Freiberg is home of the oldest still operating mining university (Technische Universität Bergakademie Freiberg) in the world founded in 1765 on the one hand and because the city was in direct vicinity to the uranium mining and milling sites of the former GDR (East Germany) which was the third biggest producer of uranium from 1947 to 1990 on the other hand. Communication during UMH I was facilitated with the help of simultaneous interpreters. The conferences to follow (UMH II to UMH VII) were solely based on English as communication language and accompanied the rehabilitation and cleanup of uranium brownfields in Germany as well as in many other countries worldwide. In many countries nowadays this job is more or less done. However, in some countries such as Kirgizstan and Kazakhstan uranium mining and milling areas are still waiting for a proper aftercare.

Past challenges were countless mistakes and carelessness related to mining, milling and processing of uranium that contaminated surface- and groundwater. Other topics had been advantages, disadvantages and needs for modeling tools, analytical problems, quality and uncertainties of thermodynamic data, risk assessment studies, and phosphate mining. Because phosphate often contains uranium, thorium, and radium phosphate fertilizers contain considerable amounts of uranium and thorium which are recklessly spread on agricultural soils and thus in the long term endanger surface- and groundwater.

Future challenges will focus on the situation after most of the uranium brownfields have been cleaned up. However, uncertainties of thermodynamic data and the lack of kinetic and sorption data will be a topic of future research. Another future topic will probably be the development of environmentally friendly technologies for nowadays uranium mining and mining of ores that contain significant amounts of radioactive elements, respectively (e.g. many REE ores and phosphate). Therefore, further efforts have to be made to develop efficient and low-cost techniques to extract uranium from phosphates and REE-containing ores. This has to be accompanied by the definition of limits by state authorities for permissible uranium and thorium concentrations in fertilizers and naturally occurring radioactive materials (NORM). Mining phosphates and REE ores will have to consider radioactivity as an additional risk. This might have the positive side-effect to trigger the development of new mining techniques based on robotics and remote-controlled operations.

Although worldwide efforts are made to expand the use of alternative energies it is rather likely that nuclear energy utilization in future will continue to play an important role worldwide for at least several decades. Thorium reactors might be the basis for a worldwide renaissance of producing electricity and heat by means of nuclear technologies because a reactor meltdown seems to be close to impossible for a thorium reactor and proliferation risk is much lower than for uranium. Furthermore, thorium reactors produce less and only short-living radioactive waste and ^{232}Th is about 500-times more abundant than ^{235}U.

Freiberg, September 2014

Broder J. Merkel
TU Bergakademie Freiberg

Contents

Uranium Boom in Namibia – Hausse or Baisse

Helmut Mischo[1], Rainer Ellmies[2]

[1]Professor Underground Mining Methods, Department of Mining Engineering and Specialized Underground Construction, Technical University Bergakademie Freiberg, Freiberg, Germany
[2]Chief Geologist, African Mining Capital, Windhoek, Namibia

Abstract. Uranium has been mined in Namibia since the 1970s. The worldwide increasing need for energy in the early 2000s has triggered an increased interest in uranium exploration, often referred to as "the Namibian Uranium Rush". All in all, five mining licenses have been granted by the Namibian Government, with currently two mines in operation, a third under construction undertaking trial mining, the fourth under care and maintenance after completing a large scale and successful trial period and the fifth in project finance negotiations. In addition, highly intensive prospecting activities at additional deposits are at an advanced stage. The finalization of the strategic environmental impact assessment (SEA) on uranium mining in 2010, the first of its kind and scale in the world, has enabled the Namibian government to assess this uranium rush and its tremendous legal, socioeconomic and environmental impacts and to prepare for different future scenarios, including both, a skyrocketing and a complete breakdown of demand scenario. The Fukushima disaster and plans of the Namibian government to significantly increase royalties and company taxes in 2011 have threatened the market situation, forcing investors to reevaluate Namibian uranium mining projects. However, with the tax and royalty increase initiative since withdrawn, most projects have soon been back on track with significant pace, when again low market prices hampered some of the projects or even completely put them on hold.

Introduction

While small-scale mining activities by the local San people on the outcrops of the copper deposits around Tsumeb and Otavi in northern Namibia have been reported since precolonial times, mining for base metals in the Central Namib Desert started only in 1901. The first uranium deposits were discovered as early as 1910. This did, however, not lead to any mining activities on those uranium deposits in the following decades (Southern African Institute for Environmental Assessment, 2010; Schneider et al., 1992).

B.J. Merkel and A. Arab (eds.), *Uranium – Past and Future Challenges*,
DOI 10.1007/978-3-319-11059-2_1

The increase in the use of nuclear energy in the 1960s and early 1970s triggered extensive uranium exploration worldwide. The first uranium mine in Namibia was commissioned in 1976. In the following three decades, the focus of prospecting activities in Namibia shifted away from uranium. However, since 2000, the forecast for a developing and growing global economy and its growing energy demand have sparked off a large number of new nuclear energy projects worldwide (Southern African Institute for Environmental Assessment, 2010; Schneider et al., 1992; Nukem).

In Namibia, this increasing demand did not only lead to the opening of a second uranium mine in 2006, but also caused an increased general interest in uranium exploration, often referred to as the Namibian Uranium Rush. Currently, five mining licenses have been granted by the Namibian Government, with two mines in operation, a third undertaking trial mining, the fourth under care and maintenance after a successful pre-production stage and the fifth in project finance negotiations. In addition, prospecting activities at three deposits are at an advanced stage. In total, 36 exploration licenses for nuclear fuels have been granted in the Erongo region and nearly the same number elsewhere in Namibia. The cumulated production figures of the existing mines and of those currently under construction could raise Namibia's position as an uranium producer from currently the fourth to the second largest supplier of nuclear fuel worldwide (Southern African Institute for Environmental Assessment, 2010; Schneider, 2009; Chamber of Mines of Namibia; Mischo 2012, Nukem).

In Namibia, the Uranium Rush, with its tremendous legal, socioeconomic and environmental impacts, forced the government in 2007 to place a moratorium on any further uranium prospecting licenses. This was done to ensure that the authorities and other stakeholders had time to consider how to manage the Uranium Rush. One of the most prominent results of this moratorium is the Strategic Environmental, Economic and Social Assessment for the region, the first of its kind in the world. Since 2011 low market prices for uranium, however, have a significant impact both on the active as well as on prospective mining operations (Southern African Institute for Environmental Assessment, 2010; Schneider, G., 2009; Chamber of Mines 2013).

The Uranium Deposits of Namibia's Central Namib Region

The economically mineable uranium deposits of Namibia's Central Namib Region have so far been mainly grouped into two types: the primary deposits, which occur and are associated with plutonic rocks, mainly leucogranite, and the secondary-type deposits, that are of sedimentary origin and mostly occur in paleo channels. A third type of deposit, in impregnated magnetite, is currently being prospected, but has not reached an economic status yet (Schneider et al., 1992; Schneider, 2009).

The primary deposits, i.e. uraniferous leucogranites, also known as alaskites, occur preferentially in and around anticlinal and dome structures along the so-called 'Alaskite Valley' to the east of Swakopmund, a NE-trending zone of the Damara orogen hosting all known economically mineralized leucogranites. The age of these uraniferous leucogranite intrusions is estimated at between 458 ± 8 million years at the Rössing Mine and up to 542 ± 33 million years in the Swakop River valley. They are confined to the areas of highest metamorphic grade. The predominant uranium mineral in the leucogranites is uraninite (UO_2); however, betafite and secondary uranium mineral might be the major mineral phase in some places (Southern African Institute for Environmental Assessment, 2010; Schneider et al., 1992).

The secondary-type uranium deposits are mainly found in calcretes and gravel deposits downstream from primary deposits. Today, placed entirely within the 100 mm isohyete, these deposits have been formed in a fluvial environment within paleo valleys of ancient rivers that flowed westwards from the Great Escarpment during the upper Cretaceous and the lower Cenozoic periods. The main uranium-bearing mineral in the calcretes is carnotite, a bright yellow potassium-uranium vanadate. While the age of the paleo channel system dates back to the middle and lower Cretaceous periods, the age of the uranium mineralization in most deposits is estimated to be only between 2 to 5 million years (Schneider et al., 1992; Mischo and Ellmies 2012).

Existing Uranium Mines

Currently, there are two large uranium mines in operation in the Erongo Region (Rössing, Langer Heinrich), and one other uranium mine is under construction (Husab Mine). The fourth mine (Trekkopje Mine), which has already produced 440 t of U_3O_8 (yellow cake) during large scale heap leach operations in its pre-production phase, has been put under care and maintenance in 2012 due to extremely low grades and insufficient market prices. The number of direct employees and contractors for the three active mining operations in the region, as well as tax income, has led to overall economic and social prosperity in the region with a direct dependency rate of 8 and the indirect dependency rate estimated in the region of 28 persons per job created at the mines (Chamber of Mines of Namibia: SOMP).

The two operating uranium mines (Rössing and Langer Heinrich) and the Husab mine under construction contribute significantly to Namibian development, not only through taxes and royalties, but also through employment, subcontracting, wages and salaries. The combined employment at these mines in 2013, of 1,975 permanent staff, 163 temporary staff and 3,264 contractors, represents almost 6.4% of the economically active working population of the Erongo region (based on 2011 census figures). Research by Ashby (2009) at the Langer Heinrich mine found that the dependency ratio for workers to dependents on the mines is approximately one-third higher than the average for the Erongo Region as determined in the 2001 census, which results in the average transfer of financial support to the rural areas of approximately 900 N$ per month. The same figure is estimated for the well-established Rössing Mine (Southern African Institute for Environmental Assessment, 2010; Chamber of Mines of Namibia).

The two operating uranium mines produced a total of 4,878 t of U_3O_8 in 2013, while additionally 440 t of U_3O_8 from pre-production tests at Trekoppie mine have been produced and shipped. A production target of 7,000 t U_3O_8 is expected in 2015, as Husab will ramp up towards full production. (Southern African Institute for Environmental Assessment, 2010; Chamber of Mines of Namibia 2009 - 2013; Rio Tinto, 2010; Paladin Energy, 2012; AREVA, 2010a).

Rössing Uranium Mine

The Rössing Mine, the only uranium mine in Namibia from 1976 until 2006, is currently the third largest open pit - hard rock uranium mine in the world. It is situated in the central Namib Desert, approximately 65 km east-north-east of Swakopmund (Fig. 4). The open pit is about 3 km long, 1.2 km wide and up to 390 m deep, mining a low-grade ore with an average uranium content of slightly over 350 ppm. It is operated by the Rössing Uranium Ltd., with its major shareholders Rio Tinto Zinc Cooperation, Iran and Namibia. The remaining shares are strewn worldwide (Rio Tinto, 2009; Langefeld, 2005).

The name Rössing derives from the Rössing Mountains that are situated approximately 20 km west of the current mine site. Radioactive minerals in the vicinity of the current mine site were first reported by Reuning in 1910. It was only the political interest in uranium deposits in the 1940s that led to a more intensive exploration, when the Geological Survey of South Africa examined the Rössing area for the occurrence of uranium ore. The uranium mineralizations were assessed as uneconomical for mining. Only in 1966, the British Rio Tinto group with its subsidiary Rio Tinto South Africa Ltd. was attracted to the Rössing deposit. Together with several international investors, the Rössing Uranium Ltd. was founded in 1970. After an intensive drilling program, the existence of a very large, low-grade deposit of uranium that could be mined by open pit methods was proved. After numerous metallurgical trials and some pilot plant test work, which also showed that the uranium could be recovered by means of conventional metallurgical processes, the development of the mine started in 1973 (Schneider et al., 1992).

Uraninite (UO_2) is the dominant primary mineral and occurs as grains ranging in size from a few microns to 0.3 mm, with the majority in the 0.05 to 0.1 mm fraction. The plant was commissioned in June 1976 and commercial production started in January 1978 (Schneider et al., 1992; Rio Tinto, 2009; Langefeld, 2005).

With the opening of the mine, the ore reserves of the deposit were calculated from the data obtained from the surface diamond-drilling program. Using computer techniques at a large scale for the first time in the then South-West Africa, an optimum 20-year plan for the mine life was obtained. With ongoing exploration and reduced production figures during the uranium bust of the 1980s and 1990s, according to recent planning, the life-of-mine span could be extended to 2024. In 2013, a total mass of 36 Mt of ore and waste have been moved. This significant earthworks supplied 11 Mt of ore to the processing plant (Schneider et al., 1992; Chamber of Mines of Namibia, 2013; Rio Tinto, 2009; Mischo, 2009; Langefeld, 2005).

In general, the mineability of the Rössing ore body, and thus its short- and mid-term planning, is complicated due to the fact that it consists of a mixture of uranium-bearing leucogranite and barren metamorphic rock. At the same time, both the uranium content of the leucogranite, as well as its spatial extension, ranging from large masses to narrow bodies intercalated with barren metasedimentary rock, are extremely variable. In addition, the acid consumption of the rocks for the metallurgical process of leaching the uranium from the ore varies greatly from low consumption for leucogranite to very high consumption for marble (Southern African Institute for Environmental Assessment, 2010; Schneider et al., 1992; Rio Tinto, 2009; Mischo, 2009; Mischo et Ellmies, 2012).

The mining method in the kidney-shaped main open pit is a typical hard rock drill-and-blast operation (Fig. 1). Bench heights have been adjusted to 15 m in order to allow a drill pattern adjustment to the highly altered ore body. After pre-split, the production boreholes (380 mm diameter and 18 m length) are drilled and charged. The use of "heavy ANFO" (60% emulsion and 40% ANFO prills) has proven to be effective and economical (Rio Tinto, 2009; Mischo, 2009; Langefeld, 2005).

Fig.1. Rössing Uranium Mine (source: Geological Survey of Namibia).

The diesel-electrical 180-t Komatsu haulage trucks drive up the main ramp with the help of a trolley system, which has proven to be cost-effective through the low electricity costs available in southern Africa. Truck utilization, position and speed are being monitored continuously by a computerized control system. Each truck load is scanned at stationary test areas by four scintillation meters to detect the intensity of the gamma radiation. Depending on the radioactive inventory of a truck, the ore is either transported to the processing plant, to the low-grade stockpile or to the waste dump (Rio Tinto, 2009).

The recovery of the leached uranium by an ammonia precipitation process led to a production of 2,469 t of uranium oxide (yellow cake, U_3O_8) in the year 2013. In order to handle high production, while at the same being able to handle lower-grade ore effectively, large column heap leach tests were conducted successfully in 2009. The construction and commissioning of a pilot plant for a heap leach process has been scheduled for the end of 2011, but is currently on hold due to production decrease since 2010 (Chamber of Mines of Namibia; Rio Tinto, 2009; Langefeld, 2005).

The daily resulting 80,000 m^3 of slime and wash water are led to a settling basin via a pipeline. The cleared water is reused and, thus, provides over 60% of the water demand of the plant (Rio Tinto, 2009, Mischo et al., 2012).

Langer Heinrich Mine

The Langer Heinrich Uranium Mine (LHU) is located 60 km directly east of Swakopmund. The Langer Heinrich deposit is placed in a large paleo channel system, represented today by the Gawib River valley between the Langer Heinrich Mountains in the North and the Schieferberge in the South and covers approximately 25 km^2. Uranium mineralization at Langer Heinrich is associated with the calcretization of valley-fill fluvial sediments in a paleo drainage system. Calcrete terraces within the deposit and the surrounding sediments indicate an original sediment thickness of 60 m and a channel width between 200 to 1,000 m. However, subsequently, 30 to 40 m of the original sediments in the central and eastern portions of the paleo channel have eroded (Southern African Institute for Environmental Assessment, 2010; Schneider et al., 1992; Paladin Energy, 2010).

The deposit occurs over a length of 15 km along the original flow line of the paleo channel, with seven different areas of high-grade mineralization currently identified. Mineralization occurs close to the surface, ranging from 1 m to 30 m in thickness and between 50 m and 1,100 m in width, depending on the former width of the paleo valley. It is confined to several thin tabular bodies along the length of the paleo channel and is generally situated a few meters above the bottom of the channel fill. In vertical section, there is only one ore-bearing zone, which has regular gradational hanging wall and footwall contacts. The grade of mineralization tends to be higher in the centre of the zone. Economic analysis on this resource has indicated a break-even cut-off grade of 250 ppm U_3O_8 (Fig. 2) (Southern African Institute for Environmental Assessment, 2010; Paladin Energy, 2010).

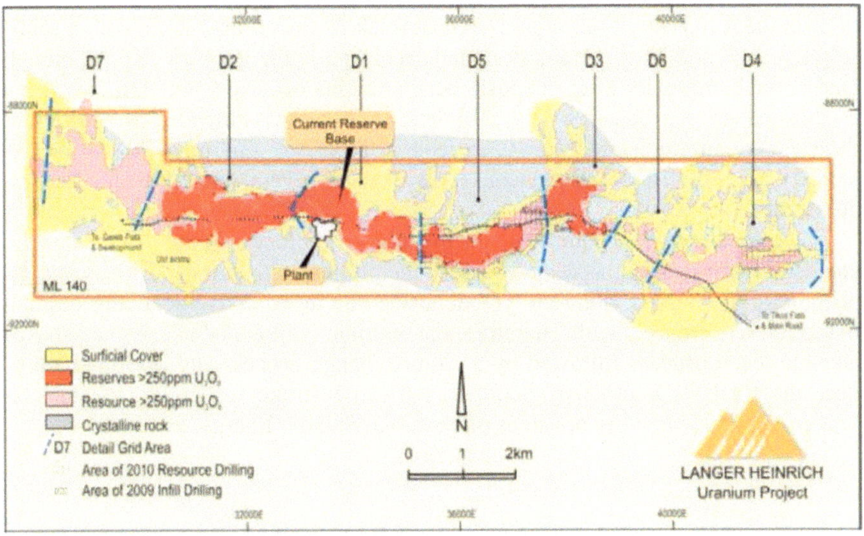

Fig.2. Langer Heinrich Deposit (source: Paladin Resources).

The ore reserve estimates for this cut of grade (250 ppm U_3O_8) are 30.0 Mt proved reserves, with an U_3O_8 content of 17,934 t, and 20.6 Mt of probable ore reserves, with a content of 11,950 t (Chamber of Mines of Namibia; Paladin Energy, 2012).

The uranium mineralization occurs as carnotite. The estimated formation of the deposit took place at least 0.5 million years ago, when the already arid environment preserved the deposit. Most geologists estimate the age of the deposit between 2 and 5 million years. Today, the deposit is located within the 70 mm isohyets (Southern African Institute for Environmental Assessment, 2010; Schneider et al., 1992; Paladin Energy, 2010).

Mining operations first started at the central high-grade zone close to the processing plant (pits A and B), followed by pit D, while the other high-grade zones along the paleo valley are to be brought into production step by step. After the Stage II expansion in 2011, which has increased lift production capacity from 760 t/a U_3O_8 to 1,665 t/a, in 2013 the Stage III expansion has been completed, which has increased annual production to 2,409 t/a U_3O_8. The original target was even higher, with a planned 2,700 t/a, but the uncertain process water supply by NamWater, the parastatal company responsible for water supply in Namibia, currently does not allow for this. Originally, a further expansion, the Stage IV expansion, had also been planned, with an annual mine production of about 4,500 t/a U_3O_8. But this would, leaning aside the low availability of process water, also require the installation of a second water pipeline and, in addition, an upgrade to the existing power supply (Southern African Institute for Environmental Assessment, 2010; Paladin Energy, 2010).

Every expansion stage increases run-of-mine production, thus leading to a sequence of parallel-operated open pits along an east-west axis. All mining operations in the different pit are organized as a truck & shovel operation, with a semi-selective drill and blast pattern, designed according to the geologic deposit model. Due to the complexity of the deposit, bench heights of only 6 and 9 m have been established at the production areas. While today material is blended from different faces for a continuous and equalized feeding the plant, this practice will be extended to parallel and simultaneous operation of several pits (Paladin Energy, 2010; Mischo, 2009).

The calcrete ore has required a new metallurgical process in the plant, since the acid tank leach process, as established for the primary uranium deposits, is not applicable to the Langer Heinrich type deposits, due to excessive acid consumption in the calcrete environment. Therefore, the uranium is liberated using a tank-based alkaline leach process followed by an ion exchange process and roasting to produce the final U_3O_8 product (Southern African Institute for Environmental Assessment, 2010; Chamber of Mines of Namibia; Paladin Energy, 2010).

Uranium Mines under Construction in Namibia

In addition to those for the two operating mines, the Namibian Ministry for Mines and Energy has issued three more mining licenses on nuclear fuel/uranium in the last ten years. One of those planned mines already started preproduction in 2009 but ceased due to low market prices, the second one is in preproduction phase while the third is currently put on hold due to financial constraints (Ministry of Mines and Energy of the Republic of Namibia, 2010).

Rössing South/Husab

Swakop Uranium has been conducting exploration work on the Husab deposit since 2008. Despite adverse conditions in the uranium market, Swakop Uranium and its Hongkong-based shareholders Taurus Investments (Pty) Ltd (90%) - owned by China General Nuclear Power Holding Company (CGNPC) as well as the Namibian state-owned Epangelo Mining Company (10%) - announced their decision to forge ahead with the construction of the mine at an investment of N$20 billion.

The Husab deposit, formerly known as Rössing South, is interpreted as being an extension of the same stratigraphy that hosts the Rössing Mine. Located 15 km south of the Rössing Mine, the ore body is buried under some 30 m of desert sands. Here again, the main uranium host rock is leucogranite. The mining license was granted in 2012 and the ground breaking ceremony took place on April 18 2013. The Husab deposit was first drilled in 2007, and the results from intensive chemical assay confirmed the discovery in February 2008. In the original exploration campaign, two major high-grade zones were discovered. Zone 1 has indicated reserves of 21 Mt of ore containing 529 ppm of uranium, which translates to

10,800 t of U_3O_8. The inferred resources at this zone were estimated at 126 Mt of ore at 436 ppm uranium, thus increasing the U_3O_8 content to 54,450 t. For the second high-grade zone, Zone 2, which is situated south of Zone 1, only inferred resources have been published, showing a uranium content of 543 ppm and an estimated U_3O_8 content of 54,900 t (Southern African Institute for Environmental Assessment, 2010; Extract Resources; Extract Resources, 2010; Spivey and Penkethman, 2010).

Both deposits seem to have a relatively simple geometry with a moderate dip. Since the rock strength of the ore, as well as the host rock, indicates it to be sufficient for opening a deep opencast mine, a traditional mine design with two large open pits has been chosen as shown in Fig. 3. A complete new infrastructure including a 22 km access road over the Khan river needed to be constructed in order to bring high performance mining equipment (e.g. 300 t Komatsu haulage trucks) on site. Designed production figures under full production are 40,000 t per day with a mine life of more than 20 years, translating into an annual production of 6,700 t U_3O_8 through an acid tank leach process. Mining operations commenced in March 2014. The plan is to ensure that a Run-of-Mine (ROM) stockpile will be ready for processing on completion of the processing plant as of end of 2014. A total investment of N$ 21 billion is foreseen until 2017, with an expected contribution up to 5 % to the Namibian Gross Domestic Product, 20% to the country's merchandise exports and generate up to N$1 700 million per year in Government revenue (Southern African Institute for Environmental Assessment, 2010; Extract Resources; Extract Resources, 2010; Spivey and Penkethman, 2010; Swakop Uranium, 2014).

Fig.3. Husab Construction site and Open Pit Model (source: Swakop Uranium).

Trekkopje Mine

The Trekkopje uranium deposit is located some 80 km northeast of Swakopmund. It occurs, similarly to the Langer Heinrich deposit further south, as sedimentary filling of a paleo channel system. While the Langer Heinrich deposit consists of spatially closely defined high-grade areas along one paleo channel, the main mineralization at the Trekkopje deposit consists of much larger ore bodies. The first published mine designs gave a possible maximum size of the future mine of 15 km by 30 km. Within that area, the operation later focused on a central part of approximately 16 km by 4 km, which has little or no overburden (Southern African Institute for Environmental Assessment, 2010; Chamber of Mines of Namibia; AREVA, 2010a).

Detected during an aerial survey in 1974, the deposit was explored only after 2005 when UraMin, a company which was established in order to acquire and develop mineral properties, predominantly uranium, took over. The prospecting program showed that measured and indicated resources could be estimated at 144.6 Mt of ore graded 146 ppm U_3O_8 (at a 100 ppm cutoff grade), thus containing an estimated 21,000 t of U_3O_8. The inferred resources have been estimated at 195.3 Mt of ore graded 138 ppm, thus containing 26,700 t U_3O_8 (Southern African Institute for Environmental Assessment, 2010; Chamber of Mines of Namibia; AREVA, 2010a).

Such high reserve volumes, even at low grades, and the favorable geology, with small or even nonexistent overburden, attracted the attention of the French Areva group, one of the world's largest nuclear companies. In early 2008, the Areva group bought UraMin and formed its own Namibian subsidiary, Areva Resources Namibia, to develop the Trekkopje project. The Trekkopje mine was designed as a shallow, open pit mining operation using conventional truck-and-shovel methods with only limited drilling and blasting. Continuously ongoing exploration has proven reserves at over 300 Mt uranium ore at an average grade of 150 ppm (UraMin, May 2007). In order to produce an estimated 3,900 t/a of uranium oxide at full production, the Trekkopje Mine had to process 100,000 t of crushed ore per day, based on the stripping ratio of 1:15. To be able to process such an amount of ore in a leaching process, Trekkopje has successfully tested a sodium carbonate/bicarbonate heap leach process in a Maxi-Pilot Plant.

Based on the favorable results of the bankable feasibility study and on the investment decision of the Areva Group, the Ministry of Mines and Energy granted the mining license in 2008. The Trekkopje mine has been operated in a preproduction stage. Figure 4 shows the heap leach pilot plant. Due to low uranium prices on the world market, Areva put all further developments on hold in 2012, and, after successfully completing the Maxi Pilot plant stage, placed the whole mine under care and maintenance effective from July 1 2013 (Southern African Institute for Environmental Assessment, 2010; Chamber of Mines of Namibia; AREVA, 2010a).

Fig.4. Trekkopje heap leach pilot plant operation (source: Geological Survey of Namibia).

Areva claims that with this heap leach process they are using far less water than with a similar tank leach process. Certain constraints with a secure water supply to its heap leach process and the location of the mine far north of any previously developed mining infrastructure have led to the decision of the Areva group to build their own desalination plant with a capacity of 20 million m^3 due west of the mine at the Atlantic coast. The Erongo Desalination Plant at Wlotskasbaken was inaugurated in October 2009 and, after entering a successful cooperation agreement with the state owned NamWater water supply entity, will keep in operation to supply the other ongoing mining activities and coastal towns (Southern African Institute for Environmental Assessment, 2010; Chamber of Mines of Namibia; AREVA, 2010a; AREVA 2012; Ministry of Mines and Energy of the Republic of Namibia, 2010).

Valencia Mine

The Valencia deposit is one of the most inland-bound uranium mining projects in the Central Namib, situated about 50 km southwest of Usakos. The Valencia radiometric anomaly was first shown on the 1968 Geological Survey radiometric map, but the significance of the anomaly was only realized in 1973 after a detailed helicopter radiometric survey was conducted across the area (Southern African Institute for Environmental Assessment, 2010; Schneider et al., 1992).

The uranium-bearing leucogranites are present as massive stock-like bodies, dykes of varying thicknesses and veins and veinlets. The owner of the deposit, Valencia Uranium (Pty) Limited, a wholly owned subsidiary of Canadian Forsys Metals Corporation, has finalized and published the mine design of the open pit of the Valencia Uranium Mine in 2010, based on an extensive drilling program (Fig. 10). The proposed Valencia Uranium Mine will apply the traditional open pit hard rock mining method of drilling and blasting with a shovel and truck operation to mine the leucogranite-type uranium ore. According to the mine design, the pit will be developed to a maximum size of approximately 1,400 m in length, 700 m in width and 360 m in depth. The modeling of the ore body has shown a probable reserve of 117 Mt of ore (at an average grade of 125 ppm) and 122.4 Mt of waste rock. The operation is currently designed to have a run-of-mine capacity of 1 Mt per month with a life-of-mine of only nine years. Due to the limited size of this operation compared to the neighboring uranium projects, construction is currently on hold pending funding, and the earliest date of commissioning is now expected to be in 2016. The Mining License (ML149) was granted in August 2008 and is valid for 25 years (Southern African Institute for Environmental Assessment, 2010; Liebenberg-Enslin, 2008; Forsys Metal, 2010; Forsys Metal, 2009; Forsys Metal 2013).

Future Uranium Mines

The database of the Namibian Ministry of Mines and Energy lists a total of 78 exclusive prospecting licenses (EPL) for nuclear fuels in Namibia. Currently 36 of them are located in the central western Namib, while a further six EPL application decisions are pending. No new EPLs have been granted by MME since 2007, as a moratorium is in place. The most advanced projects are described below (Southern African Institute for Environmental Assessment, 2010; Schneider, 2009; Ministry of Mines and Energy of the Republic of Namibia, 2010).

Bannerman / Etango Project

The Etango Project (EPL 3345), of the Australian Bannerman Resources Ltd., is situated on the south bank of the Swakop River near Goanikontes, some 35 km east of Swakopmund. Thus, it is situated in the vicinity of a well-known tourist area, the so-called Moon Landscape (Southern African Institute for Environmental Assessment, 2010; Schneider, 2009; Bannermann Resources, 2010a).

During intensive prospecting activities on eight previously identified radiometric anomalies within the EPL, two major mineralization zones, arranged in a boomerang-shaped pattern, have been identified. The total uranium content of these two high grade zones has been estimated at 56,970 t of U_3O_8, of which 40,140 t are indicted (195.5 Mt of ore with an average of 207 ppm uranium) and

16,830 t of U_3O_8 are inferred resources (87 Mt of ore with an uranium content of 195 ppm) (Southern African Institute for Environmental Assessment, 2010; Bannermann Resources, 2010a, 2010b).

During the recent enhanced prefeasibility study, the ongoing drilling has shown additional resources, even under the location originally selected for the processing plant. The uranium mineralization in the high-grade zones is found in leucogranites. Due to the comparatively low ore grades, it is expected that the development of this prospect is likely to be a large tonnage open cast hard rock operation similar to the existing Rössing mine, with a stripping ratio of 4:1. A competitive advantage of the Etango project is its location, being close to the coast and the existing infrastructure, which translates into significantly lower costs for water supply (pumping), materials and staff transport (Mischo, 2009; Bannermann Resources, 2010a, 2010b).

Due to the direct proximity to a tourist highlight of the Swakopmund region, the Etango project has been controversial to the public. Open discussions and close cooperation with the local tourism industry, ecologists and other stakeholders have led to successful compromises between the company and the local community (Schneider, 2009; Bannermann Resources, 2010a).

Reptile Uranium Namibia (Pty) Ltd. Projects

Reptile Uranium Namibia (Pty) Ltd (RUN) probably has the next most promising projects in terms of resource definition and advanced project status. RUN is a wholly owned subsidiary of the Australian Deep Yellow Ltd. It is interesting to note that the owner of the Langer Heinrich Mine, Paladin Energy Ltd., owns a 19.29% stake of Deep Yellow and, therefore, future linkages between their uranium mining projects are possible. The most advanced uranium prospecting projects of RUN are: Inca, Tubas Red Sand, Tubas, Tumas, Oryx, S-Bend, Aussinanis and Ripnes prospects, with resources having been defined for the Inca, Tumas, Tubas Red Sand and Aussinanis prospects. However, these resources are continually being refined, as drilling continues at a rate of between 12,000 and 20,000 m of drilling per month (Southern African Institute for Environmental Assessment, 2010; Schneider et al., 1992; Reptile Uranium Namibia, 2009).

For most of the projects, previously collected data from exploration campaigns in the 1970s and 1980s has been available, except for new discoveries of the Inca uraniferous magnetite. Beside its uranium activities, RUN has also evaluated a magnetite discovery named M62 near the Inca prospect for iron ore production. The two major uranium projects of RUN are briefly described below (Southern African Institute for Environmental Assessment, 2010; Reptile Uranium Namibia, 2009).

The Inca uraniferous magnetite deposit is a primary mineralization in hard rock that dips steeply underground. The resulting ore-to-overburden ratio probably requires an underground operation. The metallurgical processing of both uranium and iron ore in an acid leaching plant seems most likely (Southern African Institute for Environmental Assessment, 2010; Reptile Uranium Namibia, 2009).

The Tubas Red Sand project could be a mere soft-rock sand and gravel operation with extremely limited drilling and blasting on a secondary uranium mineralization. Depending on the geochemistry of the host rock, processing has to take place either in an acid or alkaline metallurgical process. Beside the uranium, vanadium could probably be produced as a byproduct. For the projects Inca and Tubas Red Sands, a combined annual production of 900 to 1,350 t of U_3O_8 is projected, with approximately the same amount of vanadium and an additional 100,000 to 300,000 t of iron as byproducts (Reptile Uranium Namibia, 2009).

Additional Uranium Projects

The Namibian Ministry for Mines and Energy listed, in the Strategical Environmental Assessment report and the database, another 11 companies (other than the companies discussed above) currently holding EPLs for uranium in the Central Namib (Southern African Institute for Environmental Assessment, 2010; Ministry of Mines and Energy of the Republic of Namibia, 2010).

These are the Australian Companies Erongo Energy Ltd., West Australian Metals (formerly Marenica Minerals), Toro Energy Ltd. (formerly Nova Energy), Swakop Uranium (owned by Extract Resources), Green Mineral Resources and the Canadian companies Cheetah Minerals (owned by Manica, which is 51% owned by Pitchstone Exploration), Xemplar Energy Corp. (formerly Namura) and Dunefield Mining (owned by Forsys). In addition, one company each from Russia (SWA Uranium Mines), China (Zhonghe Resources Namibia) and the British Virgin Islands (Petunia Investments 3 - 100% owned by Barlow Holdings Ltd.) are currently holding EPLs (Southern African Institute for Environmental Assessment, 2010; Ministry of Mines and Energy of the Republic of Namibia, 2010).

All of these companies are in different stages of exploration. This includes airborne and ground radiometric surveys, geological mapping, radon surveys and reconnaissance drilling with variable effort. Among those junior companies, the West Australian Metals Ltd. project is probably the most advanced, due to the fact that they started a diamond drilling program on their Marenica prospect in the area north of the Trekkopje mine in 2009 (Southern African Institute for Environmental Assessment, 2010; Schneider, 2009).

Scenarios and Reality

From the analysis of the forces and dynamics of the uranium rush in Namibia, including a consideration of the global market, four different scenarios have been derived in 2010. These form the basis for planning of all economic, social and environmental aspects of the rush through the creation of the SEMP Offices in the Directorate: Geological Survey of Namibia of the Ministry of Mines and Energy. A rating for the different projects, their ore volume and ore grade and the related uranium content are given in Fig. 5 (Southern African Institute for Environmental Assessment, 2010).

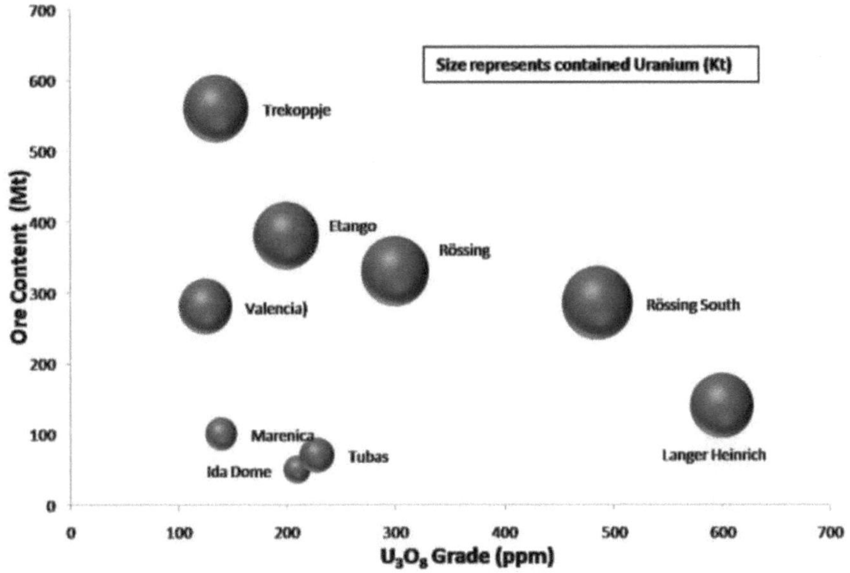

Fig.5. Uranium deposit comparison (Mischo 2009, after: Extract Resources Ltd 2009).

Scenario Development

In the first scenario, which described a situation below expectations from 2010's perspective, only those mines which are currently in production and those which already have received a mining license will be operating in the course of the coming decade (Southern African Institute for Environmental Assessment, 2010; Schneider, 2009).

The second scenario, which was called the in-line-with-expectations scenario, is based on the expectation of a slightly increasing uranium price on the world market. In this scenario, between one and three of the then mining prospects would have turned into a mining operation, in addition to those four mines with mining licenses already identified under Scenario 1 (Southern African Institute for Environmental Assessment, 2010; Schneider, 2009).

To offer the required support to all mines, major investments will need to be made in the construction of chemical and mining support facilities in Swakopmund and Walvis Bay. To meet the increased power demand, the construction of a thermal power station with a minimum capacity of 400 MW in Walvis Bay or vicinity will be necessary. The sector-wide employment would rise to a continuous 6,100 people, with a peak in 2014/2015 during the building phase of up to 8,500 employees. This substantial increase of production will stress the existing infrastructure (roads, harbor facilities) tremendously, as the increase in employment will do to the housing, health, educational and other civil services in the coastal towns. With the typical factor of eight in the SADC region, it could be assumed that the current uranium rush could lead to a total influx of people up to 50,000. In this scenario, the annual production of U_3O_8 is expected to climb to 21,200 t (Southern African Institute for Environmental Assessment, 2010; Chamber of Mines of Namibia; Schneider, 2009; Mischo, 2009).

Examples of serious constraints that might slow down or even stop the uranium rush will be attended to in the near future by the local and regional authorities and governmental institutions. This includes the inadequate supply of water, regional power shortages, deteriorated road infrastructure, port congestion and delays, overburdened health and educational facilities in the local towns, shortage of skills and, finally, inadequate government structures (Southern African Institute for Environmental Assessment, 2010; Chamber of Mines of Namibia; Schneider, 2009).

The third scenario, called above expectations is only feasible if the uranium price increases significantly. In this scenario, the production of Scenario 2 has been increased by the additional commissioning of two or even more mines before 2020, while all other existing mines are able to increase production through additional expansion projects. This would have raised the number of operating mines up to 12 in 2020 and, thus, increase the production of U_3O_8 to approximately 27,000 t/a. Scenario 3 clearly indicates the potential for future uranium projects in the Central Namib Desert (Southern African Institute for Environmental Assessment, 2010; Schneider, 2009).

Table 1 gives a detailed overview of the predicted development of the uranium U_3O_8 production in the different uranium rush development scenarios (Southern African Institute for Environmental Assessment, 2010).

Table 1. Expected increase of annual U_3O_8 production in the different uranium rush development scenarios in metric tons (BGR 2010)

	2008	2009	2010	2011	2012	2013	2014	2015	2016	2017	2018	2019	2020
SCENARIO 1	4,815	5,220	7,155	7,650	10,980	10,845	10,755	10,935	10,395	10,665	10,845	10,620	10,170
Rössing Uranium Ltd	4,050	4,050	4,050	4,050	4,050	4,050	4,050	4,050	4,050	4,050	4,050	4,050	4,050
Langer Heinrich (Stage I & II)	765	1,170	1,620	1,620	1,620	1,620	1,620	1,620	1,620	1,620	1,620	1,620	1,620
Trekkopje	0	0	1,485	1,980	3,825	3,825	3,825	3,825	3,825	3,825	3,825	3,825	3,825
Valencia	0	0	0	0	1,485	1,350	1,260	1,440	900	1,170	1,350	1,125	675
SCENARIO 2	4,815	5,220	7,155	7,650	13,950	17,640	21,690	21,870	21,330	21,600	21,780	21,555	21,105
Rössing plus expansion	4,050	4,050	4,050	4,050	4,500	4,950	4,950	4,950	4,950	4,950	4,950	4,950	4,950
Langer Heinrich (Stages I, II & III)	765	1,170	1,620	1,620	2,340	2,340	2,340	2,340	2,340	2,340	2,340	2,340	2,340
Trekkopje	0	0	1,485	1,980	3,825	3,825	3,825	3,825	3,825	3,825	3,825	3,825	3,825
Valencia	0	0	0	0	1,485	1,350	1,260	1,440	900	1,170	1,350	1,125	675
Etango	0	0	0	0	1,800	1,800	2,565	2,565	2,565	2,565	2,565	2,565	2,565
Rössing South (Zones 1 & 2)	0	0	0	0	0	3,375	6,750	6,750	6,750	6,750	6,750	6,750	6,750
SCENARIO 3	4,815	5,220	7,155	7,650	13,950	18,630	26,685	27,315	26,775	27,045	27,225	27,000	26,550
Rössing plus expansion	4,050	4,050	4,050	4,050	4,500	4,950	4,950	4,950	4,950	4,950	4,950	4,950	4,950
Langer Heinrich (Stages I to IV)	765	1,170	1,620	1,620	2,340	2,340	4,500	4,500	4,500	4,500	4,500	4,500	4,500
Trekkopje	0	0	1,485	1,980	3,825	3,825	3,825	3,825	3,825	3,825	3,825	3,825	3,825
Valencia	0	0	0	0	1,485	1,350	1,260	1,440	900	1,170	1,350	1,125	675
Etango	0	0	0	0	1,800	1,800	2,565	2,565	2,565	2,565	2,565	2,565	2,565
Rössing South (Husab) (Zones 1 & 2)	0	0	0	0	0	3,375	6,750	6,750	6,750	6,750	6,750	6,750	6,750
Omahola Project	0	0	0	0	0	990	1,485	1,485	1,485	1,485	1,485	1,485	1,485
Marenica	0	0	0	0	0	0	1,350	1,800	1,800	1,800	1,800	1,800	1,800

Finally, there was a fourth scenario to be taken into consideration: It describes a situation where, after the development of the uranium mining sector in Namibia, the uranium market collapses due to a significant drop in the uranium price. A collapsing uranium market would put an end to all new developments with all their impact on investments and infrastructure, as well as rural and town development in the area. Since all uranium projects in Namibia are situated on low-grade deposits, this scenario is a constant threat to the development of the Namibian uranium sector. The dependence of all uranium projects in Namibia on high uranium prices due to their low grade is shown in Fig. 5 (Southern African Institute for Environmental Assessment, 2010; Ministry of Mines and Energy of the Republic of Namibia, 2010; World Nuclear Association, 2011).

The disaster at the Fukushima Power Plant in Japan in March 2011 and the resulting short-term drop of the uranium prices on the international market gave a first taste of Scenario 4. Indeed, most uranium projects, even the expansion projects at the existing mines, were put on hold in the weeks following the disaster. Depending the nature of each project and its financial implications for the respective company, this situation lasted for several weeks or even months, but eventually all projects have been revived.

A much more serious impact on the future development of the Namibian Uranium Rush was created in July 2011 by the announcement of the Namibian Government to significantly increase royalties and company taxes for mining companies. This announcement has forced all mining companies active in Namibia to reassess their local mines and new projects. Subsequently, nearly all uranium mining companies declared their Namibian projects uneconomic and prepared for inevitable mine closures. After intensive negotiations with the Chamber of Mines and the whole mining industry, the government eventually withdrew the tax and royalty increase and is now investigating the possibility of implementing a windfall tax system in the future.

Environmental and Socioeconomic Issues

Namibia as a developing country needs to invest in a large variety of infrastructure and socioeconomic projects in order to meet both the requirements of the uranium mines, as well as of its national development plan. The existing and envisaged tasks are discussed below.

Logistics and Transport Infrastructure

In recent years, the port of Walvis Bay has become the central transport hub on the southwestern African coast. It is now managing a substantial import and export volume, not only for Namibia, but also for neighboring countries Angola, Zambia,

Zimbabwe, Botswana and some of South Africa. This development has stretched the existing harbor installation, resulting in continuous port development and expansion programs. Also, the main trunk road connecting this harbor to the inland is in the process of being upgraded. Nevertheless, all new uranium projects are facing the situation, in that they have to invest in the construction of heavy load tar roads. Otherwise, the transport of large and heavy equipment over the existing salt, sand and gravel roads is not only difficult and costly, but is also increasing the wear and deterioration of the existing roads at a tremendous rate. The necessary infrastructure upgrade has to meet the requirements of the import of the mining and processing equipment, as well as the operational supplies (e.g. fuel and acid), the transport of the workforce from their living surroundings to work, as well as the shipment of the product. Thus, any improvements will also beneficiate the countries' development in the whole region (Southern African Institute for Environmental Assessment, 2010; Schneider, 2009; Mischo, 2010).

The role of the existing railway line from Walvis Bay inland through Swakopmund towards the inland for the transport of liquid and bulk goods should not be underestimated. Currently, only the Rössing mine is connected to the railway, but the construction of new railway lines, especially for the transport of bulk chemicals, is under discussion. Based on the SEA uranium rush findings, the development of infrastructure corridors is recommended. Those corridors will consist not only of roads and railways, but also pipelines and power lines. This will ensure that the impact of uranium mining on other economic activities such as tourism, as well as on biodiversity and the natural beauty of the Namib Desert, is minimized (Southern African Institute for Environmental Assessment, 2010; Mischo, 2010).

Water Supply

With an average annual rainfall in the uranium province between 5 and 70 mm, and the lack of any perennial rivers, sufficient fresh water supply is a serious concern for any industrial and economic development in the Namib Desert. Before the uranium rush, the water for the coastal region was provided by the Omdel Dam in the Omaruru dry riverbed (rivier) and the Kuiseb aquifer. However, water supply for the new mines from those aquifers is not sustainable. As overpumping has to be stopped, Areva has decided to build its own desalination plant. This 20 million m^3/a plant was commissioned in 2009. All new mine developments, including Rössings and Langer Heinrich expansion programs, will require the construction of additional desalination facilities (Southern African Institute for Environmental Assessment, 2010; Chamber of Mines of Namibia; AREVA, 2010b; Heyns, 2009; NamWater, 2010).

Originally, NamWater has been expected to commission a second desalination plant in 2012/2013, with a capacity to produce 25 Mm^3/a of potable water. With some delays in the project and construction work not yet started and the Areva Trekkopje project under care and maintenance even at a pilot stage, the existing AREVA desalination plant (Fig 6) seems sufficient to fulfill the current water supply needs both of the new mining projects as well as the civil development on the coastal towns (Southern African Institute for Environmental Assessment, 2010; Heyns, 2009; NamWater, 2010, Areva 2014).

Fig.6. Areva Desalination Plant in Wlotzkasbaken (Source: World Nuclear News 2014).

Electrical Energy

Electricity is a scarce commodity in the whole Southern African developing communities (SADC) region. The Namibian peak demand of 465 MW is still not met by the installed generating capacity of 380 MW. This obvious undersupply has forced the regional energy distributor, Erongo RED, to encourage new mines to install their own basic power supply for the time being, and also to feed unused capacity into the grid (Southern African Institute for Environmental Assessment, 2010; Chamber of Mines of Namibia; NamPower, 2010).

Beside those short- to mid-term solutions, the Namibian energy producer NamPower is currently investigating the construction of a power plant at Walvis Bay in order to secure sufficient energy supply not only to the mines but also to the booming transport and industrial sector, as well as for the influx of people. From the three originally discussed options for the nuclear power plant and thermal power plants fired either with hard coal or with compressed natural gas, the Namibian government is currently the prioritizing the hard coal project, while the other two are still discussed. In addition, the country's energy supply will be stabilized with the installation of the fourth turbine at the Ruacana Hydro Power Station and the connection to the Zimbabwean grid through the Caprivi High Voltage Link (Schneider, 2009; NamPower, 2010).

Housing and Civil Development

With at least some of the currently prospected mining projects expected to become reality, the existing housing and supply infrastructure in the coastal towns is considered to be insufficient. With a typical direct dependence rate of 1:7 in the mining and support sector, an increase of the local population between 20,000 and 50,000 seems realistic, with a major impact on all community services like health, water supply, waste water, housing, shopping, household services and schools. All major social issues are followed up by an interministerial working group under the SEMP Office in the Geological Survey (Southern African Institute for Environmental Assessment, 2010; Schneider, 2009).

Supporting Industrial Development

The mining activities in the Erongo region of Namibia will inevitably trigger numerous mining-related industrial and supporting service activities. Bearing in mind that the current labor market is already stressed, the lack of qualified workforce, not only in the coastal region, but all over Namibia may hamper those developments (Southern African Institute for Environmental Assessment, 2010; Chamber of Mines of Namibia; Schneider, 2009).

Impact of Low Uranium Prices

After a decade of relatively low uranium prices on the world market, mostly due to the availability of depleted uranium fuel sourced from the decommissioning of nuclear weapons after the global political change in the early 1990s, the drying up of most of the military sources and simultaneous increase of demand for nuclear fuel for civil use let the stock market for U_3O_8 pick up again in the early 2000s. The following expected long term shortage sky-rocketed the price to an all-time high around 2007, triggering the Namibian Uranium Boom as described above.

Since then the prices have dropped again steadily, with a short living recuperation phase in early 2011, as shown in figure 7. Since the first quarter of 2014, the price has even dropped consistently below 30$ US per pound U_3O_8. This low market price since 2011 for uranium has a significant impact both on the on-going as well as on prospective mining operations, forcing investors to reevaluate Namibian uranium mining projects.

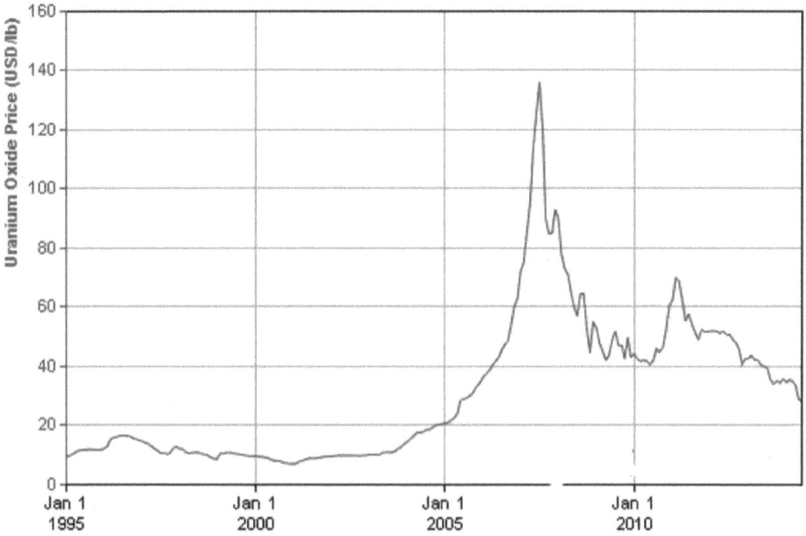

Fig.7. Uranium spot market prices 1995 – 2014 (After InfoMine 2014).

While Langer Heinrich was able to maintain the completion of its stage III extension project and even increasing its annual production to 2,469 t of U_3O_8, the company has faced a loss of N$ 336 million in 2013. While re-trenching could be avoided so far, the planned execution of the stage IV extension project has been put on hold for an undefined period. Difficult geology and a limited reserve situation together with necessary investments and maintenance work in the processing plant have shown a significantly deeper impact on the ongoing mining operation at Rössing. With a decrease in production from 4,150 t U_3O_8 in 2011 down to 2,409 t in 2013, the fate of stopping the mining and processing operation as of the end of 2015 and putting the whole operation under care and maintenance seems possible, if not even unavoidable, even in spite of a small profit of N$ 32 million in 2013, in case the uranium prices on the world market remain low (Chamber of Mines, 2013).

Also for Trekkopje, the care and maintenance period of the whole operation will be extended for an undefined period of time. The AREVA group has clearly expressed its interest in the Trekoppje mine but for the time being has decided to operate solely its water desalination plant in Wlotzkasbaken as a service provider for other operations in the region. With this the group has been able to limit its losses in 2013 to N$ 1.93 billion (Chamber of Mines, 2013).

The Husab project, by contrast, seems unimpressed by the ongoing market turmoil for uranium. Continuous high investments from its Chinese major shareholders and a highly ambitioned mine and processing plant construction well ahead of the planned schedule have turned this project into a lighthouse project not only for the Namibian uranium industry, but for the whole mining sector in the region (Swakop Uranium, 2014; Chamber of Mines, 2013).

Summary

In addition to its large-scale but low-grade uranium deposits and favorable geology (most of the recently prospected ore bodies have been found on or close to the surface, which allows opencast mining), Namibia, compared to other African countries with uranium deposits, appears to be popular among the mining and exploration companies for a range of technical, financial and regulatory reasons. The well-established Geological Survey not only provides information on all previously conducted exploration results, but also conducts regional airborne geophysical surveys to provide information for investors. The lack of skills in the region is addressed by the Namibian Institute for Mining Technology (NIMT) for artisan training, as well as by the Polytechnic of Namibia (PoN) for the formation of professional mining engineers and metallurgists.

If those uranium projects that are currently under construction or in a prospection phase materialize as new uranium mines, Namibia might increase its uranium production from the current 4,878 t/a to over 21,000 t/a or even 27,000 t/a (SEA Scenarios 2 and 3). However, in addition to the global factors regulating the uranium market, limiting factors such as inadequate infrastructure or skills may hamper fast development of the optimistic Scenario 3. On the other hand, the location of the mines close to port facilities and the relatively straightforward legal and regulatory framework and a stable government may foster further development. Also, the success of Namibia's mining industry in convincing the government to withdraw a threatening new tax regime shows the importance of a flourishing mining industry for Namibia's development and the attention the government is paying to this sector (Southern African Institute for Environmental Assessment, 2010).

Even in spite of the currently extremely low market price for U_3O_8 on the spot market, and taking all the impacts on the new and existing mines and boundary conditions into consideration, Namibia may become the second largest uranium producer in the world, as soon as the Husab mine is in full production. And there certainly is an outlook of keeping and stabilizing this position for many years to come. The 2010 SEA summarized the situation as follows: "Political will, technical capacity, enabling policies and laws, and mutually beneficial partnerships are needed to ensure that adequate capacity exists. [This] in combination with strong capacity, transparency and consistency in decision making will ensure that the Uranium Rush is a blessing and not a curse. The bottom line is the need for good governance" (Southern African Institute for Environmental Assessment, 2010).

References

AREVA, 2010a, Areva Resources Namibia: The Trekkopje Mine: Project Information on the Trekkopje Mining Project, Areva Group Company Information September 2010, www areva.com, http://www.areva.com/EN/operations-595/areva-resources-namibia-training-and-mining-projects.html, accessed 02.07.2012 (Project information currently limited).

AREVA, 2010b, Project Information on the Wlotzkasbaken Desalination Plant, Areva Group Company Information September 2010, www areva.com, http://www.areva.com/EN/operations-595/areva-resources-namibia-training-and-mining-projects.html, accessed 02.07.2012 (Project information currently limited).

AREVA 2014, Project Information on the Wlotzkasbaken Desalination Plant, http://www.areva.com/EN/operations-595/areva-resources-namibia-training-and-mining-projects.html, accessed 21.06.2014

Bannermann Resources, 2010a, Corporate Presentation July 2010; www.bannermanresources.com.

Bannermann Resources, 2010b, "Presentation to Africa Down Under Conference," September 2010.

Chamber of Mines of Namibia, Annual Reports 2006-2007, 2007-2008, 2008-2009, 2009-2010, 2010-2011, 2011, 2012, 2013 http://www.chamberofmines.org.na/main/publications/annual-reports.html; accessed 25.06.2014.

Extract Resources Ltd, "Husab Uranium Project,"http://www.extractresources.com/our-business/husab-uranium-project.html, accessed 15.06.2014, first accessed November 30, 2010.

Extract Resources Ltd, 2010, Swakop Uranium Newsletter, 06/08/2010.

Forsys Metal, 2010, "Valencia Uranium development," Forsys Metal Corp., http://forsysmetals.com/?page_id=420; accessed November 30, 2010.

Forsys Metal, 2009, 2010, 2011, 2012, 2013 Annual Report.

Heyns, P., 2009, "Uranium mining and water resource management," Presentation at the 2009 GSN Uranium Mini Conference, Swakopmund, October 2009.

Langefeld, O., ed., 2005, Excursion to Namibia and South Africa, Papierflieger Verlag.

Liebenberg-Enslin, H., 2008, Air Quality Impact Assessment For The Proposed Valencia Uranium Mine In Namibia, Valencia Uranium Limited, February 2008.

Ministry of Mines and Energy of the Republic of Namibia, 2010, 2012, 2013, 2014, Resource Information.

Mischo, H., 2009, "The nuclear cycle - to what extent is Namibia involved?" Presentation at the 2009 GSN Uranium Mini Conference, Swakopmund, October 2009.

Mischo, H., 2010, "Bulk handling technology," Presentation at the 2010 Annual Workshop of the Namibian German Centre for Logistics, Walvis Bay, October 2010.

Mischo, H and Ellmies R, "Uranium Mining in Namibia", Rise of a new Giant" SME Transactions 2012 Vol 332, PP 354 – 369, Society for Mining, Metallurgy and Exploration, Inc. ISBN 978-0-87335-271-0

NamWater, 2010, "Projects: desalination project," http://www.namwater.com.na/data/Projects_Desalination.htm.

NamPower, 2010, "Proposed coal power project at Walvis Bay," October 2010, http://www.nampower.com.na/docs/media/WW_Coal/Final/04%20Walvis%20Bay%20Power%20Station%20ESEIA%20Final%20Executive%20Summary.pdf

Nukem GmbH, "The Role of Nuclear Energy,"5 www.nukem.de, accessed November 30, 2011 (No longer available).

Paladin Energy Ltd, 2010, Annual Report.
Reptile Uranium Namibia (Pty) Ltd, 2009, "Inca Project," Presentation at the 2009 GSN Uranium Mini Conference, Swakopmund, October 2009.
Rio Tinto – Rössing Uranium Limited, 2011, Report to Stakeholders.
Schneider, G., et al., 1992, The Mineral Deposits of Namibia, Geological Survey of Namibia.
Schneider, G., 2009, "Legal aspects of the uranium rush," Presentation at the 2009 GSN Uranium Mini Conference, Swakopmund, October 2009.
Southern African Institute for Environmental Assessment, 2010, Strategic Environmental Assessment for the Central Namib Uranium Rush - Main Report; Final Report, September, Geological Survey of Namibia, Bundesanstalt für Geowissenschaften und Rohstoffe (BGR) Federal Institute for Geosciences and Natural Resources.
Spivey, M., and Penkethman, A., 2010, "Extract Resources," Presentation at RBC Metals &Mining Conference, Toronto, June 2010.
Swakop Uranium http://swakopuranium.com/index.php, accessed 15.06.2014
World Nuclear Association, 2014, www.world-nuclear.org
World Nuclear News 2014, http://www.world-nuclear-news.org, http://www.world-nuclear-news.org/uploadedImages/wnn/Images/Erongo%20desalination%20plant%20-%20460%20(Areva).jpg

Rare earth elements in Australian uranium deposits

Bernd G Lottermoser[1]

[1]Environment and Sustainability Institute, University of Exeter, Penryn Campus, Penryn, Cornwall, TR10 9FE, UK

Abstract. Rare earth elements (REE: La to Lu) are critical for high-technology applications. In Australia, undeveloped REE resources are associated with several deposit types including uranium ores. These include the Olympic Dam Cu-U-Au-Ag deposit and uranium mineralisations in the Yilgarn, Mount Painter, Olary and Mount Isa regions. At the Mary Kathleen uranium mine site, tailings are stored in a purpose-built tailings dam structure, with an estimated 7 Mt of mill tailings carrying approximately 3 wt% total REE oxides. This makes it one of the largest undeveloped REE resources in Australia.

Introduction

Rare earth elements (REE) are becoming increasingly important in high-technology devices such as rechargeable batteries, magnets and catalysts, with demand expected to increase significantly in these areas. Currently, much of the REE supply is effectively restricted to several mining districts in China, which has brought these elements to the headlines and created a critical-metals agenda (Chakhmouradian and Wall 2012; Hatch 2012).

The distribution and concentration of REE into mineral deposits is influenced by various rock-forming processes including enrichment in magmatic or hydrothermal fluids, separation into mineral species and precipitation, and subsequent redistribution and concentration through weathering and other surface processes (Hoatson et al. 2011). Using this evolutionary framework, Walters et al. (2011) proposed that REE deposits can broadly divided into: (1) primary deposits associated with igneous and hydrothermal processes (e.g. carbonatite, alkaline igneous rocks); and (2) secondary deposits concentrated by sedimentary processes and weathering (e.g. marine and alluvial placers, ion-adsorption clays, laterites). By contrast, the classification of REE deposits by Cassidy et al. (1997; cited by Hoatson et al. 2011) is based on the relative 'abundance status' of the REE component in the deposit, i.e., where the REE are either a co-product or primary product, or a by-product of the deposit. In Australia, elevated concentrations of REE are known

B.J. Merkel and A. Arab (eds.), *Uranium – Past and Future Challenges*,
DOI 10.1007/978-3-319-11059-2_2

to occur in six major deposit groups: (i) beach sands and placer deposits, (ii) carbonatites, (iii) alkaline igneous rocks, (iv) pegmatites, (v) phosphorites, and (vi) uranium mineralisations (Lottermoser 1991). This contribution documents the latter REE resources, which could potentially fulfill future economic needs.

Uranium mineralisations

There are numerous uranium ± thorium occurrences in Australia, which contain elevated REE as well as yttrium and scandium contents. These include the Olympic Dam deposit, and uranium mineralisations in the Yilgarn, Mount Painter, Olary and Mount Isa regions (Hoatson et al. 2011).

Olympic Dam

The Olympic Dam deposit in northern South Australia represents a world-class Cu-U-Au-Ag resource (Reeve et al. 1990; Oreskes and Einaudi 1992; McPhie et al. 2011). The deposit is principally hosted by a thick sequence of Proterozoic sedimentary and granitic breccias. Ore minerals are characteristically very fine-grained and intimately intergrown with gangue phases. REE and yttrium predominantly occur in REE minerals and to a lesser extent as cation substitutions within uranium minerals. The light REE (LREE: La to Eu) and heavy REE (HREE: Gd to Lu) tend to occur in different host minerals (Lottermoser 1995). Most of the LREE are present as the essential structural constituents of LREE fluorocarbonates (bastnaesite, synchysite), phosphates (monazite, britholite), and aluminophosphates (florencite). Yttrium and the HREE occur mostly as minor concentrations in the form of cation substitutions within uranium minerals (uraninite, brannerite, coffinite). Xenotime, yttrium-bearing bastnaesite and unidentified phases are rare and account only for minor yttrium and HREE concentrations (Lottermoser 1995). The deposit is LREE enriched. Submarginal and inferred resources of REE oxides amount to 53 Mt @ 0.2 % La and 0.3 % Ce (Hoatson et al. 2011).

During hydrometallurgical treatment and sulfuric acid leaching of the processed uranium ore, significant fractions of REE and yttrium (several %) are leached and liberated from uranium ore minerals and become enriched in the uranium liquors. No provision has been made for the extraction and recovery of REE as recovery is currently considered not to be economic. Hence, it is likely that Olympic Dam tailings contain significant REE concentrations.

Olary district

In the Olary district of South Australia, uranium mineralisation is related to the metamorphism and anatexis of Proterozoic volcano-sedimentary sequences (Ashley 1984; Lottermoser and Ashley 2006). REE minerals occur in pegmatites and as disseminated fracture fillings within granites and migmatites (Lottermoser and Lu 1997). REE-bearing minerals in these occurrences are allanite, betafite, brannerite, davidite, euxenite, fergusonite, florencite, monazite, polycrase, samarskite, synchysite, thorite, xenotime and yttrocrasite.

The Radium Hill uranium deposit is located within shear zones of Precambrian feldspathised gneisses, aplitic gneisses and schists, intruded by basic and acid bodies (Lottermoser and Ashley 2006). Mineralised rock at Radium Hill comprises quartzofeldspathic gneiss, schist, amphibolite and pegmatite, with the ore minerals (davidite-brannerite-uraninite) being part of a refractory Fe-Ti-U-REE oxide assemblage (Fig. 1). The Radium Hill mine operated from 1954 to 1961. Underground mining occurred with processing of ore on site, resulting in the generation of mill tailings dams and numerous dumps of waste rock material. The principal REE ore mineral is davidite. The deposit was mined mainly for uranium but a total of 136 kg of high-purity scandium oxide was produced as a by-product. Abandoned waste rock and tailings dumps are LREE enriched, with maximum concentration of 0.13 wt% La (Lottermoser and Ashley 2006).

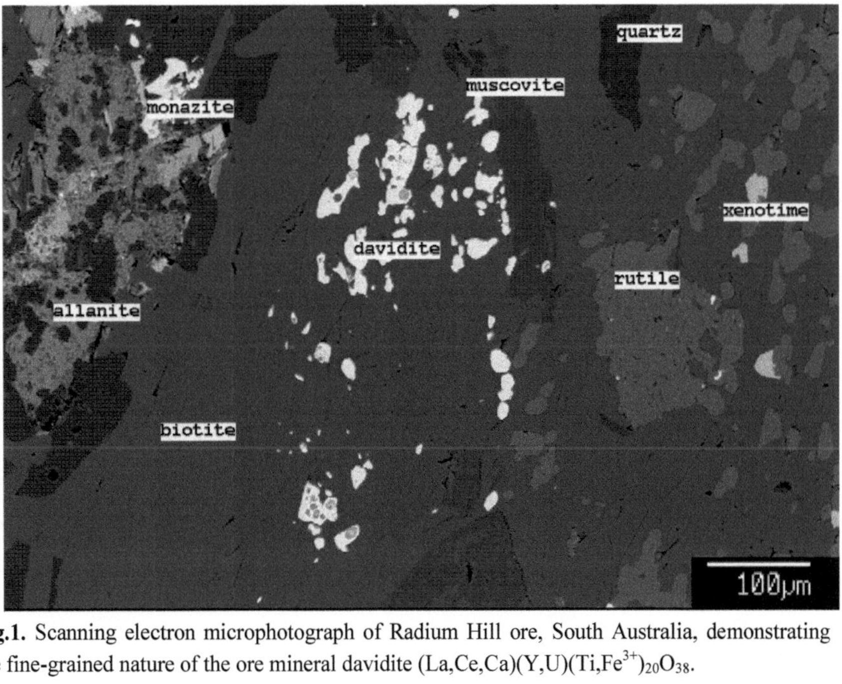

Fig.1. Scanning electron microphotograph of Radium Hill ore, South Australia, demonstrating the fine-grained nature of the ore mineral davidite $(La,Ce,Ca)(Y,U)(Ti,Fe^{3+})_{20}O_{38}$.

Mount Isa district

Several uranium-bearing skarn orebodies in the Mount Isa region contain elevated REE contents. The deposits were exploited for uranium until 1982 but REE were not recovered. The commonly occurring REE minerals are stillwellite and allanite; other REE-bearing minerals are apatite, uraninite, sphene and garnet (Oliver et al. 1999).

The Mary Kathleen uranium mine, in northwest Queensland, operated from 1956 to 1963 and again from 1976 to 1982. Mineralised rock at Mary Kathleen (ore and waste rock) is dominated by a metasomatic calc-silicate assemblage, with minor amounts of sulfide minerals, rare earth minerals and uraninite (Lottermoser et al. 2005; Lottermoser and Ashley 2005). Rehabilitation of the Mary Kathleen open pit mine, mill and tailings repository sites occurred between 1982-1985 (Fig. 2). The Mary Kathleen deposit is strongly enriched in the LREE. Tailings are stored on site in a purpose-built tailings dam structure, with an estimated 7 Mt of mill tailings carrying approximately 3 wt% total REE oxides (DNRM 2013). This makes it one of the largest REE deposits in Australia.

Fig.2. Open pit of the Mary Kathleen mine, Queensland. Disposed tailings contain approximately 0.2 Mt of REE oxides. Based on current world prices, the value of this resource would be approximately US$4000 million.

Conclusions

REE are used in a wide range of consumer products and have become indispensable in electronic, optical, magnetic and catalytic applications. As research and technology continues to advance many more applications for REE are being developed, especially in the areas of energy conservation and efficiency and environmental protection (Walter et al. 2011). Consequently, future demand for REE can be expected to increase.

In Australia, elevated concentrations of REE are known to occur in various deposit types including uranium mineralisations (Lottermoser 1991). The presence of high REE and yttrium concentrations in uranium ores is not unique to Australian uranium occurrences. There exists several other uranium deposits around the world with elevated REE and yttrium contents, and REE have been recovered from uranium operations in Canada (Elliot Lake). The elements are either present as separate REE minerals or they are incorporated as cation substitutions in other mineral phases. The presence of REE within uranium ores and minerals is due to chemical crystallographic constraints as the ionic radii of uranium are very similar to those of REE. In addition, the close mineralogical association of uranium, yttrium and REE is likely due to the transport of these elements by the very same complexing agents during the ore formation processes.

In some of these undeveloped REE resources, the unusual crystallographic siting of the REE, the exceptionally fine grain size of the REE minerals and their intense intergrowth with other phases may require new processing and extraction methods. This in turn may lead to new patents for the processing of REE minerals and for the extraction of REE.

References

Ashley PM (1984) Sodic granites and felsic gneisses associated with uranium-thorium mineralisation, Crockers Well, South Australia. Mineralium Deposita 19: 7-18

Chakhmouradian AR, Wall F (2012) Rare earth elements: minerals, mines, magnets (and more). Elements 8: 333-340

DNRM (2013) Queensland rare earth elements. Department of Natural Resources and Mines, 4pp (http://mines.industry.qld.gov.au/assets/minerals-pdf/rare-earth-elements-brochure.pdf)

Hatch GP (2012) Dynamics in the global market for rare earths. Elements 8: 341-346

Hoatson DM, Jaireth S, Miezinis Y (2011) The major rare-earth-element deposits of Australia: geological setting, exploration, and resources. Geoscience Australia, 204pp (http:// www.ga.gov.au/corporate_data/71820/Complete_Report.pdf(

Lottermoser BG (1991) Rare earth element resources and exploration in Australia. Bulletin and Proceedings of the Australasian Institute of Mining and Metallurgy 296: 49-56

Lottermoser BG (1995) Rare earth element mineralogy of the Olympic Dam Cu-U-Au-Ag deposit, South Australia: Implications for ore genesis. Neues Jahrbuch für Mineralogie, Monatshefte 8: 371-384

Lottermoser BG, Lu J (1997) Petrogenesis of rare-element pegnatites in the Olary Block, South Australia, part 1. Mineralogy and chemical evolution. Mineralogy and Petrology 59, 1-19

Lottermoser BG, Ashley PM (2006) Physical dispersion of radioactive mine waste at the Radium Hill uranium mine site, South Australia. Australian Journal of Earth Sciences 53: 485-499

Lottermoser BG, Ashley PM (2005) Tailings dam seepage at the rehabilitated Mary Kathleen uranium mine, Australia. Journal of Geochemical Exploration 85: 119-137

Lottermoser BG, Ashley PM, Costelloe MT (2005) Contaminant dispersion at the rehabilitated Mary Kathleen uranium mine, Australia. Environmental Geology 48: 748-761

McPhie J, Kamenetsky VS, Chambefort I, Ehrig K, Green N (2011) Origin of the supergiant Olympic Dam Cu-U-Au-Ag deposit, South Australia: was a sedimentary basin involved? Geology 39, 795-798

Oliver NHS, Pearson PJ, Holcombe RJ, Ord A (1999) Mary Kathleen metamorphic-hydrothermal uranium-rare-earth element deposit; ore genesis and numerical model of coupled deformation and fluid flow. Australian Journal of Earth Sciences 46:467–484

Oreskes N, Einaudi MT (1992) Origin of hydrothermal fluids at Olympic Dam: preliminary results from fluid inclusions and stable isotopes. Economic Geology 87, 64–90

Reeve JS, Cross KC, Smith RN, Oreskes N (1990) Olympic Dam copper-uranium-gold-silver deposit. Monograph Series. Australasian Institute of Mining and Metallurgy, 14, pp. 1009–1035

Walters A, Lusty P, Hill A (2011) Rare earth elements. Mineral Profile Series, British Geological Survey, United Kingdom, 54 pp (https://www.bgs.ac.uk/mineralsuk/statistics/mineralProfiles.html)

IAEA Initiatives Supporting Good Practice in Uranium Mining Worldwide

Peter Woods[1], Russel Edge[1], Martin Fairclough[1], Zhiwen Fan[1], Adrienne Hanly[1], Ib-rahim Miko Dit Angoula[1], Horst Monken-Fernandes[1], Haridasan Pappinisseri Puthanveedu[1], Marcelle Phaneuf[1], Harikrishnan Tulsidas[1], Oleg Voitsekhovych[1], Tamara Yankovich[1]

[1]International Atomic Energy Agency, PO Box 100, Vienna International Centre, A-1400 Vienna, Austria

Abstract. Recognizing the environmental impacts of mining and associated industries and their minimization has become more important over the last two or three decades. The International Atomic Energy Agency (IAEA) supports good practice in uranium mining and milling worldwide. As well as its well-known safety standards for radiation protection and waste management, it has produced guidance and acted as a gatherer and provider of information on geological, technological, environmental and regulatory aspects of the uranium production cycle. It is involved in a number of Technical Cooperation projects on this and related topics throughout the world.

Introduction and rationale

Uranium mining presents the same opportunities and challenges as comparable mining in general (c.f. Fernández Rubio 2012) and base-metal mining in particular, with the addition of necessary radiation protection measures. Worldwide guidelines are emerging for social and environmental aspects of mining in general, as well as for uranium mining specifically. The Uranium Production Cycle, (UPC) including exploration, mining and processing of the raw materials for nuclear power (uranium, and to a lesser extent thorium), is one of the themes where the International Atomic Energy Agency (IAEA) is active in promoting best practice around the world. This includes the eventual decommissioning and remediation of mining and processing facilities.

In particular, interest and expertise in these matters at the IAEA is contained in two of its Divisions:

- Nuclear Fuel Cycle and Waste Technology Division
    - Nuclear Fuel Cycle and Materials Section
    - Waste Technology Section

B.J. Merkel and A. Arab (eds.), *Uranium – Past and Future Challenges*,
DOI 10.1007/978-3-319-11059-2_3

- Radiation Transport and Waste Safety Division
    - Waste and Environmental Safety Section.

Other parts of the IAEA are also involved, including the Department of Nuclear Sciences and Applications. Perhaps the best known IAEA department is Safeguards; its primary role is to deter the proliferation of nuclear weapons, mainly through inspections of uranium enrichment facilities and nuclear reactors. Their work also includes inspections of the security and safeguards aspects of uranium mines on the ground. However, the security and safeguards aspects of IAEA work with uranium mining are not the emphasis of this paper.

Whilst nearly all environmental aspects of uranium mining are also relevant for other commodities, albeit with a special emphasis put on radiological protection in the case of uranium, it is also important to recognize that naturally occurring radioactive materials (NORMs) are relevant in many other industries, notably thorium and mineral sands mining and the oil and gas industry. Hence guidelines prepared for uranium mining have broader application, and are worth consulting for other types of mining. Further, the IAEA has published a number of guides for the management of NORM which is applicable to other industries.

Each year the IAEA organizes or participates in many activities supporting the Uranium Production Cycle. Some of its publications and activities are described below. Many activities are supported by the IAEA's Technical Cooperation Department (TC), through projects with low- to mid-income Member States.

IAEA documents hierarchy

The IAEA provides a large amount of information relevant to radiation protection in all mining, oil and gas and related industries. These are organized in a hierarchy (IAEA 2014a).

The lead requirement document, available in all official IAEA languages, is "Radiation Protection and Safety of Radiation Sources: International Basic Safety Standards" (IAEA 2011a), generally referred to as the BSS. It is one of the best known and most widely used of all the IAEA safety standards. The BSS applies to all facilities and all activities that give rise to radiation risks and lays down a consistent and harmonized system for protection of people and the environment.

Other types of documents that include material relevant to mining, moving down the IAEA hierarchy, are the:

- Safety Standards Series
- Safety Reports Series
- Nuclear Energy Series and similar
- Technical Reports Series
- Tecdoc Series (Technical Documents)
- Training Course Series
- Proceedings Series.

Ad-hoc publications of books, reports and paper collections are also issued.

The IAEA is heavily involved in the documentation of the uranium (and thorium) resources of the world, through its databases and collaboration with the Nuclear Energy Association with the joint Uranium Group and the (currently) biennial 'Red Book' publication (http://www.oecd-nea.org/ndd/uranium/). The IAEA itself produces many publications on technical aspects of exploration, geology, mining and milling of uranium; however, this paper will concentrate on initiatives supporting good practice of uranium mining.

IAEA Uranium and NORM documents

A number of higher-level guides relevant to NORM in mining and uranium mining in particular are relevant. These include:

- Radiation Protection and Management of NORM Residues in the Phosphate Industry (IAEA 2013a), including a section on the recovery of uranium from phosphoric acid
- Radiation Protection and NORM Residue Management in the Production of Rare Earths from Thorium Containing Minerals (IAEA 2011b)
- Assessing the Need for Radiation Protection Measures in Work Involving Minerals and Raw Materials (IAEA 2007)
- Occupational radiation protection in the mining and processing of raw materials (IAEA 2004)
- Monitoring and Surveillance of Residues from the Mining and Milling of Uranium and Thorium (IAEA 2002).

Broad guidance is given in the recent report "Management of NORM Residues" (IAEA 2013b).

The IAEA and environmental and social aspects of uranium mining

The IAEA, and other industry and government bodies, have long recognized the importance of environmental and social aspects of uranium mining, especially over the last three or four decades.

Some publications regarding general environmental and social aspects of uranium mining include:

- Best practice in environmental management of uranium mining (IAEA 2010)
- Establishment of uranium mining and processing operations in the context of sustainable development (IAEA 2009)
- Guidebook on environmental impact assessment for in situ leach mining projects (IAEA 2005)

- Guidebook on good practice in the management of uranium mining and mill operations and the preparation for their closure (IAEA 1998)
- Environmental impact assessment for uranium mine, mill and in situ leach projects (IAEA 1997)
- Guidebook on the development of regulations for uranium deposit development and production (IAEA 1996).

The IAEA has also organized a number of relevant conferences and open technical meetings over the last three decades including the subject. It has undertaken and continues to undertake many Technical Cooperation projects in many less-developed member states, including projects regarding uranium exploration, mining and legacy site remediation. Symposia, conferences and technical meetings are also undertaken, often with published proceedings.

Of particular relevance to good practice in uranium mining, the IAEA has become more involved in the Uranium Mining and Remediation Exchange Group or URMEG (Jakubick & Waggitt 2014). Now over 20 years old, since 2012 UMREG has been hosted by the IAEA, who also published a compilation of papers from 1995 – 2007 (IAEA 2011c). As has occurred previously, the 2014 UMREG meeting is being held in conjunction with the International conference of Uranium Mining and Hydrogeology.

Information exchanges using the modern medium of the internet are also hosted by the IAEA. One relevant forum is ENVIRONET, which aims to provide support and information exchange related to environmental management and remediation of radiologically contaminated sites including mines (IAEA 2014b).

The IAEA continues to work on a number of relevant publications, with the assistance of expert panels recruited from around the world. Amongst relevant documents that should be released over the next two years is a report on major environmental considerations associated with modern uranium mining and milling activities.

Training

Technical Meetings, workshops and symposiums organized by the IAEA may have a primary or secondary emphasis on good practice in environmental and social matters. Examples include:

- International Symposium on Uranium Raw Material for the Nuclear Fuel Cycle: Exploration, Mining, Production, Supply and Demand, Economics and Environmental Issues (URAM), Vienna, held 2005, 2009 and 2014
- Regional seminar on Good Practices in the Processing and Control of Uranium Ore Concentrate, Namibia in April 2012 (Atomic Energy Board of Namibia 2012).

In addition to technical meetings and symposiums, the IAEA undertakes many specific training meetings, usually held in a host country with either particular skills to share or a need of training. Recent examples include:

- Inter-regional Workshop on Social Licensing and Stakeholder Communication in the Uranium Exploration and Mining Industry, Turkey, February 2014 (Tulsidas and Zhang 2014)
- Regional Workshop on Sustainable Uranium Resources Development, Tanzania, September 2012
- Training Meeting on Effective Regulatory and Environmental Management of Uranium Production, Australia, August 2012

Earlier training events have been held on assessment and environmental studies, uranium mining: its operation, safety and environmental aspects, and uranium geology, exploration and environment.

Through its TC activities, the IAEA supports nominees from low- to mid-income member states for placements in other countries with more experience in a particular aspect. This includes Scientific Visits for senior professionals, typically of one to two weeks, and longer Fellowships for early and mid-career professionals and technicians. A number of these have as a main or secondary aim exposure and experience in good practice aspects of uranium mining, placing candidates with universities, government regulators and uranium miners in countries with good experience in uranium mining.

Remediation of legacy sites

Coordination and support activities

The IAEA has been active in the field of remediation of legacy sites, particularly in low- to mid-income countries. One ongoing activity is the International Forum for the Regulatory Supervision of Legacy Sites (RSLS). The overall objective of the RSLS is to promote effective and efficient regulatory supervision for the management of legacy sites, consistent with the IAEA Fundamental Principles, Safety Standards and good international practices. The scope of RSLS activities includes all types of nuclear legacy sites, and comprises support in the development of regulations and guidance through technical exchange on topics such as: licensing, monitoring, environmental impact assessment, inspection and compliance monitoring, and long term surveillance and maintenance

Activities include: professional development of regulatory staff, support of the development of international guidance, technical exchanges, scientific visits and technical meetings (usually in October, beginning in 2010).

RSLS participants are either Members, appointed from Authorities responsible for regulation, and/or supervision of radiation and environmental safety at legacy sites, or other stakeholders in legacy site management, such as international organisations, site operators, technical support organisations.

In 2012, as well as the annual technical meeting the RSLS organized a Scientific Visit to USA. 28 participants from 20 IAEA Member States visited an active uranium mill at White Mesa and the major Moab Uranium Mill Tailings Remediation Action (UMTRA) project (Karp and Metzler 2006), several remediated sites with long term surveillance and monitoring programs and a four-day workshop. In April 2014, RSLS, with the cooperation of the Canadian Nuclear Safety Commission (CNSC), organized a scientific visit and Workshop in Elliot Lake Ontario The group visited nine former mill sites and mines in the Elliot Lake area. The Workshop brought together regulators from across Canada as well as from 12 Member States who are addressing historical or legacy issues.

Another initiative is 'CGULS', the Coordination Group for Uranium Legacy Sites, primarily an EU Extra Budgetary Fund project. The group comprises national and international stakeholders to:

- Provide forum for information exchange
- Support project planning and implementation issues
- Address radioactive waste management issues
- Coordination mechanism to avoid duplication of efforts and optimize resources

Based upon recommendations in the technical baseline document developed by IAEA and EU experts in 2007, the Agency created CGULS in 2012. Its overall strategy is to leverage members' resources and plans to move forward on an integrated path with the ultimate goal of remediation of the priority legacy sites. It is comprised of national and international organizations who are addressing uranium legacy issues in Central Asia. Each organization within its own mission space will plan and implement activities/projects utilizing CGULS as a means of identifying gaps, needs, resources etc. such that there should be a cohesive overall approach. In addition to regulators and operators in Central Asia (Kyrgyzstan, Tajikistan, Uzbekistan, Kazakhstan) several international organizations are involved; the EU, EBRD, UNDP, ENVSEC, OSCE, NATO, EurAsEC, all participate in CGULS.

The group has held technical exchange meetings in Central Asia, Russian Federation and the USA. Kyrgyzstan hosted the annual CGULS Technical Meeting in June 2014 at Lake Issyk Kul.

CGULS meets to coordinate activities, which include a knowledge management platform provided by the IAEA. The group is particularly active in Central Asia, where a number of legacy sites remain from the earlier Soviet era. It has been directly

Other initiatives are part of a broader approach to the remediation of radioactively contaminated sites. An initiative directly related to legacy uranium mine remediation is the development and field testing of a Mobile Site Characterization

Unit. This was recently used during an IAEA mission in Mounana in Gabon, a former uranium mining and production site, in July 2012.

Technical cooperation activities

RSLS activities may be supported through IAEA's TC Program. Since 2005, 15 TC projects on uranium production legacy sites have been undertaken, including 1 inter-regional, 4 regional and 10 national. The overall objective of the projects has been to develop national capacity to the point such that the participating Member States can implement a risk-based remediation process consistent with international standards and good practices. To achieve these objective the projects were designed to address the following topical areas:

- Enhanced regulatory framework and monitoring programs
- Enhanced analytical capabilities (including QA/QC)
- Appropriate risk assessment methodology
- Use of Institutional Controls and appropriate stewardship measures
- Risk communication/ Public Information

These projects have included training, visits by experts, and the purchase of equipment including:

- Radiation and radionuclide measuring equipment
- Risk modelling software
- Water, soil, air sampling equipment, pumps, other field equipment
- Laboratory hardware (e.g. balances, glassware, pH meters)
- Drilling of observation wells and analysis of environmental samples.

Further, six uranium production legacy site characterization studies have been initiated and draft Safety Assessment reports were prepared.

Other relevant TC projects have included:

- Developing decontamination, recultivation and reconstruction infrastructure for existing uranium mines and former uranium production facilities – Ukraine
- Assessing radioactive contamination of surface, groundwater and other resources in mining areas – Zambia
- Building capacity for developing and implementing integrated programmes for remediation of the areas affected by uranium mining - regional, Eastern Europe and Central Asia.

Conclusions

The IAEA, as with other industry and government bodies, has emphasized the importance of environmental and social aspects of uranium mining especially over the last three or four decades. In an encouraging trend, these aspects of uranium mining are becoming 'mainstream'; that is, environmental and social aspects are increasingly becoming a normal part of planning, regulating, mining and milling of uranium, both in theory and in practice. A key lesson learned is that a life cycle planning approach can prevent future legacies from occurring. Without this approach, potentially many millions of dollars will need to be spent on remedial actions, potentially rendering a net monetary and social loss on the mining and milling operations and certainly negatively impacting the reputation of the industry.

Within the IAEA, work specific to the environmental and social aspects of uranium mining are undertaken in several departments, and this aspect of the uranium industry is a major emphasis of a significant number of workers and work there. This will continue into the future; without good practice, uranium mining will face ongoing, justified and sometimes project-stopping opposition from the society that it intends to support as the first stage of a major global electricity supply partner.

Acknowledgements

The authors of this paper are collectively the current informal members of the (also informal) Uranium Production Cycle team at IAEA headquarters in Vienna, which has met at collegially over the last several years. We acknowledge the contributions of our predecessors in raising the profile and environmental and social aspects of uranium mining and the remediation of legacy sites within and outside the IAEA and the support of middle and senior management. We must note and thank the work of large numbers of experts from around the world who support our efforts – keep up the good work!

References

Atomic Energy Board of Namibia (2012) IAEA Regional Seminar. http://aebofnamibia.org/index.php?option=com_content&view=article&id=70&Itemid=72

Fernández Rubio, R (2012) Mining: the challenge knocks on our door. Mine Water & the Environment 31(1): 69-73

Karp, KE, Metzler, DR (2006) Moab, Utah, UMTRA Site: The last large uranium mill tailings pile to be cleaned up in the United States. Uranium in the Environment (Merkel, BJ and Hasche-Berger A, Eds): Springer Berlin, Heidelberg: 671-682

IAEA (1996) Guidebook on the development of regulations for uranium deposit development and production. IAEA-Tecdoc 862, International Atomic Energy Agency, Vienna, Austria

IAEA (1997) Environmental impact assessment for uranium mine, mill and in situ leach projects. IAEA-Tecdoc 979, International Atomic Energy Agency, Vienna, Austria

IAEA (1998) Guidebook on good practice in the management of uranium mining and mill operations and the preparation for their closure. IAEA-Tecdoc 1059, International Atomic Energy Agency, Vienna, Austria

IAEA (2002) Monitoring and Surveillance of Residues from the Mining and Milling of Uranium and Thorium, Safety Reports Series 27, STI/PUB/1146, International Atomic Energy Agency, Vienna, Austria

IAEA (2004) Occupational radiation protection in the mining and processing of raw materials, Safety Guide, RS-G-1.6, International Atomic Energy Agency, Vienna, Austria.

IAEA (2005) Guidebook on environmental impact assessment for in situ leach mining projects. IAEA-Tecdoc 1428, International Atomic Energy Agency, Vienna, Austria

IAEA (2007) Assessing the Need for Radiation Protection Measures in Work Involving Minerals and Raw Materials, Safety Reports Series 49, STI/PUB/1257, International Atomic Energy Agency, Vienna, Austria

IAEA (2009) Establishment of Uranium Mining and Processing Operations in the Context of Sustainable Development. International Atomic Energy Agency Nuclear Energy Series NF-T-1.1, International Atomic Energy Agency, Vienna, Austria

IAEA (2010). Best Practice in Environmental Management of Uranium Mining. International Atomic Energy Agency Nuclear Energy Series NF-T-1.2, International Atomic Energy Agency, Vienna, Austria

IAEA (2011a) Radiation Protection and Safety of Radiation Sources: International Basic Safety Standards - Interim Edition General Safety Requirements Part 3. International Atomic Energy Agency, Vienna, Austria, STI/PUB/1531

IAEA (2011b) Radiation Protection and NORM Residue Management in the Production of Rare Earths from Thorium Containing Minerals, Safety Reports Series 68, STI/PUB/1512, International Atomic Energy Agency, Vienna, Austria

IAEA (2011c) The Uranium Mining Remediation Exchange Group (UMREG) Selected Papers 1995–2007. STI/PUB/152, International Atomic Energy Agency, Vienna, Austria

IAEA (2013a) Management of NORM Residues. IAEA-Tecdoc 11712, International Atomic Energy Agency, Vienna, Austria

IAEA (2013b) Radiation Protection and Management of NORM Residues in the Phosphate Industry. Safety Reports Series 78, STI/PUB/1582, International Atomic Energy Agency, Vienna, Austria

IAEA (2014a) IAEA Safety Standards. International Atomic Energy Agency, Vienna, Austria. http://www-ns.iaea.org/standards/default.asp?s=11&l=90

IAEA (2014b) Network of Environmental Management and Remediation (ENVIRONET) http://www.iaea.org/OurWork/ST/NE/NEFW/WTS-Networks/ENVIRONET/overview.html

Jakubick A, Waggitt P (2013) The Uranium Mining and Remediation Exchange Group. AusIMM Bulletin, Australasian Inst. Min. Metall., Melbourne, Australia, April 2014 (2): 46-47

Tulsidas H, Zhang J (2014) Breaking New Ground - Istanbul Workshop zooms in on Social Licence for Uranium Exploration and Mining. IAEA Fuel Cycle and Waste Newsletter 10(1) : 12-13 http://www-pub.iaea.org/MTCD/Publications/PDF/Newsletters/NEFW-10-01.pdf

About uranium researchers

(some of this is actually true)

Challenging Issues in Regulating Uranium Mining in Tanzania

Firmi P. Banzi[1], Peter Msaki[1], Najat Mohammed[1]

[1]Department of Physics, University of Dar es Salaam, P.O. Box 35063, Dar es Salaam, Tanzania. Email: fpbanzi@yahoo.com

Abstract. Experiences have shown that unregulated uranium mining practices have led to significant radiological risk to workers, public and the environment. Tanzania is about to start uranium mining in different parts of the country. Although the legislation and regulations to control doses to workers, public and the environment are in place, there are challenges to adequately address best practice requirements during and after the mining operations. This paper provides an overview of the challenges in addressing the regulatory compliance.

Background

Experiences from historical uranium production sites all over the world have consistently shown that unregulated uranium mining practices have led to significant damage of water, soil, put persons at risk and resulted high cost of clean up the environment (IAEA 2004). In Tanzania the primary legislation, which control practices of ionizing radiation is the Atomic Energy Act 2003 (TAEC 2003) and associated regulations: Mining (Radioactive Minerals) Regulations of 2010 (TAEC 2010) and the Atomic Energy (Radiation Safety in the Mining and Processing of Radioactive Ores) Regulations of 2011 (TAEC 2011). According to the Act an ore is classed as radioactive mineral if its total activity concentration exceeds 74 Bq/g (TAEC 2003). In this context, the activity concentration of economical uranium ores which are likely to exceed this activity are classed as radioactive ores and thus subject to stringent regulatory control. The dose limits applicable in these regulations are those recommended by the ICRP (1977) for both occupational and public exposure to ionizing radiation. The dose limits are as follows: maximum occupational dose of 50 mSv in any year with a mean of 20 mSv over any five years, and a public dose limit of 1 mSv in any single year. In addition, the regulatory authority has a clear role to enforce regulatory compliance with the national and international standards. The doses apply to both internal and external exposure exclude background radiation and medical exposures received during examination or treatment.

B.J. Merkel and A. Arab (eds.), *Uranium – Past and Future Challenges*,
DOI 10.1007/978-3-319-11059-2_4

The background radiation referred herein composes of terrestrial radiation, cosmic radiation together with the contribution of fallout from man-made radiation sources. The terrestrial radiation is mainly a result of ^{40}K, ^{232}Th and ^{238}U and associated decay products. These radionuclides are commonly found on the earth's crust (soils, oceans, food, and drinking water) and they occur with various concentrations. Implicit that depending on the altitude above sea level and radionuclides content in the environment media, the background radiation would vary from one place to another. Literature report that high doses and dose rates of background radiation occurs in soils with rich concentrations of uranium and thorium (UNSCEAR 2000). Normally igneous rocks of granitic composition are rich in thorium and uranium compared to rocks of basaltic composition. Because of the thorium and uranium concentrations, higher background radiation levels are associated with igneous rocks and lower levels with sedimentary rocks, with exception of shales and phosphate rocks (UNSCEAR 2000). Hence, extractions of uranium ore from deposits have the potential to generate wastes with substantial amount of radionuclides and heavy metals in soils, groundwater and surface water. The heavy metals associated with uranium deposits which are threat to the living organisms i.e. plants, animals and humans are: Ag, As, Ba, Cd, Hg, Pb, Sb, and Th and trace elements which are toxic to living organism above certain tolerance level such as: Co, Cr, Cu, Mn, Mo, Ni, Se, and Zn.

Airborne geophysical surveys undertaken country wide in Tanzania by Geosurvey International in 1976-1980 and follow up detailed surveys by Uranerzbergbau in 1978-82 have delineated potential uranium occurrences viable for economical exploitation. Locations of uranium occurrences are shown on blocks A-G in Figure 1 and deposits name and area coverage are indicated in Table 1. The discovery of uranium deposits in Tanzania have attracted foreign company's interests to invest on large scale mining and milling projects in some parts of the country. According to the projections, for example production of uranium in Mkuju deposit is targeted to commence very soon.

Considering that uranium occurrences in many deposits of Tanzania are on shallow deposits, heap-leaching or open cast methodologies are foreseen as economically viable methodologies that would be used to extract the uranium ores. The productions of yellowcake using these methodologies involve removing of uranium ore from the ground and extraction of the yellowcake through crushing of the ore, grinding of the ore to produce fine particles for subsequent chemical leaching (IAEA 2009). It is worth noting that all these processes produce significant amount of waste rock, tailings, and contaminated water or soils containing radionuclides and non radionuclides which potential to environmental contamination. Typical examples of expected waste products from extraction of uranium ores are: barren rocks which are uneconomical to process in the mill and tailings which retain about 5-10% of uranium and 85% of long lived radioactive radium and thorium (IAEA 2009).

While it may seem that there are proven, best available, technologies for long term management of waste rock, tailings and waste water in developed countries, in certain cases, challenging incidents were reported where large-scale discharge of wastes in the environment e.g. New Mexico, Navajo. Nabariek, Ranger, Olympic dam, Beveriey and Rum Jungle occurred (IAEA 2004). Quite often the perceived cost of managing a non-scheduled environmental release is usually very high financially, ecologically and socially. As a result, mining firms worldwide are focused on the early planning and implementation of wastes management.

It is likely that through wind and water pathways, the waste products containing radionuclides and non radionuclides from uranium mining can access the environment and become potential for radiological exposure to public. In order for the regulatory authority to be effective in assessing and mitigating the associated radiological and heavy metals risks, benchmark data of the natural distribution of these elements in the environment at the vicinity of the uranium mines are inevitably needed for managing safety and environment before the mining operations.

In recognition that the uranium mining is effectively controlled, the regulations in Tanzania require the regulatory authority and company to prepare an Environmental Impact Assessment (EIA). Knowledge of pre existing environment prior to the beginning of the mining activities is important in the uranium mining context to assess radiation doses to humans and the environment both during and after mining. This information is particularly useful in rehabilitation planning and developing closure criteria for uranium mines as only radiation doses above the natural background are usually considered 'controllable' for radiation protection purposes. The EIA also serve the regulatory authority to review the license conditions as well as indicators for monitoring and evaluation of changes on the environment as a point of comparison during and after the project operation.

Regulatory Challenges

Although a legal framework and relevant infrastructure for the management and control of occupational and public exposure, waste and the environment are in place, there are challenges for effective monitoring of the uranium mining industry in Tanzania. Finance and technical resources sounds as major limitations that may hinder the regulatory authorities to complete its primary responsibilities. Establishing baseline environmental data by the regulatory authority and the mining company prior to project development is of crucial importance and cannot be underestimated. Lack of comparable data could prevent the effective monitoring of changes in compliance with the environmental standards during and after the mining activities. Since data that established after the mining activities commences cannot be used to relate the impacts of uranium mining have on the environment and public exposure. Because the anticipated area in need of pre mining data is vast and time available to establish them before actual uranium mining commences is very short, the regulatory authority need substantial amount of resources. Big cost required establishing the baseline data and building technical capability wholly resting on the government funds are big challenges. Implicit, inadequate budget and technical capability could lead to impairment on assessing and effective controlling the uranium mining.

Conclusion

For Tanzania to effectively implement best practice principles in the uranium industry, the competent regulatory authority must have adequate financial capabilities, equipment and competent staffs for verifications of compliances with standards. This is because stringent laws and regulations alone may not guarantee the full enforcement of compliances and preventing the hazard from actually occurring. The regulatory compliance will be verified only when resources and reference data made available prior to the mining operations.

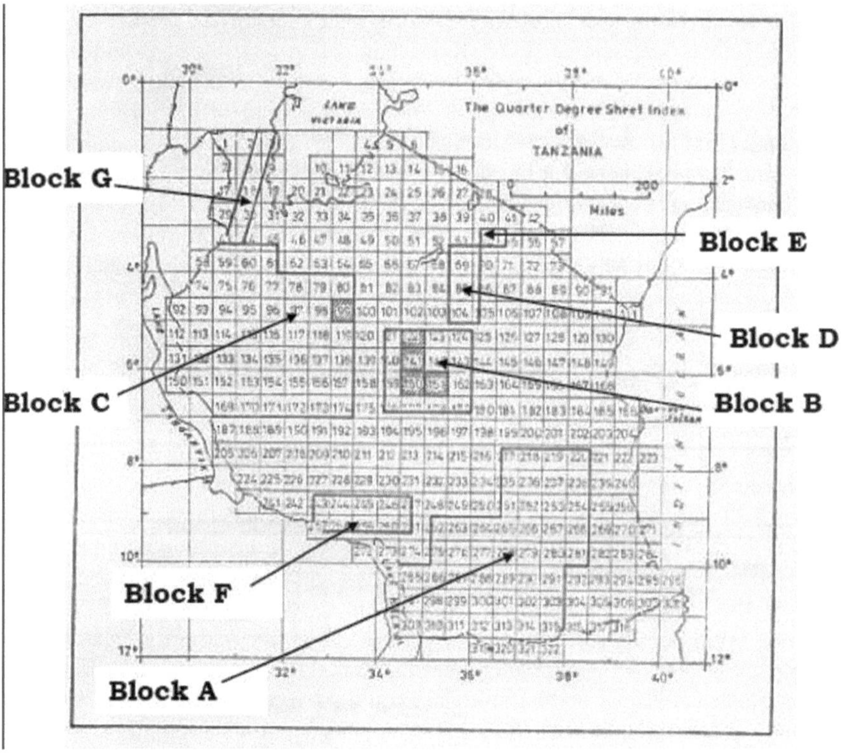

Fig.1. Blocks A – G for discovered Uranium occurrences in different geological environment in Tanzania.

Table 1. Shows block name, Locations and Area coverage of discovered Uranium occurrences in different geological environment in Tanzania.

Block Name	*Deposits Locations*	*Occurrences Area (Sq km)*
A	Mkuju, Madaba	135,000
B	Isuna, Bahi, Makutupora	37,000
C	Ndala, Igombe, Kigoma Ugalla River, Mpanda,	142,000
D	Minjingu, Gallapo	13,000
E	Monduli, Tarosero	2,000
F	Chimala, Panda, Njombe	19,000
G	Bukoba, Biharamulo	10,000

References

IAEA (2004) Environmental Contamination from Uranium Production Facilities and Their Remediation. **http://www-pub.iaea.org/MTCD/publications/PDF/Pub1228_web.pdf**

IAEA (2009) Establishment of Uranium Mining and Processing Operations in the Context of Sustainable Development. IAEA Nuclear Energy Series No. NF-T-1.1

ICRP (1977) Recommendations of the International Commission on Radiological Protection. ICRP Publication 26, Ann. ICRP 1 (3).

MEM (2010) Ministry of Energy and Minerals. The Mining (Radioactive Minerals) Regulations.

TAEC (2003) The Atomic Energy Act, 2002

TAEC (2011) The Atomic Energy (Radiation Safety in the Mining and Processing of Radioactive Ores) Regulations.

UNSCEAR (United Nations Scientific Committee on the Effects of Atomic Radtion) (2000) "Annex B". Sources and Effects of Ionizing Radiation. vol. 1. United Nations.

You have rendered great service to mankind.
Your research has helped many. Please, share with us your story.
Well...

Uranium leaching from a burning black shale deposit – present conditions and future scenarios

Mattias Bäckström[1], **Lotta Sartz**[2]

[1]Man-Technology-Environment Research Centre, Örebro University, 701 82 Örebro, Sweden
[2]Bergskraft Bergslagen AB, Harald Olsgatan 1, 714 31 Kopparberg, Sweden

Abstract. During WW2 oil was produced through pyrolysis of alum shale giving rise to waste that was deposited in the open pits and in a waste deposit. The waste deposit still today has significantly elevated temperatures (above 500°C).

Remaining pyrite in the waste material has also led to ARD with elevated trace metal concentrations.

The waste deposit is no great environmental problem today but as soon as the waste pile cools off both the volumes of drainage and concentration of uranium will increase significantly.

Introduction

During WW2 there was a fuel shortage in Sweden and it was decided to produce oil through pyrolysis of alum shale in Kvarntorp, Kumla. The alum shale contains approximately 20 % kerogene (organic matter) and 12 % pyrite. Prior to pyrolysis the alum shale was crushed and the finer fraction (below 10 mm) was discarded (approximately 20 %) to avoid sintering during the pyrolysis. Depending on the process used the waste products were either coke or ash. During the years 1942-1966 waste was deposited in the open pits but also in a waste deposit close to the industrial area. It is estimated that this waste deposit consists of finer fraction (3 Mt), coke (2 Mt) and ash (23 Mt) with a total volume of 40 million m^3 (total height around 100 m).

Combination of hot ash, pyrite and organic matter (kerogene) has led to significantly elevated temperatures (>500°C) in the waste deposit and the processes are still active 44 years after the closure of the waste pile. Oxidation of pyrite can increase the temperature to 70-100°C, followed by ignition of kerogene giving rise to temperatures close to 1 000°C (Bharati et al. 1995; Puura 1998).

The aim of the present study was to characterize the leachates from the pile, perform different leaching tests on the major waste materials and try to understand the evolution of the leachates from the pile.

B.J. Merkel and A. Arab (eds.), *Uranium – Past and Future Challenges*,
DOI 10.1007/978-3-319-11059-2_5

Methods

Sampling

Several methods, both geotechnical and chemical, were used in order to study the evolution of leachates from the waste deposit. Sampling of solid samples was performed through drilling in combination with installation of shallow groundwater wells.

Dominating materials (original shale, unprocessed fines, ash and red ash) were sampled for further leaching tests.

Measurements

Solid samples were analyzed for trace elements as well as major constituents. Uranium concentrations were determined through alkaline fusion followed by analysis by ICP-MS.

Ground and surface waters were analyzed for trace elements as well as basic chemical parameters.

Temperature measurements were performed in deep (20 m) steel pipes (n 21) on the waste deposit and through helicopter thermography (surface temperature).

Leaching

Two different leaching methods were used on the sampled material; (i) batch leaching simulating different scenarios and (ii) humidity cells for 26 weeks in order to establish weathering rates.

Leaching (L/S 20) to simulate different scenarios was performed in parallel at (i) deionized water, (ii) oxidizing conditions (H_2O_2 at 85 °C for 3 h) and (iii) reducing conditions (NH_2OH-HCl at 90 °C for 5 h) in order to determine the potential leaching in the future.

Humidity cell testing was used to predict long-term leaching from the four main constituents of the pile (red ash, unprocessed fines, shale and ash).

Humid air was obtained by deionized water heated by an immersion heater to 25°C in a 10 L container. Air was pumped into the container, where it was divided into small bubbles. The humid air was led through a tube that went into the upper part of the humidity chamber and out through the hole in the lower part. The samples were exposed for dry air at room temperature for 3 days, thereafter 3 days with dry air followed by leaching.

Samples were leached with 500 ml deionized water and the samples stayed immersed in the water for one hour. Electrical conductivity, pH and redox potential were determined immediately after sampling using relevant electrodes. Alkalinity (end-point pH 5.4) and acidity were determined through titration with HCl and NaOH, respectively. Elemental analysis was performed using ICP-MS.

Results and discussion

Solid phase

From the total concentrations of major elements a mineralogical composition was estimated for both the original shale and the ash formed after pyrolysis. The major difference between the shale and the ash is that it is assumed that all pyrite has been converted into hematite and the organic matter kerogene has been driven off or combusted. Dominating minerals are illite (31 % in shale and 42 % in ash), quartz (23 and 29 %) and K-feldspars (13 and 17 %). Pyrite (12 %) and kerogene (18 %) are present in the shale while hematite (10 %) is present in the ash.

Average concentrations (n 7) for the most common trace elements are: As 79 mg/kg dw, Mo 163 mg/kg dw, Ni 70 mg/kg dw, U 235 mg/kg dw and V 650 mg/kg dw.

Groundwater

There is a large span in chemistry in the 15 shallow groundwaters around the waste deposit (Table 1). Low pH waters are found in the areas where non processed shale has been deposited while high pH waters are found in areas where calcite has been converted to CaO.

Uranium concentrations (average concentration around 250 µg/L) have a strong relationship with pH; at low pH uranium concentrations are high (r^2 0.43). This indicates that weathering at low pH releases more uranium or that sorption is lower at low pH (indicating a cationic behaviour). However, there is also a strong relationship between uranium and molybdenum (r^2 0.71) that does not exist between uranium and nickel (r^2 0.04). Uranium probably exists as negative carbonate or sulphate complexes at pH above 7.

Table 1. Groundwater composition around the waste deposit (filtered samples). Nickel and molybdenum are included to illustrate the behavior of a cationic and anionic element, respectively.

	pH	Ca (mg/L)	Fe (mg/L)	HCO_3 (mg/L)	SO_4 (mg/L)	U (µg/L)	Mo (µg/L)	Ni (µg/L)
0401	4.6	522	628	<1.0	2 470	142	<3	613
0402	3.2	435	328	<1.0	4 260	129	<1	1 190
0403	3.6	519	23.9	<1.0	1 720	15.6	8.16	235
0404	12.2	698	<0.004	1 800	656	0.5	142	8.12
0405	-	568	0.0323	-	-	1 490	196	10.7
0406	-	511	0.0066	-	-	1 760	935	6.95
0407	4.1	495	1 130	<1.0	3 780	23	<3	173
0408	7.5	145	<0.004	220	196	5.4	12.5	7.52
0409	4.6	484	620	<1.0	6 040	30.8	6.07	271
0410	6.5	654	259	540	3 020	1.56	<1	2 420
0411	10.4	39.6	0.0494	210	23	<0.01	2.04	<0.5
0412	10.7	88.9	0.0272	170	308	<0.01	7.48	<0.5
Rb0301	3.8	483	508	<1.0	4 450	126	12.8	873
Rb0302	6.5	261	0.027	710	476	10.9	48.7	1.03
Rb0306	6.6	262	0.137	82	869	10.2	37.4	47

Leaching

Results from the leaching tests indicate an increased leaching of several trace elements if the chemical conditions are changed (Table 2). Probably pH will decrease slightly when water starts to infiltrate resulting in higher leachability for the trace elements. This behaviour is also confirmed in the groundwaters with higher concentrations in the more acidic water even though many of the oxyanions (As, U, Mo and V) were found at their highest concentrations in the neutral ground water (Table 1).

Leaching tests indicate that the leachable amounts of uranium can increase with almost a factor of 50. Even if the volumes of leachates increase with a factor 10 when the waste pile cools off the concentrations will probably increase with roughly a factor of 5. Average uranium concentrations in the surrounding groundwaters may end up around 1 000-1 500 µg/L.

Table 2. From the leaching tests estimated leachable today, leachable during the two different scenarios (mg/kg dw and tons). This is compared with the leaching today at natural pH. The numbers should be regarded as rough estimates.

		Ni	U	Mo
Leachable now	mg/kg dw	2.7	0.61	1.6
	Tons	76	17	46
Leachable reducing	mg/kg dw	13	34	26
	Tons	370	960	720
Leachable oxidizing	mg/kg dw	11	29	28
	Tons	310	820	780
Max leachable	mg/kg dw	11	29	28
	Tons	310	820	780

Results from the humidity cells can be found in fig. 1 below. Average pH is 2.8 in the shales, 3.5 in the unprocessed fines, 5.1 in the ash and 6.2 in the red ash. There is no significant change in pH during the 26 week test period.

Highest uranium concentration was noted for shale during the first leaching periods; 3 720 µg/L. The concentrations then declined to around 100 µg/L at the end of the leaching process (equal to L/S 25). Unprocessed fines had an initial uranium concentration around 320 µg/L, declining to around 10 µg/L at the end. Uranium concentrations from the humidity cells are lower than expected from the estimates from the redox leaching. However, L/S in the humidity cells is 1 at the first sampling point and around 26 at the last sampling point, while L/S through the waste pile is much lower.

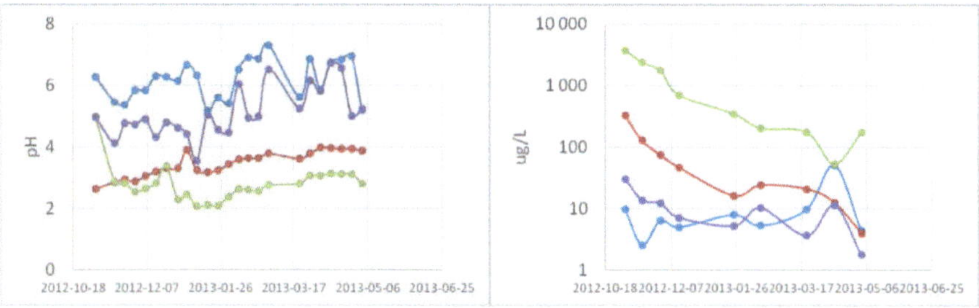

Fig.1. Left: pH. Right: uranium concentration. Blue = red ash, red = unprocessed fines, green = shale, purple = ash

Uranium concentrations as a function of pH are illustrated in fig. 2. It is quite clear that uranium behaves as a cation at low pH but there is a tendency for higher concentrations as pH increases above 6 in ash and red ash, indicating anionic complexes.

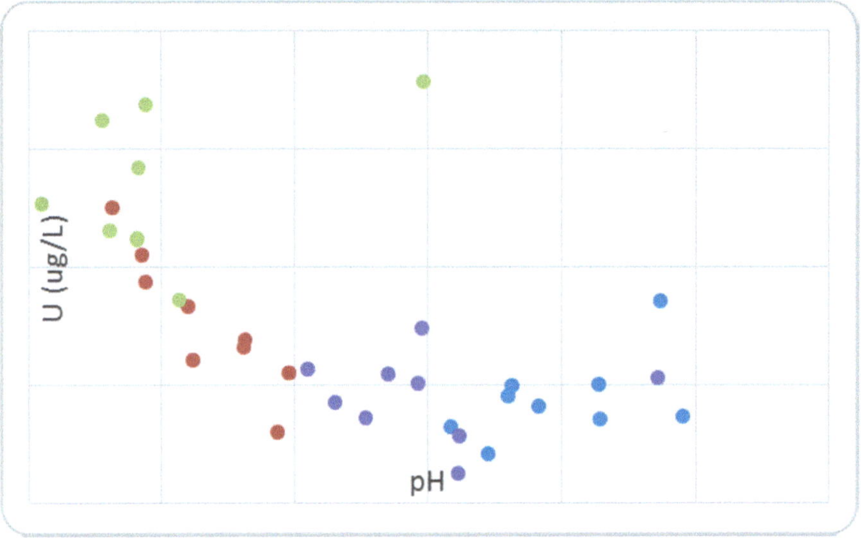

Fig.2. Uranium vs pH for humidity cells. Blue = red ash, red = unprocessed fines, green = shale, purple = ash.

Temperature

Approximately 50 % the temperature measurements performed in the waste pile had a flat profile with depth where the temperature did not exceed 50°C. The other 50 % had similar profiles, but with temperatures between 50 and 100°C. However, a couple of profiles showed temperatures exceeding 500°C (maximum range for the thermometer), indicating sustained kerogene oxidation (fire) at a depth around 15 m below the surface. These profiles agreed well with the ones performed in the early 1980s (700°C at a depth of 5 m). It is, however, possible that the zone of very high temperatures today is found at greater depths than nearly 30 years ago (15 m today compared with 5 m in the early 1980s). It is suspected that the temperature interval below 100°C is partly generated by pyrite oxidation while higher temperatures are generated by kerogene oxidation (open fire). Heat activity is still noted at the surface of the waste pile. For instance, during the spring of 2010 large areas had to be dug out and treated to lower the surface temperatures. It has earlier been shown that the worst temperatures, from a local environmental perspective, are found in the range 40-100°C, where the rates of pyrite and kerogene oxidation are high at the same time as there is plenty of water in the system (Puura 1998).

Assessment of future temperatures in the waste deposit is extremely hard since there is no good data on (i) the remaining heat in the entire deposit and (ii) the remaining chemical energy (in pyrite and kerogene) in the entire deposit. There is, in other words, no easy way to determine how long it would take for the deposit to cool down. Assuming the present heat energy (rough estimate from the temperature measurements) and no remaining chemical energy (conservative estimate) only cooling through heat conductance to the air, it would take approximately 100 years to reach below 100°C (on average) and several hundred years to reach ambient temperature. Since this simple model doesn't take into account the remaining chemical energy the cooling time is probably much longer.

Future scenarios

Today the waste deposit has an elevated temperature and minimal runoff is formed compared to the evaporation. No significant runoff will be formed until the waste deposit reaches lower temperatures (below 70°C on average) in approximately hundred years counted from 1966. It has been estimated that the volume of runoff will increase from 0.5 L/s today to 5 L/s when the deposit is cool. Chemistry will also change when the temperature has decreased. Higher rates of oxidation of pyrite and kerogene will take place (Puura 1998) in the presence of more water. This will lead to somewhat lower pH and uranium leaching will increase with a factor of 50 (Table 2). Average uranium concentrations around the waste pile will possibly increase with a factor of 5.

Conclusion

Today the environmental impact from the waste deposit is relatively small and local. However, in 100-150 years the waste deposit will cool off and start to leach significantly higher loads of uranium to the surroundings. Leaching of uranium will increase with a factor of 50 and the average concentration can possibly reach 1 000-1 500 μg/L. Further studies on both the cooling and the chemical leaching are needed in order to be able to handle the likely increased uranium loadings in the future.

References

Bharati S, Patience R, Larter S, Standen G, Poplett I (1995) Elucidation of the alum shale kerogen structure using a multi-disciplinary approach. Org Geochem 23: 1043-1058

Puura E (1998) Weathering of mining waste rock containing alum shale and limestone: A case-study of the Maardu dumps, Estonia. Dep. Chem Eng Technol, KTH, Stockholm, PhD thesis

Puura E, Neretnieks I (2000) Atmospheric oxidation of the pyritic waste rock in Maardu, Estonia, 2: an assessment of aluminosilicate buffering potential. Environ Geol 39: 560-566

Is enough information available to derive an overall EQS for uranium in French freshwaters, according to European Guidance?

Karine Beaugelin-Seiller[1], Olivier Simon[2], Rodolphe Gilbin[2], Jacqueline Garnier-Laplace[3], Laureline Février[2]

Institut de Radioprotection et de Sûreté Nucléaire, PRP-ENV, SERIS, Cadarache, France ([1]LM2E, [2]L2BT, [3]SERIS)

Abstract. An overall criterion for protecting freshwaters against uranium chemotoxicity should encompass all the compartments of the freshwater environment at risk (water column, sediment, predators, and humans). According to the latest European recommendations, we revised the interim Environmental Quality Standard currently applicable in France, to introduce physico-chemical parameters affecting uranium bioavailability in water and therefore toxicity, by using chemical speciation modelling in aqueous phase including both mineral ligands and organic matter. Additionally, we completed the poor available data set related to sediment toxicity by implementing dedicated standardized tests and we explored the question of secondary poisoning. Human health was also briefly considered, to fulfill the methodological requirements.

Introduction

Uranium is one of the substances regulated in France since 2007 for freshwater environment, by an interim Environmental Quality Standard (EQS) provisionally fixed at of 0.3 $\mu g.L^{-1}$ in increment to the natural background of the water bodies. Recently, European Guidance on the EQS derivation has evolved and requests now explicitly for metals to take into consideration their bioavailability. Three years ago, we presented our first results accounting for this new procedure in the determination of a Quality Standard (QS) for the water column (previously named predicted no-effect concentration or PNEC), following the method classically applied to other metals in Europe for regulatory purposes (Beaugelin-Seiller *et al.* 2011). From the generic value of QS equal to 0.3$\mu g.L^{-1}$ for chronic exposure, conditional QSs were derived for different representative freshwater compositions, by combining assumption on uranium bioavailable fraction and geochemical modeling approach. This modeling was firstly realized assuming simplified water compositions, without considering natural organic matter. This work highlighted the influence of pH, hardness and carbonates on uranium bioavailable fraction. Dis-

B.J. Merkel and A. Arab (eds.), *Uranium – Past and Future Challenges*,
DOI 10.1007/978-3-319-11059-2_6

solved Organic Carbon (DOC), due to its high capacity to form organic complexes with uranium (Crançon and Van der Lee 2003; Le Goff and Bonnomet 2004; Jackson *et al.* 2005, Ranville *et al.* 2007), was recently shown to decrease uranium bioavailability (Trenfield et al., 2011; van Dam et al., 2012). Therefore, the outcoming conditional QS values for the water column still included some conservatism. In this paper, we present the follow up of this study. Having previously ignored the influence of DOC on uranium bioavailability, we added this parameter in our derivation process. In parallel, we investigated the other components of interest for the determination of an overall EQS (sediment, secondary poisoning, and human health). For sediment, due to the lack of validated ecotoxicity data in the literature, we realized a basic set of toxicity tests, covering three benthic species (one plant and two invertebrates representative of different feeding habits and lifestyle) to generate a corresponding sediment QS by using the safety factor method. The question of secondary poisoning was only explored at this stage from a methodological point of view, completed with a review of available information that appears poor and fragile. Additionally, a brief review of sanitary criteria proposed by international organizations was performed to evaluate the influence of human health component in the uranium overall EQS determination.

Water Quality Standard

This first part of the study had two objectives: i) to update if possible the value of the generic QS, examining the ecotoxicity data published since its previous determination; ii) to refine and strengthen the realism of the calculated conditional QSs by including DOC, and to propose a simplified formalism, operationally applicable, while guaranteeing the necessary conservatism whatever the physico-chemical class of waters to which any conditional value applies.

Generic value

The update of ecotoxicity data aimed at integrating the most recent published knowledge about uranium ecotoxicity in freshwaters as well as, if necessary, review former information taking into account the recent changes in the European recommendations regarding the determination of generic QS. These evolutions consist in considering firstly the QS resulting from a Species Sensitivity Distribution, as long as the quality and size of the dataset allows such an approach (alternative method to that of safety factors, but requiring a more important set of data). The other newly recommended option to determine a QS is to use *in situ* data.

After review, it appeared that the complete data set, including recently published information and *in situ* data, did not allow changing the determination method of the QS. The generic value of this criterion had still to be derived applying the safety factor method, identically to what was done previously with the data identified in 2010.

Conditional QS values

Water conditional QSs were derived considering five physico-chemical classes to represent the whole range of variation of the composition of French surface waters. These classes differ by their characteristics in terms of hardness, HCO_{3total}, pH and DOC, the latter being previously ignored. The determination of these classes and their associated conditional QSs was based on the simulation of the uranium speciation in 119 French waters of contrasted physico-chemistry (originated from the FOREGS geochemical atlas (Salminen *et al.* 2005)), considering that the bioavailable concentration of uranium (Ubio) is the sum of 4 uranium bioavailable species ($Ubio=UO_2^{2+} + UO_2(OH)^+ + UO_2(OH)_2 + UO_2CO_3$). A statistical regression tree analysis (RPART library – Therneau *et al.*, 2014 - on R software – R Core Team, 2014) was performed to identify the variables (pH, hardness, HCO_3 and DOC content) explaining the variation of the maximal concentration of uranium in these waters that ensure a Ubio value below the generic QS.

The formalism finally retained is the one of a determination tree (Fig.1) built on this statistical analysis. The field of application of the tree is defined by the ranges of each of the four physico-chemical variables:

- Hardness from 11.4 to 1091 mg $CaCO_3.L^{-1}$;
- pH from 6.4 to 8.8;
- $HCO_{3\ total}$ from 8.8 to 421 $mg.L^{-1}$;
- DOC from 0.3 to 23.4 $mg.L^{-1}$.

For any surface water showing characteristics outside these ranges, the generic QS value of 0.3 $\mu g.L^{-1}$ (dissolved uranium) should be used.

A first validation step of the classes' tree was successfully performed on a set of 25 independent data, representing large French watercourses and rivers close to former uranium mining sites. This validation step is currently being implemented in depth with other data sets, in order to examine whether the tree could be applied to waters presenting characteristics outside the range of water composition used for its construction.

The water QSs tree was established from the knowledge currently available about uranium bioavailability, it would logically evolve with the knowledge improvement of the uranium environmental fate and the measurement techniques of the bioavailable fraction.

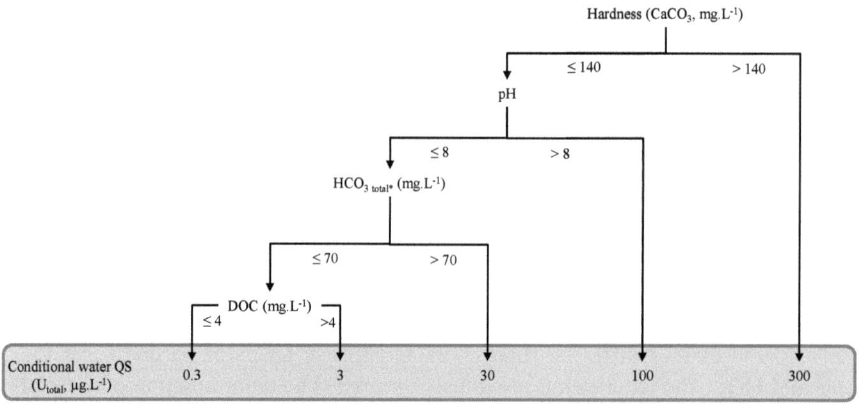

*HCO_3 total: all chemical forms of HCO_3

Fig.1. Determination tree of conditional water QSs applicable to French freshwaters.

Sediment Quality Standard

Available quality-assessed data on sediment toxicity are scarce in the literature for uranium. A review conducted in 2012 led us to identify a single value that may be considered as acceptable with regard to chronic ecotoxicity, a NOEC obtained on a diptera (growth inhibition for a larva of *Chironomus dilutus* after 10 days of exposure; Liber *et al.* 2011). Due to the paucity of the dataset, the highest safety factor (i.e. 100) should be applied to this single data (740 $mg.kg^{-1}$ dry weight) to determine the sediment QS (7.4 $mg.kg^{-1}$dry weight).

In order to consolidate the sediment QS, we decided to enlarge the dataset. On the operational point of view, obtaining additional ecotoxicity data such as NOECs or EC_{10}s for two or three benthic or epi-benthic species representative of different feeding habits and lifestyle may divide by up to 10 the safety factor to be applied to the lowest value. Three standardized tests were implemented with two animal species (the crustacean *Hyallela azteca* – ISO 2013a; the insect larva *Chironomus riparius* – OECD 2004) and one plant (*Myriophyllum aquaticum* – ISO 2013b). Despite our efforts, we were unable to obtain valid data for the latter. We finally collected 4 new data (Table 1), leading to a safety factor of 50 (two data on two different species) in the absence of complementary data published since 2012.

Results are consistent between the tested species, with a same lowest value (40 $mg.kg^{-1}$ for the lower bound of the insect EC_{10} as well as for the crustacean NOEC). The quality-assessed ecotoxicity data we obtained on invertebrates allowed us to reduce by 2 the safety factor to apply for the sediment QS derivation (from 100 to 50): the corresponding QS would be 0.8 $mg.kg^{-1}$. Adding a third data on plants would potentially authorize the use of a safety factor 5 times lower (i.e.

10), increasing moreover the confidence in the data robustness. Taking advantage of our recently acquired experience, we will renew the toxicity experiment on *Myriophyllum aquaticum* during 2014.

Table 1. Ecotoxicity data acquired through standardized tests on benthic invertebrates.

Species	Effect criterion	Endpoint	Value [confidence interval 95%] Mg.kg^{-1} dry weight
Chironomus riparius	NOEC	Emergence rate	62
	EC_{10}	Emergence rate	188 [40-885]
Hyallela azteca	NOEC	Survival rate	40
	EC_{10}	Total mass	199 [107-291]

Applying a maximum value of 1000 L.kg^{-1} for the liquid-solid partition coefficient (IAEA 2010), the sediment QS of .8 mg.kg^{-1} corresponds to an equivalent water QS of 0.4 µg.L^{-1}which is similar to the generic water QS established from water toxicity data.

Secondary Poisoning Quality Standard

Secondary poisoning is the result of a predator contamination due to the ingestion of a contaminated prey. Practically, studies related to trophic poisoning in animals are rare, and even more when considering an animal-to-animal transfer. The dedicated literature review conducted in 2013 revealed only 4 studies related to the trophic toxicity of uranium in freshwater ecosystems, among which only one was about a prey-predator interaction. Additionally, none was conceived in order to acquire data useful to a QS determination.

Aware of this type of knowledge gap for a large majority of chemicals, the European Commission recommends to base the secondary poisoning QS on chronic ecotoxicity data collated on birds and mammals (EC 2011). Biota standards derived in this way are assumed to be enough protective for benthic and pelagic predators, waiting for the development of robust guidance applicable to these predators.

From this point of view, the literature reviews conducted in 2009 by the EFSA and more recently by the ATSDR (2013) present relevant information. The ATSDR conclusions are edifying: only 7 studies were published since 1949, concerning solely 3 different mammal species and all with a statistical power too low to guaranty the robustness of the study.

In the absence of complementary information, any determination of secondary poisoning QS for uranium in freshwater ecosystems would be today so largely uncertain that its operational use would be meaningless.

Nevertheless, to obtain the order of magnitude of such QS value, an attempt was made to derive it considering the available information. Applying the European recommended approach to the lowest published relevant ecotoxicity data (2.8 mgU/kg/d after 21 days of mice exposure, Paternain *et al.* 1989), and considering a maximum concentration ratio of 560 (w.w.) published for molluscs and direct transfer from water (IAEA, to be published), the corresponding QS would be about 0.1 $\mu g.L^{-1}$, which is close to the generic water QS established from water toxicity data. The Danish government realized a similar determination using different data and parameters (Miljoministeriet, 2011) which led to a much lower secondary poisoning QS (0.015 $\mu g.L^{-1}$).

Human Health Quality Standard

It is generally admitted that, as for animals, information on the human exposure to uranium by the trophic pathway is not sufficient to derive the corresponding QS. There is consequently no specific national recommendation in France with regard to trophic contamination.

For drinking-water, the European Commission recommends firstly to refer to guidelines published in Europe or by the WHO when existing. This is the case for uranium for which the WHO guideline value for drinking-water quality was revised in 2011, raising from 15 $\mu g.L^{-1}$ to 30 $\mu g.L^{-1}$ (WHO 2012), on the basis of new epidemiological studies on populations exposed to high uranium concentrations.

Conclusions

If the water conditional QSs appear now well adapted to the state-of-the-art on uranium bioavailability in freshwaters, there is a much lower level of confidence in the QSs related to the other components of interest for the determination of an overall uranium EQS for aquatic ecosystems. New ad hoc data and establishment of European topical working groups could help in going further.

An interesting point in the various generic QSs derived until now, whatever their level of confidence, is their internal consistency when looking at ecological drivers (Fig.2). We derived QSs for the three compartments of interest (water, sediment, secondary poisoning) that are in the same order of magnitude, from 0.1 to 0.3 $\mu g.L^{-1}$, that is to say 2 orders of magnitude lower than the one related to human health.

Finally we consider that there is still a lack of data to derive an overall uranium EQS for freshwater ecosystems, with regard to European best practices in this field. We suggest using the available criteria, once reinforced by the tests in progress, to propose an interim Guidance Value (GV) which could help managing a number of French rivers impacted by uranium releases from former uranium mining activities.

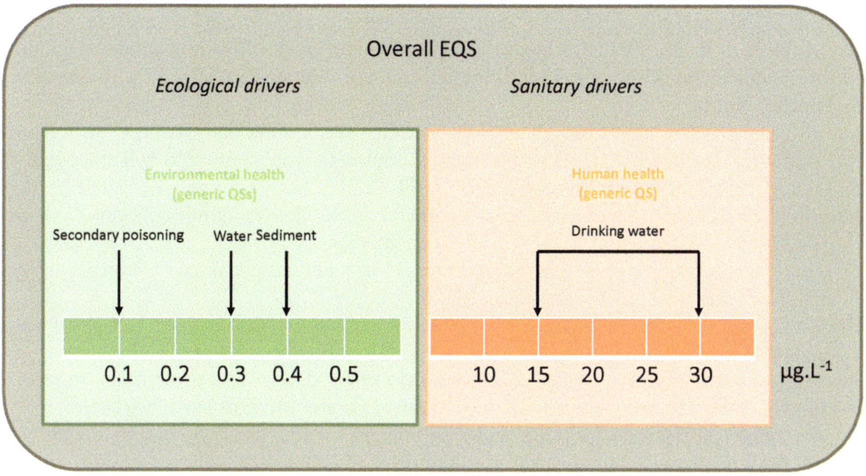

Fig.2. Relative position of the elementary generic QSs required for the derivation of an overall uranium EQS for aquatic ecosystems.

References

ATSDR (2013). Toxicological Profile for Uranium. U.S. Department of Health and Human Services. Public Health Service. U.S. DEPARTMENT OF HEALTH AND HUMAN SERVICES, Agency for Toxic Substances and Disease Registry, 526 p.

Beaugelin-Seiller K, Février L, Gilbin R, Garnier-Laplace J (2011). Ecotoxicity of Uranium in Freshwaters: Influence of the Physico-Chemical Status of the Rivers. In : The New Uranium Mining Boom. Challenge and lessons learned. Part 5 - Modelling, 507-515. Merkel and Schipek eds, Springer.

Crançon P, Van der Lee J (2003). Speciation and mobility of uranium(VI) in humic-containing soils. Radiochim Acta 91: 673-679

EC (2011). Technical Guidance for Deriving Environmental Quality Standard. Guidance Document n°27, technical report 2011-055, European Communities, 203 p.

EFSA (2009). Scientific Opinion of the Panel on Contaminants in the Food Chain on a request from German Federal Institute for Risk Assessment (BfR) on uranium in foodstuff, in particular mineral water. The EFSA Journal 1018, 1-59.

IAEA (2010). Handbook of parameter values for the prediction of radionuclide transfer in terrestrial and freshwater environments. Technical report series n°472, International Atomic Energy Agency, Vienna, Austria.

IAEA (to be published). Handbook of Parameter Values for the Prediction of Radionuclide Transfer to Wildlife. Technical Reports Series, International Atomic Energy Agency, Vienna, Austria.

ISO (2013a). ISO/DIS 16303. Water quality - Determination of toxicity of fresh water sediments using Hyalella azteca, 27 p.

ISO (2013b) ISO/DIS 16191. Water quality -- Determination of the toxic effect of sediment on the growth behaviour of Myriophyllum aquaticum, 21 p.

Jackson BP, Ranville JF, Bertsch PM, Sowder AG (2005). Characterization of colloidal and humic-bound Ni and U in the "dissolved" fraction of contaminated sediment extracts. Environ. Sci Technol 39: 2478-2485

Le Goff F, Bonnomet V (2004). Devenir et comportement des métaux dans l'eau : biodisponibilité et modèles BLM. Rapport technique, Direction des Risques Chroniques, Unité « Évaluation des Risques Écotoxicologiques », mars 2004, 85 p. INERIS, Verneuil en Halatte, France

Liber K, Doig LE, White-Sobey SL (2011). Toxicity of uranium, molybdenum, nickel, and arsenic to Hyalella azteca and Chironomus dilutus in water-only and spiked-sediment toxicity tests. Ecotoxicol Environ Saf 74, 1171-1179.

Miljoministeriet (2011). Fastsaettelse af kvalitetskriterier for vandmiljoet. Uran CAS nr. 7440-61-1 (available at http://www2.mst.dk/blst_databaser/havkvalitet/visDetaljer.asp?pWhat=MST&pStof_id=354, checked on 13th of december 2013).

OECD (2004). OECD 218/9. Test No. 219: Sediment-Water Chironomid Toxicity Using Spiked Water, 21 p.

Paternain JL, Domingo JL, Ortega A, Llobet JM (1989). The effects of uranium on reproduction, gestation and postnatal survival in mice. Ecotox Environ Saf 17:291-296.

R Core Team (2014). R: A language and environment for statistical computing. R Foundation for Statistical Computing, Vienna, Austria. URL http://www.R-project.org/.

Ranville JF, Hendry MJ, Reszat TN, Xie Q, Honeyman BD (2007). Quantifying uranium complexation by groundwater dissolved organic carbon using asymmetrical flow field-flow fractionation. J Cont Hydrol 91: 233-246

Salminen R (Chief-editor), Batista MJ, Bidovec M, Demetriades A, De Vivo B, De Vos W, Duris M,Gilucis A, Gregorauskiene V, Halamic J, Heitzmann P, Lima A, Jordan G, Klaver G, Klein P, Lis J, Locutura J, Marsina K, Mazreku A, O'Connor PJ, Olsson SÅ, Ottesen RT, Petersell V, Plant JA, Reeder S, Salpeteur I, Sandström H, Siewers U, Steenfelt A, Tarvainen T (2005). Geochemical Atlas of Europe (available at http://weppi.gtk.fi/publ/foregsatlas/index.php, checked on 24th of April 2014).

Therneau T, Atkinson B, Ripley B (2014). rpart: Recursive Partitioning and Regression Trees. R package version 4.1-8. http://CRAN.R-project.org/package=rpart

Trenfield M.A., McDonald S., Kovacs K., Lesher E.K., Pringle J.M., Markich S.J., Ng J.C., Noller B., Brown P.L., Van Dam R.A. 2011a. Dissolved organic carbon reduces uranium bioavailability and toxicity. 1. Characterization of an aquatic fulvic acid and its complexation with uranium[VI]. Environ. Sci. Technol. 45(7): 3075-3081.

van Dam, R. A., Trenfield, M. A., Markich, S. J., Harford, A. J., Humphrey, C. L., Hogan, A. C., and Stauber, J. L., 2012. Reanalysis of uranium toxicity data for selected freshwater organisms and the influence of dissolved organic carbon. Environmental Toxicology and Chemistry 31, 2606-2614.

WHO (2012). Chemical hazards in drinking-water: Uranium. (available at http://www.who.int/water_sanitation_health/dwq/chemicals/uranium/en/, checked on 3d of April 2013).

Radiation exposure and environmental remediation at the Urgeiriça mine site, Portugal

Fernando P. Carvalho[1], **João M. Oliveira**[1], **Margarida Malta**[1]

[1]Instituto Superior Técnico/Laboratório de Protecção e Segurança Radiológica, Universidade de Lisboa, Estrada Nacional 10, km 139, 2695-066 Bobadela LRS, Portugal (E-mail: carvalho@itn.pt)

Abstract. The Urgeiriça mine and milling facilities, Portugal, was in operation until 2001. Site remediation, including clean-up of milling plant facilities and terrains was implemented during 2006-8 and it was completed with the placement of a multilayer cap on milling tailings and construction of a drainage system for retrieval of mine water and waste piles seepage followed by neutralization in a mine water treatment plant. During this period the monitoring of environmental radioactivity was periodically carried out in the Urgeiriça area. Results of environmental surveys carried out during last 10 years on ambient radiation dose, radon in surface air, radionuclide concentrations in soils, agriculture products, and water from wells and from surface streams receiving mine effluents, are reviewed. Results of this environmental survey are presented and radiation exposure of the local population is discussed.

Introduction

The uranium mining and milling industry in Portugal was productive during most of the 20th century. Facilities at Urgeiriça mine, near Viseu, were the main milling site of radioactive ores and accumulated about 13 Mtonnes of tailings from chemical processing (Nero et al., 2005). The mining company ENU-SA ceased activities in 2001 and, following that decision, the government requested from national laboratories a major assessment of the environmental and public health risks posed by uranium mining and milling wastes in the region. The reports from such assessment (MinUrar project) were conclusive on the existence of a radioactive impact in the surrounding environment and public health and recommended the implementation of waste management and environmental remediation measures (Marinho Falcão et al., 2005, 2007). This work was entrusted to the EDM company under the Ministry of Economy, and the Director Plan for remediation of abandoned mine sites was submitted and approved by an inter-ministerial commission

B.J. Merkel and A. Arab (eds.), *Uranium – Past and Future Challenges*,
DOI 10.1007/978-3-319-11059-2_7

with representatives of Health, Economy, Environment, and Radiation Protection authorities (Santiago Baptista, 2005). The remediation work in the Urgeiriça area was implemented between 2006 and 2008, and consisted in cleanup of mine area and uranium chemical plant with relocation and grouping contaminated waste into the former milling tailings dump "Barragem Velha".

The coverage of milling tailings and retrieval of drainage followed by treatment of contaminated water, restrained the dispersion routes of waste and radionuclides in surface air (radon and dust), of discharge of radioactive and acid drainage in surface streams, shielded the radioactive materials responsible for enhanced radiation doses in the area, and prevented the potential misuse of radioactive materials (sands, gravel, scrape metals, woods, etc) in the future (EDM, 2011).

Radiological surveys were made over the years in this and other old uranium mining areas, as part of the duties of LPSR/IST (Carvalho, 2014). An assessment of the environmental radioactivity was made in 2008 in Urgeiriça area, after the cover of milling tailings had been concluded. This synoptic monitoring of the status of the environment is described for 2008. Temporal trends of radioactive contamination over the years are presented also and discussed.

Materials and Methods

Ambient radiation doses were determined with calibrated portable radiation monitors (Thermos, FH-40). Samples of water from irrigation wells, soil from the agriculture plots, and plants from the available horticulture production were collected on June 2008. Water samples were filtered through 142 mm diameter, 0.45 µm pore size membrane filters, and filtered water and suspended particulate matter were analyzed separately. Soil bulk samples of the top 30 cm layer were collected and sieved in the laboratory. The size fraction less than 63 µm was used for radioanalysis. Vegetables were cleaned and washed with tap water in the laboratory and tubers and fruits peeled off, as for consumption. Samples were freeze dried and aliquots of homogenized material taken for determination of radionuclides. Radon in surface air was determined with continuous radon monitors (Sarad GmbH, Scout). Surface air aerosols were sampled with high volume aerosol samplers (Andersen).

Radionuclides were determined in all samples applying radiochemical separation of radionuclides, followed by electro deposition and radiation measurement by alpha spectrometry. The analytical procedures applied and their validation through the analysis of certified reference materials and participation in international inter comparison exercises organized by the IAEA, are described in detail elsewhere (Carvalho et al., 2005, 2009a, 2009b; Carvalho and Oliveira 2007, 2009; Oliveira and Carvalho 2006).

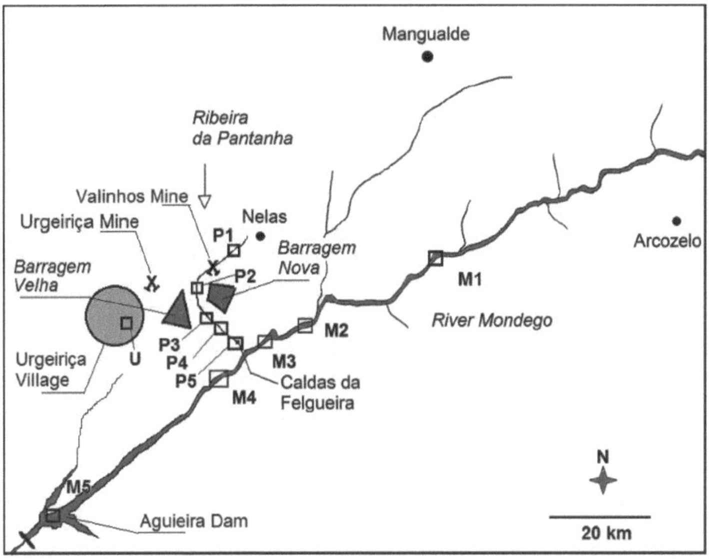

Fig.1. The region of Urgeiriça Mine and uranium milling tailings. Sampling stations are indicated with squares.

Results and Discussion

Ambient dose

Ambient radiation doses measured in 2002 on top of milling tailings "Barragem Velha" ranged from 2 to 12 μSv/h, averaging 7.5 μSv/h. These dose rates varied depending on the composition and grain size of materials disposed in the self-contained impoundments spread over a total area of about 10 hectares. The relocation of waste with transfer of contaminated materials from the mine shaft area and chemical plant area to the waste dump site "Barragem Velha", plus transfer of materials from the cleanup of industrial area onto the top and, finally, placement of additional layers of coverage materials (geotextile membrane, clay, sand, gravel, soil), allowed reducing ambient radiation dose on waste piles was to 0.35 μSv/h, i.e, about 5% of initial dose rate (EDM, 2011). Although this area remains fenced and not open to the public, the current external absorbed radiation dose in the air is of the same magnitude and even slightly lower than the regional background, 0.4-0.5 μSv/h.

Radionuclides in water

The survey of radionuclides in the stream Ribeira da Pantanha (performed in 2008), showed still high concentrations in the acidic seepage from waste piles particularly loaded with dissolved uranium, ^{230}Th and ^{226}Ra. Despite recuperation of tailings seepage and mine drainage into a water treatment plant for acid neutralization and co precipitation of radionuclides with barium sulphate, the enhancement of dissolved uranium and radium in stream water was still measurable downstream the area of waste piles. Concentrations of these radionuclides in stream water decreased downstream with increasing distance due to radionuclide sorption onto sediments and water dilution. Nevertheless, the discharge of Ribeira da Pantanha into the major River Mondego was still contributing to enhance radionuclide concentrations, particularly those of radium, above natural concentrations in this river (Table 1). Further downstream radionuclide concentrations decreased to natural levels with large dilution in the artificial lake of Aguieira Dam.

Bottom sediments of Ribeira da Pantanha stream and River Mondego showed higher radionuclide concentrations and stronger contamination downstream Urgeiriça and waste disposal areas. As sediments are more conservative of radionuclides, the concentrations in bottom sediments will remain above natural levels longer than in the flowing water (Table 2). Notwithstanding, determinations of radionuclides in suspended particulate matter and in bottom sediments showed a decrease in comparison with determinations made several years before remediation measures (Carvalho et al., 2007).

Radionuclides in vegetables

On the banks of stream Ribeira Pantanha as well as in the town of Canas de Senhorim-Urgeiriça area, there are small agriculture productions and kitchen gardens whose products are irrigated with water from wells and from Pantanha stream (Table 3). Agriculture plot # 22, in the Village Aguieira near Nelas, irrigated from a non-contaminated well is a reference area. Vegetables from plots #19-#22 located on the banks of Ribeira da Pantanha and regularly irrigated with stream water, generally showed higher radionuclide concentrations than at the reference area.

Atmospheric radioactivity

Radon in surface air was measured over the years in several locations around the Barragem Velha waste pile. Radon emanated from the uncovered waste piles was transported by the wind in any direction and rapidly diluted in the atmosphere but continuous recording averaged those effects. Measurements were made in three locations, by the nearest houses to tailings piles and at the center of the town Canas de Senhorim. Radon values in surface air at some distance from the waste piles before the placement of the tailings cover averaged 80-100 Bq/m^3, slightly higher than after placement of the cover averaging 55-65 Bq/m^3 on late 2008.

Long lived uranium series radionuclides attached to dust particles in suspension in surface air were measured also at several locations around the main waste pile Barragem Velha. U-238 concentrations in air dust were for example in the year 2006 from 200 to 400 Bq/kg, while in 2010 ranged from 70 to 190 Bq/kg. U-238 activity concentrations in atmospheric dust are an indicator of the resuspension of tailings, while ^{210}Pb and ^{210}Po in dust particles are the sum of radionuclide in minerals of tailings particles plus radon daughters scavenged in the atmosphere. Results over the years displayed fluctuations in specific activity of dust particles which indicates particles from different sources (soil particles, pollen, milling tailings particles, etc.), but a clear decrease of radioactivity in surface air occurred since the placement of the tailings cover, dropping to about half radioactivity levels of aerosols in surface air.

Compaction and coverage of waste materials reduced both the emanation of radon (^{222}Rn) into the surface air and dust re suspension, particularly during summer.

Conclusions

An abatement of ambient radiation dose from gamma radiation emitting radionuclides was obtained with the cleanup of the industrial area, relocation of residues into the "Barragem Velha" milling tailings, and placement of a multilayer cover. Cleaned areas, through removal of waste, debris, scrap materials, and top soil layer have now ambient radiation doses aligned to natural background radiation levels. These areas were opened to other uses such as construction of a new residential area, a public garden, and new roads. Some former industrial and administration buildings were released also by the company to other users and uses. The improvement of radiation protection in the area of these former uranium mining and milling activities was considerable, with radiation exposure of the members of the public being reduced, contaminated materials confined, erosion and leaching of radionuclides brought under control, and environment contamination risk and public health risk controlled, such as recommended by international standards (IAEA, 2006).

The challenge might be now the transfer of custody, maintenance, and monitoring of the remediated sites over future years in order to ensure continued radiation protection and confinement of residues.

Table 1. Activity concentrations of dissolved radionuclides in samples from surface waters (mBq/L).

Sample	Station Nr	^{238}U	^{235}U	^{234}U	^{230}Th	^{226}Ra	^{210}Pb	^{210}Po	^{232}Th
Stream Rib.ª da Pantanha	P1	81.7±2.9	3.4±0.3	82.2±3.0	2.3±0.2	33.3±2.6	11.8±0.7	3.6±0.2	0.15±0.05
Run off, B. Velha	P2	48867±2642	2387±157	47243±2556	565±41	364±49	44.1±2.8	87.7±3.6	13.1±0.3
Stream Rib.ª da Pantanha	P3	1885±100	87.2±5.65	1914±101	2.3±0.2	56.5±5.7	12.6±0.5	5.0±0.2	0.11±0.04
Stream Rib.ª da Pantanha	P4	1992±85	91.0±4.6	2009±86	8.7±0.5	59.3±3.7	7.7±0.3	4.3±0.2	0.34±0.07
Rib.ª da Pantanha, C. Felg	P5	298±9	13.8±0.7	310±9	-	69.4±7.7	12.9±0.7	8.6±0.04	-
R. Mondego, P. Cervães#9	M1	22.0±0.6	0.96±0.07	22.4±0.6	4.7±0.4	44.8±5.4	3.8±0.2	5.9±0.2	0.20±0.07
R. Mondego, C. Louça #8	M2	32.6±0.9	1.5±0.1	33.6±0.9	2.9±0.2	18.0±3.0	1.22±0.05	7.0±0.2	0.27±0.06
R. Mondego. upstream C. Felg #6	M3	28.8±0.8	1.4±0.1	29.9±0.8	2.8±0.2	1.4±0.2	3.1±0.1	5.3±0.2	0.21±0.04
R. Mondego. downstream C. Felg #7	M4	20.0±0.6	1.0±0.1	20.2±0.6	3.9±0.3	29.2±5.5	6.3±0.2	5.0±0.2	0.29±0.06
Aguieira DAM, "Lagoa Azul" #10	M5	9.2±0.3	0.51±0.06	8.8±0.3	1.5±0.1	17.7±1.9	4.3±0.3	5.3±0.2	0.08±0.04

Table 2. Activity concentrations of radionuclides in riverbed sediments (Bq/kg dw).

Sample		Station Nr	^{238}U	^{235}U	^{234}U	^{230}Th	^{226}Ra	^{210}Pb=^{210}Po	^{232}Th
Stream Rib.ª da Pantanha	Sed.#1	P1	2107±47	99±3	2136±48	923±48	403±27	312±13	171±9
Run off B.Velha, Ribª Pantanha	Sed.#2	P2	1109±26	49±2	1200±27	2752±154	2211±140	3366±107	108±7
Stream Rib.ª da Pantanha	Sed.#3	P3	1951±44	91±3	1923±43	484±33	593±48	379±13	276±19
Stream Rib.ª da Pantanha	Sed.#4	P4	4106±102	178±6	4193±104	2892±189	1141±100	1165±37	132±10
Stream Rib.ª da Pantanha	Sed.#5	P5	5422±150	244±9	5584±154	2312±127	1733±77	3736±117	169±10
R. Mondego, P. Cervães	Sed.#9	M1	414±13	20±2	431±13	298±19	366±35	495±12	162±11
R. Mondego, C. Louça	Sed.#8	M2	453±12	19±1	463±12	411±19	674±50	443±11	407±19
R. Mondego. upstream C. Felg	Sed.#6	M3	508±13	25±1	511±13	470±28	534±44	493±18	593±35
R. Mondego. downstream C. Felg	Sed.#7	M4	827±20	39±2	842±20	1302±88	797±70	665±23	440±30
Aguieira Dam	Sed.#10	M5	95±5	3.2±0.7	98±5	96±6	85±8	96±3	50±3

Table 3. Activity concentrations of radionuclides in vegetables from local production (mBq/kg ww).

Sample	Station Nr	Dry/wet weight	^{238}U	^{235}U	^{234}U	^{230}Th	^{226}Ra	^{210}Pb	^{210}Po	^{232}Th
Lettuce, Aguieira Village #22	P1	0.07	197±9	8.1±1.7	164±8	147±15	674±45	470±11	612±22	100±11
Tomato, Aguieira Village #22	P1	0.06	7.8±1.2	< 2.5	5.9±1.2	19.8±4.1	66.4±6.5	17.3±1.2	0.06±0.003	0.42±0.38
Carrot, Aguieira Village #22	P1	0.12	25.0±2.8	< 5.4	27.4±3.0	29.7±4.1	906±59	116±6	57.8±2.7	9.3±2.3
Lettuce, Urgeiriça #21	U	0.04	100±6	5.2±1.2	100±6	113±9	360±24	343±20	157±6	48.6±4.8
Tomato, Urgeiriça #21	U	0.1	1.4±0.5	< 4.0	5.2±1.5	14.8±2.6	53.3±6.9	11.9±0.8	6.8±0.3	0.34±0.17
Cucumber, Urgeiriça #21	U	0.04	2.9±0.3	0.12±0.07	1.6±0.2	22.5±2.4	17.7±1.7	16.4±2.7	18.4±5.4	3.7±1.1
Carrot, Urgeiriça #21	U	0.11	54.9±8.2	< 5.5	65.3±9.5	80.2±8.4	631±46	99.1±3.8	35.2±1.6	30.4±5.1
Lettuce, Urgeiriça #19	P3	0.13	654±24	26.6±3.7	680±25	481±41	1840±107	1770±98	1113±26	227±23
Tomato, Urgeiriça #19	P3	0.09	5.6±1.2	1.1±0.7	3.6±0.9	3.9±0.7	82.1±5.9	20.0±1.4	17±1	0.30±0.12
Corn leaf, Urgeiriça #20	P4	0.16	555±29	25.9±5.1	574±30	355±19	1768±140	970±42	177±6	65.4±5.4
Corn cob, Urgeiriça #20	P4	0.14	4.4±0.4	0.22±0.10	2.0±0.2	3.2±1.0	274±23	32.1±4.4	10.5±0.9	2.4±0.9
Pasture, Urgeiriça #20	P4	0.1	1000±26	44.0±3.4	998±26	807±40	944±52	466±21	199±7	163±10
Turnip greens, C. Felgueiras #23	P5	0.06	1270±41	58.6±5.2	1239±40	430±23	623±32	652.9±34.5	537±17	136±8

References

Carvalho FP, Madruga JM, Reis MC, Alves JG, Oliveira JM, Gouveia J, Silva L. 2005. Radioactive survey in former uranium mining areas in Portugal. In: Proceed. of an International Workshop on Environmental Contamination from Uranium Production Facilities and Remediation Measures, held in Lisbon 11-13 Feb 2004. International Atomic Energy Agency, Vienna, pp.29-40.

Carvalho FP, Oliveira JM, 2009. Performance of alpha spectrometry in the analysis of uranium isotopes in environmental and nuclear materials. J Radioanal Nucl Chem 281:591-596.

Carvalho FP, Oliveira JM, Faria I, 2009a. Alpha Emitting Radionuclides in Drainage from Quinta do Bispo and Cunha Baixa Uranium Mines (Portugal) and Associated Radiotoxicological Risk. Bull Environ Contam Toxicol 83:668-673.

Carvalho FP, Oliveira JM, Lopes I, Batista A. 2007. Radio nuclides from past uranium mining in rivers of Portugal. J Environ Radioact 98: 298-314.

Carvalho FP, Oliveira JM, Malta M, 2009b. Analyses of radionuclides in soil, water and agriculture products near the Urgeiriça uranium mine in Portugal. J Radioanal Nucl Chem 281:479-484.

Carvalho, F.P. (2014). The National Radioactivity Monitoring Program for the Regions of Uranium Mines and Uranium Legacy Sites in Portugal. Procedia Earth and Planetary Science 8: 33–37.

EDM, 2011. The Legacy of Abandoned Mines. Empresa de Desenvolvimento Mineiro, Lisboa, 2011. (ISBN: 978-972-95226-2-8)

IAEA, 2006. Release of sites from Regulatory Control on Termination of Practices. IAEA Safety Standards Series No. WS-G-5.1. AIEA, Vienna; 2006.

Marinho Falcão J, Carvalho FP, Leite MM, Alarcão M, Cordeiro E, Ribeiro J. MinUrar-Minas de Uranio e seus Resíduos. Efeitos na Saúde da População. Relatório Científico I (Julho de 2005), II (Fevereiro 2007). Publ. INSA, INETI, ITN. (available from http://www.itn.pt/docum/pt_bib_reltec.htm).

Nero JM, Dias JM, Torrinha AJ, Neves LJ, Torrinha JA., 2005. Environmental evaluation and remediation methodologies of abandoned radioactive mines in Portugal. In: Proceed. of an International Workshop on Environmental Contamination from Uranium Production Facilities and Remediation Measures, held in Lisbon 11-13 Feb 2004. International Atomic Energy Agency, Vienna, p.145-158.

Oliveira JM, Carvalho FP. 2006, A Sequential Extraction Procedure for Determination of Uranium, Thorium, Radium, Lead and Polonium Radionuclides by Alpha Spectrometry in Environmental Samples. In: Proceedings of the 15th Radiochemical Conference. Czechoslovak Journal of Physics 56(Suppl. D): 545-555.

Santiago Batista A., 2005. The programme for remediation of contaminated sites: its regulation and follow-up in Portugal. In: Proceed. of an International Workshop on Environmental Contamination from Uranium Production Facilities and Remediation Measures held in Lisbon 11-13 Feb 2004. International Atomic Energy Agency, Vienna, pp. 223-232.

Darling, you're amazing.
How did you find such a beautiful and still affordable place for our honeymoon?

The modern hydrochemical state of the Mailuu-Suuriver and radioecological problem of the Fergana Valley

Bekmamat Djenbaev[1], Umyt Karmisheva[1], Azamat Tilenbaev[1], Altinai Egemberdieva[2]

[1]Institute of Biology & Pedologyof National Academy Sciences of the Kyrgyz Republic, 265, Ave. Chui, 720071, Bishkek , Phone: 996 312 657943, Fax. 996 312 657943, E-mail: kg.bek.bm@bk.ru, bekmamat2002@mail.ru
[2]Jalalabad State University, Jalal-Abad, Street 57, Tel: 996 3722 55968

Abstract. In this region the water supply of inhabitants with drinking water is carried out from the rivers and canals originating from mountain rivers and including the river Mailuu-Suu. Therefore, the contamination of the river Mailuu-Suu and streams in the uranium province may further influence in the region on such large rivers as the Kara-Darya and Syr-Darya. Insufficient hydro isolating the bottom of the tailings could make pollution subsoil waters with radionuclide. In any case water supply from well of subsoil water may be contaminated with radioactive and other chemical elements.

Relevance

It is known that in the middle of the 20th century in the extraction of uranium ore requirements of radiation was safety hidden or not fulfilled, because the problem of increasing output capacity had priority to the detriment of the environment. Environmental protection has not been defined as a priority, and has not been identified relevant criteria safe operation. In 1943 Uranium Mining was launched in Kyrgyzstan and natural environmental issues not specifically addressed, they remained in the background.

After 1991, in Kyrgyzstan uranium production and waste left without the engineering and technical support to other independent countries in the region. Thus, the uranium industry was surprisingly opened to the world market from the beginning of the 90s in the region of Kyrgyzstan. A large number of mines in the region had stopped production, due to low profitability in the 70 years of the last century. Currently, some companies continue to pollute and tails neighborhoods.

That's way the Kyrgyzstan was the largest producer of uranium from 1946 to 1968 for the former of the Soviet Union. In our Republic the inefficient and wasteful mining minerals processing due to and the huge amount of minerals

B.J. Merkel and A. Arab (eds.), *Uranium – Past and Future Challenges*,
DOI 10.1007/978-3-319-11059-2_8

(747.22 million m^3) of waste with a high concentration of a number of potentially hazardous radionuclide's, chemical elements and their compounds was impounded in tailings and dumps. Also for storing uranium waste was imported from friendly countries, like Germany, Czech Republic, Slovakia, Bulgaria, China and Tajikistan. The condition of these dumps and storage are so abysmal that radioactive waste, heavy metals and toxic substances pollute the environment (soil, air, water) and living organisms. They are involved in biogeochemical cycles with the formation of secondary biogeochemical provinces (Aitmatov et al. 1997; Djenbaev 2009; Torgoev, Aleshin 2009).

In general, on the territory of Kyrgyzstan is a large number of radioactive sources (about 1200). Radioactive sources are stored in premises built by primitive methods (overlapping of mountain canyon). Moreover, many tailings are formed within the settlements (Mailuu-Suu, Min-Kush, Kadji -Sai, Ak- Tuz, Kan and others) in the mountain canyon and along the rivers (Aleshin et al. 2000; Djenbaev, Mursaliev 2012).

The technogenic uranium province of Mailuu-Suu is particular importance in the country and the region, because it has a cross-border value. In the zone of influence of the former enterprise of the Mailuu-Suu in Kyrgyzstan live 26,000 people, in Uzbekistan live till 2.4 million – in Tajikistan - about 0.7 million in Kazakhstan - till 0.9 million. Long Term contamination with radionuclide's will be subjected to extensive areas of Uzbekistan, Kazakhstan, Tajikistan and Kyrgyzstan; the most of them are in the zone of irrigated agriculture. Rivers and streams are contaminated, including such major rivers in the Fergana Valley, Kara Darya, Syr-Darya. Supply of drinking water is worth people take the water for drinking from rivers and canals. Even if the supply of ground water wells may be contaminated with radioactive and other chemical elements.

Materials and Methods

Sampling and analysis were performed according to conventional methods in ecology and biogeochemistry. The equipment which were used during the research is consist of a set - Sounder Model 798 ci HD, Dosimeter-radiometer DKS-96 radiometer RRA- 01M -01 with the sampling unit POU- 4 , λ- spectrometer (CAMBERRA), radiometer UMF -2000 and others.SAR measurement in water based on the use of the method of transfer of radon circulation with air sample volume of the working chamber of the PPA during bubbling. Work PPA based on the electric deposition of ionized decay products of radon in the measuring chamber to the surface of the semiconductor detector and the subsequent registration of alpha radiation Ra (^{218}Po).

To determine the radionuclide composition was used a gamma spectrometric method based on measuring the gamma radiation of the samples. Measurements were carried out on semiconductor detectors gamma spectrometer GX4019 software Genie- 2000-S 502, S501 RUS (Kuzin 1991; Radiation Safety Standards, RSS.-99; Radiation monitoring of drinking water, 2000).

Discussion of the results

Uranium deposit region of Mailuu-Suu were worked out from 1946 to 1967. Currently, the former enterprise are including in urban area, there are 23 tailings and 13 mountain damp. The total volume of uranium waste, pending the tailings is about 1.99 million m^3 and occupies 432 thousand m^2. Tailings were conserved in 1966-1973 years on existing standards. Dumps with a volume of 939.3 thousand m3 and footprint is not cultivated 114.7 thousand m^2 areas.

For a long time repair and maintenance of tailings were conducted sporadically and underfunded. At this time, the average exposure dose of gamma radiation (gamma background) on the surface of the tailings is 30-60 mR / h on local anomalous areas greater than 1000 mR / hr. However, according to scientists from the original breed of analysts produced 90-95 % uranium rock and tailings is only 5 to 10 % and therefore modern tails make a great background products of the uranium series.

Recently, the province sharply intensified landslides, mudflows, erosion phenomena on the slopes adjacent to the tailings, which may occur when an environmental catastrophe. It should be noted: the destruction of tailings will result in removal of the tail material not only in the valley of the river of the Mailuu-Suu, but in the densely populated Fergana valley, then - in the river basin of Syr-Darya.

Soil cover in the basin area of the lower reaches of the river and the middle reaches - dark gray soil, then begin in the foothills of the mountain- brown soil. General characteristics of the soil as follows: pH = 8.2 - 8.8 ; nitrate - 13.2 - 25 mg / kg dry matter ; chloride - 25 - 47 mg / kg; sulfate - 240 -895 mg / kg oil - 18 - 128 mg / kg dry substance. A more detailed study showed that not all parameters are normal, especially the level of trace elements. The results of the research indicate a relatively low level of contamination of soil trace elements in relation to the background and the MPC. Found a slight increase in concentration: Al, Mn, Se and U (2 - 3 times) in the autumn and spring , and Zn to 6 times compared with MPC background U in the sub region more 10 times ($P <0.01$) (Radiation Safety Standards,RSS.-99; IAEA, Quantification of radionuclide transfers in terrestrial and freshwater environments for radiological assessments 2009; ZSR, 2007).

We conducted a more detailed analysis of the water and in general studies have shown that the water of the river of the Mailuu-Suu is not potable. In some areas the river water found the highest concentrations of certain trace elements, for example, Se exceeding MPC 23-fold (Table 1). Fe concentration exceeds the MCL in 6 times or more, especially in the 2 and 5 points. Al Level 2 fold increase is compared with the MPC and the MPC in one sampling point of water, Hg at all points on average 2 -fold increased, Mn 2,4 and 5 points in 2-fold elevated . Elevated concentrations of total uranium mounted on section 3 compared to the other portions to 5 times apparently encounter at the site uranium rocks. Cadmium in all areas increased 2-fold compared with the MPC.

Table 1. The Microelement composition of the water in the river of the Mailuu-Suu (average years mg/L)

№	Elements	MPC	Sampling points(along the river) and the mean values					
			1	2	3	4	5	Σ
1.	Al	0,5	0,55±0,09	1,076±0,15	0,94±0,031	1,026±0,13	1,086±0,13	0,935±0,22
2.	Ba	4,0	0,068±0,01	0,009±0,001	0,024±0,003	0,088±0,012	0,102±0,012	0,074±0,068
3.	Co	1,0	0,005±0,002	0,0073±0,001	0,005±0,001	0,006±0,001	0,005±0,001	0,005±0,001
4.	Cu	1,0	0,004±0,001	0,007±0,002	0,005±0,001	0,008±0,002	0,008±0,001	0,006±0,001
5.	Fe	0,5	0,248±0,025	0,46±0,062	0,34±0,025	2,54±0,42	3,209±0,54	2,601±1,01
6.	Hg	0,005	0,01±0,003	0,01±0,001	0,01±0,002	0,01±0,002	0,01±0,002	0,01±0,001
7.	Mn	0,1	0,07±0,012	0,225±0,013	0,081±0,012	0,181±0,032	0,192±0,016	0,101±0,03
8.	Co	0,5	0,005±0,001	0,004±0,001	0,010±0,002	0,003±0,001	0,010±0,002	0,006±0,004
9.	Ni	0,1	0,026±0,006	0,032±0,001	0,025±0,003	0,028±0,004	0,025±0,004	0,22±0,14
10	Pb	0,1	0,02±0,001	0,035±0,001	0,02±0,003	0,02±0,003	0,02±0,003	0,023±0,002
11	Se	0,001	0,023±0,005	0,02±0,003	0,023±0,02	0,02±0,004	0,023±0,005	0,021±0,001
12	V	0,1	0,007±0,001	0,011±0,002	0,006±0,001	0,008±0,001	0,006±0,001	0,007±0,001
13	Zn	1,0	0,005±0,001	0,011±0,002	0,003±0,001	0,142±0,023	0,077±0,011	0,047±0,023
14	U	0,037	0,004±0,001	0,04±0,002	0,19±0,021	0,04±0,005	0,04±0,005	0,04±0,01
15	Cd	0,001	0,002±0,001	0,002±0,000	0,002±0,001	0,002±0,000	0,002±0,000	0,002±0,0002

The project «Prevention of emergency situations»had been realized in this province since 2007 years and it was funded by the World Bank. The project provides for the identification and prevention of the most significant risks from radioactive tailings in Mailuu-Suu hazards of natural origin (landslides) and improvement of emergency management . The Work is hold by VISUTEK (Germany). Previously visited the area several times by various expert missions of the IAEA, the World Bank, ADB, the Russian Federation and other international organizations.

Currently tailing number 3, which is more dangerous in this technological transferred to another province less safe place - tailing number 16, but it is located above and may spread pollution on even large areas (if the tailings are destroyed).

The results of the analysis of water from tailings and two ground rivers in this man-made uranium province were shown in the Table 2. Sufficiently high levels of accumulation of the total uranium is installed in the water 16 from the tailings number - 37050 pg/l, as compared with water from tailing at the number 3 is more than 18 times, and in the river Kulmen-Sai is compared with Kara-Zhygach up to 5 times. It should be noted that depending on the season, the water from the tailings do not always reach the river of the Mailuu-Suu.

Conclusion

In general, the soil cover of the floodplain of the Mailuu-Suu (in Kyrgyzstan), according to our data obtained is satisfactory. Specific changes in the level of the studied elements in the soil cover data were found. Naturally, the soil cover in the area of the tailings is not suitable for agricultural purposes; require specific guidelines for the region's residents. Currently, the safe storage of uranium waste in Mailuu-Suu has the following problems: disposal sites located at a distance less than 200 meters from residential city limits, waste stockpiled near river of the Mailuu-Suu . To reduce radon load to an acceptable level of the sanitary protection zone in the city should be more than 3 km. Tailings dams require continuous monitoring in the event of catastrophic floods and mud streams, as well as insufficient waterproofing bottom tailings could contaminate groundwater with radionuclide's and surface water.

Need to pay special attention and conduct regular monitoring of water pollution in the river of the Mailuu-Suu toxic elements such as - U, Se, Hg, As, Cd, Al, etc.

In these province could be formation of uranium dammed lakes and catastrophic floods. In the area of flooding may be tailing located along the river of the Mailuu-Suu , as well as residential buildings and other city and then in other countries, therefore, it need to constantly carry out geomorphological observations in this province.

Table 2. The main sources of pollution of the river of the Mailuu-Suu

№	Sources of pollution number	U [μq/L]	Discharge [L/min]
1	Seepage water from the tailings № 3	1742	Up to 0.54
2	Seepage water from the tailings № 5	8420	Up to 9
3	Seepage water from the tailings №16	37050	Up to 0.12
4	Water of the river Kara-Zhygach	38	Up to 500
5	Water of the River KulmenSai	182	Up to 100

References

Aitmatov I, Torgoev I, Aleshin Y (1999) Geoenvironmental problems in the mining industry of Kyrgyzstan. J Science and New Technologies 3: 129-137.

Aleshin Y, Torgoev I, Losev V (2000) Radiation Ecology Mailuu-Suu: 96.

Djenbaev B (2009) Geochemical ecology of terrestrial organisms: 240.

Djenbaev B, Mursaliev A (2012) Biogeochemistry of natural and manmade ecosystems Kyrgyzstan: 404.

Kuzin A.M. (1991) Natural background radiation and its implications for the Earth's biosphere: 117.

Radiation Safety Standards (RSS.-99). SP 2.6.1.758-99: 64-68.

Radiation monitoring of drinking water (2000). Guidelines. Approved by the Ministry of Health of the Russian Federation 11-2/42-09: 45-50.

Torgoev I, Aleshin Y (2009) Geoecology and waste mining complex in Kyrgyzstan: 240.

Radiation Protection Against Radon in Workplaces other than Mines (2003). Safety Report Series. IAEA, Vienna: 75-84.

International atomic energy agency, Quantification of radionuclide transfers in terrestrial and fresh water environments for radiological assessments (2009), IAEA-TECDOC-1616, Vienna: 51.

Radioecological analyses concerning the BGR project "Reduction of hazards posed by Uranium mining tailings in Mailuu-Suu, Kyrgyz Republic". Midter-Report: 111-120.

Social Licensing in Uranium Mining: Empowering Stakeholders through Information

W. Eberhard Falck[1], Joachim H. Spangenberg[2], Dominic Wittmer[3]

[1]CEARC-OVSQ, Université de Versailles St. Quentin-en-Yvelines, France
[2]Sustainable Europe Research Institute SERI, Cologne, Germany
[3]Wuppertal Institute for Environment, Climate and Energy, Germany / MINPOL - Agency for International Minerals Policy, Austria

Abstract. Modern uranium mining practices are designed to bring operations more in line with sustainable development goals, in particular with respect to environmental and socio-economic criteria. It is now widely recognised that a social license to operate is a key prerequisite for successfully establishing and running any mining project. This paper examines the use of indicators as a vehicle to convey stakeholder-relevant information about mine site development with a view to facilitate social licensing processes.

Introduction

While mining is indispensable for modern industrial economies, it comes at a price: negative environmental and social impacts can be minimised, but not entirely avoided. People living in the vicinity of mine sites are affected in a variety of ways, positively as well as negatively, and, therefore, have a natural and vested interest in the sustainable development of their regions. This is even more the case in uranium mining and milling, where there is a risk that it may stigmatise the whole region because of the massive and often unreflected fear of radioactivity. In the past, due to social and environmental indifference, mining companies have frequently exhibited a nonchalant attitude towards such impacts. Historically, these indifferences often reflect the overriding priorities of national security, when mining for peaceful uses and for nuclear weapons were not at all or not clearly separated. This is evidenced by the large number of legacy sites resulting from past uranium mining operations (IAEA 1997). Uranium mining continues to be accompanied by significant on-going and emerging social conflicts. Obtaining the support for or at least the acceptance of a mine, "the social license to operate", as Thomson and Boutilier (2001) call it, increasingly faces difficulties anywhere in the world. Obtaining the license involves not only technical measures to reduce and minimise environmental and societal impacts, thus improving the environ-

B.J. Merkel and A. Arab (eds.), *Uranium – Past and Future Challenges*,
DOI 10.1007/978-3-319-11059-2_9

mental, societal and economic performance of mining operations, but also to "enhance the participation of stakeholders [...] to play an active role in minerals, metals and mining development throughout the life cycles of mining operations, including after closure for rehabilitation purposes" (UN 2002).

Resistance against mining projects is often grounded in a lack of knowledge and/or reliable information resulting in distrust between key stakeholders. The social license is an expression of trust which in this case is based on shared knowledge. Therefore, a key issue is to make information on mine development and mine legacy management, including their socio-economic and environmental implications, accessible to all stakeholders. As trust legitimises - in a social sense - decision finding processes, it is also the basis for good governance (Braithwaite and Levi 1998). A distrust between the different stakeholders often has its origin in an unequal distribution of knowledge and/or inadequate knowledge transfer. Thus, a lack of specific knowledge is likely to seriously impede a meaningful interaction between stakeholders. Building and re-building of trust requires a shared information base to establish a rather level playing field with respect to site-specific knowledge. Some stakeholders may have a reasonably good understanding of mining in general, but lack specific knowledge about the mine operation in their vicinity; public authorities can belong to each of both camps, the fully informed or the information deficit one. Information may not be considered very trustworthy, if supplied only by the mine operators. Stakeholders will require independent and unbiased information to enter into a trustful and constructive dialogue with mine operators, and the opportunity to double-check and verify information provided by them, but they are often perplexed by the complexity of issues. Conversely, mine operators may be more willing to enter into a dialogue with knowledgeable and informed stakeholders. Compromises between representatives of different value systems can only be reached – if at all – if mutual understanding allows for rational discourses between informed discussants.

The European Commission's FP7 project EO-MINERS (2009-2013, www-eo-miners.eu) aimed *inter alia* to devise means to provide mine stakeholders with independent information on environmental and socio-economic developments at actual or projected mine sites. This paper discusses the applicability of these means and strategies to uranium mining projects.

The stakeholders in uranium mining projects

A stakeholder is any person or group who claims to have an interest in a project (IAEA 2006). The interest may be vested or not, this is not relevant for the social processes that lead to social licensing or otherwise. In the first instance it is the local-level stakeholders who are concerned directly. However, individuals or groups at national or supra-national level, who are concerned by the overall development in the uranium mining industry, can also become active stakeholders. The (political and ideological) issue of acceptability of nuclear energy systems and nuclear weapons plays a crucial role. Stakeholders can be grouped in various ways, for instance according to the interests they are likely to represent (Azapagic 2004) or whether they are internal or external to the project (Fig. 1). It is a particularity of uranium mining that it attracts significant attention of extended external stakeholders at the national and international level. This is due also to the fact that the mining companies, operate on a transnational basis within the nuclear fuel supply chain. Comprehensive social licensing can only be achieved in a multi-lateral dialogue between these groups, respecting their values and concerns.

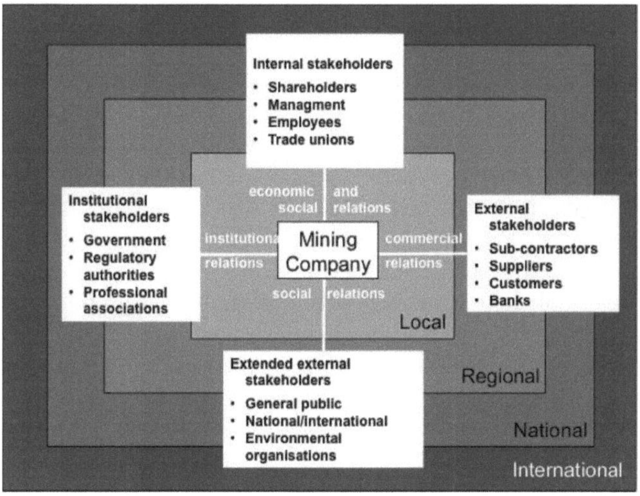

Fig.1: The stakeholders in uranium mining projects (modified after Chamaret 2007).

Framing the processes for social licensing

The framing of processes intended to lead towards social licensing of mine operations takes place within a pre-existing fabric of social relations and knowledge bases, which are shaping and shaped by the social processes themselves. The articulation of social demands often meets with difficulties due to limitations in the knowledge and understanding of stakeholders. Enhancing the stakeholders' knowledge base, therefore, appears to be a crucial task on the route to empowering stakeholders to actively take part in defining the social license.

Given the complexity of such processes and the resulting lack of transparency, applying tools for complexity reduction is recommendable, or even necessary for confidence building. The scientific information on the complex fabric of variables describing the natural, engineering and socio-economic systems of a mine site does not speak for itself; it requires a process of reducing the complexity in order to increase the comprehensibility. Based on the supposition that non-technical stakeholders typically lack the capability to properly interpret and contextualise raw technical data, i.e. giving a meaning to the data, such simplification is indispensible. It is supposed that a simplification minimising information loss would be possible by providing indicators covering issues identified as important both, from a scientific point of view, and based on stakeholder concerns.

Indicators provide a metric of the state of (complex) systems or of issues, or for trends of their development, if measurements are repeated over time. The term 'monitoring' implies repeated, often regular, measuring. Therefore, it addresses change, rather than observing a status quo. Indicators must be based on measurable or at least observable quantities in order to be useful. While indicators are useful tools to reduce a complex set of diverse data, it should be kept in mind that every process of indicator selection or aggregation inevitably entails a loss of information. Indicator sets must balance the needs for reducing complexity, being easily understandable in the prevailing socio-cultural context, resonating with a clearly defined target audience, and being limited in number (Rechatin and Theys, 1997; Spangenberg, in press). The necessary key qualities of indicators result from their purpose. They ought to enable a comparative analysis of the state and of progress and may need to be and location-specific for local stakeholders.

Processes, tools and, by implication, indicators that do not resonate with the stakeholders' need for information will not be considered useful and consequently not be used by them. Therefore, an important element of indicator development is the elucidation of stakeholder needs (Wittmer et al. 2013, 2014). Whether stakeholders consider a given indicator as useful is likely a reflection of their membership in particular groups (cf. Fig. 1). Whether stakeholders are local, regional, national, or international will also have bearing on their indicator preference. Consequently, the preference for the aggregation level built into the indicators varies. Local level actors are more likely to be interested in corporate (sustainability) reporting type indicators that reflect the local environmental and social impact. Regional or national stakeholder are expected to be interested in questions of compliance with regulations and standards. National and international actors are usually strongly interested in indicators contributing to national economic, sustainability and environmental performance indices. Indicators should not only reflect the needs of stakeholders, but also must be reproducible, and based on scientifically sound methods delivering measurable/observable quantities. Hence, once proposed based on social demand, the candidate indicators have to be assessed by experts for their quantifiability/measurability (Falck and Spangenberg, 2014).

Populating the indicators

Obtaining trustworthy, independent and unbiased information is a (time-)costly procedure and in the vast majority of cases it will be beyond the means of single stakeholders or even interested groups. Although for many environmental and socio-economic variables suitable measurement techniques can be found, it is quite possible that for a given case a technique is not available for technical, financial, or legal reasons. This can means that only a subset of the indicators desired by the stakeholders is deployable, which usually triggers another round of stakeholder consultation. To cater for these potential obstacles, the development of operational indicators should be *ex ante* conceptualised as an iterative procedure between stakeholders' expectations and deployability. A transparent selection and communicative feedback are necessary throughout this process in order to preserve and confirm the trust built up in the initial phase of indicator development.

Many stakeholders do not have the capability, nor the time, to turn data into meaningful information and knowledge through a process of contextualisation and interpretation. Conversely, it is the stakeholders who command a key resource for this process: contextual local knowledge, which external experts, but also international company management and decision makers from higher political levels are often lacking (Spangenberg, 2005).

Modes and media of communication within and between different groups of stakeholders are framed by the cultural and institutional context. Most of the suggested and technically/economically feasible indicators typically refer to spatially distributed information. The classical medium for communicating such information is thematic maps, usually a means of communication easily understood and welcome by stakeholders. This includes printed maps, which with today's printing technology can be produced almost *ad hoc* in high quality. Since much information may be either already available in digital format, or can be transformed into it, the use of Geographic Information Systems (GISs) is a natural choice for its management. GISs allow interacting with the information: once the data have been collected and processed, precisely the information the user is interested in can be extracted in real time and be presented. Different indicators can be related to each other, meaning that different thematic layers can be superimposed onto each other to form a single map, showing e.g. areas that are affected by more than one issue. In the same way cause-effect relationships that are represented by different types of indicators can be illustrated.

Experience gained within the EO-MINERS Project

The project EO-MINERS faced the challenge of soliciting stakeholder input into the development of techniques for monitoring the environmental and socio-economic development at three mine sites: the Sokolov lignite mining area in the northern Czech Republic, the eMahaleni (Witbank) coal mining area near Johannesburg in South Africa, and the Makmal gold mine near Kazarman in central Kyrgyzstan. To this end discussion fora were organised, where the stakeholders voiced their views over site and technology developments and (re-)formulated demands for information and chose indicators. Often, however, broad dialogue processes were hindered due to pre-existing and continuing social tensions between specific stakeholders. In such cases social demand-based requirements were elucidated in smaller focus groups or interviews with individual stakeholders (Wittmer et al. 2013). These events took place in stages, whereby initially a preliminary set of indicators was developed, and later the indicators and their information content, together with the supporting measuring techniques were presented to various stakeholder communities for scrutiny and commenting. Particularly in later meetings information about technologies was conveyed to audiences that comprised technical stakeholders as well as stakeholders with little or no technical background. Based on these workshops the way how the messages were conveyed was successively refined. This proved to be a challenging learning process particularly for the 'technical' project participants, who had to learn how to frame technical and scientific content in a way also accessible to non-technical audiences.

These stakeholder meetings corroborated that the level of technical knowledge and understanding can vary considerably, even within nominally 'technical' stakeholder groups (Falck and Spangenberg 2014; Spangenberg 2005). Neither stakeholders involved in regulatory, nor those involved in planning activities can be assumed to have a specific mining-related understanding. Conversely, one may encounter very knowledgeable private individuals. It will be only at the actual stakeholder engagement step that it becomes clear, which technical level of the engagement tools can satisfy the specific stakeholder information needs.

Confronted with the project results, many stakeholders immediately projected these to their own personal and/or professional situation. They also critically assessed, for instance, results of environmental analyses based upon to their own personal experience. In various instances, stakeholders clearly had the better local knowledge and insights than the project 'experts', not the least in identifying potential sources of observed issues (Wittmer et al 2014). This observation reinforces the need for iterative processes in stakeholder dialogues and for a stakeholder-need driven process of tool and service development.

It was noted that these dialogues, albeit being part of a research project, created problem solving expectations among stakeholders. Managing them to avoid disappointment requires the formulation and implementation of an 'exit strategy', in a real technology transfer situation even more than in a project.

Conclusions

Above, the proposition was put forward that mining and mining technology relevant issues can best be framed for a stakeholder deliberation process in the form of indicators. It was demonstrated at three test sites in the Czech Republic, Kyrgyzstan and South Africa respectively that indicator development is not an engineering, but a social process. Within EO-MINERS a set of candidate indicators in 11 categories was developed that is likely to cover most, if not all, issues of potential interest in a mining context (Falck and Spangenberg 2014; Table 1).

Table 1: Categories of indicators developed in EO-MINERS (Falck and Spangenberg 2014).

	Category		Category
A.	Land-use	B.	Mass and energy flows
C.	Air quality and other nuisances	D.	Soil quality
E.	Water quality	F.	Transport
G.	Geotechnical hazards and accidents	H.	Industrial and other accidents
I.	Social impact	J.	Regional development
K.	Economic vulnerability/resilience		

These categories and the 59 candidate indicators developed in EO-MINERS appear to capture most aspects that are critical in uranium mining project development. The effectiveness of the chosen iterative procedure for indicator development was demonstrated by the fact that only limited amendments were made to the original indicator set conceived by the technical experts. The actual set of indicators for a specific mine site would need to be developed with the relevant stakeholders in a deliberative process to be useful in this particular context. It would be conceivable, for instance, to have more detailed indicators pertaining to radiological contamination in environmental compartments or agricultural products, to long-term low-level exposure, or indicators that may reflect the socio-economic impact of the 'stigma' of a uranium mine in the region. Such sets of indicators covering a broad spectrum of issues would help to frame the societal dialogue during the planning as well as the implementation phase of a uranium mining project. It should be noted also that, as the deliberative process evolves, the set of indicators is likely to evolve, reflecting the heuristics of the stakeholder concerned.

Experience with stakeholder consultation gained from the EO-MINERS project shows that there is a considerable diversity in the level of technical understanding. Thus a prerequisite for stakeholder dialogue is stakeholder training and capacity building in order to enable them to enter into any meaningful dialogue.

Acknowledgements

The authors wish to acknowledge the various contributions and fruitful discussions with the whole EO-MINERS project team. The project was supported by the European Commission within its 7th Framework Programme, FP7-ENV-2009-1, Grant Agreement No. 244242.

References

Azapagic A (2004) Developing a framework for sustainable development indicators for the mining and minerals industry. J Cleaner Prod 12: 639–662

Braithwaite V A, Levi M (1998) Trust & Governance. The Russell Sage Foundation Series on Trust, 1: 386 p., New York.

Chamaret A (2007): Une démarche Top-Down/Bottom-Up pour l.'évaluation en termes multicritères et multi-acteurs des projets miniers dans l.'optique du développement durable.- Doctoral thesis, Université de Versailles-Saint Quentin-en-Yvelines, http://halshs.archives-ouvertes.fr/docs/00/19/44/45/PDF/These_A_Chamaret.pdf

Falck W E, Spangenberg J H (2014) Bases for Social Demand-Based Decision Finding in Mine Development. J Cleaner Prod, http://www.sciencedirect.com/science/article/pii/S0959652614001656.

International Atomic Energy Agency IAEA (1997) Planning for Environmental Restoration of Uranium Mining and Milling Sites in Central and Eastern Europe. IAEA-TECDOC-982, IAEA, Vienna.

International Atomic Energy Agency IAEA (2006): Management of Long-Term Radiological Liabilities: Stewardship Challenges, IAEA-TRS-450, IAEA, Vienna.

International Institute for Environment and Development IIED (2002) Breaking New Ground: Mining, Minerals and Sustainable Development. 410 p., London, IIED. http://www.iied.org/mmsd-final-report (last access 18.02.14).

Rechatin C, Theys J. (1997) Indicateurs de développement durable: Bilan des travaux étrangeres et éléments de réflexion. Coll. notes de méthode no. 8: 72 p., Orleans.

Spangenberg J H (2005) Will the information society be sustainable? Towards criteria and indicators for a sustainable knowledge society. Intern. J Innov Sust Dev 1(1/2): 85-102.

Spangenberg J H. (in press). Indicators for sustainable development. In: Redclift M, Springett D (eds.) Routledge International Handbook of Sustainable Development. Abingdon, Oxford, UK, Routledge.

Thomson I, Boutilier R G (2011) Social license to operate. In: Darling P (ed.) SME Mining Engineering Handbook, pp. 1779-1796, Littleton, CO, Society for Mining, Metallurgy and Exploration.

United Nations UN (2002) Report of the World Summit on Sustainable Development, Johannesburg, South Africa, 26 August-4 September 2002. United Nations publication A/CONF.199/20, New York.

Wittmer D, Hanise B, Coetzee, H (2013) Trialogue Workshop in South Africa: Summary and Results. CEC FP7 Project EO-MINERS, Deliverable D5.6 SA.

Wittmer D, Schepelmann P, Coetzee H (2014) Resonance Analysis of Selected Earth Observation Specifications. CEC FP7 Project EO-MINERS, Deliverable 1.6.

Social licensing and Stakeholder Communication in Uranium Exploration and Mining

W. Eberhard Falck[1], Julian Hilton[2], Henry Schnell[3], Harikrishnan Tulsidas[4]

[1]CEARC-OVSQ, Université de Versailles St. Quentin-en-Yvelines, France
[2]Aleff-Group, London, UK
[3]HA Schnell Consulting Inc., Eagle Bay BC, Canada
[4]International Atomic Energy Agency, Vienna, Austria

Abstract. Recognising how fundamental the 'social licence to operate' is to any uranium mining project, in 2012 the IAEA initiated a series of training courses and workshops on stakeholder engagement and social licensing issues. While the social licence has different features in each country, and is negotiated in different ways, without one, a project's chances of success are significantly compromised. This paper summarises the findings from a workshop held in Istanbul, Turkey, in February 2014, and offers some perspectives on likely future developments.

Introduction

Uranium production is currently taking place in some 15 countries and more than 40 further countries are engaged in exploration projects at various stages of development. Uranium is the main raw material for nuclear energy systems that will remain an essential constituent of the world energy matrix. But uranium mining and processing is widely misunderstood and feared. These concerns must be addressed, and the mineral itself must be mined and processed in an environmentally safe and socially responsible manner if its use as a fuel source for nuclear power plants is to remain assured.

The picture the public has of mining in general and uranium mining in particular has been shaped by the past. Mining companies have been at best nonchalant and at worst grossly negligent in their attitudes towards the societies and the environments in which they have operated. Historically these behaviours were often grounded in the overriding priorities of national security, as mining for peaceful uses and for nuclear weapons were not at all or not clearly separated. This resulted in large number of abandoned and costly legacy sites (IAEA 1997) the public and environmental health impact of which remains severe, and the remediation of which continues to require considerable (public) resources. The social discourse about uranium mining is also embedded in the wider social discourse about nucle-

B.J. Merkel and A. Arab (eds.), *Uranium – Past and Future Challenges*,
DOI 10.1007/978-3-319-11059-2_10

ar energy use and radioactive waste management. Hence, uranium mining continues to be accompanied by significant on-going and emerging conflicts.

Obtaining support for, or at least acceptance of, any uranium mining operation, "the social license to operate" (Thomson and Boutilier 2001), is increasingly difficult anywhere in the world. A prerequisite for this social licence are not only operations that strive to reduce and minimise environmental impacts, but that also aim to "enhance the participation of stakeholders [...] to play an active role in minerals, metals and mining development throughout the life cycles of mining operations, including after closure for rehabilitation purposes" (UN 2002). Fortunately, in the past ten years the mining industry itself has sought to address the issue from within. Starting with the seminal report 'Breaking New Ground' (2002) and now central to many mining company mission statements, the 'social licence to operate' has become the *sine qua non* of responsible mining. Leading consulting firms now include social licensing in their assessment of key business risks, citing the need for a 'community investment dividend' in any successful project. In parallel, the Extractive Industries Transparency Initiative (EITI, http://eiti.org/) has successfully put in place measures for reporting the extent to which natural resources are managed in a fair, sustainable and socially inclusive manner.

Framing the Social Processes

Obtaining a social license first of all requires that the mine operators engage with the different stakeholders, be they local, regional, national, or even international. It has been noted that a stakeholder is any person or group who claims to have an interest in a project (IAEA 2006). It is important to be 'inclusive' in order to establish trust between the various stakeholders, though this can be challenging. At the level of individual projects the social license is rooted in the beliefs, perceptions and opinions held by the local stakeholders about a project. As trust legitimises the various decision finding processes, it is also the basis for good governance (Braithwaite and Levi 1998). It takes both time and effort to create trust. True trust comes from shared experiences. The challenge for the operator is to go beyond transactions with the community and create opportunities to collaborate, work together and generate the shared experiences within which trust can grow.

Social legitimacy is based on the established norms the community, that may be legal, social and cultural and both formal and informal in nature. Companies must understand the norms of the community and be able to work with them as they represent the local 'rules of the game'. Failure to do so risks rejection. In practice, the initial basis for social legitimacy comes from engagement with all members of the community and providing information on the project, the company, and what may happen in the future, and then answering any and all questions.

The capacity to be credible is largely created by consistently providing true and clear information and by complying with any and all commitments made to the community. Credibility is often best established and maintained through the application of formal agreements where the rules, roles and responsibilities of the company and the community are negotiated, defined and consolidated. Such a framework helps manage expectations and reduces the risk losing credibility, for example from real or perceived breaches of promise.

The social licence is granted when a project has ongoing approval from local and other stakeholders. It has tangible and intangible value: effort must be made, case by case, to measure and respond to stakeholder beliefs, opinions and perceptions. It is dynamic and non-permanent. Beliefs, opinions and perceptions evolve, hence the social licence has to be both earned and maintained.

IAEA Activities in Social Licensing for Uranium Mining

The IAEA rationale and vision for the peaceful uses of nuclear energy defines that: "Any use of nuclear energy should be beneficial, responsible and sustainable, with due regard to the protection of people and the environment, non-proliferation, and security" (IAEA 2008). The basic principles on which nuclear energy systems should rest to help meet growing global energy needs include uses that are: (1) beneficial; (2) responsible and (3) sustainable. IAEA further has defined the criteria necessary to satisfy these basic principles when applied to the use of uranium resources. In particular, the criterion of transparency, is an essential precondition for meeting the requirement of beneficial use. Transparency is achieved by ensuring that: *"Information on natural uranium technologies, good practices across the uranium production cycle, and on the associated risks and benefits is distributed and discussed, engaging stakeholders and the general public"* (IAEA 2013).

Recognising the importance of social licensing of uranium mining projects, in 2012 the IAEA initiated a range of training and information exchange activities that address the direct 'technical' stakeholders, namely mine operators as well as regulators and their technical agencies. IAEA believed it was essential to all future uranium mining and processing projects that these direct stakeholders understand the importance of social licensing and subscribe to the principles of transparent governance. The workshop held in Istanbul, Turkey, February 2014 provided an opportunity to review how successful the IAEA initiative has been and to what extent the tenets of social licensing had been absorbed in the uranium mining culture. The event brought together more than 60 participants from 34 countries, representing a broad cross-section of cultures and socio-economic conditions. The authors of this paper acted as 'facilitators', seeding the discussions.

The main conclusion of the workshop was that the concept of a 'social licence', however understood, is now fundamental to mining projects around the world. Integral to the achievement and maintenance of a social licence is the generation of social capital. In some communities, creating social capital may focus on infrastructure, roads, schools, hospitals, and communications. In others, the emphasis is on technology and knowledge transfer, through training and capacity-building. In others, such as nomadic communities, the key objective for the community concerned is to preserve and enhance its traditional way of life. Common to all cultures is the wish to find a new, sustainable point of equilibrium of interest and benefit between shareholders and stakeholders that has as its outcome a true sense of partnership and complementary values. The extent to which this has now been absorbed into the business model of responsible mining may be measured by the central place given to the 'community investment dividend' by top 5 financial advisors as the key to sustaining the social licence (KPMG, 2013).

The Workshop identified a number of contributory factors in achieving a sustainable equilibrium and community dividend:

- Introduction and transparent explanation of mining proposals to indigenous communities from the moment the first exploration geologist is on the ground;
- Building an open, transparent and locally grounded narrative for explaining and justifying sustainable uranium exploration, mining and processing;
- Realistic management of expectations: mining companies have a right and an obligation to return profit to their investors, and not to be held hostage by communities demanding too much for the social licence; communities have the right to share in the benefits of such projects throughout the life-cycle of the mine, including its eventual closure and remediation;
- Periodic social and environmental impact assessments as tools to address the impact and mitigation of social and environmental issues. Of these the issues of tailings and waste management are among the most significant and arriving at 'zero waste' is a guiding principle;
- Comprehensive communications and education strategies to provide the community with sufficient understanding of the issues associated with uranium mining to enable informed decision-making and empowered consent;
- Partnering and corporate social responsibility programmes, including initiatives in local employment and sourcing, supply-chain transparency, and stimulus to local small and medium enterprises (SMEs);
- Participation in the Extractive Industries Transparency Initiative (EITI);
- Communicating health, safety and environmental (HSE) aspects of uranium mining and processing within a culture of safety; and
- Sharing of good practices, experiences and approaches for in-depth understanding of social licensing and stakeholder communications.

Uranium Mining in a Sustainable Development Context

In order to meet the criteria for sustainable development (UN 2002), uranium mining projects must meet the Triple Bottom Line (TBL) requirements. The TBL requirements were proposed by Elkington (1994) as a direct response to the sustainability agenda (Brundtland 1987). In order for an undertaking to be sustainable it must not only meet economic and financial criteria, but also social and environmental ones. The philosophy behind this is to create a win-win (‚cooperative game') situation (Nash 1950) for all three areas. It also requires a full life-cycle approach to nuclear energy systems (Fig. 1), defined by seven key stages.

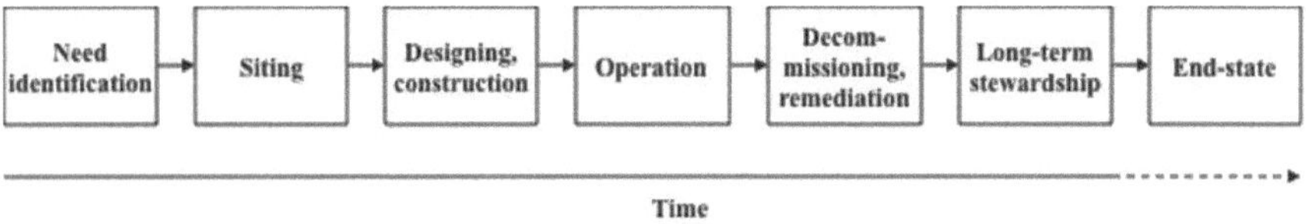

Fig.1. The life-cycle management approach to nuclear energy systems (IAEA, 2006).

The social narrative comprises three phases, beginning, middle and end:

- Investment (beginning): what are we trying to do?
- Local needs and culture (middle): what are we doing? Are we doing what we said we'd do? Better or worse?
- Outcomes (during and after, the end): what have we done? Did we do what we said we'd do? What have we left behind, what are the stewardship needs?

Investment should result in increased, self-sustaining social capital, based on capacity building, infrastructure development and long-term community/operator partnership. Success may be manifest in such outcomes as technology transfer and technology spill-over. The investment must also result in internationally recognized health and safety standards. Equitable distribution of benefits between community and operator short and longer term should reflect evolving stakeholder needs and cultures. As mines operate in a national context, the developments have to be aligned to national and local sustainable development needs and policies. The primary defensive objective is to avoid the generation of 'resource curse', i.e. the unilateral dependency on the exploitation of single natural resources. Such exploitation should not result in debts nor 'orphan' legacies.

Social Interactions and Narratives in Uranium Mining Projects

Arnstein (1969) noted that public participation without divesting (some of) the power is meaningless, "... that participation without redistribution of power is an empty and frustrating process for the powerless". However, it has puzzled social scientists and decision-makers ever since, how to actually bring about participation. For a variety of reasons, including frustration over not being able to influence or for ideological reasons, stakeholders may absolutely refuse to engage in any dialogue. The underlying reasons may be knowledge gaps (refusal to accept new/

superior knowledge or passive aggression) and ignorance (unconscious lack of knowledge). Therefore, capacity building and otherwise reducing the knowledge gaps (Falck et al., this volume) can be ways to facilitate a constructive and equitable dialogue. A challenge for all stakeholders is to 'step in the counterpart's shoes' and to try to think as the counterpart. Thus capacity building must not only address public stakeholders (and perhaps authorities), but also the industry. As noted above, trust between the various stakeholders is a prerequisite for a social licence.

In turn, the basis for building trust is a consistent and sustained narrative put forward by the initiators of a uranium mining project. The engagement narrative will begin at the very first stages of any project, exploration and feasibility studies. As soon as a mining company sets foot in a region, questions, and expectations, will be raised. Stakeholder messages have to be timely, consistent between different actors (i.e. the mining company and government authorities), and realistic. The role of the government can be ambivalent. It may promote mining with economic objectives in mind (in Malawi for example a single mine increased national GDP by 10%); but it also has the duty to manage the expectations of the community and to protect stakeholders from adverse environmental and socio-economic impacts. The open dialogue referred to should make the role and position of the government transparent. The government may also need to mediate between different development and land-use interests. The role and responsibilities of the various 'institutional' stakeholders has to be made clear, as even government authorities may have a vested interest in uranium mining projects.

It is important that thought is given to a full life-cycle management plan from the outset (Fig. 1). This includes provision (planning, financial, etc.) for the long-term stewardship of mining residues. Stakeholders have to be part of the legacy management (IAEA 2006) and they need to be part of the 'vision' for the development of the uranium-mining region. While they are being informed through the narrative about the vision of the operators and the government, they will also be given the opportunity to shape the narrative in the sense of the Triple Bottom Line that goes well beyond the life-time of the actual mining operation. Stakeholders, particularly those interested in economic development, also have to understand that the mining operation can only go ahead if the investors are satisfied that techno-economic risks are sufficiently low. Major non-technical risks include a) whether or not a social licence has been obtained, and b) and inadequate or ambiguous legal or regulatory environment. Government has a key role to play in both these risks.

Key Findings of the Workshop

Building capacity and trust was the overarching theme of the various discussion groups. It is clear, that trust can only be built if the stakeholders can communicate at the same 'eye height', which in turn depends on building capacity. Deliberative, 'good' governance will underpin this process. Clear, unbiased information about the risks associated with uranium mining projects is at the top of the trust-building agenda. It was noted that risk perception correlates to gender. Women and men place different weights on particular issues. Participants thought that communities would benefit from formal induction and training courses about the projects affecting them, aligned to project milestones across the whole life-cycle. Mentoring and knowledge and expertise transmission programmes would help to empower stakeholders. Site development beyond the active life of the mine clearly is important to stakeholders. 'End of Life' (EoL) plans for the transition from active mining to the post-mining time need to be developed and provisions made for retaining the related institutional and cultural memory.

Stakeholders wish to be able to judge for themselves and on a well-informed basis, whether or not the proposed mine developments follow the best practices available at the time, including site remediation and long-term management. This predicates the provision of standardised reference materials, in the form of a communication and training toolkit. This toolkit should contain resources on case studies, including successes, failures, lessons learned, and on conflict prevention, management and resolution. The toolkit should be tailored to different cultural and gender requirements, including different media for different communication channels (printed, broadcasts, Internet). Instruments to measure and gauge success or otherwise of the engagement strategy should also be part of the toolkit..

Conclusions

A social licence is an indispensable precondition for any mining operation, but particularly so for uranium mining and processing operations. The social license is built on trust between the different stakeholders. A stakeholder is any person or group that has voiced an interest in or concern about the operation. Trust in turn is built on a shared knowledge-base and on a mutual understanding of each other's motives and values, but these do not have to be necessarily shared. In building and retaining trust, communication and capacity-building are key activities enabling all parties to the licence to ne at the same 'eye height' and be equally empowered.

Many stakeholders are concerned about the environmental and socio-economic impacts of uranium mining project well beyond the active time of the mine. A comprehensive life-cycle management approach is needed in order to ensure the sustainability of the project.

Given how central to the mining industry the social licence has become in the past decade, there will be increasing pressure in future to formalise it as a legal requirement rather than a looser social accord. The problem this raises is that the fuzzy nature of the negotiation procedure by which a licence is first agreed, then modified and adapted as a project's life-cycle unfolds, may be its key strength. In which case, formalisation will risk causing the process to become rigid, and hence inherently conflicted – exactly the outcome that is least desired.

References

Arnstein S R (1969) A Ladder of Citizen Participation. J Amer Planning Assoc 35(4): 216–224, doi:10.1080/01944366908977225.

Braithwaite V A, Levi M (1998) Trust & Governance. The Russell Sage Foundation Series on Trust, 1: 386 p., New York.

Brundtland G (ed) (1987) Our Common Future. The World Commission on Environment and Development, Oxford University Press, Oxford.

Elkington J (1994) Towards the sustainable corporation: Win-win-win business strategies for sustainable development. California Managmt Rev 36(2): 90-100.

Ernst and Young (2012) Business Risks Facing Mining and Metals, Report 2012-13.

Falck W E, Spangenberg J H, Wittmer D (this volume) Social licensing in Uranium Mining: Empowering Stakeholders Through Information. Proc. UMH VII.

International Atomic Energy Agency IAEA (1997) Planning for Environmental Restoration of Uranium Mining and Milling Sites in Central and Eastern Europe. IAEA-TECDOC-982, IAEA, Vienna.

International Atomic Energy Agency IAEA (2006): Management of Long-Term Radiological Liabilities: Stewardship Challenges, IAEA-TRS-450, IAEA, Vienna.

International Atomic Energy Agency IAEA (2008) Nuclear energy basic principles. IAEA Nuclear Energy Series No. NE-BP, Vienna

International Atomic Energy Agency IAEA (2013) Nuclear fuel cycle objectives. IAEA Nuclear Energy Series No. NF-O, Vienna.

Mining Minerals and Sustainable Development MMSD (2002) Breaking New Ground - Mining, Minerals, and Sustainable Development. The Report of the MMSD Project, 454 p., London, Earthscan Publications Ltd.

KPMG (2013) The Community Investment Dividend, Measuring the Value of Community Investment to Support Your Social Licence to Operate.

Nash J (1950) Non-cooperative Games. Ann Math 54: 286-295.

Thomson I, Boutilier R G (2011) Social license to operate. In: Darling P (ed.) SME Mining Engineering Handbook, pp. 1779-1796, Littleton, CO, Society for Mining, Metallurgy and Exploration.

United Nations UN (2002) Report of the World Summit on Sustainable Development, Johannesburg, South Africa, 26 August-4 September 2002. United Nations publication A/CONF.199/20, New York.

Environmental radiation monitoring around uranium ore deposits and mining sites in India: An overview

Amir H. Khan[1]

[1]Ex- Scientific Officer (G) & Ex- Raja Ramanna Fellow, BARC
A-5/4-1, Millennium Towers, Sect. – 9, Sanpada,
Navi Mumbai-400705, India
Email: ahkhan1214@gmail.com

Abstract. There are many low grade uranium ore deposits identified in different states of India. Mining and ore processing operations have been undertaken at several locations in the eastern province of Jharkhand since the past several decades. Mining and processing of ore also commenced recently at a site in the southern state of Andhra Pradesh. Other deposits are in different stages of exploration and development. Monitoring for natural radiation and radon has been an integral part of the environmental surveillance at the ore deposits and mining sites. To supplement departmental environmental monitoring, the baseline radiation surveys around the proposed mining sites are sometimes entrusted to educational and research institutions of excellence in the country during the past several years. This paper aims to give an overview of the radiation monitoring in the environment around the uranium ore deposits and mining sites. Gamma radiation levels and annual exposure rates in Jaduguda area and that in different states are summarized. An overview of the studies conducted for restoration of the tailings pile and role of vegetation and plants is also given. Public awareness program undertaken in the vicinity of mining sites for information of the local stakeholders is also highlighted.

Introduction

India has moderate reserves of uranium in the form of low grade ore spread over different parts of the country. Search for uranium ore deposits began in India in early 1950s. Since then many mineable uranium ore deposits have been identified by the Atomic Minerals Directorate of Exploration and research (AMD) across the length and the breadth of the country. Mining and ore processing operations at

B.J. Merkel and A. Arab (eds.), *Uranium – Past and Future Challenges*,
DOI 10.1007/978-3-319-11059-2_11

Jadugudain Jharkhand state began in 1967on commercial scale. Subsequently, mining was undertaken at other sites in the nearby places so that the ore may be processed at the mill in Jaduguda which was expanded for this purpose. Currently, underground uranium ore mining operations are in progress at Bhatin, Narwapahar, Turamdih, Mohuldih and Bagjata and an opencast mine is in operation at Banduhurang. Another mill has been commissioned at Turamdih. Appropriately engineered tailings treatment and management facilities commensurate with the scale of operations are located at Jadugudaand Turamdih. The facilities located in the Sighbhum Thrust Belt (STB) are located within about 25 km east and west ofJaduguda as shown in Fig.-1 (Bhasin, 1998; Gupta, 2004, 2005, Sahoo et al, 2010).

Some smaller deposits were identified at Jajawal and Bodal in Chattisgarh state of central India where exploratory mining was done and the sites were closed to be taken up for development in future (Saxena and Verma, 1998).Several other workable uranium ore deposits have been identified in other states, e.g. West Khasi Hill district in the northeastern state of Meghalaya, Tumalapalle, Lambapur and other areas in Southern state of Andhra Pradesh, Gogi in Karnataka and Rohil in Rajasthan.

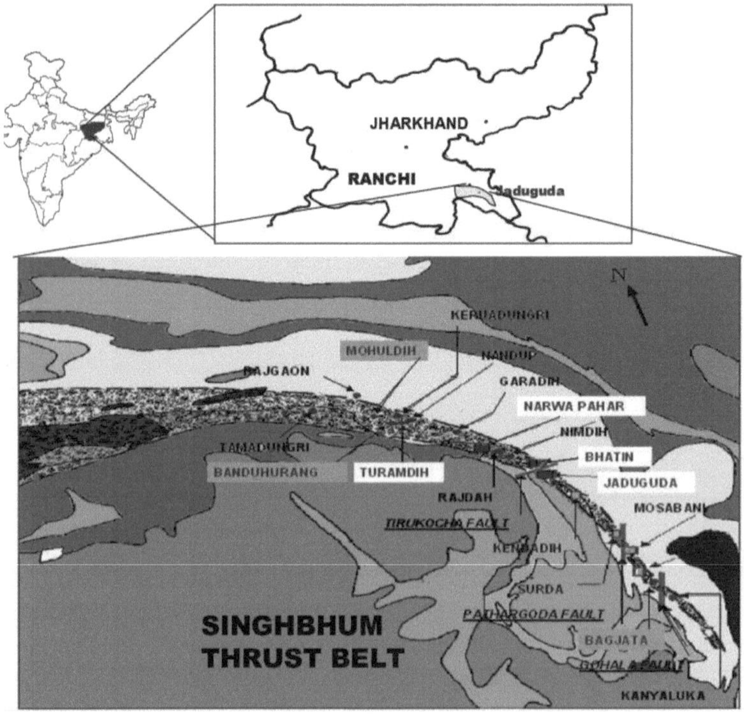

Fig.1. Uranium ore deposits and mining sites in Jharkhand state

Environmental Radiation Monitoring

Environmental and in-plant radiation monitoring at all uranium mining sites are given due importance, as in other sectors of the nuclear industry in the country. For this purpose, the Bhabha Atomic Research Centre has set up dedicated Environmental Survey Laboratories with Health Physics Units at the mining and ore processing sites (Beri, 1998; Bhasin, 1999). These laboratories are independent of the administrative control of the local mining and ore processing industry. The ESLs/HPUs are equipped with the state of the art monitoring and dosimetric equipment with qualified scientific and trained technical manpower.

The workplace monitoring comprises of evaluation of gamma radiation, radon with its progeny and the airborne long-lived alpha activity due to ore dust. On the spot gamma radiation and radon are evaluated using survey meters and scintillation cells, respectively (Raghavayya, 1981). These are supplemented with personal dosimeters comprising of thermo-luminiscent chips and SSNTD films enclosed in aluminum chambers used by workers (Raghavayya et al, 1986). Doses to the workers are computed from the workplace monitoring data (Khan et al, 2009; Khan and Puranik, 2011). Liquid and gaseous releases from the mine and mill are monitored for radioactive and chemical constituents before release to ensure compliance with regulatory limits.

The environmental radiation is monitored with survey meters as well as thermo-luminiscent dosimeters installed at difference locations around the facility. Radon is monitored on the spot with low level radon detection system (Srivastava et al, 1984) and SSNTD based radon dosimeters for the long-term measurement.

In addition to the radiation monitoring in the environment, chemical constituents of the releases from the industry to the in to the local is regularly monitored. Soil, vegetation, crop and food samples are periodically monitored for radioactivity and heavy metals etc to evaluate uptake by plants.

Results and Discussions

The workplace monitoring data for the operating facilities have been published in journals and conference proceedings during the past few decades (Khan et al, 2000; Tripathi et al 2007, Khan et al, 2009). Average annual gamma radiation levels based on TLDs in uranium mining belt around Jaduguda are shown in Fig. 2.

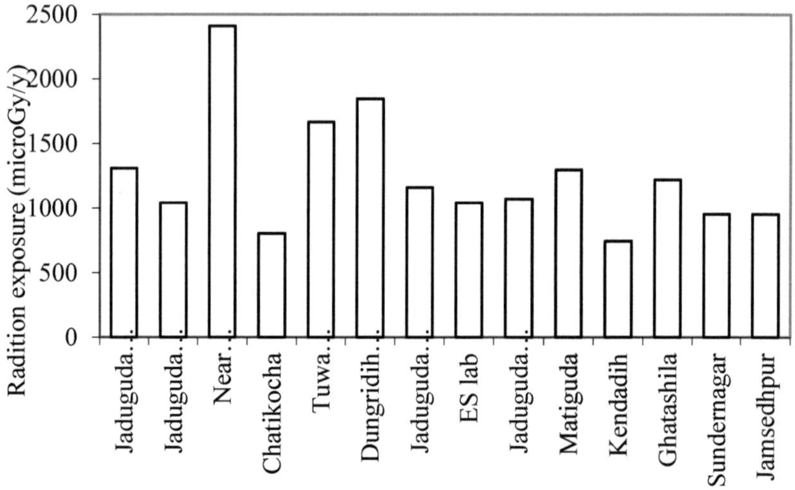

Fig.2. Mean annual gamma radiation exposure around Jaduguda

A comparison of gamma radiation levels in different states of India with that observed in uranium mining belt of Jaduguda in Jharkhand is shown in Figure-3 which indicates that the background radiation level in this region is similar to that in many other parts of the country.

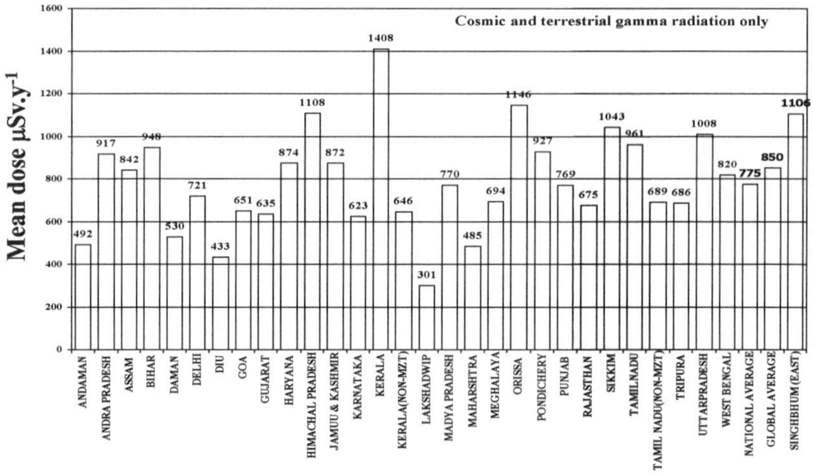

Fig.3. Mean natural radiation levels in different states, $\mu Sv.y^{-1}$

Among the many ore deposit sites, those at Bodal and Jajawal in Chhatisgarh state of central India were closed after exploratory phase for taking up mining in future. Mining and ore processing at Tumalapalle in Andhra Pradesh (A.P.) has begun a few years ago. Here the alkaline leaching process is adopted. For the relatively better grade deposits in Domiasiat and adjoining areas in Meghalaya state in the northeastern part of the country, the infrastructure facilities are being developed so that mining and ore processing operations may be undertaken after completing the regulatory process. Radiation monitoring results of some of these ore deposits sites are summarized in Table 1(Saxena and Verma, 1998; Rajeshkumaretal, 2005; UCIL, 2004; Khathing et al, 2005).

Table 1. Environmental gamma radiation at some ore deposit sites

Ore deposit site	Gamma radiation level, $\mu Gy.h^{-1}$	
	Range	Mean
Jajawal, Chattisgarh		
Near mine site	0.17 - 0.24	0.18
Near ore stack	1.10 - 2.40	1.75
Near heap leaching pad	0.30 - 0.60	0.40
3 -5 km away from mine	0.07 - 0.19	0.12
Tumalapalle, Andhra Pradesh		
Tumallapalle area (outdoor)	0.06 - 0.20	0.10
Tumalapalee area (indoor)	0.07 - 0.18	0.12
Tumalapalle mine, near ore pile	3.35 - 5.85	4.60
Near waste rock pile	0.60 - 0.90	0.75
Seripally, A. P.		
Seripally (proposed milling site)	0.16 - 0.48	0.28
Seripally (proposed T. P. Site)	0.19 - 1.08	0.42
Deverkunda area (2-6 km away)	0.23 - 0.40	0.29
Hyderabad City (100 km away)	0.21 - 1.10	0.48
Domiasiat/Killung, Meghalaya		
Geology camp site	0.04 - 0.12	0.08
Ore outcrop in Kyllung stream	1.60 - 2.00	1.80
Wahkaji, 8km from Kyllung	0.10 - 0.11	0.10
Umdohlun, (10km NE Kyllung)	0.08 - 0.10	0.09
PhudKylleng (ore outcrop)	1.13 - 1.17	1.15
Nogbahjynrin (ore block)	0.19 - 0.35	0.26
Mawathbah (proposed mill site)	0.28 - 0.53	0.38
Domiasiat, near Rangam block	0.16 - 0.31	0.23
Ranikor, 10 km SW of Kyllung	0.05 - 0.22	0.11

Radon emanation from waste rock

To evaluate the impact of waste rocks from mining operations on environmental radon, the radon emanation was measured over a pile of size about 40 m x 90 m. The whole dump was divided in to grids and radon emanation rates were measured using inverted accumulators of about 2 litre volume placed over the waste rock dumps with the edges sealed with low active soils. The radon emanating from waste rocks built up in the accumulator for 60 minutes and a sample was drawn in to a pre-evacuated scintillation cell through a swedgelok connector at the top of the accumulator. 34 samples were collected and the radon emanation rate was found to vary from 0.09 Bq m^{-2} s^{-1} to 2.13 Bq m^{-2} s^{-1} with an average of 0.45±0.40 Bq m^{-2} s^{-1}. Radon emanation rate from the waste rock dump was found much lower than the emanation rate of 1.53 Bq.m^{-2} $s^{-1}$from the uranium tailings surface (Jha et al, 2001; Khan, 2010).

Studies on restoration of tailings pond

There are three tailings ponds at Jaduguda. Two have been filled up and vegetation is grown to consolidate the surface, prevent wind erosion and control dust. A small portion has been covered with local soil and studies are underway to grow a variety of vegetation to study their growth and uptake of radio-nuclides (Khan and Puranik, 2011).

Baseline environmental studies

As a confidence building measure among the stakeholders and encourage capacity building in different parts of the country for radiological and environmental studies related to uranium industry, the educational and scientific institutions of excellence in different regions are entrusted with baseline surveys of new uranium ore deposits at proposed mining sites.

The faculty and senior level students are given appropriate training to carry the baseline environmental studies related to radioactivity and the radiation in environmental matrices, demographic and health status of the local residents, ground water modeling, radon dispersion modeling, socio-economic studies and flora fauna at the sites. Large number of such projects is awarded to different institutions and the progress is periodically evaluated by the specialists in the field. A short illustrative list of a few such research projects awarded is given below:

1. Dr. Gurdeep Singh, ISMU, Dhanbad, 'Baseline environmental studies of Bagjata and Banduhurang sites of UCIL
2. Dr. B.K. Tewari, CMRI, Dhanbad, Ground water modeling of Banduhurang mining and proposedTP area of UCIL.
3. Dr. D. Sengupta, IIT Kharagpur, 'Measurement and modeling of radon transport and distributionaround tailings pond area'.
4. Dr. V. K. Singh, Cooperative College, Jamshedpur, 'Studies on plant, animal and bird diversity at the proposed uranium mine sites of Bagjata and Banduhurang, East Singhbhum, Jharkhand.
5. Dr. PinakiSar, IIT Kharagpur, 'Baseline survey of microbial community structure present in uranium mining area of UCIL, Jaduguda

6. Dr. D. Deb, IIT, Kharagpur, ‘ Remote sensing GIS for baseline survey of propsed uranium mining sites of UCIL, Jaduguda’.
7. Dr. S. Ray Chaudhury, WBTU, Kolkata, ‘Identification and characterization of sulphate reducing bacteria for Jaduguda effluent discharge’/
8. Dr. P. Soni, FRI, Dehra Doon, ‘Identification of bushes/trees to be planted on decommissioned tailings pond’.
9. Dr. H. P. Thakur, TISS, Mumbai, ‘Epidemiological study of health status of population around Jaduguda uranium mining site’.
10. Dr. R. K. Singh, TCS/TISCO, Jamshedpur, ‘Demographic epidemiological study of Bagjata, Banduhurang and Mohuldih sites’.

Similar baselines studies have been undertaken for other uranium ore deposits and prospective mining sites.

Public Awareness program

In view of the uranium mining history in other countries, public in the vicinity of the mining industry may have concerns related to the environmental and health impact of the operations. To address the concerns of the stake holders training of workers and public awareness programs are in place.

The workers are given periodical training in the vocational training centre attached to the industry on safe handling of radioactive and toxic material, use of personal protective appliances, personal hygiene and good housekeeping. Sources of radiation exposure and control measures in place, radiation exposure limits are discussed.

For members of the public, radiological and environmental safety awareness programs are conducted in local schools, colleges and community centres. This program includes information on the safety measures adopted at workplace, treatment of tailings and effluents, control of releases from the plant and environmental monitoring systems in place. Occasionally students and community leaders are encouraged to visit the facilities. Periodic health camps are organized in the locality to evaluate the common ailments and, where appropriate, provide clinical help.

A program for societal monitoring of the environment is conceived where local students and university teachers may be trained to carry out the monitoring. This may give more confidence to stakeholders regarding the control and safety measures put in place by the facility operator.

Acknowledgement

Author wishes to thank BARC and UCIL authorities for the support extended by them.

References

Beri, K. K.,(1998) 'Uranium ore processing and environmental implications', Metals, Materials and Processess, vol.10, No. 1pp.99-108.

Bhasin, J. L.,1998, 'Mining and milling of uranium ore: Indian scenario', Impact of new environmental and safety regulations on uranium exploration, mining, milling and management of its waste, IAEA-TECDOC-1244 (2001), proceedings of a Technical Committee meeting held in Vienna, 14-17 September 1998.

Bhasin, J. L.,1999, ' Radiation Protection in uranium mining and milling' Proceedings of 24th Annual Conference of the Indian Association of Radiation Protection, Kakrapar, India, January 20 -22, 1999

Gupta, R. (2004), Uranium mining and environment management – The Indian scenario, Proceedings of Thirteenth National Symposium on Environment, June 5-7, 2004, North Eastern Hill University, Shillong, India

Gupta, R. (2005), Uranium mining: Future challenges' Invited talk at the 27th annual conference of the IARP, Mumbai, November 2005.

Khan, A. H., Puranik, V. D. and Kushwaha, H. S., 'Radiological safety in mining of low grade uranium ores: Four decades of monitoring and control in Indian mines', International Symposium on Uranium Raw Material for the Nuclear Fuel Cycle (URAM-2009), 22-26 June 2009, IAEA, Vienna, Austria.

Jha, S., Khan A. H. and Mishra, U. C., A study of the technologically modified sources of ^{222}Rn and its environmental impact in a Indian U mineralized belt, Journal of Environmental Radioactivity, 53 (2001), pp. 183-197.

Khan, A.H., Basu, S.K., Jha, V.N., Jha, S. and Kumar, R.,(2000), Assessment of Environmental Impact of Mining and Processing of Uranium Ore at Jaduguda, IAEA InternationalSymposium on the Uranium Production Cycle and the Environment at Vienna, Oct. 2 – 6, 2000.

Khan, A. H. (2010), 'Evaluation of radon emanation from uranium mine waste dumps and an assessment of it environmental impact, presented at the 3rd Asian and Oceonic Congress onRadiation Protection, IRPA Regional Congress (AOCRP-3), Tokyo University, Japan, May 24-28, 2010.

Khan, A. H. and Puranik, V. D.,(2011), 'Radiation protection and environmental safety surveillance in uranium mining and ore processing in, India' UMH-VI, Freiberg, The New Uranium Mining Boom, Springer.

Khathing, T. T.,Myrboh, B., Nongkynrih, P.,War, S. A.,Marbaniang, D.G.andIongwai, P. S. (2005), 'Baseline environmental survey of proposed uranium mining projects of Domiasiat, Meghalaya', 14th National Symposium on Environment, Hyderabad, India; published in the Special Issue of Environmental Geochemistry, vol.8, No. 1 & 2, 2005 (ISSN0972-0383)

Rajeshkumar, V. N. Jha, Sahoo,S.K. and Shukla, A.K, 2005, 'Pre-operational radiological monitoring around proposed uranium mining and ore processing, site at Tumalapalle, Andhra Pradesh', 14th National Symposium on Environment, Hyderabad, India; published in the Special Issue of Environmental Geochemistry, vol.8, No. 1 & 2, 2005 (ISSN0972-0383).

Raghavayya, M., Khan, A. H. and Jha, G., 'Dose estimate for Jaduguda mine workers', Bulletin of Radiation Protection, Vol.8 (1986).

Raghavayya, M. (1981), An inexpensive Radon scintillation cell, Health Physics, 40, 894.

Saxena, V. P. and Verma, S. C. (1998), 'Uranium mining and heap leaching in India and related safety measures – A case study of Jajawal mines', Impact of new environmental and safety regulations on uranium exploration, mining, milling and management of its waste, IAEA-TECDOC-1244 (2001), proceedings of a Technical Committee meeting held in Vienna, 14-17 September 1998.

Sahoo, S. K., Mohapatra, S.,Sethy, N. K., Patra, A.C., Kumar, A.V., Tripathi, R.M. and Puranik, V.D., 'Natural Radioactivity in roadside soil along Jamshedpur-Musabani Road: A mineralised and Mining Region, Jharkhand and associated risk', Radiation Protection Dosimetry, March 11, 2010 (doi.10.1093/rpd/neq111), Oxford University Press.

Tripathi, R. M., Sahoo, S. K., Jha, V. N., Khan, A. H. and Puranik, V. D., 'Assessment of environmental radioactivity at uraniummining, processing and tailings management facilty at Jaduguda,India', Applied Radiation and Isotopes (2008), doi:10.1016/j.aradiso.2007.12.019

UCIL (2004), 'Rapid Environmental Impact Assessment for Uranium Ore Processing Plant at Seripalli, A. P., Uranium Corporation India Ltd.

Nurse, I need more light, quick.
One minute...
Much better, thank you
But I didn't change anything yet.
What's going on here?
The baby is glowing!

Optimization of Uranium In-situ Recovery Based on Advanced Geophysical Surveying and Borehole Logging Technologies

Horst Märten[1,2], Andrea Marsland Smith[1], Jonathan Ross[1], Michael Haschke[2], Harald Kalka[2], Jens Schubert[2]

[1]Heathgate Resources Pty. Ltd. (Heathgate), Level 7, 25 Grenfell Street, Adelaide, S.A. 5000, Australia
[2]Umwelt- und Ingenieurtechnik GmbH Dresden (UIT), Zum Windkanal 21, 01109 Dresden, Germany

Abstract. Advanced geophysical methods including high-resolution seismic surveying and novel borehole logging techniques have been launched at Heathgate to better quantify both hydro(geo)logical and geochemical conditions that determine the feasibility of in-situ recovery (ISR) from sedimentary-hosted uranium. This data is the input to hydrological and reactive transport modeling to optimize well-field design and performance with regard to ore-lixiviant contact and leaching chemistry. The redox puzzle is solved for specific deposit characteristics by a new, practicable kinetic leach model providing control criteria for optimized recovery.

Conditions determining ISR feasibility

The following key factors determine the applicability of uranium ISR from sedimentary-hosted deposits:

1. Aquifer confinement within the overall sedimentary stratification,
2. Hydrology (groundwater table, porosity/permeability),
3. Mineralogy/geochemistry (abundance of reactive minerals),
4. Uranium ore morphology and grade.

In addition, technological/metallurgical, economic, environmental, market, legal, political, and social conditions apply.

This paper describes advanced methods suitable to measure the above deposit characteristics more reliably and more economically. They are used within exploration programs and for ISR mine development in the Frome Basin in South Australia, a prospective province for sedimentary hosted uranium (Märten et al. 2011; Birch et al. 2012).

B.J. Merkel and A. Arab (eds.), *Uranium – Past and Future Challenges*,
DOI 10.1007/978-3-319-11059-2_12

High-resolution shallow seismic

High-resolution shallow seismic in combination with refraction tomography (applied to the near-surface region) is used for imaging the stratigraphy of sedimentary formations, tectonic faults and other structures to improve the understanding of (hydro)geology in general and the preferred locations of the uranium deposits in particular.

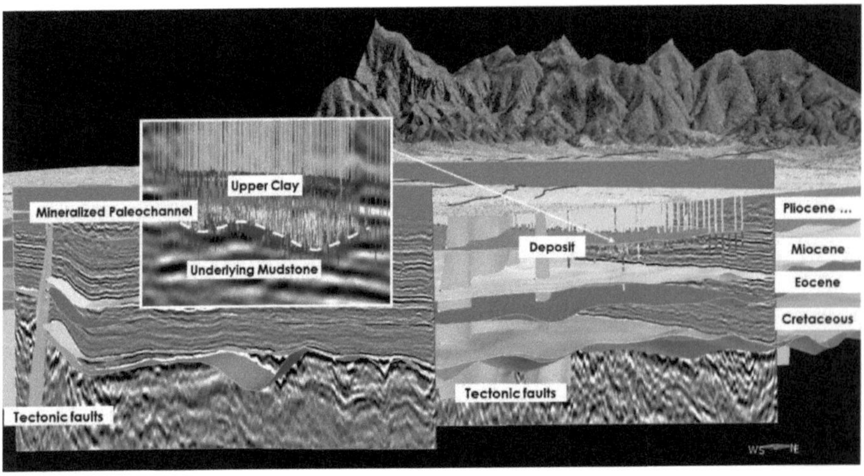

Fig.1. Illustration of seismic profiles embedded in a 3D model (gOcad) including drillhole profiles (vertical lines), example Beverley. The main picture shows the stratigraphic sections and major tectonic faults, whereas the image section indicates a mineralized paleochannel enclosed by impermeable layers.

Heathgate has acquired extensive seismic data in the Beverley and Beverley North Mining Lease (which contain the Pannikan and Pepegoona sedimentary uranium deposits). Seismic data comprised 40 lines for a total length of 318 line kilometers, with dense acquisition geometry: 5 m shot interval (Envirovibe Vibroseis, 60 kN) and 5 m geophone spacing.

After standard processing (based on post-stack time migration) all 2D seismic profiles were loaded into a 3D gOcad model to perform time-to-depth conversion (Fig.1). However, recent studies demonstrated the potential of *pre-stack* depth migration in beneficial in order to enhance the final resolution in almost all sections. The required seismic velocities were provided by refraction tomography (near-surface 2D velocity field) and by sonic borehole logging (at depths ≥ 100 m).

High-resolution seismic adapted to *shallow* sedimentary formations (in contrast to seismic surveys in gas/oil industry at depths 1-5 km) is a powerful method to image stratigraphy in detail with all its irregularities (e.g. paleochannels and faults) that can host or influence the localization of sedimentary uranium deposits. Therefore, the shallow seismic methodology (hardware, acquisition geometry/design, data processing) has been refined in this study to better identify such proxies (meandering paleochannels, etc.). Further improvements are possible when considering the upgrade in data resolution by using 2D versus 3D seismic data.

Novel borehole logging tool for U exploration and ISR wellfield development

A new-generation geophysical downhole-wireline tool with a high-performance, pulsed DT-neutron generator (1 kHz repetition rate, $>10^8$ neutrons per second) has been developed and is currently being used at Heathgate. Several conventional and new logging techniques have been combined in just one tool to enhance the efficiency (and economics) of downhole logging. Such techniques include:

1. PFN technology, i.e. counting (high-energy) prompt fission neutrons (PFN) from the interaction of thermalized neutrons with ^{235}U between the primary neutron bursts with reference to the count rate of thermal neutrons. This count rate ratio is a measure of the uranium ore grade (pU_3O_8) and subject to corrections for several systematic influences including borehole size, neutron absorption in the formation, and host-rock density. All these influencing parameters are measured by the tool itself (described below).
2. Density (ρ) measurement is carried out by counting the intensity of back-scattered neutrons, distinguishing between bulk, matrix, and dry density by considering porosity.
3. Hydrogen index (HI) is measured by the ratio of thermal neutron intensity in the center and in the periphery of the thermal-neutron 'cloud' formed after each primary neutron burst within a few μs.
4. Macroscopic thermal-neutron absorption cross section Σ_a is determined from the decline time τ of thermal-neutron intensity (multi-scaling neutron channels, i.e. time-dependent).
5. Clay content parameter ($\Delta\tau$) is determined from the difference of neutron decline times in the center and at the periphery of the thermal-neutron cloud.
6. Bore-hole radius (R) is deduced from the intensity of γ-peak caused by thermal-neutron capture by hydrogen, determined from the γ-ray spectrum measured in the main time window between primary neutron bursts.
7. Elemental and mineral abundances in the host-rock formation are measured by γ-ray spectroscopy in dedicated time windows (burst window for γ-rays from inelastic scattering of fast neutrons mainly, first intermediate window for γ-rays from thermal neutron capture, second intermediate time window for γ-rays from neutron activation).
8. Concentration of γ-emitting progenies from the natural decay of U isotopes and ^{232}Th is measured by γ-ray spectroscopy in passive mode (neutron generator

off), thus, providing an independent measure of U grade (eU_3O_8) and Th grade. The ratio pU_3O_8/ eU_3O_8 is a measure of dis-equilibrium characterizing the age and genesis of U deposition in sedimentary host-rock formations.
9. Borehole inclination is measured using an integrated (magnetic) inclinometer.

Based on dedicated algorithms to analyze the time distributions from all neutron detectors as well as γ-ray spectra from all 3 time windows, the following *deduced* parameters are calculated:

10. Porosity (ϕ) from HI and $\Delta\tau$ (the latter to quantify the abundance of clay-bound (hydrated) water).
11. (Hydraulic) permeability K from ϕ and Σ_a by applying the Kozeny-Carman equation (modified).
12. Lithological profiles from the neutron parameters ρ, HI, $\Delta\tau$, and Σ_a, forming a data quadruple that characterizes the sediment/rock type (quantitative lithology).
13. Abundance of main reactive minerals including sulfidic (e.g. pyrite), calcareous (e.g. calcite), kaolinite (dominating clay mineral), lignite a. o.

The data algorithms were developed on the basis of extensive MCNP5/6 computer simulations (LANL 2014) by using a detailed tool model inside a (water-filled) hole at given radius; all within the (saturated) rock formation at varying composition (from a consistent sediment model). The prototype tool was then intensively tested in calibration pits and test facilities in Texas, Dresden, and South Australia in 2011/2012, mainly to calibrate the tool with regard to pU_3O_8, ρ, HI, ϕ, and Σ_a. After the launch of the technically mature tool in early 2013, borehole radius (R) logging based on neutron-capture γ-ray intensity from H (H peak in γ-spectrum) was validated against caliper measurements. Core samples obtained from a recently drilled reference hole are currently being studied to provide final validation of tools' performance with respect to lithological/mineralogical logging.

Based on the comprehensive set of logging data gathered since early 2013, it has been demonstrated that the data quadruple $\{\rho,HI,\Delta\tau,\Sigma_a\}$ characterizes the lithology in a more sensitive way than neutron and resistivity logs recorded by conventional tools (Fig.2 illustrates this comparison). Fig.3 shows measured (high-resolution) lithological profiles against conventional interpretation by geologists (qualitative).

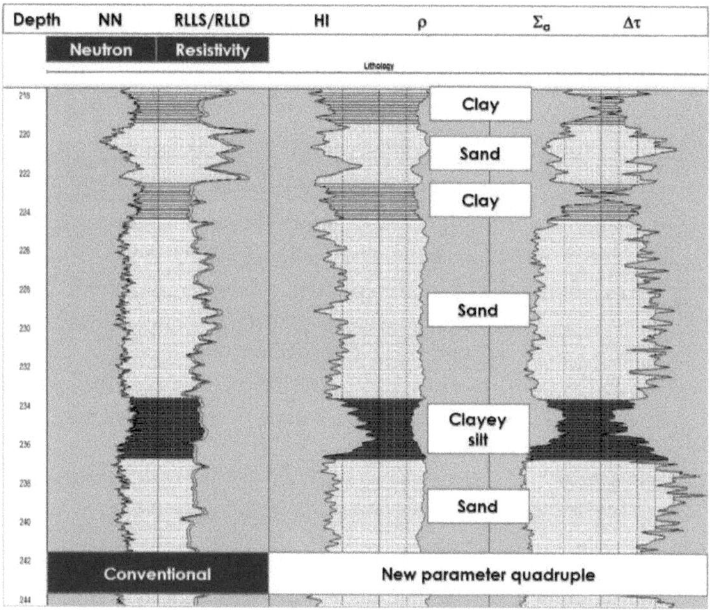

Fig.2. Conventional neutron-resistivity log versus new logging data quadruple

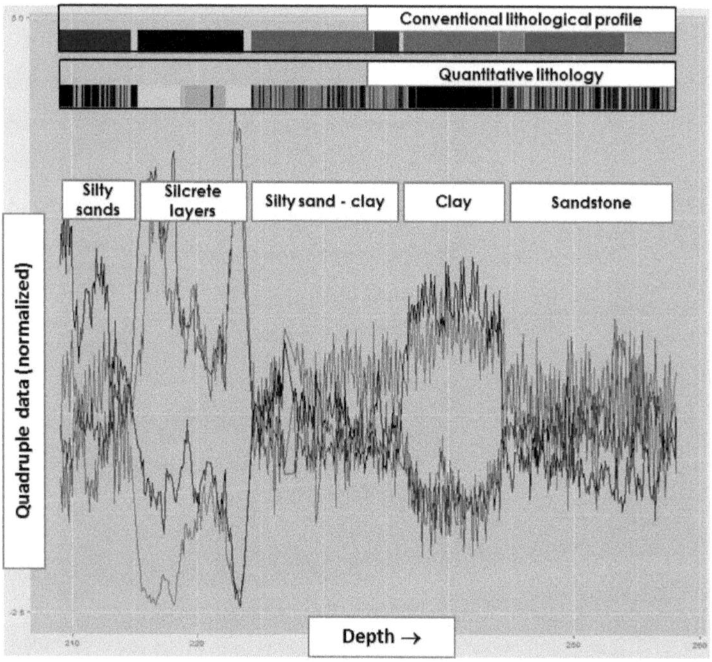

Fig.3. Lithological profiles from data quadruple $\{\rho, HI, \Delta\tau, \Sigma_a\}$ analysis (quantitative). The (actually colored) lithological sections refer to sediment-rock nomenclatures (not shown here)

The main features of the tool are summarized in Table 1.

Table 1. Main features of the geophysical wireline logging tool.

Item	Characterization/data
Neutron generator	>10^8 neutrons s^{-1}; 80 kV; pulsed; 1 kHz repetition rate; 75 μs burst width; lifetime of accelerator tube 600 h warranted (up to 1,000 h)
Neutron detectors	6 high-efficient thermal/epithermal ^3He detectors, operated in multi-scaling mode (time-resolving)
γ-spectrometer	High-performance, large-volume $CeBr_3$ scintillator; high-quantum-efficiency, stabilized photomultiplier; multi-channel analyzer operated in several time windows
In-tool processor	High-performance Blackfin processor (<1 W power consumption)
Logging speed	6 m/min in passive mode; 1 m/min in active mode
Spatial resolution	±15 cm (1σ, depth)
U grade sensitivity	Lower level of detection 0.005 wt% U_3O_8 at 1m/min logging speed; sensitivity increased in stationary mode (depending on measuring time)

Model-based wellfield design and optimized ISR performance

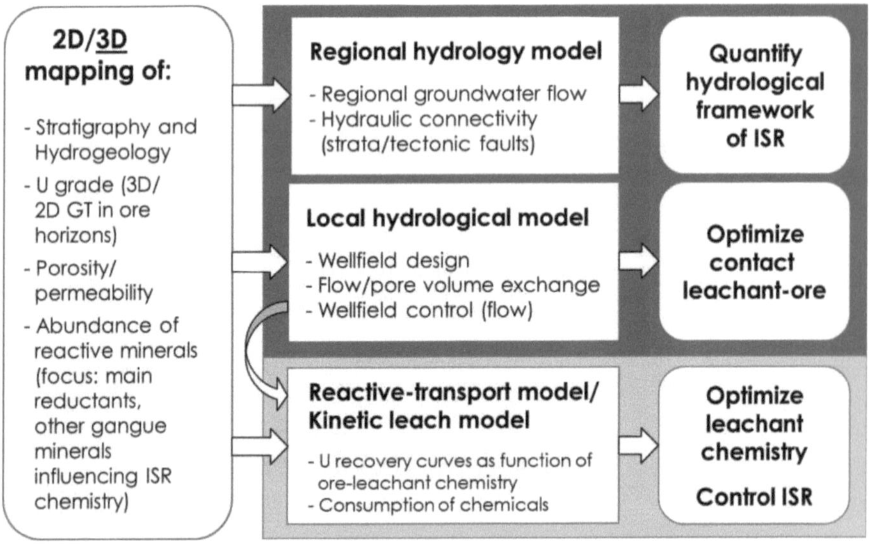

Fig.4. 2D/3D mapping of exploration data for deposit assessment and to provide input to hydrological as well as reactive-transport models for optimized wellfield design and control of leaching chemistry, respectively.

Both high-resolution seismic data and the new comprehensive information from borehole logging provide an excellent aid to the assessment of ISR feasibility, the model-based design of wellfields (hydrological setup), and planning of wellfield operation (geochemical control). In addition to hydrological software, a new kinetic leaching model (reactive transport) considers the impact of interfering reactions enabling the optimization of leaching chemistry (pH, oxidation potential) with variations in ore mineralogy. Fig. 4 illustrates the approach.

Heathgate's in-house (2D) hydrological software is used to simulate (local) wellfield hydrology and to optimize leachant flow patterns with respect to U_3O_8 grade-thickness to optimize leachant-ore contact. Other commercial 2.5/3D software tools (Modflow, FEFLOW, respectively) provides models for the (regional) hydrological framework for the area.

The dynamics of wellfield leaching is simulated by the use of a reactive-transport model code (TRN) that combines the leachant transport (advection, dispersion, dual-porosity approach) with geochemistry. PhreeqC (USGS 2014) has been embedded as an internal numerical solver into TRN. The (acidic) ISR model distinguishes between *primary* (reactive) phases and *secondary* mineral phases. Reactive phases in this case include among others: U minerals (coffinite/uraninite), pyrite, organic matter, carbonates (e.g. calcite), and ferric Fe minerals. Clays are involved in a two-fold way, as mineral phase and as ion exchanger (with specific cation-exchange capacity CEC).

The key to reliably simulate ISR performance is to consider the *kinetics* of leachant-mineral interactions as a function of the surrounding chemical conditions. In particular, it requires the solution of the so-called 'redox puzzle' where U(IV) oxidation competes with pyrite dissolution and the degradation of organics. The corresponding kinetic rates λ_i become functions of pH (or concentration of hydrogen ions $[H^+]$) and the concentration of electron acceptors $[A^{e-}]$, the absolute oxidation potential: $\lambda_i = r_{0,i} \cdot [H^+]^a \cdot [A^{e-}]^b$ where i refers to U, FeS_2, and 'CH_2O' (organics). The model parameters $r_{0,i}$, a, b have been adopted from literature data, lab tests, and most importantly, from field trials.

Chemical constraints of U leaching occur in two main cases: (i) insufficient oxidation of leachant and/or (ii) significant abundance of reductants (pyrite and/or organics, the latter expressed as total organic carbon TOC). The kinetic leach model explicitly accounts for the balance of $[A^{e-}]$ in view of competing reducing reactions during the transport of the leachant through the ore body. The mineral abundance and kinetics for neutralizing reactants (including CEC of clays) determine the amount of expected acid consumption.

As shown in Fig. 5, U recovery (expressed as net leachate grade) decreases as function of the abundance of competitive reductants. The quantitative effect (Fig. 5) is representative for reduced host-rock sediments present in the Frome Basin. Competitive reductants can be (at least partly) compensated by controlling the leaching conditions by identifying the pH and $[A^{e-}]$ range where U leaching is clearly favored against the competitors. This kinetic model approach is routinely used to optimize ISR operation which in turn maximizes production and contributes to more accurate prediction and control of costs.

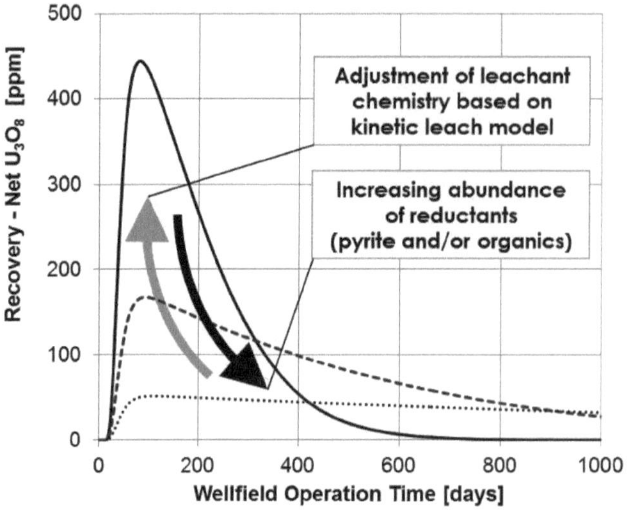

Fig.5. Influence of reducing reactants on ISR recovery and effect of model-based adjustments of leaching chemistry with regard to pH and oxidant concentration.

References

Birch GR, Every CR, Märten HG, Marsland-Smith AB, Phillips R, Woods PH (2013) Beverley Uranium Mines, Heathgate Resources, Australasian Mining and Metallurgical Operating Practices, (Australasian Int. Min. Metall., Melbourne, Australia) 2: 1799-1818

LANL: X-5 Monte Carlo Team of the Los Alamos National Laboratory (2008/2014) MCNP – A General Monte Carlo N-Particle Transport Code, Version 5 (LA-UR-03-1987), April 24, 2003 (Revised 2/1/2008), latest software update 2014

Märten H (2011), Phillips R, Woods P (2011) New Uranium ISR Satellites at Beverley North, South Australia. The New Uranium Mining Boom, Challenge and Lessons learned, Editors Merkel B, Schipek M, Springer: 23-30

USGS (2014) http://wwwbrr.cr.usgs.gov/projects/GWC_coupled/phreeqc/

Fuzzy MCDA for remediation of a uranium tailing

Danyl Pérez-Sánchez[1], Antonio Jiménez[2], Alfonso Mateos[2], Alla Dvorzhak[1]

[1] Radiological Protection of the Public and Environment, CIEMAT, Spain
[2] Decision Analysis and Statistics Group, UPM, Spain

Abstract. The Pridneprovsky Chemical Plant was a largest uranium processing enterprises, producing a huge amount of uranium residues. The Zapadnoe tailings site contains the majority of these residues. We propose a theoretical framework based on Multi-Criteria Decision Analysis and fuzzy logic to analyse different remediation alternatives for the Zapadnoe tailings, in which potentially conflicting economic, radiological, social and environmental objectives are simultaneously taken into account. An objective hierarchy is built that includes all the relevant aspects. Fuzzy rather than precise values are proposed for use to evaluate remediation alternatives against the different criteria and to quantify preferences, such as the weights representing the relative importance of criteria identified in the objective hierarchy. Finally, it is proposed that remediation alternatives should be evaluated by means of a fuzzy additive multi-attribute utility function and ranked on the basis of the respective trapezoidal fuzzy number representing their overall utility.

Introduction

The Zapadnoe uranium mill tailings site is situated in the south-western part of the main industrial site of the former Pridneprovsky Chemical Plant (PChP), located at Dneprodzerzhinsk (Ukraine). The tailings site operated from 1949 until 1954. The majority of the wastes were uranium mill tailings, disposed of using the hydraulic discharge method. The tailings site was covered in 2000 by an engineered multi-layer soil cover. As a result, the wastes are covered by a layer of non-radioactive backfill, composed from construction and industrial wastes, sand, claycy loam soils, clinker, rubbish etc. with a total thickness of 0.2 to 2.8 m. The southern parts of the tailings have been covered by a 0.3 to 1.0 m thick layer of crushed stone and asphalt layer. The slopes of the tailings pile are covered by layers of clay loam and organic soil, with a combined thickness of 0.5-1.0 m.

The tailings are situated on the slope of a sequence of the terraces of the Dnepr River. The general inclination of the ground surface is from the south to the north.

B.J. Merkel and A. Arab (eds.), *Uranium – Past and Future Challenges*,
DOI 10.1007/978-3-319-11059-2_13

The tailings themselves are located within the second terrace. The first (lower) terrace is situated to the north of the tailings. The third and fourth terraces are situated to the south of the waste site. The tailings site was surrounding by dikes that were not surfaced with protective impermeable screens and are currently buried below the layers of backfilled soil. The surface of the tailing pile is equipped with system for collecting runoff rainwater. This water goes to Konoplyanka River.

There are two aquifers at the Zapadnoe tailings site. The technogenic aquifer is a perched water horizon that is recharged by infiltration of atmospheric precipitation through the waste cover. The water from this aquifer infiltrates further down to the underlying aquifer in the alluvial deposits. The regional aquifer in the alluvial deposits is composed of alluvial sands, sandy loam and clay loam deposits, as well as loess deposits overlying the alluvial ones and the upper part of the fissured crystalline basement rocks underlying the alluvial deposits. The groundwater flow in the alluvial aquifer is directed to the north towards the Konoplyanka and the Dnepr.

In 2002-2004, as a result of a series of rainfall events, erosion of the surface and slopes of the protective dikes occurred (Economitor, 2007). Remedial works were carried out in 2005. These works included backfilling the eroded areas with clayey soil, and enforcing the slopes by a geo-technical polymer net material. The eroded surfaces were covered by an organic soil layer and planted with grasses. The surface run-off drainage system was also repaired.

The tailings site is surrounding by other industrial sites and technological communications lines that employ 2500 people. The surface of the tailing site is equipped warning signs and entry of personnel is prohibited, but the site is not fenced.

Two main sources of data currently exist regarding the physical, chemical and radiation characteristics of wastes disposed to the Zapadnoe tailings site. The first characterization studies were carried out in 2000 (Ecomonitor, 2007). Six characterization boreholes were drilled and the core material was subject to various lithological, chemical and radiometric analyses. The second characterization was carried out in 2009 in the framework of the National Program of Remediation of the PChP (UHMI, 2009). Information about radiation exposures due to contamination in soil, water and air was collected for various U-238 series radionuclides and K-40. Analyses of water samples were also performed to obtain information on contamination by chemically toxic materials.

Discrepancies between the results of inventory studies carried out in 2000 and 2009 have been identified. In particular, the 2009 studies suggest that U-238 and Ra-226 concentrations in the wastes are about a factor of two higher than previously estimated. The estimated mean Th-230 activity is increased by a factor of about 3 and discrepancies are also observed for Pb-210.

More recently, the context for a safety assessment of the Zapadnoe tailings site has been described in Bugay et al. 2012. The safety assessment itself was carried out by Ecomonitor and Geo-Eco-Consulting following the steps set out in the ENSURE project. It includes information on the operational history of the tailings site, on its engineering features, as well as on the chemical, physical and radioactive characteristics of the waste materials in the tailings. Environmental conditions (such as the geology, geomorphology and hydrogeological setting) and climate are also described.

Safety assessment can be considered as the starting point for an analysis of remediation alternatives, being equivalent to the no action alternative. The selection of a preferred remediation alternative is a complex decision-making problem in which factors additional to the radiological and chemical toxicity impacts of the wastes have to be taken into account. For example, the direct costs of the application and maintenance of remediation alternatives (manpower, consumables, equipment needed for application, management), the job creation effects and other indirect costs or benefits should be considered as economic criteria. Moreover, social impacts, as well as direct impacts on human health and safety, should be considered. These impacts include community satisfaction and the impact of remediation on the social characteristics of the neighbourhood.

A fuzzy MCDA Approach

The goal of Multi-Criteria Decision Analysis (MCDA) is to structure and simplify the task of making hard decisions to the extent that the nature of the decision permits (Belton, 1990).

What makes MCDA unique is the form in which these factors are quantified and formally incorporated into the problem analysis. Existing information, collected data, models and professional judgments are used to quantify the likelihoods of ranges of consequences, while utility theory is used to quantify preferences. The usual or traditional approach to MCDA calls for single or precise values for the different model inputs, i.e., for the weights as well as for the performances of the alternatives in terms of the identified criteria. However, we adopt a less demanding approach for the Decision-Maker (DM), who is able to provide fuzzy numbers instead of single values.

Fuzzy Logic (FL) introduced by Zadeh, 1965 is a mathematical tool for modelling using vague or imprecise measurements. In FL, a linguistic scale is usually built to characterize inputs to the model (Dursun, 2007). Each linguistic term is associated with a triangular or trapezoidal fuzzy number (Fig.1) and a fuzzy arithmetic is used to make the computations of model outputs. As shown, we consider the set of trapezoidal fuzzy numbers with support on [0, 1]. The arithmetic proposed in (Xu et al., 2010) on TF [0, 1] is used to make computations. The resulting fuzzy number from such a computation is usually translated into a linguistic term in the previously defined scale by means of a similarity function (Vicente et al., 2013).

Following the MCDA methodology, an objective hierarchy in which all relevant criteria are included is first built and attributes are then established for the lowest-level objectives of the hierarchy to indicate to what extent they are achieved.

From the results of the safety assessment and other studies, the performance of each of the options in relation to each of the considered attributes has to be determined and translated into a trapezoidal fuzzy number. Also, the relative importance of the attributes in the objective hierarchy has to be represented by means of trapezoidal fuzzy numbers. Finally, a fuzzified additive utility function can be used to derive a global utility value for each option, on the basis of which remediation alternatives can be ranked.

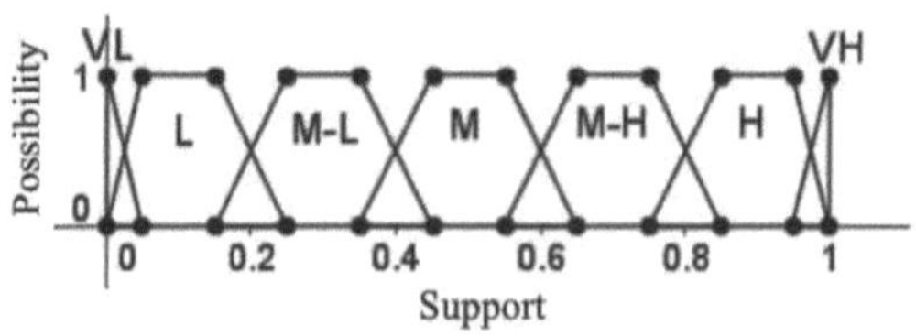

Fig.1. A fuzzy linguistic scale.

Problem structuring

To identify the criteria to be incorporated into the analysis, experts taking part in ENSURE II project were consulted and the literature on applications of MCDA for evaluating remediation alternatives (Brinkhoff, 2011) was reviewed, with a particular emphasis on applications to uranium mill tailing sites (Goldammer et al., 1999). On this basis, we built an objective hierarchy applicable to remediation options for the Zapadnoe tailings site (Fig. 2). There are four main top level criteria for the appropriate management of the Zapadnoe tailings site (global objective): Environmental impact, Radiological impact, Social impact and Economic impact.

The Environmental impact is due to contaminants discharged into surface waters that can impair the functioning of aquatic biota and the impact on groundwater bodies due to infiltration through the tailings to the underlying aquifer. Both radioactive and toxic chemical contamination are taken into account and measured in terms radiation dose or degree of chemical exposure. The doses and exposures derived from the safety assessment are adopted as reference values for the no action alternative and remediation options are evaluated in terms of differences from those values.

The Radiological impact is split into three sub-objectives. Public radiological impact refers to the doses received by the population through external exposure, inhalation (concentration in the air) and ingestion (via drinking water, food). It differentiates the doses received by the population during and after the implementation of the remediation alternative, leading to two new sub-objectives, respectively. The radiological impact on workers refers to radiation doses received by workers as a consequence of the process of implementing a remediation alternative. This objective is split into three sub-objectives accounting for the external dose (radiation exposure at the surface of the tailings site), and the doses received by inhalation and ingestion. To measure these objectives, the corresponding attributes take into account the number of workers needed to implement the remediation alternative, the number of hours each worker is exposed to the radiation and the radiation doses per hour through exposure at the surface, inhalation and ingestion, respectively.

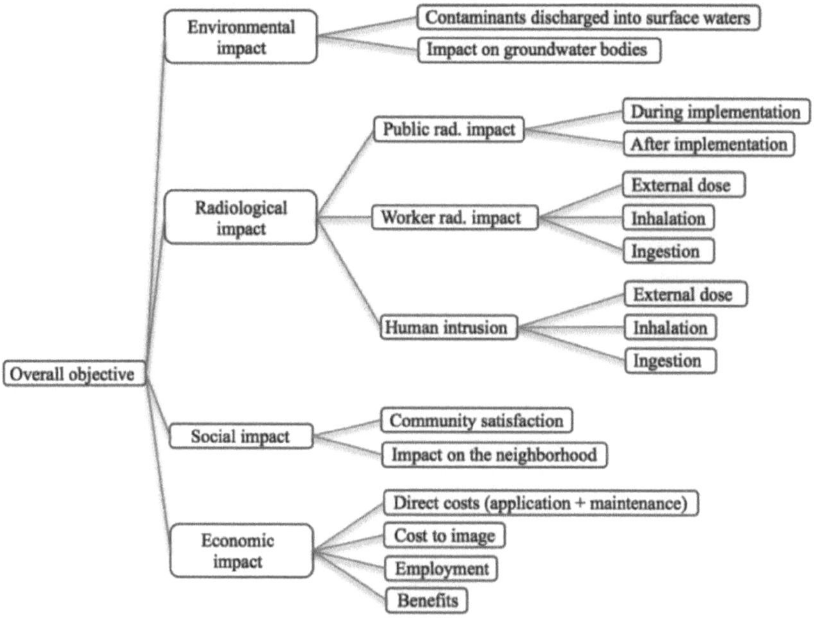

Fig.2. Objective hierarchy.

Finally, human intrusion refers to the radiation received by intruders at the Zapadnoe tailings site. The objective is again split into three sub-objectives accounting for the external dose, and the doses received by inhalation and by ingestion. The corresponding attributes take into account an estimation of the number of intruders at the Zapadnoe tailings site per year on the basis of historical data, the average number of hours each intruder spent at the site by intrusion and the radiation doses per hour through exposure at the surface, inhalation and ingestion, respectively.

Social impact is split into community satisfaction and the impact on neighbourhoods or regions. Community satisfaction refers to how a remediation alternative is perceived by individuals belonging to a critical group living in the area and the impact on the neighbourhood accounts for the impact on the local community as a whole, including dust, light, noise, odour and vibration during the remediation works and associated with traffic, including weekday and weekend day- and night time operations. The fuzzy linguistic scale is used to quantify both social objectives.

Under economic impact, direct costs refer to the costs of the implementation and maintenance of a remediation alternative (manpower, consumables, equipment needed for implementation, management requirements). A monetary attribute is used for this aspect. Cost to image comprises indirect costs associated with a remediation alternative. It relates to public perceptions, e.g., a reluctance to purchase products from the area, even if uncontaminated, or a drop in tourism. Both the no action alternative and the various remediation options may have associated indirect costs. Employment corresponds to job creation in the implementation of a remediation alternative and afterwards. Short- and long-term jobs are taken into account and the corresponding attribute is measured in person-months.

Finally, benefits refer to direct economic benefits associated with the implementation of a remediation alternative (e.g., sale of waste materials for reuse).It is measured in monetary units. Note that all the criteria apply when evaluating remediation alternatives, but some of them, such as the impact on the neighbourhood, direct costs, employment or possible benefits are directly associated to the implementation of remediation alternatives and are not considered in the no action alternative. Imprecise estimates are allowed for by means of interval values in the fuzzy logic.

Elicitation of preferences

The Generic Multi-Attribute Analysis (GMAA) decision support system provides two procedures for assessing component utilities (Jiménez et al., 2006): directly constructing a piecewise linear utility function by providing the best and the worst attribute values and up to three intermediate values with their respective imprecise utilities; or on the basis of indifference judgments between lotteries and sure amounts. In both cases, the system permits value intervals to be specified as responses to the probability questions the DM is asked, which leads to fuzzy component utilities, see Fig.3.

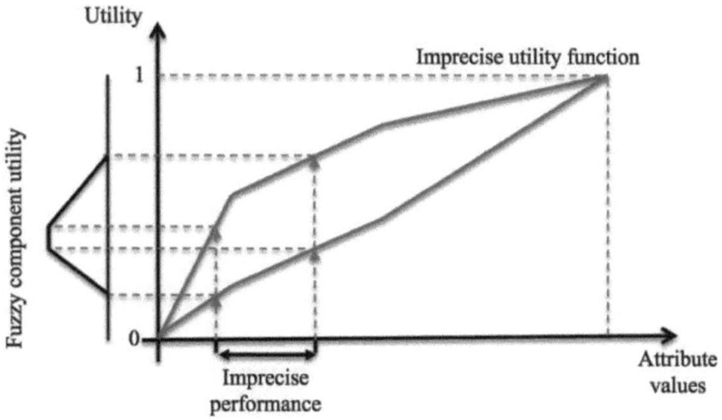

Fig.3. Fuzzy component utilities.

Weights representing the relative importance of criteria in the objective hierarchy have also to be elicited. We use a fuzzy adaptation of the procedure included in the GMAA system for eliciting weights based on trade-offs (Jiménez et al., 2006), in which the elicited individual has to make indifference judgments between lotteries and multiple sure amounts, permitting value intervals as responses. A direct assignment is also allowed by the use of weight intervals (rectangular fuzzy numbers) or using a fuzzy linguistic scale.

Once the relative importance of the objectives and attributes of those objectives has been rated along the branches of the hierarchy (Fig.2), the attribute weight can be assessed by multiplying the respective weights (represented by trapezoidal fuzzy numbers) of the objectives in the path from the root (global objective) to each leaf (attribute).

Fuzzy evaluation of remediation alternatives

Once the preferences have been quantified, the evaluation of remediation alternatives (including the no action alternative) can be performed by means of an additive multi-attribute utility function. The operators are those proposed in Xu et al. 2010. If the linguistic scale is used to value remediation alternatives in respect of a particular attribute, then the corresponding trapezoidal fuzzy numbers (Fig.1) are used as fuzzy component utilities.

Remediation alternatives are then ranked on the basis of the trapezoidal fuzzy numbers representing their overall utility. Methods for ranking fuzzy numbers are classified into four major classes according to Chen and Hwang, 1992. In this case, we propose to use a method based on a similarity function (Vicente et al.,

2013) in which the similarity of the fuzzy overall utility of each remediation alternative is computed regarding both the ideal (1, 1, 1, 1) and anti-ideal point (0, 0, 0, 0). The best-ranked alternative will be the most similar to the ideal point and, at the same time, the least similar to the anti-ideal point.

Conclusions

The evaluation of remediation alternatives in the Zapadnoe uranium mill-tailing site is a complex decision making problem involving environmental, radiological, social and economic criteria. The MCDA methodology provides a framework to structure the problem incorporating individual or group preferences. Moreover, vague or imprecise information is allowed for in the inputs to the decision-aiding process on the basis of fuzzy logic, which is less demanding for those who are elicited and makes the analysis suitable for group decision-making. Herein, we have set out a basis for such an evaluation. The evaluation itself is on-going and will be described in a subsequent publication.

Acknowledgments

This paper has been financially by the Spanish Ministry of Science and Innovation project (MTM 2011-28983-C03-03 and the Swedish International Development Cooperation Agency (SIDA) under the agreement CIEMAT-UPM-SSM for ENSURE project. The views herein shall not necessarily be taken to reflect the official opinion of SIDA.

References

Belton V (1990). Multiple criteria decision analysis practically the only way to choose. L.C. Hendry and R.W. Englese (Eds.), Operational Research Tutorial Papers 53-101, Birmingham: Operational Research Society.

Brinkhoff P (2011). Multi-criteria Analysis for Assessing Sustainability of Remedial Actions. Applications in Contaminated Land Development. Goteborg: Chalmers University of Technology.

Bugay D, Voitsekhovitch O, Lavrova A, Skalsky A, Tkachenko E (2012). Definition of the assessment context and scenarios for the selected cases. Data analyses and data collection. Case A Zapadnoe tailings. Report on the Project ENSURE-II: Work Package 2, Kiev: Ecomonitor.

Chen S.J Hwang C.L (1992). Fuzzy Multiple Attribute Decision Making: Methods and Applications. New York: Springer.

Dursun M (2007). Evaluating solid waste management scenarios using fuzzy multi-criteria decision making approaches. Thesis of Master of Science. Istanbul: Galatasaray University.

Goldammer W, Nüsser A, Bütow E, Lühr H.P (1999). Integrated assessment of radiological and non-radiological risks at contaminated sites. Mathematische Geologie, 3, 53-72.

Jiménez A, Ríos-Insua S, Mateos A (2006). A Generic Multiattribute Analysis System. Computers and Operations Research, 33, 1081-1101.

UHMI (2009). Analytical measurements of samples of material from tailings according to parameters of spectrum of priority alpha- and gamma-emitting radionuclides of U-Th series using methods of low background semiconductor alpha- and gamma-spectroscopy. Final report on contract No.31 with SE Baryer, Kiev: UHMI Institute.

Vicente E, Mateos A, Jiménez A (2013). A new similarity function for generalized trapezoidal fuzzy numbers. Lecture Notes in Computer Science, 7895, 400-411.

Zadeh L.A (1965). Fuzzy sets. Information and Control, 8, 338-353.

Xu Z, Shang S, Qian W, Shu W (2010). A method for fuzzy risk analysis based on the new similarity of trapezoidal fuzzy numbers. Expert Systems with Applications, 37, 1920-1927.

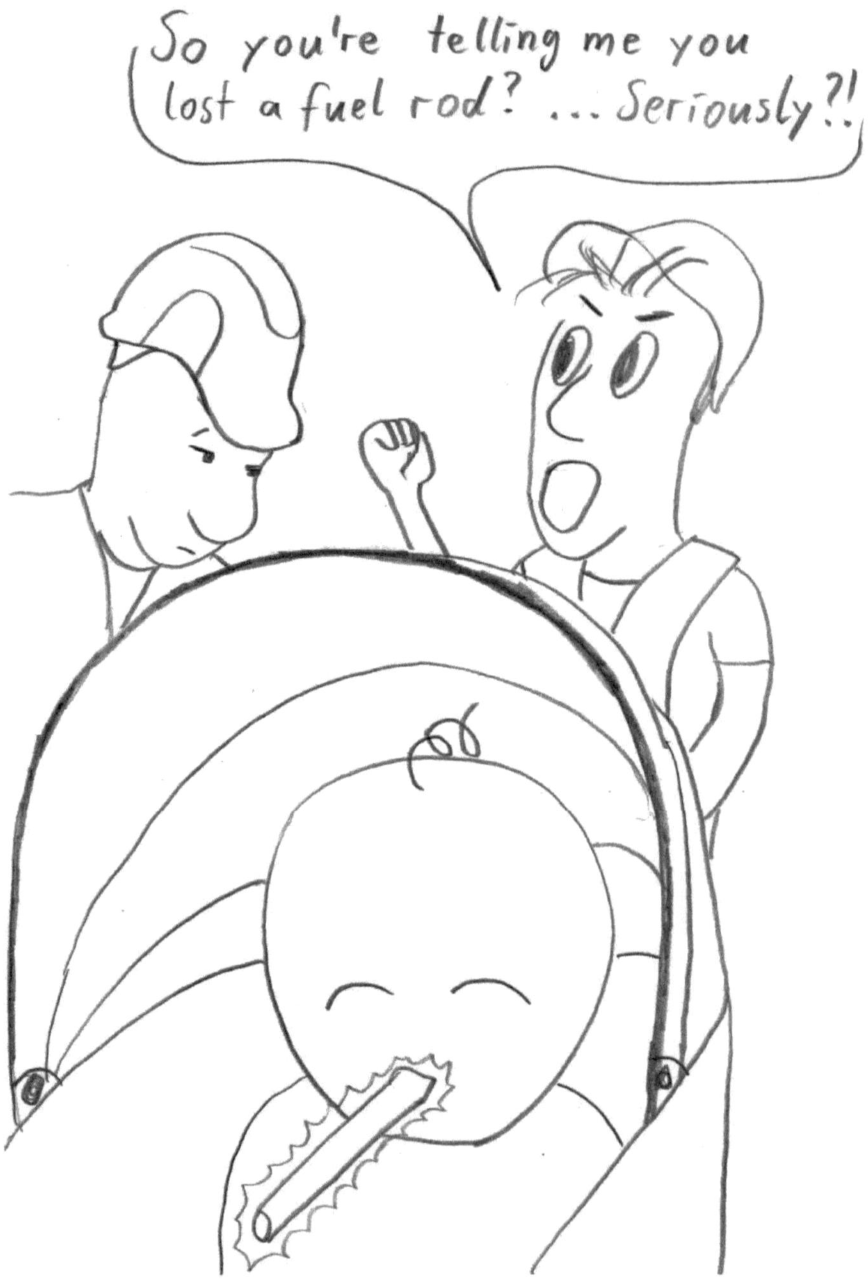
So you're telling me you lost a fuel rod? ... Seriously?!

Uranium in phosphate fertilizers – review and outlook

Ewald Schnug[1], Nils Haneklaus[2]

[1]Technical University Braunschweig - Faculty 2 Life Sciences, Pockelsstrasse 14, D-38106 Braunschweig, Germany
[2]Section of Nuclear Power Technology Development, Division of Nuclear Power, International Atomic Energy Agency (IAEA), Vienna International Centre, PO Box 100, A-1400 Vienna, Austria

Abstract. Uranium (U) can be found in concentrations of 2-200 mg/kg U in phosphate fertilizers if the rock phosphate is of sedimentary origin. About 14,000 tons of U were applied by mineral phosphate fertilizers in Germany from 1951 to 2013. This equals a cumulative load of 1 kg/ha U to agricultural soils. Actually, about 114 to 228 tons of U is distributed by fertilization.

Toxicity of uranium

U is carcinogenic and mutagenous. The chemical toxicity of U is estimated to be substantially higher than the radiological hazard (Drickhko et al. 2008, Bertell and Schmitz-Feuerhake 2008). The chemical toxicity of U ranges between that of Hg and Ni (Busby and Schnug 2008). Enhanced damages of the DNA by gamma radiation in the presence of U are reported (Smirnova et al. 2005). This suggests synergistic interactions between both modes of action. Such combination force has been attributed to the photoelectric effect: U bound to phosphate groups of the DNA absorbs ambient gamma radiation and the release of secondary electrons (beta radiation) multiplies the original alpha radiation by additional particle radiation. As a result the radiological toxicity of U increases stronger than an exclusively additive action of both radiological sources (Busby and Schnug 2008).

U concentrations of >10 µg/L (equaling the limit value for drinking water in Germany (Anonymous 2011)) induce reactions of oxidative stress in aquatic organisms and genotoxicity (Henner 2008; Beaugelin-Siller et al. 2011); leukemia may be caused (Winde 2011) and an estrogenic effect unfolded (for references see Schnug et al. 2008; Schnug and Lottermoser 2013). Important to note is that the toxicity of U is enhanced synergistically by Cd. An even significantly higher radiological and chemical toxicity of U decay products such as Po, Ra and Rn has not been considered (Thomas 2008). Moinester and Kronfeld (2014) suppose a special

B.J. Merkel and A. Arab (eds.), *Uranium – Past and Future Challenges*,
DOI 10.1007/978-3-319-11059-2_14

problem of Rn emanation from fertilizer-derived U in soils as farmlands all over the world are increasingly converted to housing. They allocate the problem not in the Rn emanation to the atmosphere where it is readily diluted, but in its leakage into buildings, driven by a pressure gradient for example through cracks in the foundation.

Uranium in fertilizers

All fertilizer products that contain mineral phosphates unfold U in varying concentrations of <1 to >200 mg/kg U (de Kok and Schnug 2008). In 1984 industry approved the augmented use of magmatic rock phosphates in order to reduce the Cd content as a measure of voluntary self-restriction, but there exists no proof for a trend towards lower Cd and coinciding U contamination of fertilizers. In contrast, fertilizers containing mineral phosphates, which were sampled from 1974 to 1984 showed by trend lower U concentrations than specimens collected from 1995 to 2005 (Sattouf et al. 2008).

The Institute for Crop and Soil Science of the Federal Research Institute for Cultivated Plants in Braunschweig, Germany maintains the worldwide largest database of elemental concentrations in phosphate fertilizers. In 303 fertilizer specimens with a P_2O_5 content of >5% and a mean P_2O_5 content of 23.1% the average U and Cd concentration was 57.9 mg/kg and 9.18 mg/kg, respectively. This equals 243 mg U/kg P_2O_5 and 35.2 mg Cd/kg P_2O_5. According to the German Fertilizer Ordinance (DümV 2012) an obligation to label the Cd content (>20 mg/kg P_2O_5) would have been necessary for 38.4% of all samples and 11.1% of the samples exceeded the limit value of 50 mg/kg P_2O_5 so that they would not be negotiable. The German Commission for Soil Protection (KBU 2012) suggested for U an obligation to label if concentrations are >20 mg U/kg P_2O_5 and a limit value of 50 mg U/kg P_2O_5. Following this approach more than 30% of the samples would not be negotiable and all other samples would require labeling of the U content.

A significant correlation between Cd and U content was determined, but the constant term of the regression is so high with values of 86 and 100 that theoretically a fertilizer product that is free of Cd will supposedly contain U in amounts, which exceed the proposed limit value by far (Schnug 2012). The results are congruent with other fertilizer surveys as that of Smidt et al. (2011a) in Brazil. Here, the mean P_2O_5, Cd and U content was 30.1%, 18.6 mg/kg (61.5 mg Cd per kg P_2O_5) and 70.6 mg/kg U (248 mg U per kg P_2O_5), respectively.

Accumulation of fertilizer-derived uranium in soils

The mean median value of U in German agricultural top soils varies between 0.5 and 3.2 mg/kg U (Utermann and Schnug 2013). Striking is the difference of 0.15 mg/kg U between median values of agricultural and forest soils (Utermann and Fuchs 2008). This value corresponds well with results determined by Salminen (2005), Huhle et al. (2008) and Takeda et al. (2006). The extractable U content of German agricultural soils is four times higher than that of forest soils (Huhle et al. 2008; Setzer et al. 2011). U is highly mobile under oxidizing conditions and in the presence of carbonate ions (Setzer et al. 2011; Baumann et al. 2008; Read et al. 2008) and thus differs from other heavy metals of environmental concern such as Cd and Zn.

The mean annual net input of U by fertilizers in Germany varies between 2.8 g/ha U if P is applied exclusively by organic fertilizers and according to the codes of Good Agricultural Practice (GAP) at a rate of 22 kg/ha P and 15 g/ha U if P is applied exclusively by mineral fertilizers (Kratz et al. 2008). Rogasik et al. (2008) analyzed soil samples from 7 long-term fertilizer experiments in Germany in order to determine the accumulation of U in soils. In all field trials the increase of the U content in the top soil correlated with the P rate. On average an accumulation rate of 3.7 µg/kg*yr U was determined with a minimum of 1 and a maximum of 15 µg/kg*yr U. The latter rate is in agreement with the values of 9.28 µg/kg*yr U from long-term studies in Japan (Takeda et al. 2006) and 14.5 µg/kg*yr U in Ireland (Tunney et al 2009). In comparison, accumulation rates of 19 to 37 µg/kg*yr U, which have been reported for New Zealand and Australia are distinctly higher (Imas and Lati 2005; Lottermoser 2009; Taylor and Kim 2008). The main reason is supposedly a slower leaching rate rather than higher phosphate fertilizer rates and a higher U content of fertilizer products.

U loads that have been calculated on basis of the official P balance yielded arithmetically an accumulation rate of 6 µg/kg*yr U equaling a mean increase of 0.32 mg/kg U in top soils within the last 60 years. This means that 5 to 15% of U in agricultural soils derives from previous fertilizations if the actual median values for the U content in top soils of Germany are used as basis (Kratz et al. 2011). This makes U the element which is accumulated in soils strongest when compared to the natural background value. Rogasik et al. (2008) determined a span of 0.13-0.20 mg/kg U which reflects the difference between calculated and actual accumulation rate of U in top soils of long-term P fertilizer experiments for the treatment where P rates equaled the off-take by harvest products. The authors identified the discharge of U by leachate as the main reason for the observed differences.

Transfer of uranium in the food chain

The mean transfer factor for U in soils to crop plants is 0.05 and thus comparable to that of As, Co, Hg and Pb (Schick et al. 2008). In general, the entry path soil/plant is evaluated as being not critical in the food chain. The U off-take by harvest products is distinctly lower than 0.5 g/ha U (Kratz et al. 2008; Gramss et al. 2011) and comparable to the annual atmospheric deposition (calculated from ATDSR (2013) and other sources). The low U uptake by plants and the fact that there exist no hyperaccumulator plants for U makes phytoremediation of contaminated soils infeasible (Haneklaus and Schnug 2008).

The by far greatest contribution to the daily intake of U by humans has U in liquids: German drinking water contributes with 65% to the daily U intake if standard consumer behavior is assumed (2000 calories per day (kcal/day), 2 liters liquids per day). An elevated intake of U was determined for 'carnivores', who prefer mineralized bottled water. Then drinking water contributes with up to 95% to the daily U intake (Hassoun and Schnug 2011). An investigation of German mineral and tap waters revealed that the relative distribution of U concentrations was similar: 1.9% and 2.3% of all samples showed an U content of > 2µg/L and 15.1% and 16.1% of > 10 µg/L, respectively (Knolle et al. 2011, Smidt et al. 2011b). The highest share of U contamination was found in tap water samples collected in Saxony-Anhalt and Thuringia, the lowest in Schleswig-Holstein (Smidt et al. 2011b).

Smidt et al. (2011b) determined not only the regional distribution of U in tap waters, but also the percentages of the population which has access to these waters. As a result up to 2 million inhabitants receive actually tap water that has a U content of >10µg/L U in Germany.

Uranium loads from agricultural soils to surface and groundwater

On a calculatory basis 420 g/ha U in German top soils that has been previously applied by fertilization cannot be traced. Higher redox potentials and higher pH values are typical for most agricultural soils and favor the mobility of U. Consequently, U is translocated into deeper soil layers and finally into surface and groundwater. Because of the distinctly higher solubility of fertilizer-derived U, this path of U contamination has to be assessed more critically than that of U from parent materials (Heshmati et al. 2008).

Conclusive evidence exists that fertilizer-derived U contaminates water bodies in Germany as it has been shown for other countries, too (Birke and Rauch 2008; Schnug et al. 2008, Garg 2012, Guttmann 1998). A close relationship between nitrate and U in shallow groundwater has been observed in intensively used agricultural areas (Smidt et al. 2011b). There exist three hypotheses for the causal relationship between nitrate and U: Firstly, U from fertilizers is translocated in the form of uranyl-carbonate complexes by percolation water faster into deeper soil layers than other heavy metals; U resembles nitrate in its transport behavior through the soil matrix. Secondly, the nitrate content of groundwater and drinking water is a direct indicator for the fertilization intensity including that of P and thus for U loads in areas without intensive livestock production. The N and P input is linked as NP and NPK fertilizers are preferred by farmers. Thirdly, immobile U (IV) in soils is oxidized to U (VI) by nitrate and then translocated by percolation water (Wu et al. 2010).

Conservative model calculations showed that the "breakthrough" of fertilizer U into deeper soil layers can be expected after 50 years assuming a mean annual U load of 9 g/ha U under climatic and soil conditions of northern Belgium. The equilibrium concentration for this scenario is 22 µg/L U (Jacques et al. 2008).

Smidt et al. (2011b) calculated conservatively that up to a quarter (U > 0.5 µg/L) to three third (U >0.1 µg/L = worst case scenario) of all drinking water samples (n=369; 11% with > 2µg/L; 0.5% with > 10 µg/L; 0.3% with >20µg/L) in the northern plains of Germany could be contaminated with U from fertilization.

The percentage of U from previous fertilizer applications in groundwater depends on regionally differing geogenous background concentrations in soils. In regions with a low natural U background more than 90% of U in groundwater can be already fertilizer-derived. It is technically no problem to extract U from drinking water. The costs for extracting 1 mg U from 1 cubic meter water are about 0.25 € cent (Riegel 2009). Smidt et al. (2011b) calculated that 2.1 t U are extracted annually from German aquifers with drinking water.

Removal of uranium from rock phosphates and the energetic and economic significance of uranium in phosphate fertilizers

The U demand for nuclear use is rapidly increasing, the supply of U from decommissioning nuclear weapons decreasing (Ragheb 2010) so that stock market prices for U can be expected to be increasing (Hu et al. 2008). This implies that investments in technology for the extraction of U from rock phosphates will not yield higher fertilizer prices.

U in rock phosphates has a high commercial value (Schnug and Haneklaus 2012). The amount of U applied annually by mineral P fertilizers during the last 10 years would have been sufficient to satisfy the energy demand of 2.4 million households of medium size based on an energy supply of 50 MW by 1 kg natural U and an electricity demand of 3.55 MW/household. Alternatively, 5.6 million hectares of forest would have been required to provide the same amount of fuel value by timber (assuming that 1 kg timber delivers 1 kW/h and a timber yield of 1.5 t/ha). Consequently, the use of mineral P fertilizers according to GAP on the basis of an U extraction process can be seen as a contribution to climate protection which adds up to 1.80 €/kg P when calculated on the figures of the CO_2 tax for cars. A monetary compensation of farmers in terms of a CO_2 (climate) bonus is expected to be motivating. Using the inherent energy in rock phosphates its natural U content may be a future option to produce higher quality phosphate fertilizers with reduced concentrations of contaminants (Haneklaus et al. 2014).

References

Anonymous (2011) Erste Verordnung zur Änderung der Trinkwasserverordnung vom 3. Mai 2011. Bundesgesetzblatt Jahrgang 2011 Teil 1 Nr. 21, ausgegeben zu Bonn 11. November 2011, 748-774

ATSDR (2013) Toxicological profile for uranium. 6. Potential for human exposure. 6.2. Releases to the environment. http://www.ncbi.nlm.nih.gov/books/NBK158802/

Baumann N, Arnold T, Read D (2008) Uranium ammunition in soil. In: de Kok LJ, Schnug E (2008.), 73-78

Beaugelin-Siller K, Février L, Gilbin R, Garnier-Laplace J (2011) Ecotoxicity of uranium in freshwaters: Influence of the physico-chemical status of the rivers. In: Merkel B, Schipek M (eds.) The New Uranium Mining Boom. Challenge and lessons learned. Springer, Berlin, 507-516

Birke M, Rauch U (2008) Uranium in stream water of Germany. In: de Kok LJ, Schnug E (2008), 79-91

Bertell R, Schmitz-Feuerhake I (2008) Radiological aspects of uranium contamination. In: de Kok LJ, Schnug E (2008), 1-10

Busby C, Schnug E (2008) Advanced biochemical and biophysical aspects of uranium contamination. In: de Kok LJ, Schnug E (2008), 11-22

de Kok LJ, Schnug E (2008.) Loads and fate of fertilizer-derived uranium. Backhuys Publishers, Leiden. 1-229

DüMV (2012) Düngemittelverordnung vom 5. Dezember 2012 (BGBl. I S. 2482). http://www.gesetze-im-internet.de/bundesrecht/d_mv_2012/gesamt.pdf

Garg B (2012) BARC: Uranium cause of Malwa water contamination. The Tribune, Candigarh, India - Main News February 19, 2012; http://www.tribuneindia.com/2012/20120219/main6.htm

Gramss G, Voigt K-D, Merten D (2011) Phytoextraction of heavy metals by dominating perennial herbs. In: Merkel B, Schipek M (eds.) The New Uranium Mining Boom. Challenge and lessons learned. Springer, Berlin, 421-431

Guttman J (1998) Defining flow systems and groundwater interactions in the multi-aquifer system of the Carmel Coast region. Ph.D. Thesis, Tel Aviv University

Haneklaus S, Schnug E (2008) A critical evaluation of phytoextraxtion on uranium-contaminated agricultural soils. In: de Kok LJ, Schnug E (2008), 111-126.

Haneklaus N, Schnug E, Tulsidas H, Reitsma F (2014) Using high temperature gas-cooled reactors for greenhouse gas reduction and energy neutral production of phosphate fertilizers, Annals of Nuclear Energy *(submitted)*

Hassoun R, Schnug E (2011) Contribution of mineral and tap water to the dietary intake of As, B, Cu, Li, Mo, Ni, Pb, U and Zn by humans. In: Merkel B, Schipek M (eds.) The New Uranium Mining Boom. Challenge and lessons learned. Springer, Berlin, 795-804

Henner P (2008) Bioaccumulation of radionuclides and induced biological effects in situations of chronic exposure of ecosystems – an uranium case study. In: de Kok LJ, Schnug E (2008), 23-32

Heshmati Rafsanjani M, Kratz S, Fleckenstein J, Schnug E (2008) Solubility of uranium in fertilizers. Landbauforschung – vTI / Agriculture and Forestry Research 58, 195-198

Hu Z, Zhang H, Wang Y, Haneklaus S, Schnug E (2008) Combining energy and fertilizer production - vision for China's future. In: de Kok LJ, Schnug E (2008), 127-134

Huhle B, Kummer S, Merkel (2008) Mobility of uranium from phosphate fertilizers in sandy soils. In: de Kok LJ, Schnug E (2008), 47-56

Imas P, Lati J (2005) Uranium in soils, fertilizers and crops. ICL Fertilizers Israel, 14.05.2005

Jaques D, Mallants D, Šimůnek J, Van Genuchten M Th (2008) Modelling the fate of uranium from inorganic phosphorus fertilizer applications. in agriculture. In: de Kok LJ, Schnug E (2008), 57-64

KBU (2012) Positionspapier der Kommission Bodenschutz beim Umweltbundesamt: Uran Einträge in landwirtschaftliche Böden durch Düngemittel Umweltbundesamt, Dessau, März 2012

Kratz S, Knappe F, Rogasik J, Schnug E (2008). Uranium balances in agroecosystems. In: de Kok LJ, Schnug E (2008), 179-190

Kratz S, Godlinski F, Schnug E (2011) Heavy metal loads to agricultural soils in Germany from the application of commercial phosphorus fertilizers and their contribution to background concentration in soils. In: Merkel B, Schipek M (eds.) The New Uranium Mining Boom. Challenge and lessons learned. Springer, Berlin, 755-762

Lottermoser B (2009) Trace metal enrichment in sugarcane soils due to the long-term application of fertilisers, North Queensland, Australia: geochemical and Pb, Sr, and U isotopic compositions. Australian Journal of Soil Research 47: 1–10.

Moinester M, Kronfeld J (2014) The radiological impact of using phosphate fertilizers. Third International Conference on Radioecology and Environmental Radioactivity, Barcelona, Sept. 7-12, 2014; http://radioactivity2014.pacifico-meetings.com/

Ragheb M (2010) Uranium resources in phosphate rocks. https://netfiles.uiuc.edu/mragheb/www/NPRE%20402%20ME%20405%20Nuclear%20 Power%20Engineering/Uranium%20Resources%20in%20Phosphate%20Rocks.pdf

Read D, Trueman E, Arnold T, Baumann N (2008) The fate of uranium in phosphate-rich soils. In: de Kok LJ, Schnug E (2008), 65-72

Riegel M (2009) Untersuchungen zur Elimination von natürlichen Uranspezies aus Wässern mit Hilfe schwach basischer Anionenaustauscher. Diss. Fakultät für Chemieingenieurwesen und Verfahrenstechnik der Universität Fridericiana Karlsruhe.

Rogasik J, Kratz S, Funder U, Panten K, Barkusky D, Baumecker M, Gutser R, Lausen P, Scherer HW (2008) Uranium in soils of German long-term fertilizer experiments. In: de Kok LJ, Schnug E (2008), 135-146

Salminen R (2005) Geochemical Atlas of Europe. Part 1: Background Information, Methodology and Maps, Geological Survey of Finland, Espoo; ISBN 951-890-921-3

Sattouf M, Kratz S, Diemer K, Fleckenstein J, Rienitz D, Schiel D, Schnug E (2008) Significance of uranium and strontium isotope ratios for retracing the fate of uranium during the processing of phosphate fertilizers from rock phosphates. In: de Kok, LJ, Schnug E (2008), 65-72

Schick J, Schroetter S, Lamas M, Rivas M, Kratz S, Schnug E (2008) Soil to plant interface of uranium. In: de Kok LJ, Schnug E (2008), 157-168

Schnug E, Birke M, Costa N, Knolle F, Fleckenstein J, Panten K, Lilienthal H, Haneklaus S (2008) Uranium in German mineral and tap waters. In: de Kok, L. J. and Schnug, E. (2008), 91-110

Schnug E (2012) Uran in Phosphor-Düngemitteln und dessen Verbleib in der Umwelt. Strahlentelex 612-613:1-8

Schnug E, Haneklaus N (2012) Environmental prospects of uranium from mineral phosphates. ATW 57: 750-752

Schnug E, Lottermoser BG (2013) Fertilizer-derived uranium and its threat to human health. Environ Sci Technol 47: 2433-2434

Setzer S, Julich D, Gäth S (2011) Sorption behavior of uranium in agricultural soils. In: Merkel B, Schipek M (eds.) The New Uranium Mining Boom. Challenge and lessons learned. Springer, Berlin, 579-584

Smidt GA, Landes FC, Machado de Carvalho L, Koschinsky A, Schnug E (2011a) Cadmium and Uranium in German and Brazilian Phosphorus Fertilizers. In: Merkel B, Schipek M (eds.) The New Uranium Mining Boom. Challenge and lessons learned. Springer, Berlin: 167-175

Smidt GA, Hassoun R, Birke M, Erdinger L, Schäf M, Knolle F, Utermnn J, Duijnisveld HM, Birke M, Schnug E (2011b) Uranium in German tap and groundwater – Occurrence and origins. In: Merkel B, Schipek M (eds.) The New Uranium Mining Boom. Challenge and lessons learned. Springer, Berlin, 807-820

Smirnova V S, Gudkov SV, Shtarkman IN, Chernikov AV, Bruskov VI (2005) The genotoxic action of uranyl ions on DNA in vitro caused by the generation of reactive oxygen species. Akademija Nauk SSSR 50: 456-463

Takeda A, Tsukada H, Takaku Y, Hisamatsu S, Nanzyo M (2006) Accumulation of uranium derived from long-term fertilizer applications in a cultivated Andisol. Science of the Total Environment 367: 924–931

Taylor M, Kim N (2008) The fate of uranium contaminants of phosphate fertilizers. In: de Kok LJ, Schnug E (2008), 147-155

Thomas PA (2008) Food chain transfer of uranium series radionuclides. In: de Kok LJ, Schnug E (2008) 169-178

Tunney H, Stojanovic M, Mrdakovic Popic J, McGrath D, Zhang C (2009) Relationship of soil phosphorus with uranium in grassland mineral soils in Ireland using soils from a long term phosphorus experiment and a National soil database. J Soil Sci Plant Nutr 172: 346-352

Utermann J, Fuchs, M. (2008) Uranium in German soils. In: de Kok LJ, Schnug E (2008), 33-46

Utermann J, Schnug E (2013) Uran - Chemisch-physikalische Eigenschaften. In: Litz N, Wilcke W and Wilke B.-M (Hrsg.) Bodengefährdende Stoffe: Bewertung - Stoffdaten - Ökotoxikologie – Sanierung. WILEY-VCH, 1-32

Winde F (2011) Challenges in assessing uranium-related health risks: two case studies for the aquatic exposure pathway from South Africa - Part I: Guideline and toxicity issues and the Pfadder case study. In: Merkel B, Schipek M (eds.) The New Uranium Mining Boom. Challenge and lessons learned. Springer, Berlin, 529-538

Wu W-M, Carley J, Green S J, Luo J, Kelly SD, van Nostrand J, Lowe K, Mehlhorn T, Caroll S, Boonchayanant B, Löffler FE, Watson D, Kemner KM, Zhou J, Kitanidis PK, Kostka JE, Jardine PM, Criddle C (2010) Effects of nitrate on the stability of uranium in a bioreduced region of the subsoil. Environ Sci Technol 44: 5104–5111

Uranium and Molybdenum transfer within the oxidized zone of uranium deposit

Irina Semenova[1], Vladislav Petrov[1], Yana Bychkova[1], Lyubov Shulik[1], Jörg Hammer[2]

[1]Institute of Geology of Ore Deposits, Petrography, Mineralogy and Geochemistry of Russian Academy of Sciences (IGEM RAS), 35 Staromonetny per., Moscow, Russia, Irina-V-Semenova@yandex.ru
[2]Bundesanstalt für Geowissenschaften und Rohstoffe, Stilleweg 2, 30655 Hannover, Germany, Joerg.Hammer@bgr.de

Abstract. Total Dissolved Solids (ion composition), concentrations of some trace elements and pH-values determination results of vein-fractured water samples from the open pit of the Tulukuevskoe U-Mo deposit in acid volcanic rocks of Transbaikalia are presented. Rest upon obtained data it is possible to assume that in the southern part of the open pit a flat-lying ore-bearing untapped structure exists. An inverse relationship between the content of hydrocarbonate ions in vein-fractured waters and concentration of uranium is presented.

Introduction

The elements and compounds uncovered and liberated through mining and processing, which are not usually part of the ecological systems (in such form or concentration) have the potential to alter the environment to its detriment. Matter of concern is that high concentrations of contaminants can be released into the surrounding aquifers. Besides environmental aspects a deep understanding concerning the migration and accumulation behavior of uranium and related elements is necessary. Consideration of these directions requires studying of the chemical composition of vein-fractured water samples. This contribution reports preliminary laboratory results, which represents trace element concentrations and ion composition in water streams around Tulukuevskoe U-Mo deposit mined by the open pit, which located within the Streltsovskoe ore field in southeastern Transbaikalia.

The Tulukuevskoe open pit (TOP) is a unique object for investigating the correlation between modern processes of the oxidation of uraniferous rocks, residence time of meteoric waters in the enclosing fractured-porous rocks, and mech-

B.J. Merkel and A. Arab (eds.), *Uranium – Past and Future Challenges*,
DOI 10.1007/978-3-319-11059-2_15

anisms of the migration of actinides at different distances from the dynamic influence zones of steeply-dipping faults (Petrov et al. 2006).

We studied five springs located at different height levels of the TOP at different distances from the steeply-dipping fault 1A with stockwork and vein type uranium mineralization.

Research object description

The Tulukuyevskoe uranium deposit is located in the central part of the Streltsovskoe ore field and controlled by northwest-trending Tulukuyevsky fault-fractured zone near its joint with submeridional Argunsky zone. Deposit occupy an area of 0,3 sq.km, it is extended in the northwest direction on 1300 meters with a width from 150 to 350 meters (Ishchukova et al. 1998).

The Tulukuevskoe U-Mo deposit is localized in acid Mesozoic (140 Ma) volcanic rocks. Late Mesozoic hydrothermal processes in the wall rocks are strongly connected with fracture network and classified (Andreeva and Golovin 1988) as pre-ore, ore-related and post-ore mineral association. In breeds hydrothermal mineral zonality (internal, intermediate and external zones) which reflects total effect of several stages of Late Mesozoic mineralogenesis is shown: pre-ore metasomatic changes (hydromicatization), the ore-related of veins-metasomatic changes and post-ore quartz-carbonate veins and argillization (kaolinite, smectite) (Andreeva and Golovin 1998; Poluektov et al. 2007). Hydrothermal mineral zonality is controlled by steeply dipping faults.

The deposit was mined by the open pit during 1972-1998. Vein stockwork ore bodies are controlled by steeply dipping fault and their close rounded systems of cracks whereas ore bodies of sheeted type in zones of flat faults are presented. Nowadays ore-bearing untapped vein structures are localized in a northwestern and southern parts of the open pit (fig. 1).In the first case the ore body is dated for a zone of northwest-trending fault 1A, and in the second is controlled by a diagonal fault 2A of submeridional direction. Vein-fractured water sources from the open pit are connected with ore-bearing fault that reflects modern hydraulic activity of these rupture structures.

The uranium mineralization on time of formation is divided (Petrov et al. 2006) into three groups: 1) the hypogene; 2) hypergene in connection with an ancient zone of oxidation and 3) hypergene in connection with modern oxidation. The hypogene mineralization is presented by two types of primary ores: mainly uranium in zones the steeply-dipping faults and uranium-molybdenum in zones of flat faults and fractures on contacts of layers. In uranium ores prevails pechblende, and also coffinite and thucholite, containing up to 50 mass.% carbon and more than 10 mass.% uranium. Primary ores are subject to processes ancient (before the open pit mining) the oxidation expressed by three subzones (from top to bottom): leaching, full oxidation and incomplete oxidation. The ancient zone of oxidation belongs to hydroxide-silicate type (Belova and Fedorov 1977) for which gradual transition of primary ores to hydroxides and silicates of uranium with preservation of morphology of allocations is characteristic. In sites of the secondary enrichment interfaced to level of the mirror of underground waters before opening of the pit, uranium blacks are formed. At the front of the modern oxidation connected with mining of the pit, are formed uranophane, calcurmolit, liebigit, etc. Level of standing of the mirror of underground waters and zone of secondary uranium enrichment corresponds to changes during the open pit mining. Hypergene changes of the rock sequence were accompanied by formation of Fe-Mn hydroxides (goethite, Fe-vernadite, hematite, ferrihydrite) which are active sorbents for uranium, and also intensive development of jordisite (MoS_2), which traces zones of flat faults and fractures (Petrov et al. 2011).

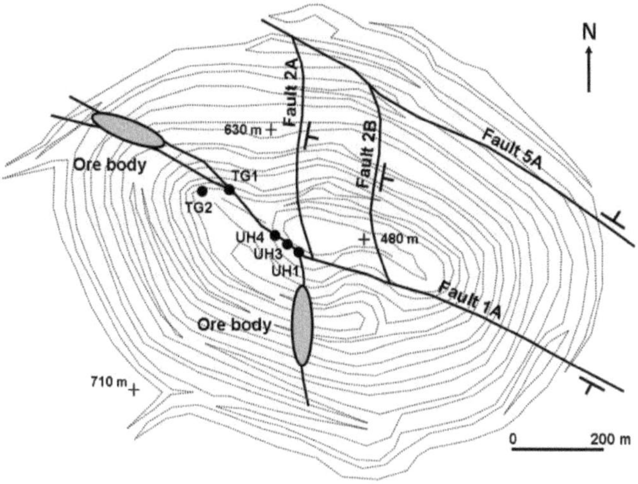

Fig.1. Arrangement and numbers of underground water sources from the open pit of the Tulukuevskoe U-Mo deposit. Tracks of the main steeply-dipping faults, projection of ore bodies to the horizontal plane, high-rise marks and sources of vein-fractured waters are shown.

Results

Total Dissolved Solids (ion composition), concentration of some trace elements and pH-values of vein-fractured water samples from the TOP are presented. Water samples were collected in north-western and central parts of the TOP in five vein-fractured water sources located in low open-pit bench along the main NW-SE fault 1A zone during 2011 fieldworks. The aliquot for analysis was placed in hermetic plastic containers with a volume of 0.5 liters. The samples poised by means of adding nitric acid at the rate of 3 ml of acid per 100 ml of water. Analysis of trace elements was carried out by inductively coupled plasma mass spectrometry (ICP-

MS) in IGEM RAS. Thus, content of thirty seven chemical elements in samples have been established.

Determination of anion-cation composition and pH-value was carried out by electrometric and titrimetric methods of the analysis. Concentrations of macro components in samples is submitted in fig. 2.

Ion composition variation

Total Dissolved Solids (ion composition) of natural water is the most important issue reflecting stability and macro element accumulation capability according to various conditions such as mineral composition of water-bearing material, geological and oxidation-reduction conditions, water exchange activity, degree of rock desalination and desorption.

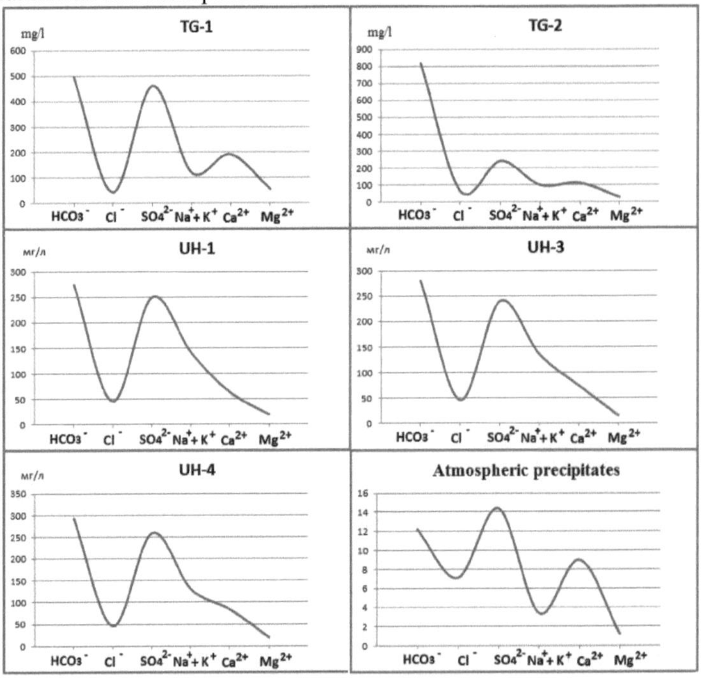

Fig.2. Total dissolved solids of vein-fractured water samples and atmospheric precipitates.

Concentration of hydrocarbonate, chloride and sulfate ions and sodium-potassium-calcium-magnesium ions in water samples of UH source group are quite close. The general mineralization varies from 796 to 832 mg/l. Basing on chemical composition they treat sulphate-hydrocarbonate-chloride magnesium-calcium-sodium waters. In samples TG-1 and TG-2 concentration of sodium- and potassium ions decreases in comparison with UH sources to 120 and 99 mg/l, respectively. In vein-fractured waters of TG-1 and TG-2 sources the contents of hydrocarbonate ions is much higher, than in UH samples and averages 42 equivalent % (494 and 817 mg/l). However on cationic composition of TG-1 source treats sulfate-hydrocarbonate-chloride water, and TG-2 source to hydrocarbonate-sulfate-chloride. According to cationic structure both sources belong to calcium-sodium-magnesium waters. The total mineralization in TG-1 and TG-2 sources as a whole is slightly higher than in UH sources and makes 1361 and 856 mg/l. pH changes slightly (8.1-8.2) except for sample TG-1 (7.6). The acid-alkali balance in atmospheric precipitation is displaced towards acidulous reaction (pH 6.2).

Thus, rest on pH values of water samples change slightly from 7.6 to 8.3, and the advance speed of a water stream is as a first approximation identical to north-east and southern parts of the open pit, it is possible to assume that high concentration of molybdenum in springs of UH group testify about much larger concentration of molybdenite as associate in the pechblende ores which are localized here. The increased concentration of manganese in waters of UH group sources indicate more intensive leaching of Fe-Mn hydroxides (goethite, Fe-vernadite, hematite, ferrihydrite) from enclosing rocks of this part of the open pit.

Analysis for Metals using ICP-MS

Concentrations of some components are illustrated in fig. 3. Results of earlier researches (Semenova et al. 2011) showed that concentration of trace elements (molybdenum and manganese are given in this work) distinctly break up to two groups. This fact is confirmed also by the new data obtained. So, considerably larger amount of molybdenum is characteristic for sources of underground waters of UH group. Thus the difference between the content of elements in TG and UH source groups varies from 339 to 4534 ppb.

Molybdenum belongs to a number of elements (U, Se, As) which are not dissolved in reducing conditions and are dissolved in the oxidizing environment. At advective and diffusive advance of the front of oxidation such elements are leached from a mineral matrix and can be transferred by the water stream to various distances. Distinctions in concentration (intensity of migration) of molybdenum in similar geological conditions depend, first of all, from: 1) the oxidation-reduction situation caused by the content of oxygen in the environment and infiltration waters; 2) distribution and reactionary ability of organic substances and other potential reducers in the water-bearing horizon; 3) speeds of circulation of underground waters; 4) form in which the element contains in the environment, i.e. mineral-chemical composition of ore bodies and sedimentary-volcanic rocks.

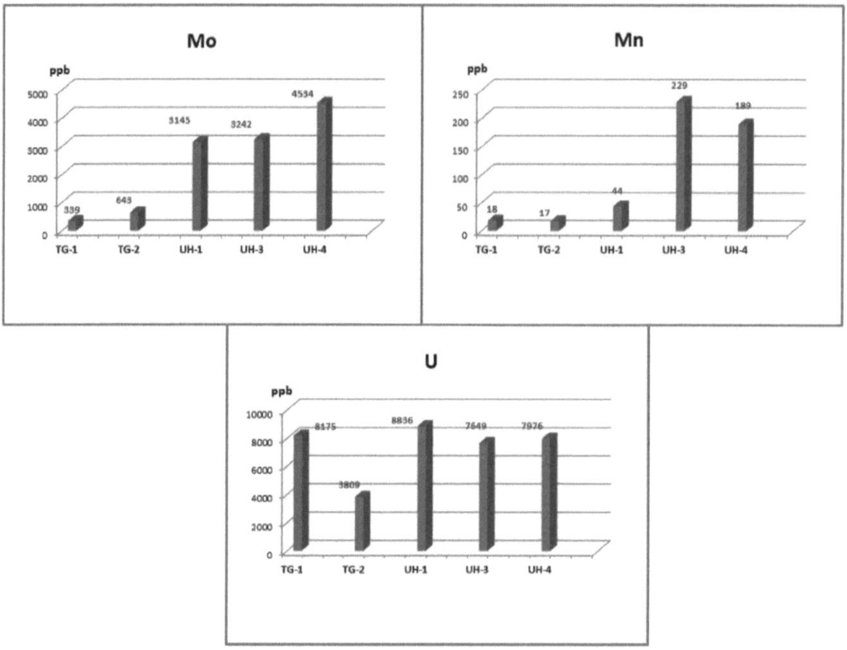

Fig.3. Molybdenum, manganese and uranium content in vein-fractured water samples.

Conclusions

Fieldworks and laboratory tests of trace element concentrations in and anionic-cationic composition of water sources of the Tulukuevskoe U-Mo deposit mined by the open pit located within the Streltsovskoe ore field in southeastern Trans-baikalia allowed us to establish the following.

Two groups of groundwater sources (TG in north-western part and UH in central part of the open pit) which are controlled by one NW-trending hydraulically active fault zone are characterized by sharply distinct concentrations of the chemical elements.

Elevated concentrations of molybdenum in the waters of central part of the pit indicate elevated content of molybdenite in ores and reflect the intensive decomposition with the formation of jordisite, accompanied by leaching and outflow of molybdenum.

As some steeply dipping U-Mo ores are controlled by zone of flat-lying faults and fractures along rock contacts, it is possible to assume that in the southern part of the open pit a flat-lying ore-bearing untapped structure exists.

The inverse relationship between the content of hydrocarbonate ions in vein-fractured waters and concentration of uranium is presented.

References

Andreeva O V, Golovin V A (1998) Metasomatic processes in uranium ore deposits of Tulukuev caldera in the East Transbaikal region (Russia). J Geology of ore deposits 40: 205-220 (in Russian)

Belova L N, Fedorov O V (1977) Some data on mineral structure of a oxidation zone of uranium deposit of the Streltsovskoe ore field. J Materials on geology of uranium deposits 45: 65-76 (in Russian)

Ishchukova L P, Igoshin Yu A, Avdeev B V (1998) Geology of the Urulyunguyevsky ore area and molybdenum-uranium deposits of the Streltsovskoe ore field: 293 (in Russian)

Petrov V A, Poluektov V V, Andreeva O V, Golubev V N, Dubinina E O, Kartashov P M, Ovseychuk V A, Schukin S I, Lespinasse M, Hammer J (2006) Migration of uranium and geochemical barriers in the Tulukuevskoe deposit aeration zone, SE Transbaikalia. Biosphere Geochemistry conference Materials: 290-292 (in Russian)

Petrov V, Poluektov V, Hammer J, Schukin S (2008) Fault-related barriers for uranium transport. Uranium Mining and Hydrogeology. B.J. Merkel and A. Hasche-Berger (eds.). Springer-Verlag Berlin Heidelberg: 779-789

Petrov V, Poluektov V, Hammer J, Schukin S (2011) Uranium mineralization in fractured welded tuffs of the Krasnokamensk Area: transfer from ancient to modern oxidizing conditions. The New Uranium Mining Boom: Challenge and lessons learned. B.J. Merkel and M. Schipek (eds.). Springer-Verlag Berlin Heidelberg: 701-710

Poluektov V V, Petrov V A, Golubev V N, Nadyarnykh G I (2007) Migration and uranium accumulation in progress of hydrothermal and hypergene transformations of acid volcanic rocks of Tulukuevskoe deposit, SE Transbaikalia. Conference Materials IGEM RAS: 157-161 (in Russian)

Semenova I V, Petrov V A, Shazzo Yu K (2011) Use of ICP-MS analysis data of water samples for an assessment of mineral structure of Tulukuevskoe uranium deposit ore bodies. Conference Materials. Minerals: structure, properties, research methods: 259-261 (in Russian)

Love is like radioactivity
Strong in the beginning but getting less and less over time.

Release of uranium from weathered black shale in meso-scale reactor systems – first year of data

Viktor Sjöberg[1], **Stefan Karlsson**[1]

[1]Man-Technology-Environment Research Centre, Örebro University, 70182 Örebro, Sweden

Abstract. In Kvarntorp, Sweden, 40 million m^3 of shale residues containing up to 150 ppm uranium is the result from some 30 years of oil extraction. This work focuses on the release of uranium from discarded unprocessed shale in contact with wood chips in 1 m^3 scale. After 15 months of incubation it was clear that shale mixed with wood chips generated leachates with a lower average content of uranium (14 µg L^{-1}) then shale covered with wood chips (1300 µg L^{-1}). This indicates that natural forestation of the pile might increase the release of uranium.

Introduction

In 1939 the Swedish government began to explore domestic sources of oil to supply the military with liquid fuels. The most promising source was the oil bearing black shale in Kvarntorp, some 160 km west of Stockholm and 15 km south of Örebro. In 1940 an oil-extraction plant was built on site and two years later started the production of oil using a pyrolysis process and this continued until the mid 60ies. Due to technical limitations shale smaller than approximately 10 mm had to be discarded in order to avoid sintering in the kilns. The discarded as well as the pyrolysed shale were piled on site in what today is known as the "Kvarntorpshögen" (the Kvarntorp pile). During the time of operation some 40 million cubic meter of discarded and pyrolysed shale were produced.

Similar to other shale deposits worldwide the shale in Kvarntorp was rich in valuable elements such as vanadium, molybdenum and uranium. Due to the high temperature process and ongoing oxidation in the pile, still 40 years after closure the temperature reached up to 800 °C a few meters below the surface (Holm 2005, in Swedish). Infiltration of rain water has therefore been effectively prevented why the element content in the discarded and pyrolysed shale is essentially the same today as when it was excavated. The inventory of valuable metals such as vanadium, molybdenum and uranium has been estimated to be at least 10'000 ton for vanadium and 4'000 ton for molybdenum and uranium, respectively.

B.J. Merkel and A. Arab (eds.), *Uranium – Past and Future Challenges*,
DOI 10.1007/978-3-319-11059-2_16

Since the closure, grass, shrubs and small trees such as birch and aspen have started to grow on the surface, thus increasing the amount of organic carbon compounds such as cellulose in the surface. By increasing the amount of metabolizable carbon the populations of heterotrophic microorganisms often also increases as a result of the increased availability of energy. Consequently increases the diversity and content of potentially metal complexing ligands produced by microorganisms (Krebs *et al.* 1997). This simple principle is widely used for leaching of several valuable elements from low grade sources such as shale (Kalinowski *et al.* 2004; Kalinowski *et al.* 2006; Edberg *et al.* 2010). However, also the opposite might occur since e.g. saw dust can act as an adsorbent for uranium in aqueous solution (Bagherifam *et al.* 2010). Immobilization might also occur due to interactions with fungal metabolites as described by Ogar *et al.* (2014).

In this work we are exploring some of the impact from organic carbon, released during decomposition of wood chips, on the mobilization of uranium from the discarded black shale. This will serve two purposes, firstly, it improves information on the behaviour of the pile when it finally cools and secondly it provides information if it is possible to use heterotrophic degradation of complex carbon sources in order to increases the leaching and recover uranium from the shale residues.

Materials and Methods

Wood chips and shale

Wood chips of *Salix viminalis* were bought locally and used as delivered why naturally occurring microorganisms were present. The average size was 50x20x5 mm and the water content approximately 25 %, according to the supplier.

From the south slope of the Kvarntorp-pile approximately 1.2 m^3 discarded shale was collected using a wheel-loader. On the test site it was unloaded on a polyethylene tarpaulin and mixed gently with a shovel. Five samples of 100 g each were collected randomly during mixing for characterization of the shale e.g. its element content. After mixing the five samples together thoroughly, one fifth was grinded until approximately 70 % passed through a 0.56 mm mesh. This material was then characterized by a slightly modified Tessier-leaching (Tessier et al. 1979). Five steps were performed but each on fresh samples and not in sequence as in the original procedure. The following steps were conducted, with their operational species given in parenthesis:

1. De-ionized water (18.2 MΩ) (water soluble elements)
2. 1 M NH_4Ac, pH 7.00 (ion exchangeable elements)
3. 1 M NH_4Ac, pH 5.00 (carbonates and amorphous hydroxides)

4. 20 mM HNO_3 in 30 % H_2O_2 followed by 3.2 M NH_4Ac in 20 % HNO_3 and double volume water (oxidizable elements)
5. Microwave assisted digestion in concentrated HNO_3 (acid extractable elements)

Step 1 to 4 were performed in duplicates at a liquid to solid ratio (L/S) of 10 using 1 gram of shale while step 5 was performed at L/S 100 using 0.1 gram of shale and five replicates. Step 1 and 2 were performed at 18 °C during constant agitation on an overhead shaker (Heidolph Reax 2) at 30 rev. min^{-1} for 24 and 4 hrs, respectively. Step 3 and 4 were performed at 90 and 85 °C, respectively, in a thermostatic water bath with manual agitation every 20 minutes. Step 5 was performed in sealed Teflon bombs in a microwave oven (CEM MarsV) for 1 hr. Before metal analysis all samples were filtered through 0.20 μm polypropylene filters.

Construction of reactors and sampling

Two reactors holding 1 m^3 each were constructed from polyethylene IBCs (Intermediate Bulk Container). Reactor 1 was filled with shale and wood chips in 27 layers containing 40 L each, with the first and final layers containing wood chips. Reactor 2 was filled with 620 L shale and then 180 L wood chips mixed with 65 kg alkaline LD-slag. After the reactors were filled and the drainage channels closed, 50 L tap water was added to each reactor. One week later the reactors were drained and the leachates were collected. Then each reactor was refilled with 10 L tap water and the procedure was repeated approximately once a week for the following 15 months.

Chemical analysis and modelling

After settling for 30 minutes, 10 ml of each leachate was decanted and acidified with HNO_3 to a final acid concentration of 2 %. Then the content of uranium was measured by ICP-MS (Agilent 7500cx) using the Merck 10580 “Multi Element Standard VI solution” for external calibration. Electrical conductivity (EC) (Radiometer CDC866T) and pH (Metrohm 6.0257.000) were measured in the remaining original samples. Calibration of the EC-probe was done as a single point calibration at 1413 μS cm^{-1} using a 0.01 M KCl solution. Calibration of the pH-probe was done using buffer solutions with pH of 4.00 and 7.00. Dissolved organic carbon (DOC) was measured using a TOC-analyser (Shimadzu TOC-V CPH). For geochemical modelling VisualMinteq v3.0 was used.

Uranium recovery

In order to recover uranium from the leachates three different methods were tested i) solvent extraction, ii) precipitation and iii) adsorption. All experiments were performed in 50 ml polypropylene test tubes and agitated with an overhead shaker (Heidolph Reax 2) at 30 rev. min^{-1}.

Solvent extraction was performed with kerosene and trioctyleamine (TOA) as the organic phase due to the estimated high content of sulphate in the leachates. TOA was added to the kerosene to give final concentrations in the kerosene of 0.1, 0.2 and 0.5 mol L^{-1}. After agitation for 12 minutes the samples were left for 5 minutes and then the uranium content in the aqueous solution was measured.

Precipitation was performed by addition of ammonia to the leachates in order to precipitate hydroxides. Final concentrations of 0.08, 0.17 and 0.33 mol L^{-1} ammonia were tested. Then the samples were agitated for 5 minutes followed by centrifugation at 3000 g for 20 minutes. After centrifugation the content of uranium in the supernatant was measured.

Adsorption was tested by addition of expanded perlite (> 70 % SiO_2, size 2-4 mm) with an estimated cation exchange capacity of over 30 milliequivalents per 100 g (c.f. Alkan and Doğan 2001) to the leachates. Three different amounts was tested, 2.5, 25 and 50 g L^{-1} and the samples were agitated for 1 hr. Then the aqueous phase was filtered through 0.20 µm polypropylene filters to remove perlite particles before measuring the uranium content.

Results and Discussion

Characterization of shale

After leaching with de-ionized water for 24 hrs the EC and pH reached 2.0 mS cm^{-1} and 2.1, respectively. This indicates that the shale, even though it has been weathered for more than 40 years, still contains substantial amounts of easily mobilized charged species as well as acid forming minerals. On average this phase contained 1 g ton^{-1} uranium. Geochemical modelling showed that the dominating aqueous uranium species were UO_2SO_4 (71 %), UO_2^{2+} (21 %) and $UO_2(SO_4)_2^{2-}$ (8 %). Saturation indices lower than 10^{-8} was estimated for all predicted uranium-bearing minerals, which proved that no secondary minerals were precipitated during the leaching. The acid extractable fraction of uranium was 30 g ton^{-1}. This was slightly lower than expected but not unreasonable since the uranium might be heterogeneous distributed in the excavated shale. The ion exchangeable fraction was almost equal to the water soluble fraction and accounted for 2.4 % of the acid extractable uranium. Approximately 20 % of the acid extractable uranium was present as carbonates, amorphous hydroxides and oxidizable species with a probable origin in secondary precipitates. It is therefore likely that substantial amounts of uranium can be released from the shale with only minor changes in the composition of the percolating water such as pH, redox potential, complexing agents e.g. organic acids etc.

Leachates from reactors

Both reactors showed seasonal variations in their leachate composition. For EC, uranium and DOC the highest readings were observed during periods with high outdoor temperatures i.e. June to August (fig.1). Similar trends were also observed for other measured elements such as aluminium, manganese and iron which will be discussed in forthcoming papers. Reactor 1 produced leachates with lower EC during 2013 than reactor 2 (fig.1). However, the correlation between EC and outdoor temperature was moderate with R-values of 0.34 and 0.37 for reactor 1 and reactor 2, respectively, probably due to a lag in the temperature distribution. By excluding outliers, R-values of 0.50 and 0.56 were obtained.

In contrast to EC and uranium, pH showed no seasonal variation but instead a pronounced non-continuous increase over time (fig.1). For reactor 2 no strong correlations was found for pH and other parameters (R = -0.49 to -0.28). No strong correlations (R = -0.42 to 0.15) except between uranium and pH were found for reactor 1. From this reactor the correlation between uranium and pH was strongly negative (R = -0.88) which could be due to an increased adsorption of mobilized uranium to the wood chips, in accordance with Bagherifam *et al.* (2010) who reported increased adsorption of uranium to saw dust as pH increases or due to immobilization by e.g. fungal metabolites (c.f. Ogar *et al.* 2014).

Reactor 2 produced leachates containing some 100 times more uranium (fig.1). For this reactor the correlation between uranium and outdoor temperature was higher (R = 0.71) than the corresponding correlation for reactor 1 (R = -0.12). In this case very few outliers were detected in the data material why the R-values for data without outliers were practically the same. This indicates that in reactor 2, uranium is mobilized by heterotrophic activity in the wood chips that releases complexing ligands that dissolves uranium by complexolysis. Because of the added LD-slag it is possible that uranium mobilization might be influenced by complexing ligands formed by alkaline hydrolysis of organic carbon. But, because of the correlation with temperature it seems more likely that microorganisms use the products from the alkaline hydrolysis as a carbon source while they produce e.g. organic acids that mobilize uranium (c.f. Sjöberg *et al.* 2011).

Modelling of the leachates was done for reactor 2 using mean values from the 15 months of sampling. The dominating uranium species were similar to the water leaching, i.e. UO_2SO_4 (66 %), UO_2^{2+} (19 %) and $UO_2(SO_4)_2^{2-}$ (8 %). In addition to the water leachates also organic uranium complexes (7 %) were present. According to the results addition of wood chips did not induce formation of secondary minerals since the saturation indices were below 10^{-7} for all predicted uranium minerals.

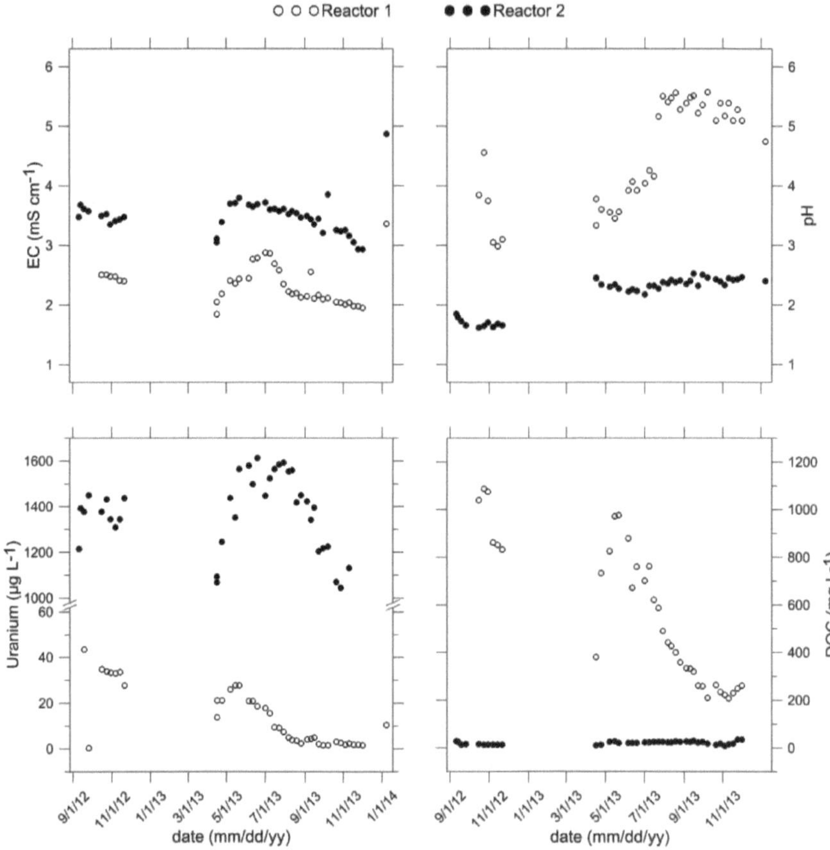

Fig.1. Seasonal variation of the leachates electrical conductivity (EC), pH and concentrations of uranium and dissolved organic carbon (DOC).

Uranium recovery

Only reactor 2 produced leachates with a content of uranium high enough to motivate recovery of the element. Also the unadjusted pH of the leachates was quite close to the optimum pH for solvent extraction using TOA. For all experiments of uranium recovery the leachate collected 10/24/12 was used (EC = 3.53 mS cm^{-1}, pH = 1.65, U = 1400 µg L^{-1} and DOC = 12.4 mg L^{-1}).

With 0.5 mol L^{-1} TOA in kerosene the recovery of uranium was approximately 10 %. Lower concentrations of TOA decreased the recovery exponentially. TOA extracts anionic species at low pH why the amount of recovered uranium corresponds with the fraction of anionic uranium predicted by the geochemical modelling. The limited extraction also indicates that the remaining dissolved uranium formed species with higher stability than $UO_2(SO_4)_2^{2-}$. According to instability constants reported by Lurie (1975) (cited in Nasab 2014) the instability of UO_2SO_4 is about five orders of magnitude lower than the instability of $UO_2(SO_4)_2^{2-}$ which prove the experimental results. Even though only a limited fraction of organic complexes were estimated by the geochemical modelling the impact of organic carbon on the speciation cannot be neglected since it has been found that the stability constant of uranium and humic substances could be as high as 10^6 (Riggle and von Wandruszka 2008), however, at a slightly higher pH.

Precipitation with ammonia removed uranium almost quantitatively from the leachate. With the lowest addition of ammonia more than 99 % of the available uranium was precipitated. Hence, it was estimated that with a final concentration of 5 to 50 mmol L^{-1} ammonia 95-99 % of the available uranium would precipitate.

Under the prevailing conditions no uranium was adsorbed by the perlite. Although perlite has been shown to have strong cation exchanging properties when it comes to e.g. copper, lead and cadmium (c.f. Alkan and Doğan 2001; Malakootian *et al.* 2011) the different ionic radius and stereochemistry of UO_2^{2+} might cause a decreased affinity between uranium and perlite. The low adsorption might also be caused by an unfavourable pH_{zpc} of the perlite. The corresponding value for quartz is 1-3 why the low pH of the leachates might have made adsorption impossible.

The low recovery of uranium by solvent extraction and adsorption is probably caused by a combination of factors such as unfavourable pH and formation of uranium-species in the leachate that has higher stability than both uranium-TOA and uranium-perlite. Since wood chips are in contact with alkaline LD-slag it is not unlikely that polyhydroxy acids such as isosaccharinic acid are involved in complexation or facilitates the mobilization of uranium (c.f. Svensson *et al.* 2007).

Conclusion

Depending on the interactions between shale and organic carbon different compositions of the leachates can be expected. In the case of degradation of organic carbon above the shale higher uranium content and lower pH in leachates can be expected than if the organic carbon is completely mixed with the shale. The uranium species in the leachates seems to have high stability and are not easily retained. However, recovery of uranium from the leachates was possible by unspecific precipitation that consequently needs to be followed by further purification.

Acknowledgements

The authors acknowledge; the municipality of Kumla for access to the Kvarntorp area and permission to collect samples, SAKAB AB for providing resources at the experimental site, the Faculty of Economy, Science and Technology at Örebro University for financial support and Prof. Bert Allard for scientific reviewing of the paper.

References

Alkan M, Doğan M (2001) Adsorption of copper(II) onto perlite. J. Colloid Interface Sci. 243:280-291

Bagherifam S, Lakzian A, Ahmadi SJ, Rahimi MF, Halajnia A (2010) Uranium removal from aqueous solutions by wood powder and wheat straw. J. Radioanal. Nucl. Chem. 283:289-296

Edberg F, Kalinowski BE, Holmstrom SJM, Holm K (2010) Mobilization of metals from uranium mine waste: the role of pyoverdines produced by Pseudomonas fluorescens. Geobiology 8(4): 278-292

Holm T (2005) Kvarntorpsområdet, studie av Kvarntorpshögen. Sveriges Geologiska Undersökning.

Kalinowski BE, Oskarsson A, Albinsson Y, Arlinger J, Odegaard-Jensen A, Andlid T, Pedersen K (2004) Microbial leaching of uranium and other trace elements from shale mine tailings at Ranstad. Geoderma 122(2-4): 177-194

Kalinowski BE, Johnsson A, Arlinger J, Pedersen K, Odeggrd-Jensen A, Edberg F (2006) Microbial mobilization of uranium from shale mine waste. Geomicrobiol. J. 23(3-4): 157-164

Krebs W, Brombacher C, Bosshard PP, Bachofen R, Brandl H (1997) Microbial recovery of metals from solids. FEMS Microbiol. Reviews 20:605-617

Makalootian M, Jaafarzadeh N, Hossaini H (2011) Efficiency of perlite as a low cost adsorbent applied to removal of Pb and Cd from paint industry effluent. Desal. Water Treat. 26:243-249

Nasab ME (2014) Solvent extraction separation of uranium(VI) and thorium(IV) with neutral organophosphorus and amine ligands. Fuel 116:595-600

Ogar A, Grandin A, Sjöberg V, Turnau K, Karlsson S (2014) Stabilization or uranium(VI) at low pH by fungal metabolites: Applications in environmental biotechnology. APCBEE Procedia In Press

Riggle J, von Wandruszka R (2008) The stability of uranium(VI) complexes of humates and fulvates in biphasic systems. Ann. Environ. Sci. 2:1-6

Sjöberg V, Grandin A, Karlsson L, Karlsson S (2011) Bioleaching of shale – Impact of carbon source. In: The New Uranium Mining Boom, Challenge and lessons learned. Merkel and Schipek (eds).

Svensson M, Berg M, Ifwer K, Sjöblom R, Ecke H (2007) The effect of isosaccharinic acid (ISA) on the mobilization of metals in municipal solid waste incineration (MSWI) dry scrubber residue. J. Hazard. Mater. 144:477-484

Tessier A, Campbell PGC, Bisson M (1979) Sequential extraction procedure for the speciation of particulate trace metals. Anal. Chem. 51(7):844-850

Environmental Issues and Proposed Assessment of Feasibility of Remediation of the Legacy Sites of Mining and Milling in the Area of Sumsar-Shekaftar in Kyrgyzstan

Isakbek Torgoev[1], **Alex Jakubick**[2]

[1]Institute of Geomechanics and Mining of the National Academy Sciences, Bishkek, Kyrgyz Republic
[2]UMREG, Uranium Mining and Remediation Exchange Group Inc., Darwin, NT, Australia

Abstract. From 1946 to 1978, extensive mining of U and Pb-Zn ores went on in the northern edges of the Fergana valley on the foothills of the Tien Shan Mountain Range. The mining and milling activities left behind large volumes of wastes in the lower section of the Sumsar-Say, which is a cross boundary river. The mill tailings of the Pb-Zn milling in Sumsar and the low-grade U ore heaps in Shekaftar need an urgent remedial action because the legacy sites became chronic sources of toxic and radioactive contaminants for the Sumsar-Say River. In the Sumsar Say valley, the river acts as a powerful transport agent of the released contaminants carrying them to the places of potential impact, which are the downstream settlements in Kyrgyzstan and Uzbekistan. For the settlements along the river, Sumsar- Say is the main source of water and the population uses it indiscriminately. Thus, the exposure of the local population to the contaminants is via a multitude of pathways. The legacy sites requiring remediation most are (a) the tailings facility #1 in Sumsar that is chronically loosing tailings into the Sumsar-Say River and (b) the low-grade U ore heaps in Shekaftar, which are within the domain of the river, particularly ore heap no. 5, which is directly located in the river. The paper discusses possibilities of remediation and, regarding the risk and feasibility assessment, arrives at the conclusion that sustainability of remediation would be best achieved by taking out the sources of contaminants from the reach of the Sumsar Say River.

Introduction to the problem

B.J. Merkel and A. Arab (eds.), *Uranium – Past and Future Challenges*,
DOI 10.1007/978-3-319-11059-2_17

The former mines of Sumsar (Pb-Zn) and Shekaftar (U) and their legacy wastes are located on the Sumsar Say River approximately 6 -10 km North of the Fergana valley in Uzbekistan (Fig. 1).

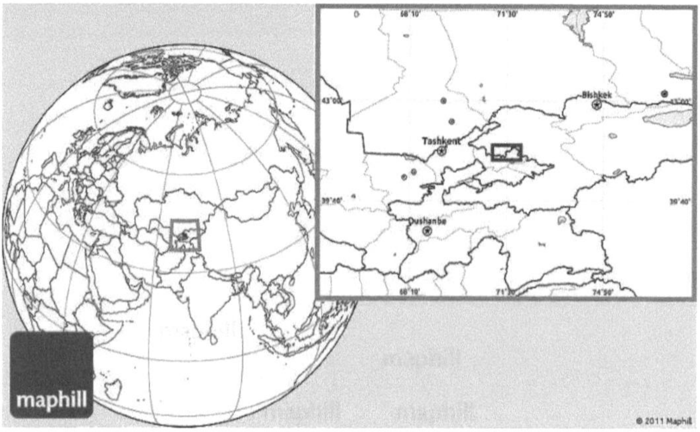

Fig.1. Location of the Sumsar – Shekaftar legacy sites in Kyrgyzstan.

At Sumsar, mining and processing went on from 1950 to 1978 and left behind a legacy of 3 tailings storage facilities, TSFs (Fig. 2).

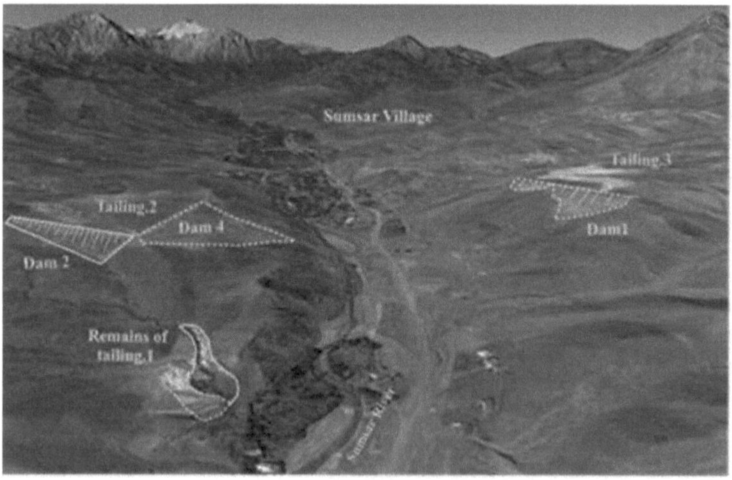

Fig.2. Location of the Tailings Storage Facilities (TSFs) of the Pb-Zn ore processing in Sumsar.

Chemical analysis showed the main components of the tails to be CaO, MgO, SiO_2, Al_2O_3, Fe_2O_3, K_2O, Na_2O, MnO and S. In addition, investigations by

Kyrzyltepegeologia identified high concentration of toxic metals and metalloids in the tails: Lead (0.1-0.5) %; Zinc (0.02-0.025) %; Manganese (0.1-0.16) %; Cadmium (9-38) mg/kg and Arsenic (10-38) mg/kg.

Mining of the Shekaftar U ore deposit went on from 1946 to 1957. Approximately two thirds of the reserves have been mined out. What remained on the surface are 110 thousand m^3 of ore refuse (low-grade ore) and waste rock.

Legacy sites of Pb-Zn ore processing at Sumsar

The containment of the Tailings Storage Facility, TSF #1, located just above the river, was to large extent destroyed by the extremely heavy rains in February/March 1994. The same rains deeply eroded the tailings dam of TSF #2. In follow up, the water diversion works and the tailings drainage became dysfunctional, which further aggrevated the erosion of the tailings dams. Without a dam, TSF #1 is today a chronic source of contamination of the Sumsar-Say River. The deterioration of the dams of TSFs #2 and #3 is progressing rapidly:

-TSF #1. From the original tailings amount (~0.3 million tonnes) approximately 70% has been lost to the river. The tailings dam has been destroyed, diversion channels eroded away and the surface water cut a deep gully (in fact a small valley, a "say") across the tailings. The little "say" is dewatered by a small creek that progressively erodes away the tailings carries them into the Sumsar-Say River.

-TSF #2. Approximately 1.3 million tonnes of tailings were stored in this facility. Upon closure in 1975 the TSF #2 has been "conserved" (preserved): The beach zone has been covered; an emergency surface water outlet constructed and a diversion channel for the storm water out in place. Today, erosion gullies, 4 m wide and 5 to 8 m deep can be observed in the foot of the dam that are places of tailings washout during the rainy season (Fig. 3). The geotechnical effect is a continuous degradation of the dam stability. The diversion channel is damaged and dysfunctional and there is no emergency floodgate.

-TSF #3. The storage facility was designed for 3.1 million m^3, however, the TSF contains only ~1.1 million m^3 of tailings (corresponding to ~1.82 million tonnes). There were no preservation measures taken at this facility and the tailings surface remained nude. The tailings on the surface of the TSF have today the consistency of very dry, very fine powder. A prominent fracture developed in the northern of the 2 dams of the TSF (Fig. 4).

According to the "Kyzyltepageologia", the amount of washed out tailings amounts to ~200 thousand tonnes. Due to weathering, a number of minerals in the tailings become soluble and release the toxic metals and metalloids they contain. The contaminants are then transported by the river and shallow ground water.

Fig.3. Erosion gullies above and below the berm of the dam of TSF #2

Fig.4. A distinct fracture in the northern dam of TSF #3 indicates structural instability.

The contaminants from the uncovered tailings surface of TSF #3 are easily re-suspended and transported by the wind to the agricultural fields in the neighborhood of the TSF (Fig. 5).

Soil sampling and measurements of toxic metals and metalloids in the Sumsar Say valley showed a directional spreading of the contaminants in the soil following the river downstream. Even at distance of 4 km from the TSFs, the toxic metal content in the soil is many times over the "norm values" used in Kyrgyzstan.

To investigate the fate of toxic contaminants in surface and ground water, samples were taken along a10 km long section of the Sumsar Say River.

As expected, the investigations showed that toxic metals concentrated preferably in the river sediments; the highest contaminants concentrations were in the river sediments (and water) in the proximity of TSF #1 and #2.

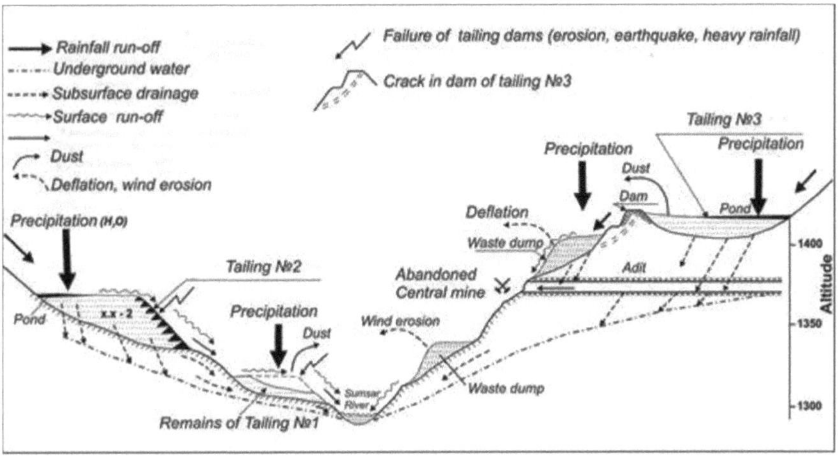

Fig.5. Pathways of contaminants from the legacy wastes of Pb-Zn ore processing into the Sumsar Say River valley and ground water.

1. Mining legacies of the Uranium deposit Shekaftar.

The Shekaftar settlement is a southern extension of the settlement of Sumsar along the Sumsar-Say River, ~ 6 km downstream of the Pb-Zn ore processing legacies.

There was no ore processing in Shekaftar and consequently there are no tailings left behind. Nonetheless, the radiometric ore separation plant left behind 113.1 thousand m^3 of ore refuse. The waste rock heaps entail 46-70 g/t of U (Lespukh et al., 2013) and, typically, up to 1.16 g/t Se. The low-grade ore is stored in form of 8 uncovered heaps.

There are legacy waste rock piles (shaft piles) around Shekaftar as well but they do not contribute to the radioactive contamination of the area (though, they contain, toxic metals and metalloids).

The measurements of gamma dose rate over the surface of the heaps range from 368 to 614 nSv/h, exceptionally reaching 921 nSv/h (Lespukh et al., 2013).

Four low-grade ore heaps (Nos.: 4, 3, 2, 5) and 2 waste rock piles (shaft piles Nos. 2 and 5) are in the domain of the Sumsar-Say River.

On the west bank of Sumsar-Say, heap No.5 is exposed to the erosion of the river throughout the year. The other heaps become flooded yearly, i.e. their feet become leached by the river at the time of high water. U contents up to 215 g/t and elevated contents of Pb, Fe, Cu, Cr, Al and other elements typical for the Sumsar-Shekaftar mining district were estimated in the Sumsar-Say sediments.

Measurements of the river water carried out upstream and downstream of Shekaftar (Fig. 6) show that erosion and leaching of the Shekaftar legacy heaps increase the content of dissolved U in the river more than 3 times even under regular flow conditions of the Sumsar Say. Investigation of the U234/U238 ratios upstream and downstream of Shekaftar provide evidence that the source of U is the leaching of the low grade ore heaps – after passing through Shekaftar the upstream U234/U238 ratio decreased from 1.77 to 1.2 (which is very significant). The U234/U238 ratio in the water supply, which comes from outside of the Sumsar-Shekaftar mining district was measured to be 1.8. The U content of the Shekaftar mine site water measures ~30 microgram/L as compared to 1.9 microgram/L in the supplied drinking water (Salbu et al.2011).

Conclusions regarding investigation and assessment of the remedial feasibility.

The presented preliminary characterization of the Sumsar Say valley legacy wastes aims to help justify systematic investigations, assessment and ultimately the remediation of the former Sumsar-Shekaftar mining district. Sumsar Say River is the most important agent of contaminants transportation in the district that also commands the capacity of transporting the contaminants to places of potential impact such as, the settlements of Sumsar/Shekatar (which are literally "next-door" to the legacy sites), to the densely populated Fergana valley (which begins ~10 km downstream) and to the Syr Darya River.

Sumsar Say River shows high seasonal variability with yearly reoccurring floods. The chronic releases of toxic metals, metalloids and Uranium (which is radiotoxic as well) from the legacy wastes introduces, even under "regular" flow conditions, an intolerably high contamination load into the river; the impact of a large-scale release could be catastrophic. Because of the poorly selected storage

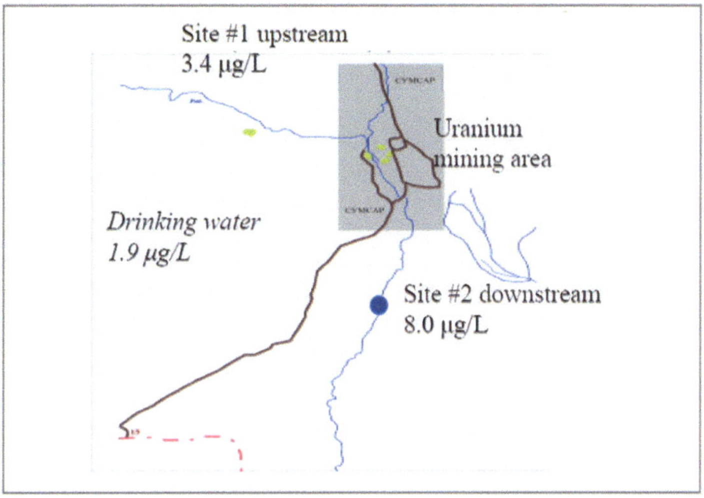

Fig.6. Total U concentrations in the Sumsar -Say River upstream (3.4 µg/L) and downstream (8.0 µg/L) of the Shekaftar low-grade U ore heaps. U content in the drinking water supplied from outside of the Sumsar-Shekaftar mining district (1.9 µg/L) is provided for reference purposes (Salbu et al.2011)

sites and poor state of the legacy waste storage facilities, a large-scale release of contaminants appears to be imminent. The remediation of the former Sumsar-Shekaftar mining district is not only justified; it is indeed, a most urgent preventive action.

Regarding the possibility of natural events prompting large-scale waste releases, suffice to say that signs of surface subsidence can be seen around the former U mine and a giant landslide (having a volume of ~200 thousand m3) is in the development on the mountain slopes of Shekaftar.

An irremediable factor in the district is that most of the legacy sites are in the domain of the Sumsar Say River, which proved to be in case of chronically released contaminants a very "effective" agent of transportation. In case of sudden, large-scale contaminants releases, it is most likely that the river would be capable of delivering the contaminants load to place of catastrophic impact.

Various projects have been proposed for remediation of the Sumsar-Shekaftar legacies, such as refortification of TSF #1, reconstruction of the dams at TSFs #2 and #3, construction of covers on all legacy wastes and rebuilding of the water works on the sites. Follow up measures would be cleanup of the contaminated neighborhood of legacy tailings sites and relocation of the houses built in front of the tailings dams. For the ore heaps of Shekaftar, fortification measures were proposed to limit river erosion and leaching of contaminants into the river, particularly from heap no. 5.

Before initiating independent remedial measures, we strongly recommend to carry out a risk based feasibility study that would include all Sumsar-Shekaftar legacy sites. This would help prioritization of the individual remedial projects/measures and finding the right remedial solution (whether to remediate in situ or relocate from the river domain) for each site. The "river domain perspective" of the assessment of feasibility would make implementation of individual site-specific projects depend on their share on minimization of the overall risk in the domain of the Sumsar Say River and thus provide a schedule for the remediation program.

For the sake of efficient use of resources, only sustainable remedial solutions should be considered, i.e. remedial measures that remain effective throughout the required lifetime of the waste facility (e.g. the lifetime recommended by IAEA for a final tailings disposal facility is 200-1000 years).

In case of the Sumsar-Shekaftar legacy sites, it is perceivable that a diversified and graded approach could provide the "right" balance of remedial measures. The number of projects implemented within the remediation program should be made dependent on the "diminishing environmental returns", i.e. when the remedial investment ceases to affect a commensurate increase of overall environmental gains further remediation is futile.

Nonetheless, further remediation may still be justified due to socio-economic requirements.

References

Salbu, B., Stegnar, P., Strømman, G., Skipperud, L., Rosseland, B. O., Heier, L.S., Lind, O.C., Oughton, D., Lespukh, L, Uralbekov B. and Kayukov, P. (2011), Legacy of Uranium Activities in Central Asia and the Joint Project between Norway, Kazakhstan, Kyrgyzstan and Tajikistan – Contamination, Impact and Risks, Summary Report of Results obtained within the NATO RESCA Project, UMB ISSN 0805 – 7214, J.Stefan Institute: IJS_DP 10733, pp. 75-76.

Lespukh, E.; Stegnar, P.; Usubalieva, A.; Solomatina, A.; Tolongutov, B.; Beishenkulova, R. Assessment of the radiological impact of gamma and radon dose rates at former U mining sites in Kyrgyzstan. *Journal of Environmental Radioactivity* **123**, 28-36 (2013).

Heavy metals and natural radionuclides in the water of Syr Darya River, Kazakhstan

Bagdat Satybaldiyev[1], Hanna Tuovinen[2], Bolat Uralbekov[1], Jukka Lehto[2], Mukhambetkali Burkitbayev[1]

[1]Al-Farabi Kazakh National University, Almaty, Kazakhstan
[2]Laboratory of Radiochemistry, Department of Chemistry, University of Helsinki, Finland

Abstract. This study deals with concentrations of heavy metals and the main uranium series radionuclide in the Syr Darya surface waters (Kazakhstan). Large proportion of studied heavy metals were transported with suspended matter (>0.45 µm), while large proportion of U (98-100 %) and Mo (81-100%) were in dissolved forms and colloidal fraction (<0.45 µm).

Introduction

The Syr Darya River is one of two major freshwater sources replenishing the Aral Sea. Today, the river crosses the territories of four central Asia republics, comprises nine major tributaries, eleven reservoirs, and numerous irrigation distribution systems and canals. The basin is now considered one of the most serious ecological disasters of modern times. As economic growth continues, there is a crucial need to improve water quality monitoring to increase environmental security through determining the extent of contaminants from agriculture, uranium mining, and municipal wastewater effluent.

Most published reports on the changes to water quality in this region have only evaluated soil salinization and estimated impact to reduced crop productivity (English et al 202; Pimentel et al 1997). The present study looks at the concentrations of heavy metals and natural radionuclides in surface waters collected during field expedition (May, 2013) to Syr Darya river basin in a watershed between cities of Turkistan and Kyzylorda.

B.J. Merkel and A. Arab (eds.), *Uranium – Past and Future Challenges*,
DOI 10.1007/978-3-319-11059-2_18

Methodology

Site description

With a length of 2,660 kilometers and a catchment area of 462,000 km^2, the Syr Darya is the largest waterway in Southern Kazakhstan. The river rises in two headstreams in the Central Tian Shan Mountains in Kyrgyzstan and eastern Uzbekistan-the Naryn River and the Kara Darya—and flows for some 2,212 kilometers west and north-west through Uzbekistan and southern Kazakhstan to the remains of the Aral Sea. Today, the river crosses the territories of four central Asia republics, comprises nine major tributaries, eleven reservoirs, and numerous irrigation distribution systems and canals.

Flow in the Syr Darya River is from mixed sources. Maximum average discharge for the lower reaches was up to 962 m^3/s at Tyumen-Aryk station, typically recorded in June, while minimum average discharge were recorded in September at 317 m^3/s (River Discharge database).

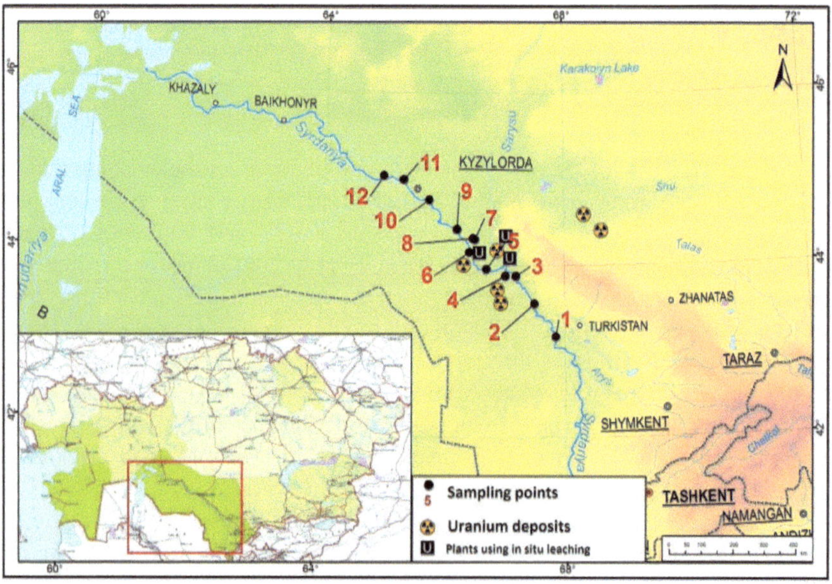

Fig. 1. Sampling points in the Syr Darya river valley

The Syr Darya River, flowing in Kazakhstan territory, runs through one of the largest uranium province worldwide, where uranium is being extracted by using in-situ leaching method. The area studied is characterized by the presence of uraniferous formations of rocks and mineral ores, pools of brown coals, granites and volcanic rocks showing enhanced concentrations of uranium and thorium series radionuclides bearing rare elements (Petrov et al 2000). Several large uranium deposits as well as uranium industrial processing plants are located within the study site (Fig. 1).

Sampling and Sample Pretreatment - Metal and Radionuclide Analysis

Surface river waters were sampled in the course of a field expedition to the Syr Darya River watershed in May 2013. The locations of the sampling points along the river are shown in Fig. 1, where position of each sampling site was determined with a portable global positioning system. Water samples of approximately 1 liter were collected at 12 stations from bridges crossing the river in clean polyethylene containers. After collection, samples were filtered using 0.45 μm membranes and immediately acidified with nitric acid to a pH of ~2.

Activity concentrations of uranium isotopes were determined by alpha spectrometry following radiochemical separation, the details of which have been fully described elsewhere (León Vintró and Mitchell 2000). Radium-226 was determined by alpha-spectrometry after an appropriate radiochemical separation.

Water samples were analyzed with respect to heavy metals and As using ICP-MS. The quality control of the data was checked by analysis of Canadian reference materials TM-27.3 and TM 64.2.

The fractions of heavy metals and As associated with suspended particulate matter was determined by comparison of the element concentrations of the unfiltered and filtered samples.

Dissolved oxygen, electrical conductivity, and pH were measured in separate samples at the time of collection using portable electrodes (Hach Corporation, Loveland, CO USA). Additional samples were collected for major dissolved ions and analyzed by standard hydrochemical procedures (State Standard 26449.1). Sodium and potassium were measured using flame photometry. Calcium, magnesium, chloride and alkalinity were determined by titration with standardized solutions. Sulphate ions were determined by gravimetric analysis.

Results and discussion

The results of the physico-chemical variables such as electrical conductivity, pH, dissolved oxygen and turbidity of waters collected at different points along the Syr Darya River are given in Table 1. The waters were found to be well oxygenated, with dissolved oxygen concentrations ranging from 8 to 12 mg L^{-1}. In all cases, the pH was slightly alkaline, being in the range from 8.1 to 8.5. Specific electric conductivity, being in the range 1335-1687 μS cm^{-1}, does not vary significantly along the river, while turbidity, being in the range 2.95-219 FNU, varies from location to location. Chemical analyses showed that calcium, magnesium and sulphate are the dominant ions in the Syr Darya River water.

Table 1. Dissolved oxygen, pH, specific electric conductivity in water samples of the Syr Darya River collected May, 2013 (±15 % uncertainty)

Sampling point	Sampling site	pH	Turbidity, FNU	O_2, mg L^{-1}	Electric conductivity µS cm^{-1}
1	Turkistan	8.27	55.2	8.31	1335
2	Besaryk	8.46	86.6	8.52	1351
3	Pervomaika	8.47	122	8.23	1365
4	Tyumen-Aryk	8.46	126	8.07	1364
5	Aktam	8.33	156	8.13	1379
6	Kyzykkain	8.29	895	8.49	1427
7	South of Baygekum	8.11	2.95	10.6	1432
8	Baygekum	8.22	7.13	12.42	1687
9	Tartogay	8.29	217	8.48	1370
10	Berkazan	8.26	219	8.58	1393
11	Karaozek	8.24	86	8.88	1400
12	Terenozek	8.43	93	9.23	1486

U concentrations together with ^{234}U and ^{238}U activity concentrations and their ratios in water samples collected in 2013 from the Syr Darya River are given in Table 2. The concentration levels of U in surface waters varied within 14-17 µg L^{-1} and in most cases within the recommendation level accepted for drinking water, 15 µgL^{-1} (WHO 2011). The $^{234}U/^{238}U$ activity ratio varied from 1.30 to 1.44 in surface waters.

In all cases, ^{226}Ra concentrations were lower than the corresponding concentrations for uranium, as evidenced by the $^{226}Ra/^{238}U$ activity ratio, which ranged from 0.17 to 0.62 (Table 2). Isotopic disequilibrium between uranium and radium can be explained by the presence of elevated concentration of sulphate and HCO_3^- ions in these waters compare to chloride and nitrate ions. In the presence of these ions, uranium forms highly mobile carbonate species ($[UO_2(CO_3)_3]^{4-}$, $[UO_2(CO_3)_2]^{2-}$), while radium has a much lower mobility and co-precipitate with carbonates and sulphates.

The concentrations of Mn, Ni, Co, As, Cd , Ba and Pb in collected water samples were lower than the Kazakhstan national regulation values for fishery water bodies, while those of V (1.7 times), Mo (7.4 times) and Zn (2.1 times) were above regulation values. Large proportion of studied heavy metals were transported with suspended matter (>0.45 µm), while large proportion of U (98-100 %) and Mo (81-100%) were in dissolved forms and colloidal fraction (<0.45 µm).

Table 2. Concentrations of uranium, radium and their activity ratios in waters sampled along the Syr Darya River in 2013

Sampling point	Sampling site	Activity, Bq L^{-1}			U, μg L^{-1}, ICP MS	$^{234}U/^{238}U$	$^{226}Ra/^{238}U$
		^{238}U	^{234}U	^{226}Ra			
1	Turkistan	0.191	0.259	0.119	17.0 ± 0.6	1.36	0.62
2	Besaryk	0.160	0.231	0.051	16.9 ± 0.8	1.44	0.32
3	Pervomaika	0.177	0.247	0.077	16.8 ± 0.5	1.40	0.44
4	Tyumen-Aryk	0.180	0.244	0.038	17.3 ± 0.8	1.36	0.21
5	Aktam	0.198	0.285	0.046	16.6 ± 0.6	1.44	0.23
6	Kyzykkain	0.204	0.275	0.044	17.1 ± 0.9	1.35	0.22
7	North Baygekum	0.188	0.257	0.037	16.0 ± 0.7	1.37	0.20
8	Baygekum	0.176	0.233	0.045	14.7 ±0.2	1.32	0.26
9	Tartogay	0.181	0.245	0.035	17.1 ± 0.3	1.35	0.19
10	Berkazan	0.183	0.251	0.032	16.9 ± 1.4	1.37	0.17
11	Karaozek	0.167	0.214	0.041	16.4 ± 0.6	1.28	0.25
12	Terenozek	0.190	0.253	0.045	17.4 ± 1.0	1.33	0.24

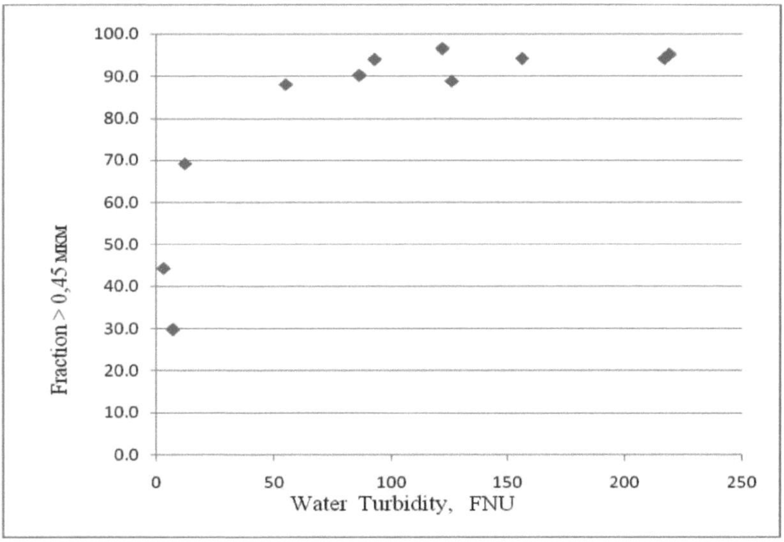

Fig. 2. Relationship between turbidity values and manganese concentration in particulate fraction (<0.45 μm) in water samples collected in May 2013 from Syr Darya river

Amount of heavy metals with suspended materials correlated with degree of water turbidity values. A plot of manganese concentrations associated with suspended particulate matter as a function of water turbidity for Syr Darya River waters is shown in Figure 2. The correlation is consistent with suspended matter as the major contributor of manganese in this watershed.

Conclusions

The concentration levels of U in surface waters varied within 14-17 μgL^{-1} and in most cases within the recommendation level accepted for drinking water, 15 μgL^{-1} (WHO 2011). The concentrations of Mn, Ni, Co, As, Cd , Ba and Pb in collected water samples were lower than the Kazakhstan national regulation values, while those of V (1.7 times), Mo (7.4 times) and Zn (2.1 times) were above regulation values. Concentration of heavy metals with suspended materials correlated with degree of water turbidity values. The fact that radionuclide concentrations are practically the same both at upstream and downstream from the uranium deposits and mining sites suggests that they do not affect the water quality of the Syr Darya River.

Acknowledgements This work is supported by project from the Ministry of education and science of the RK (Registration Number #0112RK02582).

References

English MJ, Solomon KH, and Hoffman GJ (2002) A paradigm shift in irrigation management. J Irrigation and Drainage Engineering 128: 267-277.

León Vintró L, Mitchell PI (2000) Determination of actinides and other alpha emitters. In: Encyclopedia of analytical chemistry: instrumentation and applications, Part II, Radio-chemical Analysis. John Wiley and Sons Ltd, Chichester, pp. 12848−12884.

Pimentel D, Houser J, Preiss E, White O, Fang H, Mesnick L, Barsky T, Tariche S, Schreck J, Alpert S (1997) Water resources: agriculture, the environment, and society. BioScience 47: 97-106.

Petrov NN, Yazikov VG, Berikbolov BR (2000) Uranium deposits in Kazakhstan (endogenous). Gylym, Almaty.

River Discharge database (http://www.sage.wisc.edu/riverdata/scripts/station_table.php?qual=32&filenum=2534)

WHO (2011). Guidelines for Drinking-WaterQuality. 4th edn. Geneva, Switzerland.

Establishment of a database of uranium anomalies and zones in Mongolia

Boris Vakanjac[1], Predrag Srna[2], Vesna Ristic Vakanjac[3]

[1]Faculty of Applied Ecology Futura, Singidunum University, Belgrade, Serbia
[2]EHTING, Belgrade, Serbia
[3]Faculty of Mining and Geology, University of Belgrade, Belgrade, Serbia

Abstract. Based on 26 Soviet reports, a database was compiled for uranium exploration purposes, containing information about more than 1100 anomalies of uranium and other radioactive components, and featuring some 300 zones that cover nearly all of Mongolia. The database is currently being reviewed from the perspective of relationships between recent tectonic movements, geological settings and genetic origin of anomalies. While the database was being compiled, the primary focus was on the geological properties of the reported anomalies and zones. The methods and the main results of this effort are presented in the paper.

Introduction

Soviet/Russian expeditions launched in the 1950s marked the beginning of comprehensive uranium exploration in Mongolia. Following initial regional research, comprehensive uranium exploration began in the late 1970s, in the following order: 1. airborne surveys; 2. carborne surveys; and 3. detailed geological and geochemical surveys of localities that constituted "anomalies". The outcomes of the enormous efforts expended by Soviet explorers are presented in more than 30 reports produced between 1970 and 1990. The reports were made available by the Mineral Resources Authority of Mongolia (MRAM) in 2006.

After the break-up of the Eastern Bloc, Mongolia began to open up to the rest of the world and many Western corporations and consortia of Western, Russian and Chinese companies proceeded with rigorous exploration of diverse mineral resources in that country.

Data (i.e. selected parts of Soviet reports, copies of text and scanned maps) were procured/purchased from MRAM at the then prevailing terms and conditions (June-September 2006).

B.J. Merkel and A. Arab (eds.), *Uranium – Past and Future Challenges*,
DOI 10.1007/978-3-319-11059-2_19

Soviet reports used to compile the database

The materials (photocopies of entire reports or parts thereof and scanned maps) were obtained under the terms and conditions stipulated by MRAM. The reports generally comprised four parts:

1. Narrative section, describing the study location, the geological and radiometric characteristics, and the anomalies the authors deemed significant;
2. Catalog of anomalies, generally organized as follows:
   a. Number and name of anomaly,
   b. Lat-long position of anomaly and relation to a local geographical entity (summit, spring, etc.);
   c. Date of flyover;
   d. Nature of radioactivity;
   e. Participants/team members;
   f. Supplemental research – mapping, sampling;
   g. Detailed geological description and size of anomaly;
   h. Results of spectrometric and chemical analyses;
   i. Assessment of the potential of the anomaly and proposed further action.
3. Various maps at different scales, e.g. geological maps, radiometric maps (gamma fields of uranium, thorium and potassium), detailed maps of certain anomalies and zones; and
4. Section containing a summary on field activities.

In 2003, Yu. B. Mironov published a book titled "Uranium of Mongolia". A different approach is presented in the present paper.

Database compilation problems and solutions

Given that the reports contained voluminous data, at times extremely detailed, and that different authors used different writing styles, there was the need to ensure consistency of the information and compile a homogeneous database. The process included several parallel steps.

The following activities were undertaken:

1. The narrative section was compared with the catalog of anomalies and maps.
2. The catalog of anomalies was translated from Russian into English and a system was set up in Excel and later in Access. In some cases the original data were presented in different ways, there were variations in the quality of the data and some (less than 10%) of the anomalies were not in the right place (when projected in GIS). These issues were addressed by: a. systematic reorganization and uniform presentation of anomaly data, b. checking of anomaly positions against the catalog and locations on maps, and c. compil-

ing a new catalog/database of anomalies. In the event of an easting or northing deviation of an anomaly from the relevant point on the georeferenced map, the solution was simple – printing error. However, other discrepancies had to be resolved by studying the narrative section of the report and the catalog of anomalies (there were only a few such errors).

3. Map difficulties included: varying quality, different types, varying scales, different mapping methods, positions of anomalies shown at small scale (e.g. 1:500,000), and occasionally damaged maps. The solution to these problems mainly involved careful georeferencing of the maps (at all coordinate intersections within the grid). This proved to be effective because sets of 6 or 8 maps had been joined without spaces between the individual maps. It should be noted that exact positions of mapped and cataloged anomalies are extremely important for geological exploration, especially in large countries like Mongolia, because a trip to an anomaly can be long and costly.

Organization of the database of anomalies and zones

The following reports were available and were used to compile the database of anomalies and zones (MRAM referencing system): 2447, 2462, 5251, 2411, 2448, 2449, 2427, 2419, 2520, 2452, 2426, 2442, 2437, 2414, 2453, 2455, 2428, 2429, 2432, 4289, 4358 and 4454. The catalog of anomalies was organized using the approach described below. Each anomaly was assigned the following attributes in GIS:

1. Soviet anomaly number – unique number, as stated in the respective report;
2. Soviet report number – per MRAM referencing system (the reason for this was that the original Soviet reports had several reference numbers: MGSE, VGSEI, old Mongolian nomenclature number, "top secret number" etc., while the report titles would have used up too much space in the system);
3. Location – generally distance and indicative geographical position (e.g., north of summit X, geological feature Y, human settlement Z, etc.);
4. Coordinates – latitude and longitude;
5. Comment – whether a point in the original catalog is in the right place on the georeferenced map);
6. Confirmed or corrected coordinates (errors were found in less than 10% of the anomalies);
7. Uranium range – data whose values amounted to 10^{-4}% of uranium concentration.
8. Geological settings – most important data. An attempt was made to maximize the amount of information;
9. Anomaly dimensions – if available;
10. Russian note – authors' recommendation concerning anomaly potential;

11. Associated elements in % – showing measured amounts of associated elements;
12. Background radioactivity of the environment – expressed in micro rad hours (mcr/h);
13. Radioactivity of the anomaly – expressed in mcr/h;
14. Note – brief description of geological nature of anomaly;
15. Origin: P – primary origin (related to a rock as the source of radioactivity, especially in the case of igneous rocks and processes), or S – secondary origin (sedimentary rocks and processes); and
16. Legend – color code of 12 different colors that identify the setting of the anomaly.

Examples of anomaly entries

Given below are typical examples of anomaly entries in the compiled database (Fig. 1). The anomalies are classified into 12 categories.

Ingenious rocks (intrusions and volcanics): color tan; anomaly number 273; report 2426; location – Harcit Mountain; latitude 108.939722; longitude 46.2008333; geology – erosion remnant of felsite and felsite porphyry extrusions, middle-late Jurassic age, with intrusions of trachy-dacite porphyry.

Hydrothermally altered rocks (mainly igneous): color orange; anomaly number 541; report 2453, location – Elgen occurrence, Tsagan-Suburgin plate; latitude 108.972222; longitude 43.829167; geology – middle-late Jurassic volcanics represented by basic and acidic effusives in porphyry, Silurian age. Anomaly associated with fractured and altered, lighter, carbonated porphyry, acidic composition containing epidote (veinlets in cracks and impregnations), with fractured, lighter and carbonate-altered acidic effusives.

Fluorite in igneous altered rocks: color pink; anomaly number 966; report 2426; location – Borundur deposit, latitude 109.337778; longitude 46.294444; geology – medium-fine-grained aplitoide granite, Triassic age, anomaly in set of cutting faults 20 and 600, rocks brecciated, altered and silicified, with hematite, fluorite and limonitic medium-fine-grained aplite granite. Radioactivity associated with brecciated zones with jasperoid and fluorite

Basalt: color lime green; anomaly number 439; report 2433; location – northeastern flank of paleovalley, along flank of recent valley of Sharachin-Gorki river; latitude 100.597222; longitude 48.364722; geology – granite with basalt covering layer, present are sediments - breccia built up of basalt fragments, angular pebbles and gravels. Sediments are altered, argillic, with hydro-mica and thin impregnations of autunite in cement and altered basalt fragments.

Sedimentary rocks (mainly Mesozoic): color light yellow; anomaly number 740; report 2422; location – south flank of Ulan-nuur Valley, 1.6km south of 1005.8 peak, Huroki locality; latitude 109.436111; longitude 45.640833; geology

– silt-limestone overlying limestone that belongs to effusive-sediments, Jurassic-Cretaceous age block represented by andesite-basalts, folded coarse-grained sandstones, limestone and siltstone-limestone. Uranium minerals present as banded aggregates of franklinite and clay. Thorium present as thorite.

Phosphatic concentrations in different rock types as host rocks: color light green; anomaly number 124; report 2447; location – flank of Turgen tectonic-volcanic valley; latitude 113.850556; longitude 49.563889; geology – uranium bearing phosphates as sediment cement in basal layer, containing fine-grained gravel conglomerate, gravel, arkose and polymictic sandstone and siltstone of upper Jurassic/lower Cretaceous volcano-sedimentary suite. Anomaly zone fractured, argillic, limonitic, with quartz and calcite veinlets. Valley built up of chlorite, chlorite-sericite schist and phyllite, late Proterozoic age.

Limonitic or Fe-hydroxide enriched rocks: color brown; anomaly number 561; report 2455; location – anomaly record 376; latitude 106.058889; longitude 45.079167; geology – Paleozoic bedrock uplifting at north-east, extension of Dzun-Haracav structure built up of Orcog suite, (R_1-2sv) sediments represented by limestone-siltstone, schists, brecciated and silicified sandstone cut with granite dike. Cataclastic, limonitic, weathered quartz sericite schist and sandstone.

Mn-hydroxides or Mn-hydroxide enriched rocks: color plum; anomaly number 80 (652); report 4737; location – Chirgitai depression, 950 m on azimuth 2100 from Maihan-Tolgoi Mountain; latitude 92.33060; longitude 48.518333; geology – conglomerate ($D_{1\text{-}2}$), covered with coarse-grained sediments (D_{2ef}), including pyrolusite layers. Uranium bearing horizon in sandstone-limestone under red-color horizon with pyrolusite concentrations, where gray sediments are becoming reddish-brown.

Coal, coal bearing sediments: color black; anomaly number 309; report 2452; location – 1.1 km N-W of Lake Ulaan-Honhor-nuur; latitude 115.290278; longitude 49.144444; geology – oxidized zone in coaly clays and sooty brown coals from coal-bearing terrigenous sediments in early Cretaceous Dzunbain suite, here represented by pebble conglomerates, sandstones, siltstones, coaly clays and brown coals.

Saline lake situated on different rocks: color gray; anomaly number 465; report 2439; location – eastern part of Yanachivlin, late Triassic-early Jurassic granite massif, 0.9 km south-west of 1926.6 peak; latitude 107.643056; longitude 47,680556; geology – water-salty aureoles of uranium disseminations in Quaternary sediments – terrigenous and humus accumulations, which fill the fractured zone.

Metamorphic rocks: color teal; anomaly number 247; report 2419; location – 600 m west of Dzun-Obo Mountain; latitude 109.748661; longitude 47.941667; geology – outcrop of graphite schist horizon, late Proterozoic age, covered with early Paleozoic metaeffusive. Chlorite-sericite schist under graphite schist horizon.

Contact of serpentinite and other rocks: color green; anomaly number 433; report 2433; location – Cagan-Burgas ultramafic massif; latitude 100.312500; longitude 49.987222; geology – hyperbasite formation, early Cambrian age, at contact of serpentinite peridotite and granites in zones with tectonic breccias including carbonate cement.

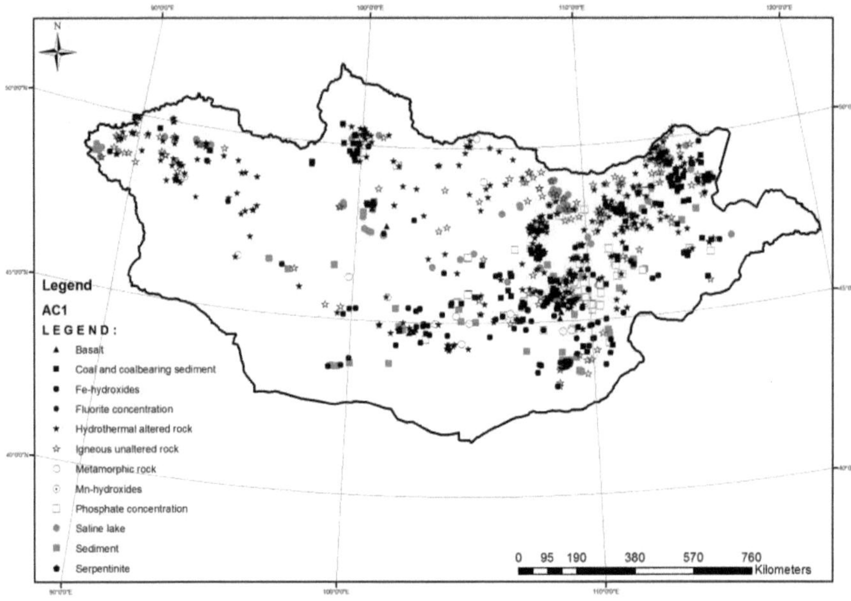

Fig.1. Locations of anomalies in Mongolia

Examples of zone entries

The zones were classified into four categories:

1. **Volcano-tectonic structures**, generally acidic vulcanites, bearers of elevated radioactivity, and ore deposits such as Dornod and Gurvan-bulak;
2. **Granite complexes**, of different ages, mainly Paleozoic, bearers of elevated radioactivity but with generally no uranium mining potential.
3. **Terrigenous sediments**, different types featuring elevated radioactivity and uranium concentrations, potential prospecting targets.
4. **Coal and coal-bearing sediments**, which constitute a reducing factor but it is questionable whether uranium deposits can be expected to exist.

Some typical examples are provided below. The stated coordinates are central coordinates of the individual zones (Fig. 2).

Volcano-tectonic Structure Zone (VTS): report 2451; name – Ugtam; central coordinates – latitude 113.825703, longitude 49.27258; geology – F-Mo-U, effusive acidic composition. Northern Choybalsan Isometric VTS built up of late Jurassic-Early Cretaceous volcano-sedimentary rocks. Knot in upper part of volcano-sedimentary batch, made up of quartz-feldspar porphyry. Ugtam ore occurrence developed in central part of volcanic structure and related to large north-west Ugtam fault in contact with north-east faults and submeridial Central Structure Zone. Several felsite and morion-sanidine rhyolite dikes intrude into porphyry body. Rocks brecciated with chlorite-quartz-fluorite-hydro-mica cement. Ore represented by coffinite and uranium-bearing sphene with molybdenite and sulfides. Exploration 1:10000 recommended.

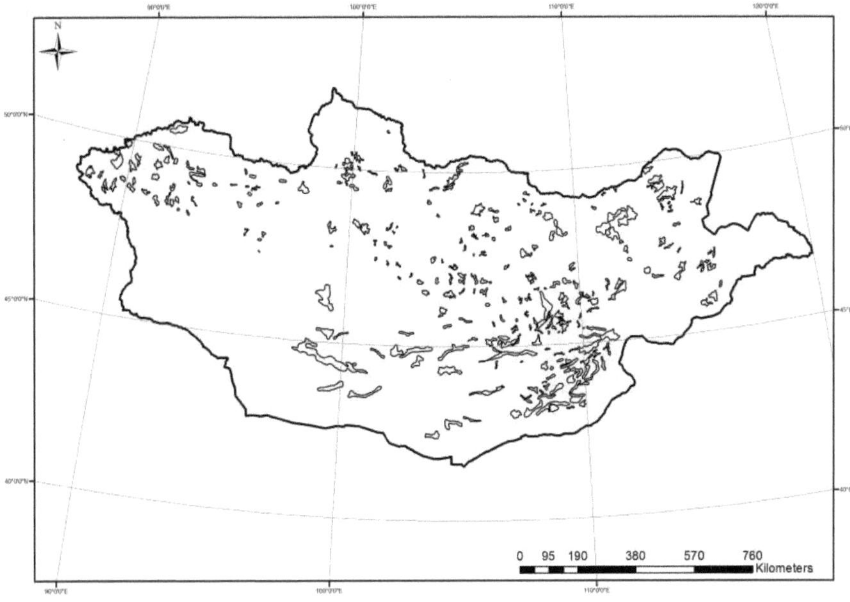

Fig.2. Locations of zones with elevated uranium concentrations in Mongolia

Granite Zone: report 2456; name Baidarak; central coordinates – latitude 99.2, longitude 49.3; geology – uranium-thorium origin of radioactivity, equally distributed. Uranium content 5-5.5x$^{10-4}$%, thorium 15-42x$^{10-4}$% and potassium 2-5%. Sources of radioactivity biotite granite and two-mica muscovite-biotite granite. Anomaly 607 in layers of granite-gneiss, in mica-quartzite and marble, Proterozoic age (thorium origin of radioactivity). Situated south-west of Hara-Uus nuur, on the right side of the Dzabhangol River, 60 km south-west of Delger-sum in granite and granite-diorite, late Carboniferous age, elevated radioactivity detected in coarse-grained pink granite.

Terrigenous Sediments Zone: report 4737; name Dzabhan; latitude 92.984; longitude 48.678; geology – area in valley of same name, middle course of the Dzabahan River, trending north-west. Anomalies at valley edges in friable sediments of Neogene age. There are 2 local fields of thorium origin in alluvial sediments. Geochemistry of uranium unexplored.

Coal and Coal-bearing Sediments Zone: reports 2451 and 2453; name Motod–Choyren; latitude 108.216; longitude 45.591; geology – zone situated at end of north-western part of Undurshil Plate, trending north-west, 65km long and 10km wide, includes south-western contact of Undursil Valley and Choyren Valley, filled with coal-bearing sediments of early Cretaceous Dzunbain suite (K_{1dz} - Hautervian – Albian). According to aerial exploration, zone features intensive gamma field – 50 micr/h, high concentrations of uranium 10-15x$^{10-4}$ %, low concentrations of thorium 4-8x$^{10-4}$%, and potassium 1-2% (3% in western part). Four areas with elevated uranium concentrations detected, at locations of anomalies 508, 522, 523 and 548, designated after ground verification as Motod occurrence. Modot occurrence related to oxidized layers of brown coal and western side of Motod. According to Mironov Y. 2003, in southern part of the Choyren depression (Modot prospect), U-bearing coal overlain by completely limonitic sand over an area of 25 × 5 km^2.

Conclusion

This paper is basically a preliminary report about ongoing activates. Each anomaly entered in the database is being re-assessed from a provenance perspective, based on analysis and generalization of physicochemical processes using Soviet reports and contemporary literature and knowledge.

Acknowledgments

The authors express their gratitude to Miles Worsley, who initiated the creation of a database of anomalies and zones in Mongolia, and to Neil Rutherford for his valuable suggestions and briefings in 2006-2009.

References

Kichman E S (1990) Internal report 4737: Report on Uranium Potential of Western Mongolia 1970-1990, Ministry of Geology, USSR

Kiselev V Ya et al. (1985) Internal report 2451: Evaluation of Uranium Bearing Potential of East Mongolia (Vols. 1 and 2), Ministry of Geology, USSR

Mironov Yu B (2003) Uranium of Mongolia. Translated by Victor S. Popov, Centre for Russian and Central Eurasian Mineral Studies (CERCAMS), Natural History Museum, London 2005: 239

Samovich D A et al. (1984) Internal report 2433: Report on Airborne Geophysical Survey of the Ho'vsgol region at a scale of 1:200,000, Ministry of Geology, USSR

Shmelev Yu. S. et al. (1984) Internal report 2456: Report on airborne geophysical and geological mapping of Altai area at a scale of 1:1,000,000, Ministry of Geology, USSR

Uranium, Rare Earths and NORM: Mining and current prospects in Australia's Northern Territory

Peter Waggitt[1]

[1]Department of Mines and Energy, Darwin, Northern Territory, Australia

Abstract. The Northern Territory of Australia (NT) has a long association with mining, including the mining of uranium. Current activity includes all aspects of the uranium production cycle including many exploration projects, the Ranger Uranium Mine and a number of remediation sites. There are also a number of rare earth projects approaching the development phase and some NORM related projects which are equally approaching development status.

The paper describes the current state of uranium exploration activities and progress on remediation works in the NT before providing an update on the ongoing development and remediation-related activities at the Ranger Uranium Mine. The paper will also provide an update on recent changes in the regulatory regime.

Introduction

Whenever any natural material or mineral is mined or processed there will always be the potential for the radiation exposure of the workforce and/or the surrounding community to increase. This is because natural raw materials all contain naturally occurring radioactive material (NORM). Admittedly it is only in a small percentage of such situations that it will be necessary to introduce changed work practices and radiation protection plans. However, it is important that the radiation risk is assessed in order to ensure the protection and of individuals and the environment as well as facilitating determination of the appropriate level of regulatory supervision to be applied. International Atomic Energy Agency (IAEA) published Safety Report Number 49 (IAEA, 2006) published in 2006 provides a complete overview of this whole issue. This report lists the most common industries and activities concerned where the matters of NORM and subsequent risk of increased radiation exposure are most likely to be encountered, with the exception of uranium mining for which there is a separate and extensive range of IAEA documents.

B.J. Merkel and A. Arab (eds.), *Uranium – Past and Future Challenges*,
DOI 10.1007/978-3-319-11059-2_20

In the Northern Territory of Australia (NT) the economy is based on three "hubs", tourism, primary industry and mining, of which mining is the most important. For this reason issue of NORM is frequently under scrutiny by the various authorities and agencies responsible for regulating the industry.

The Northern Territory and NORM

There is a significant history of NORM related industrial development in the NT; this is primarily related to the uranium mining industry since 1945. More recently other NORM-related industries have begun developing. Currently there are exploration, development and recently closed operations involving NORM. These projects include rare earths, mineral sands and phosphate deposits as well as ongoing activities related to all parts of the uranium production cycle. The major elements of the uranium related activities were described recently in a paper by Waggitt (2013) and are not discussed at length here.

The NT has an active and developing oil and gas industry with three active oil fields in the southern NT at Palm Valley and East and West Mereenie with three more not yet operational. In addition there are offshore oil and gas production fields which use Darwin as a supply base as well as the operating LNG plant at Wickham Point and the new LNG plant presently under construction for the Icthys project at Blaydin Point, both located on Darwin harbour. Offshore oil and gas activity is regulated by the Australian Government.

All of these developments and projects have the potential to produce NORM related exposures, primarily arising from the cleaning of pipes and tanks involved in the pumping of associated or formation waters. These potential exposures are managed in accordance with the requirements of the regulating authorities and taking heed of the guidance provided by the IAEA in the relevant Safety Report (IAEA, 2003). The radiation protection issues are regulated through the Radiation Protection Act, which is administered by the NT Department of Health.

Mineral Sands

The NT had only one minerals sands project which was operated by MZI Resources Limited. The company has other mineral sand interests in Western Australia and recently completed operations at the Lethbridge South site in the Tiwi Islands located in the Arafura Sea just 50 km to the north of Darwin. MZI operated in the Tiwi islands 2009-2013 and produced some 33,000t of heavy mineral concentrates which were sold mostly to China. The mines in the Lethbridge series on Melville Island ended operations in 2013 but the approvals process for a new mine at Kilimiraka on Bathurst Island has begun with a full feasibility study expected to be completed later in 2014. In recent statements the company has described an inferred resource of 56.2 Mt grading at 1.6% heavy minerals and an anticipated heavy mineral content of 894,000 t. The process would be a standard excavation and wet separation in spirals; and to date there have been no significant radiation protection issues. However, the operator is aware of the potential for such issues to arise and appropriate safety measures are in place. In early 2013 the company submitted a Notice of Intent for the new project at Kilimiraka and the environmental impact assessment process was set in train by the NT Environmental Protection Authority in March. Heritage, environmental and native title works have begun and the Scoping and Feasibility study is planned to commence later in 2014.

Rare Earth Elements

The NT has a good prospectivity for rare earth deposits and extensive exploration activities are taking place in several areas. The most advanced prospects include the Nolans Bore deposits, being developed by Arafura Resources Limited, and the Charley Creek deposits being evaluated by Crossland Strategic Metals Limited. The NT Department of Mines and Energy (DME) is responsible for regulating mines through the Mining Management Act (MMA) and refers all operators, be they explorers or miners, to the IAEA Safety Report No. 68 (IAEA, 2012) when considering the preparation of their Mining Management Plan (MMP) and the included (radiation Management Plan (RMP).

The Nolans Bore deposit has published resource information as shown below which shows that there may be possibilities for by- and co-production of other materials. However, much of that will depend on market conditions for the respective materials in the future. The original plan was for the project to produce a beneficiated concentrate on site, about 500,000 t annually, which would then be transported to the company's Rare Earths Complex to be built at Whyalla, in South Australia.

Table 1. Nolans Bore Resource Classification

	Mt	REO%	REO kt	P2O5%	U3O8 lb/t
Measured	4.3	3.3	144	13	0.57
Indicated	21	2.6	563	12	0.42
Inferred	22	2.4	511	10	0.37
Total	47	2.6	1,217	11	0.41

(Source: http://www.arafuraresources.com.au)

The project has recently been re-scoped to concentrate on on-site beneficiation and no off-site processing. The company hopes to recover 848,000 t of rare earths (lanthanum, cerium, praseodymium, neodymium, samarium, europium, gadolinium, terbium, dysprosium, andyttrium) in addition the deposit contains 3.9 Mt of phosphorus pentoxide and 13.3 M lbs (6,045t of uranium oxide. The mining operation would be regulated under the Mining Management Act which will require an annual Mining Management Plan (MMP) to be submitted for approval and that MMP will be required to include a Radiation Management Plan (RMP).

The Charley Creek project is being investigated by Crossland Strategic Metals Limited in partnership with Pancontinental Uranium Corporation of Canada with a scoping study currently being completed with hopes to commence development soon, including a degree of local processing. The project was originally exploring for uranium but it became apparent that the deposit contains significant amounts of rare earth oxides and is thought to contain 17% of the heavier, and more keenly sought after, elements amongst the rare earths, rather more than is usual. The Total Rare Earth Oxide (TREO) resource is inferred at 121.1kt from an ore mass of 418Mt with an average of 289ppm TREO In addition there are reported resources of 30.6kt Xenotine, 167.2kt Monazite and 219.9kt Zircon. To date only a small portion of the deposit has been explored and so the potential is considered to be considerable. Another interesting point is that the deposit appears to contain low levels of radionuclides associated with the rare earths and it is thought unlikely that there will be any significant radiological protection issues for the operators.

Phosphates

Phosphate deposits are being investigated at a number of locations within the NT. The evidence has been that all the deposits identified to date have been found to be low in natural radionuclides and radiation protection issues have been minimal.

One deposit is the Amaroo project being developed by Rum Jungle Resources Limited where the total of measured, indicated and inferred resources has been reported as 1.08 Bn t at an average grade of 14% P_2O_5 with a cut off at 10% P_2O_5. This is a significant increase in previous statements as it now includes the adjacent resources of the Arganara deposit, formerly owned by Central Australian Phosphate, which was taken over by Rum Jungle Resources in 2013. Throughout both deposits the uranium levels are reported as being between 19 and 31 ppm which is why the radiation risk is deemed to be low.

Rum Jungle Resources is also exploring for potash deposits at the Karinga Lakes area south west of Alice Springs. The salt lakes are a possible source of potassium and potassium magnesium sulfates. The project is in early stages and as yet no resource data are available.

Minemakers Limited are the owners of Australia's largest undeveloped rock phosphate deposit at Wonarah and so far only 15% of the deposit has been explored sufficient to make a resource statement which has 78M t of measured resources at 20.8% P_2O_5 , and inferred and indicated resources of 933M t at 13% P_2O_5 with a 10% cut-off. The ore body contains uranium at an average of 18ppm although the maximum recoded concentration was 75 ppm. Work carried out by the company as part of the preparation of the environmental impact statement have indicated that the levels of uranium are not sufficient to warrant further investigation as the ore will nether beneficiated nor concentrated. Thus the product and the residues are not considered to present a radiological risk for the proposed project with direct shipping of ore. The company also has a development plan which would employ the Improved Hard Process at the site to produce super-phosphoric acid as a higher value added product.

Thus, for both these projects the possibility exists that there may be a change from an emphasis on direct shipping of crushed ore to some form of processing which is likely to involve phosphoric acid production and possibly even further downstream processing to fertilizers or similar products. Such developments would require a re-assessment of the radiological implications, especially in relation to NORM scales building up in pipework and reaction vessels etc. Both projects could also include creation of a fertiliser plant in later stages of development which would also need to be assessed for radiological risks. However, it is unlikely that such a plant would be located on a mining tenement and would there not be subject to regulation under the MMA.

Uranium

Uranium mining activity has been a feature of the Northern Territory economy for many years. The "modern" era of uranium mining began with Rum Jungle and the South Alligator operations of the 1950s and 1960s. These have been well documented elsewhere (Waggitt, 2012, Waggitt, 2013). In the late 1960s exploration in the Alligator Rivers Region discovered several deposits of which four, Jabiluka, Ranger, Koongarra and Nabarlek were of economic significance. The political situation in Australia at the time was changing and as a result only Nabarlek and Rranger were brought into production.

Nabarlek was a small resource but of relatively high grade (2%) and operated from 1979 to 1988 producing about 10,800t of Uranium Ore Concentrates (UOC-U_3O_8) over that period The site was remote and an early example of the modern Fly in-Fly out (Fifo) style of operating. The site was remediated in 1995/6 and is currently still under assessment for completion of closure. The lease area has also been the location of a number of exploration operations many of which have extended considerable distances into the surrounding countryside and some of which are still ongoing. To date no new resource suitable for development in the current economic climate and state of the industry has been found.

Koongarra is a deposit that is located on country that has been handed back to the Kakadu National Park by the Aboriginal Traditional Owner and will not now be further exploited.

Jabiluka is a significant deposit owned by Energy Resources of Australia Limited (ERA), the owners and operators of the adjacent Ranger Uranium Mine (RUM). The deposit was explored as part of a feasibility study and EIS preparation in the 1990s but agreement could not be reached with the Aboriginal Traditional Owners and so work was halted and the sire remediated to await possible development at some time in the future. This will not be before operations at RUM have been completed

Ranger began operating in 1980 and has since produced more than 100,000t UOC from two open pits, #1 and #3. Pit #1 was mined from 1980 to 1995 and Pit #3 was operated from 1997 until November 2012 at which point mining ceased. Pit #1 has been back filled with tailings and Pit#3 has recently been partially filled with 30Mcu.m. of low grade mineralized waste rock as part of the preparations for the in-pit disposal of tailings which will commence in January 2015. This operation will also involve the relocation of some 50Mcu.m of tailings from the existing tailings storage facility (TSF) which is a 1 kilometer square dam containing about 50m depth of tailings. This operation is planned to commence later in 2015.

Exploration across the Ranger Project area (about 65 sq.km.) has continued throughout the life of the mine and a number of interesting anomalies have been found. In 2013 ERA commenced an underground exploration campaign adjacent to pit#3 to investigate a deposit named Ranger 3 Deeps (R3D). It has been stated that the initial resource at R3D is about 32,000t UOC and a $57M pre-feasibility study has been commenced. Work underground to date has included a decline and about 2000m of tunnel with many side drilling locations to increase the knowledge about the orebody. The complete campaign will extend to 2400m plus a bulk sampling exercise to enable various characteristics of the ore to be checked for future processing options. The current mining agreement with the Federal Government and the Aboriginal Traditional Owners will expire in 2021 with ERA required to have stopped mining in 2012, and have the site remediated by 2026. As the future of the operation beyond that time is not assured at present ERA is hoping to be able to exploit a good proportion of R3D within the present mining agreement time frame.

Major issues at RUM have always centered around water management and in 2013 the company commissioned a Brine (BC) at a cost of $220M as a means of rapidly reducing the inventory of process water on site. The plant is operating and reduces volume of water by 2/3 with the remainder being a waste brine that will be injected into the rockfill in the bottom of pit#3. The 1.8GL clean water condensed annually from the BC is released from the site in compliance with a strict series of protocols developed by the regulators of the operation in consultation with major stakeholders. The BC will continue to operate for the rest of the mine life.

Other Mining Operations

When one considers the list of NORM related industries in IAEA Safety Report No.49 (IAEA, 2006) it is necessary to understand the possibility that there could be a need for radiation protection issues in any potential mining operations other than uranium or the industries described above. Many of the mineralizations that exist in relation to uranium, for example, are often associated with other valuable deposits. The uranium mines at Guratba (Coronation Hill) and the other South Alligator Valley deposits also produced gold in the 1960s (Fisher, 2002); these sites were also explored in the 1990s when a gold, platinum and palladium mineralisation was identified at a number of locations. However, the sites are all located in an area that is now Stage 3 of the World Heritage listed Kakadu National Park and no further mining activity will be permitted.

The site at Rum Jungle was primarily a copper mine that also produced uranium (Barrie, 1982)], in fact several of the uranium related radiological anomalies in and around that district are now areas where gold mining has either taken place or is still in progress. For this reason all underground mines in the Pine Greek geosyncline are regarded as having the potential for exposing workers to doses in excess of 1 mSv/y and the DME recommends and advises that operators carry out suitable monitoring and screening programmes and then discuss the results with the Department of Health where appropriate. For example the Cosmo Howley operations of Crocodile Gold are on tenements which include the former Fleur-de-Lys uranium mine (Annabel, 1971). The site is known to the workforce and there are appropriate procedures in place. To date there have been no results of significance reported, although the presence of uranium in some drainage waters indicates that the matter cannot be ignored and ongoing surveillance is required. The DME's Environmental Monitoring Unit checks many legacy sites as well as active mines which are also required to submit monitoring data to DME periodically. Any anomalous results, most usually found in water chemistry data, are discussed with the operators, where applicable. Such data may also be referred to the Department of Health and/or the NT Worksafe authority as appropriate. The data are also shared with the Legacy Mines Unit of the DME so that a risk assessment can be undertaken to ensure that the site is appropriately classified in the legacy mines inventory for the NT.

Conclusion

The Northern Territory of Australia has a long and strong association with the mining industry that continues to the present time. Many of the minerals mined in earlier times contained NORM although the risks associated with these substances were not always recognised. More recently the NORM industry in the NT has been dominated by uranium mining where world class leading practices in radiation protection are a regular part of operations.

As the world has become more interested in rare earth elements and the demand for fertilisers has grown so suitable mineral deposits related to these commodities have been discovered in the NT. Most of these deposits are associated with NORM minerals that occur in varying concentrations. Several of these deposits are likely to be developed in the near to medium future and the DME are ensuring that the potential risks associated with NORM are clearly explained to operators who in turn are adopting suitable monitoring programmes and radiation protection plans where appropriate.

References

Annabel, Ross (1971).The Uranium Hunters, Rigby, Australia.

Fisher, Joe (2002). Trials and Triumphs. Pub. WJ and EJ Fisher, Darwin Australia

Barrie, D R, (1982), 'The Heart of Rum Jungle', The Dominion Press, Hedges and Bell, Sydney. p. 142.

IAEA (2003). Radiation protection and the management of radioactive waste in the oil and gas industry. Safety Report Series No. 34. IAEA, Vienna.

IAEA (2006). Assessing the need for radiation protection measures in work involving minerals and raw materials. Safety Report Series No. 49. IAEA, Vienna.

IAEA (2011). Radiation protection and NORM residue management in the production of rare earths from Thorium containing minerals. Safety Report Series No. 68. IAEA, Vienna.

Waggitt, P. (2012). Uranium mining and regulation in the Northern Territory. In Proceedings International Uranium Conference, Perth, Western - June 2012. Australia. Australasian Institute of Mining & Metallurgy, Melbourne.

Waggitt, P. (2013). The Northern Territory's Uranium Industry: Past Present and Future. In proceedings 7th International Symposium on Naturally Occurring Radioactive materials (NORM VII), 22-26 April 2103, Bejing, People's Republic of China (in press).

Technical Status of Mine Water Control in China and Its Development Strategy

Hao Wang[1,2], Shuning Dong[1,2], Rui Chai[1], Qisheng Liu[1,2]

[1]Xi'an Research Institute of China Coal Technology & Engineering Group Corp., Xi'an 710077, China;
[2]Shaanxi Key Lab of Mine Water Hazard Prevention and Control, Xi'an 710077, China

Abstract. In this paper, investigation of mine water disasters was conducted, the current situation of mine water prevention and control was analyzed, the basic theories and practical technology for mine water control were summarized, and some science and technology countermeasures were developed to prevent and control mine water disasters, which can be briefly summarized as studying basic theories, developing key technology and equipment, constructing research and development platform, and strengthening safety management.

Introduction

In China, under the effective supervision and management of state administration of Mine safety, mine water disasters decline sharply. However, shallow resource is dried up gradually, so mine companies turn to deep resource. Generally, the deeper the mines locate, the more complex the hydrogeological condition is, and the harder water hazard prevention and control work is. In such conditions, large mine water inrush disasters still occur frequently in some mine areas. In order to control mine water hazard thoroughly, mine water researchers still have a long way to go. It is urgent to develop new science, technology and equipment, as well as strengthen the research of water hazard control mechanism.

Technical status and problems of mine water hazard exploration in China

Geophysical exploration technology

B.J. Merkel and A. Arab (eds.), *Uranium – Past and Future Challenges*,
DOI 10.1007/978-3-319-11059-2_21

Geophysical exploration technology can be divided into ground geophysical exploration technology and underground geophysical exploration technology.

At present, ground geophysical exploration technology includes 2D/3D seismic exploration, transient electromagnetics, high-density electrical method, direct current method, controlled-source audio-magnetotellurics, ground penetrating radar, rayleigh wave and interborehole fluoroscopy etc.

Three-dimensional seismic exploration is the most effective geophysical prospecting technology for high-resolution exploration of geological structures. It uses sound waves to form sharp three- dimensional images of underground formations. In China, it has been successfully used to detect the faults whose throw is less than 3m. And it can also be used to prospect concealed geological structure and adverse geologic body.

Ground electromagnetic method is especially effective for detecting underground water-bearing geologic body with low resistivity, such as flooded goaf and water-bearing sink hole. For example, in March, 2010, a severe accident caused by Ordovician limestone water inrush occurred in the coal seam 16 mining roadway in the Luotuoshan Coal mine. Thirty-two coal miners died in this accident. Ground transient electromagnetic technology was applied to detect the water inrush channels. In Fig.1, the red point is water inrush point, and the blue low resistivity areas clearly show the water inrush channels of Ordovician limestone. Some boreholes were drilled at the low resistivity areas, which proved the accuracy of transient electromagnetic technology.

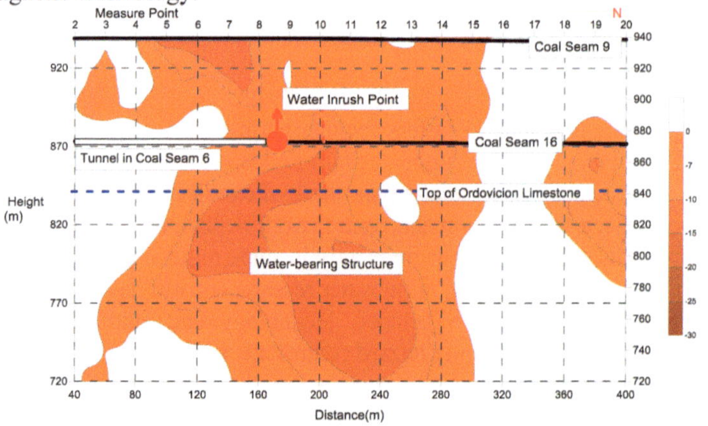

Fig.1. Diagrammatic cross-section of apparent resistivity for detecting water inrush channel

However, this technology is limited severely by geological conditions. Under the conditions of long detection distance and accidented terrain, the detection precision of ground geophysical technology is relatively low. For example, with the increase of detection depth, the detection precision of ground geophysical exploration would be reduced obviously. In addition, the detection precision is still not accurate enough to find out small faults and sink holes.

Underground geophysical exploration technology can be divided into two groups.The first group is electromagnetic wave detection technology, which includes radio wave perspective, transient electromagnetics, direct current method, high-density electrical method, ground penetrating radar, and audio frequency electric perspective, etc. The second group is elastic wave detection technology, which includes channel wave seismic, MSP (Mine Seismic Prospecting), microseismic monitoring, rayleigh wave exploration, multi-component seismic exploration, etc.

Among those methods, direct current method, transient electromagnetic, ground penetrating radar, Rayleigh wave exploration, and mine remote seismic method are applied widely and obtain good results in Chinese mines.

Although many kinds of technology and equipments for underground geophysical exploration make good effects, there are still some problems that need to be solved. For example, the basic theory research under the condition of non semi-infinite foundation and non infinite foundation is weak and there is serious electromagnetic interference, which result in interpretation ambiguity, failed reporting and misinformation.

Drilling technology

Drilling technology for mine water hazard control can be divided into two types, conventional rotary drilling and directional drilling technology, which can be applied under both ground and underground conditions. In recent years, accurate positioning technique, branch drilling technique, directional drilling technique and large diameter borehole drilling technique have made great progress. For example, directional drilling equipment came into use in 2008. Up to now, it has worked very well in China's mines. In the Hongliu mine, located in northwestern China, directional drilling technology was successfully used to explore underground water-bearing areas which were near the mining roadway. In addition, in the Zhaogu mine, located in central China, directional drilling technology was employed to drill boreholes effectively which were used for underground grouting.

However, both conventional rotary drilling technology and directional drilling technology are difficult to drill in the high pressure aquifers, as well as less effective when drilling in extremely hard(or soft) stratums.

Technical status and problems of mine water hazard monitoring and early warning technology in China

Water hazard monitoring is the precondition of safe mining above or under water. In terms of monitoring environment, roof or floor aquifer monitoring can be divided into two groups, ground monitoring and underground monitoring; In terms of monitoring object, it can be divided into dynamic groundwater properties monitoring, and water inrush monitoring; in terms of monitoring conditions, it can be divided into monitoring under natural conditions and monitoring under mining conditions.

In China, a series of water inrush monitoring, data acquisition, data processing systems and related software have been developed. In addition, some new technology and equipment have been researched. Take mine water real time monitoring system for example, it is a distributed serial digital communication network system based on Controller Area Network (CAN) technique. The monitoring system includes three parts: monitoring center (mainframe computer, printer, network communication adapter, etc), monitoring substation (sensors, data acquisition unit, network communication interface, etc) and data transmission network (Dewu J, et al., 2013).The monitoring items include water level, water pressure, water temperature, precipitation, mine water inflow, maintenance condition of mine water facilities, real time condition of mine drainage system. This system has been successfully used in the Jiaozuo, Pingdingshan, Yongcheng mine areas in China in recent years.

Fig.2 shows the roof water hazard monitoring and early warning system developed by Xi'an Research Institute, which was successfully applied in the undersea working faces of the Beizao mine in Longkou. Monitoring indexes of this monitoring system include water pressure, water quality, water temperature, and water inflow rate. Surface and underground boreholes were used to monitor water pressure, water temperature and water quality of target aquifers. Early warning thresholds of water quality indicators were determined by water matching tests.

However, there are some urgent problems that need to be solved. First, it is difficult to determine the threshold value of water hazard monitoring and early warning. Second, there is a long way to go to meet the practical demands for accuracy, validity and practicability of water hazard monitoring.

Technical status and problems of mine water hazard control technology in China

Mine water hazard control technology can be divided into two types, water hazard hidden danger control technology and post-disaster control technology.

The widely used water hazard hidden danger control technology includes floor grouting reinforcement and reconstruction, the shaft grouting, preliminary grouting for faults and sink holes, as well as the curtain grouting technology.

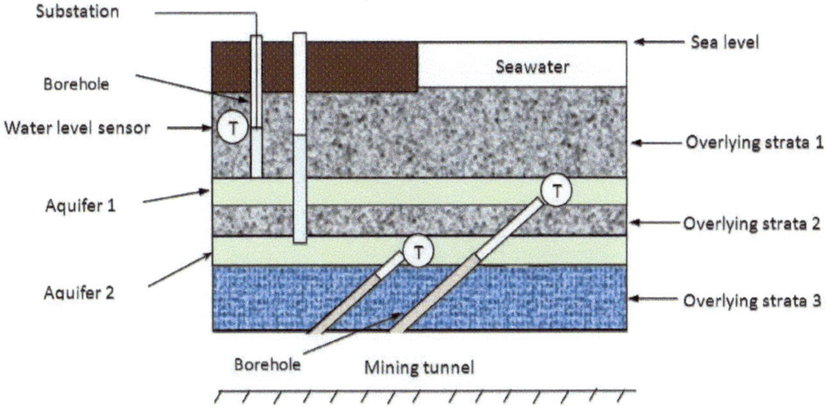

Fig.2. Roof water hazard monitoring system

Post-disaster control technology includes comprehensive geophysical exploration of water inrush channels, branch borehole drilling and directional drilling technology, water-retaining wall construction technology and sink hole water stopper construction technology. Take sink hole water stopper construction technology for example, a sink hole water stopper is constructed successfully by grouting in a proper position to block the sink hole water inrush channel in the Dongpang mine (Shenghui N, et al., 2008). Fig.3 shows the sink hole water stopper built in the Dongpang coal mine in north-eastern China. The column stopper was built by Xi'an Research Institute of China Coal Technology & Engineering Group. The total grout volume of the water stopper was 49339m^3. Grouting and monitoring boreholes were drilled with directional drilling technology to ensure designed trajectory. There are four grouting steps to construct a water stopper: regular grouting, pressure grouting, drainage grouting and reinforcement grouting. The effectiveness of water stopper can be evaluated by underground dewatering tests.

However, the present research on the regularity of diffusion and solidification of the grout under flowing water conditions is still insufficient. Furthermore, there is no systematic experimental study on the grouting technology under different hydrodynamic conditions.

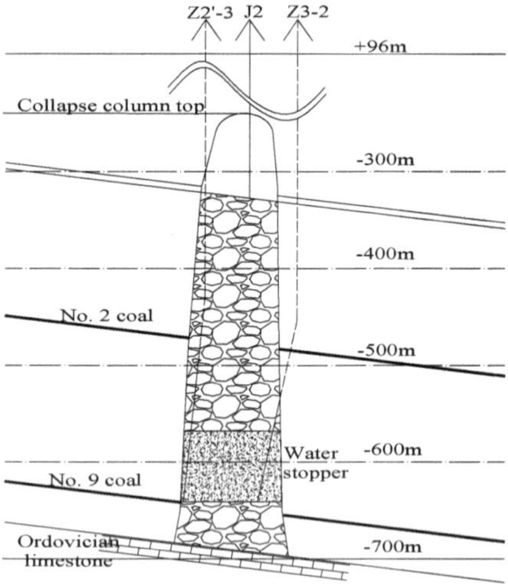

Fig.3. Water stopper in Dongpang Coal mine

Science and Technology Development Countermeasures for Mine Water Hazard Prevention and Control in China

Focusing on two key types of water hazards

First, in the terms of water source, it is urgent to take goaf water and limestone water as the key water source hazard. Second, as for water inrush channel, make permeable faults and sink holes as the key prevention and control channel. A large amount of work should be done to strengthen research on detection, monitoring, prediction and control of water hazard of goaf water, limestone water, faults water and sink hole water, which is on the basis of comprehensive water hazard prevention and control.

Strengthening two types of promotion and development technology for water hazard control

To strengthen the promotion and development of water hazard control technology, the first important point is to promote the existing technology and equipment which is advanced and well developed for water hazard prevention and control, in order to improve mine prevention and control ability comprehensively. The second point is to strengthen the prevention and control technology research of urgent need to develop, leading the development frontier of technology and equipment.

Adopting three science and technology development strategies of water hazard control

The first strategy is to develop basic theory deeply; the second one is to strengthen research and development of the key technology and equipment; the third one is to promote R&D platform construction.

Developing basic theory

The basic theories that need to be researched are as follows:

Study the basic theory of geophysical exploration for detecting goaf water, with the emphasis on the geophysical basis characteristics and the comprehensive exploration theory of goaf water;

Research the mechanism of water inrush from mine floor, especially for deep mine resource which is under high water pressure and high ground stress, in order to provide fundamental basis for water inrush risk assessment.

Strengthening research and development of the key technology and equipment for water hazard control

For water hazard detection technology and equipment, the key is to develop the technology and equipment for fine detection of goaf water, permeable faults and sink hole, as well as detection and dewatering of high pressure aquifer.

For water hazard monitoring and early warning technology and equipment, further study should be done to research the real-time monitoring and early warning technology, with emphasis on the research of threshold value of water hazard monitoring and early warning.

For water hazard evaluation and prediction technology, influence of the key water inrush factors, such as floor damage depth, lift height of fractured zone and thickness of floor aquitard, should be comprehensive analyzed based on the deeply study of mechanism of water inrush from floor.

For water hazard control technology and equipment, utilization and grouting reconstruction of the top of Ordovician limestone technology should be researched, and effective grouting process and device should be developed.

For water hazard emergency rescue technology and equipment, rapid drilling of ground large diameter borehole technology and ancillary facility should be developed, in order to provide support for scientific, timely and effective rescue of mine water disaster.

Promoting scientific and technical platform construction

In order to effectively promote technological innovation of the mine water hazards prevention and control and share the professional resources socially, the mine water hazards prevention and control engineering technology center should be established by the professional institutions as soon as possible. National key laboratories and supporting facilities, which cover water inrush mechanism research, grouting technology and process research and hydrochemical analyses, should be constructed. In addition, some typical mine should be supported to speed up the construction of water hazard prevention and control engineering demonstration base.

Suggestion

In order to prevent mine water disasters effectively, some measures should be carry out on time, such as making proper laws and policies, providing project support, offering fund guarantee and developing new industry standards. These measures can make mine safety situation improved steadily. In addition, the support of research projects on the mine water hazards prevention and control, comprehensive utilization of water resources, as well as water environment protection, should be strengthened.

References

Shuning D, Dewu J, Hong F (2008) Practical technology and equipment for mine water prevention and control. Coal Science and Technology 36(3): 8-11

Zhijun SH, Shuning D, Ningping Yao, etc (2011) The underground directional drilling technology and equipment for kilometre deep borehole with MWD in coalmine. Procedia Earth and Planetary Science 3: 17-22

Dewu J, Yingfeng L, Zaibin L, etc (2013) New progress of study on major water inrush disaster prevention and control technology in mine. Coal Science and Technology 41(1): 25-29

Shenghui N, Qinming J, Xiaoshan G, etc (2008) Construction technique of groundwater-preventing piston in Karst flow sink hole. Coal Geology and Exploration 36(4): 29-33

Revisiting a case study on uranium exposure linked to leukaemia – preliminary results

Frank Winde[1], **Ewald Erasmus**[2], **G. Geipel**[3], **AAA Osman**[3]

[1]Mine Water Re-search Group, North-West University, Vanderbijlpark, South Africa, Frank.Winde@nwu.ac.za
[2]Mine Water Re-search Group, Geotechnical Environmental Specialists, Groenkloof, Pretoria, South Africa, geotech@netactive.co.za
[3]Helmoltz Centre Dresden-Rossendorf, Institute for Resource Ecology, Bio-Geochemical laboratory, Germany, g.geipel@hzdr.de

Abstract. Triggered by observations that a disproportionally high number of leukaemia patients came from a particular area around the small town of Pofadder in the arid Northern Cape Province, a subsequently conducted research project established a geo-statistical link between naturally elevated U-levels in groundwater and haematological abnormalities. In this paper results from a follow-up study are presented indicating the water is indeed the most important pathway for U-exposure while consumption of contaminated sheep meat also contributes but to a lesser extent.

Background

Triggered by observations at the Community Health Unit of the Tygerberg Hospital in Cape Town (South Africa) that a disproportionally high number of leukaemia patients originated from a limited number of farms near the town of Pofadder the Water Research Commission of South Africa funded a study to explore a possible link to the quality of drinking water as common denominator of all patients (Toens et al. 1998). Using two large data sets on U levels in 126 boreholes in the area and results of blood samples taken in 1993 from 630 local residents a significant geo-statistical correlation between the spatial distribution of elevated U level in borehole water and abnormal haematological values of residents (used as proxy for leukaemia) was established. When the leader of the study subsequently approached the Minister of Water Affairs for rapid intervention the latter responded that no other possible causes such as contaminated dust or toxic agrochemicals had been investigated and declined to intervene.

B.J. Merkel and A. Arab (eds.), *Uranium – Past and Future Challenges*,
DOI 10.1007/978-3-319-11059-2_22

Furthermore, most affected residents live on sheep farms where they not only drink contaminated water themselves but also use it for the animals. Therefore, ingestion of possibly contaminated sheep meat, a staple food in the area, may also constitute a potential pathway that has not been considered so far. In order to address these gaps and arrive at a better understanding of the U exposure this follow-up study was designed and an associated field sampling campaign conducted. A main objective was to determine whether and if so to what extent potentially contaminated meat may contribute to the U exposure of residents.

Geological and farming conditions in the study area

The original study area is located in Kenhardt municipal district of the Northern Cape Province touching the border to Namibia in the north (Fig. 1)

Fig.1. Location of the study area shown on satellite imagery of Google Earth with the original extent of the study area indicated by a superimposed map extracted from Toens et al. (1998) depicting the spatial distribution of residents with abnormal lymphocyte counts.

Situated at the southern fringe of the Kalahari desert the area is fully arid with an average rainfall well below 100 mm/a. High interannual variability means that several years may pass without any rainfall. Consequently, there are no perennial streams and groundwater is the sole source of water supply on farms. Usage of bottled water was not reported in a single due to both, affordability and long travel distances to the next purchase point. The used groundwater occurs mainly in low yielding, fractured granitic gneiss of the Namaqualand Metamorphic Complex (NMC) at depths of 30-100 m below surface with high electrical conductivity (EC) levels of up to 30 mS/cm reflecting high evaporation and low recharge rates. Because of the tectonic stability of the area and its aridity limiting the danger of seepage the national radioactive waste disposal site is located in the vicinity of the study area at Vaalputs (Fig. 1). Uranium mainly occurs in the area in granites, gneiss and pegmatites of the NMC as well as in younger (tertiary to recent) surficial calcrete deposits in river drainage systems. Owing to elevated levels of U and other heavy metals in the host rock and prolonged contact times to the often stagnant groundwater, the latter often displays increased levels of U and other toxic trace elements leached from the host rocks (Toens et al. 1998). Agriculture in the area is confined to extensive sheep and goat farming using borehole water as sole source for animal watering, with very little house gardening taking place in a few instances. While the predominantly white farm owners often use commercially available domestic filter system to clean borehole water before consumption, the mostly mixed race farm workers (termed 'Coloured' in South Africa) mostly consume untreated groundwater as do all farm animals.

Methods

Only farms with residents being permanently on site were used (as some are only sporadically visited by owners living elsewhere) giving preference to farms with confirmed cases of leukaemia or other cancer types. Ultimately 5 farms were sampled comprehensively (including sheep tissue samples). Apart from sampling borehole water, outcropping rocks and dust also samples of fodder and sheep tissue (wool, teeth, hooves, inner organs, meat) were taken. In most cases the oldest animal available on the farm was sampled with one exception where kidney from a lamb was used). In addition, a commonly used locust pesticide was analysed for U and gamma radiation was measured inside farm houses and on rock outcrops using a hand held device (Radex, RDX 10-353). Calcite scales from water kettles and geysers have also been analysed as long-term indicators of U levels in water, as have salt crusts forming on watering troughs and groundwater storage tanks. For water samples the pH and EC was determined on site. In total 89 samples were analysed. For all samples the total U-concentration was determined using mass-spectroscopy (ICPMS) related to dry weight for biological samples. Two blind split sample analyses indicated a deviation between identical water samples of less than 1% (0.7-0.9%). Additionally, the speciation of U in all water samples was analysed using Time Resolved Laser-induced Fluorescence Spectroscopy (TRLFS).

Uranium levels in water

At all 8 sampled farms groundwater from local boreholes was used for the watering of sheep. U levels varied considerably from 13.7 to 22 100 µg/l with a median of 147 µg/l. Only 1 of the 9 sheep water samples and 2 of the 6 tap water samples complied with the raised WHO limit of 15 μg/l and only 1 sample is 'baby safe'[1] Of particular concern is the extremely high U level in water used for sheep watering at the farm Swanepoelputs which exceeds the WHO guideline value for drinking water by nearly 1500 times (Fig. 2) (WHO 2006).

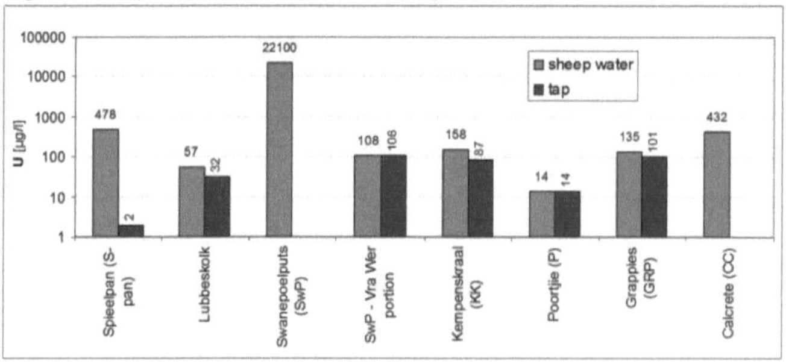

Fig.2. U levels in water used for sheep watering and human consumption ('tap')

However, no less concerning is that most tap water samples also well exceed this value by several hundred percent in cases. In the one case where this comparison was possible filtering of borehole water reduced U levels from nearly 500 μg/l to 2 μg/l suggesting that commercially available filters are able to significantly lower the U intake via drinking water (Tab. 5). Rainwater sampled from a tank that collects runoff from the roof was 0.02 μg/l indicating that dust or atmospheric wash out do not present a major source of U pollution in the area. Most of the U was found to be present in a neutral calcium-uranyl-carbonate complex of relatively high solubility (Ca2UO2[CO3]3). Only one water sample differed showing a strongly negatively charged uranyl-carbonate complex presumably due to a lack of freely available calcium ({UO2[CO3]3}4-). High to very high EC values (0.9-30 mS/cm) dominated by chloride and to a much lesser extent sulphate reflect the low groundwater recharge rate. Exceeding the target water quality range for drinking water in South Africa by 1500% to over 2000% especially high levels of chloride and nitrate respectively render the water unfit for human consumption. Uranium concentrations of nearly 50 ppm in scales of geysers and water kettles suggest that temperature driven calcite precipitation also removes parts of the dissolved U from the generally hard water. Secondary enrichment of U also takes place in evaporative salt crusts that form on watering troughs with U accumulating up to levels of almost 90 ppm. Along with commercial salt lick placed on grazing sites as mineral supplement (23 ppm U) the licking of salt by sheep may in cases add considerably to the total U intake.

---

[1] I.e. it is equal or below 2 μg/l as maximal allowable U concentration in water used for babies as stipulated in Germany (BfR 2006).

Uranium levels in sheep tissue

The U levels in sheep tissue vary widely from a minimum of 0.5 ppb in leg meat to 998 ppb in wool (all based on fresh weight). For better comparison the samples have been grouped into 3 categories of tissue: (i) (consumable) inner organs comprising of liver, kidney, brain etc. (ii) (consumable) meat from leg, neck and ribs, and (iii) (non-consumable) skeletal tissue such as bones, hooves, teeth and wool. The average U levels for the 3 categories based on fresh and dry weight are shown in Tab. 1.

Table 1. U levels in different categories of sheep tissue for animals from 5 different farms

Sampling site		**Sheep samples**							
	age	**U conc. [ppb]** (weighted averages)							
	[a]	**Meat**		**Inn. Organs**		**Skeleton**		**Total**	
Wight basis		*fresh*	*dry*	*fresh*	*dry*	*fresh*	*dry*	*fresh*	*dry*
Spieelplan (S-pan)	8	11	36	12	67	422	1960	**194**	**688**
Lubbeskolk	6	1.7	6.3	13	50	107	520	**48**	**197**
SwP – VraWer portion	?	2.7	8.9	2.8	11	127	1011	**53**	**344**
Kempenskraal (KK)	?	1.0	3.7	2.9	11	102	232	**42**	**83**
Poortije (P)	0.5			1.1				**1.1**	
n	3	8	8	16	14	16	11	**40**	**33**
n-weighted av.	**4.8**	**4.1**	**14**	**7.1**	**30**	**190**	**837**	**80**	**295**
min	0.5	0.5	2.0	0.5	2.0	6.9	11	**0.5**	**2.0**
max	8.0	18	61	37	141	998	3270	**998**	**3270**
Diff. Max-Min [%Max.]	93.8	97.2	96.7	98.6	99.6	99.3	99.7	**99.9**	**99.9**

Comparing the U levels in different tissue types per farm to those in water consumed by sheep indicates that highest U enrichment in sheep occurs where U levels in the water are also high pointing to drinking U polluted water as a major source of contamination (Fig. 3).

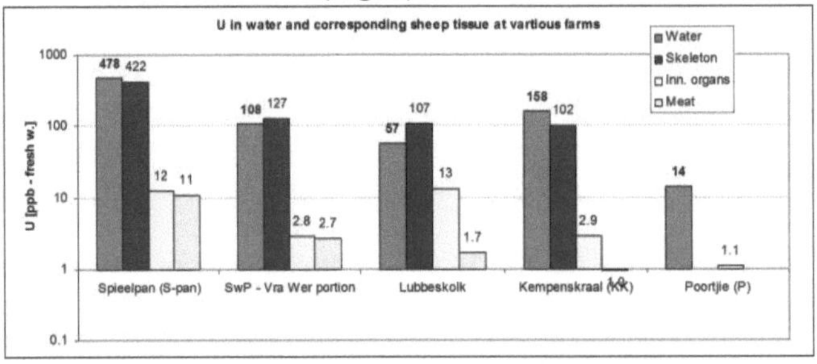

Fig.3. U concentrations in various types of sheep tissue (fresh weight) and in water consumed by the sheep at different farms

With an average of 54 ppb (dw) the U levels in kidney are significantly higher than in any other organ also displaying the highest minimum and maximum U concentrations (23 and 122 ppb dw respectively) of all analysed organs except one brain sample (141 ppb dw). Kidneys, as cleaning organs, are also the only inner organs that hint at a linear relationship to U levels in consumed water even though this relation is dominated by the maximum value and the associated correlation coefficient is reduced significantly if that value is removed (Fig. 4)

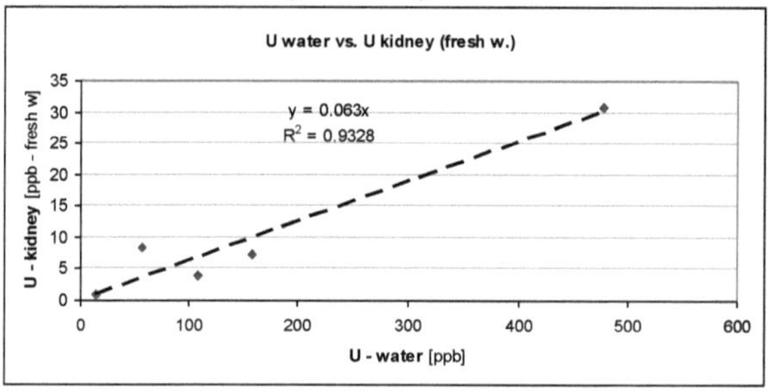

Fig.4. Relationship between U levels in kidney and drinking water of sheep

Assessing associated health risks

Given that U is radioactive as well as chemotoxic both properties need to be considered when assessing health risk possibly associated with the ingestion of polluted water and sheep meat. In order to assess the chemotoxic risk the tolerable daily intake (TDI) limit for U and the drinking water limit of the WHO are applied. Two principal scenarios are considered: one based on maximum U levels for a risk conservative approach and another based on average U levels applicable to a wider circle of users. In all calculations fresh weight U levels are used since consumption rates are also based on fresh weight.

Chemotoxic risk of meat consumption: Compared to U levels in uncontaminated meat of the European Union of 0.5 to 1.7 ppb (EFSA 2013) the sampled meat in the study area clearly shows elevated U levels exceeding the natural background by up to 70 times. Calculating the average U intake for the different age groups indicates that in a worst case scenario (i.e. only consuming inner organs with the maximum concentration of 37 ppb) account for 23% to 34% of the TDI value of the WHO (0.6 µg/kg body weight per day). The highest U-intake from regular consumption of inner organs would occur in infants and adolescents contributing up to a third of the TDI. Since many farmers use mutton, incl. inner organs, as stable food (eating sheep meat up to 3 times a day, every day) actual consumption rates may be well above the average rates used in the calculations(160 g/d for adults). Based on more realistic meat consumption rates of ca. 4 to 5 times this amount TDI for U would be exceeded by meat consumption alone.

Chemotoxic risk of drinking water: Ingesting tap water with the average U concentration found in the study area (57 ppb) exceeds the TDI limit by nearly 10 times

(1000%) for babies and 3 to 4 times for adolescents and adults without any additional U-contribution from meat taken into account. The scenario obviously worsens for farm residents relying on the worst quality tap water (106 ppb U) at which U levels babies exceed the TDI by nearly 18 times. Absolutely intolerable is the consumption of the worst quality water (221000 ppb) where the associated U intake is over 3600 times above the WHO limit. All three U levels exceed the applicable guideline value of the WHO by 3.8 to nearly 1500 times. Given the vulnerability of babies due to their high water intake relative to their (low) body weight the current WHO guideline is not applicable as it would lead to a clear exceedance of the WHO's own TDI value (250%). Applying the appropriate 'baby safe' level of 2 µg/l obviously results in another order of magnitude by which the consumed borehole water exceeds existing limits (BfR 2006).

Radiotoxic risks of meat consumption: This calculations are based on results obtained from using an online age-specific dose calculator (WISE 2013) compared to the single facility dose limit of 0.25 mSv/a, ICRP 1991). The results suggest that the consumption of meat even at maximum U levels only adds marginally to the natural background radiation.

Radiotoxic risks of drinking water: Regular consumption of the worst water quality would result in exceeding the single-facility dose limit by 3 to over 50 times. As far as could be ascertained, this water is however not consumed continuously by humans. Sheep relying on this water could not be sampled. Using the worst drinking water quality of 106 ppb U permanently results in 0.07 mSv/a for babies under 3 months (critical group) amounting to 24% of the single facility dose. For average tap water (54 ppb U) the highest effective dose is about half of this.

Conclusion and recommendation

The study determined that U levels in most sheep tissue samples are well above average values observed in uncontaminated areas with the highest accumulations occurring in skeletal parts such as wool and hooves, followed by inner organs (especially kidney) and meat showing the lowest U levels. Based on realistic consumption rates in an area where sheep is used as staple food, some farmers may exceed the TDI value for U alone by eating meat. Of greater concern is however the fact that drinking only water with the average U contents found at the sampled farms already exceeds the TDI by up to an order of magnitude, with babies being particular at risk. As U levels in drinking water are much higher at some farms exceeding WHO limits by 7 to well over 1000 times water consumption generally poses a significantly greater risks than eating sheep. This also since other water quality parameters such as nitrate, fluoride and chloride concentration all too exceed existing guideline values by considerable margins. Since both pathways, drinking water and meat consumption, are superimposed the resulting U intake for many farm residents, especially young ones, will often be well above tolerable levels. While the consumption of meat only marginally adds to the naturally occurring radiation this is different for worst quality tap water consumed in the area that accounts already for a quarter of the applicable dose limit. At the highest measured U level the effective dose associated with drinking such water (which is not known to be done by humans) would exceed the annual limit of 1 mSv/a for all pathways by over 12 times. From the above it can be concluded that water is indeed the most important pathway of exposing farm resident to U supporting the established spatial correlation between U-levels in borehole water and haematological abnormalities linked to leukaemia in farm residents using this water. Since it was recently established that U not only accumulates in the outer mineral parts of the bone where it often replaces Ca but is also able to affect the inner part with the blood producing bone marrow a causal relationship between U intake and leukaemia is in principal possible (Arrudo-Neto et al. 2004). Since the use of commercially available household level filter systems removes much of the U from tap water it is recommended to install such filters at all farms with elevated U levels in borehole water.

References

Arruda-Neto et al. (2004): The accumulation and microdistribution of uranium in the bone and marrow of beagle dog. Int J Radiat Biol, 80 (8) 567-575

BfR (Bundesanstalt für Risikobewertung) und Bundesanstalt für Strahlenschutz (BfS) (2006): Gemeinsame Stellungnahme Nr. 014/2006 des BfS und des BfR vom 16. Januar 2006. BfR korrigiert Höchstmengenempfehlung für Uran in Wässern zur Zubereitung von Säuglingsnahrung.

EFSA (2013): Uranium in foodstuffs, in particular mineral water1 Scientific Opinion of the Panel on Contaminants in the Food Chain, EFSA Journal 1018, 1-59

ICRP (1991) Recommendations of the International Commission on Radiological Protection. ICRP Publication 60. *Annals of the ICRP* 21 (1-3). Oxford, UK

Toens PD, Stadler W, Wullschleger NJ (1998): The association of groundwater chemistry and geology with atypical lymphocytes (as biological indicator) in the Poffadder area, North Western Cape, South Africa. WRC report 839/1/98

WHO (World Health Organization) (2006): Guidelines for drinking water quality [electronic resource]: incorporating first addendum. Vol. 1, Recommendations, 3rd ed. Electronic version for the web

WISE (2013): Calculator for Uranium and age specific effective dose http://www.wise-uranium.org/rdcua.html, accessed 25.05.2013)

Virtual Geographical Environments as a tool to map human exposure to mining-related radionuclides

Frank Winde[1], **Emile Hoffmann**[2]

[1]Mine Water Re-search Group, Geography and Environmental Studies, North-West University, Vanderbijlpark, South Africa, Frank.Winde@nwu.ac.za
[2]Mine Water Re-search Group, Geography and Environmental Studies, North-West University, Vanderbijlpark, South Africa, Emile.Hoffmann@nwu.ac.za

Abstract. Large volumes of uraniferous gold tailings deposited over more than a century in densely populated metropolitan areas of South Africa (SA) pose potential health hazards to millions of residents of nearby settlements, including many vulnerable, impoverished populations. In order to inform the selection of suitable sampling populations for epidemiological studies a mapping methodology based on a Virtual Geographic Environment (VGE) combined with a Geographical Information System (GIS) was developed. The structure, architecture and database of the system are presented.

Motivation: the need for assessing U related health risks in SA

The deposit of large volumes of uraniferous gold tailings in densely populated metropolitan areas has created the potential for large, widespread exposures to mining-related waste including radionuclides in South African goldfields. Most of these radionuclides are associated with uranium that accompanies the mined gold in many of the ore bodies at an average ratio of approximately 20:1. As gold is very rare with a commodity-to-waste ratio of 1:200,000 in South Africa (compared to 1:5 for iron or 1:100 for copper for example) gold mining generally produces relatively large amounts of waste. Combined with the exceptionally long period of over 120 years of mining the single largest gold deposit on Earth in the Witwatersrand basin at an industrial-scale this resulted in extraordinary amounts of (uraniferous) mining waste. By now nearly 7000 million tons (Mt) of U bearing gold tailings were generated since 1886. This is over 20 times the amount of tailings produced by uranium mines in Canada and the USA combined (300 Mt, Waggitt 1994). In 1994, the annual global production of U tailings (20 Mt, Waggit

B.J. Merkel and A. Arab (eds.), *Uranium – Past and Future Challenges*,
DOI 10.1007/978-3-319-11059-2_23

1994) equalled the tailings output of South African gold mines in 1957. Extrapolating the global tailings production rate over some 60 years of U-production (since about the mid 1940s: 1200 Mt) it is estimated that South African gold mines generated over 5 times more uraniferous tailings than the rest of the world combined. In many instances U levels in gold mine tailings are above genuine U tailings in Germany and Namibia for example where mostly low grade ore was exploited.

Compared to other major U producing countries where U was mined in remote and sparsely populated regions like the Canadian arctic circle, the Australian outback or desert-like areas in the USA, most legacy sites in South Africa are situated in very densely populated conurbations such as Johannesburg that owe their existence to mining. Many residents living closest to mine dumps are poor informal settlers that lack access to clean water. Consequently, untreated water from polluted sources is used for all domestic needs (drinking, cooking, and washing) and in some cases also for small scale food production (gardening etc.). Together with inhaling tailings dust and intentional and unintentional ingestion of tailings through hand-to-mouth activity of playing children, geophagia (especially by pregnant women) and the use of tailings as muti (traditional medicine) this has the potential to create high exposure in terms of intensity, spatial extent and number of people affected. Malnutrition, wide spread drug abuse, lack of access to health care facilities, high rates of HIV/AIDS and associated opportunistic diseases like tuberculosis as well as other poverty-related risk factors commonly render these impoverished populations very vulnerable to additional health threats.

Due to the rapid and far-reaching fluvial transport of contaminants in rivers and streams mining-related pollution also affects areas well outside the mining regions. Apart from contaminating tap water in downstream municipalities the use of polluted streams for irrigation and live stock watering also introduces U and other contaminants into commercial food production and thus the human food chain.

So far little, if any, attempts were made to scientifically investigate the possibly associated health burden in the South African goldfields. Since the absence of quantified health risks was repeatedly used in the past to justify the lack of remedial action a dedicated study into the matter is now urgent. This even more so as South Africa aspires, once again, to become a major global supplier of U as climate neutral nuclear fuel with associated plans to further expand the mining and enrichment of U. Extensive remediation programmes for U mining areas in Germany (Wismut region), the USA and even in less developed countries such as Kazakhstan are example of good international praxis illustrating the need to reduce exposure at legacy sites even if not all possibly associated health hazards are already known (William et al. 1995, Wismut-GmbH 1996, Fesenko & Monken-Fernandez 2013).

Following an initiative of the International Agency for Research on Cancer of the World Health Organisation the feasibility of conducting such much needed epidemiological study was explored together with scientists from South Africa (Schonfeld et al. 2014). It was subsequently agreed to launch a pilot project that aims to establish to what extent U indeed accumulates in exposed residents before a large scale epidemiological study can be considered. To avoid underestimating risks by sampling low exposed populations a way had to be found to screen the vast gold mining areas for hotspots of pollution. This paper presents the methodology developed for this purpose.

Mapping uranium exposure using VGE

Purpose and objectives

The main purpose of the mapping is to inform the selection of suitable sampling populations for the above mentioned pilot study. Based on a large amount of secondary data extracted from a multitude of scattered sources the mapping also aims at consolidating existing information into easy to read maps depicting the geographical distribution of mining, mine waste and associated pollution of water, air, soil, sediment, food and biota across all goldfields of the Witwatersrand basin. In addition, location, type and size of potentially affected settlements, including census data on race and gender distribution, population density are to be specified.

Operational approach, data and methodology

The general operational approach underlying the use of exposure maps is to geographically link mining-related sources with human receptors via identified pathways such as wind, water, food etc. The relative significance of each exposure pathway is characterised *inter alia* by the level of U and Ra concentrations in corresponding transport media (dust, water, soil, food etc.).

The mapping is based on an open source Geographical Information System (Quantum GIS 2.0) into which several thousands of data points extracted from a multitude of sources that include published literature, unpublished reports, monitoring programmes, research projects, academic theses etc. were imported. Most data had to be first collated in EXCEL spread sheets and manually geo-referenced before being imported into GIS. Relevant mining-related information such as the location of tailings dams, mine lease areas, shafts, mine canals etc. also had to be retrieved. Using up-to-date satellite imagery the status of tailings dams in terms of being active, decommissioned or reclaimed as well as their height, area, mass, age, name, internal code, owner, vegetation and water cover were mapped and also fed into the GIS.

In order to allow for easy interrogation of the data base and related maps by non-specialist researchers unfamiliar with operating GIS a popular and free-of-charge accessible virtual geographical environment (in this case Google Earth, GE) was employed as user interface. The use of a VGE had a number of additional advantages compared to conventional GIS including a continuously updated topographic base that is much more recent than existing maps as satellite imagery for the entire study areas generally originated from 2012 and 2013 compared to 1975 for many topographic maps. This is particular important as the reclamation of tailings dams and formation of new informal settlements is often very rapid. Using the street view function provided in GE, it was also possible to distinguish between formal and informal settlements without conducting time-consuming and costly field trips. Street view also allowed to visually asses the topographic exposure of selected areas to adjacent slimes dams. This is further aided by high-resolution elevation data underlying the offered 3D view in GE.

Lastly, based on indicated altitudes in the task bar for the cursor positions the heights of all tailings dams could be determined conveniently together with assessing the status of the structure in terms of vegetation cover, use and degree of erosion. The overlay function that allows to import and superimpose maps and aerial photos onto GE satellite imagery assisted greatly in delineating boundaries of geological provinces and other areas for which no digital data or coordinates were available.

While waste of deep level gold mining in the Witwatersrand basin often contains a range of toxic heavy metals such as Ni, Cr, Pb, Cd the mapping is confined to U (and its progeny) as it is the only contaminant in gold tailings consistently exceeding natural background levels across the Witwatersrand by at least an order of magnitude (10–100 times) and because it originates nearly exclusively from gold mining (Winde 2010). This compares to enrichment factors (i.e. the ratio between metal concentration in mine waste and natural background concentration) in average gold tailings of 1.5 to 4 for other, more ubiquitous, metals such as Cr, Pb, Cu and Ni which are much less indicative of gold mining as all are emitted from urban sources and non-mining industries too.

Complimentary hydrological data on flow volumes, water quality and water levels of streams, dams and groundwater monitoring boreholes were sourced from DWA and incorporated in the system for the improved characterisation of the water pathway. Exposure to dust plumes from tailings dams was modelled using a simple calculation of airborne transport distances for different tailings particles sizes based on Stoke's law and the different tailings dam heights above ground at a fixed critical wind speed. Combined with wind rose data for the dry winter period (June to September) from 4 airports in the area this allowed to model dust plumes for each unvegetated tailings dam and to identify potentially affected settlements. For a number of tailings dams in the KOSH goldfield the modelled plumes showed good agreement with on-site field measurements.

For characterising the recipient term (i.e. the exposed population) high-resolution digital data of the 2011 census on sub-ward level were obtained for relevant settlements. Informal settlements not covered by the census were mapped using recent satellite imagery of GE and population size estimated by multiplying the number of counted shacks with an estimated average rate of occupancy.

Data complexity was reduced by creating various folders in the VGE grouping data for selected aspects. Depending on requirements the display of sub folder data on the map can be controlled by the user allowing for the combination of different features within a layer as well as between different layers. The following main themes have been selected as layers (Table 1).

Table 1. Major map layers and the contained features used to organise data in Google Earth

Map layer	Contained features
Gold mines	mine lease areas, tailings dams/ spills, waste dumps, shafts, canals ...
Hydrology	streams, stream names, dams, wetlands, gauging stations, groundwater monitoring boreholes, water quality data, water flow data ...
Geology	U provinces, dolomites, sinkholes, faults, dykes
Settlements	name, type, number of residents, racial distribution, population density, farm boundaries ...
Contamination	concentrations of ^{238}U and ^{226}Ra extracted from 58 studies (1967-2013) in: tailings, mine water, streams, groundwater, tap water, soil, sediments, aquatic biota, foodstuff, human tissue; dose calculations...
Hot spot sites	maximum U and ^{226}Ra levels found in each study for various media
Exposure pathways	windblown tailings dust, mine water, surface water, groundwater, tap water, mineral water, food, geophagia
Worst-case sites	superimposed high-intensity pathways, extreme concentrations
Anecdotal evidence	media reports, oral reports, community contacts etc.
Sampling sites	identified populations for sampling human tissue and control sites
Hospitals	cancer treating and referring hospitals including patient catchments

Since it is humanly impossible to visually discern any spatial pattern in a map displaying thousands of single data points of different nature a way had to be found to compare the different type of data and aggregate them into a manageable number of common categories. This is also required as contaminant levels in solutions are usually orders of magnitude below solid materials even though both may reflect identical degrees of pollution. Furthermore, levels of ^{226}Ra or absorbed dose rates cannot be directly compared to U concentrations as units and associated health hazards differ.

By relating all data to their specific natural background levels and existing legislative guideline values two common benchmarks were found that allowed comparing different contaminants as well as different media with each other. To this end all data were grouped into 5 colour-coded categories with increasingly darker red colours indicating rising levels of contamination.

For defining class breaking values (i.e. the width of each class) a balance had to be found between the need to meaningfully distribute the bulk of values across the classes (i.e. avoid that all values fall into a single class only) and maintaining comparability of classes between different media and contaminants. For the latter the enrichment compared to natural background levels and to existing guideline values/ limits were used. While the class widths of the 5 categories are not identical between U and Ra and different media they are still similar enough to allow for meaningful comparison (Table 2).

Table 2. Class breaking values for the 5 categories of ^{238}U and ^{226}Ra concentrations in different media (<I–II–III–IV->V class)

Medium	^{238}**U**	^{226}**Ra**
Tailings, soil, sediment, dust	<20–100-200-500->500 ppm	<100-500-1000-2000->2000 Bq/kg
Sewage sludge, kettle scale	<2-10-50->50 ppm	no data available to be classified
Mine-, surface-, groundwater	<10-50-200-500->500 ppb	<0.02-0.2-2->2 Bq/l
Tap-, mineral water	<2-10-30-50->50 ppb	<0.02-0.2-2->2 Bq/l
Meat (beef, pork, mutton…)	<3-10-60-300->300 ppb fresh*	no data available to be classified
Vegetables	<3-30-60-300 ppb fresh*	no data available to be classified

*= fresh weight (it is important that the weight base is specified as concentrations in dried and ashed samples may be significantly higher)

Results and discussion

An example for a map generated from the VGE is shown as a screenshot of GE in Fig. 1 with folders on the left hand depicting the underlying data structure.

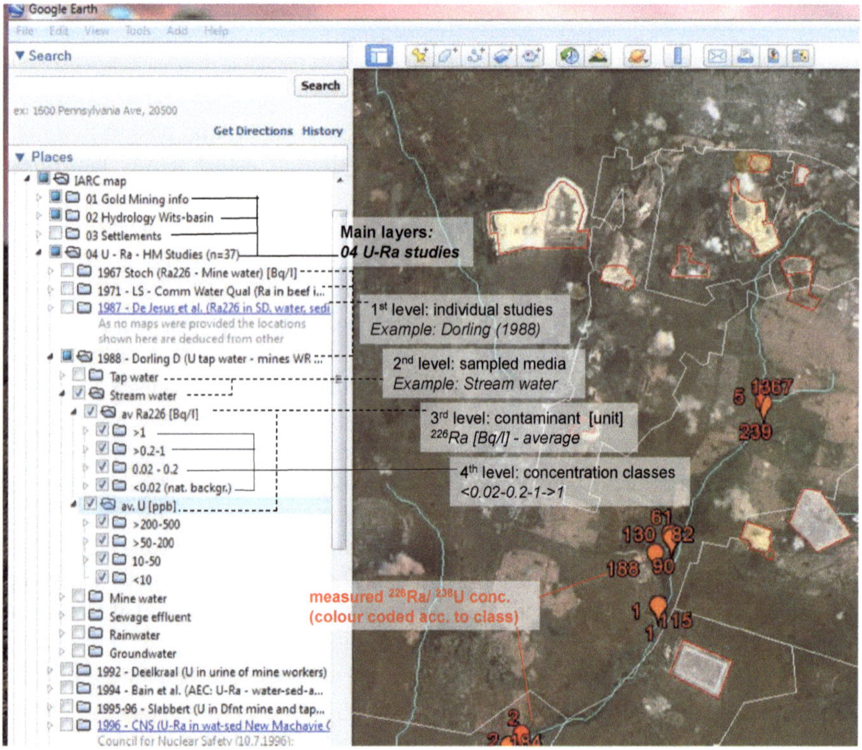

Fig. 1. Screenshot of Google Earth showing a map depicting ^{226}Ra and ^{238}U concentrations of stream water as reported by Dorling (1988) as one example of over 60 studies used for the system. The folders on the left hand side illustrate the underlying data structure (shaded comment boxes are added for explanatory purpose and not shown in GE). By ticking the various boxes different sets of data can be selected for display to superimpose different data sets for identifying possible spatial correlations and contamination patterns.

Another advantage of using a VGE for mapping pollution compared to conventional maps and GIS is the easy change of scales by using the scroll function. By superimposing the occurrence of uraniferous rocks mapped for the entire country a previously unknown spatial correlation with elevated U-levels in sewage works across SA was detected indicating the highest density of polluted sewage sludge in waste water treatment works of the Witwatersrand goldfield. The map indicates a spatial concentration of hot spots in the Witwatersrand basin suggesting that elevated U levels may also occur in drinking water of these municipalities. This too is important information to be considered in the epidemiological study. At the same time the map indicates sites where U-levels may be naturally elevated and should thus not be used as control sites.

In conclusion it was found that the combination of a GIS and a VGE proved to be a useful tool to consolidate and map a large body of fragmented and scattered point data generated over more than five decades. In many instances the functionality of the selected VGE enhanced the data base. By normalizing all data to natural background levels and legislative limits inter-comparisons between different contaminants and media are possible aiding the identification of hot spots and spatial pollution trends.

Limitations of the system relate mainly to analytical accuracy of the original lab data which could not always be ascertained, unknown weight bases for concentration values (especially in biological samples), unknown location of exact sampling sites and errors in subsequent geo-referencing based on descriptions and hand-drawn maps rather than coordinates, uncertainties introduced by converting units, and differences in the statistical handling of values below the detection resulting in variance of how means are calculated etc. Also, many studies show the tendency to follow on each other in terms of selected study areas leaving spatial gaps that are less or not at all covered by measurements.

References

Dorling D (1988): Uranium and Radium 226 concentration in the upper Wonderfonteinspruit. Monitoring results, unpublished

Fesenko S, Monken-Fernandez H (2013): Environmental remediation: from Arlington to Astana. Editorial, J Environm Radioactivity 119, 1-4.

Schonfeldt SJ, Winde F, Albrecht C, Kielkowski D, Liefferink M, Patel M, Stoch L, Whitaker C, Schüz J (2014 accepted): Health effects in populations living around the uraniferous gold mine tailings in South Africa: gaps and opportunities for research (workshop report). *Cancer Epidemiology*, pp. 10

Waggit P (1994): A review of worldwide practices for disposal of uranium mill tailings. Supervising Scientist for the Alligator Rivers Region. Technical Memorandum 48. Australian Government Publishing Service, Canberra 1994, 52pp.

William WA, Lane RG, Legator MS, Whorton EB, Wilkinson GS, Gabehart GJ (1995): Biomarker Monitoring of a Population Residing near Uranium Mining Activities. Environmental Health Perspectives, 103, 5, 466-470

Winde F (2010) Uranium pollution of the Wonderfonteinspruit, 1997-2008 Part 2: Uranium in water - concentrations, loads and associated risks. Water SA 36(3): 257-278

Wismut-GmbH (1996): Ergebnisse der Umweltüberwachung und Sanierungstätigkeit 1995. Sanierungsbetrieb Seelingstädt, Betriebsteil Seelingstädt. Chemnitz

Assessing risks associated with the flooding of mine voids on underground infrastructure and water resources in and around Johannesburg (South Africa)

Frank Winde[1], **Ewald Erasmus**[2]

[1]Mine Water Re-search Group, Geography and Environmental Studies, North-West University, Vanderbijlpark, South Africa, Frank.Winde@nwu.ac.za
[2]Mine Water Re-search Group, Geotechnical Environmental Specialists, Groenkloof, Pretoria, South Africa, geotech@netactive.co.za

Abstract. After more than a century deep level gold mining in and around Johannesburg stopped in 2008 resulting in the gradually flooding of the abandoned mine void. Triggered by extensive media reports on the risks associated with acidic mine water rising to the surface of the inner city ranging from collapse of high-rise buildings, sinkholes, water pollution and increased seismicity two major banks with headquarters in Johannesburg commissioned a desk top risk assessment from which selected results are discussed in this paper. Based on these findings low-cost, low energy alternatives to the currently adopted pump-and-treat approach are proposed.

Background

After more than a century of deep level gold in and around Johannesburg active operations finally ceased in October 2008 when an explosion disabled the last operating pumping chamber used to dewater the mine voids of the Central Rand goldfield. Known as 'central basin' this system of interconnected mine voids extends over nearly 50 km E-W strike reaching depths of up to 3600 m below surface. Following the cessation of pumping water levels in the mine void started to rise at rates of up to 60 cm per day. In its report to Cabinet the Inter-Ministerial Committee on acid mine drainage advised an appointed Team of Experts (ToE) predicted that, if no intervention takes place, mine water will breach the surface level in January 2012 resulting in an uncontrolled decant of ca. 60 Ml of acidic water (Coetzee et al. 2010). Owing to the acidic nature of the water this raised

B.J. Merkel and A. Arab (eds.), *Uranium – Past and Future Challenges*,
DOI 10.1007/978-3-319-11059-2_24

widely publicised concerns about associated risks of underground structures such as deep level basements of high rise buildings being flooded and corroded compromising their geotechnical stability in the process (McCarthy 2011, Van Vuuren 2011). In addition, various media reported on the possibility of acid mine water dissolving dolomitic bedrock resulting in potentially catastrophic sinkholes in the city centre (Eybers 2010).

This paper presents the results of a desk-based risk assessment conducted for two major banking groups occupying headquarters in the CBD (Winde et al. 2011). While the focus of the study is on the risk of mine water flooding the headquarter basements of the two banks obtained results also allowed to address a range of other pertinent aspects. Apart from the risks of sinkhole formation and pollution of receiving water bodies this includes the sources and pathways of the water entering the mine void ('ingress water'), the dynamics and factors controlling the rise of mine water in the void system as well as predictions of the date and volume of the decant differing profoundly from the ToE report (Coetzee et al. 2010) In addition, possible risks such as radon exhalation from mine shafts were identified as not having been addressed to date. Based on an improved understanding of the factors governing the mine water system the currently pump-and-treat-approach recommended by the ToE and implemented by Government is critically analysed and more sustainable alternatives based on low cost low energy approaches are presented.

Risks associated with rising mine water

Flooding of foundations of high rise buildings

In order to assess this probability of the rising mine water reaching foundations of the bank headquarter buildings ways had to be found to predict the final water in the mine void and compare this with the lowest part of the basement structures in question. In a first step the elevation of concrete piles protruding some 10 m from the bottom of the basements into solid bedrock was established (Critical Pile Base Level, CPBL). In order to reduce the uncertainty associated with elevation in topographic maps a dedicated Digital Elevation Model (DEM) of the area was developed based on data extracted from various sources including 1:50,000 digital topographic maps, high-resolution LIDAR data, Google Earth and trigonometrically surveyed mining shafts. Unexpectedly, elevation data retrieved from Google Earth corresponded best with the highly accurate survey data outperforming – at least in the open mining belt outside the build up area – all other data types. Thus the profile tool of Google Earth was used to generate cross sections of the topographic surface in order to identify low lying points where mine water could possibly decant onto surface preventing a further rise of water levels in the void. In order to assess the flooding risks for underground basement structures the highest level to which water could possibly rise inside the void needed to be determined (final mine water level, FMWL). This was based on the premise that the water table inside the void will stop rising once it reaches an outflow (decant) point that is large enough to accommodate all ingress water flowing into the mine void. Based on historical pumping figures this volume was estimated to be in the region of 35 to 40 Ml/d. With an average diameter of 8-10 m each open shaft connected to the mine void would be wide enough to accommodate this flow easily.

Given that the 50-km-long void system stretches consists of thee distinct sub-voids a single decant point will only be able to control the mine water level if the hydraulic interconnectivity between the 3 sub-voids is large enough to also accommodate the total ingress volume mentioned above. Since the sub-voids are mainly connected via holings penetrating boundary pillars between adjacent mine voids (e.g. serving as emergency escape route etc.) their partial collapse could theoretically restrict the horizontal flow between sub-voids to such an extent that water levels in the upstream sub-void would rise above the elevation of the single decant point draining the entire void system. Monthly measurements of mine water levels in each sub-void via 5 monitoring shafts indicated a nearly flat water table across all 3 sub void since 6 May 2010. At this date the water level in the Central and Eastern (ERPM) sub-voids finally reached the level of a holing through which up until then water from western sub-void (stretching from Durban Roodepoort Deep to Rand Lease Mines) was overflowing into the two neighbouring sub-voids keeping its own WL constant. After a single water table was established across all sub-voids the water levels in the western sub-void started rising again. With the same amount of ingress now filling three instead of only two sub-voids the rise of the mine water table slowed down. The largely synchronized behaviour of water levels in all 3 sub-voids for more than 2.5 years after the sub-voids merged indicates a sufficiently high interconnectivity across the entire void system to allow for a single decant point controlling the final water table elevation for the central basin. Temporal differences in water levels between individual monitoring shafts in this period were in the region of about 4 m maximum presumably reflecting differences in ingress rates that allowed for short-term rises soon to be leveled out again by horizontal flow. In a last step the future decant point was determined using the GIS-mapped compilation of all known shafts indentifying the lowest laying open shaft still connected to the mine void as future outflow point (Fig. 1).

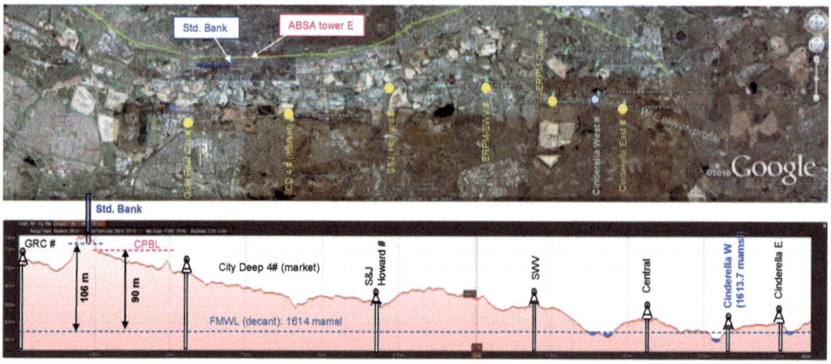

Fig.1. E-W cross section of the Central Rand indicating the elevation difference between the decant point in the East (Cinderella –West shaft) and the critical pile base level of the foundation of ABSA Bank

With 90 m difference between the collar elevation of the Cinderella W shaft and the CPBL it was concluded that no risk exists for mine water flooding the basement structures of the two Headquarter buildings. Since sub-voids are better connected the closer to surface they get as fracturing and weathering increases in vicinity of the surface interconnectivity between the sub-void is bound to increase as the water table rises adding confidence to excluding even localized flooding risks. This even more so as a number of alternative outflow points exists in form of low lying shafts that are still open as well as a total of 19 closed unlined shafts through which water could diffusely migrate out of the void. In addition low lying natural draining features such as river valleys hydraulically connected to the mine void via weathered dykes and faults would also drain water from void keeping the mine water table more or less constant.

Sinkhole formation

A GIS analysis of the geological situation indicated that no dolomite underlies central Johannesburg rendering the proposed formation of sinkholes due to dolomite dissolution irrelevant.

Pollution of surface water

Based on experiences in the West Rand where uncontrolled decant of acid mine drainage into a small, previously seasonal, stream wiped out most aquatic life and degraded the water quality due high levels of sulphate (SO_4^{2-}) and dissolved heavy metals such as uranium (U), iron (Fe) and manganese concern were raised that such pollution may reoccur once the decant in the Central Basin starts as mine water there is of similar poor quality displaying pH values of 3-5, 2000-6000 ppm

SO_4^{2-}, 100-1500 ppm Fe and over 500 ppb U (Metago 1999, Ferret Mining 2004.). However, the major difference to the Western Basin is that nearly all streams draining the Central Rand are already heavily polluted by seepage from mining residues covering large proportions of their small headwater catchments. In many cases streams, for decades, display the very same AMD signature as mine water with reddish coloured water of a pH around 3 and high levels of sulphates and heavy metals (Stander 1963, Winde & Sandham 2004). Thus, any decanting mine water is unlikely cause any further deterioration of water quality of the receiving streams. This also applies to the Vaal River into which all streams finally flow as the same loads of contaminants had been discharged for decades during active mining in the past.

Pollution of shallow groundwater

Generally the potentially affected aquifers are low yielding situated in fractured granite. In many instances the water quality of shallow groundwater is already poor as it often the medium via which seepage from mine waste deposits such as tailings dams and rock dumps migrates into nearby streams. This is particular true for groundwater in the mining belt which will be most affected by AMD diffusely leaving the mine void. Since the area generally dips towards the south groundwater to the north of the void is only affected to a very limited extend. Based on the determined decant elevation (FMWT) possible pollution of shallow groundwater is confined to a limited low lying area near the decant point, which is already affected by mining pollution on surface.

Generally it appears that neither the quality of surface water nor of groundwater would be significantly affected by allowing mine water to decant freely. Where groundwater levels increase due to the diffuse influx of mine water it must be ascertained that the geotechnical stability of structures erected in low lying areas such as slimes dams, sand and rock dumps is not compromised.

Geotechnical stability

Where groundwater levels increase due to the diffuse influx of mine water it must be ascertained that the geotechnical stability of structures erected in low lying areas such as slimes dams, sand and rock dumps is not compromised.

Increased seismicity

It has been suggested that the gradual filling of the mine void will lead to an increase in seismicity mainly due to the lubrication of flooded fault lines adjusting the void structure to the new mass distribution. While such adjustments do occur sporadically in flooded mining areas they generally do not exceed the magnitudes of events observed during active mining (Durheim et al. 2006) and neither increase in frequency as observations in the only fully flooded basin in the Wes Rand indicate. Furthermore, in case of the Central Basin no seismicity increase was observed when ca. two thirds of the mine void were flooded in the mid 1970s. It is thus believed that the drastic increase in seismicity reported in the media is unlikely to occur in the Central Rand.

Radon exhalation

One aspect not covered at all in previous reports on risks associated with AMD is the exhalation of Radon from mine shafts. Ongoingly produced by the radioactive decay of U in un-mined ore and water it is proposed that the toxic gas which is linked to lung cancer is displaced by the rising mine water and escapes via shafts into the near surface atmosphere potentially exposing residents. Rapid transport of displaced Rn to surface is promoted by shafts acting as conduits to equalize day-night atmospheric pressure differences between the underground mine void and the surface by alternating between sucking and blowing action. Especially where shacks of informal settlers are build close to old shafts which may no longer be visible on surface this may lead to an accumulation of radon in the poorly ventilated structures that lack sealed floor to minimize the concentrated influx of radon channeled into shafts from a large underground area. As many of the poor inhabitants have to sleep on the floor where Rd accumulates (as it is heavier than air) it is proposed that this pathway of exposure is to be investigated.

Governmental response: pump-and-treat

As recommended by the ToE Government installed pumps at one of the monitoring shafts in order to extract water from the mine void and discharge it into a nearby stream after neutralization in a newly erected treatment plant discharging the generated toxic sludge onto existing tailings deposits. The purpose of this pump-and-treat approach is to keep the mine water level at a fixed level (termed 'Environmental Critical Level'; ECL) citing the prevention of stream pollution as the main objective. It was also stated that the neutralized decant is required to co0mbat the water deficit and maintain the dilution capacity in the Vaal River as ultimate recipient of the semi-treated mine water. Since for the foreseeable future

the salt load of the mine water is not reduced this approach does little to increase the dilution capacity of the Vaal River but in fact reduces it.
In addition to capital costs of R 237 million this approach also requires some R 128 million per annum of operating costs with a significant proportion needed for covering electricity costs for pumping. Given the facts that pumping need to be continued indefinitely at ever increasing energy costs and that toxic sludge is generated while polluted water is discharged this approach is neither economically nor ecologically sustainable. The most important shortcoming is however that it only addresses the symptoms but not the underlying cause of AMD and therefore needs to be applied in perpetuity.

Proposed alternatives

Much of the shortcomings of the current approach are blamed on time pressure as the decant date predicted by the ToE left less than 2 years for developing an appropriate response at the time. This is unfortunate as Winde et al. (2010) did indicate that nearly double the time is actually available given the observed slow down in the rate of rise, which was unfortunately ignored by the ToE. With the pump-and-treat infrastructure recently ceremoniously opened facts were created that make the implementation of low-cost, low energy alternatives now difficult.

With regard to the latter a phased approach consisting of the following options is proposed:

(i) No pumping: Instead of pumping AMD to surface eternally at great cost let the mine water rise naturally to the lowest lying decant shaft where it is captured for further use. This is possible as none of the risks cited (e.g. basement flooding, water pollution, sinkhole formation) was found to be significant and large amounts of money could be saved. It also relieves Eskom from additional long-term pressure on its electricity supply which still operates at capacity limits illustrated by frequent load shedding and black outs. The current attempt to eternally fight a natural hydraulic equilibrium in the mine void from being established is not only expensive and unsustainable but also counterproductive. Keeping the mine water level artificially low (between 240-270 m below surface) prevents the natural reduction of water ingress that occurs when the water table comes close to the surface. This, in turn, results in maintaining unnecessarily high pumping volumes. I.e. the process of pumping itself ensures that pumping stays expensive indefinitely. Furthermore, since the void is prevented from being flooded completely atmospheric oxygen continues to access the mine void generating AMD from unoxidised pyrite underground keeping the quality of decanting water poorer than necessary. This too has costs implications as long as decant is treated.

(ii) Use AMD for tailings reclamation: Provide the freely decanting mine water to existing companies which are hydraulically reclaiming tailings deposits in the mining belt until all tailings are removed and re-deposited elsewhere. This will not only save the clean water currently imported for this purpose from being polluted but also reduce the ingress into the mine void and thus the decant volume. The significant amount of water lost to the mine void in the process is simply recycled without further increasing the volume of ingress. At an estimated demand of 20-25 Ml/d for hydraulic tailings dispersion about two thirds of the decanting water could be utilised in this way with Government stipulating that all tailings – also those of low gold and uranium content - are to be removed completely and deposited outside the mining belt in an environmentally sound manner. Government should further stipulate as a condition of granting the water fro free that the reminder of the decant (5-15 Ml/d) is to be sprayed onto unvegetated tailings in the mining belt in order to suppress the dust which currently affects hundreds of thousands of nearby residents especially during dry winter times. Since indications are that much – if not most – of the pollution of mine water is actually caused on surface by the large mine waste deposits their removal would ultimately result in improving the quality and reducing the volume of the decant. If the mine void is simultaneously allowed to flood fully, the subsequent depletion of oxygen will also reduce possible contamination from underground sources. Since a dedicated company already exists that derives an income from removing tailings from the mining belt while securing employment and generating revenue Government should explore means to increase their capacity with resources saved by abandoning the expensive, self-perpetuating and unsustainable pump-and-treat approach. In addition this would free up significant areas of land in prime inner city location for further development (that has to be mindful though of remaining pollution at tailing footprints and old shafts).

(iii) Blend AMD with municipal sewage: Once all surface tailings deposits are reclaimed the decanting water will be much better in quality and reduced in volume as major sources of pollution and ingress have been removed. No longer needed for hydraulic tailings removal and dust suppression the decanting water from the mine void can be either distributed directly to users (if quality allows) or simply blended with the large volumes of municipal sewage currently amounting to approximately 1500 Ml/d for the metropolitan areas in and around Johannesburg (incl. Ekurhuleni). Resulting in a dilution ratio of 1: 50 (at 30 Ml/d decant volume) all of the problematic contaminants in worst quality AMD (sulphate, uranium, iron) would thereby reach acceptable discharge standards without requiring any further treatment. Since municipal waste water is slightly alkaline blending at this ratio would simultaneously neutralize the AMD without any need for adding lime as is currently done preventing the generation of large amounts of toxic sludge produced by the current neutralisation. As migration to urban areas continues municipal waste water volumes will increase further allowing for even greater dilution in the future. But already now sewage volumes are sufficient to dilute the decant from all 3 basins (estimated ca. 150 Ml/d) by about ten times and meet current discharge standards.

(iv) Divert water: Given that water shortages north of Johannesburg impose severe restrictions for economic development it is proposed to divert all AMD as well as most municipal waste water away from the Vaal River System into the Crocodile West catchment to combat the acute water short. This would also relieve the Vaal River from nutrient loads which together with poorly or untreated waste water from expanding informal settlements south of Johannesburg frequently result in massive algae blooms in the river. To counterbalance the associated loss of water in the Vaal River System it is proposed to clamp down on the massive illegal water abstraction from the Lesotho Highland Water Scheme by irrigation farmers identified as early as 2007 amounting to 479 Ml/d – i.e. more than triple the volume of all AMD. Furthermore, it is recommended that measures are rapidly implemented to reduce leakage from pressured water distribution systems amounting to an estimated 506 Ml/d in urban areas of Gauteng Province alone. Lastly, water losses and inefficient water use in the downstream irrigation schemes that consume 35% of the Vaal River water are another source for recovering wasted water in the system. For the Vaalharts scheme alone as single largest user water losses from canals and inefficient irrigation methods total an estimated 1173 Ml/d, i.e. 8 times the AMD volume from all 3 basins. Given that more than 2100 Ml of water are currently lost from the system each day it is difficult to see the need for expensively treating AMD to discharge standards in order to combat the water deficit when much larger volumes can be saved by measures that are less expensive and need to be implemented in any case at some point in the future.

Conclusions and recommendations

The lack of proactive planning and preparation of mine closure in major goldfields in and around Johannesburg resulted in the uncontrolled flooding of abandoned mine voids and associated risks of adverse impacts on the environment and geotechnical stability of structures in densely populated urban areas. Driven by inflated risks reports in media and a corresponding rise in public and political pressure Government finally responded by adopting a sub-optimal pump-and treat approach that is economically unsustainable and does little to alleviate the associated burden for the receiving environment. Instead of eternally continuing with the energy and resources intense approach that tends to reinforce itself the authors propose low cost, low energy solutions that are sustainable and take advantage of existing synergies in the area simultaneously freeing up water and land resources for economic development and saving significant public funds.

References

Brouwer H (2003): pers. comm., GFL Geological Centre Oberholzer, May 2003.

Coetzee H, Hobbs PJ, Burgess JE, Thomas A, Keet M, Yibas B, van Tonder D, Netili F, Rust U, Wade P, Maree J (2010): Mine water management in the Witwatersrand gold fields with special emphasis on acid mine drainage. Report to the Inter-Ministerial Committee on acid mine drainage. Unpublished, December 2010. pp. 128

Durrheim RJ, Anderson RL, Cichowicz A, Ebrahim-Trollope R, Hubert G, Kijko, McGarr A, Ortlepp WD, van der Merwe, N. (2006): The risks to miners, mines, and the public posed by large seismic events in the gold mining districts of South Africa. In:

Hadjiggeorgiou J, Grenon M [eds.]: Proceedings of the 3rd. Internat. seminar on deep and high stress mining, 2-4 October 2006, Quebec City, Canada, pp. 14.

Eybers J (2010): Van 'n omgewingskrisis. Rapport newspaper, 15.8.2014

Ferret Mining (2004): A strategic water management plan for the prevention of water ingress into underground workings of the Witwatersrand Basin – Phase 1. pp. 413, unpublished, Ferret Mining and Environmental Services (Pty.) Ltd., Pretoria

Malan DF, Vogler UW, Drescher K (1997): Time-dependent behaviour of hard rock in deep level gold mines. Journal of the South Afric Inst of Mining and Metallurgy, May/ June 1997, 135-148

McCarthy TS (2011): The decanting of acid mine water in the Gauteng city-region – analysis, prognosis and solutions. Provocations series, Gauteng City Region Observatory. Johannesburg, South Africa, pp. 40.

Metago (1999): Report on the determination of the effects of the Amanzi water treatment venture on surface water flow rate and quality in the affected river systems. Appendix report no. 17 of: The Amanzi water treatment venture strategic environmental impact assessment. Scoping Report appendices. Volume 4. Prepared for GDACE (Gauteng Department of Agriculture, Conservation and Environment) on behalf of JCI Projects, Metago Environmental Engineers Pty Ltd. report no. 103-134, August 1999, prepared for JCI Projects (Pty) Ltd., pp. 18

Stander GJ, co-workers (1963): A survey of river pollution in the Witwatersrand catchment area of the Vaal River. National Institute for Water Research, CSIR special report W22, Pretoria, unpublished, 83 pp.

Van Vuuren L (2011): Red letter year for authorities to prevent mine-water catastrophe. *The Water Wheel*, Jan/ Feb 2011, 12-14

Winde F, Erasmus E, Stoch EJ, Hoffmann E (2011) Desktop assessment of the risk for basement structures of buildings of Standard Bank and ABSA in Central Johannesburg to be affected by rising mine water levels in the Central Basin. Final Report, 3 Volumes, unpublished, Rosebank, Johannesburg, May 2011, pp. 265

Winde F, Sandham L (2004): Uranium pollution of South African streams – n overview of the situation in gold mining areas of the Witwatersrand. *Geojournal*, 61, 131-149

Hydrogeological testing for ISL uranium mining: some Australian experience

Peter Woods[1], Ben Jeuken[2]

[1]IAEA, PO Box 100, Vienna International Centre, A-1400 Vienna, Austria
[2]Groundwater Science, Mines & Energy House, 290 Glen Osmond Road, Fullarton SA 5063, Australia

Abstract. In situ leach (ISL) has become one of the standard uranium production methods. Hydrogeology is important for all mining below the water table, but it is doubly important in ISL mining for efficient production of uranium. Prevention of unwanted groundwater contamination must be considered and controlled, and the final status of the target aquifer is also of high importance.

Hydrogeological testing has been undertaken at a number of Australian ISL sites, some of which proceeded to production. Here we consider the associated hydrogeological testing undertaken for some of these projects, with case studies and lessons learned.

Introduction

In situ leach (ISL; also called in situ leaching or in situ recovery, ISR) has become one of the standard uranium production methods, recently accounting for over 40% of world production. Hydrogeology is important for all mining below the water table, and often also for 'dry' mining; but it is doubly important in ISL mining, where the mining method relies on forcing a leaching solution through an aquifer to dissolve and recover uranium. The possibility of unwanted contamination of non-target aquifers must be considered and controlled, and the final status of the target aquifer is also of high importance.

Australia was not the first country to consider or commence the ISL mining of uranium; it was first considered there in about the 1980s, and hydrogeological testing has been undertaken at a number of sites, some of which proceeded to production in the late 1990s and early 2010s. Here we consider the associated hydrogeological testing undertaken for most of these projects, with case studies and lessons learned.

B.J. Merkel and A. Arab (eds.), *Uranium – Past and Future Challenges*,
DOI 10.1007/978-3-319-11059-2_25

ISL mining as a hydrogeological activity

ISL mining involves the dissolution of uranium under oxidizing conditions by means of the circulation of a mining solution through uranium-bearing porous sediments in an aquifer under suitable hydrogeological conditions (IAEA 2001). Uranium is subsequently recovered for further concentration and eventual packaging for use. Whilst the discovery and delineation of an ISL-amenable uranium orebody is largely a geological and geophysical exercise, the assessment of its feasibility is heavily dependent on hydrogeological information and modelling. The mining process itself and associated sub-surface environmental monitoring is a metallurgical (dissolution and recovery) and hydrogeological (extraction) activity, i.e. the construction and maintenance of groundwater wells and pumping systems. An ISL uranium deposit is only minable if the hydrogeological circumstances permit. Techdoc 1239 (IAEA 2001) provides a useful set of hydrogeological feasibility criteria for ISL uranium mining.

Brief history of ISL mining in Australia

Australia hosts a number of known sandstone-hosted uranium deposits (Penney 2012), many of which are considered ISL-amenable; others are candidates for open-pit mining. The Frome Embayment in South Australia hosts the only developed mines, but projects and prospects are known in several other sedimentary basins across the country, with a good likelihood of further discoveries.

Exploration for sandstone-hosted uranium deposits began in earnest in Australia in the 1960s, with discoveries occurring before the end of that decade and continuing into the 1970s and 1980s (McKay and Miezitis 2001).

The first discovery to proceed to testing and mining was Beverley in the Frome Embayment in South Australia (Graham 2000; Birch et al. 2013), discovered in 1969 but with the first operations in late 2000. Later the nearby Beverley North deposits, discovered in 2007, were developed as satellite operations to the central Beverley processing plant (Märten et al. 2012). The nearby Four Mile uranium project, with a different ownership structure, was discovered in 2005 and first production is expected in 2014 (Alliance Resources 2014).

Further east in the Frome Embayment several, smaller sandstone-hosted uranium deposits are known. Of these only Honeymoon, discovered in 1972, proceeded, with first production in 2011 (Anonymous 2013).

Hydrogeological testing of these deposits and some others in Australia that have not yet produced uranium is discussed below.

Case Studies

Beverley

Although the earliest considerations included open-cut mining (McKay and Miezitis 2001), ISL mining was soon after established as the preferred mining method at Beverley. Early hydrogeological studies were reported as part of the first draft environmental impact statement (South Australian Uranium Corporation 1982). However, the introduction of the Commonwealth Government's 'three mines policy' in 1983 and declining uranium market prices meant the project was again interrupted in mid-1985 (McKay and Miezitis, 2001).

A new owner, Heathgate, undertook further technical and environmental studies (Dobrzinski, 1997) and an initial field leach trial (FLT) was carried out in 1998 and a new EIS released (Heathgate, 1998). Howles (2000) summarized the findings of the studies up until 1999.

The hydrogeology of the Beverley deposits was reviewed during the approval process for the extension of mining to adjacent, hydrogeologically similar small deposits (Heathgate 2007) which refined the hydrogeological understanding without changing the basic conclusions of Howles (2000; see also Armstrong and Jeuken, 2009). Investigations included six pumping tests in small, essentially isolated aquifers containing small orebodies and additional baseline measurement in many wells (Heathgate 2007). By the time of the lease extension application, 7 years of operational experience had been gained, geological information compiled from over 9,000 geological holes, and over 10,000 monitoring data from over 100 monitoring wells and several hundred productions wells were available for review. The application document included a 131-page hydrogeological review and a separate 78-page report with hydrochemical data and graphs (ibid).

Beverley North

Following their discovery in 2007, hydrogeological investigations at the Beverley North deposits approximately 15 km NNW of Beverley preceded rapidly. Whilst superficially similar to Beverley, the Beverley North deposits are hosted in a geologically older formation (the Eyre Formation) and are part of a large, slow-moving regional aquifer system rather than hosted in isolated sandy lenses as at Beverley (Heathgate 2010; Märten et al. 2012). There is no widespread overlying aquifer, although some groundwater is encountered in minor sandy lenses in the saturated sediments overlying the orebodies. The Great Artesian Basin aquifer is absent; the underlying aquifer is in sparsely fractured basement rock (ibid.). The sequence of hydrogeological investigations was as follows:

- Establishment of the geological setting using exploration and delineation drilling and geophysics
- Establishment of investigation wells in the three aquifers; the ore-bearing Eyre Formation, the underlying basement rock and the overlying sediments; groundwater levels and groundwater quality were established and horizontal and vertical hydraulic gradients determined
- Baseline groundwater levels and (where possible) groundwater quality in nearby existing pastoral wells were obtained
- Two local pumping tests
- Hydrogeological and geochemical modelling.

An important stage was the creation of a regional groundwater conceptual model which was provided in the publicly available approval document (prepared by Sinclair Knight Merz and published in Heathgate 2010). This allowed the preparation of numerical models of groundwater flow and hydrogeochemistry.

Over 2008-11staged applications and approvals were made and received for the following stages of final testing and development:

- Hydrogeological testing of wellfield patterns (no modification of groundwater chemistry)
- Field Leach Trial (FLT) with introduction of mining solution and extraction of uranium
- Full mining approvals, with FLT infrastructure expanded to become first satellite plant (Pepegoona)
- Second satellite plant established (Pannikan).

The hydrogeology of Beverley North is summarized as (after Märten et al. 2012 and Heathgate 2010):

- The deposits are sandwiched between a mostly clayey overlying sedimentary formation and a thick underlying shale, beneath which is a highly variable fractured bedrock aquifer
- The deposits are in a slow-moving regional aquifer of water that does not meet Australian water quality guidelines for domestic, irrigation or stock use, and is not used for those purposes near the mine nor down-gradient
- Groundwater flow velocities are in the order of 10 to 25 m/a at the mining area, slowing down-gradient to 7 m/a
- The groundwater travel time to the interpreted groundwater discharge area, beneath the salt (playa) Lake Frome, was estimated to be in the order of thousands to tens of thousands of years
- Geochemical conditions are more strongly reducing that at Beverley.

Kalka et al. (2011) describe the hydrogeochemical modelling undertaken as part of the permitting process. This considered both natural attenuation (NA) and Enhanced Natural Attenuation (ENA) options for mine closure. Kalka et al. (2011) and Märten et al. (2012) conclude that:

- Range of potential impact on groundwater uranium content limited to some 200 m from the mining area.

- The future validation of the hydrogeological and hydrochemical (E)NA modelling based on real-world data will be a valuable opportunity to refine the conceptual and numerical modelling tools, as well as providing important evidence of successful closure of the Beverley North mining areas.
- Further modelling will also be used to optimize the ISR operation by balancing the mobilization of uranium against interfering leaching effects

The impact of waste solutions was not considered at the Beverley North site, as the mining proposal included further processing at the Beverley site and ongoing waste-water disposal in the small and isolated Beverley aquifers.

Four Mile

Following the Four Mile discoveries in 2005, hydrogeological and environmental studies were commenced and published in the Public Environmental Report (PER; URS 2008), concentrating on Four Mile East. Like Beverley North and in contrast to the stagnant mining aquifers at Beverley, the mining aquifer at Four Mile East is part of a slow-moving flow-through groundwater system (ibid; Jeuken et al. in press). The studies mirror the initial studies described above for Beverley North prior to the establishment of wellfields; in fact, the Four Mile studies preceded those at Beverley North and were used as a model for the work there, where ore is hosted in the same aquifer as Four Mile East. As such additional detailed description will not be given; please refer to the PER (URS 2008, full project studies) and Jeuken et al. (in press, hydrogeochemical studies), and the current environmental plan for the project that was approved in 2013 (Heathgate 2013).

It is safe to conclude that without the extensive hydrogeological and hydrogeochemical studies done at Four Mile and the various Beverley deposits, neither the required government approvals could have been obtained nor could the proponent had been able to design the ISL wellfields required for the projects.

Honeymoon

The Honeymoon deposit was Australia's first attempt at developing an ISL uranium mine (McKay and Miezitis 2001). The deposit is located within the Yarramba palaeochannel, which is infilled by Tertiary sediment of the Eyre Formation. These sediments provide the host aquifer for the uranium deposit.

Feasibility mining was assessed through a series of field trials including push-pull tests and two iterations of field leach trials. The main hydrogeological studies and some findings comprised:

- A small scale 'push-pull' test 1978
- Field Leach Trial facilities established early 1980s
- A larger FLT 1998-2000 (Reif 2002) demonstrating the recovery of uranium, using solvent extraction technology as the groundwater salinity was considered too high for ion exchange technology
- The hydrogeology is described as multiple layers within main aquifer, and surrounding aquifers were described (Southern Cross Resources Australia 2000; details on pumping tests and computer modelling are described by Rooke et al. 2002 and the general hydrogeology updated in Uranium One 2011). The Great Artesian Basin aquifer is absent.
- Additional drilling, data and computer modelling of waste disposal were undertaken for the final approval documents (Uranium One 2011), leading to a Groundwater Management Plan as part of the Mining and Rehabilitation Plan.
- Operation of the mine has entailed ongoing hydrogeological and geochemical testing and modelling comprising (ibid.):
    - 3 dimensional groundwater and solute transport modelling
    - 1 dimensional reactive transport modelling of dissolved metals including uranium.

Oban

Oban was discovered in the early 1980s 60 km north of the Honeymoon project. The deposit is described as a discontinuous roll-front type within palaeochannel sands. Oban uranium mineralization is hosted by a horizontal lignitic, pyritic and carbonaceous lower sand member of the Eocene Eyre Formation at depths of between 80–90 m (McKay and Miezitis 2001).

The assessing company decided to undertake a small Field Leach Trial (Oban Energy Pty Ltd 2008) based largely on the technology employed at Beverley. The FLT involved the circulation of acidified (using sulfuric acid) groundwater commencing in July 2010. However, although breakthrough of acidified and oxidizing solution was achieved, no significant uranium was detected in the first test. To further investigate the leach results, a series of three sonic core holes was drilled south of the initial leach pattern. Assays of the cores confirmed the presence of uranium in leachable sands but also showed a significant amount of mineralization hosted in lignitic clay which is probably not leachable. The core holes were cased and screened and further leach trials were conducted between pairs of wells. Again no significant uranium was detected and the tests suspended. As at the end of 2011 further investigations at Oban and exploration nearby were continuing (Curnamona Energy Limited 2012).

Mullaquana/Blackbush

The Mullaquana/Blackbush uranium deposits in South Australia are being evaluated by UraniumSA for possible ISL mining (UraniumSA 2011).

A significant body of work has been undertaken to develop a robust understanding of the hydrogeological system at the project site. This work has included (ibid):

- Drilling of over 500 exploration holes and interpretation of stratigraphic data
- Installation of nearly 50 monitoring wells and measurement of chemical and water level data
- Extensive aquifer pumping tests of up to 5 days duration at 4 locations including two test sites within the Retention Lease area and another two sites further to the east between the field trial site and the nearby Gulf St Vincent
- A circulation trial in ISR pattern (tracer NaBr added) in 2011
- An assessment of the regional hydrogeological structure including the continuity of aquifers and confining strata under the Gulf St Vincent
- Testing of ion exchange resins suitable for high-Cl waters (Hall 2012).

The company has continued a lower level of investigation as it further evaluates the deposit for possible future mining.

Manyingee

Manyingee was discovered in 1974 in the northern part of the Carnarvon Basin of Western Australia. The deposit is roll-front type within Cretaceous palaeochannel sandstones (Valsardieu et al. 1981).

Two pumping tests and one 5-spot Field Leach Trial were completed in the mid-1980s to evaluate ISL amenability and solution confinement. The 5-spot test utilized a solution of sodium carbonate and bicarbonate. Test results were apparently favorable, but negative government policy at both the State and Federal levels stymied further development (Bautin and Hallenstein 1997). Paladin Energy acquired the project in 1998 and is currently reviewing the potential for commercial development (Paladin Energy Ltd 2013).

Discussion and Conclusions

A robust understanding of the local and regional geology and hydrogeology is required during the investigation of an ISL-amenable uranium deposit. These studies should at least include the following in the early stages of the assessment of a deposit:

- Understand the geology of the deposit and its setting, including the lithology and permeability of ore relative to surrounding aquifer
- Thorough literature review, if data exists, from both published and unpublished sources, data bases and the like
- Obtain and compile field data (water levels and groundwater quality) for existing wells, or any exploration drillholes that are converted to observation wells
- Obtain and interpret geophysical logging results for hydrogeologically important parameters such as porosity as well as for an understanding of the geology and mineralization.

The outcomes of this work should be reviewed, together with the attractiveness of the project on other accounts, before additional, more extensive and expensive work is undertaking. Next stage testing could include:

- Undertake and interpret pumping tests, preferably with observation wells at different distances and directions in the pumped aquifer and also in any adjacent, higher or deeper aquifers. Water quality should be taken at least at the beginning and end of the test from the test production well
- Additional water level and water quality testing of existing wells in the area
- Undertake studies on cores, e.g. grain size distribution, presence of clay, lignite, potentially reactive minerals such as pyrite or carbonates.

Again, review and decide if further work is warranted. This might include:

- A groundwater circulation test, perhaps with an inert tracer added
- A Field Leach Trial with mining lixiviant (acid or alkaline) and recovery of uranium from the extracted solution
    - 'push – pull' on single wells, or preferably
    - pair(s) or pattern(s) of wells
    - monitoring of specific observation wells to check their response to the tests, regarding both hydraulic and water quality response
- It is common to establish a numerical groundwater model based on the pumping test results (aquifer parameters and calibration), that can be used for planning future production wellfields and estimating closure scenarios.

Review and decide if further work warranted – sometimes the first test is inconclusive or negative (e.g. Oban), or at the wrong scale (e.g. Honeymoon).

Where a project proceeds to production, some or all of the FLT equipment may (e.g. Beverley North) or may not (e.g. Honeymoon, Oban) be suitable to form the basis of a production phase.

From the FLT stage, the proponent and regulator must be aware of the potential environmental effects of added chemicals (acid or alkaline) that are involved. Hydrogeochemical numerical modelling, incorporating or using a separate hydraulic model, is often a necessary part of approvals and also the proponent's planning for the production stage. Contingency plans should be made for the rehabilitation of an FLT is the project does not proceed to production, and a regulator may require a bond or surety to guarantee this possible remediation.

References

Anonymous (2013) Fact Sheet Uranium One Honeymoon Mine. Australasian Mining and Metallurgical Operating Practices, (Australasian Inst. Min. Metall., Melbourne, Australia) 2: 1819-1822

Alliance Resources Limited (2014) Four Mile Project commissioning. ASX (Australian Stock Exchange) Announcement, 24 March 2014

Armstrong D, Jeuken B (2009) Management of in-situ recovery (ISR) mining fluids in a closed aquifer system. Abstracts of the International Mine Water Conference, 19-23 October 2009, Pretoria, South Africa (International Mine Water Association): 703-712

Bautin F, Hallenstein C (1997) Plans for uranium mining by COGEMA, 1997. ANA 97 Conf. Nuclear Sci. & Engineering in Aust., 16-17 Oct. 1997, Sydney, Aust. Nuclear Assoc. Inc.: 20-24

Birch GR, Every CR, Märten HG, Marsland-Smith AB, Phillips R, Woods PH (2013) Beverley Uranium Mines, Heathgate Resources. Australasian Mining and Metallurgical Operating Practices, (Australasian Inst. Min. Metall., Melbourne, Australia) 2: 1799-1818

Curnamona Energy Limited (2012) Annual Report 2011, Curnamona Energy Limited, Adelaide, Australia

Dobrzinski I (1997) Beverley and Honeymoon uranium projects, MESA Journal, Mines and Energy Department of South Australia, Adelaide, Australia, April 1997, 5: 9-11

Graham JJ (2000) Lessons learned from the start-up and operation of the world's newest and largest ISL uranium mine. Conf. Proc. Metallurgical Plant Design and Operating Strategies, 15-16 Apr. 2002, Sydney, Australia, Australasian Inst. Min. Metall., Melbourne, Australia: 8-20

Hall S (2012) Metallurgical development of the Samphire uranium project. Proc. uranium sessions at ALTA 2012, 31 May – 1 June 2012, Perth, Western Australia, 151-166

Heathgate (1998) Beverley Uranium Mine: Environmental Impact Statement – Main Report. Heathgate Resources, June 1998, Adelaide, Australia

Heathgate (2007) Mining proposal for proposed extension of Beverley mine. Prepared for Heathgate Resources Pty Ltd by URS Australia Pty Ltd

Heathgate (2010) Beverley North Project Mining Lease Proposal and Draft Public Environment Report. Prepared for Heathgate Resources Pty Ltd by URS Australia Pty Ltd

Heathgate (2013) Four Mile Uranium Mine program for environment protection and rehabilitation, and radioactive waste management plan. Heathgate Resources Pty Ltd, Adelaide, South Australia

Howles SR (2000). Beverley Uranium Project – groundwater resources, management and monitoring. MESA Journal, Mines and Energy Department of South Australia, Adelaide, Australia, April 2000, 17:4-6

IAEA (2001) Manual of acid in situ leach uranium mining technology. International Atomic Energy Agency, Vienna, IAEA-TECDOC-1239

Jeuken, B, Kalka H, Märten H, Nicolai J, Woods P (in press) Uranium ISR mine closure – general concepts and model-based simulation of natural attenuation for South-Australian mine sites. Proc. Int. Symp. on Uranium Raw Material for the Nuclear Fuel Cycle: Exploration, Mining, Production, Supply and Demand, Economics and Environmental Issues (URAM-2009), 22-26 Jun 2009, Vienna, Austria, IAEA-CN-175/15

Kalka H, Märten H, Woods P (2011) ISR mine closure concepts. Nachhaltigkeit und Langzeitaspekte bei der Sanierung von Uranbergybau- und Aufbereitungsstandorten. Proceedings des Internationalen Bergbausymposiums WISSYM_2011, Ronneburg, 25-27 May 2011. Wismut GmbH: 201-215

Märten H, Phillips R, Woods P (2012) New uranium ISR satellites at Beverley North, South Australia. The New Uranium Mining Boom, (Proc. Int. Conf. Uranium Mining and Hydrogeology VI, Freiberg, Germany, 18-22 Sep. 2011) (Merkel, B and Schipek M, Eds): Springer Berlin, Heidelberg: 23-30

McKay AD, Miezitis Y (2001) Australia's uranium resources, geology and development of deposits. AGSO - Geoscience Australia, Canberra, Mineral Resource Report 1

Oban Energy Pty Ltd (2008) Retention Lease Proposal on Mineral Claim 3777 for uranium field leach trial at Oban, South Australia

Paladin Energy Ltd (2013) Quarterly activities report for period ending – 31 December 2012.

Penney R (2012) Australian sandstone-hosted uranium deposits, Appl. Earth Sci. 121 (2): 65-75

Reif T (2002) Honeymoon uranium project update, MESA Journal, Mines and Energy Department of South Australia, Adelaide, Australia, 25:16-17

Rooke ER, Gallagher, M, Bush, PD (2002) Geohydrological characterisation of the Yarramba Palaeochannel, Honeymoon Uranium Project. International Association of Hydrogeologists Conference, Darwin, November 2002

South Australian Uranium Corporation (1982) Beverley Project Draft Environmental Impact Statement. South Australian Uranium Corporation, July 1982

Southern Cross Resources Australia Pty Ltd (2000) Honeymoon Uranium Project Environmental Impact Statement. Southern Cross Resources Australia Pty Ltd, Adelaide, Australia

UraniumSA (2011) Retention Lease Proposal on Mineral Claim 4280 for a Uranium In-situ Recovery Field Trial. Uranium SA, Adelaide, Australia

Uranium One Australia (2011) Mining and Rehabilitation Program Honeymoon Uranium Mine Uranium One Australia Pty Ltd, Adelaide, South Australia

URS Australia Pty Ltd (2008) Beverley Four Mile Project Mining Lease Proposal and draft Public Environment Report, for Heathgate Resources Pty Ltd, Adelaide, Australia

Valsardieu CA, Harrop DW, Morabito J (1981) Discovery of Uranium Mineralization in the Manyingee Channel, Onslow Region of Western Australia, Proc. Australasian Inst. Min. Metall. 279: 5-17

Note : Heathgate (2007), (2010) and (2013), Oban Energy Pty Ltd (2008), Uranium One Australia (2011), UraniumSA (2011) and URS Australia Pty Ltd (2008) can be downloaded by a search on the South Australian government web sites http://www.pir.sa.gov.au/minerals or https://sarigbasis.pir.sa.gov.au/WebtopEw/ws/samref/sarig1/cat0/MSearch

Remediation of a uranium geological exploration facility

Wei Zhang[1], Lechang Xu[1], Xueli Zhang[1], Jie Gao[1]

[1]145 Jiukeshu, Tongzhou District, Beijing, China

Abstract. The tunnels were sealed by rock walls with cut-out overflow and closed down after backfilled with clay soil or mixed soil. Ventilating shafts were backfilled with uranium waste and then closed down by backfilling clayed soil and revegetation. The waste rock piles were covered by clayed soil and revegetation.γ-rays radiation dose rate from waste rock piles reduced by 90% after remediation. Radon fluxes were reduced to 0.5Bq/m^2·s from 1.66 Bq/m^2·s after remediation.

The uranium geological prospecting work in the Northeast China Region (NCR) was conducted from 1957, and has submitted numbers of uranium deposit to the country, contributing to the China nuclear industry. With the improvement of people's environmental awareness, the decommissioning projects management was gradually conducted for the uranium exploration facilities starting from the early 1990s. The uranium exploration facility decommissioning projects management mentioned in this paper mainly refers to the exploration tunnels and mining debris in the NCR during the period of 'Eleventh Five-Year (2006-2010) Plan' of China.

General situation of the facilities

The reason for management

The uranium exploration facilities in the NCR mainly include exploration tunnels and mining debris, with the uranium tailings stacked in the open air without management. Slag heap of waste rocks were to be taken to build, paving and build walls. The remaining mining debris has been leaching and erosion, and has caused the nuclide migration and diffusion, with a certain degree of environmental pollution and the damage of the natural landscape (Fig.1, Fig.2). With the improvement of people's living standards, environmental awareness, and legal awareness, local

B.J. Merkel and A. Arab (eds.), *Uranium – Past and Future Challenges*,
DOI 10.1007/978-3-319-11059-2_26

residents and the government claimed that the relevant facilities are required to be managed.

Fig.1. Current situation of waste rock piles

Fig.2. Current situation of non-closed pithead

The goals of management

The existing uranium prospecting facilities which were not managed were tunnels and waste rock piles. Goaf and the waste rocks with high activity (the ratio between radioactivity and activity is greater than 7.4×10^4 Bq/Kg) need to be backfilled and the pithead need to be blocked off to prevent the escape of radon gas and the outflow of waste water, and then to protect the public health and safety. The pithead after blocked off was forbidden to knock down, and needs to set up warning board. The waste rock piles decommissioning after the after the shaping and stability, and then covered with a certain thickness of loess and vegetation, it basically recovered the natural landscape. γ-rays radiation dose rate, the Radon concentration, and the Radon exhalation rate almost reached the basic background value, after the remediation. The decommissioning management should keep the regulation works in good condition, take good care of environmental protection engineering without destruction, and realize the open use restrictedly.

Treatment plan for decommissioning

Treatment plan for pithead

The non-water tunnels pithead was sealed by two Cement rough walls, filled with mining debris (Fig. 3). As for the pithead with water in the tunnels, the passive filter water bed and the collecting tank were used to drain the water away. Two concrete walls were built to seal the tunnels, and the passive filter bed made by gravels was fixed between the two concrete walls. Reinforced concrete board was used as the roof cover. The water in the tunnels flows through the gravel layer and then filter pipe, went into the collecting tank thereafter, then was led-out from the cement culvert, and was drained by the concealed conduit finally (Fig. 4).

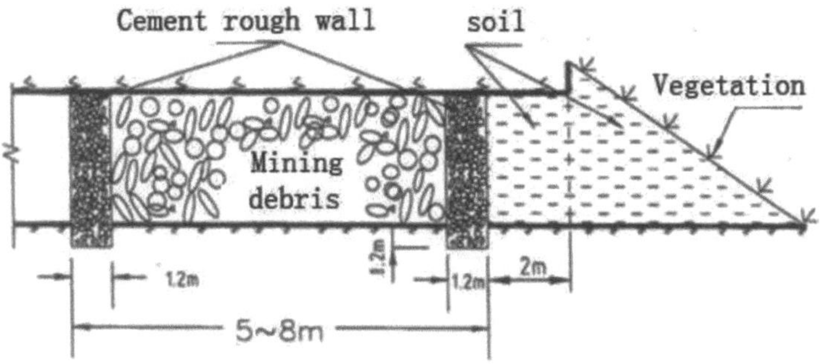

Fig.3. Treatment plan for the pithead of the non-water tunnels

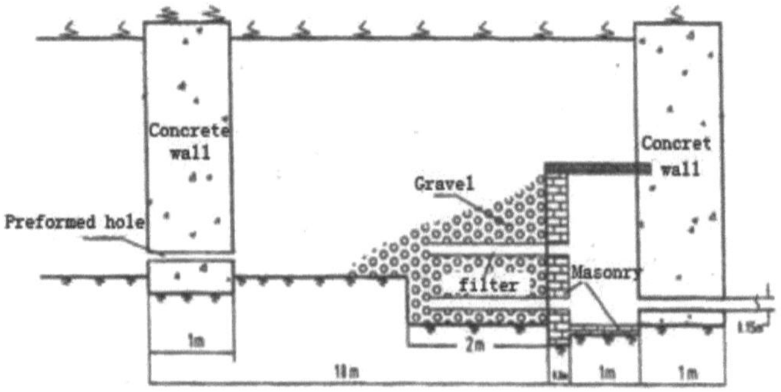

Fig.4. Treatment plan for the pithead with water in the tunnels

Treatment plan for waste rock piles

All the waste rock piles were leveled off , shaped, and smashed according to the terrain, to make the waste rock piles into the hillside stacked mullock field. Then the hillside stacked mullock field was dealt with stably through preventing the tailings by building the wall or damming. As for the steeper area, the square-shaped water retaining skeleton was introduced to protect the slope and the slag heap was buried on-site, and then covered with loess, planted vegetation on it (Fig. 5). All the methods mentioned above could make the slag heap shield the γ-rays radiation dose rate, control the release and exhalation of Radon and its progeny, to reduce the Radon exhalation rate finally.

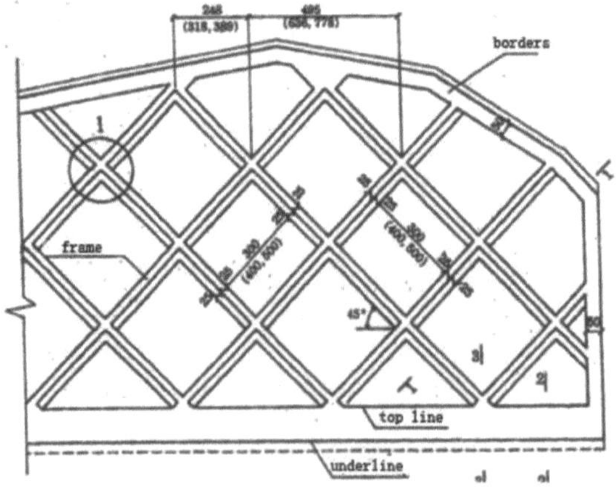

Fig.5. The sketch map of the square-shaped water retaining skeleton

Environment monitor for decommissioning management

Apparatus

FH40GX-γ dose rate instrument (FH40GX-γ DRI) and the RAD7 emanometer were mainly used for the environment monitor. The performance indexes were shown in Table 1.

Table 1. Instruments and the performance indexes used for environment monitor

items	Instruments	Model	Main performance indexes
γ-rays dose rate	γ dose rate X dose rate	FH40G -L10	range: 1nSv/h~100μSv/h relative intrinsic error: 6%
Radon exhalation rate	Emanometer	RAD7	range: 4~7.5×10^5Bq/m^3 Sensitivity: 0.4 count/min/pCi/L Built-in pump : 1L/min flux normally

3.2 Measuring method

As for the tunnels needed to be managed in the waste heaps, we set the grid points according to the site situation, with the size of about 10m×10m for each point. FH40GX-γ dose rate instrument (FH40GX-γ DRI) was introduced to measure the γ-rays radiation dose rate on each site (the height was about 100cm above the ground). In each grid, the measurement was repeated five times on five positions according to the distribution of the plum blossom. The averaged result of these five values was considered as the γ radiation dose rate of this grid.

The pattern and the parameters of the RAD7 emanometer needed to be set before the measurement of the Radon exhalation rate. According to the manual of RAD7 and the instrument measurement specification, instantaneous measurement method was conducted in this project, and the measurement cycles were set to 10 minutes, with ten circulations (Liu and Qiu 2007).

The concentration of the Radon could be measured following with the methods above, and then the Radon exhalation rate could be calculated by the Equation (1) as follows:

$$R = \frac{(C_{t2} - C_{t1})}{A \bullet \Delta t} \times V \qquad (1)$$

Where, R refers to the Radon exhalation rate, Ct1 and Ct2 refer to the Radon concentration at the time t1 and t2, respectively. V refers to the volume between the Radon collection cover and the medium surface, A refers to the superficial area of the Radon collection cover, and Δt refers to the interval between the two measurements.

3.3 The analysis of measured results

Ten waste rock piles were chosen to analyze the decommissioning results, γ-rays radiation dose rate and the Radon exhalation rate before and after the decommissioning (Table 2).

Table 2. γ-rays radiation dose rate and Radon exhalation rate before and after decommissioning

Waste rock piles	γ-rays radiation dose rate (nGy/h)		Radon exhalation rate ($Bq/m^2\cdot s$)	
	before the decommissioning	after decommissioning	before the decommissioning	after decommissioning
1	881	136	0.67	0.22
2	1130	165	0.76	0.13
3	1324	190	1.13	0.17
4	2231	197	1.38	0.22
5	1254	152	1.02	0.11
6	1446	205	1.38	0.35
7	1889	204	1.46	0.17
8	2210	210	1.22	0.44
9	1999	185	1.05	.50
10	2140	167	1.66	0.2

The environment γ-rays radiation dose rate, which was approximately 10% of that before the management, decreased significantly after the waste rock piles management, which indicated that covering with loess and planting vegetation could manage the waste rock piles effectively. According to the management goals, the target value of the air absorbed dose rate byγ-rays radiation should be the reference inventory plus by 170 nGy/h. The reference inventory raged from 82 nGy/h to 138 nGy/h, with the averaged value of 97 nGy/h. The value of the air absorbed dose rate byγ-rays radiation met the requirement of the target value after the decommissioning management (Fig.6).

The Radon exhalation rate, with the maximum value of 0.5 Bq/m^2•s, decreased significantly after the decommissioning, compared with that, with the maximum value of 1.66 Bq/m^2•s, before the decommissioning. The Radon exhalation rate of the waste rock piles surface met the requirement of the target value, which was less than 0.74 Bq/m^2•s, after the decommissioning management (Fig. 7)

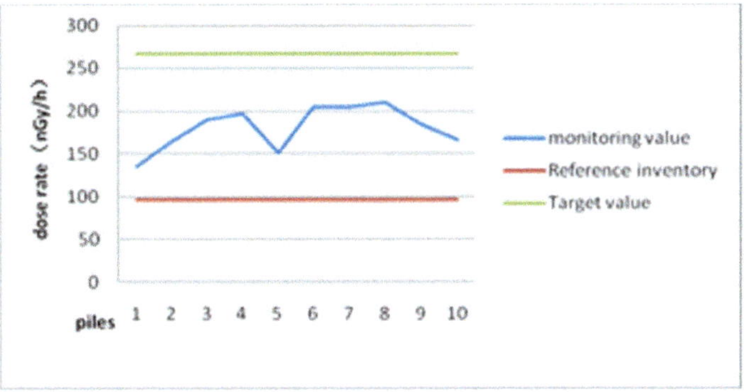

Fig.6. The comparison betweenγ-rays radiation dose rate and management value

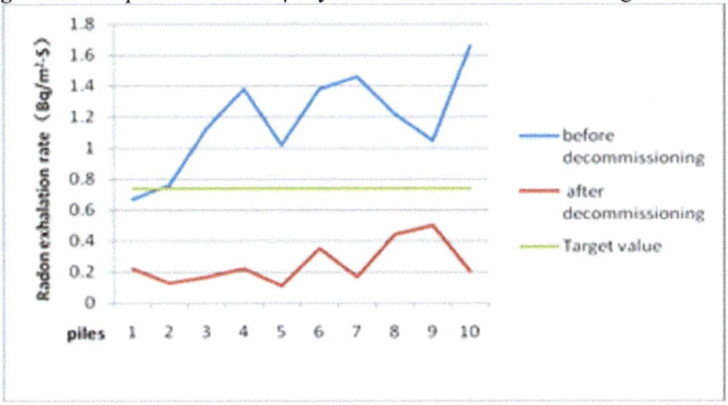

Fig.7. The change of Radon exhalation rate of the waste rock piles

Conclusions

1. The decommissioning project in this paper was mainly to deal with waste rock piles and the pithead. The main method is through the Loess covering and planting vegetation, to reduce the radon release and the radon exhalation rate, so as to realize the open use restrictedly.

2. Through the comparison of the monitoring data of ten waste rock piles, we could conclude that the methods mentioned above had obvious effect. Theγ-rays radiation dose rate and the Radon exhalation rate decreased significantly. Theγ-rays radiation dose rate reduced to approximately 10% of the reference inventory.

3. Theγ-rays radiation dose rate and the Radon exhalation rate met the management target value regulated by the decommissioning project. The decommissioning manage work was conducted quite effectively.

References

GB 14586-93 (1993) Technical regulations for the environmental management of decommissioning of uranium mining and milling facilities. Beijing: Ministry of Environmental Protection t of the People's Republic of China

Liu X, Qiu K (2007) A rapid and accurate method for measuring radon exhalation rate. Radiation Protection 27(3): 156-16

Does wind energy production cause more radioactive doses than nuclear power plants?

Gerhard Schmidt[1]

[1]Oeko-Institute, Rheinstrasse 95, D-64295 Darmstadt/Germany

Abstract. Lynas Australia plans to re-use its mill tailings from its rare earth production in Malaysia for road construction purposes. As those tailings are large masses and rich in thorium, radiological doses would result from that use. Comparing the collective doses of this re-use with collective doses from nuclear energy production shows that a wind power plant constructed with rare earth magnets so poses doses in the lower range of nuclear power production as per GWa.

Introduction

Modern electric generators can be constructed with the use of permanent magnets made from an alloy consisting of a mixture of neodymium and praseodymium (called didymium), boron, iron and a small quantity of dysprosium. Large generators, such as for wind power converters, require several tons of that material. The rare elements Nd, Pr and Dy in that alloy are produced from natural ores that also contain elevated concentrations of uranium and thorium.

If the wastes from processing of such ores are not disposed but re-used, which is the current plan of the Australian company Lynas in Malaysia, this re-use leads to doses for the general public. It is speculated, that this practice can range wind power into the same dose region than the production of nuclear energy. The following analysis therefore evaluates those doses from nuclear energy production and from wind power production and compares both.

The applied dose concept: committed collective effective doses

To compare doses on that level, a thorough selection of the dose concept is required. Individual doses, expressed as mSv per year for the most exposed individual, as usually referred to in dose calculations, are not appropriate here. Those doses would be completely misleading here because they are depending solely

B.J. Merkel and A. Arab (eds.), *Uranium – Past and Future Challenges*,
DOI 10.1007/978-3-319-11059-2_27

from the selected exposure path and would only yield the personal risk for that single individual.

Furthermore, in the case of tailings re-use, large quantities of wastes in the order of several million tons of material are involved. So, not only a few individuals but a very large number of persons can be affected. Though with small doses each, those are applied to a large number of individuals and so can cause large damages.

Instead, the committed collective effective dose, divided by the generated electricity, is used here. The collective dose in man·Sv stands for the sum of all individual doses and so is a linear measure for the complete health effects of a practice. Collective dose calculations were performed for typical exposure situations to assess their overall health effects, e.g. for above-ground nuclear testing, for the Chernobyl disaster, for health effects from natural background radiation. The collective dose for nuclear power generation was in part taken from UNSCEAR and completed by own calculations.

As will be shown, the resulting collective doses from large scale tailings waste re-use are depending from typical scenarios for that re-use. Those scenarios have to be selected realistically to reflect the typical individual dose contributions associated with that practice.

Collective doses from nuclear energy generation

Fig. 1 provides an overview over all stages of the production of nuclear energy and the associated dose contributions. The left stages have been assessed in detail in the (UNSCEAR 1993) report, so there is no need to repeat that calculation. On the right side the stages are listed for which so far no collective doses had been calculated. It had to be assessed if those stages contribute significantly to the collective dose.

The fuel reprocessing stage contributes significantly to total doses, mainly due to large emissions of carbon-14 in this stage. To assess the collective dose the liquid and gaseous emission data between 1990 and 2011 of the La Hague/France reprocessing plant were divided by the amount of spent fuel reprocessed in that year and converted to collective doses per GWa by the application of an estimate for the average fuel burn-up in those years (increasing burn-up). The results are consistent.

- The collective dose per GWa decreases from 120 man·Sv/GWa in the beginning of the Nineties due to improvements in emission reduction,
- reaches a low of 40 man·Sv per GWa in the Mid-Nineties, when the facility reprocessed large quantities of fuel, and
- increases to the average of 65 man·Sv per GWa again beginning in 2000 due to smaller processed fuel quantities.

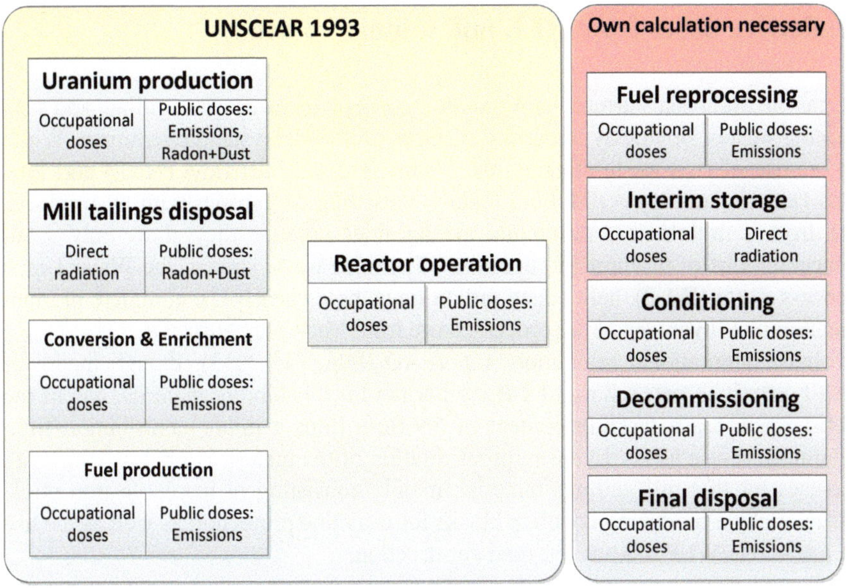

Fig.1. Stages in nuclear energy production and dose contributions

The following three stages to be assessed do not add significant contributions to the collective dose as emissions are small and the number of affected personnell in these stages is small.

To assess collective doses from final disposal the respective dose assessments for the four most advanced repository projects were analysed, the resulting individual doses were integrated over one million years, the typical timeframe for which the safety of repositories has to be demonstrated, and multiplied by the estimated number of affected persons (Yucca Mountain Project: current population of Amargosa Valley/Nevada). As performance assessments for repositories are from their nature (regulatory safety proof) based on some extreme conservative assumptions (total dissolution of fuel in contact with water, total failure of package material, etc.) the resulting values are generally on the upper edge and unrealistically high.

The resulting average collective dose of 9 man·Sv per GWa contributes to total doses of nuclear energy production.

Table 1. Collective dose assessment for the final disposal of spent nuclear fuel and High-Active-Wastes

Country	Final disposal project	Site	Host rock formation	Total dose Sv/Ma	Persons affected	Collective dose man·Sv/GWa
Belgium	(Research) SAFIR2	Mol	Boom clay	1,66	1000	22,46
Switzerland	Entsorgungsnachweis	Benken	Consolidated clay	0,20	1000	1,02
USA	Yucca M license app.	Yucca Mountain	Tuff	5,23	2400	5,17
France	Dossier Argile 2005	Bure	Consolidated clay	10,50	1000	6,51
Average				**4,40**		**8,79**

Collective doses from REE mill tailings re-use

Lynas in Malaysia currently follows the concept to re-use its milling and processing wastes, obviously under the auspices of the Malaysian government. While the re-use of two waste streams, the Neutralization Underflow (NUF) and Flue Gas Desulfurization (FGD), both mainly consisting of gypsum with by-products, are from a radiological standpoint manageable[2] because they show only small concentrations of thorium and uranium[3]. The third waste stream, the Water Leach Process waste (WLP), here called mill tailings, holds nearly the complete thorium and uranium inventory of the processed ore from Mt.Weld, Australia.

Since international regulation, e.g. (FAO 1996, EU 2013), though made for small quantities, sets a limit of 1 Bq/g thorium for this kind of material, and as the WLP waste has a thorium content of 5.9 Bq/g (plus a minor contribution from uranium), re-use as per Lynas requires dilution of the material to below that level. Re-uses for this mill tailings material, mainly consisting of insoluble iron phosphate, include producing concrete blocks for coastline protection as well as the use of a concrete/WLP mixture for road construction.

Use case road construction

In case the WLP waste is used for road construction two typical applications have to be considered: construction of a freeway, mainly frequented by cars, or a city road, mainly frequented by pedestrians. Common to both use case scenarios is the large amount of material that has to be used, 8.4 million tons of WLP mix.

In the freeway scenario, the following conditions were used to calculate dose rates for car drivers. It involves the use of 30 cm of the 1+6 WLP-concrete mix as road base material and an asphalt cover of the road base of 8 cm thickness. Additional shielding layers of 20 cm of air (underneath the car) and a 2 mm thick stainless steel layer (car bottom) have been added. The car driver's exposure point is in a distance of 70 cm from the road surface (half the body height). The calculation was performed with a well tested Java tool[4]. The resulting dose rate in this case is 64.7 nSv/h, calculated without contributions from radionuclides (uranium, thorium) in the road cover material.

---

[2] That does not imply that any re-use is recommendable, taking other environmental aspects into account (by-product content of non-radiological hazardous constituents, acidification of agricultural land by gypsum application, quality and purity implication, etc.)

[3] See (Schmidt 2013) for respective dose calculations in chapter 4.5.3, p. 74.

[4] See http://www.wise-uranium.org/rdcx.html

The calculation of the collective dose is described in Table 2. The length of the freeway that can be constructed with this material is roughly 1,200 km, a 10 m wide freeway assumed.

Table 2. Collective dose calculation for the highway road construction scenario

Highway road construction scenario		
Road construction parameters		
Total amount WLP to be reused:	1.2	Mio. tons
WLP concrete mix ratio 1 + :	6	
Total amount of WLP mix:	8.4	Mio. tons
Density of mix:	2.36	t/m³
Total mix volume to be produced:	3.56	Mio. m³
Layer thickness in road:	30	cm
Road size (width):	10	m
Amount mix per km road:	3,000	m³
km road total	1,186	km
Road use parameters		
Capacity of road (4 lane freeway):	72,000	vehicles per day
Capacity factor:	30%	
Average travel distance per vehicle:	100	km
Average travel speed:	80	km/h
Average travel time per vehicle:	1.25	h/travel
Total vehicles per day:	256,271	vehicles per day
Average persons per vehicle:	1.2	persons per vehicle
Total persons per day:	307,525	persons per day
Total exposure hours:	384,407	hours per day
	140,404,576	h/a
Dose calculation		
Dose, whole year:	566.3	µSv/a
Dose rate:	64.7	nSv/h
Dose over total exposure hours:	9.08E+09	nSv/a
Collective dose per year:	9.08	man·Sv/a
Life time of constructed road:	30	a
Total exposure:	99	man·Sv

The resulting collective dose in Table 2 is 9 man·Sv per year, adding up to 99 man·Sv if the road is used for 30 years, a realistic life-time estimate for highways. This dose stands for roughly 6 serious health cases, ICRP's dose/risk relation assumed.

A similar calculation was performed for a city street scenario. The parameters for the road construction are nearly the same, with the exposure point in a height of 75 cm above the asphalt layer. The resulting dose rates are listed in Table 3. As can be seen, the dose for an individual staying the whole year on that road is by roughly a factor of 2 below the applicable dose limits (1 mSv/a).

Table 3. Dose rate calculation for the city street scenario

Parameters		
Total amount WLP to be reused:	1.2	Mio. tons
Mix ratio, 1 + n	6	
Total amount of WLP mix	8.4	Mio. tons
Density:	2.36	t/m³
Total amount to be reused:	3.56	Mio. m³
Layer thickness:	30	cm
Road size (wide):	10	m
Amount per km road:	3,000	m³
Constructed roads total:	1,186	km
Dose rate in 70 cm distance:	65.41	nSv/h
Dose over complete year:	573	µSv/a

To calculate the collective dose, typical use profiles for such a city street were defined and the resulting contribution to the collective dose were calculated in Table 4.

Table 4. Collective dose calculation for the city street scenario

Collective dose estimate			
Duration	**Density**	**Dose**	**Coll.dose**
h/a	**Pers./km**	**µSv/Pers.**	**man·Sv**
7,000	50	457.9	27.16
5,000	100	327.1	38.80
2,000	250	130.8	38.80
1,000	500	65.4	38.80
500	1,000	32.7	38.80
100	5,000	6.5	38.80
15,600	**182.7**	**11.9**	**221.17**

As can be seen all user categories (persons with long exposure as well as persons with short stays on that road) in that case contribute roughly the same portion to the total of 221 man·Sv per years. The average dose per person is small (12 µSv), but the number of affected persons is very large (16,000 persons). The resulting dose of 221 man·Sv per year stands for more than 10 serious health damages per year, using ICRP's dose/risk factor of 18 man·Sv per case.

If the lifetime of such roads of 30 years are used, it is obvious that neither the freeway nor the city street application of WLP waste are harmless in that they would cause serious health damages to a large number of people. Again, the individual risk is small and on an acceptable level in all these cases, but the large amount of applied material increases total health damages.

Loss of control – misuse of released material

It can be doubted that the construction industry, that would accept and receive such WLP+concrete mixtures, handles those materials only as designed for. To achieve such an exclusive use for road construction a strict regulatory regime would be required that follows, balances and controls material use completely. It can be doubted that this control over the released material would work under given Malaysian framework conditions. Even if such a control would be installed, the misuse of a considerable portion of the material has to be assumed.

To identify if such misuse would increase the collective doses considerably, it was assumed that 10% of the released material is misused for building construction. The resulting dose rate in 1 m distance to a wall constructed with WLP-concrete mix 1+6 is 358 nSv/h, exceeding applicable individual dose limits if applied over 2,000 hours per year (office building) or 8,760 h/a (housing). The collective dose for a 50% mixed use in house and office buildings yields 315 man·Sv per year. Considering that buildings are constructed to last for very long times (e.g. 50 years), the collective dose from that use is extremely large and would cause more than 800 additional serious health damages in total.

The material release would still be inacceptable if only losses of 1% of the material were assumed for which extreme tight controls of material usage would be required, which are unrealistic to function in the construction sector.

Collective dose calculation for wind energy production

The following parameters and assumptions are used to calculate collective doses per GWa of wind energy production. First of all, the collective dose for wind energy generation would be near zero, if the mill tailings would not be re-used by Lynas but isolated in a permanent disposal facility (PDF). The second assumption concerns the re-use of Nd/Pr magnets, because this re-use reduces primary needs linearly. The third decision to be made concerns the attribution of collective doses to the material mix of the ore. This can be done by mass (roughly one fourth of the ore's REE content are Nd and Pr) or by the price that can be achieved in the market (Nd and Pr are in the lower price range of REEs). The collective doses for an attribution by mass is shown in Table 5.

Table 5. Collective dose attribution to wind energy production

Recycling re-use cycles	Pr/Nd need t/GWa	City street man·Sv/GWa	Freeway man·Sv/GWa	Loss-of-control man·Sv/GWa
0	22,86	3,8797	0,1598	6,9783
1	11,43	1,9399	0,0799	3,4891
2	7,62	1,2932	0,0533	2,3261
5	3,81	0,6466	0,0266	1,1630
10	2,08	0,3527	0,0145	0,6344

As can be seen from the table the collective dose is up to 7 man·Sv/GWa if only primary material is used and no recycling is performed. Recycling decreases the collective dose considerably.

Comparison of doses from wind and nuclear energy production

The results of the comparison are listed in Table 6. Doses from wind energy production are in the lower range of nuclear, assumed that no reprocessing takes place and uranium production uses high performance cover systems.

Table 6. Collective doses in nuclear and in wind energy production, in man·Sv/GWa

Nuclear	Minimum	Average	Maximum	Depending from ...
Fuel supply	9		151	Performance of the tailings cover
Reactor operation	1,8	2	2,3	C-14 emissions
Fuel reprocessing	40	60	120	Optional, not needed
Final disposal of nuclear waste	1		22	Enclosure performance
Wind energy	Minimum		Maximum	Depending from ...
Nd/Pr magnet, from Mt. Weld ore, attributed by mass	0		7	Re-use of mill tailings, recycling of Nd/Pr

References

EU (2013) Council Directive 2013/59/EURATOM of 5 December 2013 laying down basic safety standards for protection against the dangers arising from exposure to ionising radiation. - Official Journal of the European Union L 13/1, 17.1.2014

FAO (1996) FAO/IAEA/ILO/OECD/NEA/PANO/WHO: International Basic Safety Standards for Protection against Ionizing Radiation and for the Safety of Radiation Sources. – IAEA Safety Series No. 115, Vienna 1996

Schmidt, Gerhard (2013) Description and critical environmental evaluation of the REE refining plant LAMP near Kuantan/Malaysia - Radiological and non-radiological environmental consequences of the plant's operation and its wastes. – Darmstadt/Germany, January 25, 2013

UNSCEAR (1993) United Nations Scientific Committee on the Effects of Atomic Radiation (UNSCEAR): UNSCEAR report 1993 to the General Assembly - Annex B Exposures from man-made sources of radiation. – New York, 1993

Meeting of the Uranium mining and hydrogeology self-help group
My name is Bob and I work with uranium...
Welcome, Bob.

Impact of humic substances on uranium mobility in soil – A case study from the Gessenwiese test field, Germany

Stefan Karlsson[1], Viktor Sjöberg[1], Bert Allard[1]

[1]Man-Technology-Environment Research Centre, Dep. Science and Technology, Örebro University, 70182 Örebro, Sweden

Abstract. The study has shown that in a soil with a logKd of 4-5 for uranium between the aqueous and the solid phase only some 20% of the soil inventory of uranium is mobilized. Five consecutive L/S 10 extractions with alkali released some 75% of the uranium inventory but not the entire content of organic matter. The mobilization of uranium coincides with that of organic matter as evidenced by size exclusion chromatography and ICP-MS. The stationary uranium co-elutes with to a high extent with humic and fulvic substances.

Introduction

Humic substances are omnipresent in the soil environment and have a large impact on physical soil. They also influence the distribution of many elements between stationary and mobile species (Sachs and Bernhard 2011; Li et al 2013, Stockdale and Bryan 2011) which in turn will influence the mass flux and bioavailability (Bednar et al. 2007; Claveranne-Lamolère et al. 2011; Zhao et al 2011). Many microorganisms and plants can access the stationary metals through several independent processes that are both general and specific. Hence, the impact from the humic and fulvic acids (HFA) on the metal cycling in a soil system is complex and there are few theoretical approaches that can be used for either quantitative or qualitative predictions.

From a chemical perspective the HFA is best described by their functionality. In summary, they are the stable remains from bio-degradation of notably plants with molecular weights from a few hundred to several hundred thousand Daltons (D). Their backbones consist of carbon and hydrogen but they also contain varying amounts of nitrogen, oxygen and sulfur as well as minor amounts of other elements. They are polyelectrolytes that lack defined pKa but rather described by acidity functions that corresponds to their amount of titrable moieties (Marinsky and Ephraim 1986a). Depending on pH, ionic strength and ionic composition in the water phase they will flocculate and can therefore behave as dissolved species

B.J. Merkel and A. Arab (eds.), *Uranium – Past and Future Challenges*,
DOI 10.1007/978-3-319-11059-2_28

or colloidal/particulate matter (Marinsky and Ephraim 1986b). Since they are excellent complexing agents for both cat- and anions they will influence the behavior of metals. Their interactions with metal ions are best described as "stability functions" instead of constants. For practical reasons the stability functions ca be approximated as a set of intrinsic constants (Marinsky and Ephraim 1986b).

In soils, humic acids (HA) are typically associated with the stationary phase, either as surface coatings or aggregates while the fulvic acids (FA) are more mobile (Claveranne-Lamolère et al. 2011; Harguindeguy et al. 2014; Mibusa et al 2014; Stockdale and Bryan 2013). These properties are to some extent related to their molecular sizes but also to their contents and distribution of functional groups within the structure. In general the complexing capacity of HFA is around 10 meq g^{-1} but the affinity differs between metal ions and their species in solution. The composition of the solution phase has a great impact on the actual stability. Particularly pH is important since it affects the protolysis of the HFA and its quaternary structure as well as the distribution of metal species.

In boreal regions the general hydrochemical conditions in the soil environment is a function of season with respect to carbon loading/turnover, biological activity, water availability etc. In an ongoing study on the impact of low temperature on the properties of soil organic matter, notably HFA, and their impact on metal speciation some preliminary results are available. Here we report on some observations on the presence and impact from organic matter, including HFA, on the distribution of uranium between the stationary phase and the pore water after 8 months of incubation. The samples are from a control plot on the Gessenwiese test field where interactions between minerals, soil biota and metal uptake in plants are studied (Grawunder et al. 2009). The properties of the organic matter and its distribution under the hydrochemical conditions are also included.

Materials and Methods

Soil samples from a control plot at the Gessenwiese test field in Thuringia (Grawunder et al. 2009) were collected in September 2013. The samples were kept in sealed glass jars at 5°C during transport. They were then incubated in a refrigerator at 4°C with their original content of soil organic matter during the study, except for sampling. So far samples have been collected at arrival to the laboratory and after 3 and 8 months, respectively. The following characterization has been made on the material as it was sampled, i.e. without drying or sieving.

The distribution of the HFA was determined by equilibration Ω of the soil samples with 18.2 MΩ water at a liquid to solid ratio (L/S) of 10 for 24 hours on an overhead shaker (Heidolph Reax2). These solutions were also used for the determination of electrical conductivity, pH and extractable principal an- and cations. A parallel sample was extracted with alkaline medium (1.0 M NaOH) to release stationary HFA. Both extractions were repeated five times on each sample and sampling to allow for a semi quantitative estimate of the soil content. In the alkaline extracts only the apparent molecular weight (mw_{app}) of extracted constituents and metal concentrations were determined. After each extraction the samples were centrifuged at 4000 rpm for 30 minutes and then passed through a 0.20 μm polypropylene syringe filter. The alkaline extracts were mixed with the buffers (see below) to neutralize them and lower the rate of hydrolysis.

Extracted inorganic (IC) and dissolved organic carbon (DOC) in the solution phases were determined with a Shimadzu TOC-V CPH instrument. Determination of mw_{app} of extracted organic matter was made with size exclusion chromatography using an Agilent SEC-5 column. For estimations of mw_{app} a 50 mM phosphate at pH 6.8 was used with polysulfonate standards (American Polymer Standards Corporation) in the range 0f 1.2 to 60 kD. The identity of the HFA substances was determined according to their apparent molecular weight in combination with UV/Vis absorbance and fluorescence properties. Because of the high complexing capacity of phosphate ions a 50 mM NH_4NO_3 buffer at pH 6.8 was used for the size distribution of the metals. A fair correlation (r^2 0.97) between the buffer systems was established for polysulfonate references in the calibration range. The major limitation is the high absorptivity from the buffer components, particularly in the short wavelength region. Principal anions were quantified with ion chromatography using a Metrohm Metrosep A Supp 5 with a buffer of 1.0 mM $NaHCO_3$ and 3.2 mM Na_2CO_3.

Metal analysis was performed with an Agilent 7500cx ICP-MS with ^{103}Rh as internal standard. Leaching of the solid phase with concentrated nitric acid in a Mars 5 microwave oven represents the "total" concentrations.

Results and Discussion

Water extracts

The general composition of the water L/S 10 extracts shows that there is an extractable pool of ions (Table 1) but the release rate is limited. The second extraction gave significantly lower concentrations for all anions and the uranium concentrations were some 13%. Concentrations of nitrite, nitrate and phosphate ions were below their limits of detection at 0.005, 0.002 and 0.010 μg L^{-1}, respectively. The only observed increase was for pH from 4.25 to 4.68 and inorganic carbon from 0.30 mg L^{-1} to 0.63 mg L^{-1} which indicate that there had been a production of acid before to the extraction.

Table 1. Concentrations of dissolved organic carbon (DOC), inorganic carbon (IC), principal inorganic anions, uranium and pH in two consecutive water extracts (n=4).

	pH	DOC mg L^{-1}	IC mg L^{-1}	Cl^- mg L^{-1}	F^- mg L^{-1}	SO_4^{2-} mg L^{-1}	U µg L^{-1}
Extract 1	4.26	1.73	0.39	5.00	0.52	267	0.90
Extract 2	4.73	0.78	0.78	0.72	0.27	47.	0.12

Extractable DOC was lowered from 1.73 to 0.78 mg L^{-1} why there was a limited pool of readily mobile organic matter and the rate of production was insufficient to maintain the original concentrations. The DOC ratio between the first and second extraction (0.45) indicates a higher release rate than for chloride (0.18) and sulfate (0.17). It is therefore likely that at least a part of the DOC came from biological activity during the time of extraction rather than desorption from the stationary phase.

The quality of the DOC, as determined with SEC-analysis using the phosphate buffer, is rather difficult to determine because of the low concentrations (Figure 1). A peak at 11.50 minutes (mw_{app} 1.7 kD; area 11.5) is non-HFA in character according to the fluorescence spectrum while the one at 12.05 minutes (mw_{app} 1.2 kD; area 11.5) is a fulvic acid according to these criteria.

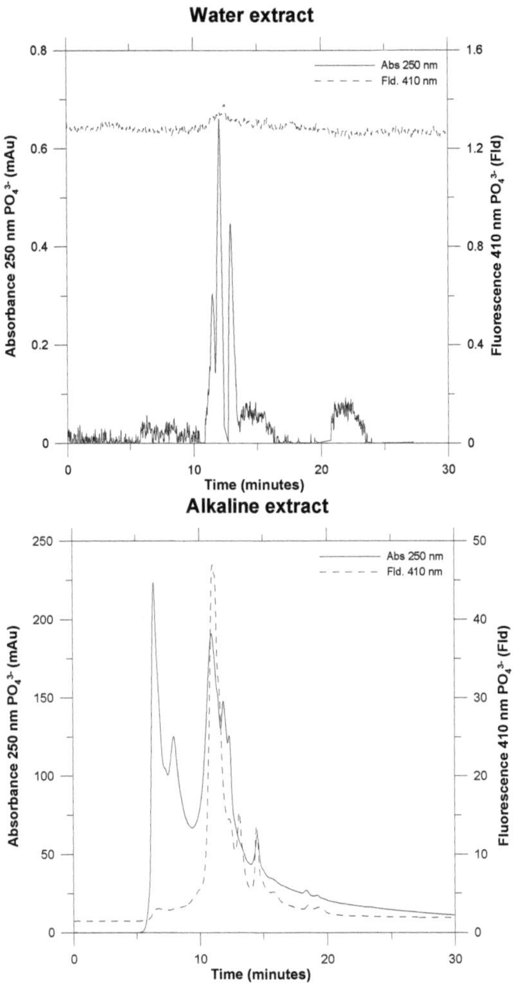

Fig.1. SEC chromatograms of water and alkaline extracts in phosphate buffer. Detections with UV at 250 nm and fluorescence at 410 nm.

There is also a highly polydisperse population of low molecular weight organic compounds appearing from 20 to 24 minutes. Although the column has a poor separation at these retention times preliminary studies have indicated mw_{app} in the range of 0.2 kD. As mentioned previously, the uranium concentrations in the water extracts went down from 0.90 µg L^{-1} to 0.12 µg L^{-1}.Since the acid extractable amount was 5.61 mg kg^{-1} the resulting logKd were 3.7 and 4.6, respectively, which is a clear indication that uranium becomes more inaccessible with increasing L/S ratio.

The size distribution of uranium in the SEC analysis is presented in Figure 2 as counts per second (cps). A small fraction has no interaction with the column at all and would represent the uranyl ions. In the high molecular weight region (3 to 12 minutes) there is no systematic relationship to retention time. In fact, the fulvic acid does not seem to carry any uranium above background. In the low molecular weight region (12 to 30 minutes) there are some fractions with slightly increased uranium concentrations at 16 and 17 minutes as well as 23-27 minutes. At these retention times it was not possible to detect any organic substances. There is, however, a fair chance that organic ligands were undetected, either because of their spectroscopic properties or too low concentrations.

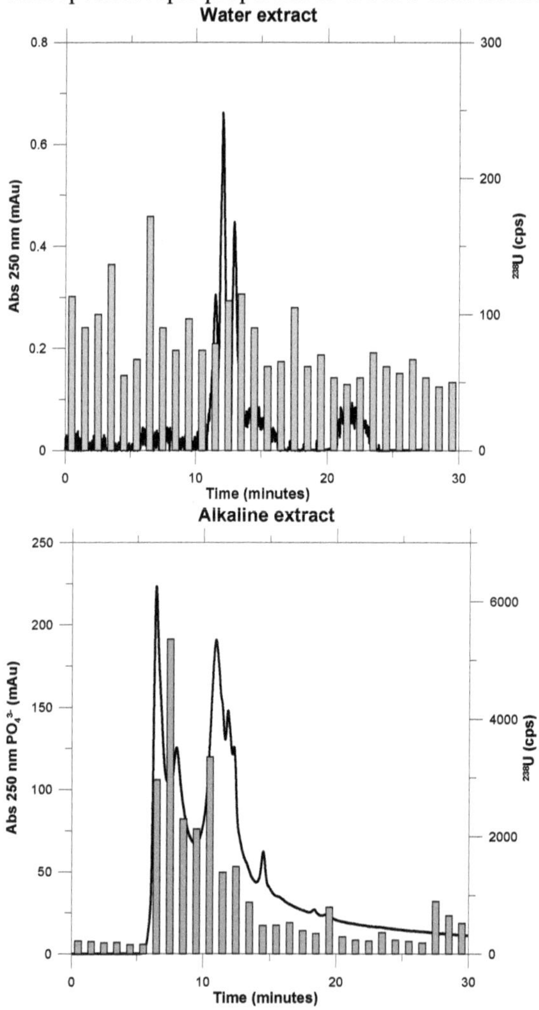

Fig.2. SEC chromatograms of water and alkaline extracts (absorbance at 250 nm, solid lines) and uranium (counts per second for ^{238}U).

Alkaline extracts

The alkaline extracts contained considerably higher DOC concentrations than the corresponding water extracts. In the four consecutive alkaline extractions there was a systematic lowering of DOC in the order 251, 71, 28 and 8 mg L^{-1}. The corresponding uranium concentrations were 380, 371, 287 and 62 µg L^{-1}.In fact, the results indicate that the number of extractions should have been extended to get an accurate estimate of the total content. This extraction behavior can be interpreted in several ways but it seems most rational that the soil DOC content is not readily available for extraction with alkali. If this is related to a kinetically constraints or just a matter of contact between the soil particles and the solution is unknown. Although incomplete the extraction indicates an inventory of extractable carbon in the range of 1.8-2.1 mg g^{-1}. Using the water extractable DOC this inventory indicates a logKd for DOC in the range of 3 to 4. Hence, in this system the metal complexing properties of the carbon fraction would result in an immobilization. The conditions also infer that metal uptake in the plants would be heavily influenced by the formation of stationary organic complexes.

In comparison with the water extracts the SEC/phosphate buffer peak pattern for the alkaline extracts is far more complex and the signals much higher. With respect to the spectroscopic properties of the components some general quality features can be discussed. It must be remembered, however, that a more precise characterization must be made in order to elucidate the true nature of the compounds. The peaks between 6 and 10 minutes are within the optimum calibration range of the column and the corresponding mw_{app} are 48.5 kD, 15.0 kD and 5.2 kD. According to mw_{app} they are all within the size range for humic acids (HA) and fulvic acids (FA) which are expected to be found in this kind of system. The spectroscopic properties suggest, however, that the first two fractions contain a high but unknown amount of non-HFA components. From this investigation it is impossible to identify any specific kind of compound but their mw_{app} would indicate proteins or bio-polymers.

In the intermediate size range with retention times from 10 to 15 minutes the accuracy and resolution of the column becomes less reliable. Calibration with a set of organic acids indicates that reasonably accurate estimates of mw_{app} can be made in this time window but the uncertainty increases to some 30%. The rather high peak at 10.9 minutes corresponds to an approximate mw_{app} of 1.6 kD. The spectroscopic characterization strongly indicates a dominance of fulvic acids in this peak. The uncertainty of the mw_{app} estimates increases with increased retention times but the three peaks that follows at retention times around 11.9, 12.3 and 14.5 minutes would represent approximate mw_{app} of 0.8, 0.6 and 0.1 kD. These substances do not show spectroscopic properties typical for HFA compounds.

At retention times above 15 minutes at least four different peaks are discernable, but at rather low signal intensities. The corresponding mw_{app} are not possible to elucidate from this study but they are definitely in the low molecular weight range, i.e. mw_{app} lower than some 0.2 kD.

Uranium enrichment in soil, as U(VI), due to the association with HFA, is common and well-known (Choppin and Allard 1983). Conditional stability constants ($logk_1$) in the range 4.5 to 9 have been reported, reflecting the origin of the HFA, different conditions, the degree of dissociation, ionic strength etc. Also the physical and chemical adsorption to soil components is highly affected by the presence of HFA, generally enhancing the adsorption at low pH and reducing it at high pH.

In conclusion, the alkaline extraction mobilized not only the expected HFA components in the soil but also a high amount of other substances, with mw_{app} both higher and lower of the HFA. Hence, the stationary organic phase in the soil can be expected to possess a wide range of metal complexing functional groups.

In the alkaline extracts some 81% of the acid leachable uranium inventory is recovered and there is a rather pronounced size distribution of the element (Fig. 4). Unfortunately, the resolution of the NH_4NO_3 buffer system does not allow for a more precise separation of the different size classes but some features can be identified. The fraction at 7 minutes contains an increased uranium concentration and as explained previously this material is a complex mixture of high molecular weight material with an identifiable fraction of HA. The majority of the uranium appears in the fractions from 8 to 12 minutes which corresponds to intermediate mw_{app}. Limited amounts are also present in the low molecular weight region of the chromatogram but without any evident relation to single size fractions.

The extractability in combination with the size distribution of uranium clearly demonstrates the impact of organic components with a low water solubility. It is therefore reasonable to assume that not only the soil-water interactions are influenced by these conditions but also the bioavailability. From a pollution risk perspective it would be essential to maintain, or increase, the retention capacity of these components in order to prevent an increased mobility of the element. In the case of phytoremediation it is more appropriate to focus on the interactions with biota. In order to improve the uptake it is not only important to consider how and to what extent a plant accumulates uranium but also on the transfer mechanisms between the soil pool and the roots. Hence, the function of mycorrhiza is probably an important key to understand and optimize phytoremediation processes.

Conclusions

The study has shown that that majority of the organic matter in this soil is not extractable in water but in alkali. The water soluble DOC is dominated by a population of fulvic acids with low molecular weight. In the alkaline extracts both humic and fulvic acids were found but also other non-identified organic constituents in the high as well as in the low molecular weight region. The uranium had a logKd of 3-4 in the water extracts. Water extractable uranium did not co-migrate with the fulvic acid but was rather homogeneously distributed in the 0.1-60 kD size range. In the alkaline extracts some 80% of the uranium inventory was retrieved, with the highest abundance in high molecular weight compounds. Hence, for a successful removal or fixation of uranium in this kind of soil the impact of stationary organic species must be considered for the treatment scheme.

Acknowledgements

The authors express their gratitude to Prof. Dr. Erika Kothe at the Friedrich Schiller University, Jena, for the samples and for discussions. The Faculty of Economics, Science and Technology, Örebro University, is acknowledged for financial support to S Karlsson.

References

Bednar AJ, Medina VF, Ulmer-Scholle DS, Frey BA, Johnson BL, Brostoff WN (2007), Larson SL. Effects of organic matter on the distribution of uranium in soil and plant matrices. Chemosphere 70: 237-247

Choppin G, Allard B (1983). Complexes of actinides with naturally occurring organic compounds. In Freeman AJ, Keller C (Eds). Handbook on the Physics and Chemistry of the Actinides, Vol. 3, Elsevier Sci. Publ., North-Holland, Amsterdam, p 407-429

Claveranne-Lamolère C, Aupiais J, Lespes G, Frayret J, Pili E, Pointurier F, Potin-Gautier M (2011) Investigation of uranium–colloid interactions in soil by dual field-flow fractionation/capillary electrophoresis hyphenated with inductively coupled plasma-mass spectrometry. Talanta 85: 2504-2510

Grawunder A, Lonschinski M, Merten D, Büchel G (2009) Distribution and bonding of residual contamination in glacial sediments at the former uranium leaching heap of Gessen/Thuringia, Germany. Chem Erde 69(2): 5-19

Harguindeguy S, Crançon P, Pointurier F, Potin-Gautier M, Lespes G (2014) Isotopic investigation of the colloidal mobility of depleted uranium in a podzolic soil. Chemosphere 103: 343-348

Li X, Wu J, Liao J, Zhang D, Yang J, Feng Y, Zeng J, Wen W, Yang Y, Tang J, Liu N (2013) Adsorption and desorption of uranium (VI) in aerated zone soil. J Environ Radioact 115: 143-150

Marinsky JA, Ephraim J (1986a) A unified physicochemical description of the protonation and metal-ion complexation equilibria of natural organic acids (humic and fulvic acids). 2. Influence of polyelectrolyte properties and functional group heterogeneity on the protonation equilibria of fulvic acid. Environ Sci Technol 20(4): 354-366

Marinsky JA, Ephraim J (1986b) A unified physicochemical description of the protonation and metal-ion complexation equilibria of natural organic acids (humic and fulvic acids). 1. Analysis of the influence of polyelectrolyte properties on protonation equilibria in ionic media – Fundamental concepts. Environ Sci Technol 20(4): 349-354

Mibusa J, Sachs S, Pfingsten W, Nebelung C, Bernhard G (2007) Migration of uranium(IV)/(VI) in the presence of humic acids in quartz sand: A laboratory column study. J Contam Hydrol 89: 199-217

Sachs S, Bernhard G (2011) Humic acid model substances with pronounced redox functionality for the study of environmentally relevant interaction processes of metal ions in the presence of humic acid. Geoderma 162: 132-140

Stockdale A, Bryan ND (2013) The influence of natural organic matter on radionuclide mobility under conditions relevant to cementitious disposal of radioactive wastes: A review of direct evidence. Earth-Sci Rev 121: 1-17

Zhao P, Zavarin M, Leif RN, Powell BA, Singleton MJ, Lindvall RE, Kersting AB (2011) Mobilization of actinides by dissolved organic compounds at the Nevada Test Site. Appl Geochem 26: 308-318

U-Th-Pb data as a tool for bordering small-scale regions of in-situ monazite mineralization

Jan Mestan[1,2], Libor Volak[2], David Sefcik[3]

[1]Faculty of Geosciences, LMU, Luisenstrasse 37, 80333 Munich, DE
[2]Faculty of Science, CUNI, Albertov 6, 12843 Prague 2, CZ
[3]Faculty of Science, USB, Branisovska 31a, 37005 Ceske Budejovice, CZ

Abstract. The aim of this paper is to give a new way of phosphate deposits prospection using in-situ monazite dating. U-Th-Pb data for a single grain of monazite from locality Kamenne Doly Quarry were combined with previously obtained monazite ages from Pisek pegmatites and their GPS positions (φ, λ) were processed by a MATLAB script to define region of monazite or phosphate occurrence. At least 3 assumptions have to be made. Monazites are of similar ages, are products of an extensive geological event affecting a significant amount of rock bodies in a region and the shallow rock environment is the same at different depths.

Introduction

Phosphate mining plays a significant role in primary sector of the economy. It includes prospecting, which is a key part of pre-mining activities. It is helpful to define borders of valuable regions for subsequent exploration. It is also desirable to get their areas for next quantitative analysis.

In this paper we deal with a requirement on finding a new eco-friendly and economically acceptable way how to border regions of in-situ phosphate deposits using radiometric dating. Because a significant quantity of them is associated with trace monazite, we decided to focus primarily on it. CHIME monazite dating has been spreading for several decades and belongs today to relatively cheap (100-200 EUR for a single grain analysis), fast (max. hours for a single grain) and reliable methods. An outline of the method was introduced by Montel et al. (1996).

There was developed a MATLAB script, because MATLAB is easy to get and is widely used in research institutions as well as industrial enterprises. Input parameters of the script are geodetic coordinates (φ, λ) of monazite localities with similar ages acc. WGS 84 and output is a plot and an area of a convex hull region determined by orthogonal multiple linear regression. The height above the reference ellipsoid is neglected (Fig.1).

B.J. Merkel and A. Arab (eds.), *Uranium – Past and Future Challenges*,
DOI 10.1007/978-3-319-11059-2_29

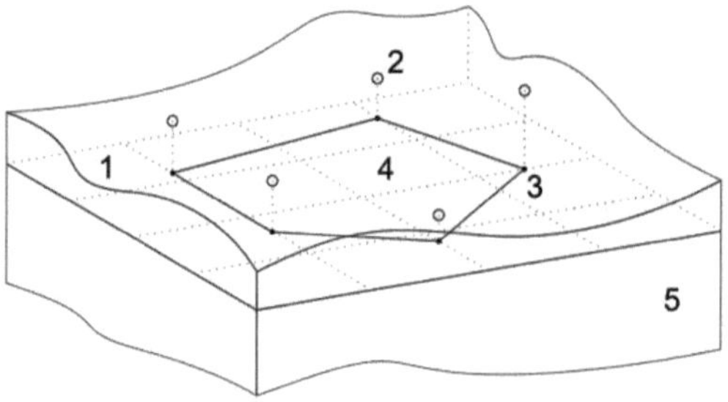

Fig.1. An outline of the procedure. 1 – ellipsoid's surface ≈ plane approximation, 2 – monazite actual position, 3 – monazite position on ellipsoid, 4 – convex hull region, 5 – rock massif.

We investigated 3 localities from Pisecke Hory, the Bohemian Massif. The first two are the granitic Pisek pegmatites U Noveho Rybnika Quarry and Udraz Quarry. The third locality is Kamenne Doly Quarry. Data from all localities were processed by the script and the results are presented and discussed.

Geological settings

The Bohemian Massif was consolidated during the Variscan orogeny (≈ 400-300 Ma), which involved several oceanic subductions and collisions of continental micro-plates (Maierova 2012). Pisecke Hory belongs to its central part, the Moldanubian Zone. It is a wooded ridge with the highest peak at 632 masl and length of approximately 20 km. The most common rock types in the Pisecke Hory area are leucocratic metagranites representing the Podolsko Complex and quartz syenite (Fig.2). They are intruded by numerous dikes of granite, pegmatite and aplite, which is well observable in Kamenne Doly Quarry.

Kamenne Doly Quarry has been an active stone quarry since 1939. It exposes mainly leucocratic metagranite of simpler mineralogy, which is sometimes accompanied by intrusive bodies of quartz syenite with abundant accessory apatite. Houzar et al. 2008 described a body of scheelite skarn with unusual mineralization with niobian titanite and Bi-tellurides in the second level of the quarry. Xenoliths of marbles and calc-silicate rocks are sometimes present.

U Noveho Rybnika Quarry and Udraz Quarry are very abundant in minerals and allow to study macroscopic monazite samples. Some of them were earlier defined as younger postectonic phases (Rajlich et al. 2010). They formed at ≈ 337 ± 2 Ma (Novak et al. 1998). Both localities belong to numerous Pisek pegmatite quarries, which were mined for feldspar and rose quartz in the late 19th century and early 20th century. Besides these minerals also macroscopic high quality samples of beryl, apatite and tourmaline are to be found (Krejci 1925).

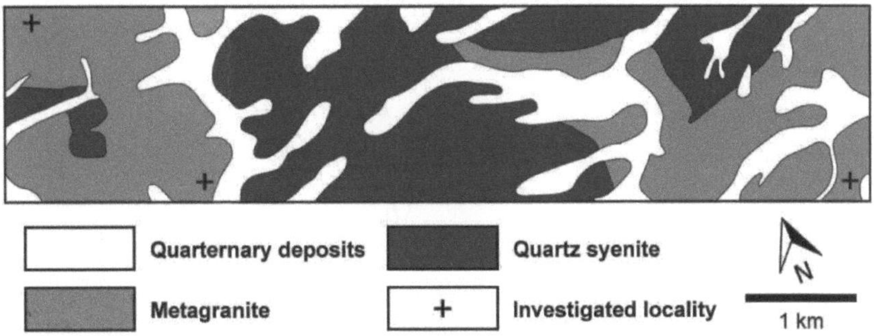

Fig.2. Geological sketch of the Pisecke Hory area (simplified after Fisera et al. 1991).

Monazite dating

A thin section of metagranite from Kamenne Doly Quarry was made and coated with carbon at the Institute of Geology, Academy of Sciences of the Czech Republic. U-Th-Pb data for 14 spots in one monazite grain (100 μm) were obtained by CAMECA SX 100 electron microprobe (WDS) at the State Geological Institute of Dionyz Stur (SGUDS) in Bratislava, Slovak Republic. Measured elemental concentrations were corrected for the interferences at an accelerated voltage 15 kV so that the counting statistics was with a minimum worsening of measured spots. The calibration was checked by monazite standards previously dated on SHRIMP. The age calculation was accomplished using MS Excel program DAMON (Konecny et al. 2004) according to the model by Montel (1996). It is assumed that monazite contains no initial lead. All errors are at the 2σ level.

Bordering procedure

The input geodetic coordinates (φ, λ) of monazite localities according to WGS 84 are transformed to Cartesian coordinates with the neglect of an ellipsoidal height h after Ligas and Banasik (2011) as follows:

$$\begin{bmatrix} x \\ y \\ z \end{bmatrix} = \begin{bmatrix} N\cos\varphi\cos\lambda \\ N\cos\varphi\sin\lambda \\ (1-e^2)N\sin\varphi \end{bmatrix}. \tag{1}$$

For the radius of curvature in the prime vertical N is

$$N = \frac{a}{\sqrt{1 - e^2 \sin^2 \varphi}}, \tag{2}$$

where

$$e^2 = 1 - \left(\frac{b}{a}\right)^2 \tag{3}$$

is the first eccentricity squared.

The Cartesian coordinates are then approximated by an orthogonal multiple linear regression using Principal Components Analysis in MATLAB. All coordinates are orthogonally projected onto a regression plane and after Kovacs (2012) their rotation into the Cartesian plane is performed. We expand an equation 2.3 in his paper by a rotational matrix R_P, which ensures that the north-pole direction on resulting plot is in the $+y$ direction:

$$\mathbf{R}_\mathrm{p}(\gamma) = \begin{bmatrix} \cos(\gamma) & -\sin(\gamma) & 0 & 0 \\ \sin(\gamma) & \cos(\gamma) & 0 & 0 \\ 0 & 0 & 1 & 0 \\ 0 & 0 & 0 & 1 \end{bmatrix}, \tag{4}$$

where γ is given as follows:

$$\gamma = j\pi \pm \arccos\left(\frac{\mathbf{u} \cdot \mathbf{v}}{\|\mathbf{u}\|\|\mathbf{v}\|}\right), \tag{5}$$

where **u** is a normal vector to a vector of an intersection of the regression plane and the Cartesian plane, **v** has a form (0, 1, 0) and j depends on quadrant.

The expanded transformation matrix **M** is then

$$\mathbf{M} = \mathbf{R}_\mathrm{p}(\gamma)\mathbf{T}^{-1}(-\mathbf{p}_0)\mathbf{R}_\mathrm{x}^{-1}(\theta_\mathrm{x})\mathbf{R}_\mathrm{y}^{-1}(-\theta_\mathrm{y})\mathbf{R}_\mathrm{z}(\theta)\mathbf{R}_\mathrm{y}(-\theta_\mathrm{y})\mathbf{R}_\mathrm{x}(\theta_\mathrm{x})\mathbf{T}(-\mathbf{p}_0). \tag{6}$$

The convex hull region is determined using MATLAB convhull function and its area A is calculated according to Cromley (1992) as follows:

$$A = \frac{1}{2}\sum_{\mathrm{i}=1}^{\mathrm{n}} (x_{\mathrm{i}+1} - x_\mathrm{i})(y_{\mathrm{i}+1} + y_\mathrm{i}). \tag{7}$$

Method limitations

The above introduced procedure is a suitable approximation for regions where the Earth's curvature is neglected. D_L to D_E ratio k is applied in order to estimate the limitations of the neglect. D_L stands for a linear distance of two points (monazite localities) and D_E for an ellipsoidal distance of them.

$$k = \frac{D_\mathrm{L}}{D_\mathrm{E}}. \tag{8}$$

Line element of D_E is as follows:

$$\mathrm{d}D_\mathrm{E} = \sqrt{1 + \left(\frac{\mathrm{d}y}{\mathrm{d}x}\right)^2}\,\mathrm{d}x. \tag{9}$$

Since the Earth's figure is as a reference frame for computations in the geosciences described by an ellipsoid of revolution, we obtain two limiting k values. For k_i of an ellipsoid with a center in (0, 0, 0) we have

$$k_i = \frac{D_\mathrm{L}}{2}\left[\int_0^{\frac{D_\mathrm{L}}{2}} \sqrt{1 + \frac{V_1^2 x^2}{V_2^4\left(1 - \frac{x^2}{V_2^2}\right)}}\right]^{-1}, \tag{10}$$

where for k_1 $V_1 = b$, $V_2 = a$ and for k_2 $V_1 = a$, $V_2 = b$. For given D_L k values are obtained numerically. The ellipsoid's flattening is very small ($\approx$ 0.003) and so it may be written $k \approx k_1 \approx k_2$.

The D_L value depends on the user himself. For D_L 400 km the k is very high (0.9998), which is a square with a size of $\approx$ 283 km. Such area should be sufficient for most of the purposes related to proposed bordering procedure (e.g. small-scale areas with a geological homogeneity, engineering goals, prospecting areas). The given reference D_L is compared with linear distances of all data points. A relevant algorithm will then point out, if the D_L was exceeded.

Method assumptions

We must make at least 3 assumptions before we start to apply bordering procedure on our data.

It is assumed that monazites have similar ages. It means that they might be put into the same geological period. The choice depends on the user himself. The shorter the period is the more similarity we get.

The second assumption is a requirement on geological homogeneity of studied region. Monazites must be products of an extensive geological event affecting a significant amount of rock bodies in the region (primary magmatic origin, late extensive hydrothermal event, other metamorphic events).

The third one is the shallow depth similarity. Since we get samples from different heights above the reference ellipsoid, it is necessary to assume that the shallow rock environment is the same at different depths.

Results

Table 1. Chemical analysis of monazite grain and calculated ages.

Spot	Pb [wt. %]	U [wt. %]	Th [wt. %]	Y [wt. %]	Th*[a]	Age [Ma]	Error [Ma]
1	0.155	0.287	9.450	2.285	10.38	334	15.5
2	0.071	0.281	3.795	1.191	4.71	339	30.7
3	0.130	0.233	8.333	1.037	9.09	320	17.1
4	0.178	1.420	7.830	2.207	12.45	321	13.0
5	0.152	0.407	8.703	1.745	10.03	339	15.9
6	0.152	0.417	9.059	1.668	10.41	327	15.2
7	0.142	0.299	8.658	0.787	9.63	330	16.4
8	0.131	0.299	7.982	1.089	8.95	327	17.3
9	0.135	0.241	8.589	1.030	9.37	323	16.6
10	0.148	0.250	8.829	1.115	9.64	343	16.3
11	0.103	0.197	6.352	1.421	6.99	329	21.5
12	0.132	0.340	7.542	1.248	8.65	342	17.8
13	0.128	0.458	7.078	1.260	8.57	336	17.9
14	0.144	0.323	8.886	3.017	9.94	325	15.9

[a] Th plus equivalent of U.

The histogram (Fig.3) shows that monazite age falls best into range 330 ± 4.5 Ma with MSWD 0.86 and 7 spot analyses. The isochron on the right plot nearly passes through the origin, which serves as an additional proof of dating reliability.

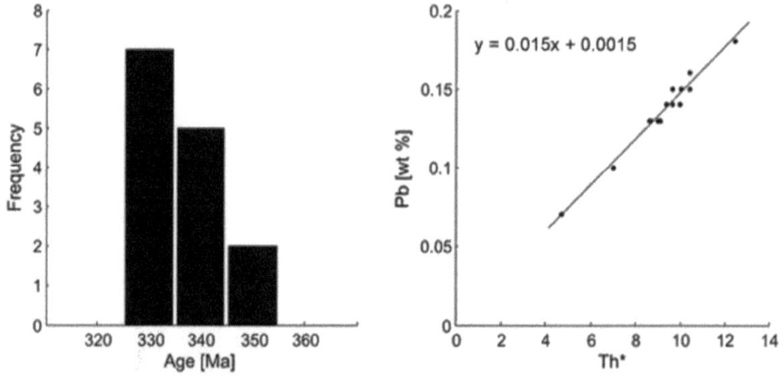

Fig.3. Results for CHIME monazite dating. Th* stands for Th plus equivalent of U.

Data from studied localities are summarized in Table 2. A suitable period for all data is an epoch Mississippian, which is an interval 358.9 ± 0.4 to 323.2 ± 0.4 Ma (ICS 2014/02). In addition our data overlap each other in applied age intervals. The final plot of a convex hull region is in Fig.4.

Table 2. Data overview.

Locality	φ [°]	λ [°]	Age [Ma]	Epoch
Kamenne Doly Quarry	49.31831111	14.18910278	330 ± 4.5	Mississippian
U Noveho Rybnika Quarry	49.30114444	14.20124444	337 ± 2.0	Mississippian
Udraz Quarry	49.28071944	14.27526111	337 ± 2.0	Mississippian

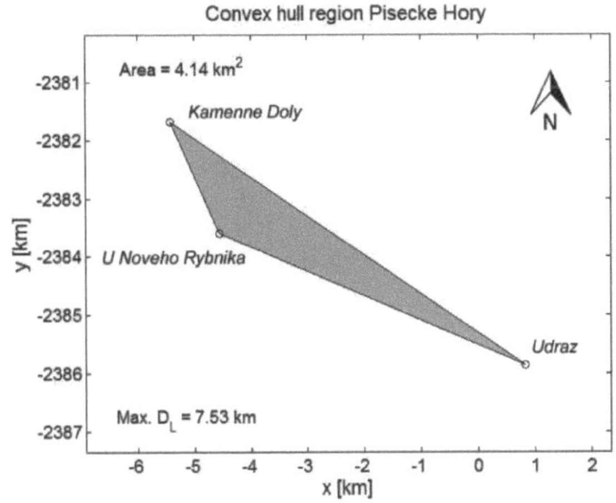

Fig.4. Convex hull region for data from Table 2.

Discussion

Our max. D_L value is at a more than sufficient level (7.53 km) as well as monazite ages. A big amount of radiometric data from the Moldanubian Zone is around 340 Ma (Maierova 2012), which is similar to the Pisecke Hory area (metagranite as well as quartz syenite). This age should date peak P–T conditions reached inside the Moldanubian crust. From this point of view the studied region has a geological homogeneity. We assume that for neglected heights (hundreds of meters) above the reference ellipsoid the rock environment is similar at different depths. We await similar plutonic and metamorphic rocks instead of sedimentary (older than quaternary) or volcanic rocks, which are not present in the map (Fig.2).

Our subsequent goal could be to estimate a quantity of monazites or phosphates according to an analysis of rock samples from the surface or boreholes. If we describe the density by a continuous function ρ_m, we have for a quantity q_m:

$$q_{\mathrm{m}} = \iint \rho_{\mathrm{m}}(x, y)\mathrm{d}S. \tag{11}$$

It is obvious that it *will be complicated to get close to* q_{m} value, because of economical and environmental reasons.

Conclusions

There was presented a method how to prospect in-situ phosphate deposits associated with monazite. It represents a single tool, which has to be combined with other data from surveyed region like previous geological mapping.

If a geologist decides to use it, he must deal with a fact that the method has its limitations connected with the Earth's curvature and at least 3 assumptions.

It would be possible to continue with exploration of bordered region in the future according to an equation (11). This should enable a quantitative characterization of the region with a help of geochemical data.

Acknowledgements

Analytical techniques were funded by the SGUDS and MSMT OP VK grant no. CZ.1.07/2.3.00/09.0034. Special thanks for support are due to Kamen a pisek, Ltd. and GFPS e.V. We gratefully thank to Patrik Konecny and Ivan Holicky for a help with monazite dating, to Lukas Ackerman for an assistance with a preparation of thin section and to Marek Svoboda for lending a GPS Tracker.

References

Cromley R G (1992) Digital cartography, Prentice Hall

Fisera M et al. (1991) Basic geological map of the CSFR, sheet 22-412, CGU (in Czech)

Houzar S, Litochleb J, Sejkora J, Cempirek J, Cicha J (2008) Unusual mineralization with niobian titanite and Bi-tellurides in scheelite skarn from Kamenne doly quarry near Pisek, Moldanubian Zone, Bohemian Massif. J Geosci 53: 1-16

Konecny P, Siman P, Holicky I, Janak M, Kollarova V (2004) Methodics of monazite dating using electron microprobe. Miner Slov 36: 225-235 (in Slovak)

Kovacs E (2012) Rotation about an arbitrary axis and reflection through an arbitrary plane. Ann Math Inform 40: 175-186

Krejci A (1925) Pisek minerals and their deposits. Cas Nar Muz Odd Prir 99: 49-65 (in Czech)

Ligas M, Banasik P (2011) Conversion between Cartesian and geodetic coordinates on a rotational ellipsoid by solving a system of nonlinear equations. Geod Cart 60: 145-159

Maierova P (2012) Evolution of the Bohemian Massif: Insights from numerical modeling. Unpublished doctoral dissertation, CUNI, Prague, the Czech Republic

Montel J M (1996) Electron microprobe dating of monazite. Chem Geol 131: 37-53

Novak M, Cerny P, Kimbrough D L, Taylor M C, Ercit T S (1998) U-Pb ages of monazite from granitic pegmatites in the Moldanubicum and their geological implications. Acta Univ Carol, Geol 42: 309-310

Rajlich P, Mestan J, Benes V (2010) Younger postectonic monazite in the Pisek pegmatite. J S Boh Mus Ces Bud Nat Sci 50: 26

Hoo Boy, isn't working with uranium bad for your health?

Sorption of U(VI) and As(V) on SiO_2, Al_2O_3, TiO_2 and FeOOH: A column experiment study

Sreejesh Nair[1], Broder J. Merkel[1]

[1]Department of Hydrogeology, Technische Universität Bergakademie Freiberg, Gustav-Zeuner Str.12, 09599 Freiberg, Germany.

Abstract. One important factor which affects migration of U(VI) in the subsurface is the sorption at the solid/solution interface. Many factors control the sorption of U(VI) on mineral surfaces and one significant candidate among them is the distribution of aqueous species. In this study, column experiments were carried out to investigate the transport of uranium, arsenate and uranium-arsenate together (0.5:0.5 μM/l) in columns packed with SiO_2, Al_2O_3, TiO_2 and FeOOH at pH 6.5. Transport behavior of U(VI) and As(V) through SiO_2 and TiO_2 packed columns are identical in the inlet solutions containing either U(VI) or As(V) separately, or both together. In the presence of equimolar U(VI) and As(V), a substantial increase in As(V) mobility and a slight decrease in U(VI) transport through Al_2O_3 were observed. When Al_2O_3 is replaced with FeOOH, a significant change in the pattern of mobility was shown by As(V); whereas U(VI) showed only minor changes. The changes in transport behavior of both elements can be attributed to the competitive sorption between uranyl and arsenate species or due to the formation of uranyl-arsenate species. The immobilization of uranyl and arsenate with the aforementioned minerals are in the order FeOOH>TiO_2>Al_2O_3>SiO_2 under our experimental conditions. This study thus gives potential information about the transport behavior of uranyl and arsenate in natural systems, especially when both elements are present.

Introduction

Uranium is a naturally occurring radioactive and chemically toxic metal, which is widespread in nature. The release of uranium to the environment is not only from natural sources but also from anthropogenic sources like uranium mill tailings, nuclear industry, coal combustion, phosphate fertilizers etc (Merkel 2002; Merkel and Hasche-Berger 2006). Under oxidizing environments, uranium is readily soluble in water in the +VI oxidation state, but as well depending on pH. In porous mediums such as soils, aquifers or mine wastes, the transportation and immobilization of U(VI) is mainly governed by sorption/desorption reactions at the solid-solution interface. Batch sorption experiments were widely carried out to study the

B.J. Merkel and A. Arab (eds.), *Uranium – Past and Future Challenges*,
DOI 10.1007/978-3-319-11059-2_30

sorption of U(VI) on different mineral surfaces like clay minerals (Pabalan and Turner 1996; Krepelova et al. 2006; Bachmaf and Merkel 2011), quartz (Lieser et al. 1992; Sylwester et al. 2000; Nair and Merkel 2011), zeolite (Aytas et al. 2004; Camacho et al. 2010), goethite (Hsi and Langmuir 1985; Missana et al. 2003) etc. under various conditions. Although batch sorption experiments are capable of providing useful information about solid-solution interaction, they poses prominent difference from the transport conditions in subsurface such as; relatively low solid-solution ratio, influence of reaction kinetics on sorption, duration of the experiment, lack of hydrodynamic mass-transport limitations which can happen in porous media, abrasion of the mineral particles etc. Hence column experiments are recommended to understand the reactive transport of radionuclides in porous media.

One of the important factors which drives the sorption behavior of U(VI) under different transport conditions is the aqueous speciation. Presence and absence of various ligands changes the U(VI) speciation in different aquatic conditions. Phosphate is one of the candidates which forms strong complexes with U(VI) and is very stable as well as insoluble in geological settings (Liu and Byrne 1997). Arsenate is analogous to phosphate and is a well-known contaminant to the environment. Long term exposure of As leads to a number of serious diseases including skin, bladder and lung cancers (Smith et al. 2000). Under oxidizing conditions, arsenic occurs arsenate with +V oxidation state [As(V)]. The formation of uranyl-phosphate complexes and its impact on the U(VI) sorption behavior have been studied extensively (Brendler et al. 1996; Payne et al. 1996; Cheng et al. 2004). On the other hand, the formation of uranyl-arsenate complex and its influence on sorption are rather less investigated (Rutsch et al. 1999; Gezahegne et al. 2012). Existence of natural minerals like Trögerite, $H_2(UO_2AsO_4)_2$. $8H_2O$ and $UO_2(HAsO_4)$. $4H_2O$ are good evidence for the affinity between uranium and arsenic to form as well aquatic species.

Reactive transport of U(VI) with various ligands and different minerals have been studied and reported elsewhere (Barnett et al. 2000; Cheng et al. 2007; Zhang et al. 2009). However, less information is available on the transportation of U(VI) and As(V) in systems containing both elements. Changes in sorption behavior of U(VI) and As(V) in columns containing bentonite (Bachmaf et al. 2009) and iron-coated sand (Schulze and Merkel 2011) were studied and reported. Less or no information is available on the transportation of U(VI) and As(V) together with minerals such as SiO_2, Al_2O_3, TiO_2 and FeOOH, which are very common in natural environments.

The objective of this study was to investigate the reactive transport of uranium (0.5 μM/l), arsenate (0.5 μM/l)) and uranium-arsenic together (0.5:0.5 μM/l) in columns packed with SiO_2, Al_2O_3, TiO_2 and FeOOH at pH 6.5. The study is also extended to investigate the influence of uranyl-arsenate species on the transport of U(VI) and As(V) with aforementioned minerals where the influent contains both elements at pH 6.5.

Materials and Methods

Materials

The stock solutions for U(VI) and As(V) (0.5 μM/l) were prepared from $UO_2(NO_3)_2.6H_2O$ (Chemapol, Czech Republic) and As_2O_5 (Alfa Aesar, Karlsruhe Germany) respectively by dissolving in deionised and purified water (TKA, Germany). To avoid precipitation or sorption of U(VI), the water was pre-acidified with HNO_3 to pH 2.5. The minerals Al_2O_3 and TiO_2 were obtained from Sigma–Aldrich (Germany). The assay, specific surface area and particle size of Al_2O_3 and TiO_2 are 99.7%, 0.6 m^2/g and 10 μm as well as ≥ 99.9%, 2.5 m^2/g and < 5 μm respectively. Goethite was prepared in the laboratory as described by Schwertmann and Cornell (2000) and the surface area of the mineral was measured using BET (44.77 m^2/g). The silica sand (SiO_2) used for the experiment (brand name F32) was obtained from Quarzwerke Frechen, Germany. The average grain size and surface area of SiO_2 are 0.24 and 102 cm^2/g respectively. XRD analysis revealed that F32 contains 98.6±0.26% quartz and 1.4±0.26% calcite. The chemical analysis shows that F32 has 99.7% SiO_2, 0.2% Al_2O_3 and 0.03% Fe_2O_3. The purification process of F32 and the removal of calcite and iron oxides were explained in Nair and Merkel (2011). All chemicals used for the column experiment were of ACS reagent grade or better.

Column Experiment

Column experiments were conducted at room temperature (22±1°C) using 12 PTFE columns with 2 cm inner diameter and 30 cm length. 0.2 μm PTFE filters were placed at both end of the column to keep the porous materials in place. 10 gm of SiO_2 and Al_2O_3, 0.5 gm of TiO_2 as well as 0.1 gm of FeOOH were mixed well with PTFE beads (grain size: 0.29-0.35 mm) and dry-packed in PTFE columns with a porosity of 0.40. The 12 columns were filled with corresponding minerals as shown below;

(1) U(VI) – SiO_2, (2) As(V) – SiO_2, (3) U(VI) – As(V) – SiO_2, (4) U(VI) – Al_2O_3, (5) As(V) – Al_2O_3, (6) U(VI) – As(V) – Al_2O_3, (7) U(VI) – TiO2, (8) As(V) – TiO2, (9) U(VI) – As(V) – TiO2, (10) U(VI) – FeOOH, (11) As(V) – FeOOH & (12) U(VI) – As(V) – FeOOH

All tools such as columns, tubes, connecting valves, collecting bottles etc. used for the experiments were made of PTFE in order to avoid the sorption of U(VI) on it especially at low U(VI) concentration and at high pH. To minimize the dead volume, short tubes with small inner diameter were used to connect the valves to the corresponding columns (Fig. 1).

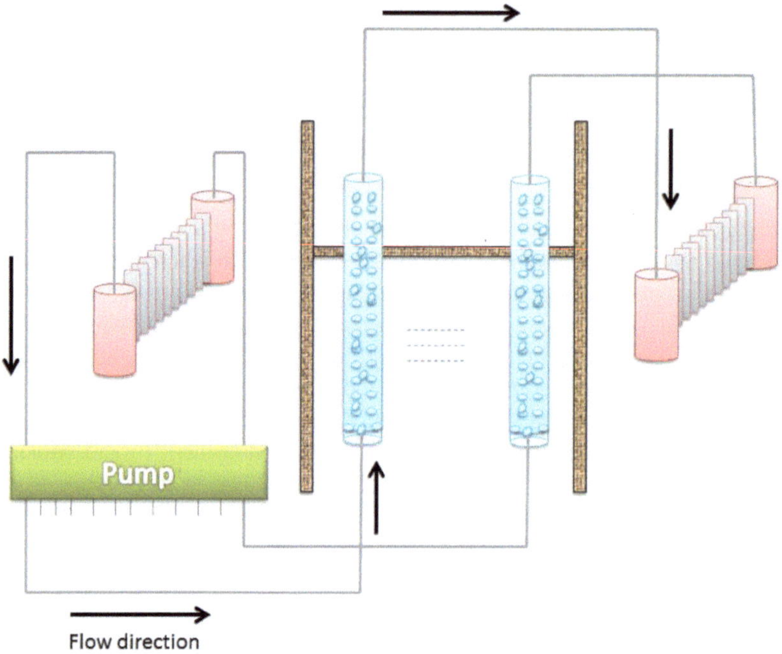

Fig.1. Schematic representation of the column experiment (see text for details and note that the size of the different parts is not to scale).

The solutions were pumped from bottom to top with a uniform flow rate of 100 μL/min using a high precision peristaltic pump ISMATEC IPC 24 (Ismatec SA, Switzerland). Due to high sorption of U(VI) and As(V) onto FeOOH, the flow rate was increased to 250 μL/min in the columns 10, 11 and 12 from 146th day of the experiment onwards. In each column experiment, the de-ionised water (DI) was pumped slowly from bottom to top for 12 hours in order to eliminate the air spaces within the porous media and to pre-condition the same. Following to this, U(VI), As(V) and U(VI):As(V) solutions were flushed through corresponding columns until the breakthrough curves (*BTC*) reached equilibrium. The pH of the inlet solution was adjusted to 6.5 using 0.1 M NaOH/HNO_3. One of the main reasons to select this pH was the assumed dominance of the proposed uranyl-arsenate complex at this pH range. After reaching the equilibrium *BTC*, the inlet solutions were changed back to DI water to study the desorption behavior. Desorption experiment using DI was only effective to certain extend with SiO_2 filled columns. In order to flush out the U(VI) and As(V) from the columns, the pH of the inlet solution was increased to 10 using LiOH. During the sorption and desorption experiments, samples were collected every day for the first week and later with an interval of three days per week. pH of the collected samples were noted (6.5 ± 0.2) and analysed for uranium and arsenic by using ICP-MS (XSeries 2, Thermo Fisher Scientific). To study the fate of sorbed U(VI) and As(V) on minerals, the mass balance calculation was carried out by calculating the amount of U(VI) and As(V) sorbed and desorbed from the columns over the sorption and desorption periods.

Results and Discussion

U(VI)/As(V)/U(VI) – As(V) – SiO_2 Column

Figure 2 shows the transport and *BTC* of U(VI) and As(V) through columns packed with SiO_2. Arsenate has been identified in the outlet solution without any delay (from the 2nd day of the experiment) and showed practically no retardation in the transport behavior. As(V) has less or no affinity towards SiO_2 and has been reported elsewhere (Darland and Inskeep 1997; Xu et al. 1988). U(VI) movement was retarded in the U(VI) – SiO_2 column. The equilibrium *BTC* for U(VI) was reached after 200 pore volumes. The affinity of the U(VI) on silica is due to the hydrolysed uranyl species which are dominant at pH 6.5. More or less similar sorption as well as *BTC* for U(VI) and As(V) were observed (Fig. 2) for the experiment U(VI) – As(V) – SiO_2 column. There was no significant indication about the influence of uranyl-arsenate species in the transport of either U(VI) or As(V). Initially, DI water was pumped through the SiO_2 columns as part of desorption experiment. More than 65% of the sorbed U(VI) came through the outlet flow and shows that the sorbed surface species are not strong enough to retain on SiO_2 when compared with the other three minerals. Almost all U(VI) was recovered (~95%) from SiO_2 after pumping the solution with higher pH. Similar sorption behaviour of U(VI) and As(V) onto SiO_2 was observed in batch experiments.

U(VI)/As(V)/U(VI) – As(V) – Al_2O_3 Column

The sorption of uranyl and arsenate on Al_2O_3 and their transport behaviour are shown in figure 3. The retardation of U(VI) with Al_2O_3 is comparable to those of SiO_2 and reached the equilibrium after 350 pore volumes. This can be explained by the high reactivity of aluminol surface sites to U(VI) than the silanol surface sites in SiO_2 (Borovec 1981; Kohler et al. 1992). The sorption retardation of arsenate is more prominent than U(VI) and the *BTC* equilibrium was achieved after 650 pore volumes. The presence of arsenic oxy-anions ($HAsO_4^{2-}$, $H_2AsO_4^-$) enhances the sorption of arsenate on aluminol sites. The strong affinity of arsenate on alumina at acidic to neutral pH range has been studied and reported (Arai et al. 2001; Goldberg and Johnston 2001).

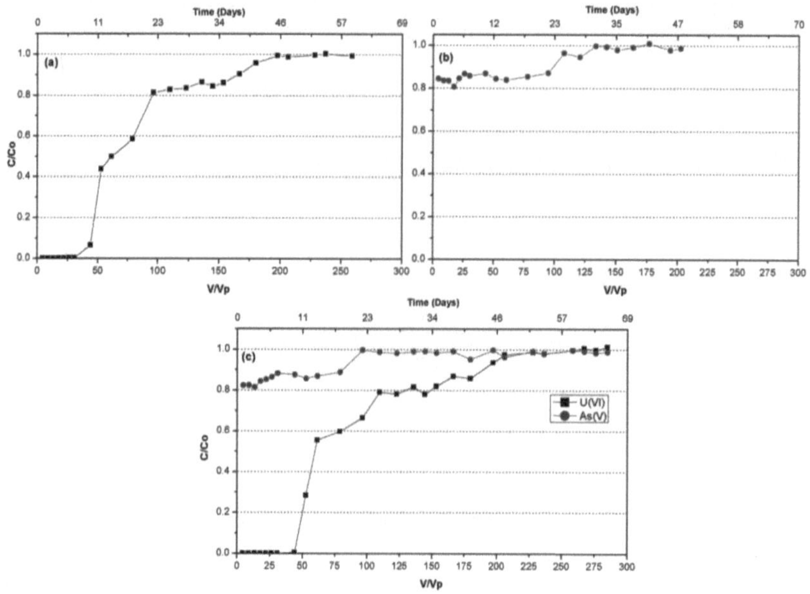

Fig.2. Experimental breakthrough curves (BTCs) of U(VI) and As(V) in SiO_2 columns. (a) U(VI) – SiO_2, (b) As(V) – SiO_2, (c) U(VI) – As(V) – SiO_2 [0.5 μM/l U, 0.5 μM/l As, 10 g SiO_2, pH 6.5, temp.:23°C].

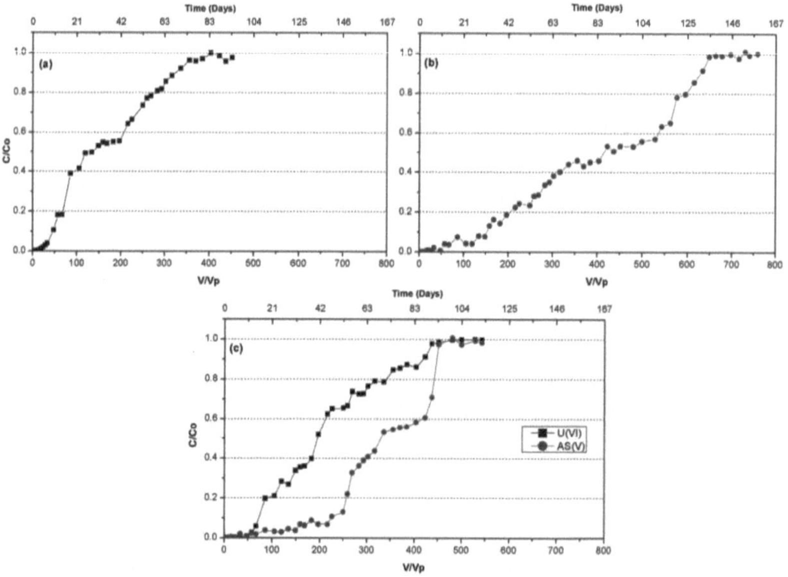

Fig.3. Experimental breakthrough curves (BTCs) of U(VI) and As(V) in Al_2O_3 columns. (a) U(VI) – Al_2O_3, (b) As(V) – Al_2O_3, (c) U(VI) – As(V) – Al_2O_3 [0.5 μM/l U, 0.5 μM/l As, 10 g Al_2O_3, pH 6.5, temp.:23°C].

The presence of both U(VI) and As(V) in the inlet solution changes the sorption behavior of both elements onto Al_2O_3. The sorption of U(VI) slightly increased and the *BTC* achieved the equilibrium after 450 pore volumes whereas the sorption of As(V) retarded and reached equilibrium similar to U(VI) (~450 pore volumes). In the initial stage of the transport, As(V) shows more retardation (up to ~250 V/Vp) and increased to reach the equilibrium. This retardation could be due to the better sorption of uranyl-arsenate species ($UO_2AsO_4^-$) compared to the $H_2AsO_4^-$ ion. The change in transport behavior of both elements is due to the competitive sorption between the uranyl and arsenate species or due to the formation of uranyl-arsenate complex. Almost 70% of the U(VI) and As(V) were desorbed from Al_2O_3. This indicates that the sorption is partially irreversible or slow desorption of both elements under the above said experimental conditions.

U(VI)/As(V)/U(VI) – As(V) – TiO_2 Column

Transport of U(VI) and As(V) through TiO_2 packed columns is shown in figure 4. The concentration of U(VI) was identified in the outlet after 100 pore volumes and reached the break through equilibrium after 250 pore volumes. U(VI) shows strong affinity to TiO_2 than SiO_2 and Al_2O_3.

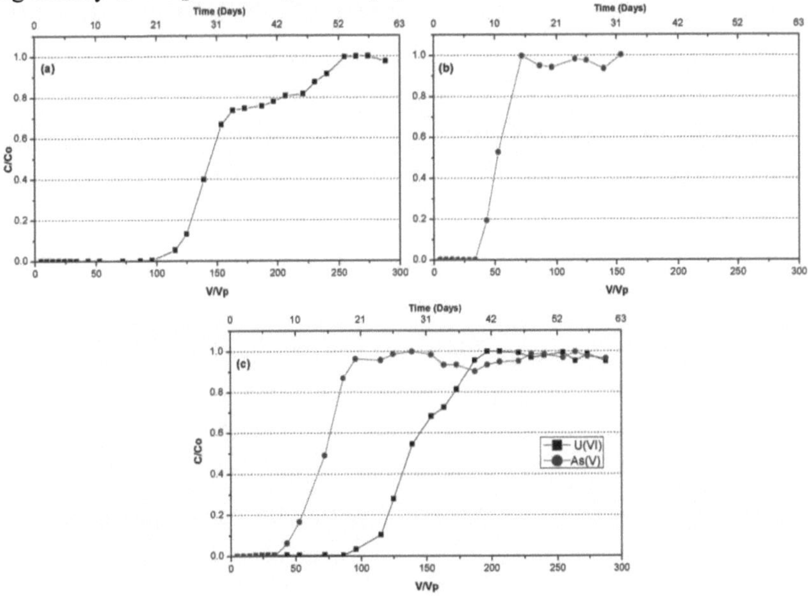

Fig.4. Experimental breakthrough curves (BTCs) of U(VI) and As(V) in TiO_2 columns. (a) U(VI) – TiO_2, (b) As(V) – TiO_2, (c) U(VI) – As(V) – TiO_2 [0.5 μM/l U, 0.5 μM/l As, 0.5 g TiO_2, pH 6.5, temp.:23°C].

The *BTC* of As(V) reached the equilibrium after 70 pore volumes and showed less affinity when compared with U(VI). Previous studies reported that As(V) shows higher sorption onto TiO_2 at higher pH and lower towards low pH values (Lee and Choi 2002; Dutta et al. 2004; Pena et al. 2005). When both elements were present in the inlet solution, the concentration of As(V) was detected initially (40 V/Vp) and is followed by U(VI) (100 V/Vp) in the outlet solution. As(V) showed *BTC* equilibrium at 70 pore volumes and at 200 pore volumes for U(VI). The sorption of As(V) on TiO_2 for both columns (As(V) – TiO_2 and U(VI) – As(V) – TiO_2) was almost similar, whereas U(VI) showed a slight decrease in the sorption behavior in U(VI) – As(V) – TiO_2 when compared with U(VI) – TiO_2 column. The recovery rate for uranyl and arsenate from all three columns were similar to that from Al_2O_3 (~70%) columns.

U(VI)/As(V)/U(VI) – As(V) – FeOOH Column

Experimental *BTC*s and the transport behavior of U(VI) and As(V) through goethite packed columns are presented in figure 5. The change in flow rate (from 100 µL/min to 250 µL/min) of the inlet solution can be seen in between ~700 V/Vp to ~1100 V/Vp in figure 5. Arsenate in the outlet solution was detected after 1500 pore volumes and reached the equilibrium at 2100 V/Vp. The retardation in transport of arsenate is due to the presence of $H_2AsO_4^-$ species, which sorbs effectively with the positively charged goethite surface. Arsenate has strong affinity to FeOOH and is previously reported elsewhere (Stollenwerk 2003). U(VI) concentration in the outlet was first detected at 1000 pore volumes. The transportation of uranyl was gradual and reached the equilibrium at 2400 pore volumes. The high sorption is attributed to the $(UO_2)_2CO_3(OH)_3^-$ species which is dominant at this pH range. Previous studies reported that U(VI) has great affinity to FeOOH over a wide range of pH (Sherman et al. 2008; Guo et al. 2009). The transport behavior of arsenate in U(VI) – As(V) – FeOOH column differs from that of As(V) – FeOOH column. As(V) started to appear in the outlet after 500 pore volumes and the *BTC* reached equilibrium at 2000 pore volumes. U(VI) has similar transport behavior as of U(VI) – FeOOH column and shows better sorption affinity to goethite than arsenate. The arsenate curve showed a sudden increase (at ~500 V/Vp) and followed a gradual movement of the element till the equilibrium. The sudden increase could be due to less sorption of arsenate species ($H_2AsO_4^-$) when competing with the uranyl species ($(UO_2)_2CO_3(OH)_3^-$). The gradual transport of arsenate after the sharp increase could be attributed to the better sorption of $UO_2AsO_4^-$ on the goethite surface. Uranyl ions showed better affinity to goethite than arsenate, even though the latter is known to be a good sorbing element. The recovered U(VI) and As(V) through the effluent solution was less than 20% and proved that the sorption is irreversible or desorption is very slow.

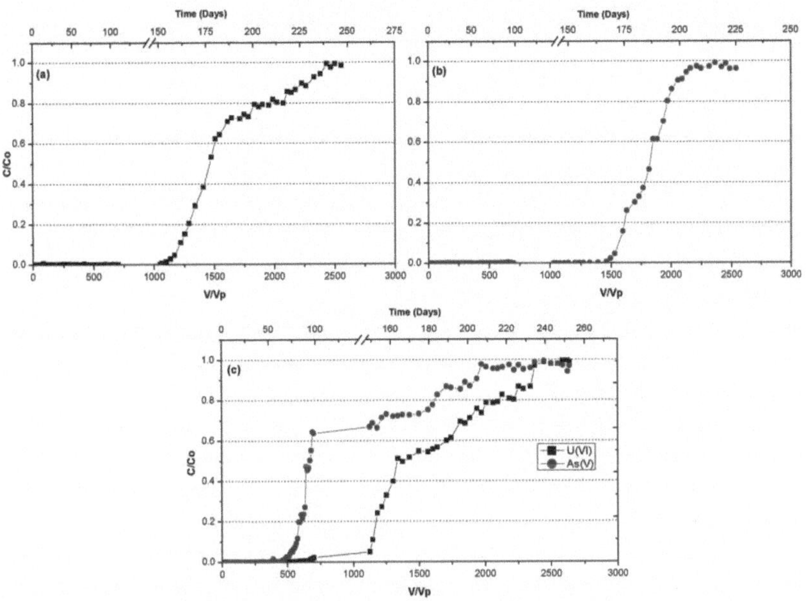

Fig.5. Experimental breakthrough curves (BTCs) of U(VI) and As(V) in FeOOH columns. (a) U(VI) – FeOOH, (b) As(V) – FeOOH, (c) U(VI) – As(V) – FeOOH [0.5 μM/l U, 0.5 μM/l As, 0.1 g FeOOH, pH 6.5, temp.:23°C].

Conclusions

Column experiments were conducted to study the transport of uranyl and arsenate, which revealed that the mobility of U(VI) and As(V) is highly dependent on the speciation of these elements as well as the sorbents. The immobilization of uranyl and arsenate with the aforementioned minerals are in the order of FeOOH>TiO_2>Al_2O_3>SiO_2. Transport behavior of U(VI) and As(V) through SiO_2 and TiO_2 packed columns are identical in the inlet solutions containing either U(VI) or As(V) separately, or both together. In the presence of equimolar U(VI) and As(V), a substantial increase in As(V) mobility and a slight decrease in U(VI) transport through Al_2O_3 were observed. While Al_2O_3 is replaced with FeOOH, a significant change in the pattern of mobility was shown by As(V); whereas U(VI) showed only minor changes. The changes in transport behavior of both elements can be attributed to the competitive sorption between uranyl and arsenate species or due to the formation of uranyl-arsenate species. Mass balance calculations indicated that the sorption of U(VI) and As(V) is strong with the above said minerals and the effectiveness of these minerals to retain the elements are in the order of FeOOH>TiO_2=Al_2O_3>SiO_2 under our experimental conditions. This study thus gives potential information about the transport behavior of uranyl and arsenate in natural systems, especially when both elements are present.

References

Arai Y, Elzinga EJ, Sparks DL (2001) X-ray absorption spectroscopic investigation of arsenite and arsenate adsorption at the aluminum oxide-water interface. Journal of Colloid and Interface Science 235:80-88

Aytas SM, Akyil S, Eral M (2004) Adsorption and thermodynamic behavior of uranium on natural zeolite. Journal of Radioanalytical and Nuclear Chemistry 260:119-125

Bachmaf S, Merkel BJ (2011) Sorption of uranium(VI) at the clay mineral-water interface. Environmental Earth Sciences 63:925-934

Bachmaf S, Planer-Friedrich B, Merkel BJ (2009) Competitive sorption and desorption of arsenate and uranium on bentonite and kaolinite. In: Podosek FA (ed) 19th Annual V.M Goldschmidt Conference. Elsevier, Davos, Switzerland, p A 67

Barnett MO, Jardine PM, Brooks SC, Selim HM (2000) Adsorption and transport of uranium(VI) in subsurface media. Soil Science Society of America Journal 64:908-917

Borovec Z (1981) The Adsorption of Uranyl Species by Fine Clay. Chemical Geology 32:45-58

Brendler V, Geipel G, Bernhard G, Nitsche H (1996) Complexation in the system $UO_2^{2+}/PO_4^{3-}/OH^-$(aq): Patentiometric and spectroscopic investigations at very low ionic strengths. Radiochimica Acta 74:75-80

Camacho LM, Deng SG, Parra RR (2010) Uranium removal from groundwater by natural clinoptilolite zeolite: Effects of pH and initial feed concentration. Journal of Hazardous Materials 175:393-398

Cheng T, Barnett MO, Roden EE, Zhuang JL (2004) Effects of phosphate on uranium(VI) adsorption to goethite-coated sand. Environmental Science & Technology 38:6059-6065

Cheng T, Barnett MO, Roden EE, Zhunag JL (2007) Reactive transport of uranium(VI) and phosphate in a goethite-coated sand column: An experimental study. Chemosphere 68:1218-1223

Darland JE, Inskeep WP (1997) Effects of pH and phosphate competition on the transport of arsenate. Journal of Environmental Quality 26 (4):1133-1139

Dutta PK, Ray AK, Sharma VK, Millero FJ (2004) Adsorption of arsenate and arsenite on titanium dioxide suspensions. Journal of Colloid and Interface Science 278:270-275

Gezahegne W, Hennig C, Tsushima S, Planer-Friedrich B, Merkel BJ (2012) EXAFS and DFT Investigations of Uranyl-Arsenate Complexes in Aqueous Solution. Environmental Science and Technology 46 (4): 2228–2233

Goldberg S, Johnston CT (2001) Mechanisms of arsenic adsorption on amorphous oxides evaluated using macroscopic measurements, vibrational spectroscopy, and surface complexation modeling. Journal of Colloid and Interface Science 234:204-216

Guo ZJ, Li Y, Wu WS (2009) Sorption of U(VI) on goethite: Effects of pH, ionic strength, phosphate, carbonate and fulvic acid. Applied Radiation and Isotopes 67:996-1000

Hsi CKD, Langmuir D (1985) Adsorption of Uranyl onto Ferric Oxyhydroxides - Application of the Surface Complexation Site-Binding Model. Geochimica Et Cosmochimica Acta 49:1931-1941

Kohler M, Wieland E, Leckie JO (1992) Metal-Ligand-Surface Interactions during Sorption of Uranyl and Neptunyl on Oxides and Silicates. In: Kharaka YK, Maest AS (eds) Water-Rock Interaction (VII). Balkema, Rotterdam, p 51–54

Krepelova A, Sachs S, Bernhard G (2006) Uranium(VI) sorption onto kaolinite in the presence and absence of humic acid. Radiochimica Acta 94:825-833

Lee H, Choi W (2002) Photocatalytic oxidation of arsenite in TiO_2 suspension: Kinetics and mechanisms. Environmental Science & Technology 36:3872-3878

Lieser KH, Quandtklenk S, Thybusch B (1992) Sorption of Uranyl Ions on Hydrous Silicon Dioxide. Radiochimica Acta 57:45-50

Liu XW, Byrne RH (1997) Rare earth and yttrium phosphate solubilities in aqueous solution. Geochimica Et Cosmochimica Acta 61:1625-1633

Merkel BJ, Hasche-Berger A (2006) Uranium in the Environment - Mining Impact and Consequences, Vol. Springer-Verlag Berlin Heidelberg

Merkel BJ, Planer-Friedrich, B., Wolkersdorfer, C. (2002) Uranium in the Aquatic Environment, Vol. Springer-Verlag, Berlin Heidelberg

Missana T, Garcia-Gutierrez M, Maffiotte C (2003) Experimental and modeling study of the uranium (VI) sorption on goethite. Journal of Colloid and Interface Science 260:291-301

Nair S, Merkel BJ (2011) Impact of Alkaline Earth Metals on Aqueous Speciation of U(VI) and Sorption on Quartz. Aquatic Geochemistry 17:209-219

Pabalan RT, Turner DR (1996) Uranium(6+) sorption on montmorillonite: Experimental and surface complexation modeling study. Aquatic Geochemistry 2:203-226

Payne TE, Davis JA, Waite TD (1996) Uranium Adsorption on Ferrihydrite - Effects of Phosphate and Humic Acid. Radiochimica Acta 74:239-243

Pena ME, Korfiatis GP, Patel M, Lippincott L, Meng XG (2005) Adsorption of As(V) and As(III) by nanocrystalline titanium dioxide. Water Research 39:2327-2337

Rutsch M, Geipel G, Brendler V, Bernhard G, Nitsche H (1999) Interaction of uranium(VI) with arsenate(V) in aqueous solution studied by time-resolved laser-induced fluorescence spectroscopy (TRLFS). Radiochimica Acta 86:135-141

Schulze R, Merkel BJ (2011) Sorption of uranium on iron coated sand in the presence of arsenate, selenate and phosphate. In: Merkel BJ, Schipek M (eds) The New Uranium Mining Boom; Challenge and Lessons Learned. Springer, Berlin Heidelberg, p 573-578

Schwertmann U, Cornell RM (2000) Iron Oxides in the Laboratory. Preparation and Characterization, Vol. WILEY-VCH, Weinheim Germany

Sherman DM, Peacock CL, Hubbard CG (2008) Surface complexation of U(VI) on goethite (alpha-FeOOH). Geochimica Et Cosmochimica Acta 72:298-310

Smith AH, Lingas EO, Rahman M (2000) Contamination of drinking-water by arsenic in Bangladesh: A public health emergency. Bulletin of the World Health Organization 78:1093-1103

Stollenwerk KG (2003) Geochemical Processes Controlling Transport of Arsenic in Groundwater:A Review of Adsorption. In: Welch AH, Stollenwerk KG (eds) Arsenic in Groundwater - Geochemistry and Occurrence. Kluwer Academic Publishers, Boston

Sylwester ER, Hudson EA, Allen PG (2000) The structure of uranium (VI) sorption complexes on silica, alumina, and montmorillonite. Geochimica Et Cosmochimica Acta 64:2431-2438

Xu H, Allard B, Grimvall A (1988) Influence of Ph and Organic-Substance on the Adsorption of As(V) on Geologic Materials. Water Air and Soil Pollution 40 (3-4):293-305

Zhang HX, Song SP, Tao ZY (2009) Effect of flow rate on the sorption breakthrough behaviors of uranium(VI), phosphate, and fulvic acid onto a silica column. Journal of Radioanalytical and Nuclear Chemistry 281:505-511

Characterization of phosphogypsum deposited in Schistos remediated waste site (Piraeus, Greece)

F. Papageorgiou[1], A. Godelitsas[1], S. Xanthos[2], N. Voulgaris[1], P. Nastos[1], T.J. Mertzimekis[1], A. Argyraki[1], G. Katsantonis[3]

[1]School of Science, University of Athens, Zografou Campus, GR-15784, Athens, Greece
[2]Department of Automation Engineering, Alexander Technological Educational Institute of Thessaloniki, GR-57400 Thessaloniki, Greece
[3]Environmental Association of Municipalities of Athens-Piraeus

Abstract. The operation of a phosphate fertilizer industry in Drapetsona, near Piraeus port (Greece), resulted in the deposition of 10 million tons of phosphogypsum(PG) into an old limestone quarry, in the period 1979-1989. The whole deposit has been recently remediated using geomembranes and thick soil cover with vegetation. The purpose of the present study was to characterize representative samples of that phosphogypsum, using diffraction (powder-XRD), microscopic (SEM-EDS), analytical (ICP-MS), and spectroscopic techniques (High-resolution γ-ray spectrometry and XRF). The material contains crystalline gypsum ($CaSO_4.2H_2O$) and Ca-Si-Al-S-F (chukhrovite-type/ meniaylovite) phases. The natural radioactivity is mainly due to the ^{238}U series and particularly ^{226}Ra (average: 462 Bq/kg), which is relatively low compared to PG from the rest of the world. Furthermore, leaching experiments using local (Attica) rainwater, together with ICP-MS, were performed to assess the potential release of elements in the environment.

Introduction

Phosphogypsum (PG) is generated during the manufacture of phosphoric acid by treating phosphate rock (phosphorite) with sulfuric acid, according to the following reaction (e.g. Fukuma et al., 2000; Haridasan et al.,2002; Abdel-Aal, 2004; Papastefanou et al., 2006; Perez-Lopez et al., 2007; Bituh et al., 2009; Silva et al., 2010; Al Hwaiti et al., 2010; Villalobos et al., 2010; Zielinsko et al., 2011; Chauhan et al., 2013; Roselli et al., 2009):

$$Ca_{10}(PO_4)_6F_2 + 10H_2SO_4 + 20H_2O \rightarrow 6H_3PO_4 + 10CaSO_4{\cdot}2H_2O + 2HF \quad (1)$$

B.J. Merkel and A. Arab (eds.), *Uranium – Past and Future Challenges*,
DOI 10.1007/978-3-319-11059-2_31

PG is a powdery material that has little or no plasticityand is mainly composed by dihydrated calcium sulfate ($CaSO_4.2H_2O$) and fluorosilicate (Na_2SiF_6) (Berish, 1990; Kacimi et al., 2006). It also contains impurities, such as H_3PO_4, $Ca(H_2PO_4)2{\cdot}H_2O$, $CaHPO_4{\cdot}2H_2O$ + $Ca_3(PO_4)_2$, residual acids, fluorides (NaF, Na_2SiF_6, Na_3AlF_6, Na_3FeF_6, CaF_2), sulphate ions, trace elements (Cd, Cr, Cu and Zn), organic matter as aliphatic compounds of carbonic acids, amines, and ketones adhered to the surface of gypsum crystals (Rutherford et al., 1996). Mineral phases remaining from phosphorite concern fluorapatite [$Ca_{10}F_2$ $(PO_4)_6{\cdot}CaCO_3$], goethite, quartz, minor amounts of Al–phosphates, anatase, magnetite, monazite and barite. The chemical composition of PG differs depending on the source of the main product. Major components are CaO, sulphates (expressed as SO_3), SiO_2, Al_2O_3, Fe_2O_3, P_2O_5, whereas trace metals include As, Ag, Cd, Ba, Cr, Pb, Hg and Se. In addition, PG has high content in Ag, Au, Cd, Sr, LREE and Y (e.g. Carbonell-Barnachina et al., 2002, Oliveira and Imbernon, 1998). The USEPA(1998) has classified phosphogypsum as a "*Technologically Enhanced Naturally Occurring Radioactive Material*" (TENORM). Up to 15% of PG is used to make building materials, as a soil amendment, and as a set controller in the manufacture of Portland cement, but its uses have been banned in most countries. (Tayibi et al., 2009). Phosphogypsum is a major environmental issue in Greece due to former and current fertilizer industry. In particular, in Schistos area (Drapetsona, near Piraeus port, Greece) 10 million tons of PG have been deposited into an old limestone quarry, in the period 1979-1989. The deposit has been recently remediated using geomembranes and thick soil cover with vegetation. The purpose of the present study was to characterize representative samples of Schistos PG using labpratory techniques. It should be no noted that no previous scientific research has been carried out, besides some short unpublished reports from the Greek Atomic Energy Commision/GAEC (private communication).

Materials and Methods

Sampling was conducted during September 2013, in collaboration with the Environmental Association of Athens-Piraeus (EAMAP), using GPS to record the exact location of each sampling site. A portion of each sample was hand-pulverized to be characterized by powder-XRD (Siemens/Bruker AXS D-5005, evaluation using Match! software). SEM-EDS (Jeol 5600 equipped with Oxford EDS) was performed on free surfaces of the particles constituting the material. Bulk analyses were performed by ICP-MS (Perkin Elmer Sciex Elan 9000 and $LiBO_2/LiB_4O_7$ fusion and HNO_3 digestion) techniques. Moreover, γ-ray spectrometry was performed using a 50% relative efficiency HPGe detector and samples packed in individual plastic beakers of about 260 mL of volume All samples as referred above were measured for about 6 hrs, at 20000–80000 counts per photopeak of interest and besides were measured for about 12 hrs each, at 80000 counts to establish the detector calibration. For each sample, a γ-ray spectrum was collected containing specific energy photo peaks, which correspond to specific radionuclides. In order to evaluate the solubility of Schistos PG, leaching experiments were carried out using local (Attica) rainwater. All the leachates were analyzed using ICP-MS.

Results and discussion

The powder-XRD patterns of the studied PG showed that the crystalline phases comprising the material are a gypsum phase ($CaSO_4{\cdot}2H_2O$), a meniaylovite (Ca4[$FSO_4SiF_6AlF_6$]$\cdot 12H_2O$)-type phase and clays (Fig.1).

The SEM-EDS investigation approved the XRD findings and indicated the existence of three different solid phases. The first one applies to the gypsum phase and it is characterized by the presence of Ca Kα and S Kα peaks in the EDS spectra. Most gypsum microcrystals exhibit a characteristic platy morphology while there are also twinned crystals (Fig.2). The second one is a more complex phase which combines Ca, S, Al, Si and F and corresponds to the aforementioned meniaylovite-type ($Ca_4[FSO_4SiF_6AlF_6]{\cdot}12H_2O$) phase related to chukhrovite (Coates & Woodard, 1966). The meniaylovite octahedral microcrystals are created during the manufacture of phosphoric acid from phosphorite attacked by sulfuric acid. Lehr et al. (1966) also reported on phases precipitated from a wet process phopshoric acid preparation, which includes a material with the chemical formula $Ca_4AlSiSO_4F_{12}{\cdot}12H_2O$. The microcrystals exhibit a distinct octahedral morphology as shown in Fig.2. Moreover, there is also a third phase referring to Fe-oxides, according to Fe Kα peak in the EDS spectra.

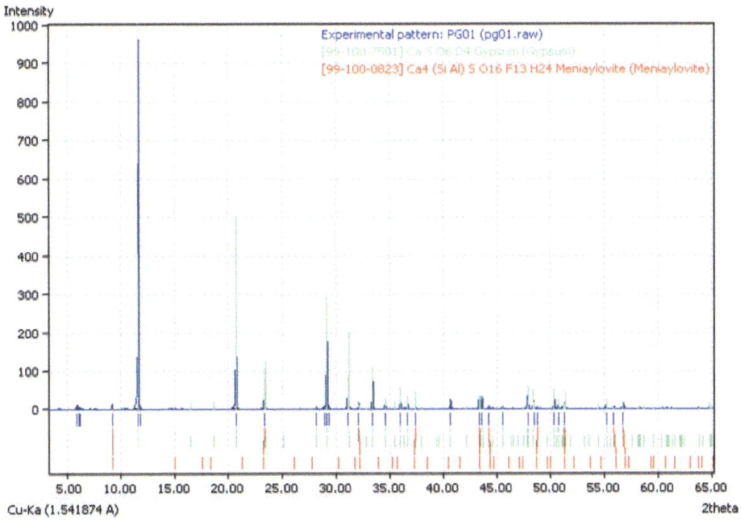

Fig.1. Powder-XRD pattern of the studied PG.

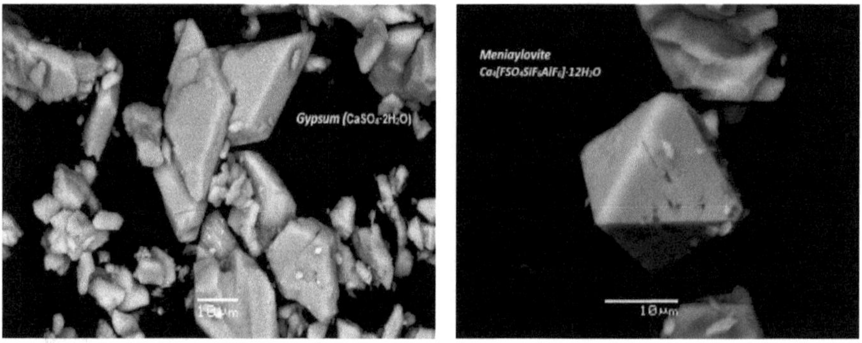

Fig.2. SEM-EDS data concerning solid phases in PG.

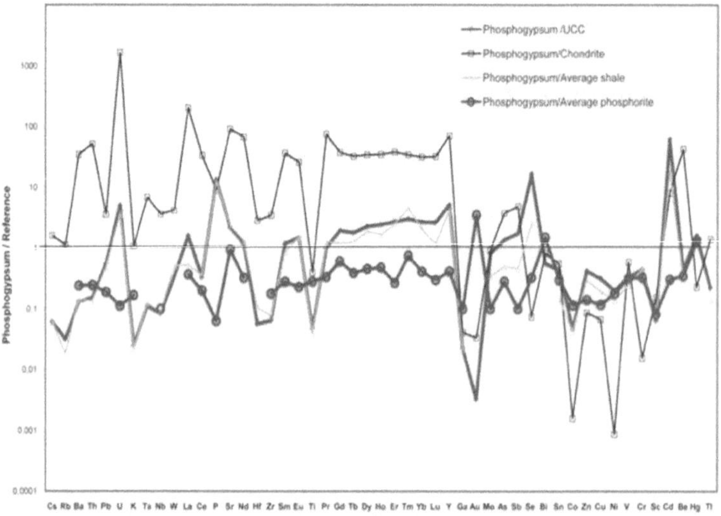

Fig.3. The relative chemical composition of the studied PG.

The results obtained from the ICP-MS analyses of PG, were on chondrite, upper continental crust/UCC, average shale and average phosphorite, representing the global geological background and the parent material (Rudnick and Gao 2003; Li and Schoonmaker, 2003). The relevant patterns are shown in Fig. 3. It is clear that Schistos PG is enriched in U, REE, Cd and Se, compared to UCC, while, based on average phosphorite, it is generally depleted for the same elements.

The HR γ-ray measurements indicated natural radioactivity mainly due to the ^{238}U series and particularly ^{226}Ra. The mean radioactivity concentration of ^{226}Ra

(average: 462 Bq/kg) was plotted together with the global stats for immediate comparison (Fig. 4; Fukuma et al., 2000; Beretka, 1990; Al-Attar et al. 2011; Okeji et al., 2011; El-Didamony et al., 2013; Bituh et al., 2009; Roessler et al., 1979; Papastefanou et al., 2006; Luther et al., 1993; Fourati and Faludi, 1988; Maiche and Scott et al., 1991; Azouazi et al., 2001; Kobal et al., 1990; Villalobos et al., 2010). It is evident that the PG deposited in Schistos waste site has typical values of radioactivity concentrations, rather low compared to the global statistics. Concerning the geological background of the Schistos waste site, represented by Mesozoic limestones, it is known that there are negligible amounts of U and Th (Godelitsas, private communication), simply meaning there is little or no interference in the measured radioactivity of the deposited phosphogypsum.

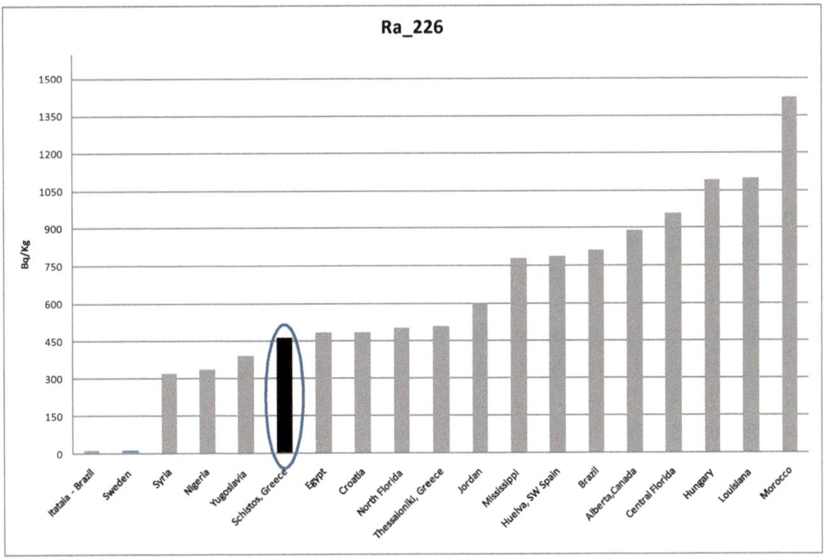

Fig.4. Average radioactivity values due to ^{226}Ra in the studied PG samples compared to the rest of the globe.

According to the leaching experiments, as shown in Fig. 5, it is obvious that there is a significant dissolution of PG in Attica rainwater, corresponding to severe Ca and S release. At the same time, a significant release of Sr, Na and K, exists. However, these elements play no critical role in environmental issues.

Element	Ca	K	Mg	Na	S	Sr	Th	U
Unit	ppm	ppm	ppm	ppm	ppm	ppb	ppb	ppb
Rainwater (diss. elements)	0.67	0.09	<0.05	0.35	<1	0.93	<0.05	<0.02
Rainwater (total elements)	2.26	0.35	0.12	1.18	<1	3.50	0.22	0.05
PG leachate	627.70	2.00	5.00	61.00	541	1838.00	<500	<3000

Fig.5. SEM-EDS and analytical (ICP-MS) data for selected elements with respect to leaching experiments using local (Attica) rainwater.

Finally, it is essential to comment on Sr^{2+} and Ra^{2+} in gypsum. The presence of Sr in PG is reasonable, due to the valence and ionic radius similarity of Sr^{2+} and Ca^{2+} in eight-fold coordination into gypsum crystal structure ($^{[8]}Sr^{2+}$ = 1.26 Å and $^{[8]}Ca^{2+}$ = 1.12 Å). Regarding actinides, related to the observed radioactivity, it has been concluded that Th release is negligible (also accordingly to its very low concentration in the studied PG: 1.6 ppm) but the release of U, though rather low (i.e., 3 ppm in solution, compared to 13 ppm in the solid), needs further research with special ICP-MS and/or α-spectrometry. Nevertheless, U, and particularly ^{238}U and its decay products, may be released in the aquatic environment and follow bioaccumulation processes.

References

Al - Attar, L., Al - Oudat, M., Kauakri, S., Budeir, Y., Khalify, H. (2011) Radiological Impacts of phosphogypsum. Journal of Environmental Management 92: 2151-2158.

Al-Hwaiti, M.S., Ranville, J.F., Ross, P.E. (2010) Bioavailability and mobility of trace metals in phosphogypsum from Aqaba and Eshidiya, Jordan.Chemie der Erde 70: 283-291.

Azouazi, M., Ouahidi, Y., Fakhi, S., Andres, Y., Abbe, J.Ch., Benmansour, M. (2001) Natural radioactivity in phosphates, phosphogypsum and natural waters in Morocco. Journal of Environmental Radioactivity 54: 231-242.

Beretka, J. (1990) The current state of utilization of phosphogypsum in Australia. In: Proceedings of the Third InternationalSymposium on Phosphogypsum, Orlando, FL, FIPR Pub. No. 01-060-083, December 1990, vol. II 394-401

Berish, C.W. (1990) Potential Environmental hazards of phosphogypsum storage in central Florida. Proceedings of the third international symposium on phosphogypsum. Orlando, FL, FIPR Pub. No.01060083;2: 1-29.

Bituh, T., Marovic, G., Franic, Z., Sencar, J., Bronzovic, M. (2009) Radioactive contamination in Croatia by phosphate fertilizer production. Journal of Hazardous Materials 162: 1199-1203.

Carbonell – Barrachina, A.,DeLaune, R.D., Jugsujinda, A. (2002)Phosphogypsum chemistry under highly anoxic conditions. Waste Management 22 (6): 657-665.

Chauhan, P., Chauhan, R.P., Gupta, M. (2013) Estimation of naturally occuring radionuclides in fertilizers using gamma spectrometry and elemental analysis by XRF and XRD techniques.Microchemical Journal 106: 73-78.

Coates, Woodward (1966) Similarity between "chukrovite" and the octahedral crystals found in the gypsum in the manufacture of phosphoric acid. Nature: 212-392

El-Didamony, H., Gado, H.S., Awwad, N.S., Fawzy, M.M., Attallah, M.F. (2013) Treatment of phosphogypsum waste produced from phosphate ore processing. Journal of Hazardous Materials: 244-245 596-602.

EPA, 1998. Code of Federal Regulations, 1998. Title 40, Vol. 7, Parts 61.202 and 61.204

(40CFR61.202 and 40CFR61.204).

Fourati, A., Faludi, G. (1988) Changes in radioactivity of phosphate rocks during the process of production. Journal of Radioanalytical and Nuclear Chemistry 125: 287-293.

Fukuma, H.T., Fernandes, E.A.N, Quinelato, A.L. (2000) Distribution of naturalradionuclides during the processing of phosphate rock from Itataia - Brazil for production of phosphoric acid and uranium concentrate. Radiochim.Acta 88: 809-812.

Haridasan, P.P., Maniyan, C.G., Pillai, P.M.B., Khan, A.H. (2002)Dissolution characteristics of 226Ra from phosphogypsum. Journal of Environmental Radioactivity 62: 287-294.Kacimi, L., Simon – Masseron, A., Ghomari, A., Derriche, Z. (2006) Reduction of clinkerization temperature by using phosphogypsum. Journal of Hazardous Materials 137 (1): 129- 137.

Kobal, I., Brajnik, D., Kaluza, F., Vengust, M. (1990) Radionuclides in effluents from coal mines, a coal-fired powerplant, and a phosphate processing plant in Zasanje, Slovenia (Yugoslavia). Health Physics 58: 8-85.

Laiche, T.P., Scott, M.L. (1991) A radiological evaluation of phosphogypsum. Health Physics 60: 691-693.

Li Y.H and Schoonmaker J. (2003) in: F.T. Mackenzie (Ed.)H.D. Holland, K.K. Turekian (Eds.), Sediments, Daigenesis, and Sedimentary Rocks, Treatise on Geochemistry, vol. 7, Elsevier-Pergamon, Oxford (2003) 1–35.

Luther, S.M., Dudas, M.J., Rutherford, P.M. (1993) Radioactivity and chemical characteristics of Alberta phosphogypsum. Water, Air, and Soil Pollution 69: 277-290.

Mullins, G.L., Mitchell Jr., C.C. (1990) Use of phosphogypsum to increase yield and quality of annual forages. FIPR Pub. No. 01-048-084, Auburn University, 56.

Okeji, M.C., Agwu, K.K., Idigo, F.U. (2012) Assessment of natural radioactivity in phosphate ore, phosphogypsum, and soil samples around a phosphate fertilizer plant in Nigeria. Bull Environ Contam Toxico 89:1078–1081

Oliveira, S.M.B., Imbernon, R.A.(1998) Weathering alteration and related REE concentration in the Catalao I carbonatite complex, central Brazil. J S Am Ear Sci 11 (4): 379-388.

Perez – Lopez, R., Alvarez – Valero, A., Nieto, J.M. (2007)Changes in mobility of toxic elements during the production of phosphoric acid in the fertilizer industry of Huelva (SW Spain) and environmental impact of phosphogypsum wastes. J Haz Mat 148: 745-750.

Papastefanou, C., Stoulas, S., Ioannidou, A., Manolopoulou, M.(2006) The application of phosphogypsum in agriculture and the radiological impact. J Env Rad 89: 188-198.

Roessler, C.E., Smith, Z.A., Bolch, W.E., Prince, R.J. (1979) Uranium and radium-226 in Florida phosphate materials. Health Physics: 37 269-277

Roselli, C.,Desideri, D., AssuntaMeli, M. (2009) Radiological characterization of phosphate fertilizers:Comparison between alpha and gamma spectrometry. Microchem J91: 181-186.

Rudnick, L.R., Gao,S., (2003) The composition of the continental crust ,in: R.L. Rudnick (Ed.)H.D. Holland, K.K. Turekian (Eds.), The Crust, Treatise on Geochemistry, vol. 3, Elsevier-Pergamon, Oxford 1–64

Rutheford, P.M., Dudas, M.J., Arocena, J.M. (1996) Heterogeneous distribution of radionuclides, barium and strontium in phosphogypsum by-product. Science of the Total Environment 180: 201-209.

Shannon, D.R. (1976) Revised Effective Ionic Radii and Systematic Studies of Interatomic Distances in Halides and Chalcogenides.ActaCryst. A32:. 751-767.

Silva, L.F.O., Hower, J. C., Izquierdo, M., Querol, X. (2010) Complex nanominerals and ultrafine particles assemblages in phosphogypsum of the fertilizer industry and implications on human exposure. Science of the Total Environment 408, pp. 5117 - 5122.

TayibiHanan, Chouva Mohamed, Lopez A. Felix, Alguacil J. Francisco, Lopez - Gelgado Aurora(2009) Environmental impact and management of Phosphogypsum. Journal of Environmental Management 90: 2377-2386.

Villalobos, M. R., Vioque, I., Mantero, J., Manjon, G. (2010) Radiological, chemical and morphological characterizations of phosphate rock and phosphogypsum from phosphoric acid factories in SW Spain. J Haz Mat 181: 193-203.
Zielinski, R.A., Al-Hwaiti, M.S (2011) Radionuclides, trace elements, and radium residenceinphosphogypsum of Jordan.EnviGeochem Health 33: 149-165.

I stopped drinking tap water completely. Only bottled water.
Why is that?
You scientist should know!
There's uranium in our tap water!

Rare earth ore refining in Kuantan/Malaysia – the next legacy ahead?

Gerhard Schmidt

[1]Öko-Institut e.V., Rheinstraße 95, D-64295 Darmstadt/Germany

Abstract. The company Lynas mills and refines the concentrated rare earth bearing ore from its Mt.Weld/Australia mine in its LAMP facility in Kuantan/Malaysia. The ore concentrate has high concentrations of thorium and small concentrations of uranium as by-products. Oeko-Institute was commissioned with an independent evaluation of the environmental aspects of the plant. Several aspects show serious deficiencies in applying emission standards. The largest radiological problem is that the operator plans to re-use its tailings as road-building material. Large collective doses would result from this practice.

Introduction

Rare Earth Elements (REE) are currently mined and processed mainly in China, which also delivers large shares of those special materials such as neodymium to the world market. Production conditions, gaseous and liquid discharges, waste management practices and radiation protection there are known as seriously deficient (Qifan 2010, Hua 2011)[5]. To diversify the sources, to avoid monopole structures and to improve the environmental records of the supply chain for REEs, alternative producers have come up that claim to implement a "cleaner production". This article analyses such an example in detail and concludes that the goal is still to be reached.

---

[5] The Chinese Nuclear Safety Administration (Hua 2011) concludes: „1 NORM radiation is the major additional dose exposure both to the public and occupational workers. NORM radiation has become an urgent problem. 2 The regulatory body should draft a list of non-uranium mine for regulatory control, and make related regulations and rules as soon as possible. […]"

B.J. Merkel and A. Arab (eds.), *Uranium – Past and Future Challenges*,
DOI 10.1007/978-3-319-11059-2_32

Production scheme of LAMP

The mining and initial part of the production scheme at LAMP in Kuantan is shown in Fig. 1. The initial REE concentration of between 5 and 10% and up to 17% is increased to roughly 35% in the beneficiation stage, then the ore concentrate is shipped to Malaysia.

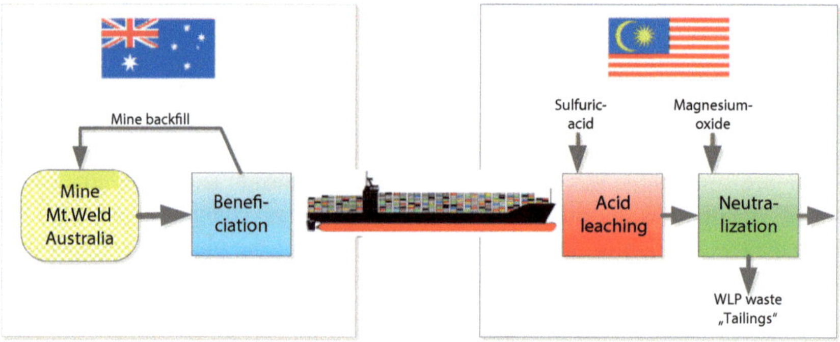

Fig. 1: Mining process and leaching process in the LAMP facility

There the concentrate is treated with boiling sulfuric acid at roughly 400°C to dissolve the REEs. The relation between ore concentrate (after beneficiation) and consumed chemicals and energy in this part is shown in Table 1.

Table 1. Consumed mass chemicals and natural gas per ton of ore concentrate in the cracking stage at LAMP

Material	used for ...	Amount t/t
Sulfuric acid	dissolving REE by cracking carbonate/phosphate	1.83
Natural gas	heating the rotary kiln to boiling point of acid	0.66 [a]
Magnesium oxide	neutralizing excess sulfuric acid	0.27

[a] in Nm^3 per ton

Chemicals added increase the material to more than double. All un-dissolvable parts of the concentrate (esp. iron phosphate) as well as all thorium and uranium in the concentrate leave the neutralization process as filtrated solids. This waste stream is called "water leach process" waste or WLP and pumped as slurry to an interim storage facility nearby called "residue storage facility" (WLP-RSF, see Picture 1).

The rest of the processes at LAMP is pure chemical technology: acidification, extraction, neutralization, precipitation, purification and product conditioning. Products and amounts are:

10,417 t/a Lanthanum-Cerium-Praseodymium-Neodymium (LCPN) carbonate, 4,919 t/a Cerium carbonate, 2,222 t/a Lanthanum-Cerium (LC) carbonate, 1,369 t/a Lanthanum oxide, 1,354 t/a Neodymium-Praeodymium („Didymium", Nd/Pr) oxide, 1,067 t/a Samarium-Europium-Gadolinium (SEG) carbonate, 113 t/a Heavy rare earth elements (HRE) carbonate (RanhillWorley 2008).

Picture.1. Distant view on the tailings impoundment for WLP wastes

In total, it is planned to import 1.27 million tons of ore concentrate, leading to roughly the same amount of WLP tailings.

Non-radiological environmental impacts of LAMP

According to our task we were to check if the applied technical, safety and radiological standards in the Lynas facility in Malaysia are meeting those that would be applicable in Europe and worldwide.

Emissions to the air

Two main emission sources are to be considered: the emission of acidic gases from the cracking stage and dust emitted from the furnace oven for the calcination of oxalate to oxide. The cracking stage emits mainly sulfuric acid, sulfur dioxide and fluoric acid. The off-gas is filtered through a scrubber system[6]. Emission reduction was compared to standard and improved filter systems in European sulfuric acid production plants[7]. The off-gas system, as fully functional, emits roughly 5 kg sulfuric acid per hour (Environ 2007) and so is a relevant acid point source in the local area. Lynas's emissions per unit sulfuric acid consumed are by a factor of up to 10 less effective than those at sulfuric acid production plants in Europe[8].

Even worse is the situation at the furnace oven. Not even the most primitive filter system against dust emissions from that oven has been chosen[9]. Not even an

6 See (Environ 2007), p. 5-73

7 See (EC 2007)

8 For a detailed comparison see (Schmidt 2013), p. 29 ff

9 (Environ 2007), l.c., p. 5-72

estimate of the discharged substances has been performed (Carbon monoxide, oxalic acid, PM10, rare earth oxides).

Wastewater discharge

Wastewater is produced in the facility mainly for washing and extraction and neutralization purposes. The wastewater is neutralized and chemically treated. The concentrations of several toxic constituents are monitored before the water is batchwise transferred to the surface water retention pond. The water is then discharged into a small canal (see Picture 2) that flows later into River Balok and to the sea.

Picture.2. Lynas's discharge point to public water channel

The species for which wastewater is tested was oriented on the Malaysian water guideline[10]. As this guideline was not made for rare earth production plants, the wastewater is probed for lead and arsenic, but not for a single REE. While the toxicity of lead is well known, that of neodymium is widely unknown, so why not control that and define restrictive discharge levels?

As had been described above the plant works with large masses of acids and neutralizes those liquids. This leads to a high salt content of the wastewater (up to 2.7% chloride in the discharge channel[11], but salt is not even listed as emitted species and the salt balance for the operation had to be calculated in an extra step.

---

[10] (Environ 2007), l.c., p. 5-18 f and Exhibit 5.2.4

[11] For a calculation of the discharged salt see (Schmidt 2013, p. 34 ff

Radiological environmental impacts of LAMP

The following radiological impacts have to be considered:

- public exposures through radon emissions over the stack in the cracking stage,
- occupational exposures by radium sulfate scales enrichment
- public exposure through the re-use of tailings waste

Radon emissions

Public exposures by emitted radon were assessed in (Bangi Ray 2012). The assessment has been confirmed in (Schmidt 2013) as accurate. The doses are small because the radon-220 generated in the thorium-232 decay chain is very short-lived and decays not to longer-lived radionuclides.

Scales enrichment

From phosphate production plants that treat apatite with sulfuric acid and filter generated gypsum it is well known that radium sulfate enriches in certain areas of the filtration stage and give rise to high gamma radiation doses. (AEC 1998) reports on dose rates in the phosphate industry of 1 to 10 mSv/h, resulting in restrictive measures to limit occupational exposures in certain work areas. As LAMP uses the same chemical process under the same conditions like the phosphate industry and with even higher concentrations of thorium in its ore concentrate scales enrichment is highly likely to at least the same extent. A realistic estimate in (Schmidt 2013) yields similar doses to be expected at LAMP.

Neither the EIA (Environ 2007) nor the RIA (Bangi Ray 2012) identify scales enrichment as a potential source for high occupational doses nor are consequences for radiological procedures drawn (monitoring of work places, change of filter clothes, repair procedures, waste storage and disposal procedures). Lynas simply denies learning from those experiences in the phosphate industry.

Waste issues at LAMP

The concentration of thorium in the solid phase of the WLP tailings waste is 5.9 Bq/g (Bangi Ray 2012), slightly above the amount of thorium in the ore concentrate. Uranium concentrations are by a factor of 20 smaller, at 0.37 Bq/g.

Typical option for materials with those concentrations is their long-term stable isolation in a permanent disposal facility. As the majority of WLP waste components, iron phosphate and silicon dioxide, are insoluble and geochemically stable, and as only thin layers of cover materials are required to reduce the direct radiation dose rates the requirements for such a disposal facility are not too sophisticated: stable site conditions (flooding, small subsidence, etc.), an engineered protective layer with long-term stability and a layout against average and extreme local weather conditions - and to identify a community willing to host such a facility on their grounds for virtually unlimited future times.

As the capacity of the WLP-RSF on the LAMP site has been designed to hold only the tailings waste produced in the first five years, the site selection, permit procedure, construction and to gain the acceptance for such a facility has only a short period.

The WLP-RSF at the Kuantan site would be unsuitable for that purpose because it is highly prone to flooding already under average weather conditions in Kuantan. Picture 3, taken December 2013, shows a nearly completely water filled RSF during a longer storm period – with a high risk of overtopping of the tailings impoundment and a catastrophic failure of the storage facility.

Picture.3. Flooding of the RSF during the monsoon season in December 2013 (courtesy SMSL)

One of the potential waste management options discussed is to send back the WLP tailings back to Australia, where it originated from, and to dispose them.

Unfortunately those cross-country shipments of such wastes require consent of the receiving country, so that option has not been followed any more.

One of the provisions in the Temporary Operating License (TOL) for LAMP is to select a site for a permanent disposal facility (PDF)[12]. Lynas has done so, but denies to name the selected site publicly – an inappropriate method to gain wide acceptance for such a sensitive facility.

In the meantime Lynas, with support of the Malaysian government, is following another path for managing its wastes: re-use. As the direct re-use of the WLP tailings is impossible due to its high direct radiation dose rates (1 to 3 μSv/h[13]), it is planned to dilute the material 1+6 and to use this mix as material for road construction.

As even the dilution leads to elevated dose rates in the order of doubling usual background radiation, three scenarios were calculated. The first scenario is a typical city street, mainly accessed by pedestrians. The second assumes a typical highway situation. In the third scenario the loss of control over 10% of the produced mix material is assumed, that is misused for house and office construction.

Table 2 lists the results of the calculation. Calculated was the committed collective effective dose of such a use. Individual doses are inappropriate for this purpose because of the large amount of material released (8.4 million tons mix 1+6) and of the short individual use, so that individual doses would not account for the total damages caused by this practice.

Table 2. Dose calculation for different waste re-use scenarios (WLP mix 1:6)

Scenario	Assumptions, conditions, dose rates calculated	Collective Effective Dose (man·Sv per year)
City street	Typical user profile for an inner city street, pedestrians, 1,186 km roads constructed from 1.2 million tons WLP, dose rate 65.41 nSv/h	221
Freeway	Typical user profile of a highly frequented freeway, 1,186 km roads, dose rate 64.7 nSv/h	99
Loss of control	10% loss, 111,000 cases, 50% housing, 50% office construction, dose rate in 1 m distance 357.6 nSv/h	318

It can be concluded that the committed collective doses and the accepted damages from such a release practice are extremely large. Even if less frequented roads or only 1% loss-of-control cases would be assumed, the doses are still in a region that are inacceptable.

[12] „III. The plan and location of the PDF have to be submitted and approved in a period not later than 10 (ten) months from the issuance of the Temporary Operating License;". The TOL has been issued 01.02.2012, no information is accessible on the named site approval.

[13] For dose rates of the materials see (Schmidt 2013), p. 73 ff

Conclusions

Environmental and waste issues in the Lynas case are to a high degree unsustainable and are in many aspects not current state-of-the-art. Open environmental issues require urgent action by the Malaysian competent authorities. Especially the still unresolved waste issue, even though the production of those wastes has already been practiced over the past two years, lets expect that we are currently witnessing the creation of a new legacy case of future post-operational clean-up.

Acknowledgements

This work was supported by SMSL, an NGO based in Kuantan/Malaysia.

References

AEC (1998) Applied Environmental Consulting (AEC): Evaluation of Exposure to Technologically Enhanced Naturally Occurring Radioactive Materials (TENORM) in the Phosphate Industry. – Bartow/Florida, July 1998

Bangi Ray (2011) Bangi Ray Services Sdn. Bhd.: Radiological Impact Assessment (RIA) of Lynas Advanced Materials Plant. – Rev. 2, December 2011

EC (2007) European Commission (EC): Reference Document on Best Available Techniques for the Manufacture of Large Volume Inorganic Chemicals - Ammonia, Acids and Fertilisers. – Brussels, August 2007

Environ Consultants (2007) Environ Consulting: Preliminary Environmental Impact Assessment and Quantitative Risk Assessment - Proposed Advanced Materials Plant Gebeng Industrial Estate, Kuantan, Pahang, Malaysia. – Selangor/Malaysia, January 2008

Hua, Liu (2011) The Situation of NORM in Non-Uranium Mining in China, Presentation of the Dept. of Nuclear Safety Management, Ministry of Environment Protection, China, (National Nuclear Safety Administration)

Qifan, Wu (2010) Enhanced Natural Radiation Exposure in China. – Presentation at the Third Symposium of Natural Radiation Exposure and Control, Aug 30- Sep 3,2010, Baotou, Inner Mongolia, China, hosted by Nuclear and Radiation Safety Center of MEP in co-operation with China Institute of Atomic Energy

RanhillWorley (2008) Lynas Advanced Materials Plant Expansion Philosophy. Prepared by Ranhill WorleyParsons Sdn Bhd, document No: 4219298-000-GE-DG-001-0F, Kuala Lumpur/Malaysia, 3 March 2008

RIA (2010) Radiological Impact Assessment of Advanced Materials Plant Gebeng Industrial Estate, Kuantan, Pahang. Prepared by Nuclear Malaysia, Kuala Lumpur/Malaysia, June 2010

Schmidt, Gerhard (2013): Gerhard Schmidt: Description and critical environmental evaluation of the REE refining plant LAMP near Kuantan/Malaysia. – Darmstadt/Germany, 25.01.2013

Uranium sorption onto the granites of Nizhnekansk massif

Anna Shiriaeva[1]

[1]Lomonosov Moscow State University, Department of Chemistry, Russian Federation, Moscow, Leninskie Gory, 119991

Abstract. Study of sorption of uranium onto mineral surfaces plays a key role in controlling its mobility in groundwater. This work focuses on sorption properties of granites of Nizhnekansk massif in order to find optimal conditions for restriction of uranium migration in the environment. Four different types of granites from Itatsky and Kamenny sites were involved in sorption experiments. Distribution coefficients were determined in different conditions and sorption kinetics was analyzed. The results of these experiments are considered to have practical application in searching for the optimal place for efficient geochemical isolation of spent nuclear fuel and high-level radioactive waste.

Introduction

Nizhnekansk granitoid massif is one of the most promising objects for construction of the storage facility for spent nuclear fuel. According to geological and geophysical research, Itatsky and Kamenny sites show the best compliance with the standarts of geological isolation of waste radioactive products. Itatsky and Kamenny sites are located 25 and 30 kilometers away from Mining-Chemical Complex in Zheleznogorsk and the RT-2 plant. Itatsky and Kamenny sites have an area of 15-20 km^2 each and are composed of granitoids from several intrusive phases. The predominant varieties, biotite granites and granodiorites, are generally viewed as providing a favorable environment for the isolation of radioactive wastes due to various attributes, such as their high hardness, low water content, poor permeability, and low porosity (Anderson et al. 2005).

B.J. Merkel and A. Arab (eds.), *Uranium – Past and Future Challenges*,
DOI 10.1007/978-3-319-11059-2_33

Sorption experiments

Samples of granitic material were obtained from the depth of 408,3 m and 501,1 m at Itatsky site (hereafter denoted by I-408,3 and I-501,1), and from the depth of 524,3 m and 689,4 m at Kamenny site (hereafter denoted by K-689,4 and K-524,3). Fine-grained granites with particle size not exceeding 1 mm were used in the experiment. The composition of the samples is presented in Table 1.

Table 1. Composition of the granite samples

I-408,3	I-501,1	K-689,4	K-524,3
plagioclase, 55%	plagioclase, 55%	hornblende, chlorite, kaolinite, 55%	plagioclase, 60%
hornblende, 30%	hornblende, 15%	potash feldspar, al-bite, 40 %	quartz, 15%
quartz, 5%	quartz, 25%	apatite, grothite, 5%	biotite, chlorite, 10%
potash feldspar, 5%	potash feldspar, 5%		potash feldspar, 5%
grothite, apatite, zir-con, 5%			

The experiments were carried out in aerobic conditions at room temperature (25°C). 2 g of each type of granite were held in contact with liquid phase containing $^{233}UO_2^{2+}$ (volume of the liquid phase was 8 ml). pH was kept at constant levels (3, 5, 7, 9) by adding solutions of HNO_3 and NaOH. Samples of the liquid phase were taken after 1, 24, 48, 144, 360, 1008 hours after adding the radioactive solution. Activity was determined by liquid scintillation spectroscopy in the alpha-ray detecting mode.

Sorption kinetics

Sorption curves are given on fig. 1-5. The results of the experiments indicate that at pH value of 3, first-order kinetics is observed predominately (Fig. 2). Efficient sorption rate constants are given in Table 2.

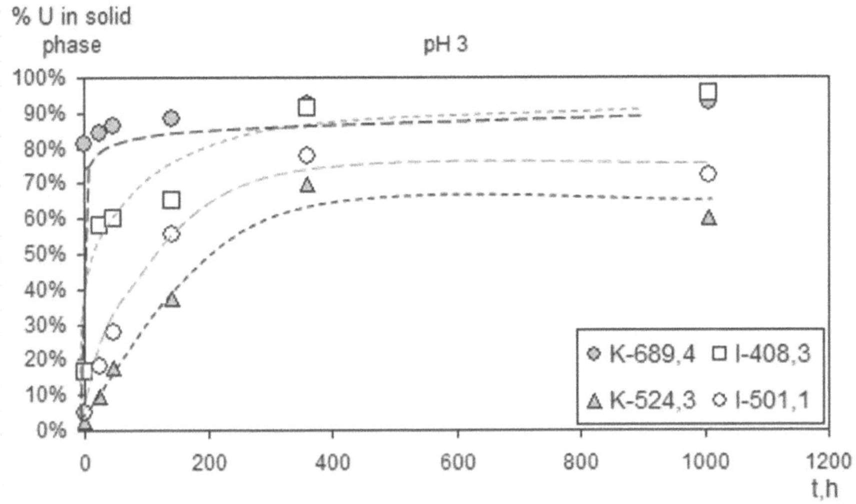

Fig.1. Dependence of the relative sorption of ^{233}U on time at pH=3.

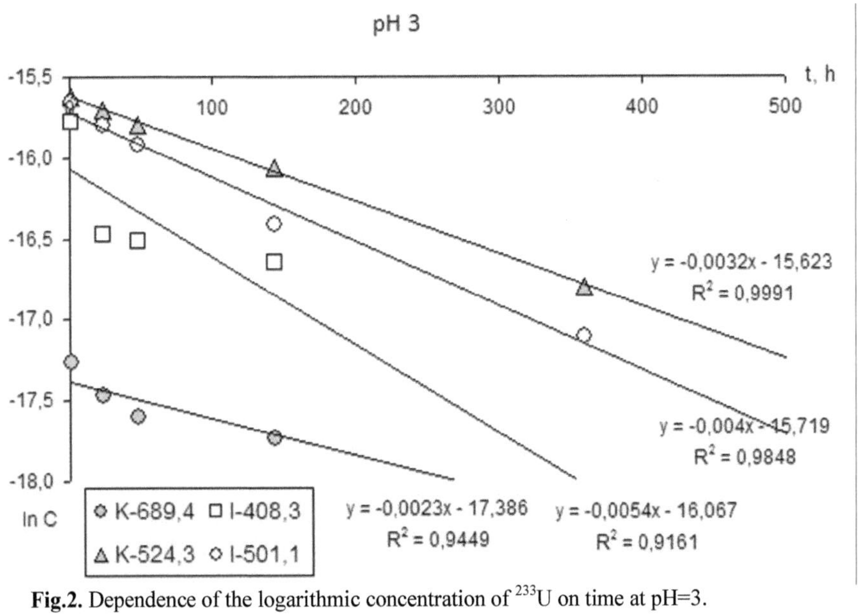

Fig.2. Dependence of the logarithmic concentration of ^{233}U on time at pH=3.

At pH value of 5, sorption on the samples I-501,1, I-408,3, and K-524,3 still corresponds with first-order kinetics (Fig. 3,4). However, sorption on the sample K-689,4 significantly declines from the first-order kinetic model. This behavior can be explained by presence of competitive sorption centers on the mineral surface.

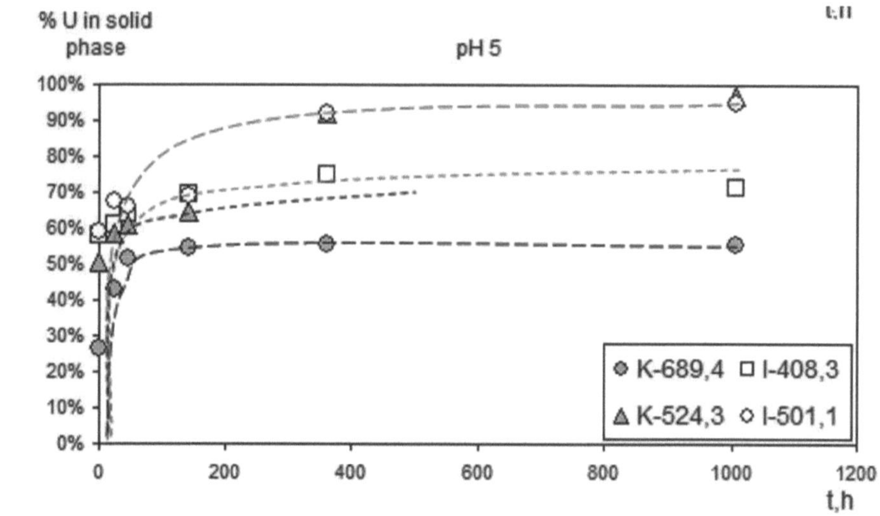

Fig.3. Dependence of relative sorption of ^{233}U on time at pH=5.

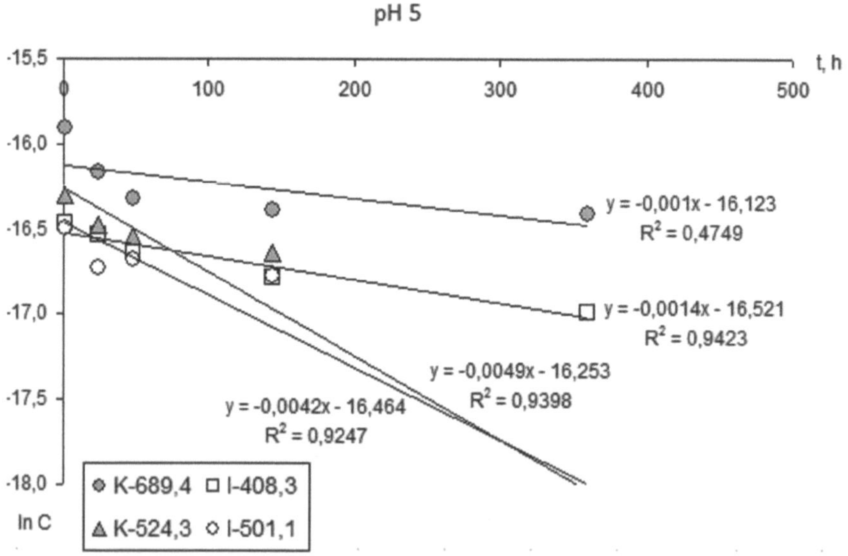

Fig.4. Dependence of the logarithmic concentration of ^{233}U on time at pH=5.

At pH value of 7, sorption on the samples I-501,1, and K-524,3 corresponds with first-order kinetic model (Fig.3,4). Sorption on the sample I-408,3, declines from the first-order model and sorption on K-689,4 does not fit any kinetic model of an integer order.

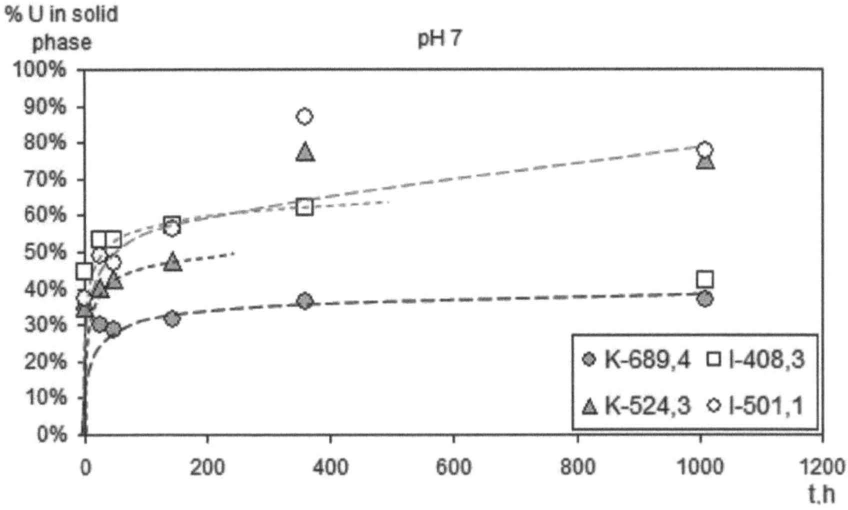

Fig.5. Dependence of relative sorption of ^{233}U on time at pH=7.

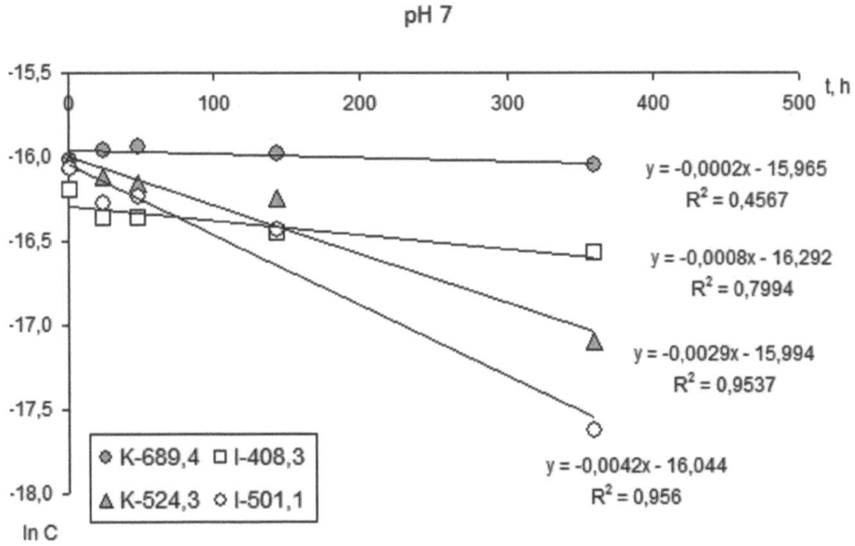

Fig.6. Dependence of the logarithmic concentration of ^{233}U on time at pH=7.

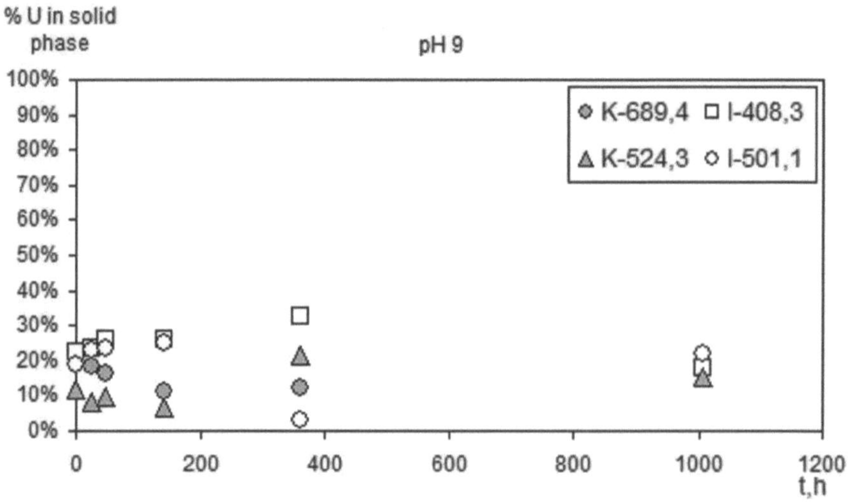

Fig.7. Dependence of relative sorption of ^{233}U at pH=9 on time.

At pH value of 9, distribution coefficients are the lowest, and sorption of UO_2^{2+} does not exceed 30%. As expected, sorption is inefficient due to the formation of complexes with carbonates and hydrocarbonates.

As follows from the efficient sorption rate constants (Table 2), sorption rate tends to decrease when passing from pH=3 to pH=5 and remains constant when passing from pH=5 to pH=7.

Table 2. Efficient sorption rate constants, min^{-1}

Sample \ pH	3	5	7
K-689,4	0,138	0,06	0,06
K-524,3	0,24	0,048	0,048
I-408,3	0,192	0,252	0,252
I-501,1	0,324	0,174	0,174

Distribution coefficients of UO_2^{2+} between solution and solid phase at the different levels of pH are given in Table 3.

Table 3. Distribution coefficients K_d^{eff} of ^{233}U in different conditions between solution and solid phase, cm^3/g

Sample \ pH	3	5	7	9
K-689,4	51,8	5,0	2,3	0,9
K-524,3	6,0	120,9	12,1	0,72
I-408,3	84,4	9,9	2,9	0,9
I-501,1	10,5	75,2	14,1	1,1

The experimental results indicate that in the considered range of pH, sorption of UO_2^{2+} on the samples I-408,3 and K-689,4 tends to decrease within the increasing of pH. Sorption on the samples K-524,3 and I-501,1 has a maximum at pH=5. Granite sample from Kamenny site (K-524,3) shows the highest sorption at pH=5. The lowest distribution coefficients are observed for sorption at pH= 9, as UO_2^{2+} tends to form soluble complex compounds at high levels of pH. Matching sorption efficiency to the composition of granites indicated that sorption onto granites with high presence of quartz is more efficient and less pH-dependent. This behavior can be explained by the possibility of uranium to form a large variety of complexes (Fig.8) with silica surface (Batuk D. et al.2011).

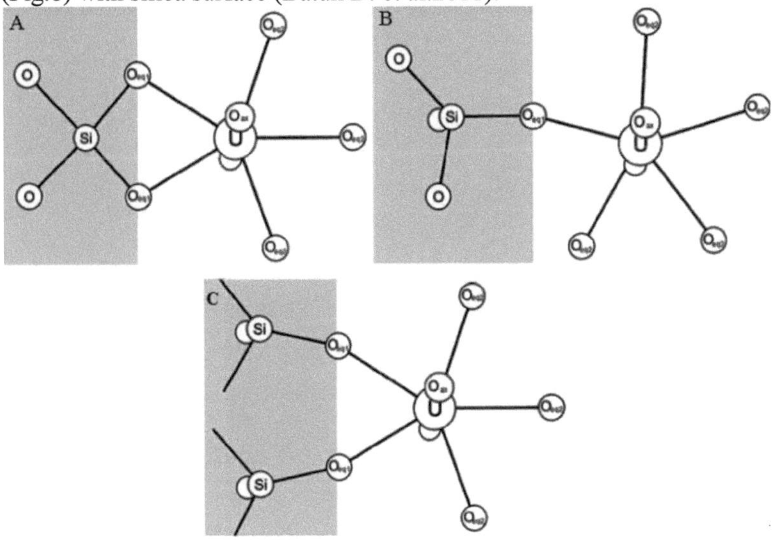

Fig.8. (a) Mononuclear bidentate uranium (VI) surface complex; (b) mononuclear monodentate uranium (VI) surface complex; (c) binuclear bidentate uranium (VI) surface complex.

Conclusion

According to the results of the sorption experiments, granites with high content of quartz are optimal for the restriction of UO_2^{2+} migration. Attention should be paid to the conditions of geochemical isolation of uranium, as different chemical nature of the sorption centers on the mineral surface can affect significantly sorption rate and efficiency.

References

Anderson Ye, Lyubtseva Ye, Savonenkov V, Shabalev S, Alekseev N, Khlopin V. (2005) Creation of an Underground Storage Facility for Spent Nuclear Fuel near the City of Zheleznogorsk. An International Spent Nuclear Fuel Storage Facility – Exploring a Russian Site as a Prototype: Proceedings of an International Workshop: 166-178.

Batuk D, Shiryaev A, Kalmykov S, Batuk O, Romanchuk A, Shirshin E, Yan V. Zubavichus Ya (2011) Sorption and Speciation of Uranium on Silica Colloids. Actinide Nanoparticle Research: 315-332.

REE fractionation and distribution of Fe, Ni and U in the soil-water-biomass system along the flow path of Gessenbach, Eastern Thuringia (Germany)

Daniela Sporleder[1], Anja Grawunder[1], Georg Büchel[1]

[1]Friedrich Schiller University Jena, Institute of Geosciences, Applied Geology, Burgweg 11, 07749 Jena, E-mail: Daniela.Sporleder@uni-jena.de

Abstract. *Fraxinus excelsior*, *Glyceria fluitans* and *Hypholoma subericaeum* were investigated as contemporary indicators for the impact of acid mine drainage (AMD) in the soil-water-biomass system. The study was carried out in a (former) U mining area with special emphasis on the elements Fe, Ni and U and the fractionation of Rare Earth Elements which can be found in higher contents in growth locations close to groundwater discharge from the underground mines. *F. excelsior* showed higher contents of these elements in sapwood, than in heartwood.

Introduction

From 1945 until the German Reunion, U was mined in the Ronneburg area in Eastern Thuringia, Germany. Groundwater was pumped to lower the groundwater level in mining days. In 1995, the resulting depression cone was at 220 m a.s.l. (Merten et al. 2005). Since 1997 with no reasonable alternative for remediation, the groundwater level is naturally rising as planned. Pyrite oxidation is one of the reactions still taking place in the underground, resulting in Acid Mine Drainage (AMD). Caused by the orographic situation and former drillings in Gessental (Gessen valley), the valley was predicted to be the main discharge area (Paul 1998). AMD- impacted groundwater exceeds the underground (groundwater level in 03/2014: 248 m a.s.l., Wismut database Al.vis 2014) and seeps aboveground into an isolated part of the creek Gessenbach since 2006, while the Gessenbach itself is redirected in a bypass to prevent contamination (Fig. 1).

This worked aimed to study the mining impact in the system soil-water-biomass based on the distribution of Fe, Ni, U and REE. Therefore, besides seepage water and soil, different biological systems were investigated: *Fraxinus excelsior* and *Glyceria fluitans* both grow along the flow path of Gessenbach and the isolated part. The wooden core of *F. excelsior* is a biogeochemical archive as the ring-porous structure reflects an age of up to 100 a. However, *G. fluitans* is a perennial grass growing on the banks of Gessenbach. The fruiting bodies of *Hy-*

B.J. Merkel and A. Arab (eds.), *Uranium – Past and Future Challenges*,
DOI 10.1007/978-3-319-11059-2_34

pholoma subericaeum were found growing on a humid substrate impacted by seepage water. The fungus' fruiting bodies are contemporary detectors, although its mycelia might be older.

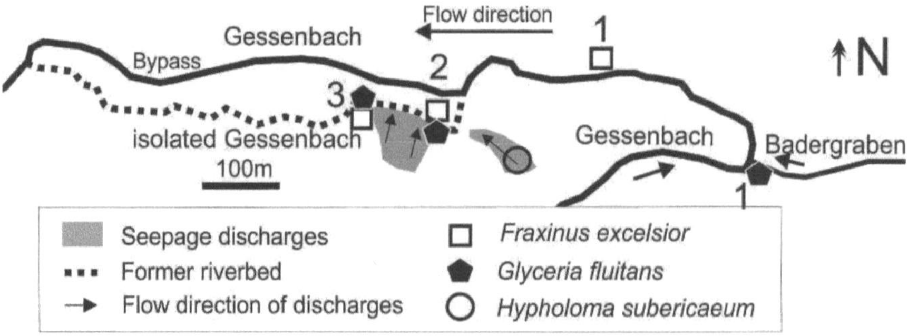

Fig.1. The creek Gessenbach is redirected in the Gessen valley, the former riverbed (isolated Gessenbach) is influenced by seepage discharges. Locations of samples along the flow path of Gessencreek are shown (1, 2, and 3).

Material and Methods

Water Samples

Water samples were taken in 2013 (Fig. 1). Hydrochemical parameters pH, redox potential and temperature (pH320, WTW) and electrical conductivity (EC, LF320, WTW) were determined in field. Samples for chemical analysis were filtered with glassfibre prefilters (Millipore, Germany) and cellulose-acetat-filter (Sartorius) to 0.45 µm. A sample aliquote for cation measurement was acidified with HNO_3 (65%, subboiled) to pH<2. HCO_3^- was determined on the same day by titration of unfiltered sample (Titrino 716 DM, Metrohm). SO_4^{2-} was determined with ion-chromatography (DX/120, Dionex). Ca, Mg, Fe were determined by inductively coupled plasma-optical emission spectrometry (ICP-OES, Varian 725 ES). Ni, U, and REE were determined by ICP-Mass Spectrometry (ICP-MS, X-Series II, Thermo Fisher Scientific). Fe (II) was measured photometrically with the 1.10 phenanthroline FerroVer® Iron reagent (Hach) method. Fe (III) was calculated as difference from Fe total and Fe(II).

Soil Samples

Soil samples were taken with a plastic shovel and dried at 40°C in an oven (Memmert). For total digestion, soil samples were ground with agate mortar. Total digestion was performed by using strong acids (HNO_3, HF, $HClO_4$) and is described elsewhere (Grawunder et al. 2014). The bioavailable fraction is assembled by the first (F1) and second fraction (F2) of sequential extraction (Zeien and Brümmer 1989) extracted by 50 ml NH_4NO_3 (F1) and 50 ml NH_4OAc (F2), respectively, and measured as described above.

Vegetation and fungal samples

Wooden and grass samples were taken along the flow path of Gessenbach: 1) in an area which is not impacted by seepage water, 2) directly at the seepage water discharges and 3) downstream the seepage discharges (Fig. 1).

Wooden samples were taken with an increment borer (SUUNTO, 400 mm, 2N, Finland) in breast height (approx. 1.4 m) vertical to the direction of growth. The samples reached the depth of approx. 35 cm which implies an age of 74 up to 100 a. The cores were transported and dried in H-Profiles and fixed with parafilm to prevent rotation during the process of drying in an oven at 40 °C. Afterwards the core was split into pieces of 8 cm length and ground with Al_2O_3 abrasive (ZirMet, grain size 78 μm). That implies a planar surface for counting of tree-rings.

Furthermore, the inner (heartwood) and outer (sapwood) 5 cm of the wooden core are abraded, ground with an agate mortar and analyzed after digestion.

Grass samples were carefully cleaned with a series of tap water, twice aqua dest. and ultrapure water (0.055 μS/cm; GenPure UV-TOC, Fa. Thermo Scientific). Afterwards they are split into root, stem and leave and dried for digestion.

Mushrooms were sampled and cleaned with aqua dest. and tissues on the same day, separated into stem, head and lamellae and dried for digestion.

For microwave digestion (MARS 5, CEM) of plant and fungal material, to approximately 0.2 g of sample, 5 ml HNO_3 (65%, sub boiled) are added in containers (HP-500 Plus) and pre-reaction occurs for minimum 20 min. Afterwards, the sample is heated to 180 °C within 15 min, left for 15 min at 180 °C and is cooled down for 30 min. The digestion solution is transferred into 25 ml volumetric flasks and filled up with ultrapure water, followed by additional centrifugation of unsolved components at 3000 U/min.

Rare Earth Elements

The group of REE constitutes of 15 elements (14 stable) from La to Lu, often subdivided into LREE (La-Nd), MREE (Sm-Dy) and HREE (Ho-Lu).As a function of the atomic number, they show smooth but continuous variation in their chemical behavior. As a result of different abundances of odd and even atomic numbers (Oddo-Harkins-Rule), they show a saw tooth pattern when plotted against atomic number. To simplify a comparison, the REE are standardized to (Post Archean Australian Shale, McLennan 1989), which represents the upper crust. The result is a flattened pattern.

Results and discussion

Water

As a consequence of pyrite oxidation in the former U mine Ronneburg, AMD-impacted water is discharging into the Gessental. The seepage water is in the range of acidic to neutral pH (pH 3.2 – 7.3) and of Ca-Mg-SO_4 water type. The EC is 4561 ± 590 µS/cm, the redox potential is slightly oxidizing (303 ± 108 mV). The Fe content in the seepage water shows peak values of 259 mg/l. Mostly Fe^{2+} (2.4 % Fe^{3+}) is discharging in the water phase in Gessental. The ferrous iron is oxidized to ferric iron on the surface, resulting in reddish precipitates. Besides Fe, also peak values of Ni (1990 µg/l) and U (399 µg/l) are observed in the seepage water.

ΣREE concentrations range from 0.4 to 194.6 µg/l. Normalized REE patterns show MREE and HREE enrichment (Fig.2). MREE are enriched in seepage water being discussed to be a result of preferential release from pyrite (Grawunder et al. 2014).

Soil

The soil sampled together with plants (locations 1, 2, and 3, Fig. 1) from the banks of the creek Gessenbach is rich in Fe (up to 99.6 mg/g) with a very low bioavailable proportion (1-6 µg/g, Table 1). Ni and U show peak contents of 127 µg/g and 106 µg/g. However, the bioavailable fractions are higher for these elements (6-17µg/g Ni and 10-22 µg/g U, respectively, Table 1).

Table 1. Contents of Fe, Ni, U and ΣREE in total and bioavailable fraction from soil samples in µg/g.

Location		Fe	Ni	U	ΣREE
1	bioavailable	6	13	10	0.9
	total	35,878	87	46	120
2	bioavailable	2	6	12	0.6
	total	62,601	67	60	154
3	bioavailable	1	17	22	0.4
	total	99,621	127	106	126
Hypholoma-	bioavailable	734	7	0.3	3.6
substrate	total	413,094	10	0.8	13

The substrate sampled together with *H. subericaem* fruiting bodies consists of dead biomass of reed on Fe-precipitates what is represented in high contents of Fe (0.4 g/g, Table 1), likewise.

Total contents of REE are except for *Hypholoma* substrate, close to the content of the upper crust standard (184.7 µg/g; PAAS, McLennan, 1989) with weak depletion of LREE (Lu/La < 1.6, Fig. 2). In comparison, the REE content of the *Hypholoma* substrate is one order of magnitude lower, therefore having a higher bioavailable content. Like the water samples, the bioavailable fractions of all soil samples showed a clear MREE enrichment reflecting the impact of the discharging groundwater and the creek water (Fig. 2).

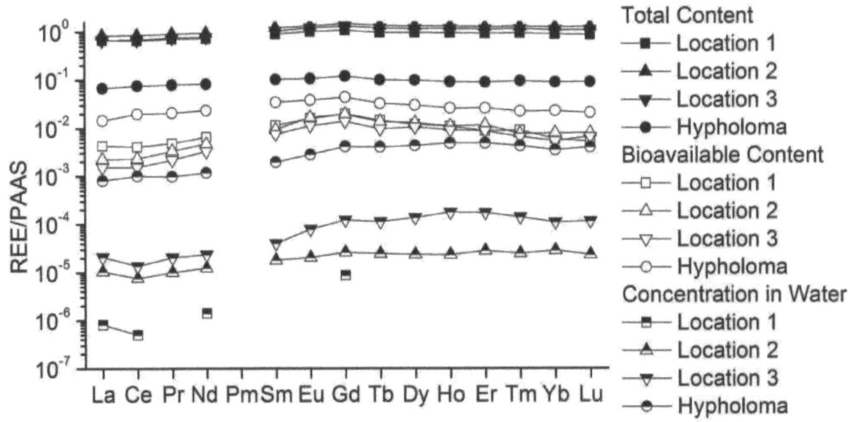

Fig.2. REE patterns of bioavailable fractions (open symbols), total content (filled symbols) and water samples (half-filled symbols) related to the different sampling locations of the vegetation.

Vegetation and fungal samples

The stem of *Fraxinus excelsior* is used as a biogeochemical archive looking back into former mining days. The older heartwood showed lower concentrations for Fe (Fig. 3), Ni, U and REE than the younger sapwood, indicating the influence of mine discharges in the Gessental. The concentration of these elements in the outermost rings is highest in the tree nearest to the seepage waters (location 2 and 3, Fig. 1). U was only detected in the sapwood. U is geogenic present, however, it is not expected to have been bioavailable in high proportions the pre-mining days. The post flooding situation causes rising of slightly acidic groundwater in Gessental, which might support or represents itself a source of bioavailable U.

REE could only be detected for locations 1 and 2 in the youngest rings (data not shown). The ΣREE is 1.55 μg/g for location 2 and 0.13 μg/g for location 3. The LREE are slightly depleted (Lu/La = 1.5). However, compared with the REE pattern of water sample, the LREE are enriched in the sapwood, a feature often found in vegetation (Liang et al. 2008).

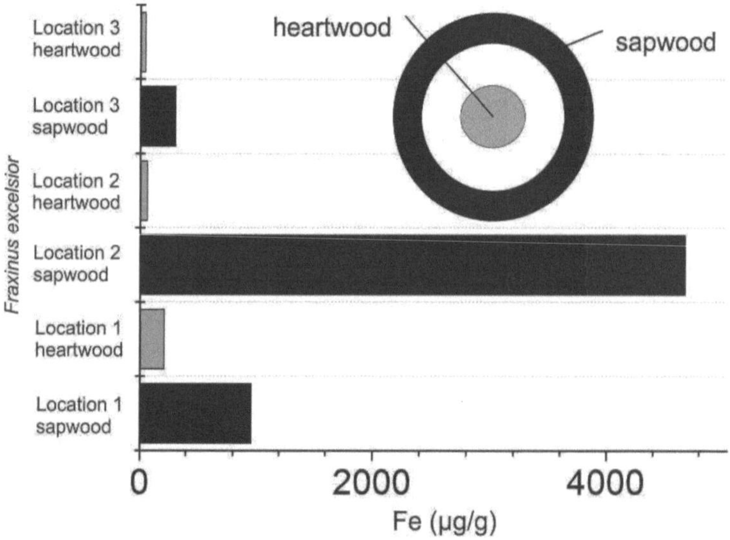

Fig.3. Fe contents in sapwood and heartwood of *Fraxinus excelsior* at locations 1, 2 and 3.

The contents of Fe, Ni and U in *G. fluitans* are compartment dependent (data not shown in detail). Highest enrichment was found in the roots, while lowest enrichment occurs in the tunneling tissue stem, coinciding with e.g. Tyler (2004). While in the root from uncontaminated location 1 0.6 mg/g Fe were measured, the concentration in seepage water influenced locations 2 and 3 is with 52.2 and 111.8 mg/g, respectively, much higher. Enrichment factors (relative to bioavailable fraction of the related location) for Fe range from 6 to 78,211, depending on the compartment. For Ni and U solely the root's enrichment factor exceeds 1.

The concentration of ΣREE in *G. fluitans* is highest in roots > stem > leave. These total contents are in agreement with concentration distribution in related water and soil samples (location 2 > 3 > 1). However, Wyttenbach et al. (1998) found no or little agreement within the distribution of REE in plants and those in soil, coinciding with Markert (1987). The MREE are enriched in all compartments (Tb/La >1.2; Lu/Tb < 0.8), again tracing the mining impact.

The contents of Fe, Ni, U and ΣREE in *H. subericaeum* fruiting body compartments do not vary significantly (Fig. 3). However, especially Fe and Ni are enriched (related to bioavailable fraction) in the compartments (Table 2).

Table 2. Average bioconcentration factor of Fe, Ni and U related to the bioavailable fraction in stipe, lamellae and head of *H. subericaeum* (n=8).

	Fe	Ni	U
Stipe	5.5	25.3	2.0
Lamellae	1.3	36.8	0.7
Head	1.5	22.9	0.6

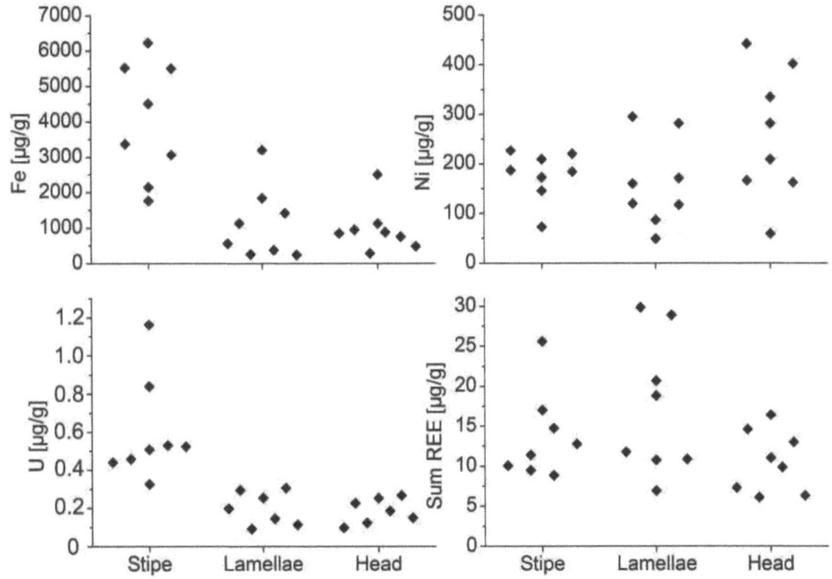

Fig.4. Point clouds of Fe, Ni, U and ΣREE in stipe, lamellae and head of *H. subericaeum* showing no significance in heavy metal uptake (n=8).

Conclusion

F. excelsior, *G. fluitans* and *H. subericaeum* are used to indicate the influence of mining for different time periods, reflecting the development of the former U mining area Ronneburg. The area is influenced by uprising of AMD-impacted groundwater. *F. excelsior*, a biogeochemical archive of up to 100 years mirrors a lower uptake of Fe, Ni, U and REE in the oldest heartwood, compared to the sapwood. *G. fluitans*, which is a perennial grass, showed compartment depending enrichment of Fe, Ni, U and REE. Within *H. subericaeum*, no significant compartment depending uptake was determined. However, the fungus enriches Fe and Ni in stipe, lamellae and head of its fruiting bodies.

Acknowledgements

We thank the DFG Research Training Group (GRK 1257/2) for funding. We acknowledge I. Kamp (ICP-OES, total digestion), G. Weinzierl (IC, photometer), D. Merten (ICP-MS) for analytics and U. Buhler for plant digestion and sequential extraction. Furthermore we thank D. Berger (all FSU Jena) for field assistance and M. Gube (Uni Göttingen, Büsgen-Institute) for identification of the fungus. We thank Wismut GmbH for enabling access to the sampling area.

References

Liang T, Ding S, Song W, Chong Z, Zhang C, Li H (2008) A review of fractionation of rare earth elements in plants. J Rare Earth 26: 7-15

Markert B (1987) The pattern of distribution of Lanthanide elements in soils and plants, Phytochem. 26: 3167-3170

McLennan S (1989) Rare earth elements in sedimentary rocks. Influence of provenance and sedimentary processes. Rev Mineral Geochem 21: 169-200

Paul M, Sänger HJ, Snagowski S, Märten H, Eckart M (1998) In: Merkel B, Helling C (Eds.), Uranium Mining and Hydrogeology II, Proceedings of the International Conference and Workshop, Freiberg, September 1998. Geocongress, Vol. 5, Köln: 401-410

Tyler G (2004) Rare earth elements in soil plant systems – A review. Plant Soil 267: 191-206

Wyttenbach A, Furrer V, Schleppi P, Tobler L (1998) Rare earth elements in soil and in soil-grown plants. Plant Soil 199: 267-273

Zeien H, Brümmer GW (1989) Chemische Extraktion zur Bestimmung von Schwermetallbindungsformen in Boden. Mitt. Dtsch. Bodenkdl. Ges. 59: 505-510

The externalized costs of uranium mining in the United States

Doug Brugge[1], Aparna Dasaraju[2], Yi Qi Lu[3], Brianna Dayer[2]

[1] Tufts University School of Medicine, 136 Harrison Ave., Boston, MA 02111
[2] Tufts University, School of Arts and Sciences, Medford, MA 02155
[3] University of Massachusetts Amherst, Amherst, MA 01003

Abstract. Uranium mining in the US spanned a period from the end of World War II into the 1980s. The legacy of this mining was illness and death for thousands of workers along with thousands of abandoned mines. Compensation programs for former workers and their heirs as well as remediation efforts for the contaminated sites appear to have cost well over $5 billion US dollars, which may exceed the value of the uranium ore extracted. Most of the cost has been born by the taxpayers to date, although a recent settlement will require one of the private companies to pay about $1 billion US dollars for uranium site remediation. The history suggests that there are likely to also be externalized costs from present-day uranium mining, especially in developing countries where regulation is less stringent.

History of uranium mining in the US

Uranium mining in the United States accelerated after World War II as the nuclear arms race picked up, peaking in the 1960s and declining by the 1980s. Today there are only a half dozen in situ leach mines in operation. Early production was for nuclear weapons with nuclear power becoming more important in the latter years. Most of the mining was in the Southwestern United States in the Colorado Plateau –a broad expanse of land in the states of New Mexico, Arizona, Utah and Colorado. From 1949 to 1989 more than 4000 uranium mines operated in the Four Corners Region (Southwestern) of the US with limited regulation of health and safety conditions or control of environmental contamination (Fettus and Mckinzie, 2012). Over 10,000 uranium workers were recruited to work in the mines, including substantial Native American populations who lived nearby. Miners uniformly state that they were not told of the hazards at the time (Brugge and Goble, 2002; Brugge et al., 2007; Ringholtz, 1989; Zoellner, 2009).

In fact, instead of regulations aimed at protecting miners, the US Public Health Service initiated a study of uranium miners to better gauge the dose response relationship between radon, the primary toxin of concern, and lung cancer (that it was

B.J. Merkel and A. Arab (eds.), *Uranium – Past and Future Challenges*,
DOI 10.1007/978-3-319-11059-2_35

a causal agent was already established). A federal standard limiting radon exposure in mines was enacted only in 1969, two decades after large numbers of miners had started receiving high doses in the mines. As a consequence the US Government failed to protect the miners from a known hazard, greatly increasing their chances of illness and death (Brugge and Goble, 2002; Committee on the Biological Effects of Ionizing Radiation, 1999).

Lagging development of environmental regulations also led to widespread contamination around the mines, mills and other uranium processing plants. The largest radioactive release in the history of the United States flowed from a uranium mill tailing pond in 1979. While little noted, but more dramatic than everyday releases and spills, the Church Rock spill exemplifies the comparatively weak level of concern (relative to today) and attention paid to the spread of uranium ore and tailings into the environment. Considerable evidence exists that environmental releases combined with family members visiting the mines also led to exposure of people living in nearby communities (Brugge et al., 2007; deLemos et al., 2008).

In recent years there has been interest in reviving mining, but little has actually started, largely due to the low price for uranium which peaked only briefly in 2007-2008. In the first quarter of 2014 there were only 6 in situ leach mines and 1 mill operating in the US (http://www.eia.gov/uranium/production/quarterly/). Expansion of uranium mining today depends heavily on expansion of nuclear power, but nuclear power generation has been almost level for decades, increasing only very slightly since 1995 (http://www.euronuclear.org/info/encyclopedia/n/nuclear-power-plant-world-wide.htm). Not surprisingly, post Fukushima prices for uranium have fallen and remained low in recent years (Koven, 2014).

Health consequences

The United States Public Health Service study showed a statistically significant association between radon exposure in the mines and lung cancer by 1959. Subsequent research refined knowledge about the dose response relationship and the interaction of radon and smoking (Archer et al., 1962). Ultimately by the 1990s there were detailed statistical models, based on the United States study together with 10 other similar studies from around the world that incorporated age at exposure and time since last exposure to calculate risk. Further, it was clearly shown that radon could cause lung cancer in non-smoking Native American miners(Committee on the Biological Effects of Ionizing Radiation, 1999).Conclusive evidence also accumulated that miners were at elevated risk of pneumoconiosis from inhaling freshly blasted rock dust (Roscoe, 1997).

Evidence for the health risks for community residents living nearby uranium mines has been slower to develop and is less definitive. Most of the community studies are ecological in design and therefore limited in terms of usefulness for causal inference. However, in recent years a small, but growing body of literature has begun to emerge suggesting that non-miners may also experience health risks from their exposures (e.g., Lourenc et al., 2013).

Economic impact

Because of the legacy of health problems experienced by US uranium miners, there was a long period of advocacy on the part of the miners and their allies to seek redress from the US Government. A campaign for compensation, which was largely led by Native American uranium workers and their communities, met with numerous roadblocks and challenges over about two decades before being partially successful. Sadly, the US law that eventually passed to provide compensation proved to be flawed and unfair in numerous ways, which took another decade of advocacy to repair (Brugge and Goble, 2002).

The US Radiation Exposure Compensation Act (RECA) was first passed in 1990 and then amended in 2000. By June 2014 $790 million had been distributed under RECA to almost 8,000 uranium workers and their survivors (another 4,000+ claims were denied). It is notable that it is the US taxpayers who have paid this restitution. This was justified based on the monopoly that the US Government had on purchase of uranium prior to 1971. Accordingly, uranium work done after 1971 is not eligible.

While RECA provided a semblance of justice for the miners, concern has more recently turned to the potential of health consequences for communities situated near the mines. Because health studies are not as well developed for community members, there is not a compensation program for people who were exposed, but did not work in the mines. One could imagine compensation for community residents in the future if the epidemiologic evidence becomes strong enough. Bu this potential future cost associated with community health problems is not possible to ascertain at this time,

Fig.1. The location of uranium mine tailings (mills) across the U.S (http://www.epa.gov/rpdweb00/docs/radwaste/402-k-94-001-umt.html).

There has been some remediation of the abandoned mills (see Figure 1, above) and mines in the US. A federal program, the Uranium Mine Tailings Radiation Control Act (UMTRCA) focused on containing contamination at mills. This program had spent about $1.5 billion by the year 2000 with long term costs, including indefinite monitoring, expected to exceed $2 billion. It has been suggested that his expense which exceeds the value of the uranium produced(http://www.sric.org/voices/2004/v5n3/umtrca.php).A similar level of remediation as was done for mills does not exist for the much larger number of mines. Even so, between 1998 and 2007 the US Government spent about $2.6 billion on clean-up of abandoned uranium mines, an expense which continues today with additional efforts to control and remediate abandoned mines (U.S. Department of Energy, 2014).

Federal government hearings in 2007 proposed $500 million to remediate dozens of abandoned mines in Native American areas and over $100 million has been spent. But these efforts fall short of full cleanup of the contamination. Often, contaminants from abandoned mines and mills have spread extensively into nearby communities due to the long period of time between mining and the beginning of remediation efforts. The US Environmental Protection Agency is heading remediation efforts and outlined a 5-year plan in 2013 that is guiding the agency.

US EPA has conducted preliminary assessments of 521 contaminated mines, structures and water sources, including finding high-priority sites in need of immediate attention. They have also set up case control studies of health risks faced by nearby community members. The US EPA is also working, as outlined in the 5-year plan, to find Potentially Responsible Parties under the Comprehensive Environmental Response, Compensation and Liability Act (or Superfund) federal hazardous waste site clean-up program. However, cleanup has started at only a few sites and the costs so far have largely accrued to taxpayer dollars. (http://www.epa.gov/region9/superfund/navajo-nation/pdf/NavajoUraniumReport2013.pdf).

A recent court decision against the former mining company Kerr-McGee (Ezell, 1979) will recover $5.15 billion in a settlement for environmental remediation of contaminated sites, including uranium sites. The Kerr-McGee decision is a landmark case as it is the largest environmental enforcement recovery by the Department of Justice to be paid by a responsible company. Kerr-McGee and the parent company Anadarko Petroleum Corp were found guilty of fraudulently conveying assets to evade the responsibilities of their industrial endeavors. Of the $5.15 billion agreed to in the settlement, $4.4 billion will fund remediation of contaminated sites across the country and approximately $1 billion will be used to cleanup abandoned mines and contaminated mills in the Navajo Nation (http://www.justice.gov/opa/pr/2014/April/14-dag-338.html).

Discussion and conclusions

This brief summary clearly shows that there are substantial externalized costs that followed in the decades after uranium mining declined in the US. It is unlikely that our effort to assess these costs, which was limited to review of easily obtainable documents, is comprehensive. Also, some of the sources produce estimates that are not current. Thus, this should be considered a floor, rather than a ceiling for the costs that have accrued to date. Further, additional costs will be incurred in the future, many of which cannot be predicted well. A few years ago the possibility of such a large settlement with Kerr-McGee seemed improbable, for example.

The lessons for the current time period seem to be the following: 1) If uranium mining is conducted today with similarly lax occupational and environmental regulation and oversight, as appears to be the case in many developing countries, it may again lead to extremely large future externalized costs, perhaps eclipsing the value of the uranium extracted and processed; 2) That there remains a risk that these costs will be borne by the taxpayers instead of the uranium extraction companies; and 3) That the largest costs are associated with remediation of sites contaminated from mining or mill operations, although a secondary cost is compensation to former uranium workers and their families.

A pressing question, which we did not attempt to address here, is what the externalized costs might be for mining conducted under the more strict regulatory regime of highly developed countries such as the US. However, given the magnitude of costs from the prior era, it is difficult to imagine that even if they were substantially reduced that there would still be externalized costs from present day uranium mining under even the most highly regulated circumstances.

References

Archer VE, Magnuson DA, Holaday DA, Lawrence PA (1962) Hazards to health of uranium mining and milling. J Occ Med, 4:55-60.

Brugge D and Goble R (2002) The history of uranium mining and the Navajo People. Am J Pub Hlth, 92:1410-1419.

Brugge D, Benally T, Yazzie-Lewis E eds. (2006) The Navajo People and Uranium Mining. University of New Mexico Press, Albuquerque, NM.

Brugge D, deLemos JL, Bui C (2007) The Sequoyah Corporation fuels release and the Church Rock spill: unpublicized nuclear releases in American Indian communities. Am J Pub Hlth. 97:1595-600.

Committee on the Biological Effects of Ionizing Radiation (1999) Health effects of exposure to radon (BIER VI), Washington, CD, National Academy Press.

deLemos JL, Bostick BC, Quicksall AN, Landis JD, George CC, Slagowski NL, Rock T, Brugge D, Lewis J, Durant JL (2008) Rapid dissolution of soluble uranyl phases in arid, mine-impacted catchments near Church Rock, NM. Env Sci& Tech, 42:3951-3957.

Ezell, JS (1979) Innovations in energy: The story of Kerr-McGee, University of Oklahoma Press, Norman, OK.

Fettus, GH.;Mckinzie MG (March 2012). "Nuclear Fuel's Dirty Beginnings: Environmental Damage and Public Health Risks From Uranium Mining in the American West". National Resources Defense Council. Retrieved 25 June 2014. http://www.nrdc.org /nuclear/files/uranium-mining-report.pdf

Koven P (2014) Uranium stocks tumble after RBC takes axe to price forecasts, Financial Post, accessed at: http://business.financialpost.com/2014/06/05/rbc-annihilates-uranium-price-outlook/.

Lourenc , Pereira R, Pinto F, Caetano T, Silva A, Carvalheiro T, Guimarães A, Goncalves F, Paiva A, Mendo S (2013) Biomonitoring a human population inhabiting nearby a deactivated uranium mine, Toxicology 305: 89–98.

Ringholtz RC (1989) Uranium frenzy: Boom and bust on the Colorado Plateau. New York: W.W. Norton & co.

Roscoe, R. J. (1997), An update of mortality from all causes among white uranium miners from the Colorado plateau study group. Am J Ind Med, 31: 211–222.

United States Announces $5.15 Billion Settlement of Litigation Against Subsidiaries of Anadarko Petroleum Corp. to Remedy Fraudulent Conveyance Designed to Evade Environmental Liabilities, accessed at http://www.justice.gov/opa/pr/2014/April/14-dag-338.html

U.S. Department of Energy (2014) Defense-Related Uranium Mines Cost and Feasibility Topic Report. Final AUM Technical Topic Reports.Doc. No. S10859, accessed at http://www.lm.doe.gov/default.aspx?id=10668.

US Environmental Protection Agency (2013) Federal Actions to Address Impacts of Uranium Contamination in the Navajo Nation: Five-Year Plan Summary Report, accessed at: http://www.epa.gov/region9/superfund/navajo-nation/pdf/NavajoUraniumReport2013.pdf

Zoellner T (2009) Uranium: War, Energy, and the Rock That Shaped the World. The Penguin Group, New York, NY.

Microbial consortia in radionuclide rich groundwater

Katja Burow[1], Sven Gärtner[1], Anja Grawunder[1], Erika Kothe[2], Georg Büchel[1]

[1]Friedrich Schiller University, Institute of Geosciences, Applied Geology, Burgweg 11, D-07749 Jena, Germany
[2]Friedrich Schiller University, Institute of Microbiology, Microbial Communication, Neugasse 25, D-07743 Jena, Germany

Abstract. In water-rock interactions, microbes play an important role in mobilization and immobilization of metals, including radionuclides. Water of three wells from a rhyolite joint aquifer known for high Rn and Ra concentrations were investigated for their microbial communities and hydrochemistry. The saline waters of Na-Cl type were used to isolate strains of the native microbe population. Both Gram-negative, like *Pseudomonas grimontii,* and Gram-positive, e.g. *Mycobacterium hodleri,* were identified. These microbes can produce siderophores, necessary to supply the cells with iron. In addition to iron, such small molecules may complex other elements, including radionuclides. Thus, they can impact the mobility of radionuclides in the groundwater. The potential of microbial processes in radionuclide reactive transfer is further discussed with respect to other potential mechanisms.

Introduction

The potential release of radionuclides (RN) into soils and groundwater poses a long-time risk to the environment. In mining and urban sewage waters, RN mobility depends on the processes of reactive transport, with mobilization as well as immobilization processes possible in groundwater aquifers. Microbes occur in different communities of bacteria and/or archaea depending on the hydrochemical composition of the groundwater. The different niches will determine microbial diversity and activity. The resulting autochthonous flora of microorganisms, in turn, may substantially alter and control these processes, e.g. by production of siderophores. These small molecules of several classes of secondary metabolites usually are excreted under Fe limiting conditions for iron supply. By complexing alternative ions also can have a strong impact on RN mobility (Gadd 1997; Gazsó 2001; Francis and Dodge 2009).

B.J. Merkel and A. Arab (eds.), *Uranium – Past and Future Challenges*,
DOI 10.1007/978-3-319-11059-2_36

Other mobilization processes including complexation, as well as enzymatic oxidation and/or reduction processes are controlled by microbial activity (Gazsó 2001; Simonoff et al. 2007). If RN are transported into the cell, bioaccumulation can result. Adsorption at the cell walls (biosorption) will result in immobilization of RN (Fig.1). All processes can contribute to reactive transport in groundwater. Thus, microbial diversity and the resulting diversity of microbial complexing agents are important factors for transport of RN in the rock-water system (Premuzic et al. 1985; Pederson 1999; Gadd 2002; Kim et al. 2013).

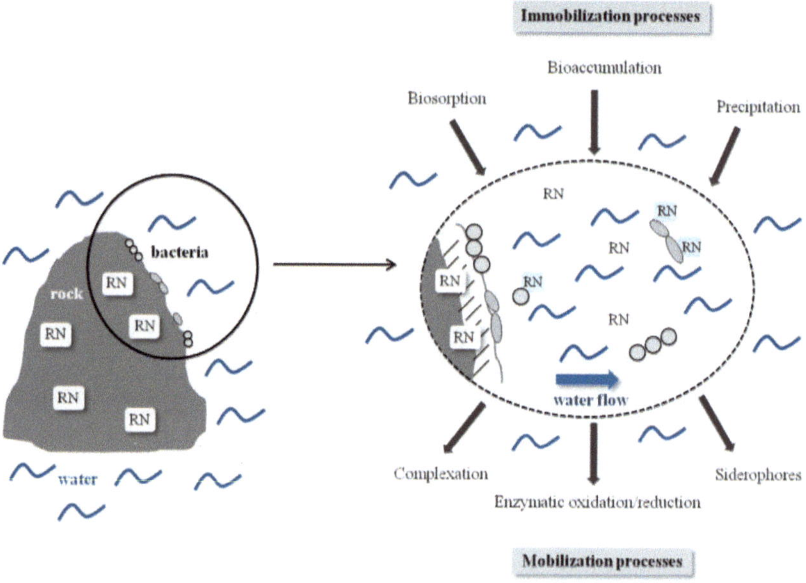

Fig.1. Rock-water-bacteria interaction processes resulting in immobilization/mobilization of radionuclides (RN-Radionuclides).

Here, we studied thermal groundwater from the inhalation spa of Bad Kreuznach, Germany. The saline water flows through a joint rhyolite aquifer, where mobilization of RN such as Ra-226, Ra-228 and Rn-222 results in high concentrations of these RN (Hemfler 1993; Hohberger 2008), but also Th and U are transported with the water after being mobilized from the rhyolite. We characterized the hydrochemistry and microbial native populations of microbes in radionuclide-bearing groundwater. We studied the strains isolated from the aquifers for their abilities to mobilize RN or to contribute to their transport in water-rock interactions.

Material and methods

Site description

The study area Bad Kreuznach is situated close to the city of Mainz, Germany. Geologically, it is situated in the Saar-Nahe basin south of the Rhenish Slate Mountains (Rheinisches Schiefergebirge). Three groundwater samples were taken from wells in different depths: Karlshaller Bäderquelle (KB, 136 m), Inselbäderquelle (IB, 75 m) and Theodorshaller Thermalbrunnen (TB, 505 m). This mineral water arises from joint aquifers of the local Permian rhyolite (Kreuznacher Rhyolith) containing high concentration of Rn (up to 163 Bq L^{-1}, Hohberger 2008) and Ra (14.8 Bq L^{-1}, Hohberger 2008).

Water sampling

Water samples were taken in October 2013. The parameters electric conductivity (EC), pH, Eh and temperature were measured in field using a portable instrument Multi 340i (WTW) with TetraCon 325 (EC), Schott BlueLine 31 Rx (3 M KCl, Eh), WTW SenTix 41 (3 M KCl, pH and temperature). Eh was corrected to standard hydrogen electrode. Samples for element and anion (except for HCO_3^-) measurements were filtered using glass fiber prefilters (Sartorius) and cellulose acetate filters (0.45 µm, Sartorius), acidified with HNO_3 (65%, sub boiled) to pH < 2 and subsequently kept at 6°C until analysis. HCO_3^- was determined at the day after sampling by titration (Titrino 716 DM, Metrohm). Cl^-, Br^-, F^- and SO_4^{2-} were measured by means of ion chromatography (DX-120, Dionex). The elements Ca, Fe, K, Mg, and Na were determined using ICP-OES (Varian 725 ES). ICP-MS (X-Series II, ThermoFisher Scientific) was used for analyzing the elements Th and U.

Microbiological investigations

To investigate the microbial community, 2.5 L were filtered through an autoclaved membrane filter (0.22 µm PTFE filter, Millipore) using a filtration unit (Steritop-GV, Millipore) and low-pressure vacuum. The filters were incubated on StdI medium (Merck, Darmstadt, Germany) and on R2A medium (Reasoner et al. 1985). Strains isolated were characterized by 16S rDNA sequencing. DNA was extracted using 5% Chelex Solution (Chelex® 100 Resin Pulver, Bio-Rad). 16S rRNA was amplified by PCR (50 µL reaction volume, bacterial primers 27F and 1492r (Lane 1991) and 0.2 µL Taq polymerase (Mango *Taq*™ DNA Polymerase; Bioline). The

PCR products were sequenced (GATC, Konstanz, Germany) and the sequences analyzed using gene bank entries for comparison.

Production of siderophores was investigated using two-partite plates containing Chrom-Azurol S media (Sigma Aldrich, Germany; Schwyn and Neilands 1987) and nutrient agar (Difco, USA).

Results

Hydrogeochemistry

All samples were of a Na-Cl water type, with pH of 6.6 to 6.9 and high EC of 24.7 (KB) to 31.2 (IB) mS cm^{-1}. The Eh was low with 230 mV for IB and KB and 210 mV for TB. The temperature was highest in TB with 25.0°C. The Fe concentrations were in the range of 5.8 to 13.3 mg L^{-1}, Th was below detection limit (<0.05 µg L^{-1}) for all samples and U was measured only at very low concentrations of up to 0.23 µg L^{-1}. Among the anions, F^- and SO_4^{2-} were below detection limit, while besides Cl^- (up to 11,722 mg L^{-1}), Br^- (up to 70 mg L^{-1}) and HCO_3^- (up to 386 mg L^{-1}) were detected.

Microbiology

The wells showed different bacterial diversities, with 36 isolates belonging to the phyla Firmicutes, Actinobacteria and Proteobacteria. Most isolates from IB belonged to the genus *Bacillus* sp., whereas *Pseudomonas* sp. dominated at KB. Isolates from TB showed the highest bacterial diversity, comprising bacteria like *Mycobacterium* sp., *Methylobacterium* sp., *Enhydrobacter* sp., *Erythrobacter* sp. and *Sphingomonas* sp. Firmicutes were only detected from TB.

Regarding the production of siderophores, *Mycobacterium* sp., *Bacillus* sp. and *Pseudomonas* sp. were able to secrete these complexing agents, and producers were detected in all three water samples.

Discussion

Processes of RN mobilization and transport are important for groundwater, and the importance of microorganisms in these mechanisms are not well resolved. Radionuclide mobilization/immobilization by microbial activity from rocks into the groundwater thus needs to be investigated with more emphasis. Here, we identified different microbial communities in three different wells at different depths, even if all three wells are based in rhyolite.

The hydrochemistry was very similar and comparable to former investigations (Hemfler 1993; Hohberger 2008), including U levels at a maximum of 0.23 $\mu g\ L^{-1}$ well comparable to data obtained from other rhyolite aquifers (e.g. Olguin et al. 1990).

The residential microbes need to be able to cope with high concentrations of Na and Cl, while differences determining the niche would be present with different depths (Itävaara et al. 2011) or temperatures. The isolated strains belong to known groundwater phyla (e.g., *Pseudomonas Microbacterium*, *Kocuria* and *Sphingomonas*) and even have been obtained from groundwater of a low-level waste repository, where especially the pseudomonades showed a high sorption capacity for RN (Luk'yanova et al. 2008). Another study proved, also for pseudomonads (*P. fluorescens*, *P. stutzerii*), the production of metabolites such as siderophores in the presence of ^{59}Fe(III), ^{147}Pm(III), ^{234}Th(IV) and ^{241}Am(III) (Arlinger et al. 2003).

All pseudomonades identified in our study were producing siderophores, with *Pseudomonas grimontii* known to be a producer of the fluorescent siderophore pyoverdin (Meyer 2000).

Future work will cover the geochemical and mineralogical composition of the rhyolite including a more precise composition of RN, as well as experiments to evaluate the impact of bacteria, and in particular siderophores, on mobilization and immobilization of RN in groundwater.

Acknowledgments

This work was supported by the BMBF (F.-Nr.: 02NUK030C). We want to thank the 'Kurverwaltung Bad Kreuznach' providing hydrochemical data, and Dirk Merten, Gerit Weinzierl and Ines Kamp for the hydrochemical analyses.

References

Arlinger, J., Oskarsson, A., Albinsson, Y., Andlid, T., Pedersen, K. (2003) Mobilisation of radionuclides by ligands produced by bacteria from the deep subsurface. MRS Fall Meeting.

Itävaara, M., Nyyssönen, M., Bomberg, M., Kapanen, A., Nousiainen, A., Ahonen, L., Hultman, J., Paulin, L., Auvinen, P., Kukkonen, I. T. (2011) Microbiological sampling and analysis of the outokumpu deep drill hole biosphere in 2007-2009. Geological Survey of Finland, Special Paper 51, 199–206, 2011

Francis, A. J., Dodge, C. J. (2009) Microbial transformation of actinides and other radionuclides. Submitted to the 9th Biennial DAE-BRNS Symposium on Nuclear and Radiochemistry (NUCAR02009)

Gadd, G. M. (1997) Roles of micro-organisms in the environmental fate of radionuclides. Novartis Foundation Symposium: 94-104.

Gadd, G. M. (2002) Microbial interactions with metals/radionuclides: the basis of bioremediation. In: Keith-Roach, M. J., Livens, F. R.: Interactions of microorganisms with radionuclides 2: 179-203

Gaszó, L. G. (2001) The key microbial processes in the removal of toxic metals and radionuclides from the environment. CEJOEM, 7: 178-185

Hemfler, M. (1993) Untersuchungen zur Hydrochemie, Isotopie, Hydraulik und Strukturgeologie der Mineralwässer von Bad Kreuznach. PhD thesis, Johannes Gutenberg-Universität Mainz

Hohberger, K.-H. (2008) Bad Kreuznach; Stadt Bad Kreuznach, Lkr. Bad Kreuznach, Reinland-Pfalz. In: Vereinigung für Bäder- und Klimakunde e.V., Käß W and Käß H: Deutsches Bäderbuch. 2nd Edition, Schweizerbart'sche Verlagsbuchhandlung, Stuttgart

Kim, J.-W., Baik, M.-H., Jung, H., Jeong, J.-T. (2013) Reactive transport of uranium with bacteria in fractured rock: Model development and sensitivity analysis. J Contam Hydrol 152: 82-96

Lane, D. J. (1991) 16S/23S rRNA sequencing. In: Stackebrandt, E., Doodfellow, M.: Nucleic acid techniques in bacterial systematic. John Wiley and Sons, New York

Luk'yanova, E. A., Zahkarova, E. V., Konstantinova, L. I., Nazina, T. N. (2008) Sorption of radionuclides by microorganisms from a deep repository of liquid low-level waste. Radiochem 50: 85-90

Meyer, J.M. (2000) Pyoverdines, pigments, siderophores and potential taxanomic markers of fluorescent *Pseudomonas* species. Arch Microbiol 174: 135-142

Olguin, M.T., Segovia, N., Carrillo, J., Ordoñez, E., Iturbe, J.L., Bulbulian, S. (1990) ^{222}Rn content and $^{234}U/^{238}U$ activity ratio in groundwaters. J Radioanal Nucl Chem 141:17-23

Pederson, K. (1999) Subterranean microorganisms and radioactive waste disposal in Sweden. Eng Geol 52: 163-176

Premuzic, E. T., Francis, A. J., Lin, M., Schubert, J. (1985) Induced formation of chelating agents by *Pseudomonas aeruginosa* grown in the presence of thorium and uranium. Appl Environ Microbiol 14: 759-768

Reasoner, D. J. and Geldreich, E. E. (1985) A New Medium for the Enumeration and Subculture of Bacteria from Potable Water. Appl Environ Microbiol 49: 1-7

Schwyn, B. and Neilands, J. (1987) Universal chemical assay for the detection and determination of siderophores. Analyt Biochem: 160.

Simonoff, M., Sergeant C., Poulain, S., Pravikoff, M. S. (2007) Microorganisms and migration of radionuclides in environment. C. R. Chimie 10: 1092-1107

Flooding of the underground mine workings of the old Witwatersrand gold/uranium mining areas: acid mine drainage generation and long term options for water quality management

Henk Coetzee[1], **Supi Tlowana**[1], **Mosidi Makgae**[1]

[1]Environmental Geoscience Unit, Council for Geoscience, Pretoria, South Africa, henkc@geoscience.org.za

Abstract. The underground workings in the older parts of the Witwatersrand Gold Fields of South Africa have been abandoned and allowed to flood. Pyrite in the ore and host rock leads to the generation of acid mine drainage in the flooding underground workings as well as in surface residue deposits. The difference in topographic elevation across these goldfields creates a driving hydrostatic head which has caused the discharge of acid mine drainage in the West Rand Gold Field and threatens to cause similar problems in the Central and East Rand. Laboratory simulations and simple models have been used to try to understand the hydraulic and chemical dynamics of the flooding mines with the aim of optimising solutions to this problem in the medium- to long-term.

Introduction

South Africa's Witwatersrand Gold Fields extend along an arc of more than 300km from the Free State in the South West, to the Evander area in the North East, passing through the City of Johannesburg. Gold and co-occurring uranium are extracted from narrow dipping pyritic conglomerate bands, extending to depths of more than 4km in places. After the discovery of Gold in 1886 in what would soon become the City of Johannesburg, mining commenced in the Central, West and East Rand Gold Fields, with surface operations soon giving way to deeper underground operations.

The continuous nature of the ore bodies has led to the excavation of large complexes of hydraulically interconnected underground mine workings. In the Johannesburg area, 3 large gold fields – the East Rand, the Central Rand and the West Rand developed, separated by geological structures resulting in discontinuities in the ore bodies and hence breaks in the hydraulic interconnection of the mines. Deep underground mining ceased in these three gold fields between 1998 and

B.J. Merkel and A. Arab (eds.), *Uranium – Past and Future Challenges*,
DOI 10.1007/978-3-319-11059-2_37

2011 and the workings were allowed to flood. Gold and, in some cases uranium, are still extracted from mines in other gold fields, accessing the Witwatersrand Supergroup ores to the east, west and southwest of Johannesburg, along a strike length of approximately 300km.

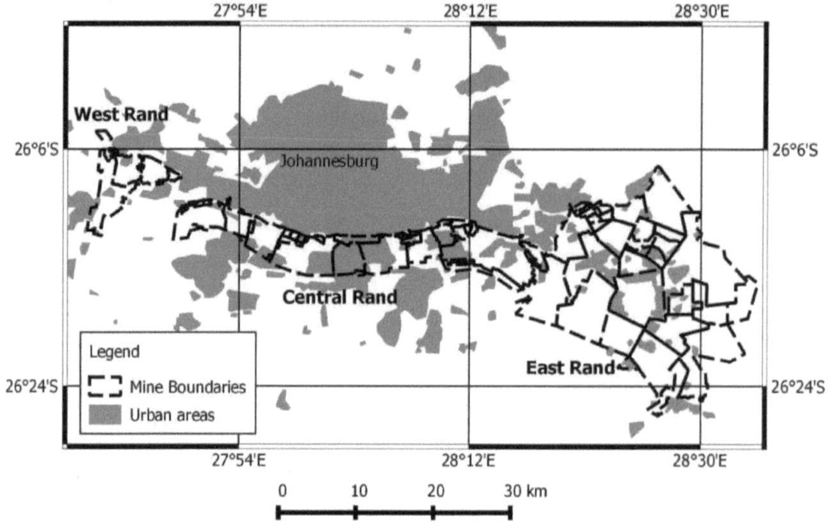

Fig.1. Location of the gold fields in the vicinity of Johannesburg

Anecdotal reports of acid water in the Witwatersrand mines can be traced back for decades (Hocking 1986). During mining operations, water was pumped from the voids and neutralised before being discharged to surface streams. After the cessation of underground operations in the West Rand in 1998, the underground workings were allowed to flood, with water first decanting to surface streams in 2002 (Hobbs and Cobbing 2007). In 2010, an Inter-Ministerial Committee was formed to address this situation and a technical proposal adopted to manage AMD in the three affected gold fields by pumping and treatment (Ramontja et al. 2011).

Development of solutions to high-priority mining areas

The solutions proposed for the three affected gold fields are based on similar principles. The water level should be maintained at or below an Environmental Critical Level (ECL), defined to be the minimum depth which will ensure that no water decants to surface or flows into surrounding aquifers. In the case of the West Rand, where AMD has already discharged to surface, this will require pumping to lower the water level, while in the Central Rand and East Rand, pumps are needed to maintain the water level and stop it from rising. The proposed ECLs have been established using conservative assumptions, and may be brought closer to the surface when control over the water level has been established by pumping.

It is anticipated that the water quality will be poor, with the West and Central Rand Gold Fields having produced acidic water in the past, while the East Rand is expected to produce circumneutral water. All of these will require treatment. Regional considerations make the desalination and re-use of the mine water the most environmentally attractive option (Department of Water Affairs and Forestry 2009). The options for the different Gold Fields are summarised in Table 1.

Table 1. Information required for the design and implementation of the short-term solution for acid mine drainage in the Witwatersrand.

Information required	West Rand	Central Rand	East Rand
Environmental Critical Level	Maximum depth to ECL well constrained due to good conceptual hydrogeological model.	ECL poorly defined using a simple conceptual understanding of groundwater levels.	ECL conservatively defined based on possible interactions with near-surface dolomitic aquifer.
Required pumping rate	Difficult to determine due to a lack of historical pumping data and poor flow measurements under decant conditions	Based on reported pumping rates.	Based on reported pumping rates.
Expected water quality	Based on intensive monitoring of decant and decant affected systems over several years.	Based on modelled and reported water quality.	Based on measurements on pumped water (circumneutral)

Anticipated water quality

Topographic controls over the level of flooding in the Witwatersrand mine voids

The surface topography of the Witwatersrand undulates, with the elevation along the trace of the Main Reef varying between around 1650m and 1750m above mean sea level in the Central Rand and 1700m and 1800m in the West Rand. The surface discharge points for both of these mine voids is a few tens of metres below the lowest point on the reef outcrop at or close to lower lying shafts accessing the void. In both cases, the environmental critical level has been determined to be around 150m below this (Ramontja et al. 2011). As a result, at the topographic highs within each gold field, as much as 300m of vertical extent of the mine void will not be flooded, exposing sulphide-rich ore and host rock to air, while water entering this zone will flow though the underground workings towards the flooded zone below this level. Even if the voids were allowed to flood completely and discharge to surface, underground workings with a vertical extent of up to 150m would remain open.

Laboratory simulation of the saturated and unsaturated zones in the underground mine voids

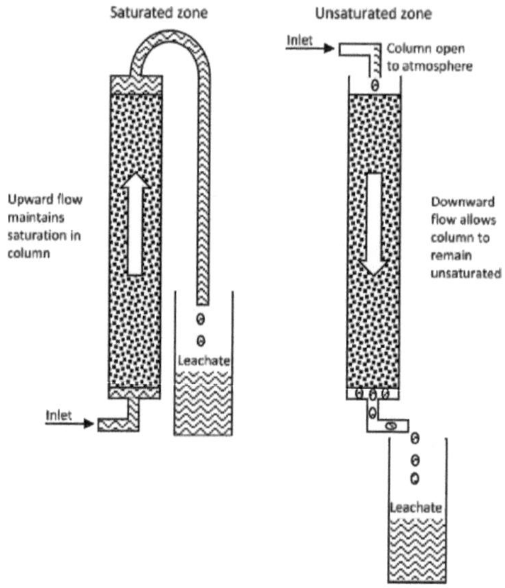

Fig.2. Column setup for the comparison of AMD generation in the saturated and unsaturated zones of the underground mine voids of the Witwatersrand.

The impact of this unsaturated surface zone in the underground workings has been investigated experimentally, using different column configurations, as illustrated in Figure 2. A sample of pyritic conglomerate was collected from an abandoned open cast operation in the West Rand Gold Field. The sample was crushed and a <4mm fraction collected. Two subsamples with masses of approximately 1.3kg each were placed in columns 5cm in diameter and 40cm long, and inoculated with acid mine water collected from a shaft in the vicinity of the sampling site. One column was configured for upward flow, ensuring complete saturation with water and the other left open at the top with water percolating downwards through it.

The water flow rate in both columns was adjusted to allow a flow of approximately 1L/kg every three weeks. At the time of writing, this experiment had been running for 71 weeks, giving a cumulative liquid to solid ratio of approximately 23L/kg.

pH and electrical conductivity (EC) values for the two column experiments are presented on Figures 3 and 4 respectively. The unsaturated column shows a sharp deterioration in water quality at a liquid to solid ratio of around 8L/kg, most likely indicating the consumption of all available buffer capacity in the sample. In the saturated column, water quality improves sharply soon after the start of the measurement period, representing the washout of already oxidised material and any residual acidic water remaining from the inoculation of the columns.

The implication for long-term water management is that flooding to higher levels would reduce the oxygen availability in the flow path from the surface and near-surface ingress of water to the underground workings, potentially resulting in better quality water reporting at the pump stations, with a consequent reduction in treatment cost. The cost of pumping from shallower depths would also reduce the total cost of water management, while the possibility of a gravity-driven decant tunnel (Ramontja et al. 2011) may become technically and economically feasible.

The implications of an upward adjustment of the current determinations of the environmental critical levels would need to be fully assessed, particularly where this could have a negative effect on groundwater quality. It must be remembered that the currently proposed environmental critical levels have been determined conservatively, based on limited information. The current short-term initiative to pump and treat water to regain the pre-flooding status quo will provide valuable information with regard to the hydrodynamics of the underground workings and possible interactions with shallow groundwater.

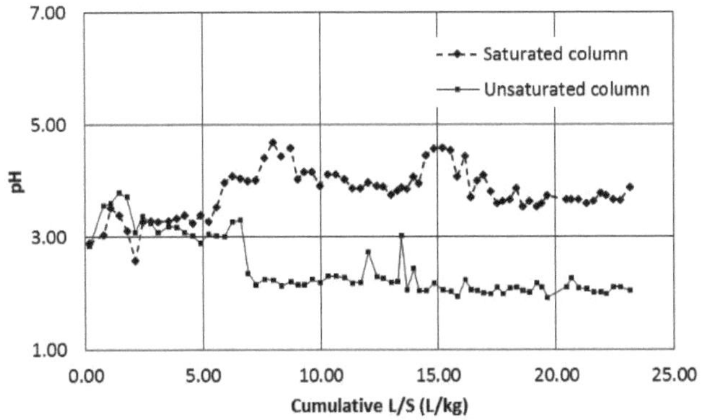

Fig.3. pH values recorded for the saturated and unsaturated columns used to simulate water flow through the saturated and unsaturated zones in the Witwatersrand mine voids.

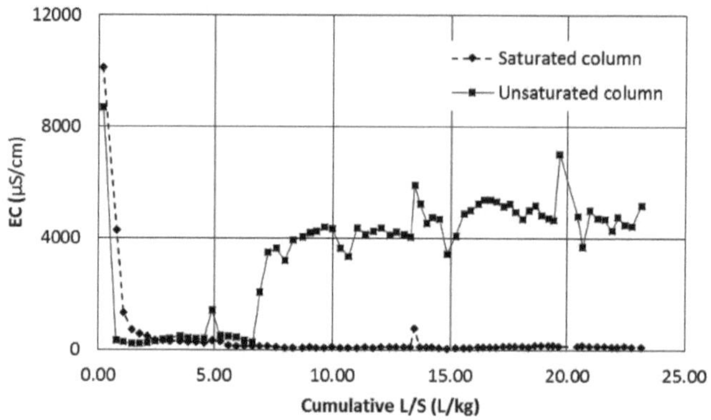

Fig.4. Electrical conductivity values recorded for the saturated and unsaturated columns used to simulate water flow through the saturated and unsaturated zones in the Witwatersrand mine voids.

Anticipated water volume

For planning of the required pump and treat facilities in the Witwatersrand it was essential to predict the ultimate volume of water which would require treatment. At the times that mines closed, no reliable measurements of water volume were available, complicating the selection of pumps and the design of treatment facilities. In the Central- and East Rand Gold Fields, summaries of historical pumping data could be used, presented by Scott (1995), based on historical data from earlier periods when multiple mines were operating and pumping. These values were used as a basis for design.

Prediction of the required pumping rate for the West Rand

In the West Rand Gold Field, no reliable and comprehensive record of historical pumping data could be identified. Initial planning after the flooding of the underground workings and discharge to surface in 2002 was based on a modeled discharge rate of between 7 and 12ML/d (pers comm. M van Biljon), but designs based on this proved inadequate for the volumes of water discharging.

Data collected by the mines in the area between January 2010 and June 2011 show a reported discharge (pumped or decanted to surface) volume between 12.5 and 67.2ML/d, with strong seasonal variation. Furthermore, the water level was monitored throughout the flooding of the underground mine void (pers comm. M van Biljon). This provides sufficient data to estimate the recharge to the mine void and, in combination with the rate of rise during mine flooding, can be used to estimate the volume of the mine void for any specified depth interval. It should be noted that the data available covered only slightly more than a single hydrological year, making a robust prediction, incorporating year-to-year variability in discharge rate impossible. The quality of the measured flow data could also not be verified.

Despite the inherent uncertainty in the available data, a simple model was constructed (Coetzee, 2011) using the basic assumption that, recharge to the underground workings is equal to the discharge from the underground workings, when flooded to surface decant level. A specified pumping volume could then be subtracted from this recharge value and the difference be used to predict the change in water level, based on the water level curve measured during flooding of the underground workings. Using this model it was recommended that a pumping volume of 40ML/d would dewater the underground workings and lower the water level to the ECL within a reasonable period of time.

Pumping throughout the 2013 dry season lowered the water level in the mine void from decant level by more than 3 meters. Unfortunately, technical problems with the pumps, together with above average rainfall recorded in early 2014 caused the discharge of AMD to the surface to start again in March 2014 (Kolver, 2014).

Conclusions

The acknowledgement of acid mine drainage as a problem by the South African government in 2010 facilitated the development and start of implementation of practical solutions to the problem in areas where deep underground mining has ceased and the underground voids have been allowed to flood. Because of insufficient pre-closure planning, very little rigorous prediction of the rate and consequences of flooding was done and the flooding process was not managed.

Limited data were available for the prediction of water volumes and flow rates and the consequences of flooding, creating a requirement for the proposal of solutions based on conservative conceptual models of the flooding process and AMD generation. In some cases, it will only be possible to properly validate these models during the flooding and solution implementation processes.

Laboratory simulations have indicated that in mining areas such as the Central- and West Rand Gold Fields, where the discharge level is lower in elevation than a large portion of the underground workings, AMD generation is likely to continue due to the flow of water through air-filled voids, while it may be expected to be limited in the lower levels which are fully flooded.

Estimates of water volume have been difficult due to a lack of reliable historical data on pumping volumes. It is nevertheless possible to predict appropriate pumping volumes using limited historical data.

These results underline the importance of comprehensive planning for long-term water management before the closure of complex underground mining operations with the potential to generate acid mine drainage. In areas where multiple mines are interconnected, such as South Africa's Witwatersrand Gold Fields, closure planning must be undertaken at a regional scale as well as the scale of individual mines.

References

Coetzee, H. (2011). A flooding and pumping model for the Western Basin of the Witwatersrand Gold Fields, based on empirical data. Council for Geoscience Report No. 2011-0176, Pretoria, Council for Geoscience: 10pp.

Department of Water Affairs and Forestry (2009). Vaal River System: Large bulk water supply reconciliation strategy: second stage reconciliation strategy (March 2009). Report Number P RSA C000/00/4406/08, Pretoria, Department of Water Affairs and Forestry: 130pp.

Hobbs, P. J. and Cobbing, J. E. (2007). A hydrogeological assessment of acid mine drainage impacts in the West Rand Basin, Gauteng Province. CSIR/NRE/WR/ER/2007/0097/C, Pretoria, CSIR/THRIP: 59pp.

Hocking, A. (1986). Randfontein Estates: The first hundred years. Bethulie, Hollards.

Kolver, L. (2014). Govt lauded for AMD paradigm shift, but complexity not fully grasped. Mining Weekly, 18 April 2014. Creamer Media.

Ramontja, T., Coetzee, H., Hobbs, P. J., Burgess, J. E., Thomas, A., Keet, M., Yibas, B., van Tonder, D., Netili, F., Rust, U. A., Wade, P. and Maree, J. (2011). Mine Water Management in the Witwatersrand Gold Fields With Special Emphasis on Acid Mine Drainage Pretoria, Inter-Ministerial-Committee on Acid Mine Drainage. Report of the Team of Experts for the Inter-Ministerial Committee on Acid Mine Drainage: 146.

Scott, R. (1995). Flooding of Central and East Rand Gold Mines; An investigation into controls over the inflow rate, water quality and the predicted impacts of flooded mines, Pretoria, Water Research Commission: 238pp.

Assessment of the success of rehabilitation at waste rock piles of the former uranium mining from the supervisory authority's perspective by the example of Schlema-Alberoda (Germany)

Klaus Flesch[1], **Andrea Sperrhacke**[1]

[1]Saxon State Office for Environment, Agriculture and Geology (Germany)

Abstract. The Wismut GmbH has the obligation to rehabilitate the damages at their legacies resulting from the former uranium ore mining. According to the legislations in Germany, the handling with radioactive material is subject to the regulatory control through licensing. As a result of the rehabilitation measures, the limit of 1 mSv per year on the effective doses for public exposure must not be exceeded. In this article, the success of rehabilitation from the supervisory authority's perspective is debated.

Introduction

On 21 Dec 1991 the former SDAG Wismut in the GDR was transformed into the Wismut GmbH due to the unification of Germany. This transformation act was connected with a restructuring process of a uranium mining operation into a rehabilitation company. The main issues of the Wismut GmbH are to rehabilitate the damages at their legacies resulting from the former uranium ore mining.

These rehabilitation issues were characterized as intervention measures by the German Commission on Radiation Protection at the beginning of the 1990s. According to the legislation in Germany, the handling with radioactive substances is subject to the regulatory control through licensing. As a result of the rehabilitation measures, the limit of 1 mSv per year on the effective doses for public exposure must not be exceeded. In the past 20 years numerous regulatory approvals on radiation protection for the rehabilitation measures on industrial tailings ponds, waste rock piles and on operational sites have been granted.

Especially for the Schlema-Alberoda site, a solid experience on the success of the rehabilitation process at waste rock piles can be concluded because the most remediation projects have been finished or will be finished in the next time. One of the most important rehabilitation measures is to attain sustainable effects with regard to minimizing the releases of radon via the air path and the releases of radi-

B.J. Merkel and A. Arab (eds.), *Uranium – Past and Future Challenges*,
DOI 10.1007/978-3-319-11059-2_38

onuclides via the water path out of the waste rock piles. After finishing the rehabilitation measures, their success of rehabilitation is controlled by monitoring the air path and the water path. These monitoring issues are supervised by the Saxon State Office for Environment, Agriculture and Geology (LfULG). In this article, the success of rehabilitation from the supervisory authority perspective is debated. Especially the results to the release of radon via the air path out of the rehabilitated waste rock piles are sensitive with respect to the current annual effective doses for public exposure.

Historical background or uranium ore mining in the GDR

After 1945, the uranium ore mining started in Saxony and Thuringia. Therefore, the Soviet occupation forces established the so-called "Sowjetische Aktiengesellschaft Wismut" (SAG Wismut). At the beginning, the uranium ore mining was characterized by a rigorous production philosophy with inefficient processes and an excessive use of land. In 1954, the "Sowjetisch-Deutsche Aktiengesellschaft" (SDAG Wismut) was founded and the GDR was implemented as a shareholder within the company. This company had about 45,000 employees from the end of the 1950's until 1990. A total mass of 231,000 tons of uranium were produced in the time period between 1945 and 1990. Together with this considerable amount of the uranium ore production, large-scaled contaminated areas especially with tailings ponds and waste rock piles were generated.

The uranium ore mining was terminated on 31 December 1990. Due to the German reunification, the Soviet Republic disclaimed its shares and Germany became a sole shareholder. The so-called Wismut GmbH was founded and this company changed from a uranium ore producer into a rehabilitation company by enacting the Law to the Treaty on 12 December 1991 ("Wismut Law").

The rehabilitation issues of the Wismut GmbH

The Wismut GmbH has to rehabilitate the large-scaled radioactive contaminated sites originated from the former uranium ore mining in the GDR since 1991. These legacies include more than 1,000 objects in Saxony and Thuringia (Fig. 1). Some of the important issues of the Wismut GmbH are the safe encapsulation of tailings ponds and of waste rock piles. And the company has also to abandon and to flood the underground mines or to dismantle and to demolish contaminated buildings.

Fig.1. Wismut-sites in Saxony and Thuringia.

The legal bases for the rehabilitation of the Wismut legacies

The obligation of the rehabilitation measures were characterized as intervention for existing situations by the German Commission on Radiation Protection at the beginning of the 1990s. The Unification Treaty of 6 September 1990 and furthermore the provisions of § 118 of the German Radiation Protection ordinance of 2001 defined in regard to the decommissioning and the rehabilitation of Wismut legacies that the provisions of the former GDR concerning radiation protection shall continue to apply. These provisions are the

5. Ordinance on the Guarantee of Nuclear Safety and Radiation Protection of 11 October 1984, abbreviated VOAS,
   and the
6. Order for the Guarantee of Radiation Protection in Dumps and Industrial Sedimentation Installations and for the Use of Materials Stored therein of 17 November 1980 (GBl. I, No. 34 p. 347), abbreviated HaldAO.

In the VOAS is determined in § 4 that an authorization is required for the use of radioactive materials with activity values above 0.2 Bq/g. And according to the §§ 9 to 12 of the VOAS, it is required that for the protection of life and health of the population the radiation protection principles "justification" and "optimization" have to be fulfilled, and the adherence of the dose limits shall be verified. With regard to the dose limit is to ensure that the average value of the effective dose equivalent per year will be limited for a period of 50 years to 1 mSv. This dose criterion as a primary guidance level is one of the most important objectives, to assess the need of rehabilitation measures and its success.

In addition to the VOAS, the HaldAO regulates in § 4 more detailed aspects on the requirement of the authorization for the rehabilitation work at waste rock piles and at tailings ponds that have an effect on the environment and for the re-use of rehabilitated areas. And, according to § 9 of the HaldAO the effects of rehabilitation measures have to be controlled by a monitoring system.

Regulatory control on radiation protection for rehabilitation measures at Wismut legacies

If rehabilitation measures are executed, a handling of radioactive materials occurs. This is a subject to the regulatory control through licensing according to the legislations in Germany.

The competent authority

For carrying out the approval process the radiation protection regulations, the LfULG is the competent authority in Saxony (Germany). The LfULG is a subordinate official authority of the Saxon State Ministry of Environment and Agriculture (SMUL) as the supreme authority.

The LfULG granted the approvals to radiation protection on the rehabilitation of each individual object of the Wismut legacies in Saxony. In the past 20 years about 430 approvals on radiation protection were granted for the remediation measures on Wismut legacies.

Application for authorization

For the granting of an approval application documents must be submitted for each object for which the rehabilitation measures to be realized. These documents are prepared by the Wismut GmbH as the applicant and they include at least information on

- a preliminary and an approval planning,
- a radiological and an environmental assessment and
- a radiation protection instruction.

In each individual approval, the objectives of the rehabilitation measures are defined. However, these objectives depend on the kind of the object and they are very different. Any subsequent programs of measures must have a rehabilitation objective which is achievable at reasonable efforts. The relevant rehabilitation objectives cannot be discussed comprehensively for all object types (tailings ponds, waste rock piles, underground mine etc.) in this article. For that reason the relevant criteria from the point of view of radiation protection for the waste rock piles on the Schlema-Alberoda site as a case example will be presented (Fig. 2). At this site about 120 approvals have been granted for some 15 objects in the last 20 years.

Fig.2. Waste rock piles on the Schlema-Alberoda site.

Relevant objectives for rehabilitation of waste rock piles (example from the Schlema-Alberoda site)

Especially for the Schlema-Alberoda site, a solid experience is available to assess the rehabilitation process at waste rock piles because the most rehabilitation projects have been finished or will be finished in the next time. Due to the rehabilitation measures, sustainable effects should be attained with regard to minimizing the external gamma radiation, the release of radon via the air path and the release of radionuclides via the water path out of the waste rock piles as well as the prevention of a direct contact to the waste rock material and the elimination of dusting from exposed waste rock surfaces. But also the slope stability and the long-term effects are relevant objectives. The waste rock piles at the Schlema-Alberoda site were covered with an insulation layer (standard cover system, Fig. 3) that was designed taking into account the radiological and geotechnical aspects listed above.

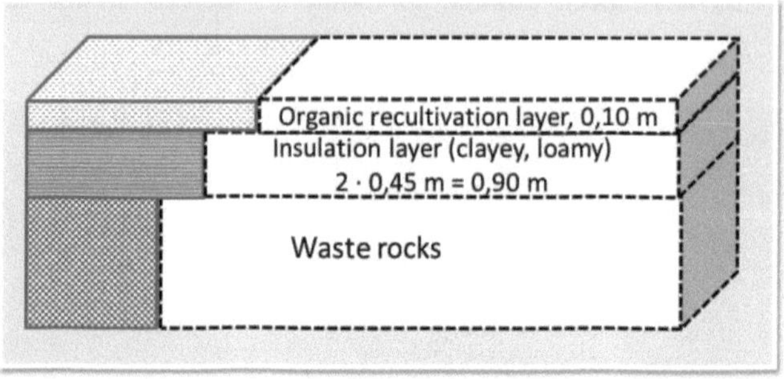

Fig.3. Standard cover system at the rehabilitated waste rock piles on the Schlema-Alberoda site.

Monitoring program to assess the rehabilitations success

After finishing the cover system on the waste rock piles, the success of the rehabilitation measures are controlled by monitoring the air path and the water path. The Wismut GmbH as the authorization holder has to manage the monitoring programs for each rehabilitated object at the Schlema-Alberoda site.

These monitoring issues are supervised by the LfULG. Especially the results to the release of radon via the air path out of the rehabilitated waste rock piles are sensitive with respect to the current annual effective doses for public exposure. As mentioned above, the limit of 1 mSv per year on the effective doses for public exposure must not be exceeded. This criterion is based on the limit value of 1 mSv per year for public exposure averaged over a time period of 50 years according to the VOAS.

The success of rehabilitation from the supervisory authority's perspective

The first rehabilitation measures at the waste rock piles were finished about ten years ago. At this time the monitoring programs to control the success of rehabilitation began at these objects. For this reason, we are able to evaluate some results with regard to the success of rehabilitation. The most relevant criteria with regard to radiation protection (external gamma radiation, dust, seepage water) do not play any significant role. However, the data that were sampled over the ten-year period for observing the release of radon via the air path show an inhomogeneous picture.

On the one hand, the essential requirement with compliance of the limit of 1 mSv per year on the effective doses for public exposure is fulfilled in the most rehabilitated waste rock piles at the Schlema-Alberoda site. But on the other hand, a significantly increased radon exhalation was locally observed in some objects. Especially during the summer, in individual cases at the lower parts of the waste rock piles nearby the dwelling houses, the magnitude of release of radon is comparable with the situation before the rehabilitation measure. The effective doses for public exposure attain in these locally restricted areas values between 3 to 5 mSv per year. This size cannot be neglected.

One conclusion is that the standard cover system does not generally fulfill the intended function to retain the release of radon. From the supervisory authority's perspective, there is a need for a new basic strategy for optimizing the radon situation at the waste rock piles nearby the dwelling houses. However, also the legal situation should be reviewed to determine to what extent it can be assessed sensible and reasonably, whether an annual exposure from 3 to 5 mSv can be accepted in locally restricted areas. In accordance with the Alara-principle ("As low as reasonable achievable"), the situation within the relevant areas at the foot of the embankment of the waste rock piles has to be examined, if the radon situation could still be improved under technical, economic, but also social aspects.

Basic strategy for optimizing the radon situation

Till the beginning of 2014, the Wismut GmbH has investigated several approaches to solve the problem. Based on the international state of the art, the company examined alternative solutions applying non-invasive or less-invasive methods. These methods include among others investigations to regulate the water balance in the cover, to install gas wells/gas drainages or to control the vegetation. All measure did not lead to the desired result to achieve an annual exposure below 1 mSv per year for the public exposure.

In conclusion, the LfULG agreed that the radon situation nearby the dwelling houses could be optimized but therefore methods with a more-invasive technology have to be taken into account. Depending on the suitable options, the technical, economic and the social aspects have to be analyzed within an optimizing process that has recently started.

Alternative view with regard to the legal situation

With regard to the recommendations of the International Commission on Radiological Protection (ICRP) and also to the Directive 2013/59/EURATOM (Basic Safety Standards (EU BSS)) published in January 2014) is to consider, whether and to what extent the now existing exposure situation of 3 to 5 mSv per year for public exposure is allowed to be accepted at the waste rock piles nearby the dwelling houses permanently.

The ICRP recommends in their publication no. 103 that national authorities should set reference levels as an aid to optimization of protection against radon exposures. For the sake of continuity and practicability, the ICRP retains the upper value of 10 mSv for the annual dose reference level.

If the Wismut legacies can be interpreted as existing situations according to article 73 (2) and Annex XVII of the new EU BSS, then reference levels expressed in effective doses should be set in the range of 1 to 20 mSv per year. The site specific reference levels should take account of the features of prevailing situations as well as societal criteria. Having regard to the general principle of justification, an existing exposure situation warrants no consideration of protective or remedial measures.

Assessment of the radiological situation at the rehabilitated waste rock piles

As mentioned above the limit of 1 mSv per year on the effective doses for public exposure is fulfilled in the most rehabilitated waste rock piles at the Schlema-Alberoda site. A first result can be summarized that the realized rehabilitation measures are in principle successful.

The locally existing situation with an annual exposure of 3 to 5 mSv per year cannot be assessed in a conclusive way at this stage. The technical, economic and the social aspects for optimization have yet to be analyzed. For transparency this process has to be developed with a maintaining close stakeholder involvement and with access to high-level expertise. At this stage, the LfULG is not only the supervisory authority. The LfULG also has the responsibility to communicate with the local representatives at the Schlema-Alberoda site as well as residents affected by a higher exposure due to the release of radon out of the waste rock piles.

Microbes affect the speciation of various uranium compounds in wastes and soils

A. J. Francis[1]

[1]Division of Advanced Nuclear Engineering, Pohang University of Science and Technology, Pohang, Korea; Environmental Sciences Department, Brookhaven National Laboratory, Upton, New York 11973, USA

Abstract. Large volumes of uranium wastes are generated by nuclear- fuel production, nuclear power plants, and by facilities producing nuclear weapons. Uranium is present in various chemical forms such as elemental, oxide, sulfide, ionic, inorganic complex, organic complex, and co-precipitated with iron-, manganese-oxides, and carbonate. The mineralogical association and the oxidation states of uranium affect solubility, stability, bioavailability, and environmental mobility. Microorganisms affect the speciation of various chemical forms of uranium present in wastes and contaminated soils. The speciation of uranium affected by microbial action such as bioreduction, biosorption, biotransformation of uranium complexed with naturally occurring organic ligands, and solubility are reviewed.

Introduction

Contamination of soils, sediments, and materials with uranium from mining and milling activities, fabricating the nuclear fuel, nuclear-power plants, nuclear accidents, disposal of nuclear wastes (low- and intermediate-level waste (LILW), transuranic (TRU) wastes), reprocessing of the spent nuclear fuel, nuclear weapons testing, phosphate mining wastes and disposal of depleted uranium is a major environmental concern. In addition to the radionuclides the LLW, ILW, and TRU wastes contain absorbed liquids, sludges, processed nitrate waste, organic compounds, cellulose, plastics, rubber materials, and chelating agents from decontamiantion. Unlike organic contaminants, the uranium cannot be destroyed to innocuous compounds, but must be converted to a stable form or removed for proper disposal or recycled.

Uranium in contaminated environments may be present in one or more chemical forms depending on the process and waste stream. Microbial activity could affect the chemical nature of the uranium by altering the oxidation state and speciation, the solubility and sorption properties and thus could increase or decrease the

B.J. Merkel and A. Arab (eds.), *Uranium – Past and Future Challenges*,
DOI 10.1007/978-3-319-11059-2_39

concentrations of uranium in solution. Initially uranium may be present as insoluble form and may be converted to soluble form after disposal by the actions of microorganisms. The chemical form, the availability of electron donors, nutrients (nitrogen, phosphorus), the presence of electron acceptors (Fe^{3+}, Mn^{4+}, NO_3^-, SO_4^{2-}, organic compounds), and the environmental factors (pH, Eh, temperature, moisture) affect the type, rate, and extent of microbial activity, and hence, uranium transformation.

The direct implication of microorganisms in the biotransformation of various uranium compounds is of considerable interest because of its potential application in bioremediation of contaminated sites, in pre-treating radioactive wastes, and in processes critical to the performance of nuclear waste repositories. The *in-situ* or *ex-situ* bioremediation strategies of uranium contaminated soils, groundwater, and waste streams include one or more of the microbially mediated processes such as bioaccumulation, bioreduction and biodegradation of uranium-organic complexes. However, one of the major concerns is that uranium stabilized by microbial treatment is not very stable and the potential for remobilization of biorduced uranium are very high, for example, via re-oxidation by oxygen and other oxidants or complexation with soluble organic ligands.

Chemical speciation of uranium

Uranium exists in several oxidation states U(III), U(IV), U(V), and U(VI). Environmentally relevant ones are U(IV) and U(VI). The former is relatively insoluble while the latter is highly soluble. Uranium in contaminated environments and wastes may be present as inorganic salts, oxides, coprecipitates with calcium carbonate, iron and manganese oxides, inorganic-, organic- complexes and naturally occurring minerals.

Dissolution of uranium

Oxidizing and reducing conditions affect the mobilization and immobilization of uranium. Inorganic and organic complexes of uranium affect its mobility in the environment. Uranium is predominantly present in the natural environment as soluble uranyl carbonate species. Many organic compounds form stable complexes with uranium, and increase its solubilization and leaching. Likewise, microbial metabolites and waste degradation products or intermediates may be an important source of agents that affect uranium solubility. Aerobic and anaerobic microorganisms are directly or indirectly involved in the mobilization and immobilization of various chemical forms of uranium.

Microbial leaching of uranium from ores. Uranium in ores is present as uraninite and pitchblende and in secondary mineral phases associated with silicates, phosphates, carbonates, and vanadates. The concentration of uranium can vary between 0.5 and 20%, with the highest amount occurring in Canadian ores. Mill tailings, a by-product of the mineral extraction process, contain up to 2% uranium. The residual uranium that has not been extracted may be present as a result of newly formed insoluble mineral phases (e.g. $CaSO_4$, $MgCO_3$, $Fe(OH)_3$) which provide surface sites for uranium adsorption. Both autotrophic and heterotrophic microorganisms solubilize uranium.

Autotrophic microbial activity. The iron and sulfur oxidizing bacteria play a significant role in the solubilization of uranium from ores and in mill tailings. The biogeochemistry of uranium recovery from ores and bacterial leaching has been extensively studied. The role of autotrophic bacteria *Thiobacillus ferrooxidans* in the extraction of uranium from ore is primarily indirect action due to generation of the oxidizing agent ferric sulfate (indirect oxidation by Fe^{2+} / Fe^{3+} cycle as the chemical electron carrier) and the solvent sulfuric acid from sulfur oxidation (Francis 1990). Direct oxidation of uranous sulfate and UO_2 by *T. ferrooxidans* has also been reported

$$UO_2 + O_2 + 4H^+ \rightarrow 2UO_2^{2+} + 2H_2O$$
$$UO_2 + O_2 + 2Fe^{3+} \rightarrow 2UO_2^{2+} + 2Fe^{2+}$$
$$4Fe^{2+} + O_2 + 4H^+ \rightarrow 2Fe^{3+} + 2H_2O$$

Heterotrophic microbial activity. Aerobic and anaerobic microorganisms in the presence of electron donors such as metabolizable organic compounds and electron acceptors such as NO_3^-, Fe(III), Mn(IV), SO_4^{2-}, and CO_2 affect uranium oxidation state and solubility. Organic carbon is degraded under aerobic conditions to intermediary products, CO_2 and H_2O; while under anaerobic conditions it is degraded to intermediate products which accumulate or completely degraded to CH_4 and or CO_2, H_2, with concomitant release of NH_3, H_2S, PO_3^4. A *Halomonas* sp when grown in WIPP brine under denitrifying conditions solubilized uranyl hydrogen phosphate precipitate to soluble uranyl carbonate due to metabolic activity (Fig 1). The organism metabolized the carbon source succinate for growth under anaerobic (denitrifying) conditions using nitrate as an alternate electron acceptor. The carbon dioxide produced from the metabolism of the carbon source accumulated in the medium causing the dissolution of uranyl hydrogen phosphate to uranyl carbonate (Francis et al. 2000).

Dissolution of uranium by organic compounds. Dissolution of uranium is due to organic acid metabolites produced by heterotrophic microorganisms as well as lowering of the pH of the medium from the metabolism of organic compounds (Berthelin and Munier-Lamy 1983; Bloomfield and Pruden, 1975). In many cases, a combined effect is important. For example, organic acids produced by microorganisms may have a dual effect in increasing U dissolution by lowering pH, and

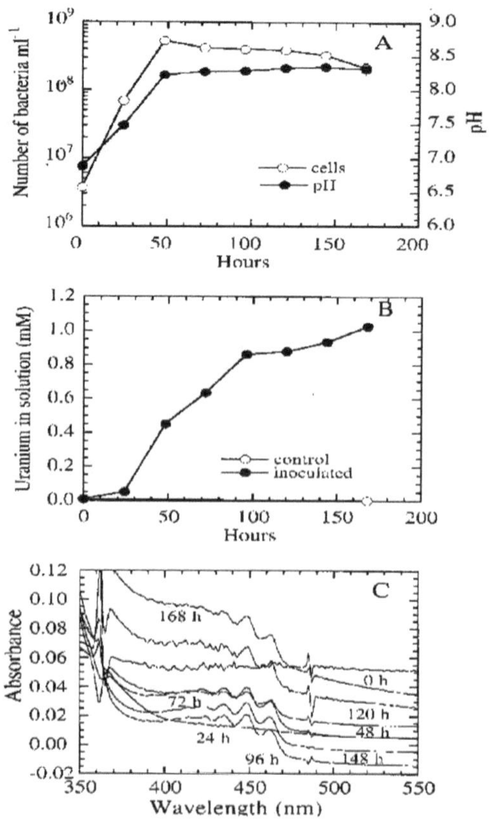

Fig.1. Dissolution of uranylhydrogen phosphate precipitate by *Halomonas* sp. (Francis et al. 2000). (A), Growth of bacteria and changes in pH of the medium. (B), Dissolution of $UO_2HPO_4 \cdot 2H_2O$ to uranyl dicarbonate as function of bacterial activity. (C), UV-vis spectra of the culture medium showed the formation of uranyl carbonate as a function of bacterial growth.

by complexation. Heterotrophic bacteria and fungi are not only known to solubilize various minerals including silicates (quartz, feldspar, mica) but also release metals associated with them, including Cu and Ni from copper-nickel concentrates, Cu from low-grade copper ore, uranium from granites, and potassium from leucite.

A wide variety of heterotrophic microorganisms, such as *Bacillus sp., B. luteus, B. subtilis, B. cereus, B. pumilis, Pseudomonas striata, P. viscosa, P. perolens, P. choloroaphis, Achromobacter xerosis, A. stoloniferum, and A. healii* may be involved in solubilizing uranium from granitic rock where uranium is generally present as an oxide. Such solubilization is due to the production of organic-acid metabolites, such as oxalic, isocitric, citric, succinic, hydroxybenzoic, and coumaric acids via their carboxylic and phenolic groups (Bertelin and Munier-Lamy,

1983). When microorganisms are grown in an iron-deficient medium, they elaborate specific iron chelators, such as siderophores (Brainard et al. 1992). Iron-sequestering agents such as siderophores elaborated by microorganisms could play an important role in the complexation of radionuclides and so increase their solubility. *Pseudomonas aeruginosa*, grown in the presence of uranium elaborated several metabolic products which complexed uranium (Premuzic et al.1985). Microbially produced dicarboxylic acids, oxalic, isocitric, citric, succinic, ketogluconic acid, polyhydroxy acids, and phenolic compounds such as protocatechuic acid, and salicylic acid are effective chelating agents but their ability to extract uranium from ores has not been fully explored.

The complexation of uranium with organic ligands such as ketogluconic, oxalic, malic, citric, protocatechuic, salicylic, phthalic, and fulvic acids and catechol were investigated (*Dodge, Vasquez, and Francis, unpublished results*). Potentiometric titration of uranium with the organic ligands confirmed complex formation. EXAFS analysis and electrospray ionization-mass spectrometry (ESI-MS) showed that ketogluconic acid formed a mononuclear complex with uranium involving the carboxylate group, while malic acid, citric acid, and catechol formed binuclear complexes. Phthalic acid formed a bidentate complex involving the two carboxylate groups, while catechol bonded to uranium through the two hydroxyl groups. The hydroxycarboxylic acids were bound in a tridentate fashion to uranium through two carboxylates and the hydroxyl group.

Immobilization of uranium

The immobilization of uranium is brought about by bioaccumulation, bioreduction and bioprecipitation reactions. Uranium is reduced by a wide variety of facultative and strict anaerobic bacteria under anaerobic conditions in the presence of suitable electron donor. Consequently, the potential exists for the use of anaerobic bacteria to concentrate, contain and stabilize uranium in contaminated groundwaters and in waste with concurrent reduction in waste volume. However, the long-term stability of bacterially reduced uranium in the natural environment is not known.

Biosorption and bioaccumulation of uranium. Biosorption and bioaccumulation of uranium has been observed in a wide range of microorganism (Francis et al. 1998, 2004; Gillow et al.2001). It is one of the intensely investigated areas of research because of the potential use of biomass to remove uranium from waste streams. Bacterial cell walls, exopolymers, proteins, and lipids contain carboxylate, phosphate, amino, and hydroxyl functional groups which bind to uranium. Extracellular and intracellular association of uranium with bacteria was observed but the extent of its accumulation differs greatly with the species of bacteria.Nuclear magnetic resonance spectroscopy (NMR), time resolved laser fluorescence spectroscopy (TRLFS), and extended X-ray absorption fine structure (EXAFS) have been used to determine the functional groups involved in the complexation of uranium with bacteria. In *Halomonas* sp. uranium accumulated as electron-dense intracellular granules and was also bound to the cell surface (Fig 2).

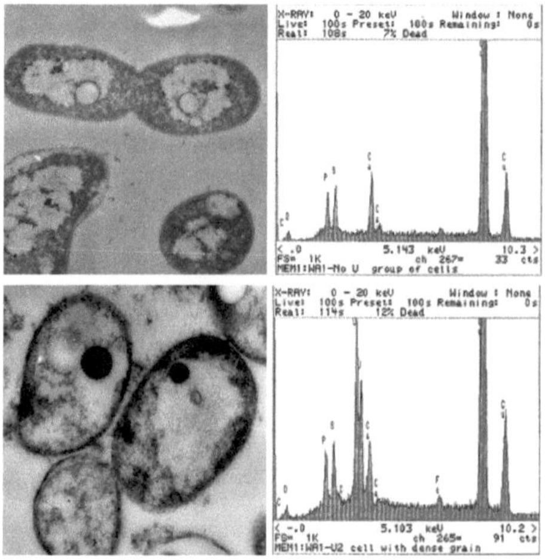

Fig.2. Intra- and extra- cellular accumulation of uranium by *Halomonas* sp. (Francis et al 2004). EDS shows U and P as the major constituents of the intracellular granules. (A) *Halomonas* sp. whole cells (not exposed to U shows poly beta hydroxybutyric acid (PHB) and polyphosphate granules; EDS shows almost equal amounts of Ca and P. (B) After exposure, U is accumulated in the polyphosphate granules.

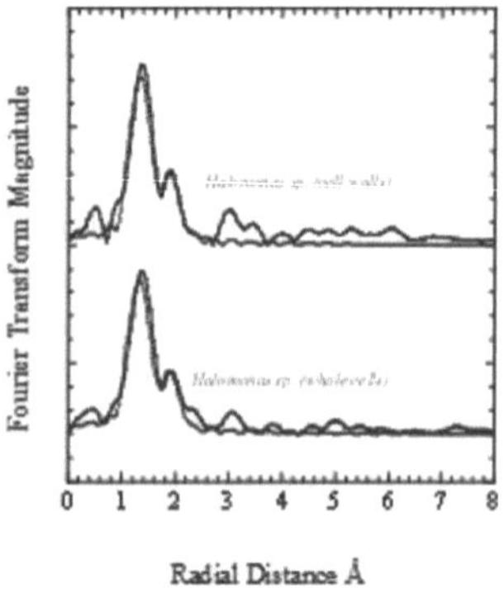

Fig.3. EXAFS analysis show uranium is complexed to diphosphate with whole cells; whereas the lysed cells show bidentate carboxylate bonding in addition to uranyl hydroxide (Francis et al 2004).

Extracellular association of uranium with bacterial cell surfaces is primarily due to physical- and chemical- interactions involving adsorption, ion exchange, and complexation and does not depend on metabolism. Intracellular accumulation is due to binding to anionic sites or precipitating as dense deposits. Intracellular accumulation involves transporting the metal across the cell membrane, which depends on the cell's metabolism. The intracellular uranium transport system into the cell has not yet been unidentified.

EXAFS analysis of the association of U with halophilic bacteria showed that it was associated predominantly with phosphate as uranyl hydrogen phosphate and additional forms of phosphate such as hydroxophosphato or polyphosphate complexes as well as other ligands such as carboxyl species (Figs 3 and 4) (Francis et al. 2004). These studies demonstrate that phosphate, including the polyphosphates, bind significant amounts of uranium in bacteria. Uranium associated with the bacteria is not very stable, as it was desorbed completely by Na_2CO_3, $NaHCO_3$ or 0.1M EDTA.

Reductive precipitation of uranium. In anaerobic environments, certain metals can be reduced enzymatically from a higher to a lower oxidation state which affects their solubility and bioavailability. For example, reduction of Fe(III) → Fe(II), Mn(IV) → Mn(II) increases the solubility, while reduction of Cr(VI) → Cr(III), U(VI) → U(IV) decreases the solubility. The reduction of uranium was reported in axenic cultures of iron-reducing bacteria, fermentative bacteria, sulfate-reducing bacteria, and cell-free extracts of *Micrococcus lactilyticus,* and in uranium wastes by clostridia.

Bioreduction is the reduction of of uranium from higher to lower oxidation state resulting in its precipitation. Although it is a two electrons transfer process, however, it has been proposed that it occurs as one electron transfer reaction with the reduction of U(VI) to U(V) by bacteria and disproportionation of U(V) to U(IV) and U(VI). A wide variety of facultative and strict anaerobic bacteria reduced uranyl-nitrate or uranyl carbonate to U(IV) under anaerobic conditions.

$$U(VI)_{aq} \xrightarrow[\text{fermenters, sulfate reducers}]{\text{Dissimilatory metal reducers}} U(IV)_{S}$$

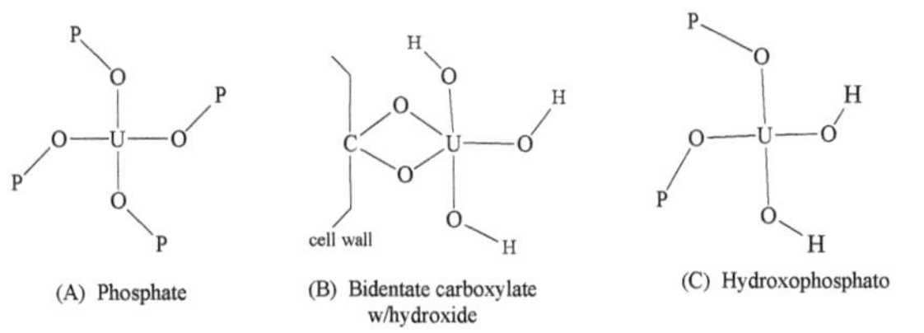

Fig.4. Proposed structures showing uranium association with bacteria (Francis et al 2004).

Sulfate-reducing bacteria catalyze uranium reduction in the presence of electron donor and electron acceptor sulfate with the generation of hydrogen sulfide.

$$SO_4^{2-} + CH_3COOH + 2H^+ \rightarrow HS^- + 2HCO_3^- + 3H^+$$

$$UO_2^{2+} + H_2S \rightarrow UO_2 + 2H^+ + S^0$$

Enzymatic reduction is due to reductases and that cytochrome C_3 is involved in the reduction of U(VI) by *Desulfovibrio vulgaris*. The mechanisms of uranium reduction by various microbes are not fully understood (Wall and Krumholz, 2006).

Uranium reduction by clostridia. Clostridia are strict anaerobic spore-forming fermentative bacteria ubiquitous in soils, sediments, and wastes, reduce uranium from higher to lower oxidation state (Fig 5). The bacterium also reduces iron, manganese, technetium, and plutonium. Reactive barrier technology is based on the activities of the anaerobic bacteria. Reduction of soluble uranyl carbonate ($UO_2(CO_3)_2^{2-}$), uranyl nitrate (UO_2 $NO^{-}_3)_2$, uranyl acetate to insoluble U(IV) by *Clostridium sp.* in culture medium was confirmed by x-ray absorption near edge spectroscopy (XANES) (Fig 6).

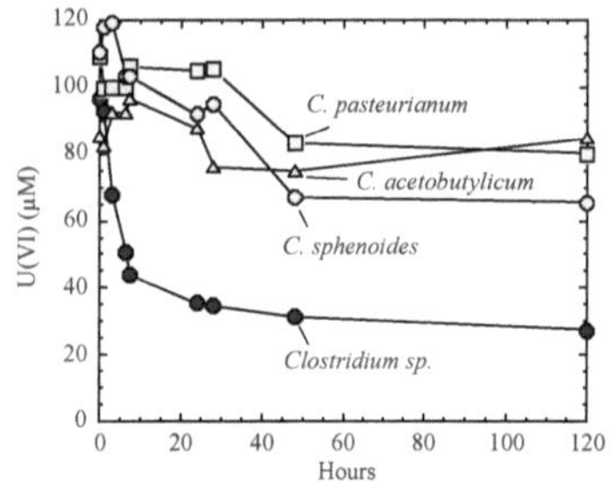

Fig.5. Reduction of U(VI) by Clostridia (Gao and Francis,2008).

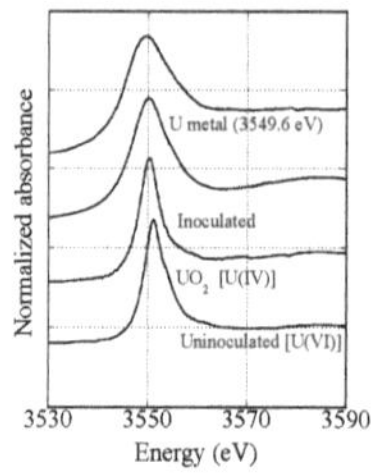

Fig.6. XANES spectra of U reduction by *Clostridium* sp. (Francis et al 1994).

Analysis of uranium contaminated sediment and sludge from Y-12 plant, Oak Ridge, TN showed that the uranium was predominantly in the hexavalent form. Anaerobic bacterial treatment of the sludge and sediment sample showed that uranium was reduced to the tetravalent form. XANES analyses of the samples treated with *Clostridium* sp showed a shift (decrease) in the position of the absorption peak maximum indicating reduction of U(VI) to U(IV) (Fig 7). Complete reduction of U(VI) to U(IV) in the sludge and only partial reduction in sediment were observed. The lack of complete reduction is due to the nature of the mineralogical association of uranium in the sediment (Francis, 1994, 1999).

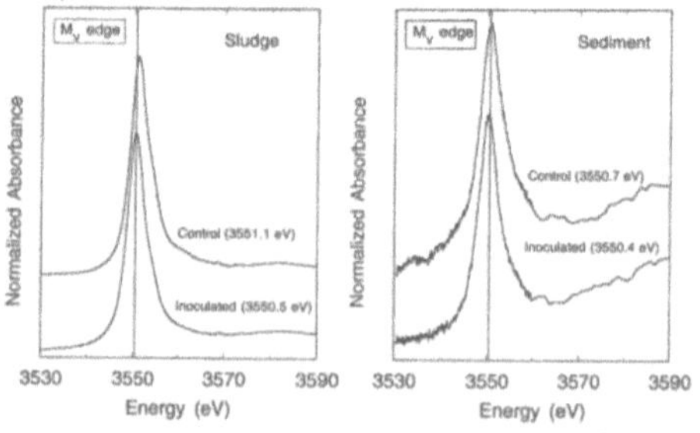

Fig.7. Comparison of XANES spectra of uranium contaminated sludge and sediment before and after anaerobic microbial treatment show reduction of U(VI) to U(IV) (Dodge et al. 1966).

The bioreduced uranium is not stable over the long-term because it can be readily reoxidized by the oxidants such as O_2, Fe(III), hydrogen peroxide or it can be mobilized if soluble organic ligands are present. Therefore it is important that the bioreduced uranium is converted to a more stable form by some additives.

Biotransformation of uranium associated with organic ligands

Naturally occurring soluble organic complexing agents present at the uranium-contaminated sites may not only affect the mobility of uranium but also affect the microbial transformation and reductive precipitation of uranium. Biotransformation of the complexed uranium should result in its precipitation and retard migration. There is a paucity of information on the mechanisms of microbial transformations of uranium complexed with naturally occurring low molecular weight soluble organic ligands.

Complexation and biotransformation of uranyl citrate under aerobic conditions. Citric acid is a naturally occurring, multidentate ligand which forms stable complexes with various metal ions. It forms stable complexes with transition metals and actinides and can involve formation of bidentate, tridentate, binuclear, or polynuclear complex species (Fig 8). Calcium, ferric iron and nickel formed bidendate, mononuclear complexes with two carboxylic acid groups of the citric acid molecule. Copper, ferrous iron, cadmium and lead formed tridentate, mononuclear complexes with citric acid involving two carboxylic acid groups and the hydroxyl group. Uranium has been shown to form a predominantly binuclear complex with two uranyl ions and two citric acid molecules involving four carboxylic groups and two hydroxyl groups. However, the proposed basic structure can vary as a result of changes in solution pH, the ratio of uranium to citrate, temperature and presence of other metals.

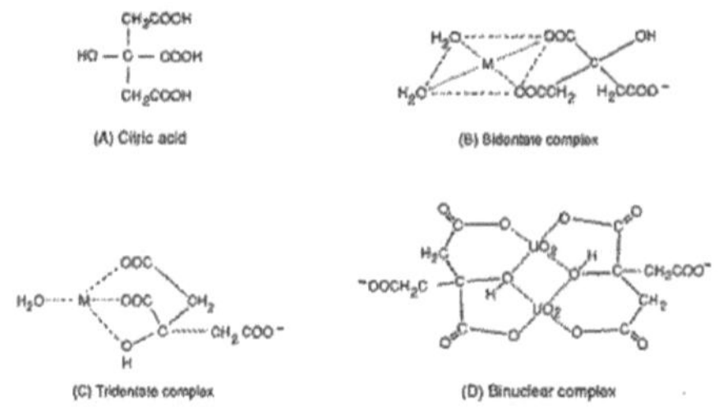

Fig.8. Citric acid forms different types of complexes with metals (Francis et al 1992).

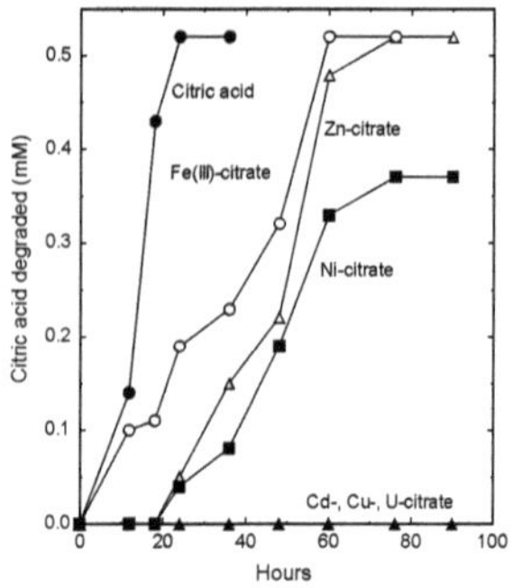

Fig.9. Biodegradation of metal citrates by *Pseudomonas fluorescens* (Francis et al. 1992, Joshi-Tope and Francis 1995).

The type of complex formed plays an important role in determining its biodegradability (Francis et al. 1992, Joshi-Tope and Francis 1995). The rate and extent of biodegradation of several metal-citrate complexes by microorganisms varies. For example, *Pseudomonas pseudoalcaligenes* degraded Mg-citrate at a much lower rate than Ca-, Fe(III)-, and Al(III)-citrate. Studies with a *Klebsiella sp*. showed that citric acid and Mg-citrate were readily degraded, whereas Cd-, Cu-, and Zn-citrate were resistant. Both studies also showed that metal toxicity was not responsible for the lack of or the lower rate of degradation of certain metal-citrate complexes but gave no other explanation. Biodegradation studies with *Pseudomonas fluorescens* showed that bidentate complexes of Fe(III)-, Ni-, and Zn-citrate were readily biodegraded, whereas complexes involving the hydroxyl group of citric acid, the tridentate Al-, Cd- and Cu-citrate complexes, and the binuclear U-citrate complex were not (Fig 9). The presence of the free hydroxyl group of citric acid is the key determinant in effecting biodegradation of the metal complex. The lack of degradation was not due to their toxicity, but was limited by the transport and/or metabolism of the complex by the bacteria (Francis et al. 1992, Joshi-Tope and Francis 1995). No relationship was observed between biodegradability and stability of the complexes. The tridentate Fe(II)-citrate complex, although recalcitrant, was readily biodegraded after oxidation and hydrolysis to the bidentate Fe(III)-citrate form, denoting a structure-function relationship in the metabolism of the complex (Francis and Dodge 1993). Biodegradation of Fe(III)-citrate complex resulted in the formation of carbon dioxide and ferrihydrite. Uranyl-citrate however, is not biodegraded and remains in solution relatively in a pure form which can be further processed by photodegradation.

Biotransformation of uranyl citrate under anaerobic conditions. The presence of organic ligands affected the extent of precipitation of reduced uranium

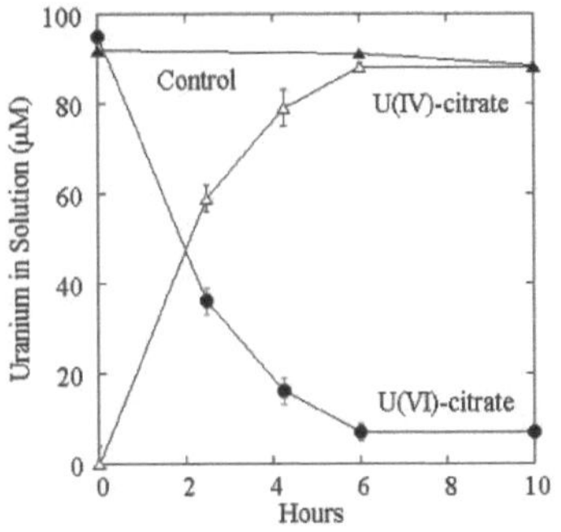

Fig.10. Bioreduction of U(VI)-citrate complex by Clostridia (Francis and Dodge 2008).

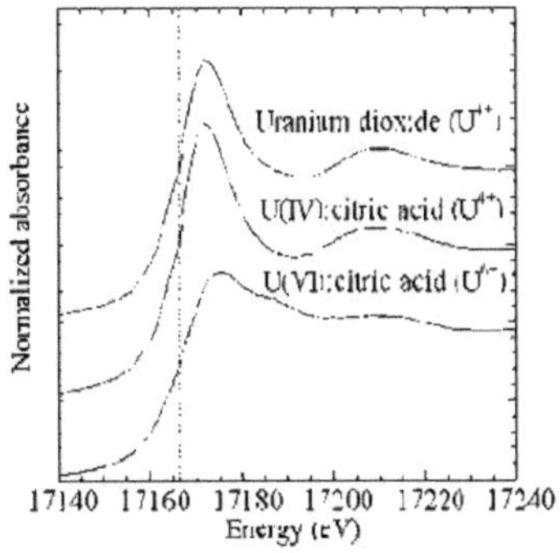

Fig.11. XANES spectra of U(VI)-citrate before and after bacterial reduction (Francis and Dodge 2008).

under anaerobic conditions. Gu et al. (2005) reported that, in the presence of humic materials, the bioreduction of U(VI) did not result in its precipitation; but, the uranium remained in the solution phase as the U(IV)-humic acid complex. The sulfate-reducing bacteria *Desulfovibrio desulfuricans* and the facultative iron-reducing bacteria *Shewanella halotolerans* reduced U(VI) complexed with oxalate or citrate to U(IV) under anaerobic conditions with little precipitation of uranium (Ganesh et al. 1997). Studies with anaerobic bacteria *Clostridium* sp. (ATCC 53464) and *C. sphenoides* (ATCC 53464) showed U(VI) complexed with organic ligands was reduced to U(IV) under anaerobic conditions with little precipitation of uranium (Francis and Dodge 2008). The reduction of U(VI)-citrate to U(IV)-citrate occurred only when supplied with an electron donor glucose or citrate (Fig 10). The bacteria did not metabolize the citrate complexed to the uranium. XANES analysis showed that the reduced form of uranium was present in solution (Fig 11), while EXAFS analysis (Fig 12) showed that the U(IV) was bonded to citric acid as a mononuclear biligand complex (Fig 13).

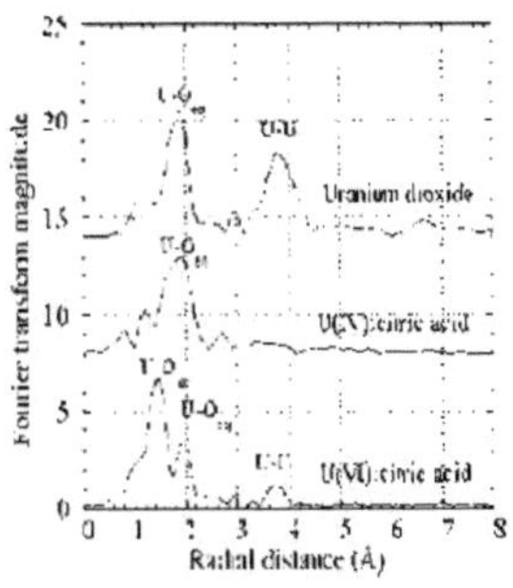

Fig.12. EXAFS analysis of U-citrate complexes before and after bacterial action (Francis and Dodge 2008).

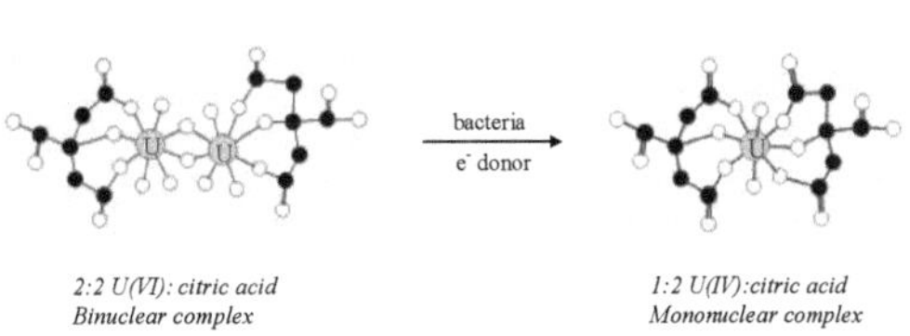

Fig.13. Reduction of biligand U(VI)-citrate to monoligand U(IV)-citrate complex (Francis and Dodge 2008).

Biotransformation of U(VI)-phthalate. Phthalic acid, a common environmental organic ligand, forms strong complexes with the uranyl ion. It formed with uranyl ion bidentate-, mononuclear-, and biligand complexes involving both carboxylate groups. Anaerobic bacterium *Clostridium* species reduced U(VI)-phthalate to a U(IV)-phthalate under anaerobic conditions. EXAFS analysis revealed the association of the reduced uranium with the phthalic acid as a biligand 1:2 U(IV):phthalic acid and finally existing as a repeating biligand 1:2 U(IV):phthalic acid colloidal polymeric U(IV)-phthalate biligand complex (Vazquez, et al. 2009).

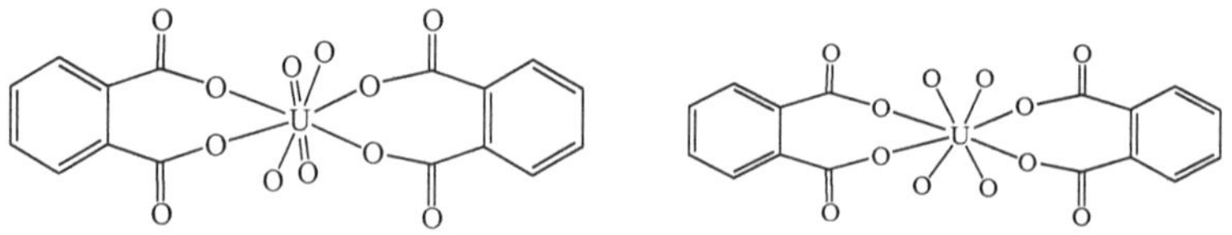

1:2 U(VI)-phthalate   1:2 U(IV)-phthalate

Fig.14. Reduction of U(VI)-phthalate to U(IV)-phthalate complex by *Clostridium* sp under anaerobic conditions (Vazquez, et al. 2009).

Although the bioreduction and the precipitation of uranium is an attractive strategy to immobilize uranium in contaminated environments, the presence of naturally occurring organic ligands such as, oxalic, citric, pthathlic, and humic acids might prevent the precipitation of the bioreduced uranium under anaerobic conditions and enhance its mobility as uranium organic complexes or colloids (Ganesh et al. 1997; Gu et al. 2005; Francis and Dodge, 2008; Vazquez et al. 2009). These results show that the complexed uranyl ion is readily accessible as an electron acceptor despite the inability of the bacterium to metabolize the organic ligand. These results also suggest that reduced uranium, when complexed with an organic ligand, can remain in solution; and this finding is contrary to the conventional belief that reduced uranium will precipitate from solution.

Summary

Microorganisms play a major role in the transformations of uranium and regulate its mobility and stability in the environment. The key microbial processes of interest include oxidation-reduction reactions, dissolution, precipitation, biosorption and colloid formation. Fundamental understanding of the mechanisms of microbial transformations of several chemical forms of the uranium in the presence of electron donors and acceptors under various environmental conditions such as aerobic and anaerobic (denitrifying, fermentative, sulfate-reducing, and methanogenic) will be useful in assessing the microbial impact on the long-term behavior of uranium in contaminated environments.

Acknowledgments. This research was in part supported by BK-21 plus program through the National Research Foundation of Korea funded by the Ministry of Education, Science and Technology (R31 – 30005) and by the Office of Biological and Environmental Research, Office of Science, U.S. Department of Energy, under contract No. DE-AC02-98CH10886.

References

Berthelin, J. and C. Munier-Lamy. 1983. Microbial mobilization and preconcentration of uranium from various rock materials by fungi. *In* R. Hallberg, (ed.) Environmental Biogeochemistry. *Ecol. Bull. Stockholm* 35:395-401.

Bloomfield, C. and G. Pruden. 1975. The effects of aerobic and anaerobic incubation on the extractabilities of heavy metals in digested sewage sludge. *Environ. Pollut.* 8:217-232.

Brainard, J. R., B. A. Strietelmeier, P. H. Smith, P. J. Langston-Unkefer, M. E. Barr and R. R. Ryan. 1992. Actinide binding and solubilization by microbial siderophores. *Radiochim. Acta* 58-59:357-363.

Dodge, C. J., Francis, A. J., and Clayton, C. R. 1966. X-ray spectroscopic studies of microbial transformations of uranium. pp. 159-168. *In* Synchrotron Radiation Techniques in Industrial, Chemical, and Materials Science, K. L. D'Amico, L. J. Terminello, and D. K. Shuh, Editors, Plenum Publishing, NY. 1996.

Francis, A.J. 1990. Microbial dissolution and stabilization of toxic metals and radionuclides in mixed wastes. *Experientia* 46:840-851.

Francis, A.J. 1994. Biotransformation of uranium and other actinides in radioactive wastes. *J. Alloys Compds.* 271/273:78-84.

Francis, A.J. 1999. Bioremedition of Radionuclide and Toxic Metal Contaminated Soils and Wastes. *In Agronomy Monograph 37; SSA, Madison WI, pp* 239-271.

Francis, A.J. and C.J. Dodge. 1993. Influence of complex structure on the biodegradation of iron citrate complexes. *Appl. Environ. Microbiol.* 59:109-113.

Francis, A.J., C.J. Dodge and J.B. Gillow. 1992. Biodegradation of metal-citrate complexes and implication of for toxic metal mobility. *Nature* 356:14-142.

Francis, A.J., C.J. Dodge, F. Lu G. Halada and C.R. Clayton. 1994. XPS and XANES studies of uranium reduction by *Clostridium* sp. *Environ. Sci. Technol.* 28:636-639.

Francis, A.J., J. B. Gillow, C.J. Dodge, M. Dunn, K. Mantione, B.A.Strietelmeier, M.E. Pansoy-Hjelvik and H.W. Papenguth. 1998. Role of microbes as biocolloids in the transport of actinides from a deep underground radioactive waste repository. *Radiochim. Acta* 82:347-354.

Francis A.J., C.J. Dodge, J.B. Gillow and H.W. Papenguth. 2000. Biotransformation of uranium compounds in high ionic strength brine by a halophilic bacterium under denitrifying conditions. *Environ. Sci. Technol.* 34: 2311-2317.

Francis A.J, J.B Gillow, C.J Dodge, R Harris, T.J Beveridge and H.W Papenguth. 2004. Association of uranium with halophilic and nonhalophilic bacteria and archaea. *Radiochim. Acta* 92:481-488.

Francis, A.J. and C.J. Dodge. 2008. Bioreduction of uranium (VI) complexed with citric acid by Clostridia affects its structure and solubility. *Environ. Sci. Technol.* 42:8277-8282.

Ganesh, R., K.G. Robinson, G.R. Reed and G. S. Saylor. 1997. Reduction of hexavlaent uranium from organic complexes by sulfate- and iron-reducing bacteria. *Appl. Environ. Microbiol.* 63:4385-4391.

Gao, W. and A.J. Francis. 2008. Reduction of uranium (VI) to uranium (IV) by *Clostridia. Appl. Environ. Microbiol.* 74:4580-4584.

Gillow, J.B., Dunn, M., Francis, A.J., and Papenguth, H.W. 2001. The role of subterranean microbes as biocolloids in the transport of actinides at the Waste Isolation Pilot Plant and Grimsel Test Site. *Radiochim Acta.* 88:769-774.

Gu, B.; Yan, H.; Zhou, P.; Watson, D.; Park, M.; Istok, J. D. 2005. Natural humics impact uranium bioreduction and oxidation. *Environ. Sci. Technol.* 39:5268-5275.

Joshi-Tope, G. and Francis, A. J. 1995. Mechanisms of biodegradation of metal-citrate complexes by *Pseudomonas fluorescens. J. Bacteriol.* 177:1989-1993.

Premuzic, E.T., A.J. Francis, M. Lin and J. Schubert. 1985. Induced formation of chelating agents by *Pseudomonas aeruginosa* grown in presence of thorium and uranium. Arch. *Environ. Contam. Toxicol.* 14:759-768.

Strandberg, G.W., S.E. Shumate II and J. R. Parrot Jr. 1981. Microbial cells as biosorbents for heavy metals: accumulation of uranium by *Saccharomyces cerevisiae* and *Pseudomonas aeruginosa*. *Appl. Environ. Microbiol.*41:237-245.

Vazquez, G.J., C.J. Dodge and A.J. Francis. 2009. Bioreduction of U(VI)-phthalate to a polymeric U(IV)-phthalate colloid. *Inorganic Chemistry*. 48: 9485–9490.

Wall, J. D. and L. R. Krumholz. 2006. Uranium reduction. *Ann. Rev. Microbiol.* 60:149-166.

Uranium induced stress promotes fungal excretion of uranium/metal stabilizing ligands: Analysis of metal-organic compounds with Size Exclusion Chromatography and Inductively Coupled Plasma-Mass Spectroscopy

Anna Grandin[1], Anna Ogar[2], Viktor Sjöberg[1], Stefan Karlsson[1]

[1]Man-Technology-Environment Research Centre, Örebro University, SE-70182 Örebro, Sweden
[2]Institute of Environmental Science of the Jagiellonian University, Gronostajowa 7, 30-387 Krakow, Poland

Abstract. Weathering of pyrite rich alum shale processing waste has led to metal pollution in Kvarntorp, Sweden. Here we use a fungal strain isolated from the site to monitor the excretion of uranium/metal stabilizing ligands under uranium induced stress. After two weeks 91 % was lost from a 10 mg L^{-1} solution but 57 % already within ten minutes. The formation of colloidal/particulate uranium is mainly controlled by organic exudates phosphorus excreted by the fungus. Most likely, the change in solution properties from metabolic processes resulted in the formation species through adsorption and precipitation.

Introduction

Alum shale was mined in open pits at Kvarntorp, Sweden, some 200 km WSW of Stockholm, between 1942 and 1966 for oil production by pyrolysis. This alum shale contains some 15-18 % kerogen and 6-9% sulfur and is rich in uranium, molybdenum and aluminum (Allard et al. 2011). Weathering of the exposed pyrite rich waste has given elevated concentrations of both major and trace elements in the area (Allard et al. 2011). The uranyl ion, UO_2^{2+} which is thermodynamically stable under oxic conditions at low to intermediate pH (Rai et al. 2003), can form complexes with a number of ligands released from microbes. They include for instance low molecular weight organic acids (LMWOAs) such as citric and oxalic acid (Renshaw et al. 2003) but also polysaccharides (Huang et al 2004, Sharma et al 2009). In addition, uranium also forms complexes with many other organic ligands and compounds from different sources (Lenhart et al, 2000). Metal stress of-

B.J. Merkel and A. Arab (eds.), *Uranium – Past and Future Challenges*,
DOI 10.1007/978-3-319-11059-2_40

ten cause an increased production of fungal exudates as shown by Green and Clausen (2003). Besides oxalate and citrate, numerous other LMWOAs have been identified (Takao, 1965). However, LMWOAs only represent a minority of fungal exudates (Ogar et al. 2014). Results from a study of metabolite composition from *Rhizoctonia solani* by Aliferos and Jabadi (2010) revealed a mixture of 109 different organic compounds where carboxylic acids, amino acids and fatty acids were the dominant groups. The exudate composition can vary greatly among fungal species and also depends on environmental conditions (Plassard and Fransson 2009). Many fungal species tolerate high levels of toxic metals, and microorganisms such as bacteria, fungi and algae are being used in large scale removal of metals from industrial and domestic effluents (Zapotoczny et al. 2007) but also for metal recovery from ores (Orłowska et al 2011; Mulligan et al. 2003). Here we have screened for fungi adapted to an extreme alum shale environment with low pH, high levels of heavy metals and poor water holding capacity to be used for element control. Since little or no biosorption of metal ions to fungal biomass can be expected at pH less than 3 (Kapoor et al. 1999; Fomina et al. 2007) we focus mainly on the redistribution of uranium in the aqueous phase by organic ligands.

Materials and Methods

Fungal cultures

Unprocessed weathered alum shale particles from Kvarntorp, Sweden, approx. size 0.5 x 3 mm, were spread on solid malt extract agar plates (Sigma, Germany) with pH 3.5, adjusted by addition of an autoclaved lactic acid solution according to protocol (www.scharlab.com). Over 30 fungal species were isolated and tested for uranium tolerance on uranium supplemented agar plates by addition of a separately autoclaved solution of 20 mg L^{-1} uranyl nitrate to final concentrations of 10, 20 and 30 mg L^{-1}. One fungal strain, BS16, was inoculated to liquid growth medium in sterile 120 mL (Sarstedt®) polypropylene containers on an orbital shaker at 150 rpm at 22 ±1 °C for 5 weeks. Adjustment of pH to 3.5 was made by HCl. After one week of incubation, uranium was added (10 mg L^{-1}) as autoclaved uranyl nitrate to fungal cultures (MFU). Fungi controls (MFC) without uranium were also included as well as malt extract controls (MC). A first sample was taken 10 minutes after the addition of uranium and the repeated weekly for 4 weeks. All samples were filtered through 0.20 μm polypropylene syringe filters (VWR International, USA) immediately after sampling and the filtrates were stored at -20 °C until analysis. After 5 weeks of incubation the biomass was recovered by filtration through a 32 μm polyester mesh (Sefar LFM, Switzerland), washed with sterile 18.2 MΩ water, dried and weighed. The dry biomass was microwave digested in concentrated HNO_3 before analysis of uranium. Malt extract without uranium (<0.01 μg L^{-1}; MU0), 10 mg L^{-1} (MU10) and 100 mg L^{-1} (MU100) uranium, were included. MU10 was sampled each minute after addition of uranium during 10 minutes and analyzed for dissolved uranium. MU0, MU10 and MU100 were analyzed with size exclusion chromatography for organic compounds.

Analytical procedure

Electrical conductivity and pH were measured with standard electrodes (Radiometer CDC866T and Metrohm 6.0257.000, respectively). The size distribution of dissolved organic compounds was evaluated with size exclusion chromatography (SEC) using an Agilent SEC-5 column. The mobile phase for fraction collection and analysis of metals was a 50 mM ammonium nitrate buffer at pH 6.8 while a 50 mM phosphate buffer at the same pH was used for size determinations. Absorbance was measured 210, 225, 250, 365 and 600 nm and fluorescence at 370 and 410 nm. Excitation was made at 250 nm. Calibration was made with polyvinyl sulphonate standards (American Polymer Standards Corp.) in the range 1.2 to 60 kD. Metal analysis was performed with an Agilent 7500cx ICP-MS using the Merck VI multi element standard solution for calibration and ^{103}Rh as internal standard. The solution speciation of U(VI) was estimated with Visual MINTEQ 3.0 (KTH). Total phosphorus was analyzed with Agilent 7500cx ICP-MS, with/without Octopole Reaction System (ORS).

Results and Discussion

Uranium in fungal cultures and biomass

After only 10 minutes of incubation, substantially lowered concentrations of dissolved uranium was noted that corresponded to some 57 % of the total inventory. After 2 weeks of incubation the decrease of dissolved uranium had reached 91 %. The dry biomass contained only some 1.7 mg g^{-1} why uptake in and surface adsorption to the hyphae were limited. Hence, the majority of the uranium species in the systems remained in the aqueous phase but was retained by the filter. Most likely, the change in solution composition by excretion of metabolites and/or presence of phosphate in the culture medium resulted in the formation of colloidal/particulate uranium species. It was possibly caused by the combined action of adsorption, precipitation and agglomeration. According to the chemical speciation of U(VI), the dominant solution species are uranyl hydrogen phosphate

(OU_2HPO_4) and uranyl ions (UO_2^{2+}). A rapid formation of uranyl phosphates over the uranyl binding to organic ligands is suggested to be favored at pH 3.5. As demonstrated by Kelly et al. (2002), the binding of uranyl to phosphoryl groups was favored over binding to carboxylate ligands on bacterial cell surface at pH 1.67 whereas at pH 3.22 and 4.8 increased complexation by carboxylates was observed. Dissolved uranium concentration in malt extract controls (MU10) decreases readily within one minute after addition (Fig.1) and remains quite stable over time with a uranium concentration of 2.9 mg L^{-1} during four weeks of incubation. In the fungal samples (MFU) the uranium concentration decreased during the four weeks (Fig.1). The results suggest that a major part of the uranium inventory is initially immobilized by compounds in the malt extract medium and retained by the filters.

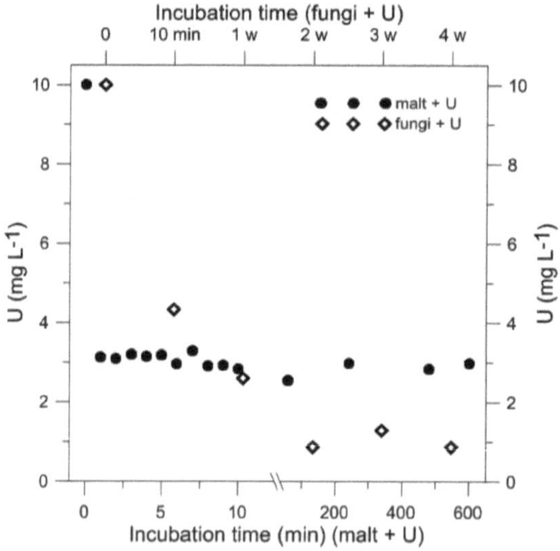

Fig.1. Uranium concentration in aqueous phase during incubation of MFU (fungi+U) and MU10 (malt+U).

SEC analysis of organic compounds

Analysis with size exclusion chromatography (SEC) indicate timed dependent and stress induced production of organic compounds in the low- and high molecular weight range that corresponds qualitatively with the loss of dissolved uranium. The characterization of these organic compounds is in progress. Chromatograms from absorption at 250 nm (Fig.2) and fluorescence at 370 nm (Fig.2) show the molecular weight distribution of dissolved compounds in the aqueous phase 10 min and 4 weeks after addition of 20 ppm uranyl nitrate to fungal cultures (MFU). The calibration range is 1.2 kD to 60kD but the column separates reasonably well down to 150 D but at lower accuracy. Peaks (P) in the chromatograms are denoted by their retention time. In MFU, there is an increased production of compounds at P13.5, P14.5, P15.5, P21 (Fig.2) and P22-24 (Fig.2) after four weeks. There are large differences in the signal intensity compared to the fungal control (MFC) (Fig.2), suggesting that the addition of uranium induces a stress related excretion of organic compounds in MFU. The SEC chromatograms will only reflect the change in concentration of compounds that are detectable at the selected wavelengths and reasonably well separated by the column. There is an obvious risk that other processes affect the apparent molecular size in these systems. One of these would be the change in size as a function of the changing conditions in the solution phase as a response to the bio-processes. This is suggested in figure 2 where there is a shift in the retention time around 10 minutes as a function of uranyl concentration. The change is systematic with lowered apparent molecular size as the uranyl concentration increases from no addition at all to 100 mg L^{-1} and would reflect a contraction of the complexing compound(s). Such a contraction will occur in many macromolecules as a response to the charge density introduced by coordinating ions, particularly for molecules where weak forces of attraction control their structure. For the particular medium this observation is important since unnoticed it might be taken for a production or consumption by the organisms under evaluation.

ICP-MS analysis of metals

Results show that at incubation week 4 many metals, including U, Cu (Fig.3), Al, Co, V, As, Mn, Pb and Cd are mainly associated to high molecular weight range (retention time up to 15 min). The highest concentration is retrieved in the 12 minutes fraction. However, the distribution of metals in the fractions varies during incubation. For example Mg, where at incubation week 1, the highest concentrations are found in the low molecular weight range (retention time 25-30 minutes) shifts towards the high molecular weight range at incubation week 3. There are some exceptions where the highest metal concentrations are mostly found in the low molecular weight range at week 4 (retention time 15-30 minutes) such as Zn, Fe (Fig.4), K, Li and Rb and Li. Some elements though (U, Cu, Cr, Ni, Li, K and Rb) are consistently found in the same fractions during the 4 weeks of incubation. In contradiction to what is generally proposed regarding LMWOAs as the major metal complexing agents, it seems like the complexation of many metals in the studied systems is heavily influenced by compounds in the high molecular weight range. These findings suggest that other candidates for uranium/metal complexation than LMWOAs, e.g. oxalic and citric acid which are generally regarded as the most important ligands produced by fungi, could be of equal or greater importance. It is also important to monitor which metal-organic compounds that are favored under certain conditions in a certain microbial-waste system, since there is a change over time possibly related to media composition, fungal exudates and growth conditions.

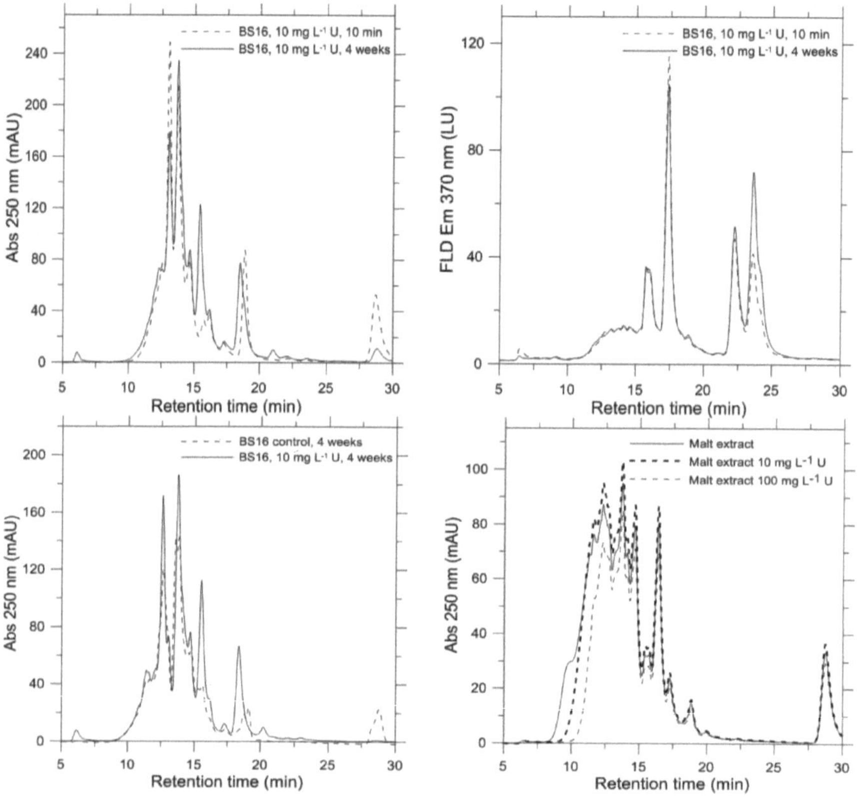

Fig.2. SEC chromatograms at 250 nm and with fluorescense detection at 370 nm.

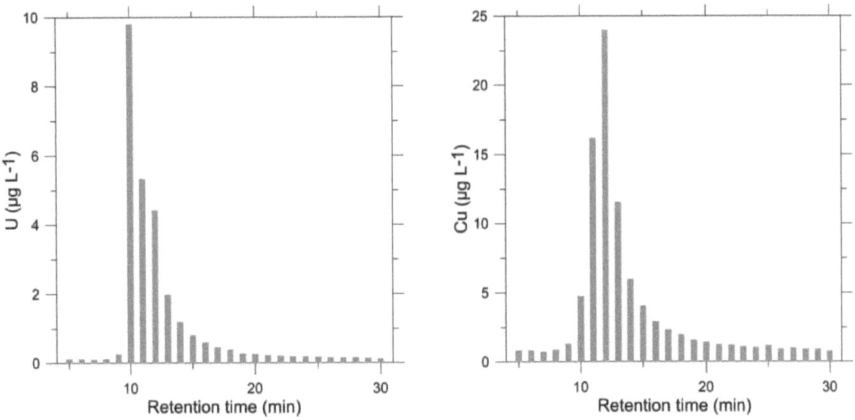

Fig.3. Metal concentrations in fractions collected during SEC analysis.

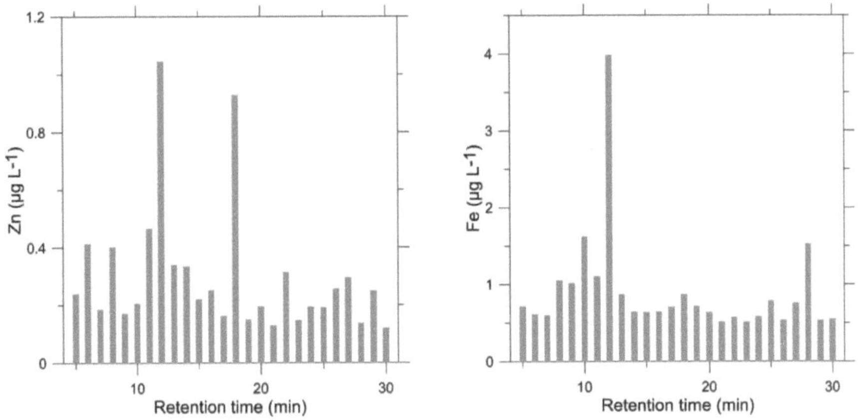

Fig.4. Metal concentrations in fractions collected during SEC analysis.

Conclusions

Uranium induced stress increased production of organic compounds in fungal cultures compared to controls. Additional uranium in the cultures (10 mg L^{-1}) resulted in a 57 % retention by the 0.20 µm filter already within ten minutes and increased to 91 % after four weeks of incubation. Uranium concentration in the solution phase from culture medium controls decreased with some 70 % within one minute but remained stable during the incubation. Size distribution of dissolved uranium in systems with fungi was dominated by an apparent molecular weight around 4 kD.

Acknowledgements

The authors acknowledge the Faculty of Economy, Science and Technology at Örebro University for financial support to S Karlsson.

References

Aliferis KA, Jabadi S (2010) Metabolite Composition and Bioactivity of Rhizoctonia solani Sclerotial Exudates. J Agric Food Chem 58: 7604-7615

Fomina M, Charnock JM, Hillier S, Alvarez R, Gadd GM (2007) Fungal transformations of uranium oxides. Environ Microbiol 9(7): 1969-1710

Gadd GM (2007) Geomycology: biogeochemical transformation of rocks, minerals, metals and radionuclides by fungi, bioweathering and bioremediation. Mycol Res 111:3-49

Green F, Clausen CA (2003) Copper tolerance of brown-rot fungi: time course of oxalic acid production. Int Biodeter Biodegr 51: 145-149

Huang H, Yang X (2004) Synthesis of Polysaccharide-Stabilized Gold and Silver Nanoparticles: A Green Method. Carbohyd Res 339(15): 2627-2631

Kapoor A, Viraraghavan D, Cullimore R (1999) Removal of heavy metals using the fungus Aspergillus niger. Bioresource Technol 70: 95-104

Kelly SD, Kemner KM, Fein JB, Fowle DA, Boyanov MI, Bunker BA, Yee N (2002) X-ray absorption fine structure determination of pH-dependent U-bacterial cell wall interactions. Geochim Cosmochim Acta 66: 3855-3871

Lenhart JJ, Cabaniss SE, MacCarthy P, Honeyman BD (2000) Uranium (VI) complexation with citric, humic and fulvic acids. Radiochim Acta 88: 345-353

Mulligan CN, Kamali M (2003) Bioleaching of Copper and Other Metals from Low-Grade Oxidized Mining Ores Byaspergillus Niger. J Chem Technol Biotechnol 78(5): 497-503

Ogar A, Grandin A, Sjöberg V, Turnau K, Karlsson S (2014) Stabilization of uranium at low pH by fungal metabolites: applications in environmental biotechnology. APCBEE Procedia, in press.

Orłowska E, Orłowski D, Mesjasz-Przybyłowicz J, Turnau K (2011) Role of Mycorrhizal Colonization in Plant Establishment on an Alkaline Gold Mine Tailing. Int J Phytorem 13(2): 185-205

Plassard C, Fransson P (2009) Regulation of low-molecular weight organic acid production in fungi. Fungal Biol Rev 23: 30-39

Rai D, Yui M, Moore DA (2003) Solubility and solubility product at 22°C of UO_2 (c) precipitated from aqueous (VI) solutions. J Sol Chem 32: 1-17

Renshaw JC, Halliday V, Robson GD, Trinci APJ, Wiebe MG, Livens FR (2003) Development and application of an assay for uranyl complexation by fungal metabolites, including siderophores. Appl Environ Microbiol 69: 3600-3606

Sharma VK, Yngard RA, Lin Y (2009) Silver Nanoparticles: Green Synthesis and Their Antimicrobial Activities. Adv Colloid Interfac 145(1): 83-96

Takao S (1965) Organic acid production by Basidiomycetes I. Screening acid-producing strains. Appl Microbiol 13: 732-737

Zapotoczny S, Jurkiewicz A, Tylko G, Anielska T, Turnau K (2007) Accumulation of copper by Acremonium pinkertoniae, a fungus isolated from industrial wastes. Microbiol Res 162: 219-228

Passive treatment of heavily polluted drainage waters in a uranium deposit

Stoyan Groudev[1], Irena Spasova[1], Plamen Georgiev[1], Marina Nicolova[1]

[1]University of Mining and Geology "Saint Ivan Rilski", Sofia 1700, Bulgaria

Abstract. The treatment of acid mine drainage in the uranium deposit Curilo, Bulgaria, was very efficient for a long period of time. However, the treatment of such waters containing high concentrations of iron, including a portion in the ferrous state, caused serious problem on the efficiency of the cleaning process. These problems were avoided by the pretreatment of these waters by a system for iron removal consisting of a unit for bacterial oxidation of the ferrous ions to the ferric state and a unit for precipitation of the ferric iron as a result of chemical neutralization.

Introduction

The acid mine drainage is considered to be a major environmental problem associated with mining activities. This phenomenon is connected with the oxidation of pyrite and other sulphide minerals as a result of which acidic waters containing sulphuric acid, dissolved heavy metals and solid iron precipitates are released to the environment. The acid mine drainage from the uranium mines contain, apart from the iron and other heavy metals, also radioactive elements such as uranium and radium (Johnson and Hallberg 2005).

The uranium deposit Curilo, Bulgaria, was for a long period of time a site of intensive mining activities including both open-pit and underground mining techniques as well as *in situ* leaching of uranium. The main uranium-bearing minerals in the ore were uraninite, torbernite, metatorbernite, pitch-blende, metaotunite and bassetite. The ore was rich-in-pyrite and, apart from uranium, contained some non-ferrous metals such as copper, zinc and lead, present mainly in the form of sulphide minerals. Quartz and feldspars were the main minerals in the host rock. Clay minerals and some iron hydroxides were also present.

The mining operations in the deposit were ended in 1990 but since that time and until recently the deposit was a permanent source of acid drainage waters. The generation of these waters was connected mainly with the bacterial oxidation of pyrite, other sulphide minerals and tetravalent uranium present in the ore (Groudev et al. 2003). Different methods to prevent the generation or to clean

B.J. Merkel and A. Arab (eds.), *Uranium – Past and Future Challenges*,
DOI 10.1007/978-3-319-11059-2_41

such waters were tested under field conditions. The treatment of acid drainage containing relatively low concentrations of iron (usually less than 1.0 – 1.5 g/l) by some passive systems was efficient for long periods of time, even during the cold winter months and the best results were achieved by systems consisting of alkalizing limestone drains, permeable reactive multibarriers for microbial sulphate reduction and biosorption, and aerobic wetlands, arranged in a series (Groudev et al. 2008). However, the treatment of acid drainage containing high concentrations of iron (over 5 g/l) was a more difficult problem, especially when a considerable portion of iron was in the ferrous state. The removal of ferrous iron in the permeable multibarrier at pH about the neutral point and prevalent anoxic conditions (the drainage waters entering in the multibarrier contained some dissolved oxygen) was connected with their slow oxidation to the ferric state, followed by precipitation of the ferric ions to the insoluble ferric hydroxide. This slow oxidation was mainly a chemical process carried out by the residual amounts of oxygen dissolved in the drainage waters being treated. Some heterotrophic bacteria able to oxidize the ferrous ions at slightly acidic pH (within the range of about 4.5 – 6.5) were present but their role in this process was negligible. The large amounts of iron precipitates formed in the multibarrier caused a strong negative effect on its permeability and on the rates of the other cleaning processes in this system such as the microbial dissimilatory sulphate reduction and biosorption.

The problems above-mentioned were avoided by the prior removal of most of the iron from the relevant drainage waters by a system consisting of a unit for the bacterial oxidation of the ferrous ions to the ferric state followed by a unit for precipitation of the ferric ions by means of chemical neutralization. Some data about the efficient work of a system of this type constructed in the Curilo deposit are shown in this paper.

Materials and methods

The passive system was constructed in a ravine collecting a considerable portion of the acid drainage waters generated in the deposit. The multibarrier was a pond dug into the ground and its bottom and walls were isolated by impermeable plastic sheets. The multibarrier consisted of two sections: an alkalizing limestone drain and an anoxic section for microbial dissimilatory sulphate reduction, biosorption and additional chemical neutralization. The alkalizing drain had a volume of about 2.5 m^3 and was filled by a mixture of crushed limestone and gravel pieces (in a ratio of about 1:2 as dry weight) with a particle size less than 12 mm. The second section of the multibarrier had a volume of about 20.4 m^3 (8.0 m long, 1.7 m wide and 1.5 m deep) and was filled by a mixture of biodegradable solid organic substrates (cow manure, plant compost, straw) and crushed limestone. This section of the multibarrier was inhabited by a microbial community consisting of sulphate-reducing bacteria and other metabolically interdependent microorganisms.

The constructed wetland was a pond dug into the ground and its bottom and walls were also isolated by impermeable plastic sheets. The wetland was 14 m long, 3.2 m wide and 0.6 m deep. Its bottom was covered by a 0.3 m layer consisting of soil rich-in-organics and mineral nutrients. *Typha latifolia* and *Typha angustifolia* were the main plant species in the wetland but *Phragmites australis* and representatives of the genera *Juncus, Eleocharis, Potamogeton, Carex* and *Poa* as well as different algae were also present.

The unit intended for a prior decreasing of the iron concentrations in the acid mine drainage consisted of two sections. The first section was intended for bacterial oxidation of Fe^{2+} ions present in the acid drainage. The section was a concrete canal of the cascade-type construction consisting of five steps. Each of the steps was 2 m long, 3.2 m wide and 0.4 m high. The bottoms of the steps were covered by a 20 cm layer consisting of crushed gravel with a particle size less than 12 mm. The surface of the mineral grains was covered by biofilms consisting of a matrix of jarosites, secreted bacterial metabolites (mainly exopolysaccharides) and active Fe^{2+}-oxidizing chemolithotrophic bacteria related to the species *Leptospirillum ferrooxidans* and *Acidithiobacillus ferrooxidans*. This section was connected with the second section which was an alkalizing limestone drain (4.0 m long, 3.2 m wide and 0.4 m high) filled by a mixture of crushed limestone and gravel (in a ratio of 1:2 as dry weight) with a particle size less than 12 mm.

The quality of the waters was monitored at different sampling points located at the inlet and outlet of all units included in this passive system. Sampling points were located also at different depths within the alkalizing drain and in the multibarrier and the wetland.

The parameters measured in situ included: pH, Eh, dissolved oxygen, total dissolved solids and temperature. Elemental analysis was done by atomic absorption spectrometry and induced coupled plasma spectrometry in the laboratory. The radioactivity of the samples was measured, using solid residues remaining after their evaporation, by a low background gamma-spectrometer ORTEC (HpGe-detector with a high distinguishing ability). The specific activity of Ra-226 was measured using a 10 l ionization chamber. Mineralogical analysis was carried out by X-ray diffraction techniques. The mobility of the pollutants was determined by the sequential extraction procedure (Tessier et al. 1979).

The fraction analysis of the solid organic components in the multibarrier was performed using air dried samples after their prior washing with distilled water to remove the carbonates and the water soluble organic compounds. The determination of the different organic fractions (cellulose, hemicellulose, lignin) were carried out by the methods described earlier (Groudev et al. 2008).

The isolation, identification and detection of the acidophilic chemolithotrohic bacteria present in the acid mine drainage was carried out by both cultivation tests at different conditions as well as by the PCR 16S rRNA sequencing (Rawlings 2001; Sanz and Köchling 2007; Escobar et al. 2008).

Results and Discussion

The passive system for treatment of acid mine drainage used in this study initially consisted of a permeable multibarrier (with an alkalizing limestone drain and an anoxic section for microbial dissimilatory sulphate reduction, biosorption and additional chemical neutralization) and a constructed wetland arranged in a series. This passive system for a relatively long period of time (about four years) worked efficiently at flow rates of about 2 – 10 m^3/24 h and residence times of about 24 – 4.8 hours, with periodic replacements of organic substrates, by fresh batches of these materials. It must be notes that this efficient work was connected with treatment of acid drainage waters containing relatively low concentrations of iron (usually less than 1 g/l). However, the treatment of waters containing high concentrations of iron (in some cases over 5 g/l) caused considerable problems. It was impossible to decrease the residual concentrations of some pollutants (sulphates, iron, manganese, dissolved organic carbon) below the relevant permissible levels for waters intended for use in agriculture and/or industry (Table 1). Furthermore, the amounts of iron and other pollutants retained in the multibarrier and in the wetland rapidly increased and caused a serious negative effect on the permeability of the multibarrier, the size of the free exposed surface of the organic substrates in the multibarrier and in the wetland, as well as on the microbial activity in this two units of the passive system. It was possible to decrease the residual concentrations of most of the pollutants in the wetland effluents below the relevant permissible levels by limiting the flow rates of the polluted waters being treated to about 2 m^3/24 h.

The construction of the unit for the prior removal of iron from the acid mine drainage had a strong positive effect on the treatment of these waters. It was found that the continuous bacterial oxidation of the ferrous ions in the acid waters being treated proceeded at high rates and depended on the temperature of these waters (Table 2). The CO_2 fixation ability of these bacteria was also increased at higher temperatures and resulted in increasing of their number. In most cases the residual concentrations of ferrous ions were as low as about 5 – 10 % from the relevant concentrations before the bacterial oxidation.

Experiments performed with samples of the polluted waters subjected to the prior iron oxidation, as well as with samples from the biofilms formed in the unit for this oxidation in the deposit, revealed that the species composition of the relevant microflora was relatively stable during the treatment but the concentrations of the different microorganisms in some cases changed to a considerable extent.

Table 1. Treatment by the passive system of acid mine drainage non-pretreated by a prior iron oxidation

Parameters	Acid mine drainage	Alkalizing drain effluents	Multibarrier effluents	Wetland effluents
pH	1.90-3.25	4.04-5.54	6.40-7.35	6.80-7.40
Eh, mV	(+312) - (+554)	(+280) - (+415)	(-185) - (-240)	(+370) - (+495)
DO_2, mg/l	1.4 - 4.6	1.4 - 4.4	0.1 - 0.4	2.8 - 5.7
DOC, mg/l	0.6 – 2.1	0.8 – 2.1	59 – 170	17 – 44
U, mg/l	0.68 – 4.15	0.27 – 3.14	0.07 – 0.84	< 0.05 – 0.35
Ra, Bq/l	0.12 – 0.80	0.07 – 0.55	0.02 – 0.20	< 0.02
Cu, mg/l	0.71 – 8.75	0.62 – 3.27	0.12 – 0.79	< 0.05 – 0.35
Zn, mg/l	1.25 – 20.3	0.80 – 15.2	0.21 – 3.25	<0.05 – 0.75
Cd, mg/l	0.02 – 0.12	0.01 – 0.07	< 0.01 – 0.02	< 0.01
Pb, mg/l	0.04 – 0.32	0.03 – 0.28	< 0.02 – 0.14	< 0.02 – 0.09
Ni, mg/l	0.21 – 1.50	0.18 – 1.20	< 0.05 – 0.57	< 0.05 – 0.23
Co, mg/l	0.15 – 1.25	0.10 – 0.92	< 0.02 – 0.41	< 0.05 – 0.21
As, mg/l	0.12 – 0.59	0.10 – 0.37	< 0.02 – 0.12	< 0.01
Mn, mg/l	10.2 – 82.4	7.10 – 73.4	0.84 – 11.3	0.14 – 2.03
Fe, mg/l	2282 – 5726	925 – 3290	212 – 1403	60 - 790
SO_4^{2-}, mg/l	3755 – 10645	3591 – 8890	1870 – 6842	1382 - 5394

Table 2. Bacterial oxidation of Fe^{2+} ions in the acid mine drainage

Parameters	Temperature of the acid mine drainage		
	8 – 11 °C	11 – 14 °C	14 - 19°C
pH	1.7 – 3.5	1.8 – 3.2	1.5 – 3.5
Total iron, g/l	2.51 – 5.45	2.97 – 572	3.29 – 5.51
Fe^{2+} iron, g/l	1.04 – 3.25	1.16 – 3.14	1.14 – 3.20
Fe^{2+} oxidation rate, mg/l.h	14 – 35	34 – 68	59 – 107
$^{14}CO_2$ fixed for 24 h, counts/min.ml	820 – 2500	2500 – 3700	3500 – 4600
Fe^{2+} - oxidizers, cell/ml	$10^3 – 10^6$	$10^4 – 10^7$	$10^6 – 10^8$
S – oxidizers, cell/ml	$10^2 – 10^5$	$10^3 – 10^6$	$10^5 – 10^7$

Apart from the temperature, factors such as the chemical composition, pH, Eh and contents of dissolved oxygen and CO_2 of the polluted waters were essential in this respect. Regardless of this, *L. ferrooxidans* and *At. ferrooxidans* were always the prevalent microbial species, although present in different concentrations and ratios. In general, the maximum ferrous-oxidizing ability of these two species was similar. They were the prevalent microorganisms participating also in the leaching of the uranium ores and in the generation of polluted waters in the deposit. *At. ferrooxidans* was present in the ores usually in higher concentrations than *L. ferrooxidans*. This was probably due to the ability of *At. ferrooxidans* to oxidize, apart from iron, also S^0 and different sulphur compounds. However, *L. ferrooxidans* was present in the biofilms in the unit for iron oxidation usually in higher concentrations than *At. ferrooxidans*. This was mainly due to the considerably lower Michaelis constant (Km) during the oxidation of Fe^{2+} by *L. ferrooxidans* than that during the oxidation by *At. ferrooxidans*.

The efficiency of the iron oxidation in the unit mentioned above depended also on the chemical composition and structure of the inorganic matrix associated with the biofilms. This matrix consisted mainly of jarosites. However, in some cases, at relatively higher pH values (> 3.2) of the polluted waters, ferric hydroxides were also present. The secondary precipitation of these hydroxides decreased temporarily the oxidizing ability of the biofilms isolating them from the efficient direct contact with the ferrous ions present in the polluted waters. However, bacteria soon formed new layers of biofilms on the hydroxides surface.

The waters containing almost the whole dissolved iron as ferric ions were efficiently treated in the new constructed alkalizing drain. Most of these ions were precipitated in this drain and the concentrations of most of the other pollutants were also decreased as a result of their co-precipitation by the gelatinous iron compounds (jarosites and ferric hydroxides).

The further cleaning of the waters pretreated in this way was very efficient and proceeded within the initial range of the flow rates of about 2 – 10 m^3/24 h, reflecting water residence times of about 22 – 4.4 hours in the multibarrier (Table 3).

The removal of pollutants in the multibarrier was due to different mechanisms. The microbial dissimilatory sulphate reduction played the main role in this removal, especially during the warmer months of the year. The sulphate-reducing bacteria inhabiting the multibarrier were a mixed population, in which the representatives of the genera *Desulfovibrio* (mainly *D. desulfuricans*), *Desulfobulbus* (mainly *D. elongatus*) and *Desulfosarcina* (mainly *D. variabilis*) were the prevalent microorganisms. Their total number in the waters being treated usually was about 10^6 – 10^7 cells/ml. However, most of these bacteria were firmly attached to the solid organic substrates in the multibarrier and usually exceeded 10^8 cells/g.

The sulphate-reducing bacteria precipitated the dissolved heavy metals and arsenic as the relevant sulphides by means of the hydrogen sulphide generated in the process of microbial sulphate reduction. The hexavalent uranium was reduced to the tetravalent state and was precipitated as uraninite (UO_2). The role of the sorption mechanisms by the dead plant biomass present in the multibarrier was essential mainly as an initial step in the removal of pollutants. The adsorbed heavy metals were then transferred into the relevant sulphides and the adsorbed hexavalent uranium was reduced to UO_2. Portions of the pollutants were further encapsulated by the ferric hydroxides formed in the multibarrier. Radium was removed mainly by sorption by the dead plant biomass and the iron hydroxides present in the multibarrier.

The microbial sulphate reduction was a function of the concentration of organic monomers dissolved in the waters. These monomers were generated as a result of the biodegradation of the solid biopolymers by the different heterotrophs possessing hydrolythic enzymatic activity. This process resulted in a steadily decrease in the concentration of the easily degradable solid biopolymers (cellulose and hemicellulose) present in the multibarrier. It must be noted that at air temperatures about 0 °C and water temperatures close to the freezing point, the temperature inside the multibarrier, within the deeply located layers, usually was in the range of about 3 – 5 °C. Under such conditions the microbial sulphate reduction still proceeded, although at much lower rates. At the same time, the role of the sorption of pollutants by the dead plant biomass and of the chemical neutralization by the crushed limestone present in the multibarrier was essential during such climatic conditions when the growth and activity of the microbial community in the multibarrier were strongly decreased. The sorption capacity of the dead biomass as well as the alkalization ability of the limestone steadily decreased in the course of time. This imposed the periodic substitution of the content of the multibarrier by fresh batches of the relevant components.

Table 3. Treatment by the passive system of acid mine drainage pretreated by a prior iron oxidation.

Parameters	Pretreated acid mine drainage	Multibarrier effluents	Wetland effluents	Permissible levels for waters used in industry
pH	4.10 – 5.15	6.84 – 7.45	6.80 – 7.35	6 – 9
Eh, mV	(+302) – (+486)	(-195) – (-260)	(+350) – (+440)	-
DO_2, mg/l	1.8 – 4.8	0.1 – 0.4	2.6 – 5.1	2
DOC, mg/l	1.0 – 2.5	68 – 154	16 – 28	20
U, mg/l	0.23 – 3.17	< 0.05	< 0.05	0.6
Ra, Bq/l	0.03 – 0.45	< 0.02	< 0.02	0.15
Cu, mg/l	0.51 – 2.95	< 0.10 – 0.35	<0.10	0.5
Zn, mg/l	0.80 – 13.7	< 0.20 – 0.80	<0.20	10
Cd, mg/l	0.01 – 0.05	<0.01	<0.01	0.02
Pb, mg/l	0.03 – 0.25	<0.02 – 0.08	<0.02	0.2
Ni, mg/l	0.14 – 1.04	<0.03 – 0.10	<0.03	0.5
Co, mg/l	0.10 – 0.82	<0.03 – 0.08	<0.03	0.5
As, mg/l	0.06 - 0.23	<0.01	<0.01	0.2
Mn, mg/l	4.98 – 60.4	0.70 – 5.14	0.10 – 0.80	0.8
Fe, mg/l	78 – 401	0.80 – 19.4	1.4 – 3.2	5
SO_4^{2-}, mg/l	1702 – 3470	352 – 1465	305 – 1142	400

The effluents from the multibarrier were enriched in dissolved organic compounds and in some cases still contained sulphates, iron and manganese in concentrations higher than the relevant permissible levels. These effluents were treated in the constructed wetland where the Fe^{2+} and Mn^{2+} were oxidized to Fe^{3+} and Mn^{4+}, respectively, by some heterotrophic bacteria producing peroxide compounds from the organic substrates present in the wetland. The Fe^{3+} ions precipitated as $Fe(OH)_3$ and the Mn^{4+} ions precipitated as MnO_2. The dissolved organic compounds [resent in the waters were efficiently degraded by the different heterotrophs inhabiting the wetland.

References

Groudev S N, Spasova I I, Komnitsas K, Paspaliaris I (2003) Microbial generation of acid drainage in a uranium deposit, in Mineral Processing in the 21st Century (eds: L Kuzev, I Nishkov, A Boteva and D Mochev) pp. 728-732 (Djiev Trade Ltd: Sofia).

Groudev S N, Georgiev P S, Spasova I I, Nicolova M V (2008) Bioremediation of acid mine drainage in a uranium deposit, Hydrometallurgy, 94: 93-99.

Escobar B, Bustos K, Morales G, Salazar O (2008) Rapid and specific detection of *Acidithiobacillus ferroxidans* and *Leptospirillum ferrooxidans* by PCR, Hydrometallurgy, 92: 102-106.

Johnson D B, Hallberg K B (2005) Acid mine drainage remediation options: a review, Science of the Total Environment, 338: 3-14.

Rawlings D (2001) The molecular genetics of *Thiobacillus ferrooxidans* and other mesophilic, acidophilic, chemolithotrophic, iron or sulphur-oxidizing bacteria, Hydrometallurgy, 59: 187-201.

Sanz J L, Köchling T (2007) Molecular biology techniques used in wastewater treatment: An overview, Process biochemistry, 42: 119-133.

Tessier A, Campbell P G C, Bisson M (1979) Sequential extraction procedure for speciation of particulate trace metals, Analytical Chemistry, 51 (7): 844-851.

At the crossroads: Flooding of the underground uranium leach operation at Königstein (Germany) – A 2014 status brief

Ulf Jenk[1], **Micheal Paul**[1]

[1]Wismut GmbH, Jagdschänkenstraße 29, 09117 Chemnitz, Germany

Abstract. Controlled Flooding of section I of the uranium leach operation at Königstein has been finished by January 2013 with about two third of mine volume being filled. The paper provides an extensive compilation of the general technological, hydraulical and geochemical findings of the flooding process. Furthermore the permit application for final flooding up to the natural water level and the approval procedure regarding the final flooding step are been discussed.

Introduction

The Königstein Uranium mine is situated in an ecologically sensitive and densely populated area near Dresden, Germany (Fig.1). From the early sixties through 1990, approximately 19,000 t of uranium were produced. Remediation of the Königstein mine is a very special case within the WISMUT rehabilitation project due to the regional setting and the leaching mining method.

The ore body is located in the 4th sandstone aquifer, the deepest of four hydraulically isolated aquifers in a Cretaceous basin. The 3rd aquifer is an important water reservoir for the Dresden region and is environmentally very sensitive. The uranium was extracted from the 4th sandstone aquifer initially using conventional mining methods, but later an underground block leaching method using sulphuric acid supplanted conventional mining. Development of the mine site extended over a surface area of ca. 6.5 km^2 and comprised 4 main levels (Fig. 1).

B.J. Merkel and A. Arab (eds.), *Uranium – Past and Future Challenges*,
DOI 10.1007/978-3-319-11059-2_42

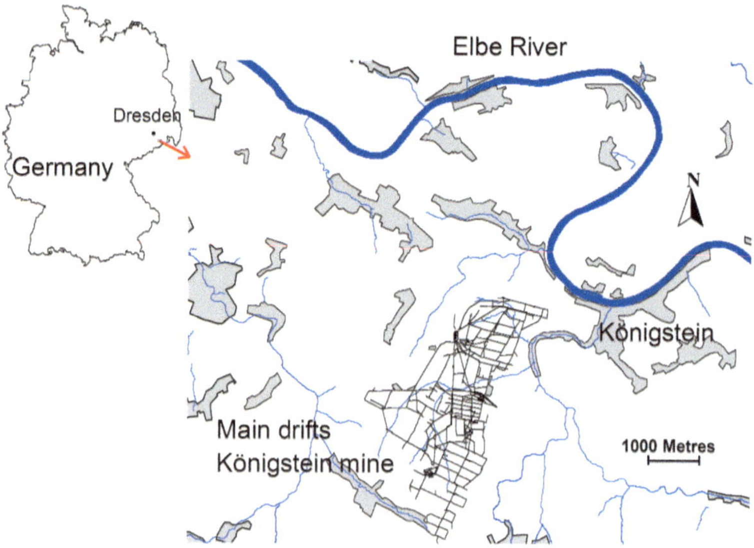

Fig.1. Location and schematic map of main drifts of the Königstein mine

Mining and the associated acidification processes but more particularly leach mining using sulphuric acid have fundamentally altered geochemical conditions within the developed section of the fourth aquifer. Pre-remediation investigations documented elevated levels of sulphuric acid, dissolved metals, and uranium.

Results of controlled flooding of section I

Rehabilitation of the former Königstein uranium mine was carried out by way of flooding. Flooding of mine section I (TB I) was initiated in January 2001 when the flood water table was at a level of ca. 24 m ASL and was terminated in January 2013 upon rising to the target water level of ca. 139.5 m ASL, were discharge of contaminated flooding water into surroundings is been ensured. This brought a complex remediation project involving orderly closeout to a successful conclusion.

Due to the specific site conditions and to permit requirements, preparations for flooding took several years to complete and included a number of large-scale flooding experiments. As a result of these preliminary studies, the concept of controlled mine flooding was devised (Jenk 2001). This concept, in essence, consists in a stepwise flooding of the mine and providing protection for the nearby aquifers. For that purpose, the mine was equipped with a specific underground water catchment system (control drifts (Fig. 1, red lines), pumping wells) to ensure the controlled withdrawal of flooding water. Basically, control of flood water rise was based on the reduction and scheduled shutdown of pumping capacities inherited from mining as well as on the caption of outflowing contaminated flooding water by means of the control drift system. Pumped to the surface, the flooding water is treated prior to discharge into the Elbe River.

The entire flooding process was followed by comprehensive and partly automated monitoring of the atmospheric and hydraulic exposure pathways as well as of a range of geomechanical topics. In addition to conventional equipment for flooding and groundwater monitoring, a number of technological innovations were put into practice to obtain representative water samples from the aquifers. Among those innovations was a mobile pump-based sampler unit for groundwater sampling down to a depth of 300 m.

Flooding of the mine took place in several stages in order to ensure required underground stabilisation efforts to proceed. In the flooding scenario per se, not only the mine voids volume, but also the readily fillable pore volume is an operative issue. As a consequence of the length of the flooding process, slowly-replenishable pore volumes also gradually filled up.

Table 1. Königstein mine: Main parameters of flooding of section I (water and compound balance).

	Königstein mine: Main parameters of flooding of section I
	Water storage (total)
Mine voids	2,386,000 [m³]
Pore volume	4,500,000 [m³]
Total volume	6,986,000 [m³]
	Release (total)
Uranium	940 t
Zinc	500 t
Iron	9,100 t
Sulphate	50.000 t

When the flooding water table reached the maximum permitted level at the end of January, 2013, the total fill volume has been identified amounting to a total of ca. 7 million m³ of water. This figure, minus the mine voids volume considered to be fully flooded, meant that ca. 4.5 million m³ pore porosity were replenished, which is some 10 % above the pore volume rated as readily fillable but significantly smaller than the total fillable volume (Table 1). In this way, predictions have been confirmed by the observed filling volume.

Particular importance was attached to the description of how a heavily contaminated flood water body would evolve. For that purpose, a model concept capable of describing aquifer recharge behaviour in the mine environment as well as hydraulic and hydrochemical conditions within the flood water body was developed at an early project stage in order to establish credible predictions of the flooding scenario (Metschies 2011). Tailor-made and adapted model tools were used to predict the flooding process. Flooding water quality predictions had been proven true in all their major aspects by the actual flooding process.

As expected, the evolution of flooding water quality had been characterised by highly elevated acid and contaminant levels (pH ~ 2, uranium ~ 260 mg/L) during the early stages of flooding. In the further course of flooding, contaminant levels diminished as a consequence of dilution with groundwater and chemical processes. Initially elevated local differences began to fade away. Even though the flooding water table has risen to its target level, the water remains acidic (pH ~ 3) and contains uranium and heavy metals in a concentration range of some mg/L (Fig. 2).

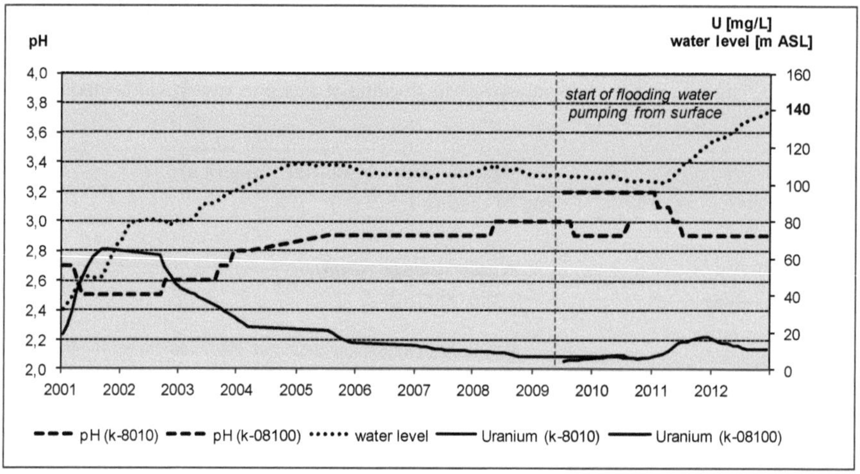

Fig.2. Mine water level, pH and uranium concentration of flooding water over 12 years.

During the flooding of TB I, approx. 35 million m³ of water were pumped to the surface as a flood control measure by which a significant amount of acid and dissolved contaminants was removed from the void subject to flooding and taken to safe disposal. Within that 12-year-period, approx. 940 t of uranium and approx. 500 t of zinc were removed from underground for commercial use (U) or proper disposal (Zn) (Table 1). This corresponds approximately to half of the estimated total mobile potential. In this way, the mining-induced impact to the 4th aquifer was significantly mitigated and a partial cleanup of the contaminant source (underground mine) achieved.

Due to the flooding of the mine and of the control drifts, the depression cone at the mine's northern and western edges (4th aquifer) began to recede as predicted. Sitting on top of the mine and also depleted due to mining activities, the third aquifer groundwater table is recharging at a slow rate. This gets us closer to hydraulic conditions which prevailed prior to mine development. Mine flooding had no impact on hydrochemical conditions in the 3rd aquifer. Hydrochemical impacts to the 4th aquifer were confined to a small area on the edges of the mined area.

Environmental monitoring of the atmospheric pathway was carried out at 8 cast-up locations (mine shafts and ventilation boreholes) to determine air volumes, radon and dust-borne radon decay products. 37 immission measurements in the near-ground air were performed at yet another set of 37 measurement locations. Underground occupational surveillance involved measurements of radioactive and non-radioactive components at 4 stationary and 10 mobile measurement points in connection with ambient dose measurements at working places as well as personal dosimetry. All concerning this matter boundary limits have been kept.

Geomechanical surveillance demonstrated that ground subsidence was minimal (maximum settlement 52 mm). Subsidence slopes derived from measurement data are within detection limit (< 1mm/m). Ground movements, on the whole, were marginal and have faded away. There is no risk of surface subsidence whatsoever.

Final flooding – status of approval procedure

Since flooding of TB I reached the target level of ca. 139.5 m ASL in January, 2013, while maintaining an elevated water throughput (rinsing of the flooding water body) the flooding water table is kept and maintained at this level. The further rise of the flooding water table to a natural stable level of approx. 200 m ASL as outlined in the mine closure plan and for which a permit application was submitted in 2011 has not yet received regulatory approval.

Continued flooding water pumping via pumping wells is imperative to prevent the rise of the water table in the mine beyond the level of 140 m ASL owing to the natural inflow of groundwater. Further continuation of the flooding process notwithstanding, rehabilitation of the site's surface structures and of areas no longer used for mining or water treatment purposes shall be continued.

In parallel, the information gained and, in particular, the modelling tools developed were used for design planning and permit application regarding the final flooding step, focussing of the further mine water rise to a natural water level (Section II), which is in accordance with the approved closure plan. Flooding of Section II would enable a complete and conclusive remediation of the site but is unavoidably linked with limited emissions of pollutants into the surrounding aquifers (Table 2).

Table 2. Main features of final flooding and long term pump and treat (Königstein mine)

Flooding up to natural water level	Long term pump and treat
▪ Long term stable hydraulic and geochemical improved status inside mine ▪ Avoidance of long term additional contaminant delivery ▪ Sustainable remediation if source (mine) ▪ Complete and finished remediation of Königstein minig site → conformance of basics of German mining law (BBergG) ▪ Costs limited ▪ Not available but limited impacts of partial areas of aquifers 3 und 4	▪ instable hydraulic und chemical status ▪ Long term contaminant release, radioactive waste ▪ Remaining of not remediated parts of mine ▪ No finishing of remediation of Königstein mining site ▪ Long term demand of technical facilities ▪ Long term costs ▪ Transmission of tasks and costs to following generations ▪ But: nor influence of downstream areas of aquifers during pumping

Due to those emissions, which were predicted based on worst case assumptions; the approval of flooding of Section II has been dismissed by the regulatory bodies with reference to the strict rules of German water legislation. Basically both options are connected with ecological impacts and risks: there is no "ideal solution".

WISMUT as the operator, however, considers the given situation and the absence of a sustainable, alternative technical solution as sufficient to apply a possible exception from this "zero emission policy", which is in principle possible according to the relevant legislation. Therefore, Wismut filed an objection, since keeping the water at the present level of 140 m a.s.l. implies technical efforts and costs regarding mine water management in perpetuity, which might not be justified.

References

Jenk U, Schreyer J (2001) Pollutant Release Level Prognosis – A Major Input into the Flooding Concept for the Former ISL Uranium Mine at Königstein (WISMUT Germany). Proceedings 8th International Conference on Environmental Management, 2001, Bruges, Belgium

Metschies T, Jenk U (2011) Implementation of a modelling concept to predict hydraulic and geochemical conditions during flooding of a deep mine, The New Uranium Mining Boom, Challenges and Lessons learned. Springer Verlag Berlin Heidelberg 2011, ISBN 978-3-642-22121-7

Do macrofungi accumulate uranium?

Jaroslava Kubrová[1,2], Jan Borovička[2,3]

[1]Institute of Geochemistry, Mineralogy and Mineral Resources, Faculty of Sci ence, Charles University, Albertov 6, CZ-12843 Prague 2, Czech Republic
[2]Nuclear Physic Institute, v.v.i., Academy of Science of the Czech Republic, CZ-25068 Řež near Prague, Czech Republic
[3]Institute of Geology, v.v.i., Academy of Science of the Czech Republic, CZ-16500 Prague 6, Czech Republic

Abstract. Samples of saprotrophic (24 species) and ectomycorrhizal (26 species) macrofungi were collected from a U-polluted Norway spruce plantation and analyzed for the metals content. Concentrations of metals (Ag, Pb, Th and U) were also analyzed in ectomycorrhizal tips, non-mycorrhizal spruce fine roots and underlying soil. The concentration of U in macrofungi from the polluted area were circa 4x higher than those reported for macrofungi from pristine sites but generally did not exceed 3 mg/kg (dry weight).

Introduction

The geochemical behavior of many heavy metals in soil is affected by microbial and fungal activity (Gadd 2010, Gadd et al. 2012), and U is no exception. It has been demonstrated in in vitro experiments that fungi are able to solubilize U oxides and accumulate U within their mycelia (Gadd and Fomina 2011). However, our knowledge on environmental interactions between fungi and U is limited as no field studies have been conducted. Both saprotrophic and symbiotic macrofungi are known for their ability to accumulate various metals in their fruiting bodies (Falandysz and Borovička 2013), but there is a lack of data on interactions between macrofungi and U in polluted environments.

To assess the possible role of macrofungi in the biogeochemical cycling of U in polluted forest soils, we analyzed the fruiting bodies of macrofungi and ectomycorrhizal tips (EM tips). Furthermore, Th and Pb, which are excluded by macrofungi (Borovička et al. 2011), and Ag, which is efficiently accumulated in macrofungi (Borovička et al. 2010) were analyzed.

B.J. Merkel and A. Arab (eds.), *Uranium – Past and Future Challenges*,
DOI 10.1007/978-3-319-11059-2_43

Experimental

Selected site

For the purpose of this study, we selected the former U mining district of Příbram in Central Bohemia (Czech Republic), which is a region in the Bohemian Massif with perigranitic monometallic veins (Ruzicka 1993). Macrofungi, EM tips, fine spruce roots and organic soils were sampled throughout the selected part of the spruce forest plantation (mixed with birch, pine, poplar and lime trees on its margins) on granitic bedrock near the village of Bytíz.

Macrofungi, EM tips and fine roots samples

Samples of macrofungi were collected, identified and processed as described elsewhere (Borovička et al. 2011). Spruce fine roots were collected with EM tips from the Oe horizon, transported to the laboratory, stored in a fridge at 4°C and processed within 24 hours. Samples were washed thoroughly in tap water. EM tips were cleaned with tweezers and stainless steel needles in a Petri dish under a stereomicroscope. A miniscule piece of each EM sample was separated and stored at -20°C for DNA extraction. EM samples were air-dried on filter paper and later in an oven at 50°C to a constant weight. A portion of non-mycorrhizal fine root tissue was also taken from each sample, cleaned thoroughly with tap water and subjected to chemical analysis. EM tips (0.40-8.08 mg) and spruce fine roots (23.6-110 mg) were heat-sealed into polyethylene capsules with a diameter of 16 mm for EM tips and 25 mm for roots.

DNA was extracted from EM tips using the NucleoSpin Plant II DNA isolation kit. To amplify fungal DNA, we performed PCR with the primer pair ITS1F-ITS4, as described elsewhere (Nieto and Carbone 2000), or semi-nested PCR was used with the primer pairs ITS1F-ITS4B and ITS1F-ITS4 (Borovička et al. 2010). All sequences were identified to the species/genus/family level by querying the GenBank database, using the nucleotide-nucleotide (blastn) BLAST search option, available through the National Centre for Biotechnology Information (http://blast.ncbi.nlm.nih.gov).

Soil samples

Based on a preliminary survey of 16 points, a representative place for digging a representative soil profile was chosen (GPS coordinates N49 40.708, E14 04.426).

To assess the extent of pollution, 23 samples of the Oa horizon were collected throughout the locality. Soil samples were air-dried and sieved through a 2 mm sieve. A representative portion of each sample was milled in an agate mill and subjected to chemical analysis.

Chemical analyses

The concentrations of Ag, Pb, Th and U in macrofungi were determined using the Element2 sector field mass spectrometer with inductively coupled plasma (ICP-SF-MS) according to Borovička et al. (2011). The concentration of U in EM tips and roots was determined by instrumental neutron activation analysis (INAA) using the radioisotope ^{239}U (74.7 keV, $t_{1/2}$ = 23.45 min) according to Řanda et al. (2005).

The total concentrations of U, Th and Ag in soil samples (weight ~250 mg) were determined non-destructively by INAA using the radioisotopes ^{239}Np (106.13 keV $t_{1/2}$ = 2.36 d) for the determination of U, ^{233}Pa (311.9 keV, $t_{1/2}$ = 27.4 d) for the determination of Th and ^{110m}Ag (657.75 keV, $t_{1/2}$= 249.88 d) for the determination of Ag.

As Pb concentrations cannot be determined by INAA, the samples were analyzed by instrumental photon activation analysis (IPAA) using the radioisotope ^{203}Pb (279.19 keV, $t_{1/2}$ = 51.873 h) according to Řanda et al. (2007).

In order to determine the mobility of metals in the soil profile, BCR sequential extraction was performed according Rauret et al. (1999). The concentrations of metals in the soil extracts were analyzed by ICP-SF-MS and the quality of the BCR sequential extraction was tested by the use of the certified reference material BCR 483 (Sewage Sludge Amended Soil).

Results and discussion

Soils

The concentrations of U in 23 soil samples (Oa horizon) varied in the range of 6-75 mg/kg. According to Kabata-Pendias (2001), such concentrations indicate pollution by this element. The site was also contaminated by Pb, which originated from long-term deposition due to smelting (Ettler et al. 2004). The median concentration of Pb in the Oa horizon was 558 mg/kg. The Ag concentrations in the Oa horizon were slightly elevated (median 0.77 mg/kg), but still can be considered as background (Evans and Barabash 2010). The concentrations of Th were similar to those found in pristine environments (Kabata-Pendias 2001).

As shown in Table 1, the highest concentrations of Ag, Pb and U were detected in the organic horizons (Oa > Oe > Oi) and rapidly decreased with depth (Ah> Bw> Cr). This enrichment in the organic layers can be explained by the sorption of metals to organic matter. The only element for which the highest concentrations were found in the lower horizons (mineral horizons Bw and Cr) was Th; this can be explained by the natural distribution of this element in the soil profile. These results are consistent with the findings of other authors (Komárek et al. 2007, Ribera et al. 1996).

Table 1. Total metal concentrations (dry matter) in particular soil horizons in the soil profile at the polluted site.

	concentration (mg/kg)			
soil horizon	U	Th	Pb	Ag
Oi	0.43	0.18	28.5	0.17
Oe	14.2	3.11	382	0.45
Oa	23.2	5.78	789	0.99
Ah	7.16	12.2	388	0.52
Bw	4.09	15.3	38.4	0.07
Cr	5.41	21.8	42.5	0.06
granite	6.59	21.4	< 25.0	< 0.09

Statistical uncertainty of INAA/IPAA determinations of U, Th, Pb and Ag was mostly below 1%, 1%, 5% and 7%, respectively.

Our knowledge of the biochemical behavior of U in soils is rather limited. Several authors have studied the behavior of depleted U in polluted areas (Graham et al. 2008, 2011), but not in forest soils. Virtanen et al. (2013) published several results for sequential extraction of U in forest soils, but their study was focused on an unpolluted environment.

The fractionation of U by BCR found in this study was very similar to the results observed Oliver et al. (2008). Most of the U was associated with the oxidizable fraction (Oe, Oa and Ah horizons) and U was highly immobile in the Bw and Cr horizons, with nearly 90% in the residuum.

The majority of the Th (~90%) was present as firmly bound in the residuum, which is in agreement with Virtanen et al. (2013).

Lead was the most mobile element, especially in the organic horizons. Similar results have been published from various polluted areas in the Czech Republic, but in some studies, higher percentages for the oxidizable fraction were reported (Ash et al. 2013, Ettler et al. 2005, Komárek et al. 2007).

Most of the Ag was associated with the reducible fraction, followed by somewhat lower percentages for the oxidizable and residual fractions, respectively. Very few data have been published for the Ag fraction in soils (Jacobson et al. 2005, Jones et al. 1984).

Comparing the total concentration and fractionation of the investigated metals, Pb may be considered the most bioavailable, followed by U, Th and Ag, respectively (data not shown).

Macrofungi

Concentrations of U in the macrofungal fruiting bodies reported from unpolluted sites are very low (units or lower dozens of µg/kg); the concentrations in saprotrophic species are slightly higher than those in the ectomycorrhizal ones (Borovička et al. 2011).

At the polluted area near Bytíz, the U concentrations in the fruiting bodies were elevated, but did not exceed 3 mg/kg; the highest value was measured in the saprotrophic *Helvella lacunosa* (2.61 mg/kg). In ectomycorrhizal species, the highest concentration was found in *Inocybe dulcamara* (1.98 mg/kg). Similar results were published by Baumann et al. (2014) for macrofungi collected from a U-contaminated area in Germany.

Thorium concentrations in macrofungi were very low, with a maximum value of 78.6 µg/kg found in *Clitocybe costata*. This indicates effective exclusion of this element (Borovička et al. 2011).

The concentrations of Pb in macrofungi were much higher than those of Th. The highest concentration of 72.0 mg/kg was found in *Lycoperdon foetidum*.

Despite the fact that Ag is an element that occurs in soils at the lowest concentrations of the elements studied (and was the least mobile element), relatively high concentrations (median 7.32 mg/kg and the maximum value of 53.9 mg/kg) were found in macrofungi, with the highest value in *Agaricus arvensis.*

A statistical summary of the metal content in macrofungi is shown in Table 2. These findings suggest that the uptake of elements by macrofungi is not primarily driven by the content or mobility of particular elements in soils.

Table 2. Statistical summary for ectomycorrhizal and saprotrophic macrofungi collected from a U-polluted site: element concentrations (dry matter).

	concentration (µg/kg)							
	ectomycorrhizal macrofungi				saprotrophic macrofungi			
	U	Th	Pb	Ag	U	Th	Pb	Ag
minimum	3.27	3.00	84.5	58.5	5.28	2.81	195	59.6
median	63.0	7.09	764	1065	82.4	5.93	7381	7318
mean	157	12.0	2285	2599	216	12.9	13100	9833
maximum	1978	49.3	10181	48157	2611	78.6	72009	53942

Ectomycorrhizae

The concentrations of U in ectomycorrhizal tips and fine roots were very similar (median 2.33 mg/kg and 2.03 mg/kg, respectively). Among the EM tips, the highest concentration of U was detected in *Russula ochroleuca* (6.95 mg/kg).

Uranium does not accumulate in ectomycorrhizae, which is in contrast with Cd, Cu, Pb and Zn (Berthelsen et al. 1995, Krupa and Kozdrój 2004). It can be concluded that the environmental role of EM macrofungi in the biogeochemical cycling of U in forest soils is considerably limited, which is in contrast to what was observed in arbuscular mycorrhizal symbiosis (Dupré de Boulois et al. 2008).

Conclusions

Although the concentrations of U in the macrofungal fruiting bodies from the polluted area were higher than those observed in unpolluted environments, they did not exceed 3 mg/kg. These values as well as the low concentrations of U found in the EM tips indicate that the role of macrofungi in the biogeochemical fate of U in polluted forest soils is rather limited.

Acknowledgements

This research was supported by projects GAUK 535112 (The Charles University Grant Agency) and GAČR 504/11/0484 (Czech Science Foundation).

References

Ash C, Tejnecký V, Šebek O, Němeček K, Žahourová-Dubová L, Bakardjieva S, Drahota P, Drábek O (2013) Fractionation and distribution of risk elements in soil profiles at a Czech shooting range. Plant Soil Environ 59: 121-129

Baumann N, Arnold T, Haferburg G (2014) Uranium contents in plants and mushrooms grown on a uranium-contaminated site near Ronnenburg in Eastern Thuringia/Germany. Environ Sci Pollut Res 11: 6921-6929

Berthelsen BO, Olsen RA, Steinnes E (1995) Ectomycorrhizal heavy metal accumulation as a contributing factor to heavy metal levels in organic surface soils. Sci Total Environ 170: 141-149

Borovička J, Kotrba P, Gryndler M, Mihaljevič M, Řanda Z, Rohovec J, Cajthaml T, Stijve T, Dunn CE (2010) Bioaccumulation of silver in ectomycorrhizal and saprobic macrofungi from pristine and polluted areas. Sci Total Environ 408: 2733-2744

Borovička J, Kubrová J, Rohovec J, Řanda Z, Dunn CE (2011) Uranium, thorium and rare earth elements in macrofungi: what are the genuine concentrations? Biometals 24: 837-854

Dupré de Boulois H, Joner EJ, Leyval C, Jakobsen I, Chen BD, Roos P, Thiry Y, Rufyikiri G, Delvaux B, Declerck S (2008) Impact of arbuscular mycorrhizal fungi on uranium accumulation by plants. J Environ Radioact 99: 775–784

Ettler V, Mihaljevič M, Komárek M (2004) ICP-MS measurements of lead isotopic ratios in soil heavily contaminated by lead smelting: tracing the sources of pollution. Anal Bioanal Chem 378: 311-317

Ettler V, Vaněk A, Mihaljevič M, Bezdička P (2005) Contrasting lead speciation in forest and tilled soils heavily polluted by lead metallurgy. Chemosphere 58: 1449-1459

Evans LJ, Barabash J (2010) Molybdenum, silver, thallium and vanadium. in: Hooda PS (Ed.) Trace elements in soil. Wiley, Chichister: 515-549

Falandysz J, Borovička J (2013) Macro and trace mineral constituents and radionuclides in mushrooms: health benefits and risks. Appl Microbiol Biotechnol 97: 477-501

Gadd GM (2010) Metals, minerals and microbes: geomicrobiology and bioremediation. Microbiology 156: 609-643

Gadd GM, Fomina M (2011) Uranium and fungi. Geomicrobiol J 28: 471-482

Gadd GM, Rhee YJ, Stephenson K, Wei Z (2012) Geomycology: metals, actinides and biominerals. Environ Microbiol Rep 4: 270-296

Graham MC, Oliver IW, MacKenzie AB, Ellam RM, Farmer JG (2008) An integrated colloid fractionation approach applied to the characterisation of porewater uranium-humic interactions at a depleted uranium contaminates site. Sci Total Environ 404: 207-217

Graham MC, Oliver IW, MacKenzie AB, Ellam RM, Farmer JG (2011) Mechanisms controlling laternal and vertical porewater migration of depleted uranium (DU) at two UK weapons testing sites. Sci Total Environ 409: 1854-1866

Jacobson AR, McBride MB, Baveye P, Steenhuis TS (2005) Environmental factors determining the trace-level sorption of silver and thallium to soils. Sci Total Environ 345: 191-205

Jones KC, Peterson PJ, Davies BE (1984) Extraction of silver from soils and its determination by atomic-absorption spectrometry. Geoderma 33: 157-168

Kabata-Pendias A (2001) Trace elements in soil and plants. Fourth ed. CRC Press, Boca Raton

Komárek M, Chrastný V, Štíchová J (2007) Metal/metalloid contamination and isotopic composition of lead in edible mushrooms and forest soils originating from a smelting area. Env Int 33: 677-684

Krupa P, Kozdrój J (2004) Accumulation of heavy metals by ectomycorrhizal fungi colonizing birch trees growing in an industrial desert soil. World J Microbiol Biotechnol 20: 427-430

Nieto PM, Carbone SS (2000) Characterization of juvenile maritime pine (*Pinus pinastre* Ait.) ectomycorrhizal fungal community using morphotyping, direct sequencing and fruitbodies sampling. Mycorrhiza 19: 91-98

Oliver IW, Graham MC, MacKenzie AB, Ellama RM, Farmer JG (2008) Distribution and partitioning of depleted uranium (DU) in soils at weapons test ranges – Investigations combining the BCR extraction scheme and isotopic analysis. Chemosphere 72: 932-939

Rauret G, López-Sánchez JF, Sahuquillo A, Rubio R, Davidson C, Ure AM, Quevauviller Ph (1999) Improvement of the BCR three step sequential extraction procedure prior to the certification of new sediment and soil reference materials. J Environ Monit 1: 57-61

Ribera D, Labrot F, Tisnerat G, Narbonne JF (1996) Uranium in the environment: occurrence, transfer, and biological effects. Rev Environ Contam Toxicol 146: 53-89

Ruzicka V (1993) Vein uranium deposits. Ore Geol Rev 8: 247-276

Řanda Z, Soukal L, Mizera J (2005) Possibilities of the short-term thermal and epithermal neutron activation for analysis of macromycetes (mushrooms). J Radioanal Nucl Chem 264: 67-76

Řanda Z, Kučera J, Mizera J, Frána J (2007) Comparison of the role of photon and neutron activation analyses for elemental characterization of geological, biological and environmnetal materials. J Radioanal Nucl Chem 271: 589-596

Virtanen S, Vaaramaa K, Lehto J (2013) Fractionation of U, Th, Ra and Pb from boreal forest soils by sequential extractions. Appl Geochem 38: 1-9

Treatment of Acid Wastewater Containing Uranium by Sulfate Reducing Bacteria

Jie Gao[1], Lechang Xu[1], Yalan Wang[1], Wei Zhang[1]

[1]Beijing Research Institute of Chemical Engineering and Metallurgy, Beijing, 101149

Abstract. Sulfate Reducing Bacteria (SRB) strain form and its sulfate-reducibility was identified after enrichment, separation and purification of SRB from some anaerobic sludge of a uranium mine. The growth of SRB was stabilized 72 hours after inoculation. The influence of inoculation quantity, pH, temperature, SO_4^{2-} concentration on the growth of SRB was discussed. Sulfate treatment effect with 6.5g/L SO_4^{2-} concentration was 51.8% in 15 days and 69.1% in 30 days by SRB, respectively. Treatment effect of sulfate was 64% and 83%, the one of uranium was 60% and 82.3% from simulated wastewater by SRB and SRB+Fe^0, respectively.

Introduction

Nowadays sulfuric acid leaching is mainly applied for uranium extraction techniques, including in-situ leaching, heap leaching and stope leaching. However, some heavy metal and non-radioactive heavy metal will unexpectedly obtained during uranium extraction process, which lead to contamination of surface water and groundwater, especially dissolved radioactive nuclides will transport for in-situ leaching and stope leaching. Currently the main methods for treatment of acid uranium containing wastewater include chemical precipitation, ion exchange, evaporation, absorption and microbiological technology (Yang et al. 1994; Yu et al. 2002; Pan 1984; Hou 2003; Junta-Ross and Hochella 1996; Tang et al. 2003; Huang and Zheng 2002) among which microbiological technology is gradually being paid attention to due to advantages of low cost and decreased secondary pollutant. Hence Sulfate-Reducing Bacteria (SRB) is used for treatment of acid uranium containing wastewater in microbiological process.

B.J. Merkel and A. Arab (eds.), *Uranium – Past and Future Challenges*,
DOI 10.1007/978-3-319-11059-2_44

Principal of treatment of acid uranium containing wastewater by SBR

SRB is kind of anaerobic bacteria which reduce sulfates through dissimilation, and normally SRB, with diverse shape and nutrition, use sulfates serving as electron acceptor during organic dissimilation (Hu and Gu 2003; Janssen et al. 1995).

Under initial anaerobic condition, SO_4^{2-} is reduced to H_2S by SRB in sulfate dissimilation process, only if electron supplier i.e. organic carbon source add to wastewater could SO_4^{2} biological reduction happens.

Reactions in initial anaerobic stage (Su and Li 2005):

(1) Fermentation

Orgnic molecule i.e. sucrose transforms to lactic acid.

$$2CH_3CHOHCOOH+SO_4^{2-}\rightarrow 2CH_3COO^{-}+2HCO_3^{-}+2H_2S \qquad (1)$$

It is necessary for the following SO_4^{2-} reduction, for only fermentation product i.e. pyruvate and lactic acid can be used as carbon source for SBR.

(2) Bacteria respiration

SBR use lactic acid to produce acetic acid.

$$2CH_3CHOHCOOH+ SO_4^{2-}\rightarrow 2CH_3COO\text{-}+2HCO_3\text{-}+2H^2S \qquad (2)$$

Reaction (2) could also be shown as below.

$$2C+SO_4^{2-}+H_2O\rightarrow H_2S+2HCO_3^{-}$$

There are mainly four pathways for SO_4^{2-} biological reduction on heavy metal treatment: (1) According to reduction by catabolized sulfate, SBR reduce SO_4^{2-} to H_2S, U (VI) in wastewater to U (IV) sediment (Bertine 1970; Bonatti et al. 1971; Lovley et al. 1993; Tebo and Obrazsova 1998; Lovley 2003), heavy metal with high valence to low valence state sediment; (2) Heavy metal ion in wastewater will be attenuated by reaction with H_2S from pathway above, which is a significant pathway for heavy metal treatment by SO_4^{2-} biological reduction; (3) Reduction from SO_4^{2-} to S^{2-} will increase pH of wastewater and further be favorable to eliminate heavy metal ion in the form of hydroxide sediment; (4) Certain heavy metal will be removed by reaction with CO_3^{2-} that obtained from CO_2 in SBR metabolism process (Ma et al. 2003).

Sample Collection, Cultivation, Isolation, Purification

Anaerobic sludge collected from certain uranium tailing pond was put into Hungate anaerobic tube containing Germany No.DSMZ2075 culture medium under 35°C (Su and Li 2005) in constant temperature incubator. After 7 days, sludge in Hungate anaerobic tube turned black with strong rotten egg odor, which showed well-grown of SBR. Then 1ml supernatant was transferred to newly prepare anaerobic culture medium for second purified cultivation, after that, Hungate roll tube isolation purification (Pan et al. 2007) was carried out until purified bacteria strain was obtained, as showed in Fig.1.

Hungate roll tube isolation → Culture bacteria → choosing single colony → Second roll tube

Fig.1. Isolation & Purification of SRB

After purification, there were single bacterial colony with scattered distribution and uniform shape at tube wall, which showed small black globular, raised in surface and stretched all around. The bacterial strain was identified to be Gram-negative, spore staining negative, sing polar flagellum and the cells were observed to be swinging moved, tiny rod but curved shape, as showed inFig.2.

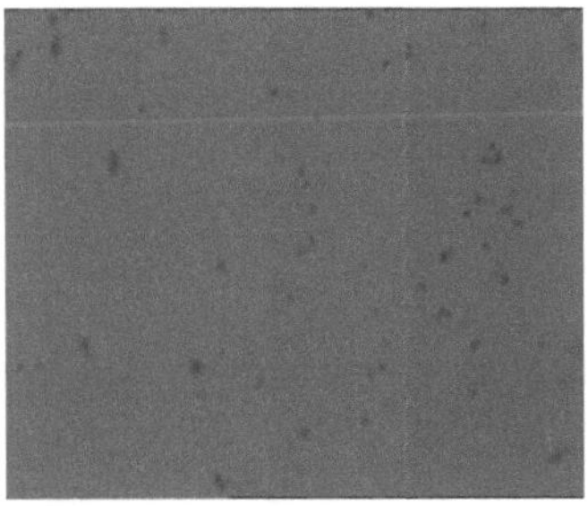

Fig.2. Isolated SRB

Being injected into liquid culture medium, the bacterial strain from above experiment turn PbAc test paper black at its logarithmic growth phase, which illustrated strong capacity in sulfate reduction, therefore, the bacterial strain was proved to be SRB (Wan et al. 2003).

Study on growth factor

Character of SRB growth

Measurement of OD_{605} of bacterial suspension by spectrophotometer demonstrated positive correlation (Shen et al. 1998) between OD_{605} and bacterial growth. Prior to measurement, bacterial suspension was diluted to 100 times as before to avoid effect of black sediment.

Growth curve

Considering the inconvenience brought by preparation of Germany No.DSMZ 2075 culture medium, formula (Chen et al. 2006) from Chengdu methane bureau under national agricultural department was adopted for growth factor study. With certain bacterial strain had inoculated into fresh uniform liquid culture medium, growth of bacterial strain experienced lag phase, logarithmic phase, steady phase and dissipation phase under adequate condition. Thus there achieved growth curve of bacterial strain with horizontal axis incubation time and vertical axis OD_{605}, which reflected variation tendency of bacterial strain.

Fig.3 showed the growth curve of isolated SRB, it is obvious that SRB went into logarithmic phase rapidly, and steady phase in 64h culure.

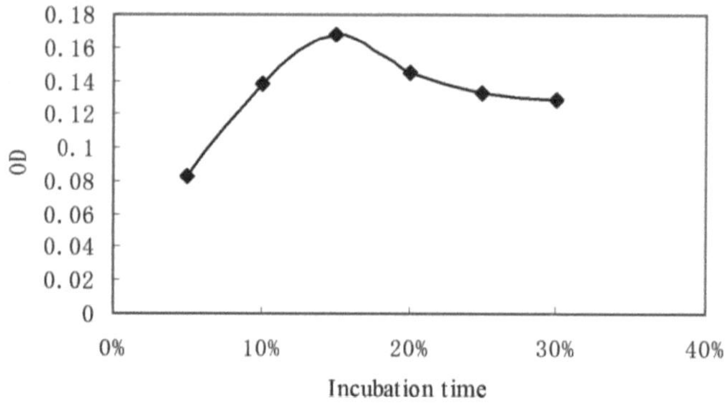

Fig.3. Growth curve of isolated SRB

Effect of inoculation quantity on SRB growth

Inoculation quantity was another factor affecting SRB growth in duration of logarithmic phase. Large inoculation quantity led to shorter logarithmic phase and competition between bacterial strain. Conversely, low inoculation quantity led to less stronger competition.

SRB was inoculated into 50ml liquid culture medium with 5%, 10%, 15%、20%, 25%, 30%inoculation quantity respectively, OD605nm was measured after 7 days culture under temperature 35°C. Results in Fig.4 indicated a maximum OD605nm at inoculation quantity 15%, which was be adopted in the following study.

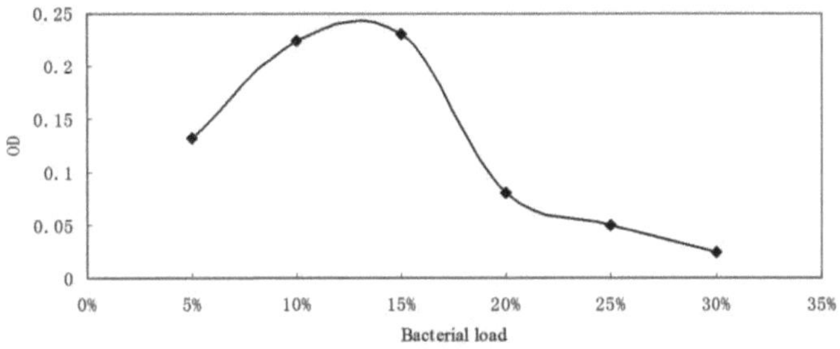

Fig.4. Effect of inoculation quantity on SRB growth

Effect of pH on SRB growth

In order to identify optimum pH of SRB growth, formula from Chengdu methane bureau was improved by adjusting pH at 2.0、2.5、3.5、4.5、5.5、6.5、7.5 with HCl (0.5mol/L) . OD605nm measurement was carried out every 24h 7 days later to identify optimum pH for SRB growth. Fig.5 showed optimum pH for SRB was 6.5.

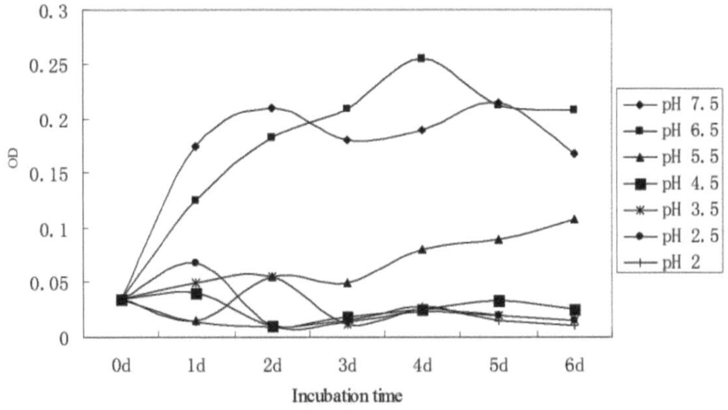

Fig.5. Effect of pH on SRB growth

Effect of temperature on SRB growth

Temperature is an important parameter for anaerobic sulfates reduction that will affect the metabolic activity and growth rate of SRB. SRB include two types, that are majority of mesophilic and the rest of thermophilic, of which mesophilic SRB are likely to achieve optimum growth between 28~38°C ,and thermophilic is 54~70°C, maximum of 56~85°C. Related research (Li and Su 2000; Laryr and

Barton 1995) revealed that activity of SRB would suffer nearly no inhibition and observable inhibition at temperature of 31~35°C and below 30°C respectively.

Effect of SO_4^{2-} concentration on SRB growth

In order to identify relationship between SO_4^{2-} concentration and SRB growth, liquid culture medium was adjusted by setting SO_4^{2-} concentration at 1.5、2、3、4、5、6、7、8、9、10g/L level with unchangeable proportion of $MgSO_4$: Na_2SO_4=1 :2. Accordingly optimum SO_4^{2-} concentration on SRB growth was identified based on OD605nm measurement taken every 24h after 7 days growth. Results showed optimum SO_4^{2-} concentration on SRB growth was 4~6g/L, see Fig.6.

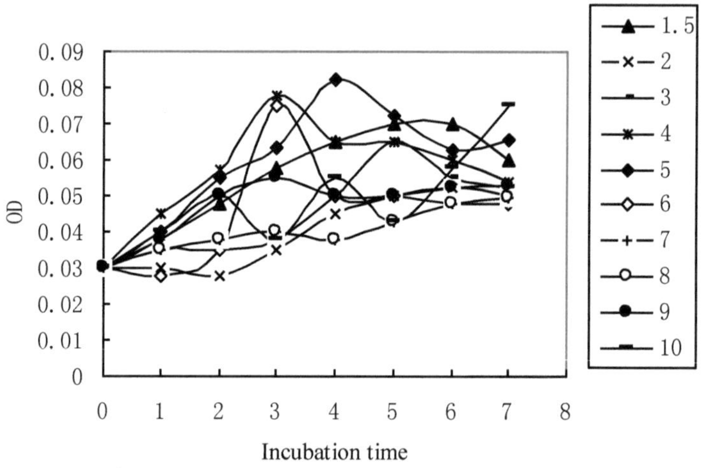

Fig.6. Effect of SO_4^{2-} concentration on SRB growth

Effect of SO_4^{2-} concentration on sulfates reduction rate

In order to identify effect of SO_4^{2-} concentration on sulfates reduction rate, liquid culture medium was adjusted by setting SO_4^{2}- concentration at 3.0、5.7、6.5、7.4、11.1、16.4、17.0g/L level with unchangeable proportion of $MgSO_4$: Na_2SO_4=1 :2. Sulfates reduction rate was calculated according to analysis on SO_4^{2-} concentration 15days and 30 days later. Effect of SO_4^{2-} concentration on sulfates reduction rate was shown in Fig.7.

As displayed in Fig.7,sulfates reduction rate was between 41.4~51.8% for 15 days culture and 43.7~69.1% for 30 days culture. Moreover, sulfates reduction rate reached maximum at SO42- concentration 6.5g/L for both 15 day and 30 days culture. This was in line with the conclusion that SRB was at its optimum growth with SO_4^{2-} concentration of 4~6g/L.

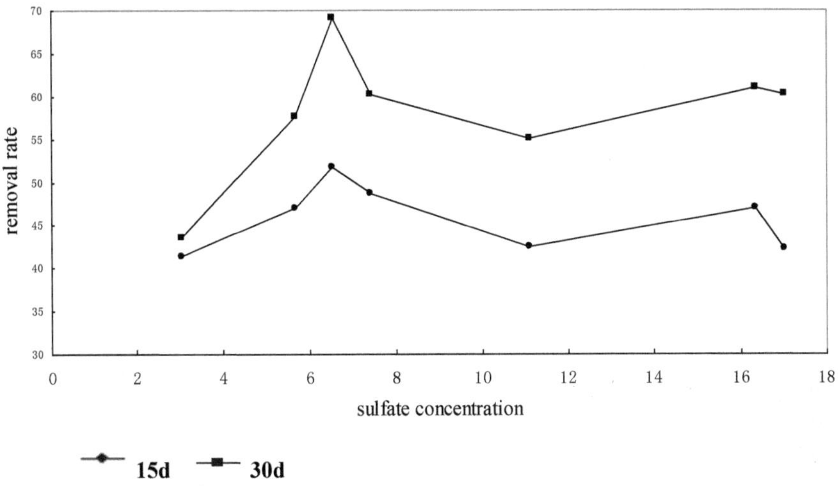

Fig.7. Effect of SO_4^{2-} concentration on sulfates reduction rate

SRB on uranium wastewater treatment

In order to explore SRB and coupling effect of both SRB and zero-valent iron(ZVI) on treatment efficiency of SO_4^{2-} and U, two set of wastewater simulation was conducted with one set 15% SRB inoculation and another set 15% SRB plus 0.3g ZVI. Uranium wastewater in this trial was prepared in lab with pH value 6.5, SO_4^{2-} concentration 2g/L and U concentration 3mg/L, culture medium was still follow the formula of Chengdu methane bureau. Results can be seen in Fig.8,9.

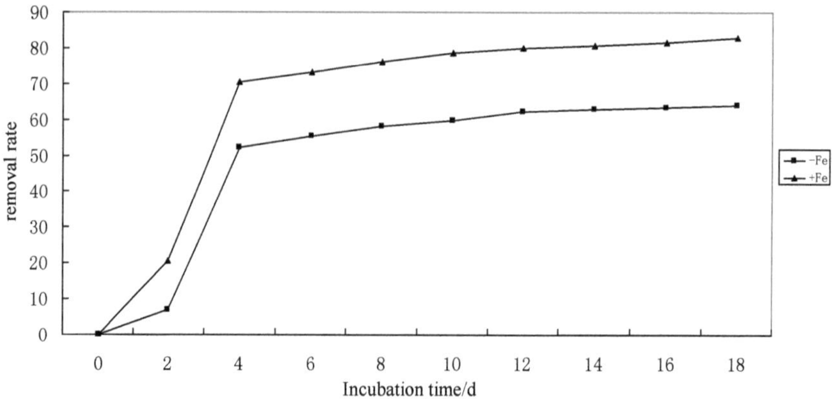

-Fe:SRB + Fe:SRB+Fe^0

Fig.8. SRB and coupling effect of both SRB and ZVI on treatment efficiency of SO_4^{2-} in wastewater

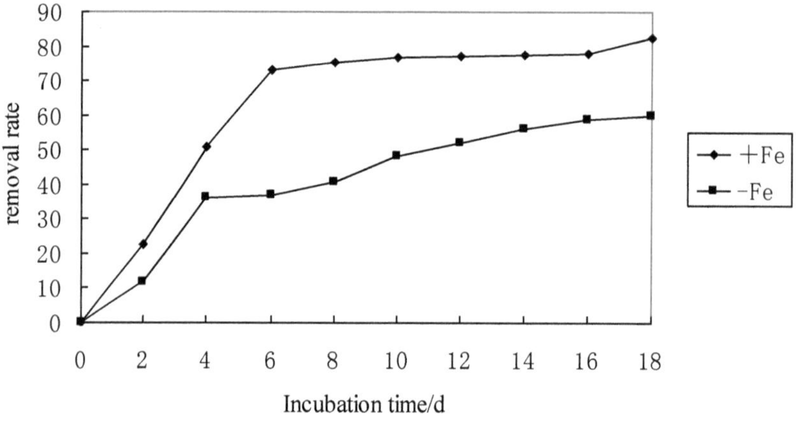

-Fe:SRB + Fe:SRB+Fe^0

Fig.9. SRB and coupling effect of both SRB and ZVI on treatment efficiency of U in wastewater

Results indicated that SO_4^{2-} concentration was 640mg/L, treatment efficiency was 64% and U concentration 1.2mg/L, treatment efficiency 60% for 15% SRB culture 18 days later; SO_4^{2-} concentration was 340mg/L, treatment efficiency was 83% and U concentration 0.53mg/L, treatment efficiency 82.3% for 15% SRB plus ZVI culture 18 days later.

In summary, ZVI promotes SRB growth in following aspects:

ZVI's consumption of H^+ under acid condition increased pH of wastewater, which accelerated SRB growth;

Hydrogen served as energy source of SRB and further promoted SRB growth;

Fe^{2+}、Fe^{3+} generated from ZVI reduction, not only promoted SRB growth, but also eliminated inhibition on SRB by sediment formation with H_2S;

ZVI weakened toxicity of SRB by decreasing heavy metal concentration in reduction process. Therefore, ZVI promoted SRB growth and strengthened its capacity on wastewater treatment and heavy metal removal.

Conclusions

SRB is widely spread, easily isolated, simply culture method, steps into steady phase in only 64h. Uranium wastewater treatment by SRB is a prospect technique with desirable efficiency of 64% and 60% for SO_4^{2-} and U;

Factors affecting SRB growth include inoculation quantity, pH, temperature, SO42- concentration. Study show optimum growth for SRB under condition of inoculation quantity 15%, pH value 6.5, temperature 35, SO_4^{2-} concentration of 4~6g/L.

Treatment of uranium wastewater by coupling effect of both SRB and ZVI achieved removal efficiency of 83% and 82.3% for SO_4^{2-} and U, which revealed ZVI promoted SRB growth and strengthened its capacity on wastewater treatment and heavy metal removal.

References

Yang C,Wang B, Ding T (1994) Barium chloride - Recycling waste slag - step-by - step and 711 by mines in acid mine drainage . Uranium Mining and Metallurgy, 13(3):172~179

Yu H, Liu S, Luo M (2002) Study on magnesium hydroxide treatment of wastewater containing uranium radioactive. TECHNOLOGY OF WATER TREATMENT, 28(5):274~277

Pan Y (1984) Uranium Mine water management experience. Uranium Mining and Metallurgy, 3(4):67~68

Hou L (2003) Special example of wastewater treatment technology and engineering,.Beijing: Chemical Industry Press,36~37

Junta-Ross J L and Hochella M F Jr.(1996) The chemistry of hematite surface. Geochim. Cosmochim.Acta, 60:305~314

Tang Z, Zhang P, Zuo S (2003) Development of Study on Treatment Technology for Wastewater with Low-Level Uranium Content. INDUSTRIAL WATER & WASTEWATER, 34(4):9~13

Huang M, Zheng L (2002) Microbial accumulation and reduction of uranium in water . nuclear technology, 25(2):123~132

Hu J,Gu X (2003) Theory and Technology of anaerobic biological treatment of wastewater. Beijing: China Architecture & Building Press, 40

Janssen A J H, Sleyster R, ochensen A, et al.(1995) Biological sulfate oxidation in a fed-batch reactor. Biotechnol. Bioeng., 47(3):327-333

Su B, Li Y (2005) Biological sulfate reduction and removal of heavy metals. industrial water conditioning, (25)9:1~4

Bertine,K.K.,Chan,L.H.&Turekian,K.K.(1970) Uranium determinations in deep-sea sediments and natural waters using fission tracks. Geochimica etCosmochimica Acta, 34:641~648

Bonatti,E.,Fisher,D.E.,Joensuu,O.&Rydell,H.S.(1971) Postdepositional mobility of some transition elements,phosphorus,uranium and thorium in deep sea sediments. Geochimica et Cosmochimica Acta, 35:189~201

Lovley,D.R.,Roden,E.E.,Phillips,E.J.P.&Woodward,J.C.(1993) Enzymatic iron and uranium reduction by sulfate reducing bacteria. Marine Geology, 13:41~53

Lovley,D.R.,Widman,P.K.,Woodward,J.C.&Phillips,E.L.P.(1993) Reduction of uranium by cytochrome C3 of Desulfovibrio vulgaris. Applied and Environmental Microbiology, 59: 3572~3576

Tebo, B.M.& Obraztsova,A.Y.(1998) Sulfate reducing bacterium grows with Cr(VI), U(VI), Mn(VI), and Fe(III)as electron acceptors[J].FEMS Microbiology Letters, 162,193~198

Lovley, D.R.(2003) Bacteria Remove Uranium from Groundwater.Industrial Bioprocessing. 25(12):8~9

Ma X, Jia X, Zhao Z (2003) Research and Application of the Processes of Disposal of Wastewater Containing Heavy Metals by Sulfate Reducing Bacteria. JOURNAL OF MICROBIOLOGY, 23(1):36~40

Pan J, Saho Z, Cao H (2007) Separation and Purification of sulfate - reducing bacteria. JOURNAL OF MICROBIOLOGY, 27(5):79~83

Wan H, Su S, Ge C et al. (2003) An Improved Method for Isolation and Culture of sulfate - reducing bacteria. Chinese journal of applied and environmental biology. 9 (5) :561~562

Shen P, Fan X, Li G (1998) microbiology experiment. Higher Education Press

Chen T, Cao Y, Yin F et al. (2006) Isolation and Study of Physiology of a Strain Acid-Resistant Sulfate-Reducing Bacterium Journal of Sichuan University (Natural Science Edition), 43(2):451~456

Li Y, Su B (2000) Study on Treatment of Acid Mine Wastewater by Sulfate-Reducing Bacteria, CHINA WATER & WASTEWATER, 16(2):13-17

Laryr L, Barton. (1995) Sulatfe-reducing bacteria, Plenum Press. New York and London.228-230

Current reclamation of historical uraniferous tailings dams and sand dumps – exacerbating the mess or minimizing the mining footprint? Case studies within the Witwatersrand goldfields.

Mariette Liefferink [1], **Simone L. Liefferink** [1]

[1]Federation for a Sustainable Environment, 8 Palladio, corner of Ryk Str. and Roux Ave., Beverley Gardens, Randburg 2194

Abstract. Since the discovery of gold in the Witwatersrand Goldfields in 1886, gold mines produced more than 270 tailings storage facilities (TSFs). In terms of their spatial dimensions tailings are by far the most important source of pollution associated with gold mining. In recent years there has been an increase in reclamation of tailings within the Central-, West, East and Klerksdorp, Orkney, Stilfontein and Hartbeesfontein (KOSH) goldfields. In a number of cases it has been identified that reclamation has been incomplete and without proper remediation. This article will highlight case studies of inappropriate reclamation activities within the Witwatersrand Goldfields and the associated environmental and social impacts.

Introduction

Since the discovery of gold in the Witwatersrand Goldfields in 1886, gold mining resulted in the establishment of more than 270 tailings dams, containing more than 6 billion tons of tailings and some 600 000 tons of uranium, and covering 400km^2. (Bambas-Nolen et al. 2013). In terms of their spatial dimension tailings are an important source of pollution associated with gold mining (Chevrel 2008; Winde 2013). With the decline of the deep level gold mining within the Witwatersrand goldfields, there has been a significant increase in surface reclamation of uraniferous tailings storage facilities and sand dumps in recent years (Bambas-Nolen et al. 2013). Tailings reclamation is particular pronounced in the Central-, West and KOSH goldfields where many old slimes dams contain relatively high gold grades (Winde 2013).

In the dump reclamation activities, a number of cases have been identified by the Department of Mineral Resources and the Council of GeoScience, where the the re-mining of dumps was not completed, either re-mining of dumps was not

B.J. Merkel and A. Arab (eds.), *Uranium – Past and Future Challenges*,
DOI 10.1007/978-3-319-11059-2_45

completed, either due to a lack of funding on the part of the miner or due to heterogeneity in the dumps which were mined (van Tonder et al. 2008). The failure by some mining companies to remove the entire residue deposits and rehabilitate the remaining footprints may result in the creation of new contaminated sites. During the reclamation process the fallout of toxic and radioactive dust and water pollution are often exacerbated (van Tonder et al. 2008). The selective extraction of gold and not uranium from the tailings in the absence of any rehabilitation measure of the tailings footprint area is a matter of concern.

The possible contribution of surface reclamation activities to acid mine drainage, including the release of waterborne radionuclides, has furthermore not been fully investigated (Winde 2009, Winde 2010). The associated contribution to ingress is likely to be considerable as old tailings are hydraulically mined using high-pressure cannons containing partially treated acid mine drainage water (Winde et al. 2011). This practice introduces air and water into anaerobic tailings, which not only contributes to acid mine drainage formation but there is also evidence for the remobilization of contaminants such as uranium and cyanides during disturbance of old tailings deposits. (Sutton & Weiersbye 2007; Winde et al. 2011).

Radiometric surveys over previously reprocessed mine residue deposit footprints have, in some cases, shown elevated levels of residual radioactivity in the soils (van Tonder et al. 2008). In these cases, it must be accepted that some areas will never be suitable for unrestricted development and that these areas will need to be demarcated as such and appropriate land-uses should be proposed and implemented. In many cases, however, new settlements have been developed on unrehabilitated footprints of tailings dams (van Tonder et al. 2008).

In this paper a civil society approach is applied that aims to understand whether the current reclamation of tailings storage facilities is delivering value to the environment and communities. It is the objective of this paper to analyse the effects, socially and environmentally, that inappropriate reclamation activities have had by reviewing several case studies. The important role of civil society is emphasised in the 2014 National Development Plan (NDP) of South Africa, which states that *'active citizenry and social activism is necessary for democracy and development to flourish, to raise the concerns of the voiceless and marginalised and hold government, business and all leaders in society accountable for their actions"* (NPC 2011).

Background to gold mining in study area

In 1885 news arrived in Pretoria of a major gold discovery on the eastern border of the Transvaal. The editor of the Pretoria Press, Leo Winthal, recorded President Kruger's reaction. After remaining silent, lost in thought, Kruger said:

"*Do not talk to me of gold, the element which brings more dissension, misfortune and unexpected plagues in its trails than benefits. Pray to God, as I am doing, that the curse connected with its coming may not overshadow our dear land ... for I tell you today that every ounce of gold taken from the bowels of our soil will yet have to be weighed up with rivers of tears*" (Meredith 2007).

This has indeed proved true because some degradation associated with mining is irreversible from an ecological perspective (Sutton & Weiersbye 2007).

Gold deposits within the Witwatersrand goldfields naturally co-occur with uranium and its progeny, and metals and metalloids such as magnesium (Mg), copper (Cu), zinc (Zn), manganese (Mn), arsenic (As), nickel (Ni), chromium (Cr), cobalt (Co) and lead (Pb). Biota, including humans, suffer from latent impacts due to bioaccumulation and exposure to naturally occurring radioactive materials and the presence of elevated toxic metals in the environment that are made bioavailable due to mining. These impacts on biota have been supported by numerous national and international literature (Sutton & Weiersbye 2007; Coetzee 2008; Coetzee et al. 2008). Gold and uranium mining in the Witwatersrand is currently in decline because of the resources exhaustion or depletion, or due to high costs or declining grades and the fall in the gold- and uranium price (Seccombe 2014). However, since 2003 many uraniferous slimes dams are reworked by local as well as foreign mining companies (Winde 2013). In most of the reclamation operations, only gold is reclaimed leaving U and other toxic metals in the redeposited tailings.

Case Studies

Stilfontein

The reclamation facility located near Stilfontein is a gold and uranium tailings recovery operation; it is approximately 160 km west of Johannesburg, South Africa. The operation processes multiple tailings dumps (15) in the area through three production modules, the last of which was commissioned in 2011. It also includes an unlined Centralised Tailings Storage Facility (CTSF) approximately 15 km from the gold plant and 2 km upstream of the Vaal River, a national water resource (Retief 2009).

The main motivation for the project in terms of Environmental Impact Assessment (EIA) of the reclamation facility was argued as follows:

"The reclamation operations will result in:

- Removal of existing TSFs which are located on potential unstable surfaces
- Clearing up of existing TSF footprints;

- The establishment of a CTSF which could be manage, maintained and rehabilitated;
- The resulting inert material due to the processing procedure (construction of sulphuric acid and uranium plants to remove minerals which are prone to pollution);
- The processing of uranium to be used for national power supply purposes."

While the motivation for the project is laudable some concerns should be highlighted. Firstly, tailings are hydraulically reclaimed from historical tailings storage facilities and the reclaimed residue is pumped in a pipeline route, crossing the Koekemoerspruit, for a number of kilometres to the CTSF. The Koekemoerspruit is a tributary of the Vaal River, a national water resource (Winde & Van Der Walt 2002). Gold mining activities in the area do not only affect the water quality of the Koekemoerspruit, but also modify hydraulic conditions. Mining related modifications include the alteration of the natural flow regime in the Koekemoerspruit to a completely artificial one. In addition to impacts on the stream the artificial flow regime also contaminates adjacent floodplain sediments (Winde & Van Der Walt 2002). The Stilfontein operation's frequent tailings spillages and wall failure in one of the tailings storage facilities impact adversely upon the Koekemoerspruit, private farmland and public areas (The FSE v DWA and others 2011; Appelgryn Properties CC v First Uranium (Pty) Limited and others 2011; NNR 2013). The geochemical analysis by DD Science Environmental Monitoring in 2012 of the affected farmland, showed the soil to possess a similar geochemical makeup as that of gold mine tailings (Table 1). The absence of national heavy metal guideline levels and remediation procedures for polluted mine soils is problematic (Van der Waals et al. 2007). Table 1 highlights some soil quality guidelines and it is evident that aluminium, arsenic, chromium, cobalt, copper, lead, manganese, nickel and uranium all exceeded these quality guidelines. These minerals have been found to have negative impacts on human, wildlife and plant health (CCME 1997).

Table 1. Geochemical analysis of contaminated soils on private farmland (Dorling 2012).

Sample ID	Mine Waste Solutions Soil Samples (mg/kg)	Guideline A (mg/kg) [a]	Guideline B (mg/kg) [b]	Guideline C (mg/kg) [c]
Aluminium	3948	0.39		
Arsenic	49	0.43	12	2
Calcium	5079 - 15372			
Chromium	89	4.7	64	80
Cobalt	46	6.9	40	
Copper	165	0.1	63	100
Iron	17605 –			

	27367			
Lead	70	0.1	70	56
Magnesium	1780 - 3321			
Manganese	2304 - 8253	0.3		
Nickel	165	1.14	50	50
Sulphur	937 - 9556			
Uranium	27		23	

[a] Guideline A shows the acceptable environmental risk concentrations in waste material from Minimum requirements for the Handling Classification and Disposal of Hazardous Waste (DWAF 1998). The values must be used with total mass of waste and area of land
[b] Guideline B shows the Canadian Soil Guidelines for Agricultural use (CCME 1997)
[c] Guideline C shows the limits for metals found in soil according to Snyman and Herselman (2006).

The frequent spillages that occurred during the tailings reclamation impact adversely on the livelihood of farmers as well as on the health of the surrounding environment. These spillages are even more concerning as large tracts of contaminated farmland remain un-rehabilitated.

The second concern with the Stilfontein reclamation operations relates to the number of significant radioactive spillages as recorded by the National Nuclear Regulator (NNR) Many spillages exceeded the radiological regulatory limits of 250μSv/a for a single facility dose or 500Bq/kg as set by the NNR (van Veelen et al. 2009; Winde 2010). The NNR is the organ of state, which is mandated in terms of the National Nuclear Regulator Act (No 47 of 1999) to protect the public, property and the environment against nuclear damage. Such damage includes protection from disease, injury, damage, loss of property or environmental damage resulting from a "facility specifically designed to temporarily store any radioactive material which is intended to be disposed of as waste material" (RSA 1999).

As a result of the Stilfontein reclamation operations (NNR Status Report/Inspection Report 2013) the North West Department of Economic Development, Environment, Conservation and Tourism (NW DEDECT) has a record of 12 spillages since 2010. These are only incidents reported and identified as NEMA Section 30 incidents (RSA 1998). The ongoing spillages resulted in the NW DEDECT issuing a directive against the mine's operations (Abader 2013). Furthermore, a criminal charge was laid against the mining operation by the FSE for the ongoing environmental incidents in 2012 (case number: CAS 147/12 /2012) (Crowley 2014).

The unlined CTSF has created a new focus of water and dust pollution in close proximity to the Vaal River.

Finally, the fact that the Vaal River's sulphate levels already exceed 600mg/l does not allow for further pollution (DWAF 2009; Holden 2012; Moolman 2012). Since 1948 concerns have been expressed by experts regarding obnoxious effluents, the most objectionable feature being the addition of sulphate, to national water resources (Reichardt 2012). The water in the Lower Vaal and Orange River is utilised mainly for irrigation and should not exceed Total Dissolved Solids (TDS) levels of <150mg/L. The plume that will reach the Vaal River, even with a sulphuric plant in operation, exceeds the regulatory TDS limit (Retief 2009).

Randfontein Cluster

A mining operation within the Randfontein Cluster in the West Rand motivated its reclamation operations, in terms of its consolidated Environmental Management Programme Report (EMPR), as follows: "The operations... have the potential to contribute meaningfully to the social, economic and environmental aspects of the Krugersdorp/Kagiso and Randfontein areas. From an environmental perspective, the reclamation and subsequent removal of dumps from the Randfontein Cluster will release tracts of land previously used for mining between Mogale City, Kagiso and Randfontein allowing for consolidation of the urban hub and alleviating peripheral urban sprawl into agricultural areas" (Nesbitt 2013).

However, the reality on the ground appears to be different as this mining operation is currently in non-compliance with its legally binding EMPR and its Water Use Licence (WUL) (Malebe 2013; Turton 2014; DWA 2014). Numerous news media articles have highlighted the issues (Bega 2014, Faku 2014, Matthews 2014, Myburgh 2014, SAPA 2014a, SAPA, 2014b, Wolmarans 2014).

Tailings within the Randfontein Cluster are hydraulically reclaimed from historical tailings storage facilities and the slurry is pumped in pipeline structures to the gold treatment plant which is located a number of kilometres from the historical tailings storage facilities (Nesbitt 2013). The residue is deposited in the unlined West Wits Pit, which has holings as a result of historical mining (Van Tonder & Coetzee 2008). Partially treated AMD from the West Wits Pit is used in the reclamation operations. Pipeline spillages frequently occur along the entire pipeline route (Turton 2014). The spillages impact upon public areas, the headwaters of the Wonderfonteinspruit, a tributary of the Vaal River, which flows into the Boskop Dam, Potchefstroom's main water reservoir, and wetlands, of ecological importance (Turton 2014).

Furthermore, the now desiccated Tudor Dam (located directly in the course of the Wonderfonteinspruit as main stream draining the area) was mined by the above-mentioned mining company to recover gold from the sediments that have accumulated as a result of past mining practices considered inefficient by today's standards (Van Veelen et al. 2009). The soils and sediments at the site, after re-mining operations stopped, contain a concentration of 8 000 – 10 000 Bq/kg U^{238} and 1700 – 2800 Bq/kg Ra^{226}. The regulatory limit is 500 Bq/kg (Van Veelen et al. 2009). The mine was directed by the Department of Water Affairs (DWA) to immediately remediate the Tudor dam site or alternatively fence off all licensed areas and make it off limits to the public until such time as the site is rehabilitated. This did not happen (Turton 2014). The wetlands downstream of Tudor Dam also contain high activity concentrations of U and Ra, at 2 000Bq/kg for U^{238} and 1200 Bq/kg for Ra^{226} (Van Veelen et al. 2009). The sediments of the wetlands have been disturbed as a result of the mining practices at Tudor Dam. During mining the Mine was responsible to contain this material but failed to do so (Van Veelen et al. 2009).

The Mine is in violation of its Water Use License at the Lancaster Dam. The Lancaster Dam is considered to be an acutely toxic acidic dam. There is visible seepage of slimes from the Lancaster Dam into the stream which forms the upper Wonderfonteinspruit (Van Veelen et al. 2009).

The hydraulic remining of tailings using high pressure water cannons introduces air and water into anaerobic tailings, which contributes to acid mine drainage formation. The possible contribution of the tailings reclamation activities within the headwaters and Upper Wonderfonteinspruit to acid mine drainage, which contain radionuclides, and the resultant mobilisation and solubilisation of U should therefore not be excluded (Winde et al 2011). Results indicate that U levels in water resources of the Wonderfonteinspruit Catchment area increased markedly since 1997 even though U-loads emitted by some large gold mines in the Far West Rand were reduced (Winde 2009). Winde (2009) recorded an average of some 3,5t of dissolved U that was released into the fluvial system from monitored discharge points alone with some 800 kg of U per year flowing into the Boskop Dam.

Reclamation operations have, furthermore, resulted in an increase in toxic and potential radioactive dust fallout as a result of the removal of the grassy vegetation from tailings dams and sand dumps or the crust surfaces, which provide moderate resistance to wind and a reduction in the frequency and severity of dust fallout (Reichardt 2012). The West Rand District Municipality's Environmental Management Framework (2013) recorded dust fallout of 42.24 t per day within West Rand goldfields (Liversage 2013). The reclamation operations, in the short and medium term, have exacerbated the dust fallout because of the failure to implement adequate amelioration measures (Barthel et al. 2007). Significant radiation exposure can occur due to the inhalation of contaminated dust and the contamination of agricultural crop (pasture, vegetables) by the deposition of radioactive dust particles, which can cause considerable dose contributions via ingestion (Barthel et al 2007).

Blyvooruitzicht

The Blyvooruitzicht Gold Mine Company (BGMC) holds an old order mining right to mine gold at the Blyvoor/Doornfontein mine complex and is also a freehold owner of the land. From 1997 Durban Roodepoort Deep Gold Ltd (DRD Gold) was the majority shareholder in the BGMC. In June 2011, the BGMC placed itself under supervision and business rescue in terms of the Companies Act 71 of 2008 (Humby 2014). On 6 August 2013 a provisional winding-up order was granted. BGMC left in its wake a number of un-rehabilitated footprints of reclaimed tailings storage facilities, containing toxic and radioactive water and soil, radioactive infrastructure, tailings storage facilities without vegetation, retainer walls and functional toe paddocks and penstocks, and a total environmental liabilities of R891 million. R35 million is held in trust for rehabilitation (Village Main Reef 2012). Recent news media reports, such as "Blyvoor: Legal minefield is a deadly hazard"; "Defunct mine gets up Blyvooruitzicht's nose"; "Aurora meets Marikana at bloody Blyvoor"; "Wrangling leaves miners in limbo"; "Savagery governs Wild West Mine"; adduces documentary evidence of environmental infractions and failure of duty of care (Kings 2013a,b; Evans 2014; Falanga 2014; Timse 2014). A criminal case was opened against the directors of DRD Gold and its operator, Village Main Reef Mine (VMRM) by the FSE at the Carletonville Police Station on the 19th of February 2014 (CAS 347/2/2014) for the alleged violations of the Environmental Management Programme Report, including the Slime Spill Contingency Plan and the Blyvoor Waste Management Strategy (Fraser Alexander Tailings Technical Division 2002) (Louw 2000, 2006, 2010a,b,c, 2014; Fraser Alexander Tailings Technical Division 2002). Pursuant to the criminal charge, the DEA: Inspectorate in collaboration with the Department of Mineral Resources (DMR), the NNR, the DWA, provincial and local municipalities has launched a criminal investigation against DRD Gold and Village Main Reef Mine.

An alternative approach proposed by a mining company operational within the West and Far West Rand

The approach of a mining company operational within the West and Far West Rand involves the construction of a large-scale Central Processing Plant (CPP) for the responsible recovery of gold, uranium and sulphur from tailings deposits and current mine waste arising within the region (Sibanye Gold (Pty) Ltd 2011). This approach offers a long-term solution in the mitigation of the risk of harmful metals including uranium and sulphur emanating from mining wastes and impacted water. The aim is to develop advanced metallurgical processing plants and water treatment facilities to remove residual gold, uranium and sulphur, thereby ensuring residual levels are maintained within limits set by the World Health Organisation (Sibanye Gold (Pty) Ltd 2011).

The resulting mine residue will be deposited onto a Regional Tailings Storage Facility ("RTSF") The proposed RTSF will be located on competent geological sub-strata where there is no risk associated with dolomitic formations and will centralise tailings from the region which will benefit the regional environment and communities (Sibanye Gold (Pty) Ltd 2011).

In terms of the Environmental Impact Assessment and Environmental Management Programme Reports the project proposes responsible remediation and the following benefits (Sibanye Gold (Pty) Ltd 2011):

- Significant regional investment by the mining company into the declining West Rand and surrounds promises business opportunities for Small, Medium and Micro Enterprises (SMMEs) in the area and sustainable jobs for local communities
- Removal of tailings dams off permeable dolomite sites reducing future environmental liabilities
- Relocation of existing Tailings Storage Facilities ("TSFs') away from urban areas, thereby reducing community exposure to mine residue sites and releasing land for development within established infrastructure
- Reducing the potential of illegal mining through removal and reprocessing of gold-bearing tailings with socially and environmentally responsible deposition in a secure and centrally managed RTSF
- Deposition of reprocessed residue on a legally compliant well managed RTSF which will limit dust and ground water pollution
- Development of regional tailings deposition and water management strategies

Recommendations and conclusion

The rehabilitation of footprints of reclaimed TSFs can be a complex endeavour, with misleading results if considered only in the short term. It is vital that the rehabilitation is relevant to the end land-use objectives as stated in closure plans. It is impossible to determine the rehabilitation measures or objectives with the aim of achieving mine closure unless the future land use has been determined in the context of societal and economic expectations (Bosman & Kotze 2005; Haagner et al 2008).

Radiometric surveys over previously reprocessed mine residue deposit footprints have, in some cases, shown elevated levels of residual radioactivity in the soils. Metals such as manganese, cobalt, copper, zinc, the metalloid arsenic, lead, magnesium and long-lived cyanide metal complexes persist in unrehabilitated footprints of reclaimed tailings storage facilities (Sutton 2007). It is therefore necessary to understand the biochemical limitations for the development of a self-sustaining vegetation community and viable end land uses on the footprints of reclaimed TSFs. It must be furthermore accepted that some areas will never be suitable for unrestricted development and that these areas will need to be demarcated as such and appropriate land-uses should be proposed and implemented.

Persistently weak gold prices are forcing marginal mines to continue cutting not only operational costs but cutting on concurrent rehabilitation. Mining companies also reduce costs by producing more, which they could do by remining higher-grade ore, but that often results in the selected reclamation of tailings dams with higher Au/U-grades while leaving uneconomical, lower grade deposits in place.

The failure by some mining companies to remove the entire residue deposits and to conduct concurrent rehabilitation results frequently in the creation of newly contaminated sites since, rather than consolidating contaminated sites, the reprocessing activities resulted in the creation of a number of contaminated sites, where one previously existed. It may also result in the exacerbation of toxic and radioactive dust fallout and water pollution (Lambeck 2009). Applications for the reclamation of TSFs should thus only be approved if stipulated reclamation operations of the affected area are adhered to and the TSF is removed completely. Ideally mining companies involved in reclamation activities should endeavour to go beyond the minimum legal requirements aiming for environmental and social, as well as economical improvements of the affected area (Lambeck 2009). Reclamation activities provide excellent opportunities for the mining sector to demonstrate a new mind set by not perpetuating the irresponsible practices of the past but to, as good corporate citizens, seek maximal environmental and social, as well as economic, benefit (Lambeck 2009).

References

Abader I (2013) Recollection of meeting with the DDG of the Department of Environmental Affairs and other officials of the relevant organs of state and the said parties' responses. As recorded by Wessels C (30 September 2013).

Appelgryn JJ v First Uranium (Pty) Ltd, Mine Waste Solutions (Pty) Ltd, CHEMWES The Department of Mineral Resources, the Department of Environmental Affairs, the Department of Agriculture, Conservation, Environment and Rural Development, North West Province, the Department of Water Affairs and Forestry, the South African Heritage Resources Agency, the National Nuclear Regulator, City of Motlosana Municipality, City of Tlokwe Local Municipality, The Department, North West Provincial Administration, Development Local Government and Housing (2011) Pretoria : North Gauteng High Court, 24 November, 2011. Case number : 66043/11.

Barthel R, Williams LA, Deissman G, Bunzmann C, Funke L, Schulz H, Ross R, Rikhosto R, Moeketsi SR, Cramer T and Brenk HD (2007) Assessment of the radiological impact of the mine water discharge to members of the public living around Wonderfonteinspruit Catchment Area. Prepared for the National Nuclear Regulator. BSA-Project-No. 0607-03. Centurion: National Nuclear Regulator.

Bega S (2014) Gold-mining outfit's woes deepen as water affairs acts on waste. *Saturday Star*, January 25.

Bosman C and Kotze LJ (2005) Responsibilities, Liabilities and Duties for Remediation and Mine Closure under the MPRDA and NWA. Paper delivered at the International Association for Impact Assessment of Southern Africa (IAIAsa) Annual Conference, August 2005, ThabaNchu, South Africa.

Bambas-Nolen L, Birn AE, Cairncross E, Kisting S, Liefferink M, Mukhopadhyay B, Shroff FM and van Wyk D (2013) Case study on extractive industries prepared for the Lancer Commission on global governance: Report from South Africa. *The Lancet* 383 (9917): 630-667.

CCME (Canadian Council of Ministers of the Environment) (1997) Recommended Canadian soil quality guidelines. CCME, Winnipeg.

Chevrel S, Croukamp L, Bourguignon A, and Cottard T (2008) A remote-sensing and GIS-based integrated approach for risk-based prioritization of gold tailings facilities – Witwatersrand, South Africa. In: Fourie AB, Tibbett M and Weiersbye IM (eds), Proceedings of the Third International Seminar on Mine Closure, 14-17 October, Johannesburg, South Africa: Australian Centre for Geomechanics. pp 639-650.

Coetzee H (2008) Radiometric surveying in the vicinity of Witwatersrand gold mines. In: Fourie AB, Tibbett M and Weiersbye IM (eds), Proceedings of the Third International Seminar on Mine Closure, 14-17 October, Johannesburg, South Africa: Australian Centre for Geomechanics. pp 617-626.

Coetzee H, Nengobela NR, Vorster C, Sebake D and Mudau S (2008) South Africa strategy for the management of derelict and ownerless mines. In: Fourie AB, Tibbett M and Weiersbye IM (eds), Proceedings of the Third International Seminar on Mine Closure, 14-17 October, Johannesburg, South Africa: Australian Centre for Geomechanics. pp 113-118.

Crowley K (2014) Water pollution near mines prompts South African ombudsman probe. *Bloomberg News,* May 22.

Dorling D (compiler) (2013) Geochemical analysis of Stilfontein soil samples. Reference: 260612/11695. The Federation for a Sustainable Environment: Johannesburg.

DWA (Department of Water Affairs) (2014) Notice of intention to issue directive in terms of section 19 of the National Water Act, 1998 (Act 36 of 1998) (NWA) to Mintails SA (PTY) LTD. Reference: 16/2/7/C231/C040. Pretoria: Department of Water Affairs.

DWAF (Department of Water Affairs and Forestry) (1998) Waste Management Series. Minimum requirements for handling, classification and disposal of hazardous waste. ISBN 0-620-22995-0. Pretoria: Department of Water Affairs and Forestry.

DWAF (Department of Water Affairs and Forestry) (2009) Vaal River system: Large bulk water supply reconciliation strategy: Executive summary. Prepared by: DMM Development Consultants, Golder Associates Africa, SRK, WRP Consulting Engineers and Zitholele Consulting. DWAF report no. P RSA C000/00/4406/09. Pretoria: Department of Water Affairs and Forestry.

Evans S, Kavahematui J and Timse T (2014) Aurora meets Marikana at bloody Blyvoor. *Mail and Guardian,* March 7.

Falanga G, Mongudhi T and Kavahematui J (2014) Wrangling leaves miners in limbo. *Mail and Guardian*, January 24.

Faku D (2014) Anger at Mintails boils over in Kagiso. *Business Report*, January 14.

Fraser Alexander Tailings Technical Division (2002) Environmental Management Program for the activities of Blyvooruizicht gold mining company Ltd in accordance with section 39 of the Mineral and Petroleum Resources Development Act 2002, Act 28 of 2002. (2007-10-02). Pretoria: Department of Minerals and Energy.

FSE (Federation for a Sustainable Environment) versus DWA (Department of Water affairs, CHEMWES (Pty) Ltd, Mine Waste Solutions (Pty) Ltd and First Uranium (Pty) Ltd (2011) Johannesburg: South Gauteng High Court, 4 March, 2011.

Matthews C (2014) Mintails maintains it can make money from muck. *Business Day*, March 28.

Haagner ASH, Kellner K and Tongway DJ (2008) Enhancing conventional rehabilitation monitoring in South Africa by adding landscape function characteristics. In: Fourie AB, Tibbett M and Weiersbye IM (eds), Proceedings of the Third International Seminar on Mine Closure, 14-17 October, Johannesburg, South Africa: Australian Centre for Geomechanics. pp 809-820.

Holden R (2012) The impact of acid mine drainage in the Witwatersrand on the mining industry in Mpumalanga, Free State, Limpopo, North West and Northern Cape Provinces. Discussion document addressed to the Chamber of Mines. Johannesburg.

Humby T (2014) Environmental obligations associated with the Blyvooruitzicht Gold Mine. Presentation delivered at the Meeting to determine the environmental obligations associated with the Blyvooruitzicht Gold Mine, February 2014, Johannesburg, South Africa as recorded in minutes by S. Liefferink.

Kings S (2013a) Blyvoor: Legal minefield is deadly hazard. *Mail and Guardian*, November 22.

Kings S (2013b) Defunct mine gets up Blyvooruizicht'snose. *Mail and Guardian*, November 15.

Lambeck RJ (2009) Mine closure or mind closure – are mining companies meeting their whole of life cycle, triple bottom line obligations? In: Fourie AB, Tibbett M and Weiersbye IM (eds), Proceedings of the Fourth International Seminar on Mine Closure, 9-11 September, Perth, Australia: Australian Centre for Geomechanics. pp 13-19.

Louw A (2000) Spill from Blyvooruitzicht's slimes dam. Carletonville Herald, May 12.

Louw A (2006) Slimes water reaches Carletonville. Carletonville Herald, February 3.

Louw A (2010a) Besoedelde mynwater in kanaal. Carletonville Herald, July 23.

Louw A (2010b) Besoedelde water droog skielik op. Carletonville Herald, September 3.

Louw A (2010c) Welverdiend water ook besoedel. Carletonville Herald, August 20.

Louw A (2014) Slime Spill at Blyvoor within days of takeover. Carletonville Herald, February Malebe S (2013) Re : Complaint regarding the tailing spillages on the pipelines in respect of the mining operation by Mogale Gold (Pty) Ltd covering on portions 209, 136 of the farm Luipaardsvlei 246 IQ, Portions 66 and 99 of the farm Waterval. Addressed to the Chief Executive Officer of the Federation for a Sustainable Environment. Johannesburg : Department of Mineral Resources.

Meredith M (2007) Diamonds, gold and war: the British, the Boers and the making of South Africa. Public Affairs: New York.

Moolman S (2012) Vaal water deficit expected by 2015. *Creamer Media Engineering News,* October 12.

Myburgh J (2014) Mintails Operations Suspended. *Krugersdorp News*, January 20.

National Nuclear Regulator (NNR) (2013) NNR inspection reports/status reports for 2009-2013. Provided by the NNR to the Federation for a Sustainable Environment on 13 November, 2013.

Nesbitt K (compiler) (2013) EMPR for Mogale Gold (Pty) Limited: Randfontein Cluster – Gauteng. DMR order reference number: GP 30/5/1/1/2/206 MR. Pretoria: Department of Mineral Resources.

NPC (National Planning Commission) (2011) National Development Plan: Vision for 2030 (11 November 2011).

Reichardt M (2012) A history of mine wastes rehabilitation techniques in South Africa: A multidisciplinary overview of mine waste rehabilitation and the non-scientific drivers for its implementation 1950s-1980s. University of Witwatersrand: Botany Department.

Retief E (2009) Mine Waste Solutions: Reworking of TSFs. Environmental Impact Assessment Report (Version 1). Client Name: Mine Waste Solutions (Pty) Ltd.

RSA (Republic of South Africa) (1999) National Nuclear Regulator Act (Act No. 47 of 1999). Government Gazette, South Africa 927(34735).

RSA (Republic of South Africa) (1998) National Environmental Management Act (Act No. 107 of 1998). Government Gazette, South Africa 401(19519).

SAPA (South African Press Association) (2014a) Shabangu suspends Mintails Mine. *Business Report*, January 17.

SAPA (South African Press Association) (2014b) Premier welcomes Mintails order. *Business Report*, January 22.

Seccombe A. (2014) Platinum mine strikes: Man versus machine. *Financial Mail* April 11-16: 20-22.

Sibanye Gold (Pty) Ltd (2011) Integrates Environmental Authorization. Authorization no. 12/12/20/1451. Gauteng. Issued in terms of the National Environmental Management Act, 1998, the Environmental Impact Assessment Regulations 2010, the National Environmental Management: Waste Act, 1998 and Government Notice 718 of 2009 for the proposed Driefontein Environmental Project, Gauteng Province.

Snyman HG and Herselman JE (2006) Guidelines for the utilisation and disposal of wastewater sludge: Volume 2 of 5: Requirements for the agricultural use of wastewater sludge. WRC report number: TT 262/06. Pretoria: South Africa.

Sutton MW and Weiersbye IM (2007) South African legislation pertinent to gold mine closure and residual risk. In: Fourie AB, Tibbett M and Wiertz J (eds), Proceedings of the Second International Seminar on Mine Closure, 14-17 October, Santiago, Chile: Australian Centre for Geomechanics. pp 89-102.

Timse T and Evans S (2014) Savagery governs Wild West Mine. *Mail & Guardian,* April 4.

Village Main Reef (2012) Village Main Reef Integrated Annual Report (30 June 2012). Johannesburg: Village Main Reef.

Turton AR (2014) External audit report : Mintails Mining SA (Pty) Ltd : Mogale Gold. DWA File Reference No: 27/2/2/C423/1/1. Pretoria: Department of Water Affairs.

Van Der Waals JH (2007) Detail assessment of metal distribution in polluted soil – South African case study. In: Fourie AB, Tibbett M and Wiertz J (eds), Proceedings of the Second International Seminar on Mine Closure, 14-17 October, Santiago, Chile: Australian Centre for Geomechanics. pp 815-820.

Van Tonder D and Coetzee H (2008) Regional closure strategy for the West Rand Goldfield. Report for the Department of Minerals and Energy. Council for Geoscience report no. 2008-0175. Pretoria: Department of Minerals and Energy.

Van Tonder DM, Coetzee H, Esterhuyse S, Msezane N, Strachan L, Wade P, Mafanya T and Madau S (2008) South Africa's challenges pertaining to mine closure – The concept of regional mining and closure strategies. In: Fourie AB, Tibbett M and Weiersbye IM (eds), Proceedings of the Third International Seminar on Mine Closure, 14-17 October, Johannesburg, South Africa: Australian Centre for Geomechanics. pp 87-98.

Van Veelen M, de Beer G, Winde F, Lush D, Cohen R, Khoathane M and Mlangeni G (2009) Wonderfonteinspruit Catchment area remediation plan. Radioactive contamination specialist task team report on site visits and recommended actions. Final Draft Report (2009-04-03). Prepared for the Department of Water Affairs and Forestry and the National Nuclear Regulator.

Winde F (2009) Uranium pollution of water resources in mined-out and active goldfields of South Africa – A case Study in the Wonderfonteinspruit Catchment on extent and source of U-contamination and associated health risks. In: Abstracts of the International Mine Water Conference, 19-23 October, Pretoria, South Africa: Cilla Taylor Conferences. pp 772-781.

Winde F (2010) Uranium pollution of the Wonderfonteinspruit, 1997-2008. Part 1: Uranium toxicity, regional background and mining-related sources of uranium pollution. Water SA 36(3): 240-256.

Winde F (2013) Uranium pollution on water – a global perspective on the situation in South Africa. Vaal Triangle Occasional Papers: Inaugural Lecture 10/2013.

Winde F, Erasmus E, Stoch EJ and Hoffman (2011) Desktop assessment of the risk for basement structures of buildings of Standard Bank and ABSA in Central Johannesburg to be affected by rising mine water levels in the Central Basin. Final Report, 3 Volumes, unpublished, Rosebank, Johannesburg, May 2011, 265 pp.

Winde F and Van Der Walt (2002) Uranium contamination of fluvial systems – Mechanisms and processes. Part II: Dynamics of groundwater-stream interaction – a case study from the Koekemoer Spruit (South Africa). Water SA 30 (2): 219-225.
Wolmarans E (2014) Kagiso: from bad to worse. *Citizen*, January 14.

Challenges of water management during tailings remediation – Site and catchment-specific focus

Thomas Metschies[1], Jan Laubrich[1], Jürgen Müller[1], Manja Haupt[1]

[1]Wismut GmbH, Chemnitz, Germany

Abstract. Management of emerging water volumes and their chemical composition is an important part of the remediation of a former mill site with 2 tailings storage facilities. A vast number of constraints are to be considered resulting from available technical installations as well as requirements for environmental protection on the site but also catchment scale. Based on a case study lessons learned for similar remediation projects are presented which may allow to consider relevant aspects of water management already at an early stage of a mining legacy remediation planning.

Site characterisation

Uranium mining and milling by the former joint Soviet-German Wismut Company started in Eastern Germany right after the end of WW II and continued till 1990 with a total uranium-production of about 230.000 t. A significant part of the mined out material was processed at the Seelingstädt milling site in Thuringia, Germany. At this site uranium ore was milled and a total of about 110,000 t of uranium processed from 1960 to 1990 (Barnekow et al. 2012). The residues were dumped in 2 tailings management facilities (TMF) located at sites of former open cast mines.

As part of its mine closure program the state-owned Wismut GmbH is remediating the Trünzig and Culmitzsch uranium TMF covering 1.2 km² and 2.4 km² resp.. In total 105 million m³ of tailings sludge were deposited with about 80% of this volume at the Culmitzsch TMF. Both TMF consist of 2 ponds separated by a central dam where the residues of the acidic (ponds A) and alkaline (ponds B) processing scheme were dumped after neutralization.

There was no sealing placed at the bottom of the former pits before the start of dumping the tailings material. Therefore, seepage waters can enter the local aquifers below and around the facilities. This led to a considerable influence of the TMF on the groundwater in the surrounding (Table 1). After a short passage a portion of the affected groundwater infiltrates into the local receiving streams draining the area around the facilities. The main components of concern beyond uranium are hardness, sulfate and chloride. As Table 1 shows the tailings contain a

B.J. Merkel and A. Arab (eds.), *Uranium – Past and Future Challenges*,
DOI 10.1007/978-3-319-11059-2_46

considerable amount of uranium which is continuously released by infiltrating waters. Pore water with high concentrations seeps out of the TMF influencing the groundwater downstream of the TMF.

Table 1. Estimated U- concentrations in source material and ranges of the resulting concentrations in seepage as well as groundwaters downstream of the TP

Tailings impoundment	Culmitzsch A	Culmitzsch B	Trünzig A	Trünzig B
U_{nat} in solids (t)	4800	2200	1500	700
U_{nat} in pore water (mg/l)	0.3 .. 3.9	1.0 .. 16.5	1 .. 19	1 .. 20
U_{nat} in groundwater (mg/l)	0.04 .. 2.8	0.03 .. 5.3	0,01 .. 2.5	

Water management at the Seelingstädt site includes the collection and treatment of surface, ground- and seepage waters at the TMF and in their surroundings. Various drainage systems and pumping wells were built to contain contaminated waters. However, there is a small proportion of seepage water which still infiltrates into the local receiving streams influencing the chemical composition of the surface waters. Apart from the remediation objects managed under the Wismut remediation program there are additional sources close-by also influencing the water quality of the receiving streams.

Remediation of the site started in the 1990s and will last until 2022. Apart from the geotechnical stabilization of the tailings the reduction of water infiltrating the dumped material is a major objective of the remediation works. The geotechnical stabilization of the TMF however temporarily generates additional contaminated water volumes by releasing pore waters which is in case of the Culmitzsch TMF in the order of nearly 5 million m³ over a 10- to 15-year period.

Scope of water management during remediation

Site specific aspects

While remediation works are in progress the conditions for the water management continuously develop facing new challenges which result from the changing volumes and chemical composition of the collected waters as well as more restrictive limit values for the release of the treated waters into the receiving streams.

From the perspective of water management the site remediation is a transition period. Due to the discharge water limits to be met and the long-term perspective of generation of contaminated seepage waters a collection and treatment will be required for a period of several decades even after the end of the active remediation. As consequence of the remediation the volume of water to be handled will decrease. At the same time the geochemical composition of the water changes.

Therefore, the long-term requirements for the water management could be reliably judged on only after the end of remediation. But during the ongoing works itself the volume and composition of the waters vary considerably. Especially climatic conditions have a considerable effect.

Because of the mostly limited extent of certain conditions major technological changes for this interim period are not feasible requiring that during remediation the existing technologies and facilities have to be continuously adapted to the changing situation to meet the discharge limits.

Changing water quantities

So far the storage of surplus water in a supernatant water pond at Culmitzsch (pond A) was important for the water management. The capacity of the treatment plant of 300 m³/h was sufficient to cope with the average climatic conditions (Fig. 1). However, the collected surface run-off from the site strongly depends on the climatic conditions. The annual average precipitation varying between 450 and 800 mm during the past 10 years dramatically influences the amount of water to be treated. During wet periods the additional water temporarily stored reached volumes of up to 800.000 m³ (Fig.1).

In addition to the storage of water this procedure also resulted in a homogenization of the water ensuring a quite constant composition of the water inflow to the treatment plant.

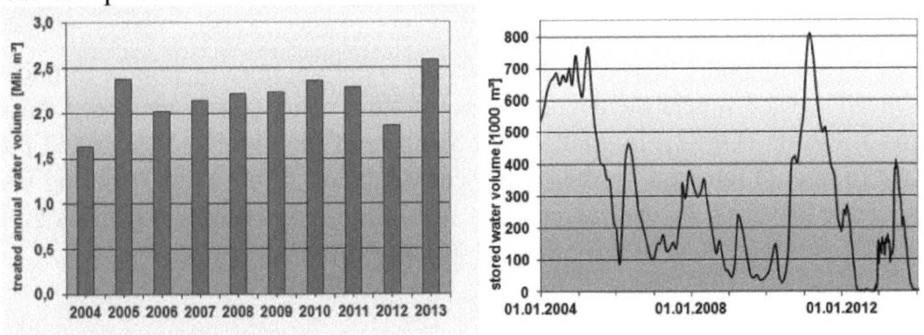

Fig.1. Treated annual water volumes (left) and water volume of the supernatant water body stored on pond A of the Culmitzsch TMF.

Remediation of the Culmitzsch TMF however required stopping the use of the pond for further storage. Erecting a new storage reservoir of a similar size was not feasible due to 1) the limited space available, 2) the necessary effort but 3) mainly because it would have been required just for a relatively short period of time. On the other hand an extension of the treatment plant capacity was also not economic because the additional capacity would have been needed just during periods of higher rainfall while an additional water release to the receiving stream is constrained by the hydrochemical conditions.

Instead fractions of the surface waters from remediated working areas and already covered parts of the Trünzig TMF meeting the set discharge limits were identified and applications made for a direct discharge into the receiving streams at an early stage. The support of this approach by the licensing authorities allowed reducing the area where surface waters are to be collected by about 30 %.

In addition storage capacity for about 100.000 m³ is provided now in separate basins to balance fluctuations due to climatic conditions. Excess water can safely overflow these storage basins into the depression of the former pond for temporary storage but thereby influencing the progress of remediation work there.

The storage volume needed was based on an optimization to statistically allow an overflow just less than once in a 4 years period. With the progressing remediation additional areas for a direct discharge of the surface waters are prepared further reducing the amount of water to be handled.

For a direct remediation of the contaminant source wells were drilled into the sandy beaches of Culmitzsch pond A where contaminated pore water is pumped. This pumping reduces the hydraulic gradient between the tailings and the surrounding groundwater and decreasing the seepage of pore waters. These wells are operated based on the available surplus capacity of the water treatment additionally balancing the seasonal changes of the emerging water volumes.

Chemical composition

With the consequent reduction of the collected water volumes by release of uncontaminated surface waters the average concentration continuously increased during the past 5 years (Fig. 2). In addition the fraction of the highly concentrated pore waters (Table 2) released by the forced consolidation of the tailings during contouring and covering as well as pumped from the sandy tailings beaches significantly increased the load of the collected waters.

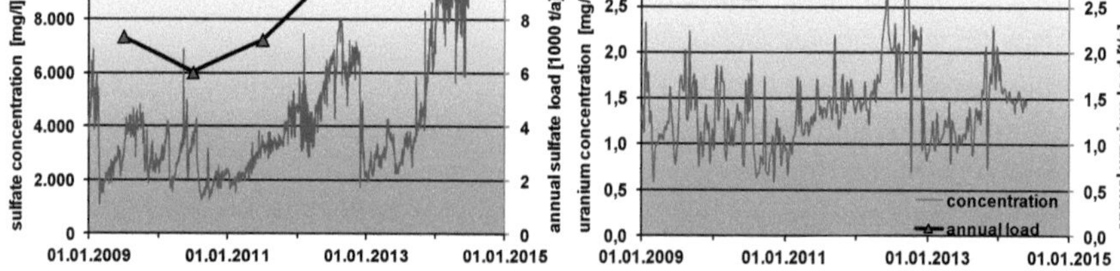

Fig.2. Concentration and annual loads of sulfate and uranium in the collected waters prior to treatment.

Table 2. Water concentrations in sampling wells at the beach and transition zone of the Culmitzsch TMF

2000 -2013	U [mg/l]	Mg [mg/l]	Ca [mg/l]	SO4 [mg/l]	Cl [mg/l]	Fe [mg/l]	Ni [µg/l]
Pond A							
Median	0.94	1310	381	11700	1530	0.67	28
10th percentile	0.25	589	335	7230	856	<0.08	<15
90th percentile	2.21	2280	420	15100	2160	104	473
Pond B							
Median	6.21	113	27	8190	1110	0.18	<20
10th percentile	0.94	66	11	6190	754	<0.04	<10
90th percentile	11.20	648	379	9480	1460	1.15	40

The contaminated waters are treated by lime addition thus reducing the concentration of metals such as uranium below the discharge limits (average annual release limit for uranium of 0.3 mg/l with maximum concentrations 0.5 mg/l). With high bicarbonate concentrations in the water uranium is present as highly soluble carbonate complex. Therefore, before the lime treatment the pH of the water is lowered to strip out carbon dioxide in stripping columns to reduce the uranium complexes.

Optimization and adaption of the treatment process is required to ensure the safe function of all treatment installations as well as to meet the limit values set for the discharge even while input water composition is changing. An implemented optimization of the stripping process will enhance the effectiveness of the treatment process increasing the uranium demobilization and reducing its dependency on the temperature conditions.

With additional pore waters to be treated the increasing iron concentration requires an implementation of a pre-treatment precipitation to assure a stable stripping and treatment process without iron clogging the available installations. Construction of an iron precipitation is in progress which will allow a pre-treatment of up to 100 m^3/h of consolidation waters from the pond A of Culmitzsch TMF having concentrations of up to 250 mg/l. The precipitation will be supported by additional aeration and lime dosage.

Limit concentrations have to be met for salt concentrations and the German Hardness downstream of the tailings site in the receiving Culmitzsch creek as well as in the Weiße Elster river where the Culmitzsch creek discharges into. These limits restrict the discharge volume of treated waters and strictly confine the water management operations (Metschies et al. 2012). As a result the full technical capacity of the water treatment plant might not be available throughout the year depending on the discharge conditions in the influenced water bodies. To assure that the installed treatment could be used even under low flow conditions in the receiving stream, processing water from the Weiße Elster (E-312 in Fig. 3) is diverted through an existing pipeline into the Culmitzsch creek. As a result the salt concentrations in the creek could be controlled without interrupting the remediation works at the site.

Catchment specific aspects

The Trünzig and Culmitzsch TMF are situated in the Weiße Elster river catchment area. Limits for several parameters in the discharge waters have to be met in the outflow from the water treatment plant and partly in the receiving stream. For the management of the Weiße Elster river catchment limits for hardness, sulfate and chloride are set for a sampling point (e-423; Fig. 2) as far as about 25 km downstream of the discharge point.

Within this catchment area a number of concurring water uses and additional contamination sources are situated. Another Wismut remediation site discharges treated waters into a tributary of the Weiße Elster river (Fig. 2) close to the town of Ronneburg having elevated concentration of salts and hardness too.

Especially under low flow conditions concentrations of the regulated parameters come close to the set limits. Additional management measures were required including:

- Integrated release control procedure for hardness and sulfate from both Wismut remediation sites
- Enhancement of the minimum river discharge in the Weiße Elster river
- Temporary increase of the limit for hardness as the most critical parameter

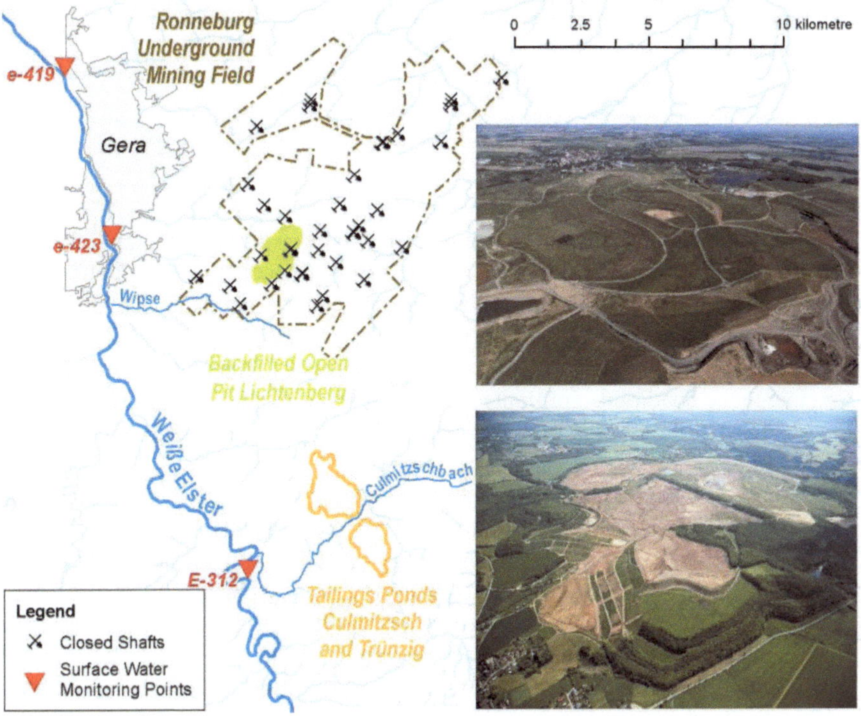

Fig.3. Weiße Elster catchment area with the location of the remediation sites as a source of salt, hardness and uranium; monitoring points for surface water quality.

The integrated control of the release from the Wismut remediation sites was implemented leading to a reduction of the discharge from the tailings remediation site when the river flow rate decreases under a critical value. Because of the on-site conditions the release from the Ronneburg site had the highest priority affecting the remediation progress at the Seelingstädt site.

To further reduce the limitations the enhancement of minimum river water discharge was negotiated with the operator of upstream water retaining reservoirs. The controlled release of water from these storage reservoirs allowed to increase the minimum discharge in the Weiße Elster river. Between 5 and 10 % of the average flow rate in the Weiße Elster were discharged additionally in the low flow periods (2012: 19 million m^3, 2013: 10.6 million m^3).

With all the additional technical measures implemented the regulating authorities allowed a temporary increase of the limit for hardness from 19°dH to 24 °dH for the remediation period. As a condition an extensive biological monitoring program to identify potential effects on special areas of conservation according to the Habitats Directive (Council Directive 92/43/EEC on the Conservation of natural habitats and of wild fauna and flora) had to be implemented downstream in the

Weiße Elster river. As expected before the monitoring has not shown any adverse effects yet.

Conclusions

Remediation of a former tailings site represents a transition period with changing emergence of volumes as well as the composition of waters to be treated. The safe management of these waters has to take into account the available technical installations for water collection, storage and treatment as well as the changes in the water management requirements due to the progress of remediation. Specific water emergence conditions are limited in time making major investments not feasible. A high flexibility of water management is required to meet the limits as well as not to interfere with the remediation progress.

During the remediation additional contaminated waters are released as a result of consolidation requiring special solutions for their treatment and release. Continuous adaptions of the treatment technology as well as appropriate conditions for the release of the treated waters into the receiving streams have to be assured. Variable discharge conditions in the receiving stream influence the water release from the remediation site. Therefore, water management has not just a site but also catchment specific focus requiring additional effort to make sure that not only the release limits at the discharge point but also favorable conditions downstream in the receiving stream are met.

Covering of the TMF results in changes of the water balance reducing the infiltration of rain water and thereby the volume of contaminated water to be treated. However, due to the extent of the contamination source and the requirements in the receiving streams a management of the seepage and influenced groundwater will be required for a long-term.

References

Barnekow U, Müller J, Metschies T, Paul M. (2012) Experiences gained and major challenges for the remediation of Wismut's Culmitzsch tailings pond. In: Forrie A B, Tibbett (2012) Conference Proceedings Mine Closure 2012, Australian Centre for Geomechanics: 209 – 222.

Metschies T, Paul M (2012) Management of mine and tailings seepagewater discharge in an intensively used watershed. In: McCullough CD, Lund MA, Wyse, L (Eds), International Mine Water Association Symposium, Bunbury Australia: p 409 – 416.

Thirteen Years Later: Status of the Moab UMTRA Project Long-term Remedial Action

Donald Metzler[1]

[1]U.S. Department of Energy, 200 Grand Avenue, Suite 500, Grand Junction, CO 81501

Abstract. For the past 13 years, the U.S. Department of Energy (DOE) has managed the Moab Uranium Mill Tailings Remedial Action (UMTRA) Project in Utah. As DOE designed and constructed tailings-removal and disposal systems, and now relocates the 14 million tonnes of uranium mill tailings away from the Colorado River, DOE has continued to place its highest priority on workplace safety. The project is ahead of the schedule and below the cost at completion, based on the approved lifecycle estimate.

Background

The Moab UMTRA Project site is a former uranium ore-processing facility located about 5 kilometers (km) northwest of the city of Moab in Grand County, Utah (Fig.1). The Moab mill operated from 1956 to 1984. When the processing operations ceased, an estimated 14 million tonnes (9.2 million cubic meters) of uranium mill tailings and other contaminated materials, collectively known as residual radioactive material (RRM), had accumulated in an unlined impoundment covering 53 hectares.

Material in the tailings pile ranges from outer compacted, course sands to inner fine clays, and an interim cover of nominally contaminated soils taken from the Moab site outside the pile, known as "off-pile areas." Debris, including asbestos, from dismantled mill buildings and associated structures was buried in the southeast corner of the pile. Some off-pile soils contain radioactive contaminants that exceed cleanup standards.

In 2001, ownership of the site was transferred to DOE. In 2005, DOE issued the Record of Decision (DOE 2005), which documented the selected alternative of relocation of the RRM primarily by rail to a permanent disposal cell 48 kilometers north near Crescent Junction, Utah.

Site-related contaminants, including ammonia and uranium, have leached from the tailings pile into the shallow ground water and are discharging to the Colorado

B.J. Merkel and A. Arab (eds.), *Uranium – Past and Future Challenges*,
DOI 10.1007/978-3-319-11059-2_47

River adjacent to the Moab site. The project scope includes active remediation of the contaminated ground water.

Also the scope includes remediating vicinity properties that exceed cleanup standards. Contaminated vicinity properties are not numerous because access to the Moab site has always been controlled.

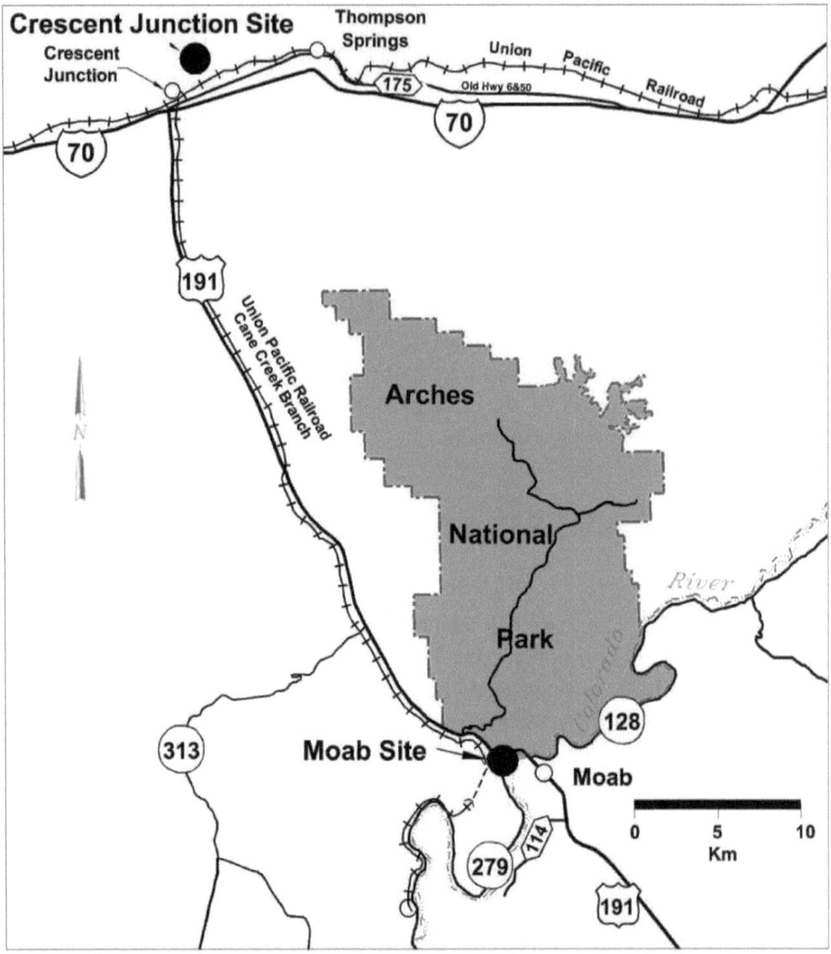

Fig.1. Location of Moab tailings pile and Crescent Junction disposal cell sites.

Status of RRM Haul and Disposal

Periodically, the project obtains aerial photographs and topographic contours of the tailings to confirm the total estimated 14 million tonnes remains the same. To date, the project's estimate of the pile has remained unchanged.

The Crescent Junction site was selected as the off-site disposal location largely because of its ideal geological isolation. In 2008, the U.S. Nuclear Regulatory Commission (NRC) conditionally concurred with the project's Remedial Action Plan (DOE 2008), which presents the basis for constructing the disposal cell near Crescent Junction.

Infrastructure construction at the Moab and Crescent Junction site began in 2008, including handling and loading facilities and haul roads at both sites. The cell is being excavated in phases. The first two phases, each about 16 hectares, have been constructed. The disposal cell is rectangular and will occupy approximately 100 hectares when complete. The unlined cell is excavated 7 ½ meters below grade into weathered Mancos Shale. RRM is placed 7 ½ meters above grade.

Relocation of the RRM began in April 2009 and as of June 2014, almost 6.2 million tonnes, or about 43 percent of the total, had been relocated. The project is currently shipping one trainload of RRM daily, four days per week. Fig.2 shows the Moab site in 2001, shortly after DOE assumed ownership. Fig.3 is a similar view of the site 13 years later.

An interim cover is placed on RRM that has reached design height in the disposal cell. The 3-meter-thick final cover consists of multiple layers of soil and rock. Material from the cell excavation is used as interim cover, radon barrier, and the frost-protection layer. Excess is used on the upslope side of the disposal cell as a protective barrier from runoff (Fig.4). Fig.5 shows the status of the disposal cell as of April 2014. The rock for the biointrusion layer and the uppermost rock armor layer of basalt is being quarried to meet NRC specifications for durability, and is being hauled from 130 kilometers away.

Status of Ground Water Cleanup

The Ground Water Program at the Moab site is designed to limit ecological risk from contaminated ground water discharging to potential endangered fish species habitat areas along the Colorado River. This protection is accomplished by removing contaminant mass in the shallow plume from eight extraction wells; the extracted water is pumped to an evaporation pond or through evaporation units on top of the tailings pile. In addition, freshwater is injected into the underlying alluvium between the river and the tailings pile to create a hydraulic barrier that reduces discharge of contaminated water to suitable habitat areas. Fig.6 shows the uranium and ammonia plumes and the location of well configurations. If necessary, the ground water interim action system allows for the diversion of surface

Fig.2. Moab site, 2001

Fig.3. Moab site, 2014

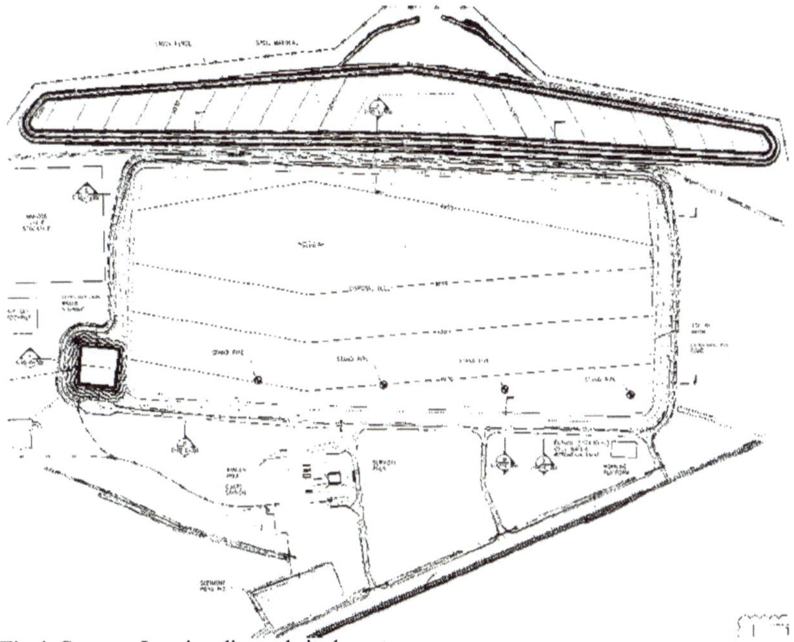

Fig.4. Crescent Junction disposal site layout

Fig.5. Crescent Junction disposal site, 2014. From left, final cover (dark grey), interim cover placed (light tan), RRM being placed (red), and excavated cell unfilled.

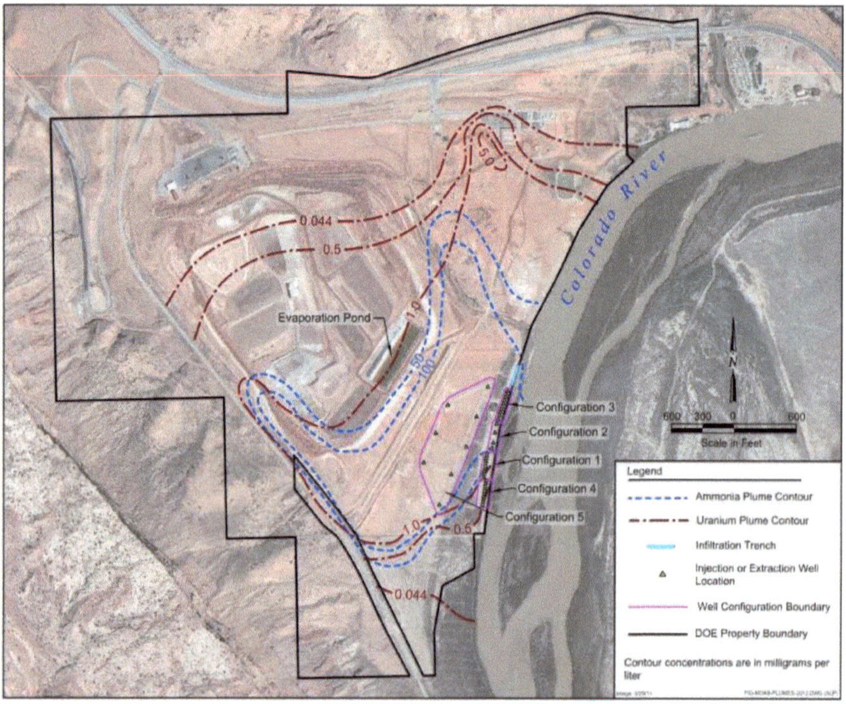

Fig.6. Ammonia and uranium plumes at the Moab site, 2014

water (freshwater) into the side channels to assist in reducing ammonia concentrations in suitable habitat areas.

Well field performance is assessed by measuring extraction/injection rates of remediation wells, measuring water levels, and sampling surface water locations, extraction wells, and monitoring wells. Ground water and surface water monitoring data will also be used in implementing a final ground water remedy. As of early June 2014, more than 790 million liters of contaminated ground water has been extracted, including 370,000 kilograms of ammonia and 1,860 kilograms of uranium. Fig.7 shows the volume of extracted water and ammonia and uranium mass removed between 2003, when the interim action system was initiated, and 2013, the last complete year. The figure indicates that although the volume of ground water extracted has significantly decreased since 2009, there was not a corresponding reduction in mass removal because the extraction wells were moved closer to the tailings pile (the source).

The Colorado River can strongly affect ground water elevations and contaminant concentrations in a portion of the well field near the river. Normally, a gaining stream near the site, it changes to losing when flows reach about 280 cubic meters per second. Depending on the runoff each year, a lens of freshwater can expand in the soils at the Moab site along the river. Early each spring Moab Project staff monitor the snowpack in the Upper Colorado River Basin as an indication of the potential flow.

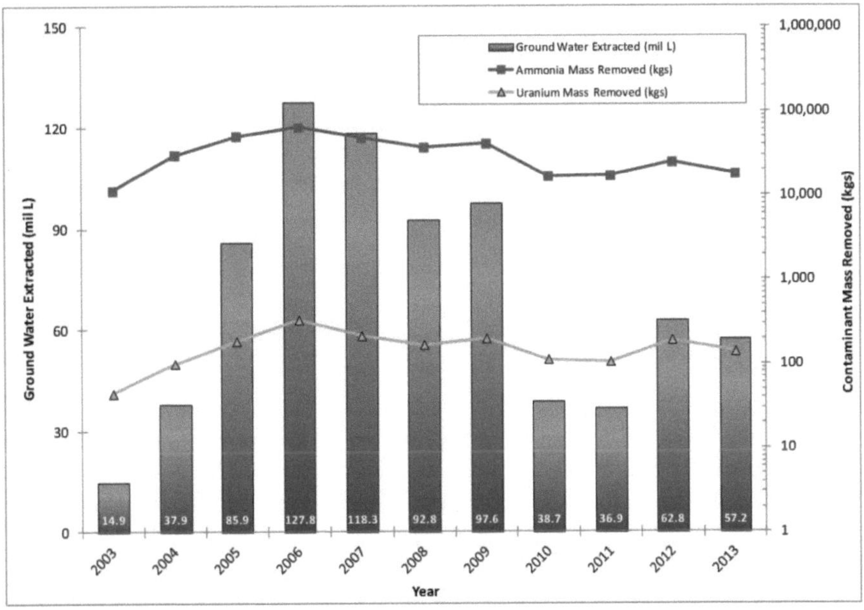

Fig.7. Volume of ground water extracted and ammonia and uranium mass removal, 2003-2013

Importance of Safety

From the time DOE began conducting activities at the project sites, an emphasis was placed on safety and a positive safety culture began to develop early on. The project held safety luncheons at the work sites where awards were presented to employees who had exhibited a positive safety attitude, and a Contractor Safety Committee, which consists of non-management employees from a variety of disciplines, was created. When the Moab Project received additional money in 2009 through the American Recovery and Reinvestment Act, the project suffered a setback in what it considered a healthy safety culture that was brought on by a large influx of employees who had not yet embraced the existing safety culture.

More recently, DOE has renewed its emphasis on safety across its complex of sites through the development of a Safety Conscious Work Environment, or SCWE. A SCWE is a work environment in which employees feel free to raise safety concerns to management without fear of retaliation. Attributes of a strong SCWE include such behaviors as frequent management engagement and time spent in the field; open communication and fostering an environment free from retribution; credibility, trust, and reporting errors and problems; and a questioning attitude.

Efforts have been made to enhance the Moab Project's SCWE through the following means: conducting employee safety questionnaires to solicit worker feedback; using the Safety Committee to track suggestions and implement corrective actions; management safety walkdowns; implementing an Employee Concerns Program where workers can anonymously report concerns; and revitalizing the Safety Incentive Program. Awareness of the SCWE has been promoted through posters at the worksites and at daily safety meetings.

The project has maintained a good safety record within the DOE complex, which sets higher standards than comparable private industries. Last March, the project celebrated reaching 2 million work hours without a lost-time injury or illness.

Cost and Schedule

In 2008, the Moab Project received approval of its lifecycle estimate at a cost of approximately $1 billion to be completed in 2028. As work has progressed on the Moab Project, and with funding received from the American Recovery and Reinvestment Act of 2009, the new lifecycle estimate is now $910 million to be completed in 2025.

Challenges Ahead

Current project funding supports moving about 816,000 tonnes this fiscal year. Other activities, such as equipment maintenance, excavation of the remainder of the disposal cell, and placement of interim cover and final cover are required to be completed while continuing to excavate, transport, and dispose of RRM. As the project continues as an operating project and activities become routine, it strives to combat complacency and improve the SCWE. Despite these challenges, the project plans to be complete on schedule and within budget.

References

U.S. Department of Energy, Record of Decision for the Remediation of the Moab Uranium Mill Tailings, Grand and San Juan Counties, Utah, September 2005.

U.S. Department of Energy, Final Remedial Action Plan and Site Design for Stabilization of Moab Title I Uranium Mill Tailings at the Crescent Junction, Utah, Disposal Site, July 2008.

Overcoming the barriers to implementation of decommissioning and environmental remediation projects - a focus on uranium mining legacy sites. The CIDER Project.

Horst Monken-Fernandes[1], **Patrick O'Sullivan**[1]

[1]Waste Technology Section, International Atomic Energy Agency, Wagramer Strasse 5, A-1400 - Vienna -Austria

Abstract. Much remains to be done in terms of addressing the decommissioning of facilities and remediation of sites affected by past uranium mining and processing operations. The Constraints to Implementing Decommissioning and Environmental Remediation (CIDER) project was launched by the IAEA to contribute to improving the current levels of performance on Decommissioning and Environmental Remediation (D&ER) programmes through promoting greater cooperation amongst IAEA Member States and relevant international organizations. This paper analyses the major constraining factors that affect the implementation of D&ER programmes and how these impediments might be overcome.

Introduction

Mining and processing of uranium ores give rise to a variety of wastes and residual materials, such as top soils, overburden material and non-mineralized rock (removed to reach the ore body). In addition, waste rock that contains sub economic levels of mineralization, and is thus not milled for extraction of uranium, may contain environmentally significant levels of radionuclides and other minerals. Mill tailings (consisting of the ground ore from which the uranium has been extracted) and sludge (containing metals and radionuclides from water treatment plants) are also relevant types of waste. In addition, there are also different types of processing waste.

Many uranium mining sites throughout the world have become orphaned sites, and are waiting for remediation. However, little progress has been made in the management of some, particularly in the understanding of the associated environmental and health risks (IAEA, 2014).

B.J. Merkel and A. Arab (eds.), *Uranium – Past and Future Challenges*,
DOI 10.1007/978-3-319-11059-2_48

The legacies from past use of nuclear energy is often a state responsibility, e.g., in countries where uranium production was/is undertaken by state-owned organizations. Therefore, in those countries in which the economy is not strong enough to afford the costs of decommissioning and environmental remediation (D&ER), these activities are severely restricted unless resources are made available from sources outside the country (e.g. from bilateral or multilateral sources). In other cases funds were typically set aside by the operating organization to defray some or all of the cost of decommissioning the facility or remediating the site. But such provisions have often proved inadequate or, in some cases, the funds have been reallocated for other purposes (IAEA, 2005).

It is of paramount importance that either the contamination is reduced to non-significant levels, or that arrangements are made to maintain institutional control of the site until the contamination is removed or attenuates naturally. In some countries it is intended that such control will be maintained in perpetuity for certain sites. This approach is generally adopted only in cases where the remediation cost is so high that it would represent a disproportionate use of state resources, given the level of risk and that the site is remote from populated areas. The corollary of this approach is that the site will not be made available for other uses; therefore, this option is adopted only rarely. There are significant costs associated with maintaining ongoing public safety regardless of whether this approach is adopted or a full D&ER programme is initiated.

It is evident that the levels of progress being achieved by different countries in implementing D&ER programmes varies widely and the reasons for these differences are considered by CIDERype here the text belonging to your first header. The first line of a paragraph is not indented after a heading (format "p1a"). This format will appear automatically if you press return at the end of a heading.

In all other paragraphs of text, the first line is indented. This format will also appear automatically if you press return at the end of the previous paragraph. Please do NOT insert additional returns between paragraphs.

Drivers for Undertaking Decommissioning and Remediation Projects Header first order

It is the responsibility of governments to put in place a clear legal framework for management of contaminated land. In particular, the absence of a mechanism to identify responsible parties for managing legacy sites and to determine appropriate end states may delay or prevent clean-up, even when the need for this is widely acknowledged. In some Member States, an extensive legislative basis and regulatory framework to address issues of decommissioning and contaminated site remediation is already in place. In others, the legislative basis may be incomplete or even non-existent. These situations offer significant challenges to programme implementation, regardless of the availability of funds. The political environment in which decisions are made concerning the implementation of environmental remediation projects is complex and the many non-technical factors bearing on the decision-making processes are discussed in (IAEA, 2002).

Four main ethical principles are related to the sustainability of nuclear activities, and that should underpin the practical arrangements put in place for funding decommissioning and environmental remediation activities (NEA/OECD, 2006). These principles are: (1) the safety of current and future generations; (2) avoiding assigning undue financial burdens to future generations; (3) ensuring assembly and preservation of financial, technical, and scientific resources for the decommissioning of nuclear facilities and remediation of contaminated sites (i.e. establishing the principle of 'the polluter pays'); and (4) intergenerational equity, according to which the generation that incurs long-term liabilities should take responsibility, and provide appropriate resources, for the management of these liabilities in a way that will not impose undue burdens on future generations. Achieving equity between generations requires that the generation which incurs the long-term liabilities selects technologies and strategies which minimize the resource and risk burdens passed to future generations.

In addition to ethical considerations, there are important economic drivers for early implementation of D&ER programmes, e.g., investment in D&ER can be good for the local economy and/or may benefit the site owner financially. Inter alia, investment in D&ER activities can create job opportunities, spur development of new businesses, facilitate sector growth, increase property values and provide potential redevelopment opportunities. Conversely, delaying D&ER projects may be costly due to potential degradation of structures, spread of contamination, loss of institutional knowledge, and uncertainty of future impacts on factors that will influence remediation efforts, such as availability and cost of waste management services and availability of a trained workforce to conduct D&ER activities.

Where state-owned commercially-operated uranium production centres are approaching the end of their lifecycles they will soon become a liability of the state in which they are located. In many other cases, commercial entities are fully liable for the final stages of a facility's lifecycle. Depending on the situation, different remediation strategies may be chosen. Timely planning and implementation of D&ER projects may increase societal confidence in the capability of the uranium production sector to deal with the long-term consequences of uranium mining and facilitate public acceptance of future developments and/or new projects. Conversely, retaining the remediation of sites for many years with no decision, plan, or effort to address restoration of the site will likely cause unnecessary increased costs and adversely impact public perception and attitude.

Applying a facility lifecycle approach with clear and transparent technical, financial, and (where practicable) job relocation plans for implementing D&ER projects can be beneficial for both the industry and the surrounding area, because it enables cost optimization and increases public awareness.

Overview of Constraints that Pose Barriers to Decommissioning and Remediation

Political commitment is generally a major driving force for implementation of D&ER and, without this, significant progress is unlikely to occur. Even with strong political motivation, constraints of different types may cause delays or impede project implementation and, therefore, these constraints should be identified and available lessons learned should be examined to facilitate more effective planning.

National policy generally provides the basis for development of relevant regulations, and is usually an important precursor for implementation of D&ER programmes. Even in the presence of a fully defined framework including a waste management system and other relevant infrastructure, progress may not be possible without strong political commitment. Countries with large numbers of contaminated sites should define prioritization mechanisms in their national policy. Prioritization needs to take account of many factors, including:

- Comprehensive risk assessment;
- Availability of resources (both existing and anticipated);
- Decisions on the scope of the D&ER programmes, optimizing benefits and risk reduction; and
- Cost.

Undertaking D&ER generally requires the use of significant resources and is therefore expensive. The cost of a D&ER programme depends on the size of the project, the technology to be used, the project management strategy adopted, and the definition of the end state. Stakeholder demands may cause escalation of costs, e.g. in cases where lower residual levels of contamination or other adjustments are required beyond those envisaged in the original plan. The cost of implementing a D&ER project may increase over time, in real terms, and this needs to be a consideration in lifecycle planning (IAEA, 2005) .

Regarding technology for D&ER, there will be a need for decontamination procedures, demolition equipment, waste collection and disposal methods, radiation detection instrumentation, and laboratory facilities, among others. In addition, for environmental remediation, technology selection may need to address site characterization and monitoring, hydrogeological modelling, cover design, and water treatment.

Moreover, an appropriate, well-defined, infrastructure has to be established in a timely manner so that a D&ER project can be successfully carried out, including providing training of personnel to execute and regulate the activities, as well as establishment of auxiliary tools. For example, regarding radioactive waste, it will be very important to establish an infrastructure for handling, conditioning, transporting, and disposing of such materials.

Stakeholder opinion/attitude is an important element of the decision-making process for D&ER projects and can represent a significant barrier to their implementation, e.g. lack of knowledge or education may drive an excessive demand for a potentially unnecessary level of clean-up. Exaggerated perceptions of risk by involved communities can negatively influence decision-making; conversely, the lack of public engagement and proper channels of communication can damage risk perception. Stakeholders with different degrees of understanding of technical and programmatic issues can strongly influence project implementation, both positively and negatively. It is clear that a lack of mutual trust can disrupt any process of negotiation.

Strategies to Overcome Barriers and Promote Decommissioning and Environmental Remediation

It is evident that some critical elements can become important constraints in the implementation of D&ER projects. Careful consideration needs to be given to how these are applied in any particular situation.

Roles and responsibilities of national entities dealing with D&ER should be clearly assigned in accordance with national legislation, and in such a way that there will be no duplication or void.

Cost benefit analyses should be conducted and supported by risk-informed technical decisions in determining the strategy for D&ER project optimization. This is particularly important when D&ER projects face important funding constraints.

D&ER projects have to be planned and implemented in such a way that their end-state enables exclusion or at least minimization of future long-term service costs. Monitoring probable contamination left on a non-operational site is generally not as costly as construction, licensing, and operation of a new onsite waste storage facility for D&ER waste.

In situations of large legacy sites with high levels of contamination, where clean-up to unrestricted use may be extremely expensive and environmentally detrimental, it may sometimes be more environmentally sustainable to construct a waste disposal facility on the site, e.g. as part of an entombment strategy for compact radioactively contaminated structures and/or former waste disposal pits. Properly designed and constructed protective engineering barriers to prevent intrusion and radionuclide migration may achieve the required level of public and environment protection for a long time period, though it should be borne in mind that this option will require long-term institutional control to ensure safety. An analysis of risk and associated cost may be helpful in evaluating the appropriateness of this option.

D&ER projects typically have long timeframes and high costs. Thus, it is of utmost importance that the generation that benefits from a nuclear activity establish arrangements for the appropriate funding of D&ER projects with the main objective of having adequate funding available during the applicable lifecycle phase. The ultimate responsibility for legacies rests within the state. Thus, the principal or primary source of finance should be national funding, which in certain cases could be complemented with international resources. Funding must, at a minimum, comply with the criteria of the polluter pays principle (which enhances the responsibility of the operator and does not provoke imbalances for free competition), sufficiency (the funds must be enough to complete the D&ER tasks), availability (the funds must be available at the appropriate times), and transparency (the funds must be used only for D&ER, and their management must be clear, auditable, and transparent). Generally, a legal framework is established to ensure and enforce these requirements. Three main types of funding models have proved effective and are in use in different countries: i) direct funding from government, ii) internal segregated or non-segregated funds, iii) and external segregated funds.

Regardless the specific advantages and disadvantages of each of the funding models and mechanisms to raise the funds (for instance, an external segregated fund could have advantages with regard to transparency), it is important to emphasize the requirements and risks for adequate funding. Without being exhaustive, to ensure appropriate funding, it is essential to have a complete inventory and an accurate estimation of the cost of the D&ER activities. This depends on a number of factors of which the D&ER strategy is one of the most important. Accurate assumptions about relevant future price inflation, discount rate, value of the asset, closure date, etc., are also significant factors to consider when designing efficient and successful funding arrangements.

The project lifecycle refers to a logical sequence of activities from the conception through to the completion of a project, including dealing with any required post-project remediation. Any project, regardless of scope or complexity, will include a series of stages. Those stages are generally described as initiation, planning, execution, and closure. As described in (IAEA, 2013) lifecycle management approaches consider each stage of a project not as an isolated event, but as one phase in an overall lifecycle. Robust and thoughtful lifecycle planning is one of the primary elements that will determine the success of a D&ER project, and transparent and formalized project planning must be undertaken at the outset. By applying good lifecycle planning, there is greater assurance that potential risks that could delay or even prevent the successful completion of the project are identified and addressed prior to project execution.

A critical element of lifecycle planning is the early engagement with stakeholders (including regulatory authorities, local government officials, and interested members of the public). Frequent and transparent communications with stakeholders may help avoid an overly burdensome bureaucracy that can present a barrier to project execution.

The management and organizational culture required for decommissioning of nuclear facilities and remediation of contaminated sites is different from that for operation of such facilities and sites. Operation is essentially a process based on a reasonably standard routine and training can easily be planned and tested. In D&ER projects, the nature of tasks is constantly changing and as a result more flexibility is required to adapt to unexpected situations. A cooperative attitude is required by management and workers of the organization which is responsible for development and implementation of such projects, as well as by contractors and regulatory authorities, to ensure that lessons learned are openly discussed and rapidly fed back into the system.

Good communication strategies will establish trust, cooperation and understanding between different interested parties in D&ER projects. Involvement of affected or interested persons can prevent fear-driven reactions, potentially damaging public response, and the creation of undue expectations or unnecessary anxiety. For all D&ER cases there is a risk that the process will fail if it does not respect the local social, environmental, political, and economic dimensions. This requires open, clear, and agreed-upon lines of communication among stakeholders within a well-defined legal framework. A general recommendation is to involve them at a very early point in the process and throughout the lifecycle of the project. In some cases, public participation in decision-making processes regarding the living environment is mandated by regulatory requirements, environmental policies of international organizations, and more broadly, even by international conventions.

Conclusions

This paper presents a summary of the fundamental requirements necessary for the successful implementation of D&ER projects, as well as the factors that may constrain progress, and discusses options for overcoming these constraints and thereby facilitate better implementation. The analysis concludes that fundamental requirements for implementation of these programmes include:

- The existence of an adequate legal, regulatory and funding framework;
- Availability of sufficient funds; and
- Having access to appropriate technologies and the associated human resources necessary for utilization and oversight of these technologies.

Approaches that will serve to mitigate the identified constraining factors include:

- Undertaking project planning in such a way that takes into consideration the entire lifecycle of the project.;
- Employing risk-based approaches to decision-making, thus ensuring that the available resources are utilized in the most efficient manner.;
- Adopting a graded approach in which the institutional and legal framework is commensurate with the scale of the problems to be addressed;
- Identifying clear roles and responsibilities of the relevant stakeholders (including site/facility owners, implementers, regulators and funding agencies).;
- Sharing experiences, good practices, and lessons learned in comparable programmes through independent reviews; and

- Giving consideration to the potential for bilateral or multilateral support may improve the availability of funds in certain situations. The international donor organizations have an important role in such considerations.

Large D&ER programmes are expensive to undertake and implement and may have significant socioeconomic impacts, both positive and negative, on local communities. Political considerations will therefore often play an important role in determining how quickly a particular programme will proceed. Significant attention therefore needs to be given to addressing these considerations

References

IAEA (2005) Financial Aspects of Decommissioning, IAEA-NE-Tecdoc 1476. International Atomic Energy Agency, Vienna, Austria

IAEA (2002) Non-technical factors impacting on the decision making processes in environmental remediation, IAEA Tecdoc 1279. International Atomic Energy Agency, Vienna, Austria.

IAEA (2013). Overcoming Barriers in the Implementation of Environmental Remediation Projects, IAEA NE-Series No. NW-T.3.4. International Atomic Energy Agency, Vienna. Austria.

IAEA (2014) Lessons Learned from Environmental Remediation Programmes. NE-Series No. NW-T-3.6. International Atomic Energy Agency. Vienna, Austria

NEA/OECD (2006) Decommissioning Funding: Ethics, Implementation, Uncertainties. A Status Report, NEA No. 5996 Nuclear Energy Agency/Organisation for Economic Co-operation and Development, Paris, France

Phytostabilization of uranium-containing shale residues using *Hieracium pilosella*

Anna Ogar[1], Viktor Sjöberg[2], Stefan Karlsson[2]

[1]Institute of Environmental Science of the Jagiellonian University, Gronostajowa 7, 30-387 Krakow, Poland
[2]Man-Technology-Environment Research Centre, Örebro University, SE-70182 Örebro, Sweden

Abstract. A greenhouse pot experiment was conducted to evaluate the feasibility of using *Hieracium pilosella* and soil microorganisms for phytostabilization of uranium-containing shale residues. Conductivity of leachates significantly decreased and pH increased when plants were grown on the substratum. *H. pilosella* has ability to change the hydrochemical parameters and to decrease the mobilization of uranium. Moreover, *H. pilosella* is able to accumulate significant amounts of uranium in the shoots.

Introduction

Phytostabilization involves establishment and maintenance of plant cover on the surface of heavy metal rich sites. Stabilization of pollutants can be an important remediation option for large areas with high and multi-elemental contamination. Phytostabilization prevents erosion, reduces leaching, increases root growth that immobilizes metals by adsorption or accumulation, and provides a rhizosphere wherein metals adsorb and precipitate (Bradl 2004). Increased heterotrophic activity as a response to increased carbon loading will also impact the hydrochemical conditions, particularly at depths where oxygen availability is limited. Increased partial pressure of carbon dioxide has a large impact on pH and increased electron activity will lower the redox potential. Hence, below the root zone sulfide forming metals will scavenge many metals by formation of sparingly soluble sulfides. Plant roots are naturally associated with microorganisms, and these plant-microbial consortia can have direct or indirect effects on the mobility, availability and uptake of elements by plants. However, plant establishment and growth on metal contaminated sites might be inhibited not just due to the toxic metals, but also by adverse soil factors, such as acidity, poor physical structure, and nutrient deficiencies (Ebbs et al. 1998). Remediation strategies for metal contaminated soils, including uranium, have been based on physical and chemical procedures such as:

B.J. Merkel and A. Arab (eds.), *Uranium – Past and Future Challenges*,
DOI 10.1007/978-3-319-11059-2_49

redox, flushing and phosphate precipitation technologies, which may be beneficial, but are costly and will in many cases lead to further environmental degradation (Dupré de Boulois et al. 2008). Phytoremediation has been proposed as a promising alternative for such sites which rely on the capacity of plants and their associated microorganisms to stabilize or extract toxic metals from soils. The *Hieracium* taxon is often chosen as a model since it occurs in a large variety of habitats and has a world-wide distribution. On open sites *H. pilosella* shows vigorous clonal growth resulting in formation of dense mats. The study was performed to investigate the influence of a biological system on metal distribution and to determine the efficiency of metal immobilization by plant-microbial system.

Materials and Methods

Site characterization

Black shale (alum shale) of Late Cambrium age is abundant in southern Sweden at the Kvarntorp area (N 59°08', E 15°17'). The shale consist of K-feldspars (3-8%), illite (24-41%), chlorite (1-5%), calcite (1-4%), quartz (18-34%), pyrite (5-17%) and organic carbon (5-18%), as well as other minerals (up to 8%) (Allard et al. 2011). During 1942-66 the black shale was mined and processed for recovery of hydrocarbons. Mixed refuse from the process is deposited in a pile that is roughly 100 m high, 700 m in diameter and contains some 40 Mm^3 waste that cover about 50 hectares. Moreover the shale is also rich in several elements such as vanadium, molybdenum, uranium, rare earth elements as well as base metals such as manganese, aluminum and iron. The uranium content in the shale (uranium-rich zone) is generally in the range 100-150 g ton^{-1} and the concentration in red processed shale (RPS, rödfyr) is about the same. Very little, if any, of the uranium is lost in the pyrolysis process but the mineralogical composition is changed. There is in the RPS initially some CaO and with time $Ca(OH)_2$ as well as $CaCO_3$. Essentially no Fe(II) is present after pyrolysis, only Fe(III) oxides. This means that pH of leachates would be near neutral, although initially high (maximum around pH 12.5). The weathered shale fines (WF, stybb) is fine-grained shale that has been exposed to the atmosphere for 50-60 years. The material has weathered continuously due to the oxidation of mainly pyrite and other sulphides (still in progress), generating a low pH and continuous release of metals, including uranium. This is probably the main reason for the high metal levels in the leachates from the Kvarntorp deposit. Uranium concentration in the WF has gone down; some 50-75% has been leached over the years and remaining concentration is generally of the order 30-50 g ton^{-1}, or even lower. The main reason for the low pH in leachates from shale, as well as from WF, (pH initially in the range 2.5-3.5) is the oxidation of Fe(II), present in pyrite, and the subsequent oxidation to Fe(III) and the hydrolysis of Fe(III).

Plant materials and preparation of substrates

The experiment consisted of a 2 factorial randomized design with four replicates. Plastic pots with 750 g WF or RPS (Kvarntorp, Sweden) and bark compost (Econova Garden AB, Åby, Sweden) that was mixed in a ratio of 5:1 (v:v) was used as substrate after sieving through a 125 μm mesh. The content and leaching of relevant elements in the bark compost was reported by Sjöberg et al. (2014). To each pot three seedlings of *Hieracium pilosella* from the ecotype collected at Kvarntorp were transferred. Healthy plants of similar size were selected, directly transferred into the pots. As controls, pots without plants were used. Twice a week each pot was given 250 ml artificial rain water (0.01 mM Na, 0.01 mM Mg, 0.2 mM K, 0.025 mM Ca, 0.035 mM Cl^-, 0.025 mM SO_4^{2-}, 0.01 mM NO_3^-, 0.005 mM F^-) but no nutrients. The experiment was conducted in a controlled greenhouse with artificial light 16 hours per day, relative humidity around 30 % and temperature at 25 to 35°C. The plants grew for 21 weeks from May 17 to October 7, 2013

Sampling of percolation water leachates

After watering the pots were left for 30 min to equilibrate after which the leachates were collected, once a week for 21 weeks. Electrical conductivity and pH were measured after settling for 24 hours. The samples were then filtered through 0.20 μm polypropylene syringe filters (VWR International, USA) and stored in a refrigerator at 4°C until metal analysis. After appropriate dilution with 1% nitric acid, element concentrations were determined by ICP-MS (Agilent 7500cx).

Harvest and sample preparation

Each plant was carefully lifted from the substrate, loosely adhering solids were removed from the roots and larger root pieces remaining in the substrate were manually picked out and washed carefully with 18.2 MΩ water. Dry weight of shoots and roots were determined after drying at 80°C for 72 h. Dried sub-samples of shoots were homogenized with an agate mortar and 0.1 g was dissolved in 10 ml concentrated nitric acid in a microwave oven (CEM MarsV).

Results and Discussion

Electrical conductivity and pH

After the initial sampling, the conductivity in the leachates from 1st to 21st week was in a range of 0.7-1.9 mS cm^{-1} for both WF and RPS but with explicit time trends. As can be seen in Fig. 1 there was a substantial lowering of the E.C. with time in both the control and the test systems but largest in the plant systems. Although the compost itself will contribute with dissolved ions the differences between the controls and plant systems clearly show a high impact from the plants on the total ionic concentrations in the solution phases. Similar observations were reported by Sjöberg et al. (2014) when historic sulphidic mine waste was conditioned with bark compost and vegetated with *Agrostis capillaris*. In the plant systems the pH would be influenced by the activity of microbes and plant roots in relation to the properties of the test materials and the release of organic acids from the compost. Oxidation of sulfides (i.e. pyrite) in combination with the oxidation and hydrolysis of particularly iron would lower pH to levels below 3 in the WF system. This was counteracted by the natural content of carbonates in the shale (i.e. calcite and traces of dolomite). The resulting pH for the WF was rather stable at 4.0-4.2 and according to the data plant growth did not have any significant (p 0.05) impact on pH.

The pH control in the RPS system was more complex. The carbonates in the shale were transformed into oxides during the pyrolysis, and the oxides (calcium and magnesium) form hydroxides when in contact with water, resulting in a pH of 12-12.5 in the micro environment. Neither of these processes seemed to dominate at present since pH in the control systems was 5.5 to 6. There might be some pyrite remaining after pyrolysis but the majority would have been lost. It is more likely that the pH is influenced by carbonized calcium hydroxide (Karlsson et al. 2013). Plant growth resulted in a significant pH increase. The exact reason is not known at present but the increase in biological activity, particularly respiration by roots and microbial communities would work in this direction. This is to some extent corroborated by the increase of pH with time. In addition, the release of organic compounds from roots, microbes and the compost would also impact pH, particularly in the root zone but evidently also in the run-off.

In both systems their pH-ranges would impact the distribution of metals between the solid phase and the solution by two in principal different processes; precipitation and adsorption. Precipitation is a rather simple and predictable mechanism that depends on the activity of the ions in a solution. Adsorption of heavy metals ions on soil components, on the other hand, is a highly complex process that in a well-defined solid-solution system and is influenced by physical as well as chemical parameters, notably pH and presence of complexing agents (Bradl 2004; Schwab et al. 2008).

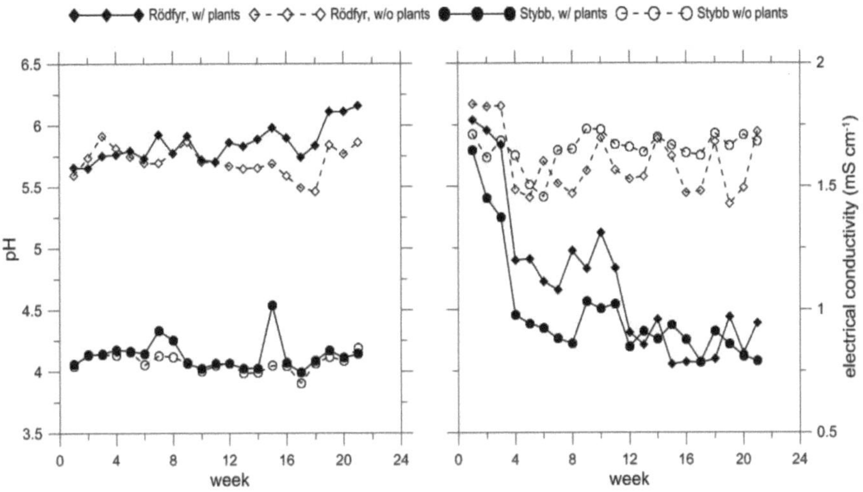

Fig.1. pH and electrical conductivity in leachates from weathered shale fines (stybb) and red processed shale (rödfyr) substratum

Metals

The uranium release from WF and RPS was different as well as the impact of the plants (Fig. 2). The uranium concentration in solid WF is around 30-50 mg kg^{-1}. A pH between 4 and 4.3 was obtained in the aqueous phase, and concentrations of uranium were in the range 0.4 to 0.8 µg L^{-1}, where uranyl and uranyl sulfate ions would dominate. There was no difference between the control and the plant system why there is no apparent impact from the plants.

The uranium concentration in solid RPS is around 100-150 mg kg^{-1} and they gave a pH of 5.5- 6.2, possibly somewhat higher in the systems with plants than without plants (6.2 and 5.8, respectively, after 21 weeks). Concentrations of uranium were 1.8-3.4 µg L^{-1} and significantly lower in the systems with plants than in the systems without (3 µg L^{-1} in the control) after 21 weeks. The adsorption to solids should have an influence on uranium distribution around pH 6. It is possible that the release has been underestimated in these experiments since it was estimated from the aqueous concentrations. It is therefore interesting that the aqueous uranium concentration was significantly lower (p 0.05) when the plants were present (Fig. 2). This indicates a stabilization of the waste or an uptake in the plant or a combination of them. If the difference in pH between the systems with and without plants is significant this would also lead to a somewhat higher adsorption in the plant systems. There is no support for accumulation of uranium in the plants why stabilization is an alternative.

One well known response to metal stress is the ability of roots and associated microbial consortia to release molecules in the rhizosphere that might both enhance bioaccumulation and bioprecipitation (McGrath et al. 2001). The resulting complex mixture of potential metal ligands and adsorbents include low molecular weight organic acids, short chain carboxylic acids, fatty acids, amino acids, inorganic ions as well as a wide range of enzymes (Tobin et al. 1994; Jones 1998; Ogar et al. 2014). It was shown that such common compounds as citric acid can greatly increase uranium bioavailability through complexation rather than the change in pH (Ebbs et al. 1998). The roots exudate a variety of extracellular compounds that serve as ligands e.g. polysaccharides, lipopolysaccharides (LPS) and a number of extracellular melanins which might be involved in surface complexation formation of metal-organic complexes in the solution and surface precipitation (McColl and Pohlman 1986; McGrath et al. 2001; Dakora and Phillips 2002). The results obtained by Rufyikiri et al. (2004) demonstrated the role of an arbuscular mycorhizal fungus in the uptake, translocation and immobilization of uranium. It was proposed that immobilization of uranium probably occurs in fungal structures such as vesicles as an effect of competition between uranium and other cations such as Ca^{2+}, Mg^{2+} or K^{+} for sites of adsorption, for uptake through cation transporters and for complexion with polyphosphates. Consequently, there are a multitude of potential mechanisms that might influence the uranium concentrations in the RPS systems, and a closer examination is needed in order to elucidate the major ones.

Concentrations of uranium in the plant shoots grown on WF ranged from 0.1 to 0.25 mg kg^{-1}, whereas for RPS between 0.55 and 1.6 mg kg^{-1}. According to Baker and Brooks (1989) hyperaccumulators of Co, Cu, Cr, Pb and Ni are defined as plants containing over 0.1% where for Mn and Zn, the criterion is 1% of any of these elements in the dry matter, irrespective of the metal concentration in the soil *H. pilosella* is able to grow on red processed shale (RPS) and accumulate high amounts of uranium, corresponding to some 0.6-0.8 % just in the aerial organs.

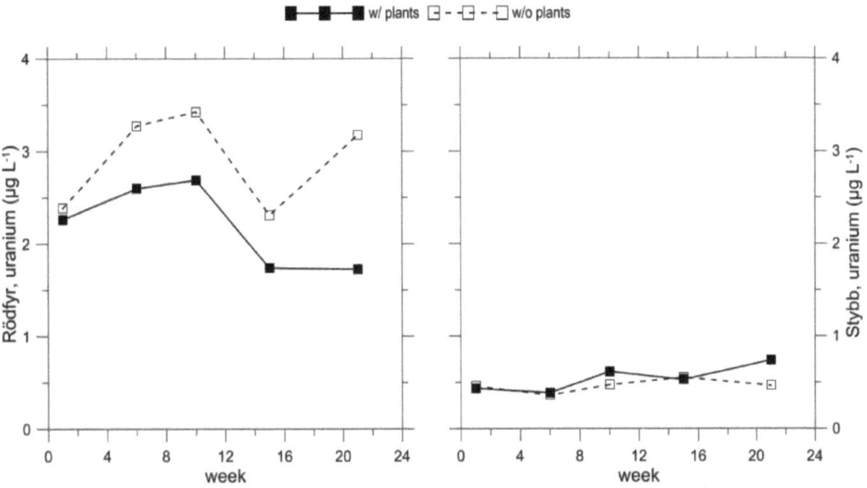

Fig.2. Uranium concentration in leachates from weathered shale fines (stybb) and red processed shale (rödfyr) substratum after 1, 6, 10, 15 and 21 weeks of experiment

Conclusions

By growing *H. pilosella* on red processed shale it was possible to reduce the release of uranium from the material by approximately 50 % whereas no impact was found for weathered shale fines. However, the lowered electrical conductivity for systems where plants grow on WF indicated lowered mobilization of other elements. However, due to the change in electrical conductivity plant growth on WF must have resulted in decreased mobilization of other elements. Hence, *H. pilosella* and its associated microorganisms might be a good option for phytostabilization of both wastes not by uptake but its ability to alter the hydrochemistry. A closer examination of *Hieracium* and associated microorganisms is needed in order to evaluate the impact on Cd, Co, Ni, V and Zn.

Acknowledgements

Funding was provided by the Foundation for Polish Science, International PhD Projects Program financed by the EU European Regional Development Fund (MPD/2009-3/5) and the Faculty of Economy, Science and Technology at Örebro University. The authors would like to thank Bert Allard and Katarzyna Turnau for constructive comments. We also thank anonymous reviewers for their assistance.

References

Allard B, Bäckström M, Häller S, Karlsson S, Panova E, Grawunder A (2011) Water chemistry and trace metal concentrations in an acidic alum shale pit lake - effects of liming. Proc. Internat. Mine Water Association Symp p 503-508

Baker AJM, Brooks RR (1989) Terrestrial Higher Plants which Hyperaccumulate Metallic Elements - A Review of their Distribution. Ecol Phytochem Biorecover 1: 81-126

Bradl HB (2004) Adsorption of heavy metal ions on soils and soils constituents. J Colloid Interface Sci 277(1): 1-18

Dakora FD, Phillips DA. (2002) Root Exudates as Mediators of Mineral Acquisition in Low-Nutrient Environments. Plant Soil 245(1): 35-47

Dupré de Boulois HD, Declerck S, Joner EJ, Leyval C, Jakobsen I, Chen BD, Roos P (2008) Impact of arbuscular mycorrhizal fungi on uranium accumulation by plants. J Environ Radioact 99(5): 775-784

Ebbs SD, Brady DJ, Kochian LV (1998) Role of uranium speciation in the uptake and translocation of uranium by plants. J Exp Bot 49(324): 1183-1190

Jones DL (1998) Organic acids in the rhizosphere – a critical review. Plant Soil 205(1): 25-44

Karlsson L, Karlsson S, Allard B, Sjöberg V, Bäckström M (2013) Release of metals from unprocessed and processed black shale due to natural weathering. Reliable Mine Water Technology. Proc. Internat. Mine Water Association Conf p 391-396

McColl JG, Pohlman AA (1986) Soluble organic acids and their chelating influence on Al and other metal dissolution from forest soils. Water Air Soil Pollut 313(4): 917-927

McGrath SP, Zhao FJ, Lombi E (2001) Plant and Rhizosphere Processes Involved in Phytoremediation of Metal-Contaminated Soils. Plant Soil 232(1): 207-214

Ogar A, Grandin A, Sjöberg V, Turnau K, Karlsson S (2014) Stabilization of uranium (VI) at low pH by fungal metabolites: applications in environmental biotechnology. APCBEE Procedia in press

Rufyikiri G, Declerck S, Thiry Y (2004) Comparison of ^{233}U and ^{33}P Uptake and translocation by the arbuscular mycorrhizal fungus *Glomus Intraradices* in root organ culture conditions. Mycorrhiza 14(3): 203-207

Schwab P, Zhu D, Banks MK (2007) Heavy metal leaching from mine tailings as affected by organic amendments. Bioresource technol 98(15): 2935-2941

Sjöberg V, Karlsson S, Grandin A, Allard B (2014) Conditioning sulfidic mine waste for growth of *Agrostis Capillaris* - Impact on solution chemistry. Environ Sci Pollut R 21(11): 6888-6904

Tobin JM, White C, Gadd GM (1994) Metal accumulation by fungi - applications in environmental biotechnology. J Ind Microbiol 13(2): 126-130

Reliable water management as key success factor for the remediation of uranium production sites under humid conditions

Michael Paul[1]

[1]Wismut GmbH, Jagdschänkenstraße 29, 09117 Chemnitz

Abstract. Successful implementation of mine remediation projects depends on a wide range of key aspects, one of the most critical is water management. Irrespective of both its importance regarding remedial success and its relevance to remediation cost, water management issues are frequently underestimated in the design and execution of remedial work. In this regard, the paper summarizes some of the most relevant findings from more than 20 years of implementing the WISMUT programme, one of the biggest mine reclamation projects worldwide.

Introduction

East German uranium industry, one of the world's premium historical uranium producers, was decommissioned in 1990. A huge environmental remediation programme, known as the WISMUT programme and sometimes dedicated to be the world's biggest mine closure project, was launched in 1991. The focus of this programme was on the dismantling of all former uranium production complexes, the decontamination of operational areas, the safe enclosure of all solid radioactive residues such as waste rock and tailings, and the sound management of remaining liquid and gaseous emissions. WISMUT's seven individual remediation sites are located in the humid Central European climate zone (Leupold & Paul 2007), with a mean annual precipitation between 700 and 1150 mm/yr, a mean annual temperature of 6...9 °C and a positive climatic water balance.

By 2014, WISMUT's remediation work is far advanced, about 75 % of the formerly contaminated mine land have been finally remediated. Remaining radioactive and other emissions have been essentially diminished and are generally in compliance with given standards. In summary, the programme receives favourable recognition and is often referred as a success story and international benchmarking project for mine closure and remediation of radioactive wastes.

The WISMUT experience, however, does also demonstrate the relevance of water management for the performance of huge mine remediation projects. Due to

B.J. Merkel and A. Arab (eds.), *Uranium – Past and Future Challenges*,
DOI 10.1007/978-3-319-11059-2_50

a large number of potential pitfalls associated to water management, especially related to predictive modeling and the approval application and permitting procedures, the remediation programme was delayed at several sites, suffered setbacks and had to be partially revised. From almost 25 years of project experience, the paper summarizes some of the most prominent conclusions for a sound and reliable water management strategy. The discussion here will be focused on three key activities: (i) mine flooding and mine water management, (ii) water treatment, and (iii) run-off management and flood protection.

Mine flooding and mine water management

The experiences and lessons learned regarding the closure and flooding of WISMUT's five underground uranium mines at Ronneburg, Schlema, Königstein Pöhla, and Dresden-Gittersee have been repeatedly reported, most recently by Paul et al. (2013). Given the prevailing climatic conditions, the mines were flooded due to natural groundwater inflow after extensive permitting approval procedures, following detailed site specific flooding concepts. The preparation activities for the decommissioning of the mine dewatering equipment had to be coordinated with a huge variety of other mine closure activities in conjunction with the stepwise withdrawal from the mine, including dismantling of the mine infrastructure, backfilling of mine openings, etc. Due to the complexity of the entire closure plan for the two complex underground mines at Ronneburg and Königstein, specific project management units were installed, to centralize responsibilities and to ensure a systematic and holistic approach. By 2014, all underground mines are completely abandoned, with the exception of galleries necessary for dewatering (Schlema, Gittersee), or exhaust ventilation in order to suppress upward radon migration in inhabited areas (Schlema).

It was a basic approach, that groundwater rebound should proceed stepwise and in a controlled manner, in order to ensure proper mine abandonment, but also to gain hands-on experience with the flooding process. Since mine flooding was initially characterized by significant uncertainties, flooding strategies had to be laid out with flexibility including various back-up options. Conceptual site models were found to be key tools for predictive modelling and decision making.

Although extensive data were collected prior to flooding and state-of-the-art modelling techniques were used by experienced mine water professionals familiar with the site conditions, some of the predictions failed. In consequence, amendments of the technical arrangements were necessary, to ensure the remediation success. Some important findings will be discussed in the following paragraphs, without claiming to deliver a complete account.

At the Schlema and Ronneburg mines, mean mine water inflow had been significantly underestimated by the original predictions. This was compounded by the fact, that the mean-to-peak-flow ratios, characterizing the relative flow increase during wet periods, were also higher than expected. As a consequence, flood water rise was faster, and water treatment capacities had to be augmented. At Pöhla, on the contrary, real post flooding mine water flow occurred lower than predicted. The reasons for those misjudgements are diverse. At Ronneburg, for instance, long-term mine water make was hard to derive from pumping rates during production times due to immense amounts of technical waters introduced. Therefore, the side wide water balance used for predictive modelling was established based on a data set of the early 1990ies, a period which was chosen since it was free of major anthropogenic input, but became evident as a dry one in retrospect. Moreover, the additional amount of surface waters to be treated together with the mine waters had been insufficiently addressed. The erroneous post-flooding flow predictions for Schlema and Pöhla were basically related to the lack of conceptual understanding regarding the head dependency of mine water inflow to the mine's depression cone during ground water rebound.

In order to prevent adverse effects of emerging mine waters suitable w**ater collection and treatment facilities had to be installed at any site in due time.** With the exception of the Pöhla mine, which is dewatered by its access tunnel, mine water abstraction concepts relied basically on the use of submersible pumps, installed in shafts or extraction wells, or shallow collection systems. At two mine sites (Gittersee, Ronneburg) the original dewatering concepts had to be changed significantly, as a result of the observations and findings as flooding became complete.

The Gittersee uraniferous coal mine, which had been originally allowed to flood up to the natural water level, caused sudden water logging in the Freital urban area. Contrary to all predictions, sufficient subsurface runoff either to a historic dewatering adit or to the local receiving stream did not materialise, obviously due to an overestimation of the hydraulic conductivity of the mined ground. Therefore, the original dewatering concept had to be discarded, and preference was given to extend a historic drainage gallery by some 3 kilometres, to ensure long-term runoff while safely precluding any surface water emergence. In consequence, completion of the site remediation was delayed by more than 5 years.

According to the original flooding concept for the Ronneburg mine site, an elevated flood water level was tantamount to minimizing the long-term contaminant supply from the unsaturated zone and to reducing volumes of groundwater needing treatment by cutting off as far as possible the inflow from the underground catchment area. According to that strategy, the deepest regional valley cut was equipped with a basic water collection system for contaminated groundwater discharge, backed up by several action plans to intervene in the flooding process via an extraction well or other measures, according to the results of an intensive environmental monitoring of the critical valley areas. After mine water emergence started in 2006 the installations proved not to be fully sufficient in their effectiveness and were, moreover, quite vulnerable to iron clogging and other disturbances. Necessary reworking measures were quite substantial. The volume of groundwater to be treated had to be adjusted from 300 m^3/h to 500 m^3/h, with weather-induced peaks up to 650 m^3/h. Optimization and increase in capacity turned out to be partially implementable only: lead time required for planning, permitting, contracting and construction was longer than expected, and emergency releases of contaminated waters could not be prevented: Further rise of mine water heads did finally cause significant uncontrolled discharges to the local creek, leading to a pollution of downstream water courses in 2010/11. Water management strategy is now being adapted for more flexibility in order to cope with inflow peaks during stormwater events. To reinforce the collection systems, a partial and temporary mine water re-drawdown is necessary (Paul et al. 2013, Baacke et al. 2014).

From a decision making standpoint, the post-flooding mine water qualities were observed chiefly in line with predictions. The well-known "*first flush*" phenomenon was observed at all mines. Concentration decrease vs. time was typically retarded, due to contaminant's discharge from saturated pore volume or various other sources (Paul et al. 2011, 2013). An unexpected self-purification effect was observed at Pöhla, where sulphate and uranium showed a spontaneous decrease shortly after flooding was completed, driven by a microbiologically catalysed sulphate-uranium-reduction process chain. Arsenic and Ra-226, on the other hand, reveal for both the Schlema and Pöhla mines a significant mobilization within the flooded mine water pool, clearly over-compensating the dilution by meteoric waters. Dissolution of native arsenic and its very common oxidation product arsenolite (As_2O_3) seems to be the key driver for As-mobilization (Paul et al 2010), whereas Ra mobility is increasing when sulfate concentrations drop.

Consequently, water treatment units had to be commissioned at any mine site, with key target parameters including radionuclides (U, Ra-226), Fe and Mn, As (most relevant at Schlema and Pöhla), and base metals like Zn, Cd, Cu, and Ni (crucial at Ronneburg and Königstein). Forecasts of the first flush peak heights as a necessary input parameter for the design of water treatment plants turned out to be a sophisticated task, since reliable full-scale estimates regarding the quantity of accumulated secondary leachable substances is usually not available. Interpretation of large scale experiments has proven as a valuable tool (Baacke et al. 2014).

In order to shorten the need for water treatment, a lot of research has been conducted to understand, use and enhance possible natural attenuation processes in flooded mine water reservoirs (Paul et al. 2006, Jenk et al. 2004, 2013). One key finding is, that full scale applications of in-situ-approaches are hard to implement and can, if at all, only be conceived as supporting measures to conventional technologies. As a matter of fact, they are hampered by a multitude of scientific and technical difficulties (Paul et al 2013). In this respect, the R&D work was focused on the Königstein mine. Based on the discoveries made during a field experiment, a fluid injection technology for full-scale application was designed and conceived as a supportive measure to enhance further mine flooding (Jenk et al. 2014b).

Considering the problem of mine flooding as a whole experience has shown, that the systems uncertainties are huge, due to (i) highly variable and poorly defined systems parameters including problems of scale, (ii) analytical complexity and insufficient knowledge of key processes, (iii) inadequate data base, especially poor quality of pre-mining data, (iv) limited possibilities for model calibration prior to flooding, but also (v) diverging interests of the parties involved when it comes to interpretation and decision-making (Younger & Robins 2002, Brown 2010, Paul et al 2013). With reasonable means, the majority of those uncertainties can only be diminished to a certain degree by data collection and exploration work prior to flooding. Therefore, a step-by-step approach for mine closure and flooding based on a conceptual site model seems to be indispensable for most mines.

Nevertheless, reaching a high correlation between model prediction and reality is possible which should be demonstrated by the Königstein case (Metschies & Jenk 2011, Jenk et al. 2014a): Monitoring data reveal that the impoundment process of the double-porosity system did fairly match the forecasts. The discharge of water contaminants occurred like expected, with a fast wash-out of the open mine voids, accompanied by a slow re-delivery from pore space. The total amount of soluble contaminants predicted has been confirmed by measurements. Also, the predictions regarding rock stability and surface subsidence were confirmed by reality. The excellent correspondence between forecast and reality is attributable to the high exploration level of the mined ground in conjunction with the implementation of an underground leach technology formerly used at Königstein, but also to very high R&D expenses during the preparation of the flooding concept due to permit requirements. Those included a number of large-scale flooding experiments, and preparations for flooding took a decade to complete.

Planning, permitting and implementation of mine flooding concepts take decades to complete. Authority requirements and expectations can change substantially in between: In contrast to the complementary situation in radiation protection legislation for instance, German water legislation has become significantly more stringent over the lifetime of the WISMUT programme, which is being reflected in remedial efforts being generally higher than originally expected.

Water treatment

Commissioning dates of the water treatment units at WISMUT's five mine sites (Pöhla 1995, Gittersee 1997, Schlema 1999, Königstein 2001, Ronneburg 2006) as well as at the former mill sites of Crossen/Helmsdorf (1995) and Seelingstädt (2001) were important milestones of the remediation programme. Thereafter, water-born emissions of radionuclides and trace metals decreased dramatically, and downstream water quality in the catchments of Weiße Elster and Mulde river has improved (Paul et al 2008). Since 2004, all conventional treatment facilities are being operated according to lime precipitation technologies laid out in different modifications. Dosing of $FeCl_3$/$FeClSO_4$ and $BaCl_2$ (for Ra-removal) is optional depending on the raw water quality. Lime precipitation stood up to alternative technologies mainly due to its robustness and the relatively low operating costs. Disposal of the treatment residues is crucial and a key cost driver, the sludges must be immobilized and are deposited at specific in-house dry disposal areas.

Solely at Königstein mine water uranium concentrations still justify an upstream ion exchange step for uranium recovery, with subsequent uranium concentrate fabrication. At any other site, uranium is precipitated together with the iron and metals.

Since sulfate, chloride and hardness are not being removed, the discharge of the treated waters has to be controlled based on the flow characteristics of the tailwaters, which is particularly challenging at the Weiße Elster river catchment, receiving the very hard and sulfate-rich discharges from both the Ronneburg and Seelingstädt treatment plants. To avoid bottleneck situations regarding discharge, a salinity management algorithm is in place, including process water supply from upstream storage reservoirs (Metschies & Paul 2012).

Mine water treatment is and will remain the most cost-intensive long-term task related to the entire WISMUT programme: Between 2008 and 2012, the total annual throughput was between 17 and 22 Mm^3 of mine and seepage water, comparing to operating cost between 33 and 35 M€/yr. According to WISMUT's 2010 work program, more than 70 % of the after-care budget will be dedicated to water treatment. Therefore, substantial resources must to be spent for process optimization, but also for investigation and design of alternative, low cost technologies.

In this regard, high expectations were put in a semi-passive treatment unit which replaced the original conventional plant at Pöhla. The unit was designed and constructed to treat a maximum of 20 m^3/h of mine effluent quite low in total mineralization and uranium, but still high in Ra-226 and arsenic. The bioengineering approach was based on co-precipitation of arsenic with iron after aeration and sedimentation, followed by ponds containing macrophytic algae as hyper-accumulators for radium. After a successful pilot test phase, the full scale unit became operational in 2003. Discharge limits for arsenic and radium, however, could not be permanently achieved, so the facility had to be upgraded by downstream reactive filters for arsenic and radium adsorption. In its final configuration, the entire installation worked satisfactorily, however, the aspired significant cost reduction did not materialize, due to high expenses for the filter material and significant monitoring and maintenance expenditures (Kunze et al. 2007). In consequence the Pöhla semi-passive unit is scheduled for decommissioning in 2015, after a modern remote control aeration-precipitation plant has started regular operation in 2014.

With remedial progress, WISMUT's first generation treatment plants will have to be either adjusted or even replaced, due to changes in boundary conditions. For mine effluents, input concentrations for most contaminants typically decrease, whereas the throughput requirements stay more or less the same. Process adjustment has to focus on permanent cost optimization, but also on the perpetuation of the treatment reliability, which can be threatened due to water quality changes. Uranium precipitation in modified lime treatment, as an example, can be inhibited if hydrogen carbonate concentrations increase as a consequence of mine water dilution and pH increase. Likewise, the evolution of iron is critical, since it is regulated for discharge on the one hand, but acts as sludge creating substance on the other, which is important for co-precipitation of arsenic and base metals. Iron clogging can also hamper CO_2-stripping as a critical upstream process for uranium precipitation.

The treatment plants at the Seelingstädt and Crossen mill sites had primarily to ensure the freewater removal step as a precursor for dry tailings remediation (Barnekow et al. 2012). As such, the total throughput was high and relatively constant over a 10 to 15 years period. After freewater removal was complete, plant operation had to cope with greater throughput fluctuations on the one hand, and increasing contaminant concentrations on the other, due to the relative increase of pore water intake. Specific treatment costs increased, however, the plant's capabilities to handle short-term inflow peaks are critical for the entire remediation progress in this phase. As an example, finishing the interim cover at the Culmitzsch A tailings management facility has been significantly delayed due to the 2010 summer and 2012/13 winter wet seasons. On the long term, after the tailings ponds will have been covered completely, water treatment will be dedicated to the tailings seepage only. For that future scenario, the existing plants will be definitely oversized and cannot be operated cost-effectively. Conceptual planning for new and adapted treatment units must take into account alternative technologies and installation sites. Enhancing energy efficiency will also be of high priority.

Since design planning and construction of new treatment plants are typically contracted out to external service providers, those activities are of high relevance regarding scheduling and budgeting. A multitude of problems can impede the procedures, and delays can endanger the remediation progress. WISMUT GmbH as a public entity has to strictly comply with all relevant regulations for public procurement. Permitting procedures are comprehensive, including various legal areas including mining, water and radiation protection legislation. After implementing the EU Mine waste directive (2006/21/EC) into German law, disposal areas for treatment residues are formally classified as waste management facilities, and if the residues are characterized as being hazardous, construction and operation of those facilities are subject to time consuming plan approval procedures.

In the context of European Water Legislation (2000/60/EC) and its further implementation at national levels, the so-called "good status" should be achieved for all ground and surface water reservoirs by 2027 at the latest. For water bodies affected by mining legacies, target achievement can be unrealistic, despite of all efforts for a sound water management. Less stringent requirements can be accepted by the permitting authority, if the target cannot be reached for technical reasons or only with disproportionate means. Generally, however, it is to be assumed that stricter environmental standards may come into force in the longer term.

As a further challenge WISMUT's long term water management approach will also have to address the regional impacts of climate change, including potential conflicts with third parties in the management of intensively used watersheds.

Run-off management and flood protection

Re-integration of remediated former mine land into the natural watershed is one of the final aims of remedial work. One fundamental requirement for that is to guarantee flood protection of downstream riparian areas. In the WISMUT programme, runoff management from remediated areas is to preserve the hydrological situation in the receiving streams at the status before remediation (1990) with respect to flood water flow resulting from the 100 year return event calculated based on the critical precipitation event in the respective catchment area. Therefore, implementation of hydrological models is a prerequisite for design planning.

Engineering complexity and necessary lead time are significant, especially for remedial activities which completely transform or reshape the former mine land such as the Lichtenberg open pit backfill project at Ronneburg or the tailings remediation activities at Seelingstädt and Crossen. Regularly, key problems arise from the superposition of the following circumstances: (i) The hydraulic capabilities of the receiving streams are relatively low in most cases, in conjunction with the fact, that the existing flood prevention at many cross-sections downstream does not comply with current requirements, even prior to the integration of remediated sub-catchments. (ii) Assessment base, boundary conditions, input parameters and permitting requirements did change several times over the project life time. (iii) Frequently, flood prevention cannot be managed without constructing storm water retention reservoirs, which themselves, however, are viewed critically by the permitting authorities, due to their inherent risk potential and maintenance requirements. (iv) The risk perception by the local public is very sensitive, also due to the increasing frequency of major floods in Central Europe and Germany over the last 20 years. (v) The construction of dewatering ditches and flood control reservoirs is not restricted to land owned by the company, which causes conflicts with private or public landowners. (vi) Very often, decision making requires planning approval procedures with multiple authorities being involved and public participation, leading to process durations of sometimes more than 10 years.

Due to these challenges, the completion of remedial works has been substantially delayed in several cases, so for instance at the Trünzig tailings management facility, the Crossen waste dump and several smaller area clean-up projects. Those delays were mostly unexpected and hard to handle from a project management perspective. Targeted and pro-active stakeholder involvement is of critical importance to keep the project work on track.

Conclusions

Sound water management is a crucial prerequisite for proper and on-schedule implementation of mine reclamation projects. Irrespective of both its importance regarding remedial success and its relevance to remediation cost, water management issues are frequently underestimated in the design and execution of remedial work. A whole variety of potential pitfalls of technical and non-technical nature must be properly managed. Referring to this, the paper documents major findings from the WISMUT environmental remediation programme regarding mine flooding, water treatment and surface water management. Many of the observations made may be transferred to other mine remediation projects in humid environments as well.

Several lessons must be learned from the experiences made under the WISMUT project. Misjudgments in the prediction of hydrological processes and decision making can lead to substantial delays and cost increase. Adverse environmental impacts as a result of emergency situations can also lead to loss of trust and prestige. The reasons for those misjudgments are manifold, including problems of an inadequate data base and poor understanding of key processes. The majority of those technical and scientific uncertainties are inherent to the problem and can only be minimized, not excluded. Therefore, a step-by-step approach based on a conceptual site model seems to be indispensable for decision making, and technological arrangements must be flexible and robust.

On the other hand, non-technical aspects do also contribute to the list of potential errors and mistakes, including a vague perception of remediation requirements and authority expectations as well as insufficient communication between mine water specialists and key decision makers at the operator's side. As in many other mining projects (Vink 2012), the role of variability and extreme events turned out to be a challenge for long term rational management and decision making also under the WISMUT programme.

Several examples do also show the importance of a realistic assessment of the lead time required for planning, permitting, contracting, construction and commissioning of facilities and structures. Since this process chain applies to many structural subdivisions of the mining company, water management must be understood as a company-wide task, in order to make it sound and reliable.

References

Baacke D, Jenk U, Paul M (2014): Flooding of the Underground Uranium Mines at Ronneburg and Königstein – a drawn-out finish.- Markscheidewesen 121 (2014) in print

Barnekow U, Müller J, Metschies T, Paul M (2012): Experiences gained and major challenges for the remediation of Wismut's Culmitzsch tailings pond.- In: Mine Closure 2012, Proceedings of Conference Mine Closure, Brisbane, Australia, 14 pp.

Brown A (2010) Reliable Mine Water technology. Mine Water Environ (2010) 29: 85-91

Jenk U, Frenzel M, Metschies M, Paul M (2014a): Flooding of the underground uranium leach operation at Königstein (Germany) – A multidisciplinary report.- Proc. IMWA 2014 Annual Congress, in print

Jenk U, Paul M, Ziegenbalg G, Klinger C. (2004) Alternative methods of mine water treatment - Feasibility and technical limitations for a full-scale application at WISMUT's

Königstein mine site (Germany). Proc. 8th Int. Mine Water Association Congress, Newcastle, UK, Vol. 1, pp. 245-251

Jenk U, Zimmermann U, Uhlig U, Schöpke R, Paul M (2014b): In Situ Mine Water Treatment: Field Experiment at the Flooded Königstein Uranium Mine (Germany).- Mine Water Environ (2014) 33: 39-47

Kunze C, Kießig G, Küchler A (2007): Management for passive biological water treatment systems for mine effluents.- In: Marmiroli et al. (eds.): Advanced science and technology for biological decontamination of sites affected by chemicals and nuclear agents, Springer, 177-195

Leupold D, Paul M (2007): Das Großprojekt WISMUT – Nachhaltige Sanierung und Revitalisierung von Uranerzbergbaustandorten in Sachsen und Thüringen.- Bergbau, 10 (2007), 438-444

Metschies T, Jenk U (2011) Implementation of a modelling concept to predict hydraulic and geochemical conditions during flooding of a deep mine.- The New Uranium Mining Boom, Challenges and Lessons learned. Springer Berlin Heidelberg 2011, 673-680

Metschies M, Paul M (2012): Management of mine and tailings seepage water discharge in an intensively used watershed. In: Proc. IMWA Ann. Conference, Bunbury, WA, September 30- October 4, 2012, 409-416

Paul M, Kreyßig E, Meyer J, Sporbert U (2008): Remediation effects of the WISMUT project to surface waters in the Elbe watershed: An overview.- In: Uranium, Mining and Hydrogeology, Proceedings of the Conference Uranium Mining and Hydrogeology V, Freiberg, September 2008, Springer, 437-450

Paul M, Metschies T, Frenzel M, Meyer J (2011) The Mean Hydraulic Residence Time and its Use for Assessing the Longevity of Mine Water Pollution from Flooded Underground Mines. Proc. 6th Conf. Uranium Mining and Hydrogeology, Freiberg, 689-699

Paul M, Meyer J, Jenk U, Baacke D, Schramm A, Metschies T (2013): Mine Flooding and Water Management at Underground Uranium Mines two Decades after Decommissioning.- Proc. IMWA Conference 2013, pp. 1081-1087

Paul M, Meyer J, Jenk U, Gengnagel M (2006) Source manipulation in water bodies of flooded underground mines - Experiences from the WISMUT Remediation Program. Proc. 7th ICARD, St. Louis, 1514-1530

Paul M, Meyer J, Jenk U, Jahn S, Klemm W (2010) Investigation of arsenic emissions from flooded ore mines of the Westerzgebirge region, Saxony, Germany. In: Proc. IMWA Sympos. Sydney, Nova Scotia, Canada, September 5-9, 2010, 387-390

Vink S (2012) Water conflicts, causes and solutions in the Australian resource sector. http://alhsudchile.files.wordpress.com/, accessed 15/06/2014

Younger PL, Robins NS (2002) Challenges in the characterization and prediction of the hydrogeology and geochemistry of mined ground. Geol. Society, London, Special publications, v. 198, 1-16

Mine Water Quality Evolution at Abandoned Uranium Mines in the Czech Republic

Nada Rapantova[1], Monika Licbinska[1], Pavel Pospisil[1], Karel Lusk[2]

[1]VSB - Technical University of Ostrava, Institute of clean technologies for mining and utilization of raw materials for energy use, 17. listopadu 15, 708 33 Ostrava, Czech Republic
[2]DIAMO, s.p. - Straz pod Ralskem, Machova 201, 471 27 Straz pod Ralskem, Czech Republic

Abstract. The objective of the presented study is to assess the evolving mine water quality of uranium mines abandoned between 1958 and 1992 in the Czech Republic. The sampling proved that actual uranium concentrations in mine waters did not in most cases exceed 0.45 mg/L. Uranium concentrations in the discharges from the adits abandoned more than 40 years ago were below MCL of US EPA for drinking water. Special attention is given to the most recently abandoned Mine Olší. Upon the geochemical modelling results using Geochemist's Workbench, three time phases of mine water evolution can be defined at this deposit.

Introduction

Czech uranium deposits were abandoned with the exception of Rožná deposit between 1958 and 1992. When uranium ore exploitation was finished and the mine dewatering stopped the process of spontaneous mine flooding started.

In the case of deposits abandoned in the 1990's, groundwater is pumped from the shafts in order to preserve groundwater level below decant level to surface water and after treatment it is discharged to surface streams. This ensures that shallow groundwater and surface water is not threatened by uncontrolled discharges of contaminated water from flooded mines. All these deposits are systematically monitored and as soon as water quality corresponds with the limits approved by the Czech Inspectorate of Environment for mining site under closure the groundwater pumping can be stopped.

The quality of mine water is impacted as soon as water during flooding enters mine spaces and under aerobic conditions (with the access of oxygen) comes in contact with primary and secondary minerals, which start to dissolve. Most of the processes are driven by Eh-pH conditions but also by temperature and concentration of some constituents.

B.J. Merkel and A. Arab (eds.), *Uranium – Past and Future Challenges*,
DOI 10.1007/978-3-319-11059-2_51

The strongest acid-forming process of all known oxidation processes that occurs in natural environment (Stumm and Morgan 1996) is sulphide (pyrite, marcasite) weathering. The generated acidity may then dissolve other constituents. Initial phases of sulphide weathering are bacteria catalyzed. Due to the pyrite paragenesis, prevailingly with sulphides but also with oxides, with compounds of vanadium and arsenic, with phosphates and carbonates, the mine water is enriched by number of elements. Their concentration depends mainly on mine water pH; maximum elements concentration is found at lower pH values.

After a mine is abandoned, the recovered groundwater is therefore enriched by dissolved substances from rock environment that was exposed to long-term oxidation and also by precipitated salts deposited in mine workings. As the flooding of uranium mine subsequently continues, overall concentration of the dissolved substances in mine waters rapidly increases which leads to increased concentrations (by several orders of magnitude) of uranium and other related metals (so-called "First Flush" in Younger et al. 2002; Gzyl and Banks 2007). Subsequently, conditions in the deposit stabilize and typical stratification appears which relates to oxidation conditions changing into reduction ones in deeper levels of the mine.

Mine waters evolution predictions require use of geochemical and reactive transport models. Utilization of such models is in practice limited by the availability of geochemical parameters, knowledge of rock environment heterogeneity and variability of transport parameters (Bain et al. 2001; Kerry et al. 2005).

It is therefore obvious that water management in abandoned mines requires the combination of theoretic model studies and long-term monitoring of abandoned uranium mining sites, which can provide empirical experience applicable for verification of conceptual and numerical geochemical and transport models.

Methods

The modelling was executed by hydrogeochemical modelling software The Geochemist's Workbench (Bethke 2007). Its database comprises 5 partial programs. Program Act2 helped to construct stability diagrams, modelling of mine water development was performed by program React and Gtplot program was used for graphic presentation of results.

In 2010 systematic monitoring of mine water quality at all available locations of previous uranium exploitation was performed (mine adits, uncontrolled discharges, shafts) - Rapantova et al. 2013. Analyses of mine waters physical parameters and chemical composition were the input data of the model. For the Olší Mine site, data from long-term monitoring of physical & chemical parameters taken by personnel of Diamo, s.p. in the period of 1987 – 2013 were used.

Results and discussion

The sampling (sampling campaign in 2010) proved that uranium concentrations in mine waters at most of the abandoned uranium mining sites in the Czech Republic did not exceed 0.45 mg/L (Rapantova et al. 2013). Uranium concentrations in mine water discharges from adits abandoned more than 40 years ago were below or slightly above the maximum uranium concentration limits (MCL) in drinking water (0.03 mg/l) of US EPA. U concentrations between 0.45 and 1.23 mg/l were detected only at water treatment stations of the Rožná and Olší mines. Rožná is an active uranium mine and at the Olší deposit, water was in 2010 pumped from a deep well and recharged to deposit after treatment (Rapantová et al. 2007; Rapantová et al. 2008). Only three samples showed abnormally high uranium concentrations. These samples were taken from pumped shafts J-19 (5.61 mg/l) and J-11A (5.44 mg/l) at the Příbram location and sample no. 5 from the deep well in the Olší mine (6.43 mg/l). All these samples quite clearly represent waters of deeper circulations. Increased values of uranium content in these waters correspond well with conclusions relating to mine waters stratification.

Volume activity of ^{226}Ra at most of the locations varies between 0.03 and 1.85 Bq/l; increased values (max. 17.1 Bq/l) were detected at two locations: Vítkov and Okrouhlá Radouň. These values are related to naturally increased background concentrations of Ra where radium is a product of uranium decay in uranium-bearing granite rocks. In the case of radium content in mine waters, complex trends depending on time can be detected.

In this paper the studies performed on the the Olší deposit will be discussed. The deposit exploited from 1959 was completely flooded in 1996 when mine water level reached the decant point – dewatering adit. The mine water level is maintained just below the bottom of the adit by pumping and mine water is treated. The information on the quality of input and output from the treatment station is available. At present, the underground spaces cannot be monitored anymore.

During the deposit exploitation, mine water was treated in the average amount of 1426 m^3/day, i.e. 16.5 l/s. Mine water quality in the uranium parameter fluctuated in the range of 1.5–2.5 mg/l. Shortly after the flooding, high concentrations of dissolved constituents, especially iron, manganese, sulphates, uranium and radium, appeared in the Olší deposit mine waters.

Uranium without the presence of other components is very little mobile even at rather low concentrations (0.01 mg/l) and in wide temperature range. Figures 1 and 2 show uranium stability in water during the mine operation (Fig. 1) and after flooding in 2010 (Fig. 2) in the Olší mine site; the monitored uranium concentrations achieved approximately the values of 2.23 (Fig. 1) or 2.48 mg/l (Fig. 2) but temperature conditions in the mine changed.

Uranylic ion is the mobile constituent but only in very acid pH. In oxidation conditions, schoepite is the stable mineral; in reduction conditions then uraninite while stability of these solid phases grows with the growing uranium concentration. It means that in the case that there are no other constituents in the system, uranium will not freely migrate. However, it needs to be mentioned that uranium mobility depends on temperature. Results of modelling at temperatures lower than 25 °C showed higher uranium mobility (Fig. 2).

At low sulphate concentrations (~30 mg/l), sulphates have the ability to stabilize uranylic ion due to the formation of insoluble phase of $UO_2SO_4.3H_2O$ (Fig. 3). Only with higher sulphate concentrations in the solution (≤80 mg/l), mobile complex $UO_2(SO_4)_2^{2-}$ (Fig. 4) appears. Higher sulphate concentrations will therefore increase uranium solubility.

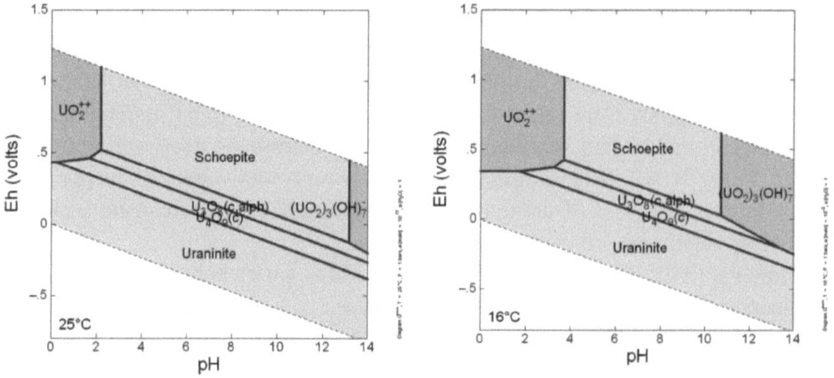

Fig.1 and 2. Eh-pH uranium diagrams with uranium concentrations 2.23 and 2.48 mg/l at different temperatures in the Olší-Drahonín location in time of active operation and in 2010.

On the other hand, the same does not apply when carbonates participate in the reactions with uranium, as the sulphate concentration is low. Then insoluble phases are dominant in wide pH spectrum and only in alkaline pH and oxidation environment uranium complexation with carbonates applies. Carbonates presence in the system therefore decreases uranium solubility (Fig. 3).

Within long-term mine waters monitoring at the Olší mine site, it can be observed that waters in first five years after the deposit abandonment have the highest concentration of uranium as well as sulphates (Fig. 5 and 6) and bicarbonates. In these waters, uranium is dissolved in form of a complex with bicarbonates $UO_2(CO_3)_3^{4-}$ (Fig. 4).

Geochemical model of mine waters evolution was created for initial phases of the mine abandonment, for the period when the mine was being flooded and finally for the situation when mine waters reached contact with external atmosphere at the discharge of these waters from the mine.

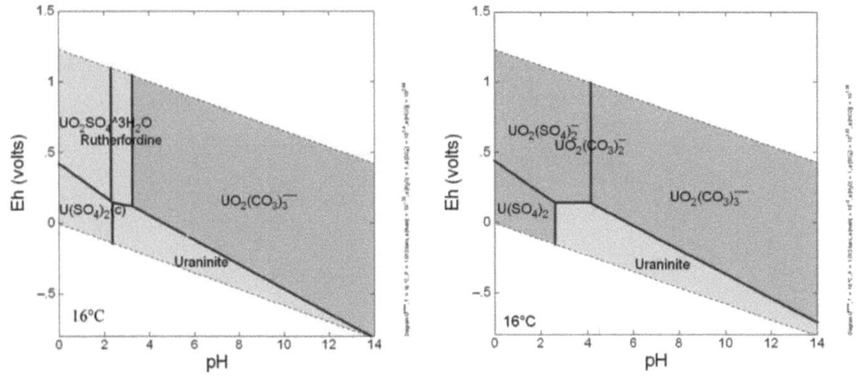

Fig. 3 and 4. Eh-pH diagrams of uranium stability for concentrations of uranium 2.48 mg/l; sulphates 30 mg/l (Fig. 3) or 80 mg/l (Fig. 4); carbonates 130 mg/l.

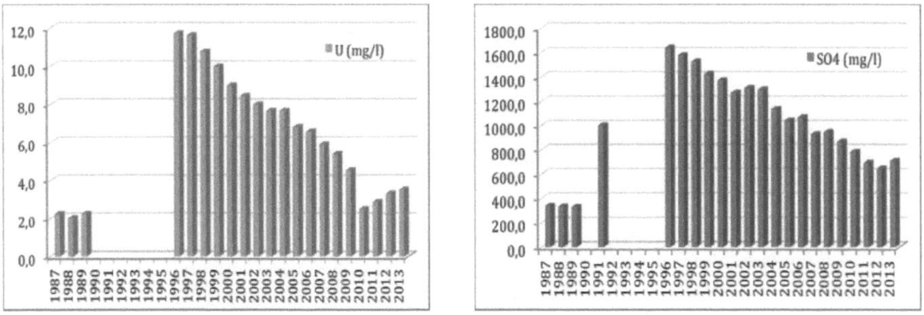

Fig. 5 and 6. Development of uranium and sulphates concentration in mg/l in the Olší mine water during mine operation (1987-1989), flooding (1990-1995) and in the period when the mine was flooded (1996-2013).

The process of mine water level rebound during the mine flooding represents the situation when the water saturated against the surrounding rock gets to reduction environment where dissolved oxygen is lacking and to the environment with higher pressure. Higher pressure at deposit bottom is characterized by increased CO_2 partial pressure. Significant decrease in oxygen is manifested in the oxidation-reduction potential decreased down to -0.076 V (Fig. 7). Total amount of dissolved substances grows approximately ten times while pH decreases to the value of 5.2. On the other hand, sulphates concentration decreases back to its original value of 1 mg/l due to their complete reduction to pyrite. Large amount of organic mass allows sorption of part of the metals contained in mine water and their concentration declines. Uranium concentration decreases prevailingly as a result of uraninite precipitation to $0.1.10^{-3}$ mg/l. Iron and especially manganese become mobile in the reduction environment and their contents in mine waters grow. Course of the precipitation and dissolving of individual minerals, Fe, Mn, U and Ra concentration development and pH, Eh and total water mineralization in course of the reaction are shown in Fig. 7 – 12. Though pH decreases in course of nearly whole reaction, it increases at the very end.

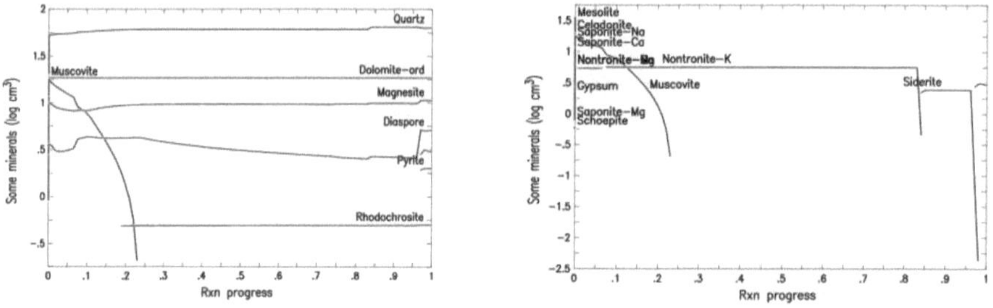

Fig. 7 and 8. Precipitation and dissolving of mineral phases in the course of mine water rebound during the mine flooding. Stage of the reaction is shown on x-axis; y-axis shows volume of mineral phase [log cm^3]; $p_{CO_2}(g) = 60$ atm; $p_{O_2}(g) = 1.10^{-66}$atm

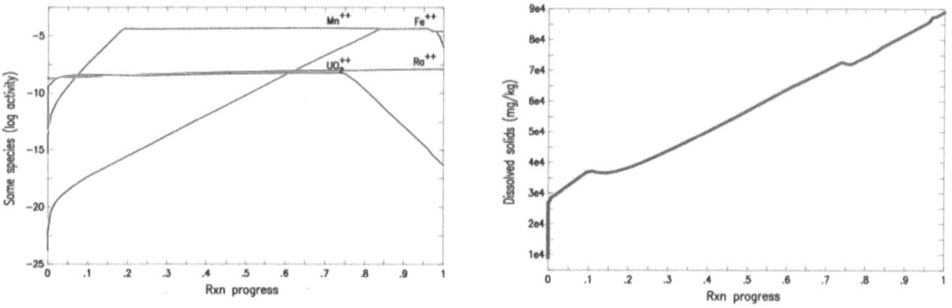

Fig. 9 and 10. Concentration evolution of Fe, Mn, U, Ra and dissolved substances in course of mine water rebound during the mine flooding. Stage of the reaction is shown on x-axis.

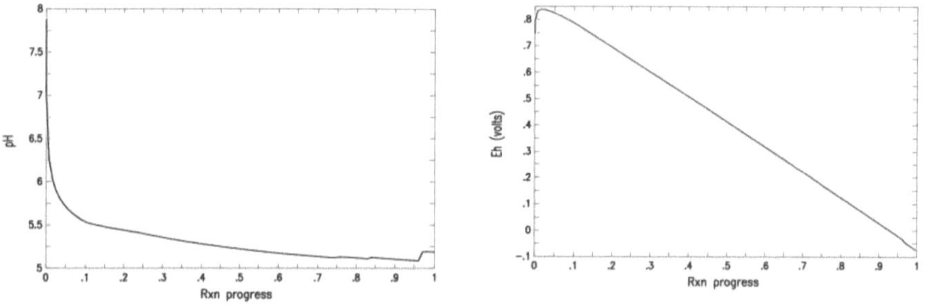

Fig. 11 and 12. pH and Eh development in course of mine water level increase during the mine flooding. Stage of the reaction is shown on x-axis.

Mine water reaching contact with the atmosphere means that the usual partial pressures of oxygen and carbon dioxide are achieved. The transition from reduction to oxidation environment is documented by the change of oxidation-reduction potential from -0.076 V to 0.657 V. In these conditions, manganese and iron oxidize and precipitate in the form of pyrolusite and hematite, or insoluble hydro-oxides. It is why the resulting Mn and Fe concentration in mine waters is low. Apart from that, dolomite, magnesite and K-saponite also precipitate. The content of U, Ra and sulphates remains more or less stable. However, the overall water mineralization decreases. Precipitation of mineral phases and development of pH, Eh and mineralization in course of the reaction are shown in figures 13 – 16.

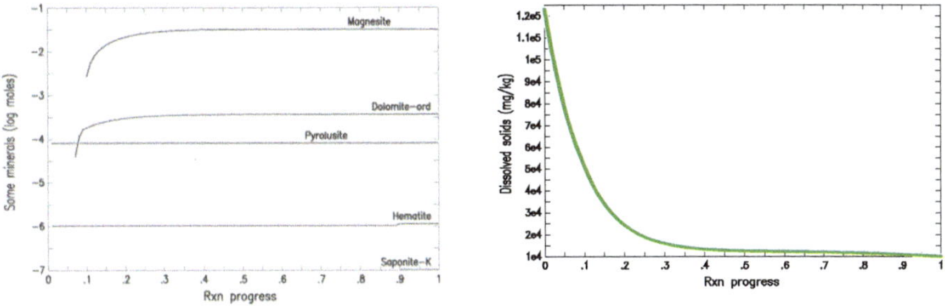

Fig. 13 and 14. Precipitation of mineral phases and mineralization development in course of mine water discharge to the surface. Stage of the reaction is shown on x-axis, y-axis shows volume of mineral phase [log moles]; $p_{O_2}(g) = 0{,}23$ atm; $p_{CO_2}(g) = 0{,}5.10^{-3}$atm.

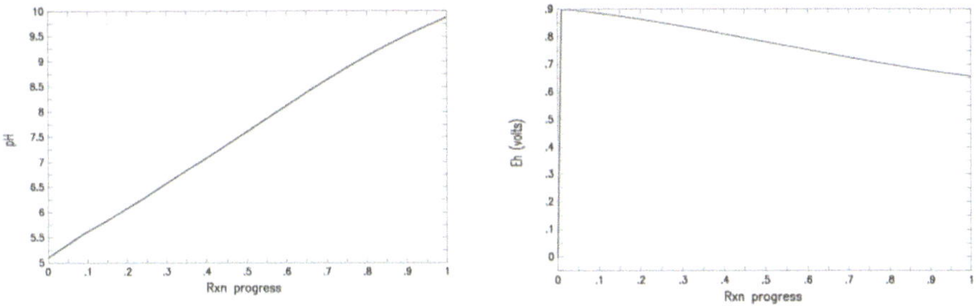

Fig. 15 and 16. pH and Eh development in course of mine water discharge to the surface. Stage of the reaction is shown on x-axis.

Mine water discharge to the surface and contact with the atmosphere result in precipitation of carbonates, manganese oxides and iron. Considerable decrease of Fe and Mn contents, maintaining of the originally low contents of U, Ra and sulphates and increase of pH contribute to rather favourable parameters of the mine water. The persisting problem however is the rather high TDS of mine water (10 335 mg/l). Results of modelling only describe the optimal situation when water reaches balance after the mine is flooded. It can only happen in broader time horizon, though. Immediately after the mine is flooded, water shows higher contents of Fe, Mn, U, Ra and sulphates and lower pH.

Conclusion

Upon the geochemical modelling results, three time phases of deposit evolution can be defined. In the first phase when the mine is in active operation or shortly after its abandonment, uranium in water is not very mobile depending on concentration of other constituents, especially sulphates and bicarbonates, and Eh-pH conditions. In the second phase, uranium is washed out of the deposit. In the third phase, approximately 40 years after the deposit abandonment, uranium is not washed out of the deposit anymore and poses no hazard to surface watercourses.

Acknowledgments

The article has been made in connection with project ICT CZ.1.05/2.1.00/03.0082 (Institute of Clean Technologies for Mining and Utilization of Raw Materials for Energy Use) supportedby the European Union and from the means of state budget by the Ministry of Education, Youth and Sports. This research was also financially supported by the Technological Agency project TA02021132.

References

Bain JG, Mayer KU, Blowes DW, Frind EO, Molson JWH, Kahnt R, Jenk U (2001) Modelling the closure-related geochemical evolution of groundwater at a former uranium mine. J Cont Hydr 52(1–4): 109–135

Bethke CM (2007) Geochemical and biogeochemical reaction modeling, 2nd edn. Cambridge University Press, Cambridge

Gzyl G, Banks D (2007) Verification of the "first flush" phenomenon in mine water from coal mines in the Upper Silesian Coal Basin, Poland. J Cont Hydr 92(1–2): 66–86

Kerry TB, MacQuarrie K, Mayer U (2005) Reactive transport modelin in fractured rock: a state-of-the-science review. Earth Sci Rev 72(3–4):189–227

Rapantova N, Grmela A, Vojtek D, Halir J, Michalek B (2007) Groundwater flow modelling applications in mining hydrogeology. J Mine Water Environ 26(4):264–271

Rapantová N, Grmela A, Michálek B, Hájek A, Zábojník P, Zeman J (2008) Utilization of Olší-Drahonín uranium deposit after mine closure. 10th IMWA Congress. Mine Water and the Environment. Karlovy Vary. Proceedings ed. Rapantova & Hrkal. Printed Esmedia DTP s.r.o. Olomouc: 221–224.

Rapantova N, Licbinska M, Babka O, Grmela A, Pospisil P. (2013) Impact of uranium mines closure and abandonment on groundwater quality. Environ Sci Pollut Res Int. 20(11): 7590-602

Stumm W, Morgan JJ (1996) Aquatic Chemistry: Chemical Equilibria and Rates in Natural Waters. 3rd edn. Wiley, New York

Younger PL, Banwart SA, Hedin RS (2002) Mine water—hydrology, pollution, remediation. Kluwer, Dordrecht

Glass Bead Filter Packs in Water Wells for Higher Efficiency and Reduced O & M costs

Reinhard Klaus[1]

[1]Sigmund Lindner GmbH, Oberwarmensteinacher Str., 95485 Warmensteinach, Germany, klaus@sigmund-lindner.com

Abstract. The selection of poor quality natural sand and gravel filter pack media for water supply wells leads to insufficient hydraulics, increased well clogging, higher electrical energy demand, reduced life-cycle, and increased Operations & Maintenance costs. Extensive comparative field and laboratory studies since 2008 proved, glass bead filter pack wells can achieve two figure savings for O & M at enhanced well performance and lifetime cycles.

Keywords Alternative filter pack media, physical, hydrological properties, well performance, sustainability, cost savings

Introduction

Until late 2007 gravel and sand were exclusively used as filter pack media in water wells. Gravel and sand are natural minerals, their availability and quality is rapidly declining in the last years. This phenomenon can be detected globally. Apart from that, even material in accordance with industry norms causes a lot of problems in well construction and functionality. For instance the German industry norm DIN 4924 which determines the specifications of mineral sands and gravel for filter packs in water wells accepts 1% of unclarified particles, 10 % of undersized and 15 % of oversized particles.

The amount of undersized particles is growing during transport of the material to the construction site due to disintegration because of insufficient crushing strength. A summary of negative effects on well construction and performance is given by Hermann and Stiegler (2008). Among others the main problems are:

- Jamming and bridging because of angular and edged grain
- High share of undersized particles and fines
- Cost intensive development work with limited effects
- Reduced porosity and permeability of the filter pack
- Clogged filter packs and well screens with gravel debris

B.J. Merkel and A. Arab (eds.), *Uranium – Past and Future Challenges*,
DOI 10.1007/978-3-319-11059-2_52

Figure 1 shows a clogged continuous wire wrapped screen after development.

Fig.1. Clogged well screen with gravel debris after development pumping

In 2010 laboratory tests were conducted by an independent consultant (Paul, K.) in order to qualify and quantify the effects physical properties of natural gravel has on the hydrologic performance of filter packs. For instance a 10 % share of undersized particles reduces the permeability of a filter pack to 55 – 75 % compared to a pack free of undersized grain. (Table1.)

Table 1. Effects physical properties of natural gravel has on the hydrologic performance of filter packs

Gravel	Share of undersized grain [%]	Kf [%]
3.5 x 8	0	100
3.5 x 8	10	75 - 55

As the equation in the figure below shows, the inferior sphericity of natural gravel causes 36% higher head loss resp. reduced permeability compared to perfect rounded glass beads

$$\frac{k_{Glas}}{k_{Sand}} = \frac{\frac{1}{3^2}}{\frac{1}{3{,}5^2}} = 1{,}36 = \frac{\Delta p_{Sand}}{\Delta p_{Glas}}$$

Examinations in the Netherlands first proved fines in gravel packs are a major source for irrelevant well aging by clogging and enhancing microbiological scaling with iron and manganese (Van Beek & Kooper 1980, Van Beek 1995, DeZwart 2007). Further indications on influencing or promoting factors of pack media for well aging or scaling are given by Treskatis & Houben (2003). Precisely they name grain shape, inner surface (coarseness), size, geometry and volume of pore channels and fines (from formation and filter pack).

With average operation times of more than 40 years, operation and maintenance costs for frequent well rehabilitation to restore capacity loss by scaling are a major financial burden in total lifetime costs of a well. In addition there are added investment costs for the substitution of irreversible damaged wells.

High performance wells, like ASR wells with their bidirectional water flow, huge water volumes transported and higher amount of fines due to the nature of the source water for infiltration created further needs to raise the standards for filter pack media. Otherwise they are impaired by irreversible clogging, due to disintegration and compaction of gravel packs which means significantly reduced well performance and lifetime.

Alternative filter pack media which will avoid these problems were in high demand. First quality demands were named by Treskatis, Hein, Peiffer, & Herrmann (2009).

Table 2. Quality characteristics for filter pack media (Treskatis, Hein, Peiffer, & Herrmann 2009)

Characteristic of the material	Quality goals
Washed and free from 'undersized particles'	Low material losses free from 'undersized particles' and compaction when developing the well; reduction of the development time
Well- rounded gravel grains	Increasing the porosity and hydraulic permeability compared with the aquifer; reduction of the lowering and pressure losses; improvement of the development ability and yield
High quartz share	Avoidance of volume changes through swellable or broken minerals
Smooth surface	Minimising deposits
Low irregular form	Low demixing when filling; avoidance of pressure losses through clogging

With regard to these demands, glass beads seemed to be a natural choice for a test.

Alternative filter pack media

In late 2007 soda lime glass beads from Sigmund Lindner were first applied in a 150m deep well in the Frankonian Keupersandstone near Nuremberg. Wells in that formation have to cope with severe and fast well aging by iron and manganese encrustation. Gravel filter packs in former wells were irreversibly destroyed after some rehab cycles with high impact hydro mechanical cleaning technics. Promising results from handling and well performance gave way to a series of comparative R & D projects.

Physical properties

A major R & D project, funded by the German Federal Ministry of Economics and Technology was conducted from 2008 – 2009. The authors, Treskatis et al. (2010) performed comparative laboratory tests of several sizes and variations of natural gravel and glass beads for the parameters:

Table 2 Tested parameters

roundness	peak-to-valley heights
specific weight	surface relief
bulk density	surface profile
grading	specific surface
breaking load during static stress	abrasion resistivity
breaking properties during static stressbreaking properties during dynamic stress	chemical resistance to rehabilitation solvents
	abrasion resistance

With the result: "Glass beads have mechanical, physical and chemical advantages compared to natural filter gravels and can make an important contribution to avoid clogging and to reduce incrustations when used in suitable unconsolidated sediments and bedrock, and thereby to an overall reduction in development and rehabilitation expenses" (Treskatis et al. 2010).

Figures 2 and 3 of this publication give a clear indication about the amount of differences in relevant properties between gravel and glass beads.

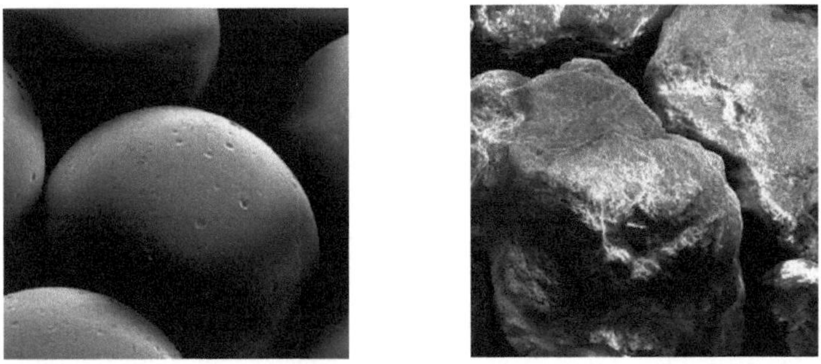

Fig.2. REM image of a glass bead compared to a filter gravel grain of the same grain size. The "smooth" surface of the glass bead prevents the formation of tensile stress when load is applied and reduces the agglomeration of incrustations. (Treskatis et al. 2010)

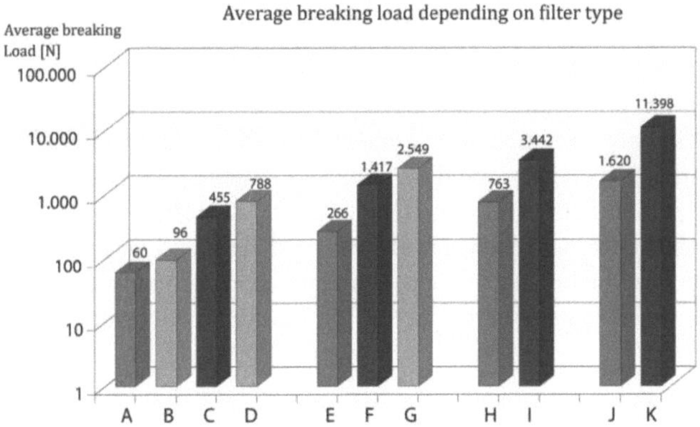

Fig.3. Magnitudes of breaking loads for filter gravel and glass beads at different granulations and bead sizes and mixtures at static load handling (Treskatis et al. 2010)

A = Filter gravel no.1 (1.4-2.2 mm)
B = filter gravel no.2 (1-2 mm)
C = glass bead type S (1.25-1.65 mm)
part no.: 4505 #923033
D = glass bead type S (1.50+-0.2)
part no.: 4505-A #820029-1
E = filter gravel no.3 (2.0-3.15 mm);
F = glass bead type S (2.85-3.45 mm)
part no.: 4511 #920032
G = glass bead type S (3.00+-0.3)
part no.: 4511-A #820022)
H = filter gravel no. 4 (5.6-8 mm)
I = glass bead type S (5-6 mm)
J = filter gravel no. 5 (8-12 mm)
K = glass bead type M (12 mm)
part no.: 5018-99-24 #855057-20

Hydraulic and Hydrodynamic properties

Comparative tests in the laboratory of Bau ABC (Federal Academy for Construction Professions) showed also better properties of glass beads for:

- Bedding properties
- Porosity
- Permeability

as well as better capabilities for sand discharge in the well development process at significantly higher efficiency. Glass beads generate a faster and, with regard to soil fines, more efficient sand discharge, while the limit of sand breakthrough, especially in uniform soils is already at a low leakage size. The grain smaller than the characteristic grain is already mobilized on glass bead packs. Thus a comparatively rapid development is possible. (Treskatis, Tholen, Klaus 2011, 2012).

Well aging caused by scaling

First column tests with gravel and glass beads in 2008 showed that in natural gravel, approx. 40 % more iron mass was embedded than in glass beads. Thus a clearly lower incrustation tendency could be expected in actual wells when using glass beads as filter pack media (Treskatis, Hein, Peiffer, & Herrmann 2009). Recent tests by the author et al. with actual wells in a test field and an extended laboratory set up with real heavy ferrous and manganiferous groundwater proved that scaling of glass beads is delayed by factor 2 – 3 compared to natural gravel. See (Klaus et al. 2013).

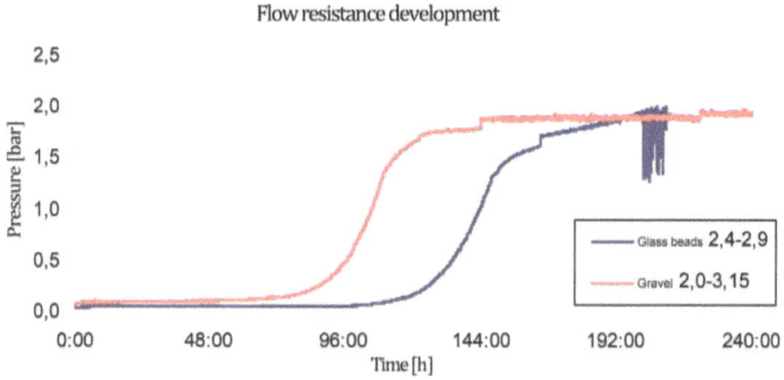

Fig.4. Pressure development in gravel packs 2.0 – 3.15 mm grading and glass beads packs grading 2.4 – 2.9 mm under continuous perfusion with iron and manganese containing groundwater

Field results

To date more than 4,000 metric tons of glass beads were used in more than 100 water wells in Germany, Italy and the USA, covering the whole hydrogeological spectrum from alluvial to bed rock and various groundwater chemistry, proving the laboratory results described above. Just for the USA 14 new wells are expected to be constructed with glass bead packs this year.
Further observations from contractors, technical consultants and well owners are:

- Easy application, no bridging or jamming during filling process
- Consolidated bedding after filling, no subsidence compared to gravel
- Time and volume for sand removal and development is down to 50 % compared to gravel
- Reduced drawdown of water table compared to former well layout
- Higher specific capacity
- Lower tendency of scaling in filter packs
- Intervals between rehabilitation can be stretched, which means lower expenses for O & M (The first water well equipped with glass beads in the town of Rosstal, near Nuremberg, still has not to be rehabilitated. The predecessor well in the same geologic setting had to be rehabilitated between every two years to once a year).

In 2013 also 2 dewatering wells for mines in Colorado, one of them app 1.400m deep, were equipped with glass bead packs. Even under these demanding conditions the positive experience with handling, better filling and yield could be proved.

Economic aspects

Based on local conditions, the investment costs for glass beads are between 2 and 5 times higher than for gravel. Regarding total costs of wells the surplus is between 0.5 % and 5 % depending on depth, diameter, screened area, etc. But material price alone is no indicator for the efficiency of a well. Wells are long term investments with lifetime cycles far beyond 40 years. Operating costs, primarily for electrical energy and rehabilitation after iron and manganese scaling are the essential factor. Due to higher specific capacity and delayed scaling, glass bead wells imply cost saving potential for O & M which will more than compensate the initial higher investment.
Klaus & Walter (2011) did a first cost benefit analysis based on the then known facts. Even this first tentative approach on the base of 1% savings for electric energy and 25 % for rehabilitation costs produced a total benefit of 8 % after 40 years considering interest and inflation.
Actual wells showed an increase in performance between 20% and 300 % with corresponding savings for pumping energy. An updated calculation by Klaus & Walter (2012) brought savings for pumping costs between 50 and 80 % per year,

which means a ROI in 3.5/8 years just on the base of cost savings for water pumping. A first extrapolation of the potential savings for rehabilitation based on the results of the recent scaling tests brings total lifetime savings up to more than 20%.

Conclusions

Glass beads as substitute for mineral gravel in filter packs of water wells are successfully applied since 7years. The field and laboratory results show, this application is a progress in well construction and shifts the state of the art to a higher level.
For the first time, physical, hydrological and chemical properties of a filter pack can stay consistent for the entire well lifetime cycle. Savings of electrical energy and O & M costs for rehabilitation are a major step towards real sustainability.
Promising results and positive feedback are coming also from the wastewater and water treatment sector.

References

Van Beek, C. G. E. M. & Kooper, W.F. (1980): The clogging of shallow discharge wells in the Netherlands river region.- Ground Water 18 (6): 578-586.

Van Beek, C. G. E. M. (1995): Brunnenalterung und Brunnenregenerierung in den Niederlanden.- gwf Wasser/Abwasser 136 (3): S. 128-137.

Herrmann, F. & Stiegler, X. (2008): Einsatz von Glaskugeln als Ersatz für Filterkies in Brunnen. – in: bbr 05/2008: S. 48-53; Bonn (wvgw).

Paul, K. (2010): Bedeutung und praxisgerechte Bestimmung von Bodenkennwerten
für den Brunnenbau. Oral presentation at: figawa geofora Fachkongress für Bohrtechnik, Brunnenbau und Geothermie 17. / 18.06.2010, Hof.

DeZwart, B.-R. (2007): Investigation of Clogging Process in Unconsolidated Aquifers near Water Supply Wells. – 200 S., Dissertation TU Delft.

Houben, G. & Treskatis, C. (2003): Regenerierung und Sanierung von Brunnen– 280 S., 111 Abb., 32 Tab., Anhang und CD-ROM; München (Oldenbourg) (ISBN: 3-486-26545-8).

Klaus, R & Walter, P. (2011): Wirtschaftlichkeit von Glaskugeln im Brunnenbau – in: bbr 08/2011; Bonn (wvgw).

Klaus, R & Walter, P. (2012): Neubau von Brunnen mit Glaskugeln-Ergiebigkeiten/Einsparpotenzial – in: Energie / Wasser – Praxis (ewp) 04/2012: S. 30 – 33; Bonn ISSN 1436-6134.

Treskatis,C., Hein., C., Peiffer, S. & Hermann, F. (2009): Brunnenalterung: Sind Glaskugeln eine Alternative zum Filterkies nach DIN 4924?.- in: bbr 04/2009: S. 36-44; Bonn (wvgw).

Treskatis, C., Danhof, M., Dressler, M. & Herrmann, F. (2010): Vergleich ausgewählter Materialcharakteristiken von Glaskugeln und Filterkiesen für den Einsatz in Trinkwasserbrunnen. DVGW energie|wasser-praxis 1/2010: S. 26 –32; Bonn (wvgw).

Treskatis, C., Tholen, L. & Klaus, R. (2011): Hydraulische Merkmale von Filterkies und Glaskugelschüttungen im Brunnenbau – Teil 1. – Energie / Wasser – Praxis (ewp) 12/2011: S. 58 – 65, 11 Abb., 6 Tab.; Bonn; ISSN 1436-6134.

Treskatis, C., Tholen, L. & Klaus, R. (2012): Hydraulische Merkmale von Filterkies und Glaskugelschüttungen im Brunnenbau – Teil 2. – Energie / Wasser – Praxis (ewp) 01/2012: S. 40 – 43, 3 Abb.; Bonn; ISSN 1436-6134.

Klaus, R., Treskatis, C., Tholen, L. (2013):Development and scaling characteristics of glass bead and gravel packs – Findings for practical well design.- in: bbr 11/2013: Bonn (wvgw).

Reduce,
Reuse,
Recycle...
PAPER
PLASTIC

Soil hydrological monitoring in the framework of the remediation and long-term safeguard of uranium ore mining residues of the Wismut GmbH

Katja Richter[2], Marcel Roscher[1], Ulf Barnekow[1], Gert Neubert[1], Manfred Seyfarth[2]

[1]Wismut GmbH, Jagdschänkenstraße 29, 09117 Chemnitz, Germany
[2]Umwelt-Geräte-Technik GmbH, Eberswalder Straße 58, 15374 Müncheberg, Germany

Abstract. This article provides an insight into the extent of remediation of residues of the uranium mining of the Wismut GmbH. Using the example of the waste rock dump Beerwalde in the Ronneburg region the layout of the measurement systems and the measurement technology used will be described. The results of the long-term measurements show the importance of this measurement campaigns for this remediation project in particular, and future remediation projects in general.

Motivation

The SDAG Wismut, following UdSSR, USA and Canada the fourth largest uranium producer in the world has mined a total of 216,000 tons of uranium by 1990. With the abandoning of the active mining the Wismut GmbH faced the remediation of, int. al., about 1500 km open mine workings, 311 million m^3 of mine dump material and 160 million m^3 of radioactive sludge in Thuringia and Saxony. This unprecedented and also world`s largest remediation project has its particular specificity in the complexity of different environmental impacts, the immediate proximity to the population and the demand for very long-term stable remediation solutions. These remediation concepts and projects have largely been developed and implemented in-house effort and in close cooperation with national and international engineering firms, consultants and government agencies.

B.J. Merkel and A. Arab (eds.), *Uranium – Past and Future Challenges*,
DOI 10.1007/978-3-319-11059-2_53

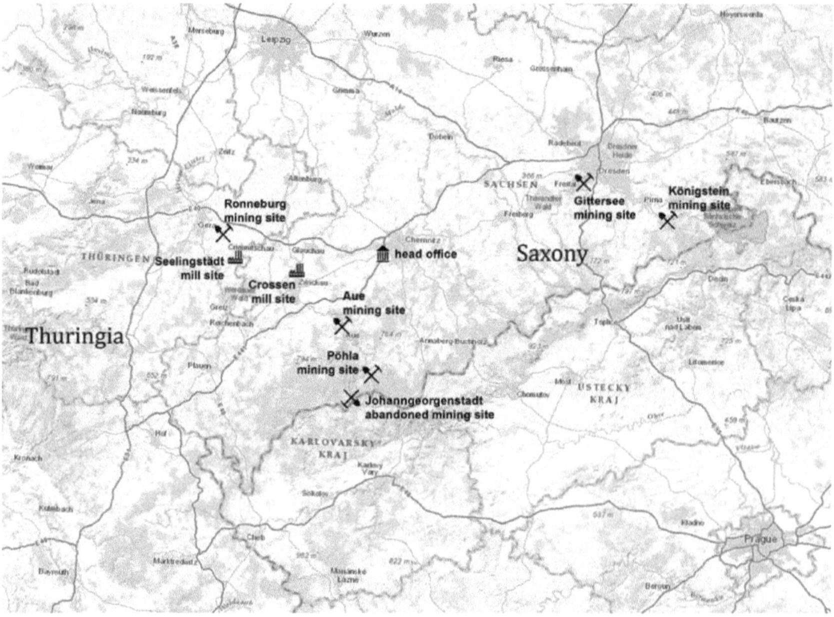

Fig.1. Map of the Wismut´s uranium mining and milling areas (Barnekow & Paul 2013)

To proof the remediation results since 1994 test fields with different cover systems (combinations of materials, thicknesses and installation parameters) were built and equipped with hydrological instrumentation.In 2000 the Wismut GmbH and UGT GmbH have started an extensive field test and monitoring program to observe the water and gas balance of the different cover systems. However, the results of those measurement campaigns are not only used for securing the above mentioned residues, but also for the effective implementation of this and other remedial measures. They also provide fundamental insights for future remediation projects.

Demands for cover systems

Based on the precursory environmental assessments object-specific demands have been developed for the respective cover systems.
They have to meet the following fundamental criteria:

- Reduction of the infiltration of precipitation and the oxygen influx as well as minimization of radon exhalation
- Compliance with the regulations concerning the radiological impact on the general public through the air path respectively the direct radiation –
- Providing sufficient root and nutrient conditions for the vegetal cover with respect to long-term utilization concepts (modelled >1000 years!)
- Guaranteeing stability and erosion protection

- If sealing layers are used, prevention of its drying (cracking) and root penetration as well as provision of a frost-free area of retreat for fossorial animals by a sufficiently heavy storage layer

Methods

To be able to compare different types of cover systems measurement sites have been installed at the former mining area. Cover systems according to storage and evapotranspiration concepts, with capillary barriers and with sealing layers have been compared.

The water balance of the cover layer here is a precondition for their functionality. On the other hand the components of the water balance proof how well the cover layer meets the demands. Therefore the measurement sites focus on hydrologic measurements. Essential measurement methods are:

- Free-drainage lysimeters; catch basins with a surface of 1 m² bis 300 m² lined with foil
- Lysimeters with controlled tension; ceramic plates with a 30 cm diameter
- Drainage and runoff catchments capturing the interflow
- Pressure transducer tensiometers and TDR-probes
- Gas lances to monitor the oxygen content in the soil

In the Ronneburg region with the open pit mining site Lichtenberg, the waste rock dump (WRD) Beerwalde and the industrial tailings ponds at Seelingstädt the measurement sites were mainly planed and installed in cooperation with the UGT GmbH.

Example WRD Beerwalde in the Ronneburg region

The WRD Beerwalde is a mining tailings pile of the Wismut in the Ronneburg region. Heaping together the tailings from Drosen, Korbußen and Beerwalde formed a terraced heap with four plateaus and a total height of about 40 m.

In large parts the stockpile is covered with a two-layer cover consisting of a 1,5 m heavy storage layer and an underlying 0,4 m heavy sealing layer out of clayey loam. Other than this standard assembly there is a small forest area at the foot of the stockpile with a cover of a 1,2 m heavy cover layer directly on the compacted tailings material.

The climatic conditions at WRD Beerwalde are temperate humid with an average annual precipitation rate of 715 mm (452 mm to 1086 mm).

Table 1. Characteristics of the Beerwalde WRD

Fact	Value
Location	Ronneburg / Thuringia
Mean Precipitation	715 mm
Altitude m asl	318...360 m
Volume	9,6 Mio m³
Footprint	33,4 ha
Slope Ratio	1:3
Exposition	In all directions
Remediation work	1991 - 2003
Cover type	Two-layer cover
Cover thickness	1,9 m
Vegetation	Grass and forest

Fig.2. Aerial photo of the Beerwalde WRD with the locations of the hydrological monitoring stations and cover types placed

The Beerwalde WRD is the first stockpile with a completed cover in the Ronneburg area. The chosen cover type therefore has pilot characteristics for following remediation projects. To test and verify the effectivity of this cover system seven monitoring stations for the determination of soil hydrological and meteorological parameters have been installed.

The configuration of each measurement station is adapted to its location. Especially site exposition and inclination of the cover have been considered.

Measurement technology and sensor systems

To monitor the water balance preferably sensor systems and measurement technologies according to GDA-guidelines have been used. For meteorologic measurement stations these are rain gauges, air temperature and humidity sensors, wind direction and wind speed sensors and global radiation sensors. For soil hydrologic measurement sites the installed technology comprises tensiometers and equitensiometers for tension measurement, TDR-sensors for the volumetric water content and soil temperature sensors. Further three different types of lysimeters are used according to the components to be measured. Surface lysimeters capture the specific surface runoff of an inclined surface with a defined area of 1 m². Soil lysimeters capture the lateral hypodermic flow of an inclined sealing or storage layer. The inlet of the lysimeter has a width of 1 m and a height of 0,3 m. The lateral flow from the direction of flow is not limited. The measured flow rate is therefore valid for the given height of the slope position. Large scale lysimeters capture the vertical seepage flow within the tailings material. The catchment area is according to GDA-guidelines 50 m². Those lysimeters reach up to 0,3 m underneath the lower boundary of the cover and are overlain by the cover layer and the vegetation according to their location. The runoff from the lysimeters is captured in one joint gauge well using tipping counters.

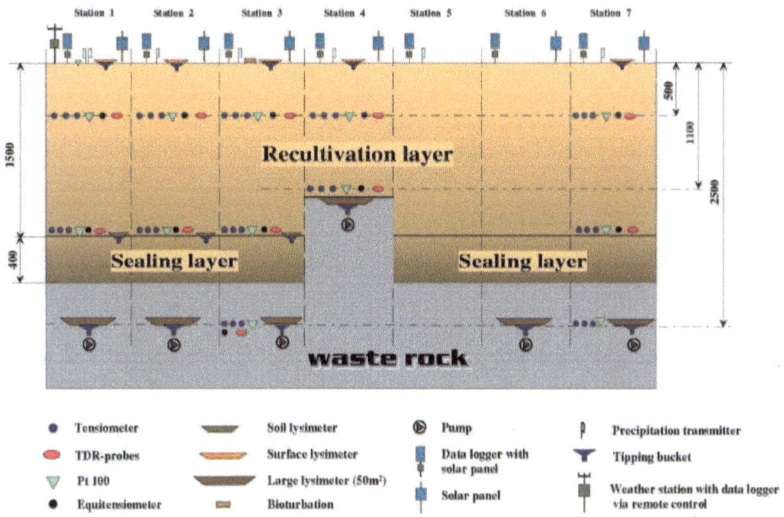

Fig.3. Graphic presentation of the instrumentation of a measurement network using the example of the Beerwalde WRD

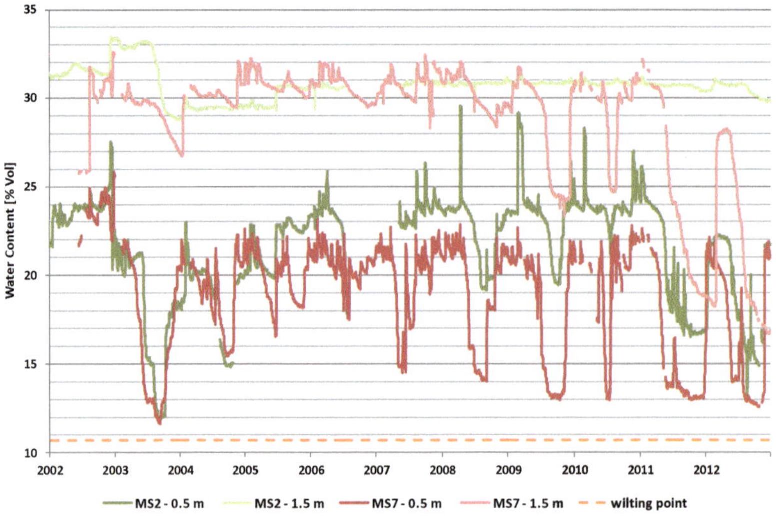

Fig.4. Development of the water content at station MS2 and MS7

Results for the Beerwalde WRD

Fig.4 shows the development of the water content at the station MS2 and MS7 in 0,5 m and 1,5 m depth. Just one year after the completion of the cover there is a clear annual cycle, which expresses itself even more strongly about the shown 10 years developed. Especially for MS7the water content gets very close to the wilting point during the summer months. Since 2009, this annual cycle also gets more pronounced in 1.5 m. This shows the increasing drainage caused by the growth of the vegetation since 2006. This shows the increasing drainage caused by the vegetation growth since 2006 and the associated root growth. Additionally the influence of the different exposition and vegetation can be seen. In contrast to MS7 the development of the water content in 1,5 m at MS2 is almost constant. The comparison of the different measurement sites allows to observe the influence of vegetation and exposition.

Due to the different water contents in the cover systems, different exposition and state of the establishment of vegetation the local annual percolation rate varies typically. At station MS 4 with a 1.2 m single storage layer and established mature forest vegetation the percolation varied from 3 mm in 2012 to 230 mm in 2002 (percolation rate of 0.4% to 28.6%) due to the annual precipitation rate during the reported period. The annual percolation rate at the stations in the two-layer-cover-system is usually lower than at MS 4. But there are also differences between the measured percolation rates at the MS 2 and MS 7. At station MS 2 exposed to the north the annual percolation rate did not rise above 1.4% of the annual precipitation rate during the entire reported period from 2002 to 2010. Since 2011 the percolation rate has increased to a maximum value of 43 mm per year (percolation rate of 6.9%). In contrast the annual percolation rate of MS 7 was increasing with time until 2010. The maximum percolation rate of 173 mm was measured at the MS 7 in 2010. In 2011 and 2012 the percolation decreased here to a minimum. The reasons are the minor and equal distributed precipitation combined with the high evapotranspiration of the developed forest vegetation. The percolation rates measured at the other monitoring stations are comparable with the results of MS 2 and MS 7 but influenced by variant exposition.

Table 2. Measurement Results for percolation (RU) of selected Lysimeters at Beerwalde

Year	2002	2003	2004	2005	2006	2007	2008	2009	2010	2011	2012	Ø
P [mm]	804	450	550	540	543	860	624	875	950	544	628	670
RU MS 2 [mm]	4	4	2	1	6	0	8	8	9	19	43	10
[%]	0,5	0,9	0,4	0,2	1,1	0	1,3	0,9	0,9	3,4	6,9	1,5
RU MS 7 [mm]		10	8	25	24	73	31	119	173	43	3	517,6
[%]		2,2	1,5	4,6	4,4	8,5	5,0	13,6	18,2	7,9	0,5	
RU MS 4 [mm]	230	107	48	78	77	63	100	85	204	98	3	86
[%]	28,6	23,8	8,7	14,4	14,2	7,3	16,0	9,7	21,4	17,1	0,5	12,8

The average percolation rate for the cover system at Beerwalde WRD amounts 8% of precipitation for the reported period including the results of all monitoring stations. So the design criterion of the two-layer cover to reduce the infiltration into the mine wastes to less than 5% of incident rainfall isn`t met.

The surface runoff RO and the lateral interflow RH is decreasing continuously during the reported period. The measurements directly after cover construction works (2002/2003) showed an average RO and RH of about 20% of precipitation. Only 1%...9% RO and RH were measured at the end of the reported period.

Summary of Results / Conclusions

From the continuous measurements since 2000 on the heaps of the Wismut GmbH in Ronneburg important conclusions regarding the water balance processes of the cover systems could be drawn.

Essential results are:

- Interflow can also occur in recultivation layers of one or two layer cover systems
- A notable percentage of the flow is partially due to preferential flow paths

- The ET evapotranspiration cover concept can provide an alternative to conventional cover concepts if a deep seepage of 10% to 20 % is admissible
- A crucial aspect of the methodology of lysimeter measurements is the extend of the catchment area
- The exposition and the vegetation status significantly influences the water balance of the cover and hence the percolation rate

The results gained were and are partially integrated in approval procedures for different remediation projects of the Wismut GmbH. The experiences with the measurement methodology are used for the long term monitoring of the remediated mining sites.The methods used, especially capturing the water flow with drainages and lysimeters, play an important role in the long term monitoring of the remediation results.

References

Schramm A., Roscher M. (2013) Two-Layer Soil Covers on Selected Radioactive Waste Rock Dumps at Wismut – Results of more than 10 years of hydrological monitoring. Mine Closure 2013

Barnekow U., Paul M. (2013) 15 Years of Design, Construction and Monitoring of Soil Covers on Wismut´s Uranium mining legacy sites – A Synopsis. Mine Closure 2013

Seyfarth M., Hoepfner U., Neubert G., (2006) Rehabilitation of uranium mining sites of WISMUT: Lysimeter measurements. Poster presentation 18th WCSS

UGT GmbH. (2002) Projekt-Dokumentation zur Instrumetierung eines Monitoringsystems auf der Halde Beerwalde

The New European Radiation Protection Safety Standards as Basis to Asses the Radiological State Achieved at Remediated Uranium Legacy Sites (WISMUT Sites) in Germany

Peter Schmidt[1], Jens Regner[1]

[1]Wismut GmbH, Jagdschänkenstr. 29, D-09117 Chemnitz, Germany

Abstract. In December 2013, the Council of the European Union adopted new Basic Safety Standards (BSS) for radiation protection (RP) which Member States are required to transpose into national law by 6 February 2018. Given that the remediation of uranium mining legacy sites in Germany (WISMUT sites) will need to continue well beyond 2018, the questions arises whether the new standards will have implications for WISMUT in terms of practical radiation protection and radiological assessment of the remedial success. Therefore, the paper contrasts current radiation protection concepts with the BSS and comments to case examples.

Introduction

In the light of new scientific information on the effects of ionizing radiation on human beings and on the environment, the European Union has revised previous radiation protection directives and adopted new Basic Safety Standards for radiation protection (EC 2013). Member States are required to bring into force new national laws and regulations to comply with this directive by 6 February 2018.

The new BSS follow the exposure situation based concept introduced by the ICRP in its Publication 103 (ICRP 2007). This approach distinguishes between existing, planned and emergency exposure situations. In addition to new regulations regarding exposures by artificial radiation sources, the management of natural radiation sources, i.e. radionuclides from the Uranium and Thorium series, K-40, and cosmic radiation as well as the related exposures are most notably regulated.

As early as 1991, the legacies of uranium mining and processing operations in eastern Germany had been classified by the German Commission on Radiological Protection as an existing exposure situation (SSK 1992). In the absence of applicable regulations at the time of German reunification, it was decreed that the regulations on radiological protection of the former GDR would continue to be applied to the remediation of the WISMUT sites. The respective ordinances (VOAS 1984,

B.J. Merkel and A. Arab (eds.), *Uranium – Past and Future Challenges*,
DOI 10.1007/978-3-319-11059-2_54

HaldAO 1980) remain in force even almost 25 years after German reunification. Admittedly, they fall out of compliance with current radiological protection concepts regarding the management of naturally occurring radioactivity.

RP regulations governing WISMUT compared to the EU BSS

The subsequent comparison highlights those radiation protection regulations introduced by the new EU Basis Safety Standards which deviate relevantly from the current regulations applicable to the WISMUT Project. Direct comparison is by the enumeration order and the link at the respective item number „#", i.e. regulation WI-# applicable to the WISMUT Project is related to BSS regulation BSS-#.

RP regulations under the WISMUT Project

The continued application of the regulations of the former GDR is stipulated by §118 of the Federal German Radiation Protection Ordinance (StrlSchV 2012). However, §118 also stipulates that the respective paragraphs of the actual RP Ordinance (StrlSchV 2012) shall apply to occupational radiation protection as well as to emission and pollution control. In addition to the statutory rules there are recommendations issued by the German Commission on Radiation Protection (SSK 1992) as well as guidelines drawn up by the Federal Ministry of the Environment (e.g. BMU 1999) which complete the set of rules and regulations governing the remediation of uranium mining legacies in eastern Germany.

WI-1: Management of U mining legacies is regulated by separate laws and regulations. To some extent, these deviate from regulations under the federal Radiation Protection Ordinance (e.g. in terms of exemption levels for discharges and of clearance values for the reuse of naturally-occurring radioactive materials).

WI-2: Having regard to the nature of an existing radiation exposure, the SSK recommended a primary reference value of 1 mSv per year for the use of surface areas, buildings, and mine dumps contaminated by uranium mining operations (SSK 1992). The level of 1-mSv/a is considered as a mining-induced effective dose to the population, i.e. as dose in addition to the natural background.

WI-3: Under the WISMUT Project, by now more than 1,300 object and project-specific radiation protection licensing procedures referred to the 1 mSv/a level as criterion for the justification of the remediation action. At the same time the level of 1 mSv/a was also taken as the remediation goal.

WI-4: The remediation goal of < 1 mSv/a embraces all exposure pathways. Likewise it is to be understood as the sum of all effective doses caused by all sources at the site. For this reason, also effective dose contributions by radon and its progenies from the various legacies at a site have to be taken into account upon approaching the remediation goal of „1 mSv/a".

WI-5: No reference values for indoor radon concentration levels are set. Mining-related outdoor radon concentration levels are added to indoor radon concentration levels. In the same way as doses from all other exposure pathways, effective doses resulting from indoor and outdoor radon concentration levels, as a total, are subject to comparison with the reference value of 1 mSv/a.

WI-6: According to (VOAS 1984), solids with a specific activity of an individual nuclide in excess of 0.2 Bq/g are radioactive materials. Management of radioactive materials from uranium mining and uranium ore processing is subject to licensing under radiation protection regulations.

WI-7: There are no separate regulations concerning the specific activity of construction materials. Potentially existing elevated radioactivity in buildings is evaluated by means of exposure pathway analysis and comparison of the findings with the 1 mSv/a reference value.

WI-8: A reference value of 1,000 Bq/m^3 is applicable to workplaces of the general public exhibiting elevated radon concentration levels. Radon concentration levels at the workplace of occupationally exposed staff of Wismut GmbH are evaluated by the assessment of the effective dose. Total dose contributions including those due to radon progeny must not exceed the dose limit value of 20 mSv/a.

WI-9: Overall responsibility for all RP issues within the environmental restoration company rests with the radiation protection supervisor. As a rule, this is the top manager of the company (director, chief executive officer, etc.). The RP supervisor shall appoint in written form radiation protection officers to supervise and perform relevant RP tasks and notify the authorities of that appointment. In accordance with §§ 33-35 of the Radiation Protection Ordinance (StrlSchV 2012), RP officers have extensive legal powers to enforce high radiation protection standards.

RP regulations provided by the EU Basic Safety Standards

With regard to items **WI-1** to **WI-9**, the new EU BSS stipulate:

BSS-1: Protection against natural radiation sources is no longer addressed separately in a specific title, but fully integrated within the overall RP requirements. Industries processing materials containing naturally occurring radionuclides should be managed within the same regulatory framework as other practices.

BSS-2: Member States shall establish reference values for existing exposure situations; their choice shall take RP requirements as well as societal criteria into account. EU BSS set reference levels in the range of 1 to 20 mSv/a for existing exposure situations.

BSS-3: Interpretation of reference values as remediation goals is not conclusively derivable from the European Basic Safety Standards. Rather, exposures above the reference level shall be given priority in optimization of radiation protection. Optimization is to be continued below the reference level.

BSS-4: For existing exposure situations involving exposure to radon, reference levels shall be set in terms of radon activity concentrations in air both for members of the public and for workers.

BSS-5: For indoor exposure, whether in new or existing buildings, the reference level to be established by Member States for the annual average radon concentration in air shall not exceed 300 Bq/m^{-3}. Under the national action plan to be established by Member States, identification of dwellings with radon concentrations exceeding the reference level of 300 Bq/m^3 and reduction of encountered concentrations by technical or other means shall be promoted.

BSS-6: EU BSS decree equal treatment (and thereby an equal footing) of exemption and release values. Values for exemption or clearance for naturally occurring radionuclides in solid materials in secular equilibrium with their progeny are:

- 1 Bq/g for natural radionuclides from the U-238 series;
- 1 Bq/g for natural radionuclides from the Th-232 series;
- 10 Bq/g for K-40.

For the clearance of materials containing naturally occurring radionuclides and which result from practices in which radionuclides were processed for their radioactive or fissile properties, the establishment of clearance values shall comply with the dose criteria for clearance of materials containing artificial radionuclides. In regard to the recycling of residues from industries processing natural radioactive materials into building materials, specific requirements shall be introduced.

BSS-7: Member States shall ensure that the selection of building materials will keep the annual exposure due to gamma radiation below the level of 1 mSv/a. Types of building materials to be examined are specified in an indicative list.

BSS-8: Member States shall establish national reference levels for indoor radon concentration in workplaces. The reference level must not exceed 300 Bq/m^3. Workplaces where the radon concentration cannot be reduced to below the reference level shall be managed as planned exposure and shall involve corresponding consequences for the radiological monitoring of workers.

BSS-9: Undertakings where planned or existing exposures to ionizing radiation occur shall in the future seek advice from a radiation protection expert. The latter shall develop strategies and measures for the undertaking on how to ensure radiation protection or check the existing radiation protection system for compliance and efficiency. The undertaking shall be responsible for the implementation of strategies and measures proposed by the radiation protection expert.

Member States shall bring into force the laws, regulations and administrative provisions necessary to comply with the European BSS by 6 February 2018. Member States may establish more stringent regulations on a case-specific basis and with regard to national prevailing circumstances.

Basic commentary

Integration of protection against natural radiation sources into a coherent RP concept and the alignment of that concept on exposure situations are a laudable intention. In that context, the new EU BSS provide Member States with sufficient flexibility in taking national prevailing circumstances into account during translation into national radiation protection legislation.

In several items, the new EU BSS are less stringent than currently applicable regulations in the framework of the WISMUT Project. That pertains to the:

- exemption or clearance values for naturally occurring radionuclides in solids;
- reference values for the effective dose to members of the public in existing exposure situations;
- separate assessment of indoor and outdoor radon occurrence;
- recycling of materials from industries processing materials containing natural radionuclides for use among others as construction materials;
- role of radiation protection officers.

In some items, more stringent benchmarks are established which concern for example:

- radon levels at workplaces of the general public;
- the required competence standards in the undertakings and the role of radiation protection experts.

In this regard, it has to be emphasized that the European BSS can reflect the specific problems facing a large-scale project like the WISMUT environmental restoration operation only in general terms and that their implementation can only be accomplished in a modified way. By way of example, direct transfer of EU clearance values for naturally occurring radionuclides to the assessment of legacies for the purpose of deriving remediation justification would be not very helpful. Almost all large mine dumps at WISMUT sites on average exhibit specific activities of individual nuclides of the uranium decay series significantly below 1 Bq/g. Anyhow, these mine dumps, when left in an un-remediated state, generate effective doses to the general public significantly in excess of 1 mSv/a.

Case-related Commentary

Use of Limit Values and Reference Values

Both regulatory frameworks permit to classify the legacies left behind by uranium mining operations in eastern Germany as an existing exposure situation. In accordance with applicable ICRP principles of radiological protection, reference values supersede limit values in such existing exposure situation. Compliance with reference values has to be pursued, also with due regard to weighing economic and social aspects. In the sense of the ALARA principle (As Low as Reasonably Achievable), RP measures have to be optimized both above and below reference values.

It is current practice of the WISMUT Project to take reference values as remediation targets. Until now, failure to comply with remediation targets had not to be appraised on a large scale. Disputes about elevated indoor radon concentrations demonstrate however that in everyday life, reference values are taken as limit values, right up to legal battles. This is in contradiction to modern optimization approaches as outlined among others in the EU Basic Safety Standards.

Management of materials with elevated natural radioactivity

At the time of uranium production by SDAG Wismut, the RP legislation of the former GDR facilitated the use of waste rock and tailings materials with specific activities of > 0.2 Bq/g. Examples of such uses include the placement of waste rock in subbase applications for roads or the use of sandy tailings as construction aggregates. Typically, the materials used contained specific U-238/Ra-226 activities of < 1 Bq/g. As a consequence of these applications, unanticipated materials are again and again encountered during road work activities. According to current legislation these materials have to be classified as radioactive materials.

Management of materials containing specific activities of > 0.2 Bq/g, such as the the replacement or disposal comes within the provisions of RP legislation. This involves investigations to determine the specific activity and to proof in specific cases compliance with the 1 mSv/a level. Construction companies, municipalities but also private property owners are hence confronted with problems which they can only solve by appealing to external expertise. While this is a cost factor on the one hand, the disquiet of the persons concerned to be exposed to „radioactivity“ must not be underestimated, on the other. Commenting such incidences, the media follow suit. In a bid to promote objectification and provide guidance, the Free State of Saxony has edited a special publication (LfULG 2013).

The circumstances portrayed above are in contradiction to EU BSS establishing exemption or clearance values of 1 Bq/g for the nuclides of the U-/Th decay series.

Outdoor and Indoor Radon

From a radiation protection perspective, inhalation of radon and its progeny plays a prominent role at former uranium mining sites in Saxony and Thuringia. This is due to the close proximity of large mine dumps to residential estates on the one hand. On the other there are near-surface non-flooded mine workings which have a negative impact on the radon situation at the ground surface (both indoor and outdoor). Findings of exposure pathway analyses indicate that radon is the main contributor to the effective dose incurred by members of the public at virtually all the WISMUT sites.

With a view to attenuating the radon situation, large mine dumps in close vicinity to the Schlema-Alberoda site were capped under the WISMUT Project with a 1 m thick standard cover, and vegetation of the cover was favored. Vegetation was intended to control the cover's water balance and to blend the remediated mine dump into the surrounding landscape. Post-care measurements performed only a couple of years after remediation revealed that the radon attenuation effect of the covers which had been good (as predicted) at the time of their placement was increasingly deteriorating. Bioturbation (root penetration, impact of burrowing animals) and water consumption by root systems expanding within the cover were identified as causes for the deterioration. Increased rates of radon exhalation are again observed at the toes of the mine dumps which cause average annual radon concentrations of up to 200 Bq/m³. In pursuance of (BMU 1999) and proceeding on the assumption that the measured outdoor radon concentration is also taking effect indoors (see above: WISMUT ruleWI-5), the effective annual dose to occupants of dwellings at the toe of the mine dump would be in the order of ca. 4 mSv.

Taking the current RP regulatory framework as an assessment basis, WISMUT has failed to achieve the remedial goal. If, however, the new EU BSS were applied, an outdoor concentration of 200 Bq/m³ would be judged tolerable on condition that efforts to improve the situation would turn out to be out of proportion.

In recent years, WISMUT has conducted comprehensive investigations and looked for ways to improve the radon situation in the vicinity of capped mine dumps. The type of solutions WISMUT experts were looking for should avoid massive perturbation of the cover in place (known as low-invasive solutions). Since that search was not successful, Wismut GmbH now has to decide whether invasive approaches would have to be implemented (e.g. construction of radon barriers in the surroundings of mine dumps and the subsequent reinforcement of the cover). Strictly speaking, what has to be evaluated is (as in the past) an existing exposure situation. Decision-making requires an input from stakeholders and will have to apply optimization principles. In doing so, evaluation approaches may be derived from the EU BSS without shelving current evaluation principles.

Conclusion and Outlook

The new EU Basic Safety Standards provide Member States with a framework for action that grants them sufficient flexibility in taking their national specifics into consideration when implementing modern radiological protection. By its sheer size, the WISMUT Project has to be considered as a national particularity of Germany. Schematic transposition of EU BSS on the WISMUT Project, however, would erode radiological regulations applicable to the remediation of uranium mining legacies in Germany.

As follows from the foregoing, once the EU BSS will have been translated into new German RP rules and regulations, there will be need for a subordinate set of rules and regulations to meet the specific conditions of the WISMUT Project. A conceivable approach would be, first and foremost, the continued application of current rules and regulations and in particular of the RP legislation of the former GDR. In the light of current planning, physical work under the WISMUT Project will be completed in 2025. Even after that deadline, radiological protection rules and regulations will have to be applied, e.g. with regard to the long-term required discharge into receiving streams of treated effluents containing residual contamination levels. The need for RP regulations governing such discharges is currently still derived from (VOAS 84), that is to say derived from the radiation protection legislation of a country that has ceased to exist some 25 years ago.

As an alternative to the continued application of the GDR regulations, Germany might adopt a radiological legacy ordinance. Such legacy ordinance would have the advantage of not only covering the existing exposure situation of WISMUT, but in addition also other radioactive legacies generated by the management of materials with elevated naturally occurring radioactivity (e.g. from activities involving radium/thorium, from the mining and processing of raw materials other than uranium ores, etc.). On the basis of the new EU BBS, such legacy ordinance might lead to the harmonization of German RP legislation. During the forthcoming three years up to the translation of the EU BBS into national law, the pros and cons of both approaches would have to be weighted.

References

BMU (1999) Calculation Bases for the Determination of Radiation Exposure due to Mining-caused Environmental Radioactivity, German Federal Ministry for the Environment, Nature Conservation and Nuclear Safety (BMU), Bonn 1999

EC (2013) Council Directive 2013/59/EURTOM laying down the basic safety standards against the dangers arising from exposure to ionizing radiation, Brussels, 5 Dec. 2013

HaldAO (1980) Ordinance on the Safety of Waste Dumps and Tailings Ponds of the former German Democratic Republic, GBl, 34 (1) der DDR, 17. Dez. 1980 (in German)

ICRP (2007) Recommendations of the International Commission on Radiological Protection, ICRP Publication 103, 2007

LfULG (2013) Radioactive Materials in Building Operations, Saxony Office for Environment, Agriculture and Geology (LfULG), Schriftenreihe Heft 13/2013 (in German)

SSK (1992) Radiological Protection principles concerning the safeguard, use or release of contaminated materials, buildings, areas or dumps from uranium mining, German Radiation Protection Commission (SSK), SSK Publication 23, G. Fischer Verlag, 1992

StrSchV (2012) Ordinance on Protection Against Damages from Ionising Radiation of the FRG (Strahlenschutzverordnung - StrSchV), BGBl. I, S. 212 (2012) (in German)

VOAS (1984) Ordinance on Atomic Safety and Radiation Protection of the former German Democratic Republic, GBl., Nr.30 (1) der DDR, 21. Nov. 1984 (in German)

Geochemical controls on U immobilization in the subsurface

Malgorzata Stylo[1], Daniel Alessi[1], Shao Paul[1], John Bargar[2], Rizlan Bernier-Latmani[1]

[1]Environmental Microbiology Laboratory, École Polytechnique Fédérale de Lausanne, CH-1015, Lausanne, Switzerland
[2]Chemistry and Catalysis Division, Stanford Synchrotron Radiation Lightsource, SLAC National Accelerator Laboratory, Menlo Park, CA 94025, USA

Abstract. The implementation of engineered solutions to U contamination in the subsurface results in reductive immobilization of uranium in the form of U(IV) precipitates, including mineral phases, i.e., uraninite (UO_2), or non-crystalline biomass-associated U(IV) agglomerates. Crystalline uraninite is a preferable end product for U bioremediation due to its greater resistance to re-oxidation and remobilization processes in comparison to non-crystalline biomass-associated U(IV) species. Nevertheless, under field conditions, the latter species was reported to be predominant. According to our findings, the groundwater composition exerts significant control over the product of non-crystalline U(IV) species in the field.

Introduction

Uranium, like other radionuclides, is not biodegradable; therefore engineered solutions to U contamination in the subsurface include methods that decrease its mobility and bioavailability in the environment. One of the proposed approaches involves the stimulation of indigenous microorganisms by the addition of an electron donor to the subsurface (Anderson et al. 2003). This promotes the microbially driven reduction of highly soluble U(VI) to sparingly soluble U(IV) (Lovley et al. 1991), and thus decreases its mobility and bioavailability in the aquifer (Borch et al. 2010). The biostimulation process results in the immobilization of U(IV) species, including mineral phases, i.e., uraninite (UO_2), or non-crystalline biomass-associated U(IV) agglomerates.

In the literature, what is referred to as "bio-UO_2" (microbially produced uraninite), is actually a mixture of both species: crystalline uraninite and amorphous non-uraninite U(IV) (Alessi et al. 2012). Laboratory-scale research demonstrated that it is possible to control the ratio of the two U(IV) species (Bernier-Latmani et

B.J. Merkel and A. Arab (eds.), *Uranium – Past and Future Challenges*,
DOI 10.1007/978-3-319-11059-2_55

al. 2010). In a simple medium (containing only bicarbonate and PIPES buffer, referred to as BP) an approximately 50/50 mixture of the two species results from U(VI) reduction mediated by the metal-reducing microorganism *Shewanella oneidensis* MR-1. On the other hand, biomass associated non-uraninite U(IV) species are formed predominantly by the same microorganism in a more complex salt-rich medium (referred to as WLP). Non-uraninite species, as the name suggests, are the U(IV) species that lack the crystalline form of the mineral uraninite. That includes amorphous U(IV) phases, characterized by a decrease in order or the total absence of the U-U pair correlation at 3.8 Å, as observed in Fourier transformation of Extended X-ray Absorption Fine Structure Spectra (EXAFS) (Bernier-Latmani et al. 2010, Boyanov et al. 2011). In addition, previous studies implied that non-uraninite U(IV) species form U(IV) complexes bound to biomass, through P-containing ligands (Bernier-Latmani et al. 2010).

When effectiveness of the bioremediation strategy applied to the contaminated site is considered, crystalline uraninite is a preferable product of U reduction in the field due to its higher resistance to the re-oxidation and re-mobilization. Nevertheless, the studies conducted at the U-contaminated sites (Kelly et al. 2008, Bargar et al. 2013) report the predominant formation of non-crystalline biomass-associated U(IV) as a result of applied biostimulation. Therefore, the main goal of this study was to determine the geochemical conditions that lead to formation of non-uraninite U(IV) species in environmental settings and to identify controls on its formation. To better characterize these controls, we used controlled laboratory settings, to test the impact of various geochemical conditions (modulated by addition of several groundwater relevant solutes) applied during microbial U reduction on the final product of its enzymatic reduction. The motivation for this study was to acquire insight into U(IV) speciation during bioremediation of contaminated field sites. The ultimate goal was to modulate the final U(IV) product by altering geochemical conditions.

Material and Methods

Microbial growth and Uranium reduction conditions

Biomass associated uranium U(IV) was produced as previously described in Bernier-Latmani *et al* (2010). *Shewanella oneidensis* MR-1 cultures were grown in sterile LB medium until they reached mid-exponential phase. Cells were harvested by centrifugation at 8,000 g for 10 minutes and washed in simple BP medium, composed of 30 mM $NaHCO_3$ and 20 mM 1,4-piperazinediethanesulfonic acid (PIPES buffer) adjusted to pH 6.8. Washed cells were suspended to an optical density (OD_{600}) of 1.0 in BP medium, amended with corresponding solutes as described in Table 1, or in filter-sterilized groundwater from a U-contaminated site at Old Rifle, Colorado (RGW) (Table 1).

Design of experiments

In order to identify conditions that lead predominantly to the formation of non-uraninite U(IV) species, we conducted U(VI) bioreduction experiments in BP medium amended with individual solutes or mixtures thereof. Duplicate reduction batches with the appropriate solute composition were amended with 20 mM lactate as an electron donor and varying concentrations of U(VI) acetate (100 and 400 μM) under anoxic conditions. In addition, batch duplicates containing RGW were prepared. The initial U(VI) concentration was lowered in all batches containing phosphate, in order to avoid U(VI) precipitation. The details of all the experimental batches are listed in Table 1.

Uranium quantification

Subsamples were taken over time from each batch to quantify dissolved uranium. Samples were filtered through 0.22 μm membranes and retained solids were refrigerated until analysis. Afterwards subsamples were diluted in 0.1 M nitric acid and measured by inductively coupled plasma optical emission spectrometry (ICP-OES).

X-ray Absorption spectroscopy (XAS)

Uranium X-ray absorption spectroscopy measurements of samples were conducted at beamline 4-1 of the Stanford Radiation Lightsource (SSRL) at the Stanford Accelerator Laboratory (SLAC). Each sample was maintained in a liquid N_2 cryostat during analysis, and kept under a vacuum of approximately 10^{-6} Torr in order to preserve anoxic condtions. X-ray absorption spectra were collected at uranium at the U L_{III}-edge (around 17,166 eV), in fluorescence and transmission modes. A double-crystal Si (220) monochromator was used to select the energy over each sweep, and was slightly detuned to reject higher energy harmonics. An yttrium metal foil was used for the initial and internal calibration, with its first derivative set to 17,038.4 eV. EXAFS spectra were processed using the SixPACK and Athena analysis packages. Backscattering phase and amplitude functions used to fit the spectra were generated in Artemis using FEFF6L. The EXAFS spectra were also fit by linear combination fitting (LCF) using previously characterized compounds (reference samples): crystalline uraninite species, non-uraninite U(IV) and U(VI) adsorbed.

Scanning Transmission X-ray Microscopy (STXM)

Carbon speciation and uranium localization analyses by scanning transmission X-ray microscopy (STXM) were conducted on the STXM end station 11.0.2.2 of the Molecular Environmental Science beam line 11.0.2 at the Advanced Light Source (ALS) with the synchrotron storage ring operating at 1.9 GeV and 500mA multibunch mode storage current. Carbon speciation stacks were collected through serial image collection along C K-edge energies (280 – 300 eV). Uranium localization maps were collected at 725 and 738 eV below and at the edge, respectively. Carbon reference spectra of albumin (protein), dipalmitoyl-sn-glycero-3-phosphocholine (DPPC) (lipid) and alginic acid (polysaccharide) were recorded along the C K-edge (280-300 eV). The uranium reference spectrum was collected along its $4d_{5/2}$ edges (720 – 780 eV), capturing both peaks in that range. All the spectra were normalized to an optical density (OD) corresponding to a 1-nm layer.

Table 1. Sample list, describing geochemical conditions of each sample (presence/absence of specific solutes) and concentration of added uranium.

Sample name	*U concentration (μM)*	*$SO4^{2-}$ (mM)*	*$SiO4^{4-}$ (mM)*	*Mg^{2+} (mM)*	*Ca^{2+} (mM)*	*$P04^{3-}$ (mM)*	*Rifle Ground Water*
BP 100	100	-	-	-	-	-	-
BP 400	400	-	-	-	-	-	-
Ca	400	-	-	-	7*	-	-
Mg	400	-	-	5*	-	-	-
S	400	10*	-	-	-	-	-
Si-1	400	-	0.3*	-	-	-	-
Si-2	400	-	1	-	-	-	-
P	100	-	-	-	-	0.2	-
Ca/P	100	-	-	-	-	2.1	-
RGW	400	-	-	-	-	-	+

Results and Discussion

Role of specific solutes

The five chosen solutes for our experiments (calcium, silicate, magnesium, sulfate and phosphate) were selected either due to their environmental relevance to the U-contaminated site at Old Rifle, in Colorado (USA), or due to their importance based on previous laboratory experiments. Calcium, silicate, magnesium and sulfate are present in Old Rifle groundwater (RGW, Table 1) at concentrations of 7 mM, 0.3 mM, 5mM and 10 mM, respectively (Campbell et al. 2011). Additionally, calcium is well known for its property to inhibit or decrease the rate of micro-

* Environmentally relevant concentration, based on Rifle Ground Water (Colorado, USA, Campbell et al, 2011)

bial U(VI) reduction (Brooks et al. 2003). Phosphate, on the other hand, has been shown to be important in non-uraninite U(IV) formation. A recent report showed that the formation of non-uraninite U(IV) [also referred to mononuclear U(IV)] was dependent on the presence of phosphate - for several strains (*Desulfitobacterium* spp. and *S. putrefaciens*) (Boyanov et al. 2011).

The outcome of our experiments is summarized in Table 2., where we estimated the contribution of uraninite and non-uraninite U(IV) fraction to the products of microbial U reduction under varying geochemical conditions, by fitting the EXAFS spectra to the standard spectra (LCF). We show that the presence of **calcium** and **magnesium** during microbial U reduction does not alter the U(IV) product composition (a similar ratio of uraninite vs. non-uraninite U(IV) to the baseline product bio-UO_2 formed in the absence of any solute). In the presence of **silicate** and lower concentration of initially present U, an increase in the non-uraninite U(IV) fraction is observed. The presence of **sulfate** or **phosphate** results in the formation of almost pure non-crystalline U(IV) species following microbial U reduction.

Table 2. Linear combination fits (LCF) of EXAFS spectra for the studied conditions: with included values of each uranium fraction formed under varying geochemical conditions. Errors for each value is presented in the brackets (last digit variation) and represent the variation calculated by the fitting program.

		uraninite	*non-uraninite*	*U(VI) adsorbed*
mixture of UO_2 and non-uraninite	**BP 400**	0.46 (2)	0.53 (3)	0 (+/- 0.03)
	Mg	0.46 (2)	0.52 (3)	0.02 (3)
	Ca	0.44 (2)	0.53 (3)	0.03 (4)
growing fraction of non-uraninite	**Si-1**	0.28 (1)	0.69 (3)	0.03 (3)
	BP 100	0.15 (1)	0.85 (3)	0 (+/- 0.03)
	Si-2	0.16 (2)	0.84 (3)	0 (+/- 0.04)
	RGW	0.12 (3)	0.85 (3)	0.03 (3
almost pure non-uraninite	**S**	0.09 (2	0.90 (3)	0 (+/- 0.04)
	P	0	0.95 (0)	0.05 (0)
	Ca/P	0.01 (2)	0.01 (2)	0.01 (2)

Tested solutes were shown to influence the U(IV) product to a different degree. Environmentally-relevant concentrations of **cations** – calcium and magnesium - were shown to have no effect on the U(IV) product composition. In contrast, the presence of **anions** was revealed to have a significant influence on the U(IV) product composition. The environmentally-relevant concentration of sulfate (10 mM) increased the contribution of non-uraninite U(IV) from 55% to 90%. The presence of 0.3 mM silicate, on the other hand, results in a relatively small increase of non-uraninite U(IV) species (from 55% to 70%). However, a higher concentration (1mM) impacted the fraction of non-uraninite U(IV) significantly (increasing it from 55% to 85%). Despite its low concentration in the Rifle groundwater, silicate is of interest due to its ability to diminish uraninite corrosion by slowing down the rate of uranium loss (Campbell et al. 2011), and therefore it should be taken into account when considering environmental conditions at U-contaminated sites. Additionally, even low concentrations of phosphate promote the predominant formation of non-crystalline U(IV) species (95%).

Mechanism of preferential non-crystalline U(IV) formation

Despite the clear relationship between the presence of solutes and the preferential formation of non-crystalline U(IV) species in the subsurface, the mechanism of this interaction remained unclear. An initial hypothesis suggested direct involvement of solutes by association with bacterial cell wall, and alteration of their surface properties, precluding uraninite formation. Similarly, association of phosphate with U(IV) may preclude the precipitation of UO_2 in a manner akin to the delayed transformation of less structured ferrihydrite into crystalline phases due to phosphate (Paige et al. 1997). In order to test these hypotheses, we tested the behavior of dissolved solutes during the course of U reduction. Results, based on the measurements of solution concentrations of selected solutes during U(VI) bioreduction, revealed that the concentrations remained constant over time (results not shown, more details in Stylo et al., 2013) suggesting that there is limited association of these ions with biomass and/or U(VI) and the formed U(IV) species.

The alternative hypothesis pointed to the presence of certain solutes as controls on the U(IV) end product by imposing an indirect influence on the biomass. As a result, we used Scanning Transmission X-ray Microscopy (STXM) to monitor biomass components (carbon species) and at the same time localize U species in a given sample. We sought evidence of extracellular polymeric substances (EPS) production and found little EPS production by microorganism when the solutes were absent (BP background) and extended EPS production when impactful solutes were present (for example phosphate) (Fig. 1). Moreover, the formed non-crystalline U(IV) species were co-localized with EPS when solutes were present, resulting in enhanced cell viability. In contrast, in the absence of solutes, U(IV) accumulated on the bacterial cell walls, resulting in diminished cell survival rates.

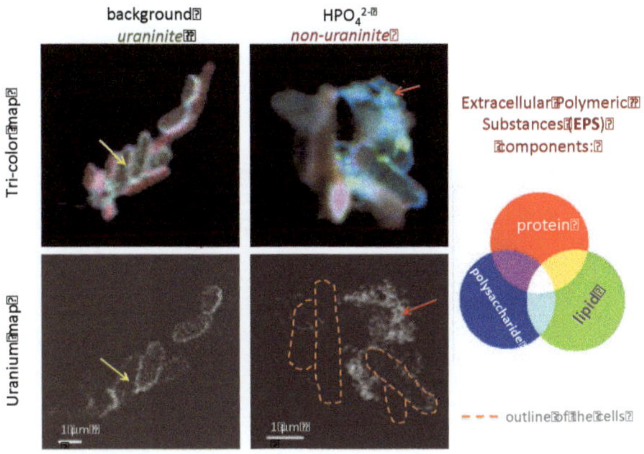

Fig.1. STXM images showing uranium map and tri-color map (indicating three main components of EPS: protein in red, lipid in green and polysaccharide in blue) for U(IV) product (modified from Stylo et al., 2013)

In conclusion, the geochemical composition of the aqueous phase (i.e., the presence of specific solutes) impacts biomass by modulating EPS production, which, in turn, impacts U(IV) speciation and cell survival. For that reason, it is not surprising that solutes typically associated with biological activity (sulfate and phosphate) impact non-uraninite U(IV) species formation. Because silicate is considered a non-essential element for microbial nutrition, it is uncertain why its presence would result in an improved physiological state that allows for EPS production.

Environmental Implications

Before planning bioremediation interventions at U-contaminated sites it is essential to understand the geochemical and biological properties that control the preferential formation of non-crystalline U(IV) over crystalline uraninite under specific field conditions. The conclusions from this laboratory-scale study enabled us to identify the role of particular solutes in the formation of non-crystalline U(IV) species. This knowledge can be further applied to field investigations, allowing for a more accurate assessment of environmental conditions and ongoing processes promoting bioremediation as well as challenges faced at specific U-contaminated sites. For example, bioremediation interventions applied at a U-contaminated site where limestone is the predominant bedrock, technically should yield better results due to preferential uraninite formation in the presence of calcium and magnesium solutes (originate from dissolution of calcite and dolomite).

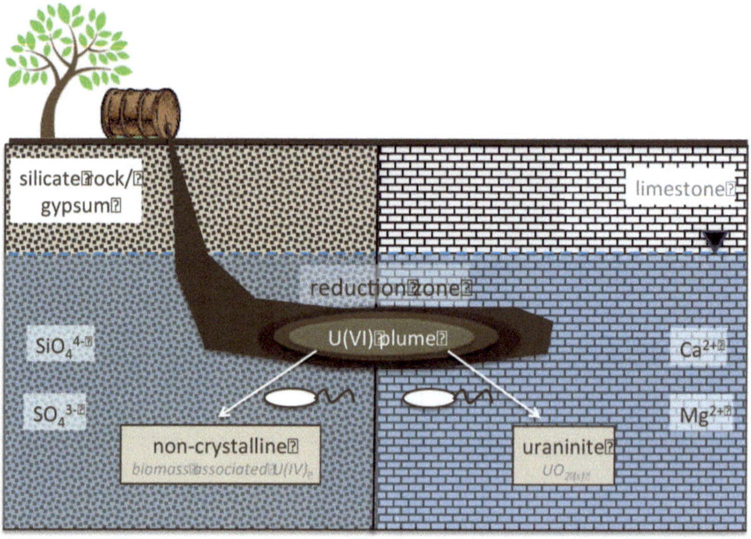

Fig.2. Schematic diagram summarizing environmental dependence of geochemical conditions at the field site and the end product of microbial U reduction (modified from Stylo et al. 2013)

On the other hand, in U-contaminated systems dominated by silicate minerals or gypsum (calcium sulfate), resulting in elevated concentrations of silicate and sulfate in the aquifer, the predominant formation of non-uraninite U(IV) species is to be expected. This environmental dependency exposes a fundamental weakness of the bioremediation approach in natural systems and underscores the need for extended research to further characterize non-uraninite U(IV) species, in particular their formation and transformation mechanisms and resistance to re-oxidation.

References

Anderson R. T., Vrionis, H. A. Ortiz-Bernard I., Resch C. T., Long P. E., Dayvault R., Karp K., Marutzky S., Metzler D. R., Peacock A. D., White D. C., Lowe M., Lovley D. R. (2003) Stimulating the *in situ* activity of Geobacter species to remove uranium from the groundwater of a uranium-contaminated aquifer. Appl. Environ. Microb. 69: 5884-5891

Lovley D. R., Phillips E. J. P., Gorby Y. A., Landa E. R. (1991) Microbial reduction of uranium. Nature 350: 413 – 416

Borch T., Kretzschmar R., Kappler A., Van Cappellen P., Ginder-Vogel M., Voegelin A., Campbell K. (2010) Biogeochemical Redox Processes and their Impact on Contaminant Dynamics, *Environ.* Sci. Technol 44: 15-23

Bernier-Latmani R., Veeramani H., Dalla Vecchia E., Junier P., Lezama-Pacheco, J. S., Suvorova E. I., Sharp J. O., Wigginton N. S., Bargar J. R. (2010) Non-uraninite products of microbial U(VI) reduction. Environ. Sci. Technol 44: 9456 – 9462

Kelly S.D., Kemner K.M., Carley J., Criddle C., Jardine P.M., Marsh T.L., Phillips D., Watson D., Wu W.M. (2008) Speciation of Uranium *in* Sediments before and after *In* situ Biostimulation. Environ. Sci. Technol. 42:1558-1564

Bargar J. R., Williams K. H., Campbell K. M., Long P. E., Stubbs J. E et al. (2013) Uranium redox transition pathways in acetate-amended sediments, PNAS 110: 4506-4511, 2013

Alessi D. S., Uster B., Veeramani H., Suvorova E., Lezama-Pacheco J. S., Stubbs J. E., Brgar J. R., Bernier-Latmani R. (2012) Quantitative separation of monomeric U(IV) from UO_2 in products of U(VI) reduction. Environ. Sci. Technol 46: 6150-6157

Boyanov M. I., Fletcher K. E., Kwon M. J., Rui X., O'Loughlin E. J., Löffler F. E., Kemner K. M. (2011) Solution and microbial controls on the formation of reduced U(IV) species. Environ. Sci. Technol 45: 8336 – 8344

Campbell K. M., Veeramani H., Ulrich K. U., Blue L. Y., Giammar D. E., Bernier-Latmani R., Stubbs J. E., Suvorova E., Yabusaki S., Lezama-Pacheco J. S., Mehta A., Long P. E., Bargar J. R. (2011) Oxidative dissolution of biogenic uraninite in groundwater at Old Rifle, CO. Environ. Sci. Technol 45: 8748 – 8754

Paige C. R., Snodgrass W.J., Nicholson R. V., Scharer J. M., He Q.H. (1997) The effect of phosphate on the transformation of ferrihydrite into crystalline products in alkaline media. Water, Air and Soil Pollution 97: 397-412

Stylo M., Alessi D. S., Shao P. P., Lezama-Pacheco J. S., Bargar J. R. (2013) Biogeochemical controls on the product of microbial U(VI) reduction. Environ. Sci. Technol 47: 12351-12358

Oooh, that paper fits exactly the scope of my research. I need to use it to improve my work.
Paper A
5 months later
Great! I finally found a paper that is perfect for my research!
Paper A
This is something I should investigate further. I put it in the "important" box

Impact of Uranium Mill Tailings on Water Resources in Mailuu Suu, Kyrgyzstan

Frank Wagner[1], Hagen Jung[2], Thomas Himmelsbach[1], Arthur Meleshyn[3]

[1]Federal Institute for Geosciences and Natural Resources, Stilleweg 2, 30655 Hannover, Germany
[2]NUKEM Technologies GmbH, Industriestrasse 13, 63755 Alzenau, Germany
[3]Gesellschaft für Anlagen- und Reaktorsicherheit (GRS) mbH, Theodor-Heuss-Straße 4, 38122 Braunschweig

Abstract. Residual waste dumps and tailings from previous U-mining activities in Mailuu-Suu represent a potential risk on local water resources. In 2006, a monitoring program has been initiated to determine the contamination status of local water resources, taken as a baseline to assess the impact of coming remediation activities. Field data supplemented by laboratory experiments shed light on uranium mobilization and transport path. The presented findings conclude into simple mitigation measures which have been introduced to local authorities and citizens.

Introduction

From 1946 to 1968, uranium ore has been mined and processed in Mailuu Suu (Kyrgyzstan). The resulting tailings and waste rocks have been deposited in nearby dumps together with ore material of inefficient low uranium content. Altogether, around 3 Million m^3 waste material (TACIS 2003) has been deposited in morphologic depressions and covered provisionally.

Due to their location in a tectonically active region, the stability of dumps and tailings is threatened by landslides, triggered by seismic events or seasonal heavy rains. Their radioactive inventory might be eroded and transferred into the river. Furthermore, landslides may block and retain the local river Mailuu Say and its tributaries such as Kulmin-Say - as already happened in 1992 and 2002 -, resulting in flooding of nearby dumps and tailing impoundments. Both scenarios are combined with erosion and solution processes and, thus, mobilization of the radioactive inventory. Beside affecting local water resources, mobilized radionuclides might be transported downstram beyond the Uzbek border.

Against this background, the World Bank has initiated a 17 Mio $ disaster hazard mitigation project (DHMP). In this context, the presented study was carried out in cooperation with the Kyrgyz Ministry of Emergency Situations (MOE) in

B.J. Merkel and A. Arab (eds.), *Uranium – Past and Future Challenges*,
DOI 10.1007/978-3-319-11059-2_56

order to provide a baseline to assess the contamination status of water resources. Major activities have been carried out 2006 to 2009, comprising implementation of a monitoring network as well as laboratory experiments. The remediation activities of DHMP project recently have been completed. Nevertheless, the threat from uranium mining residues is still reported to be high (Blacksmith Institute 2013). Therefore, the results of this study are still of actual relevance and useful for designing future remediation activities as well as assessing their impact.

Uranium mining in Mailuu Suu area

Predominating hard rocks in Mailuu-Suu valley are Cretaceous and Tertiary strata, generally consisting of fine grained sandstone and limestone sequences with intercalated marl layers. Holocene unconsolidated sediments of alluvial origin have been deposited in river valleys with a maximum thickness of up to 30 m. At outcropping Cretaceous rocks, crude oil has been observed to leak locally into the Mailuu-Say River ("Black Water").

Uranium mineralization is mainly developed within the limy strata of the upper Cretaceous and - to a lesser extend - in the lower tertiary strata. The dominating uranium ore mineral is *Uraninite* (UO_2), *Brannerite* (UTi_2O_6) and its alteration products *Carnotite* ($K_2(UO_2)_2V_2O_8{\cdot}3H_2O$), *Tyuyamunnite* ($Ca(UO_2)2V_2O_8{\cdot}5$-$8H_2O$). The uranium content of ore bearing rocks is reported to be relatively low with >0.5 % down to 0.03 %. Totally, roughly 10,000 tons of uranium was produced from mining activities in Mailuu-Suu. The grinded ore material has been processed applying both acidic as well as alkaline leaching techniques. SCHMIDT (1997) as well as local reports state that not only local uranium ores has been processed in Mailuu-Suu, but also material transported from other mining districts, such as Kasachstan or as far as from the german Ore Mountains.

23 tailing impoundments and 13 waste rock dumps have been deposited provisionally within morphologic depressions (location see Fig. 1). Generally, the design of tailing impoundments as well as dumps was carried out without consideration of a base sealing (UN 2001). Dewatering of the tailing impoundments is commonly realized by drainage tubes within the gravel front at the foot of the tailing impoundments. The collected seepage water is either directed into nearby rivers or infiltrates into the subsurface. Some of the drainage systems are observed to be out of function due to clogging (UN 2001). Furthermore, Waste rock and tailing material has been re-deposited in flooded mining excavations (ASSMANN & THOSTE, 2004).

Impact of mining residues on water resources

Hereafter, the presented findings are based on 50 sampling stations (Fig. 1) in context with scarce hydrogeological data about Mailuu-Suu area (Ostroborodov et al. 2003). Sampled water types comprise seepage from tailings, rivers and creeks as well as groundwater (deep and shallow aquifers) sampled from springs, dug wells, artesian wells and 11 new observation wells. The artesian wells are screened within Cretaceous or Tertiary hard rocks. Shallow wells tap groundwater from Holocene alluvial sediments. Reliable well data are restricted to the new observation wells drilled in frame of this study (M1-M11, s. Fig. 1). However, even in case of uncertain origin of a water sample, hydrochemical fingerprinting hints to a genetical interpretation and specific interrelations as described hereafter (Fig. 2).

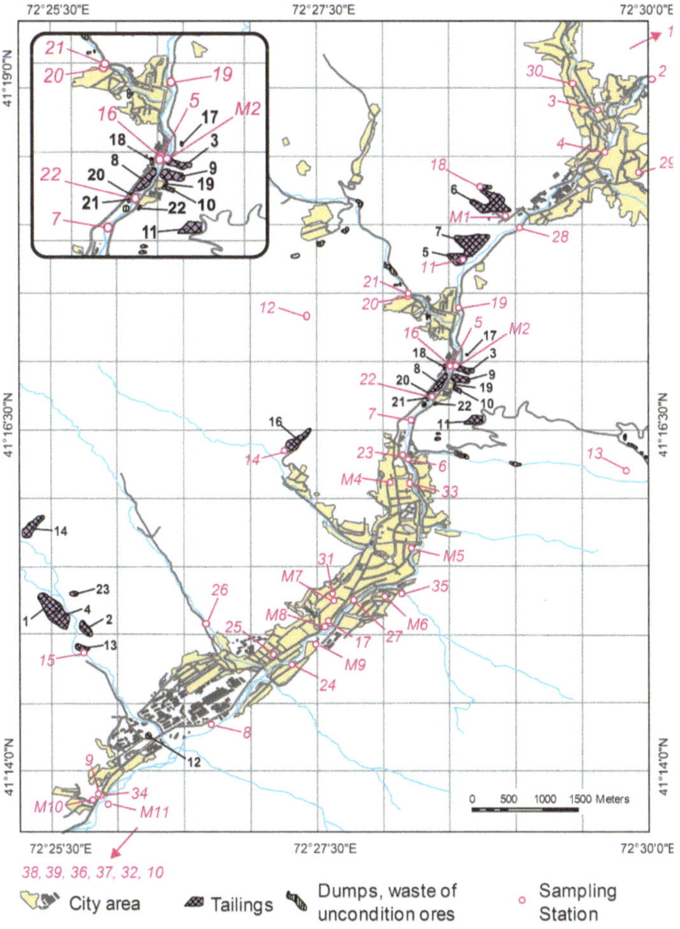

Fig.1. Sketch map of Mailuu Suu valley with locations of dumps, tailings as well the surface and groundwater sampling stations which have been integrated into the monitoring program.

The Mailuu-Say River, flowing NE to SW, represent a Ca-Mg-HCO_3 water type with low solutes and, thus, a typical meteoric composition. In contrast, the major tributary Kulmin-Say with a Na-Mg-SO_4 composition indicates a dominating impact of other sources than meteoric. In Figure 2, river samples plot along a mixing line bordered by the northernmost Mailuu-Say water on one end and Kulmin-Say on the other. This indicates a southward increasing impact of tributaries which draining water sources rich in sulphate and sodium fraction.

The sampled Holocene alluvial aquifer shows a remarkable resemblance to the Mailuu-Say water with respect to their major composition (Ca-Mg-HCO_3). Similarly to Mailuu-Say river water, shallow groundwater has a southward increasing impact of sodium and sulfate dominated fluids. Thus, the associated water composition is arranged along the mixing line discussed above (see "GW-qH" in Fig. 2). This indicates a hydraulic interconnection between shallow aquifer and Mailuu-Say river. Therefore, the shallow aquifer seems to be subject to locally influx of deep groundwater and contaminated seepage water as well as tributaries, additionally to a continuous exchange with Mailuu-Say water.

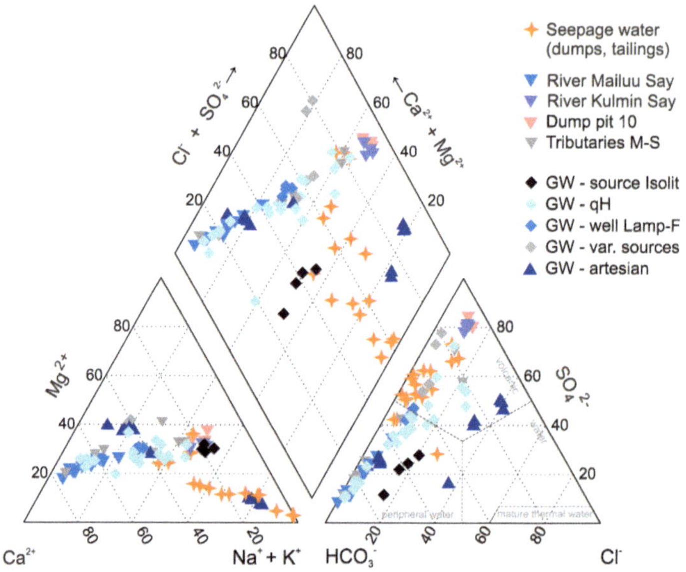

Fig.2. Major water composition of sampled water resources (n=93). Major water chemistry indicates genetic relationship of ground and surface water.

In contrast to natural water resources, the term "technogene" water is applied when influence from mining residues is likely such as tailings or dumps. The predominating water composition (Na, SO_4, HCO_3) in this group might be explained by dissolution of sulfide and carbonate minerals, commonly present within the tailings together with a high fraction of gypsum precipitated during the leaching with sulfuric acid. Especially, the sampled seepage water leaking from tailings have a remarkably high fraction of dissolved solids (TDS up to 10 g/L).

Uranium in natural water resources

Analyses of water samples points out that most water resources in Mailuu-suu area are at least locally affected by elevated dissolved uranium contents. From a chemotoxic point of view, uranium is the most problematic parameter. In more than 50% of all water samples, the threshold value for drinking water recommended by WHO (15 µg U/L) has been exceeded with uranium concentrations up to 36 mg/L. Further threshold values respective solutes such as sulfate (max. 5 g/L), fluoride (max. 10 mg/L) and arsenic (max. 1.8 mg/L) were locally exceeded and therefore increase the risk for adverse health effects in case of consumption. Repeated monitoring revealed that contaminants load of a water body varies seasonally mainly depending on precipitation and inflow of other water sources. The observed variability was within 30%, while being highest in case of surface waters (rivers, seepage water).

Hereafter, the sampled water types are discussed with respect to the observed dissolved uranium content (Fig. 3).

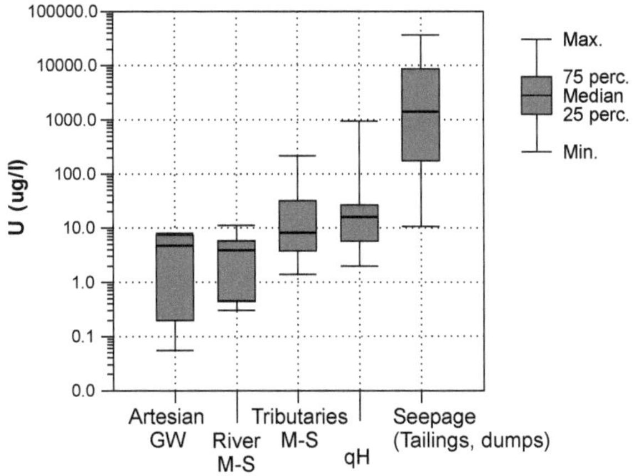

Fig.3. Distribution of dissolved uranium in 108 water samples, subdivided into clusters: "artesian" hardrock groundwater, "Mailuu-Say River", its "tributaries" incl. Kulmin-Say, Holocene aquifer ("qH") and technogene affected water, such as seepage water from dumps and tailings.

The seepage from tailings is generally dominated by high dissolved uranium (up to 36 mg/L), as a result of the tailing materials fine grain size combined with high residence time and evaporation effects (Fig. 4). Seepage discharges into receiving creeks and rivers or directly infiltrating into the subsurface.

North of Mailuu Suu valley, the main river Mailuu-Say is low in dissolved uranium (0,4 µg/L). In contrast, significant dissolved uranium has been observed in its tributaries (e.g., Kulmin-Say 170-220 µg/L). Consequently, the dissolved uranium content of the Mailuu-Say is increasing downstream but still remains below the WHO guideline for drinking water quality. The dissolved solute concentration varies in dependency of seasonal flow rate fluctuations of the Mailuu-Say and its tributaries. Consequently, the maximum uranium concentration has been observed in the south of Mailuu-Suu valley at the end of the dry season in October 2006 (up to 11 µg/L, Kok-Tash area).

In Holocene alluvial sediments, lowest dissolved uranium levels have been identified north of Mailuu-Suu valley (~3 µg/L). Further downstream, uranium content increases uot to a maximum within the city area (30 µg/L), exceeding the WHO drinking guideline threshold. Nevertheless, this aquifer is utilized by numerous domestic household wells. In the southernmost sampling station (Kok-Tash area) uranium has been observed to be still elevated (~7 µg/L).

Artesian groundwater of tertiary and Cretaceous aquifers can reach high uranium levels (up to 140 µg/L), probably associated with uranium ore mineralization in the host rocks. An impact of flooded mining excavations cannot be confirmed with the available data. In contrast, a deep well in the very northern area (Sarabiya) tap an unspecified possibly tertiary hardrock aquifer which bears very low dissolved uranium (0.05 µg/L).

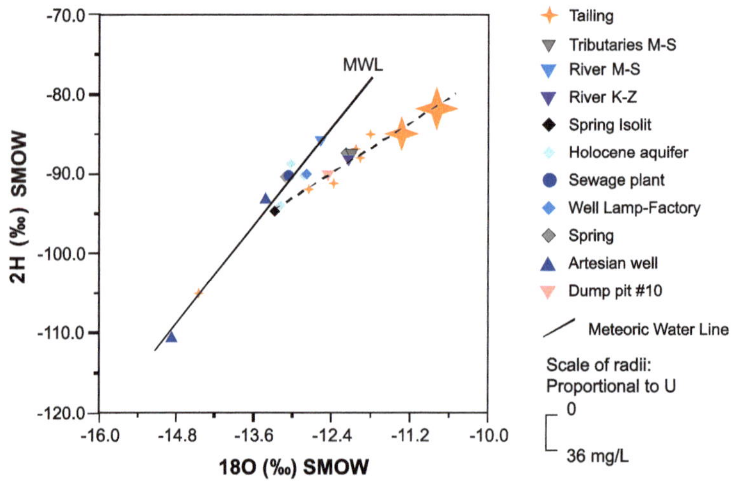

Fig.4. Fraction of the stable water isotopes δ^2H and $\delta^{18}O$ in selected water samples. Lowest values mark artesian groundwater from Sarabiya. U-rich seepage waters and receiving tributaries plot along a dashed line diverging from the local meteoric water line (MWL).

Assessment of contamination sources

Superficial mining residues such as dumps and tailing impoundments obviously represent contamination sources. However, the contamination path for specific natural water resources is quite complex since an interaction of different sources should be considered. In addition to seepage discharge from numerous dumps and tailing impoundments, artesian groundwater from Cretaceous aquifer containing dissolved uranium and metal sulfides and/or crude oil might be a significant impact when seeping into alluvial aquifers and rivers (as observed at Cretaceous outcrops north Mailuu Suu city). Moreover, uranium oxidation and dissolution in flooded mining excavations might enhance dissolved uranium levels.

Another perspective to the specific relevance of contamination sources for superficial water resources in Mailuu Suu area is provided by analysis of stable hydrogen and oxygen isotope data (Fig. 4). While the natural water samples generally plot along the local meteoric water line, high-uranium seepage waters as well as the sampled tributaries (incl. Kulmin-Say) are characterized by an increased $\delta^{18}O$ content. This hints to evaporation effects, which take place in poorly capped tailings and dumps and resulting in accumulation of heavy oxygen isotopes as well as nonvolatile solutes ($r_{(\delta 18O,TDS)} = 0.62$). Based on the available stable isotope data, the water composition of sampled tributaries including Kulmin-Say is dominated by uranium rich seepage water from dumps and tailings. This should be confirmed in future monitoring campaigns including stable isotope analysis.

Radiochemical leaching experiments with local tailing material (tailing 3) as well as thermodynamic calculations have been carried out to get insight into the uranium mobilization process within the tailing impoundments (Azeroual 2010). In summary, formation of dissolved uranium is generally controlled by pH/Eh-conditions as well as the availability of the ligands bicarbonate and calcium (Azeroual et al. 2010, Meleshyn et al. 2009). The dominating U-species are Calcium-Uranyl-Carbonates ($Ca_2UO_2(CO_3)_3$ and $CaUO_2(CO_3)_3^{2-}$). These are mobile species in sediments with neutral or negatively charged mineral surfaces. Moreover, a significant proportion of uranium has been mobilized as Uraninite colloides (<200 nm) with 20 ± 5% of total mobilized uranium.

Vandenhove et al. (2002) identified tailing 3 as the dominating environmental hazard with a total radiation of 650 TBq as high as 60% of the total radiation of all tailings impoundments. Consequently, recent remediation activities relocated the radioactive inventory of tailing 3 to a safe disposal site. However, considering the water transport path other sources are found to be more relevant as shown in table 1. Based on our field observations, seepage from tailing 5 represent a major contaminant for Mailuu-Say river with an input of up to 122 g uranium per day. Therefore, relocation of tailing 3 has limited impact on water quality due to the comparatively low daily uranium discharge into the Mailuu-Say.

Table 1. Approximate uranium release of selected sources discharging into the Mailuu-Say river, calculated from observations during field campaign in 10/2006.

Source	Dissolved U [mg/L]	Discharge [L/min]	U release [g/d]
seepage water tailing 3	1,76	0,5	1
seepage water tailing 5	8,50	10	122
seepage water tailing 16	36,00	0,1	5
Kara-Agach river	0,039	500	28
Kulmin-Say river	0,171	100	25

Recommended Mitigation Measures

A local water supply facility was designed under Soviet governance with the aim to provide drinking water for 20.000 people and, therefore, never achieved to serve all residents of Mailuu Suu. The supply facility providing uncontaminated river water upstream the mining area degraded during the last decades. Consequently, an increasing number of households directly utilize local water resources from deep artesian wells, simple dug wells and local rivers. Responsible authorities are strongly recommended to modernize and enhance the central water supply.

Furthermore, seepage water catchment and disposal systems need to be designed to reduce environmental impact of identified contamination sources. The quality of water resources should be observed by continuous monitoring approach distinguishing between surveillance, operational and investigative monitoring. This can only be realized by a long-term allocation of funds and expertise as well as sufficient laboratory capacities. Finally, all mitigation measures need to be combined with systematic information campaigns with the aim to improve awareness and reduce risks of affected local population.

References

Azeroual M, Bister S, Bunka M, Meleshyn A, Michel R (2010) Bestimmung von Radionuklidgehalten, die Speziation von Uranverbindungen und ihrer Bindungsformen in festen Proben aus Absetzhalden in Mailuu-Suu (Kirgistan). Final report 17.02.2010: 82 pp

Blacksmith Institute (2013) The World's Worst 2013: The top ten toxic threats - cleanup, progress, ongoing challenges. Blacksmith Inst. & Green Cross Switzerland: 27-28

Jung H & Wagner F (2008) Hydrogeochemical Groundwater Monitoring in Mailuu-Suu, Kyrgyz Republic. Final Report of Technical Corporation Project Reduction of Hazards posed by Uranium Mining Tailings in Mailuu-Suu, Kyrgyz Republic. Federal Institute for Geosciences and Natural Resources, Hanover: 81 pp

Meleshyn A, Azeroual M, Reeck T, Houben G, Riebe B, Bunnenberg C (2009) Influence of (Calcium-)Uranyl-Carbonate Complexation on U(VI) Sorption on Ca- and Na- Bentonites. Environ. Sci. Technology 43: 4896-4901

Ostroborodov V, Erohin S, Chutkin B, Knapp R (2003) The Water Component to Risk, Monitoring, and Rehabilitation. World Bank Disaster Hazard Mitigation Project, preparation team report

Schmidt G (1997) Anlagenzustand und radiologische Bewertung der früheren Urangewinnungsanlagen am Standort Mailii Su (Kirgistan). Report, Öko-Institut e.V. Friedrich-Ebert-Stiftung Büro Bishkek: 19 pp

TACIS (2003) Remediation of Uranium Mining and Milling Tailing in Mailuu-Suu District of Kyrgyzstan. Final report EC-TACIS Project. R-3721, SCK CEN: 679 pp

UN (2001) Kyrgyzstan - Management of radioactive and other wastes. – Environmental performance reviews, chapter 5.

Vandenhove H, Sweeck L, Mallants D, Vanmarcke H, Aitkulov A, Sadyrov O, Savosin M, Mirzachev M, Clerc J, Quarch H, Aitaliev A (2006) Assessment of radiation exposure in the uranium mining and milling area of Mailuu Suu, Kyrgyz. J Env. Rad. 88 : 118-139.

I've been thinking about this question now for 30 min straight. It was tough and I had some good ideas. So, a 5 min break should be fine.
It's late and I need to finish by tomorrow. So I better have an efficient break on the sofa is better than on the chair.
next morning:
oh... no.

Rhizofiltration of U by plant root surfaces in a tailing wetland

Weiquing Q. Wang[1], Carsten Brackhage[1], Ernst Bäuker[2], E. Gert Dudel[1]

[1]Technische Universität Dresden, Institute of General Ecology and Environmental Protection,
[2]Institute of Forest Utilization, Pienner Strasse 19, D-01737 Tharandt, Germany.

Abstract. It is well known that element transfer into plant shoots is highly variable and often very low in comparison with the accumulation in roots. On the other side, the role of Fe plaque on metalloid (P, As) and heavy metal stabilization on root surfaces has been widely researched on. In a reed dominated wetland along a flow gradient of seepage water plant and related water samples were taken to determine the relation between Fe plaque formation and U immobilization on the root surface of common reed (*Phragmites australis* TRIN. ex Steudt.). Fe plaques were observed in all sampling sites with a significant increase from near mine water inflow (Fe: 5449 mg·kg^{-1})to the near outlet (Fe: 16189 mg·kg^{-1}), while an average of 87.1% U was found fixed on root surfaces. The result indicated that an enhanced Fe plaque formation in the aeration zone of mine water inflow may inhibit more U from root inner ad- and absorption (accumulation) and/or competition with other elements (e.g. As). Both the increase of Fe plaque and the related root surface U coating is associated with a decreased redox potential in the surrounding interstitial water indicating that a reducing environment induced by organic matter decay (plant litter) could be an important factor. However, EDAX analysis showed that the U complex forming particles were not tightly adhered to the root surface but surround the root loosely. U is accumulated only to a small extent within roots (rhizofiltration *in sensu strictu*). This is different to other metalloids (like As) and heavy elements (ions) which fully bind to root surfaces and inside the rhizodermis.

Introduction

In the course of U mining and ore processing, U is still released in the environment for many years after decommissioning of mines and processing plants (Carvalho et al., 2011). This may imply a potential hazard to the environment and even may strongly affect human health (Carvalho et al., 2005; Lapham et al., 1989). Apart from engineered physico-chemical methods sun driven passive water treatment technologies using algae and macrophytes (submerse and emerse plants

B.J. Merkel and A. Arab (eds.), *Uranium – Past and Future Challenges*,
DOI 10.1007/978-3-319-11059-2_57

and algae) were proposed for cleaning such contaminated waters (Dudel et al. 2001; Kalin et al. 2005, Pratas et al. 2014). Constructed wetlands have been tested in the last years to remove U by adsorption and accumulation on/in living plants and plant litter (Johnson and Hallberg, 2005). Recent studies on different aquatic plant species showed that U removal and even accumulation mainly occurs in belowground plant parts including rhizomes and fine roots (Baumgartner et al., 2000; Boileau et al., 1985; Zafrir et al., 1992).

U occurs mainly in two oxidation states, U (IV) and U (VI). Under oxidizing conditions, U (VI) prevails as highly soluble uranyl ion (U (VI) O_2^{2+})(Finch and Murakami, 1999; Jackson et al., 2005). More soluble forms include uranyl sulfates, uranyl nitrates, uranyl acetates where in slide acidic, neutral and alkaline waters soluble carbonate complexes prevail (Mkandawire et al. 2005).

As a common micro-nutrient of plants, iron has been named as key factor when considering U immobility (Hobbs and Karraker, 1996; Hsi and Langmuir, 1985; Read et al., 1993). Many studies showed that soluble U (VI) readily adsorbs to iron oxides such as hematite (α-Fe_2O_3) and goethite (α-FeOOH forming reduced layers of insoluble U (IV) (Frederickson et al. 2000, Moyes et al. 2000).

The rhizosphere of plants with associated microbes and the surrounding soil/substrate is physically, chemically and biologically active (Anderson et al., 1993; Curl and Truelove, 1986; Hinsinger et al., 2003). Many species of wetland plants are able to oxidize the rhizosphere and hence prevent reducing conditions and a related release of potentially toxic ions like Mn^{2+}, Fe^{2+} and S^{2-} (Taylor et al., 1984). This active plant driven process can cause the oxidation of ferrous iron to a ferric state like goethite (α-FeOOH), coating the surface of roots known as Fe plaque. Fe plaques may interfere strongly with soluble element (metal) transfer within the soil (water)-plant-atmosphere continuum (Greipsson and Crowder, 1992; Ye et al., 2001). However, the role of Fe plaques in the course of metal immobilization especially U is not yet fully understood. Many studies indicate an increase in element accumulation by plants in the presence of Fe plaque (Xu et al., 2009; Ye et al., 2001). In other studies Fe plaques have been considered as a physical or chemical barrier to inhibit uptake of heavy metals by roots (Batty et al., 2000; Christensen and Sand-Jensen, 1998). Former studies indicate that these controversial results might be explained by the ratio between iron and different target ionic species (Otte et al., 1989; Zhang et al., 1998). The wetland plant species and environmental conditions like irradiation, temperature, evapo-transpiration, photosynthesis and root release of organic carbon may also interfere with plaque formation and uptake of heavy metal ions (Liu et al., 2004).

Due to the different chemical characteristics compared to other heavy metals and metalloids (like As, P) and the special relationship with iron in the process of immobilization, it is important to assess the role of iron plaques in U uptake and immobilization by plant belowground tissues. It should lead to an estimation of the removal capacity and stability of immobilization of U in natural and constructed wetlands. To elucidate this in further detail reed roots and rhizome samples were taken and analyzed from a U mine tailing to investigate the specific iron plaques and related distribution of U adsorbed/accumulated on belowground plant surfaces and also inside the plant tissues.

Material and methods

Field sampling and sample pretreatment

The sampling site is located in the decommissioned mining site Neuensalz-Zobes, Western Ore Mountains, Saxony. Due to natural succession since 1977 in the tailing pond a wetland habitat matured. It is dominated for about 10 years by common reed (*Phragmites australis* L.). An inner stream in the wetland gathers water from the flooded mine (main input P1) down to the resting part of the tailing pond close to the outflow (P5, see Fig. 1). P1, P2 and P3 are all located in the main seepage runoff, with high flow rate and low water level while P4 and P5 are sampling points surrounding the pond. The root samples from each sampling point were randomly harvested and surrounding organic rich sediment was removed and then rinsed twice with de-ionized water. Filtered and unfiltered samples of *in situ* seepage stream water and interstitial water were filled in PE-bottles. All samples were transferred into the laboratory and stored acidified at 6℃ for further analyses.

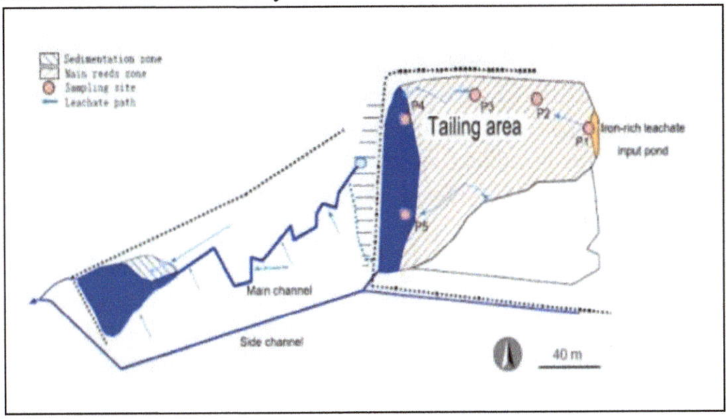

Fig.1. Sketch map of water pathway above (right site) and below tailing damm (left site) and sampling points of water, and plant sampels (P1...P5) in September 2012 near Neuensalz (Location of study site in former U mining and ore processing area of Zobes-Mechelgrün) in Western Erzgebirge mountain in Saxony, Germany.

Fresh roots were rinsed twice with de-ionized water then separated into fine roots and rhizomes. Each tissue was further divided into suitable size portions with roughly equal weight (approx. 0.1 g DW) and randomly assigned to Fe plaque extraction as is described below. The rest of the root samples were dried to constant weight at 60℃.

Fe plaque extraction

A modified dithionite-citrate-bicarbonate (DCB) extraction method was used for Fe plaque extraction (Taylor et al., 1984). Fresh samples were incubated in a solution of 40 ml of 0.3 M $Na_3C_6H_5O_7{\cdot}2H_2O$, 5 ml of 1.0 M $NaHCO_3$ and 3 g of $Na_2S_2O_4$ at room temperature for 16h. The resulting solution was filled to 100ml with de-ionized water. The residue was then rinsed with 5ml of de-ionized water, dried to constant weight at 60℃ in finally digested as described below.

Chemical analysis

The oven dried samples were weighed into digestion tubes, spiked with 3 ml HNO_3 and 2 ml H_2O_2 (DIN-EN-13805, 2002), and digested at 180℃ in a microwave heated system (Mars 5, CEM Corporation, Matthews, NC, USA.) for 2h. The resulting solution was mixed with 20ml of de-ionized water and then transferred to labeled PE bottles for storage for ICP-MS measurement. The total content Fe, U and other elements in all fractions were determined by Inductively Coupled Plasma Mass Spectrometry (X-series instrument, Thermo Fisher Scientific Inc.) according to (DIN-EN-ISO-17294-2, 2004). An internal standard was included. ICP-MS calibration functions were recorded from mixed calibration samples, which were prepared from multi-element solutions (Bernd Kraft, Duisburg, Germany). LOD was calculated as the threefold standard deviation of instrument blank (acidified water). All chemicals were of analytical grade.

Mapping analysis

Parts of fresh root samples were examined by scanning electron microscopy (SEM). The microscope (JEOL T 330A) was equipped with an element detector (EDR 288, Bruker Röntec GmbH, Berlin, Germany) and run at 15 kV. Plant samples for electron microscopy were critical-point dried in liquid N and subsequently coated with carbon (Zimmermann et al. 2000). An area of about 0.09 mm^2 in the middle of the roots was scanned. All scans were run at a magnification of 350x.

Results and discussion

Formation of Fe plaque

The data from all sampling sites indicated that Fe plaque widely deposited on the reed root surface. However, a quite different distribution pattern comparing fine roots and rhizomes can be seen. Abundant Fe was detected in all fine roots, indicated by high concentrations in the DCB extraction while only 6.8±2% of total Fe was detected for rhizomes (Fig. 2).

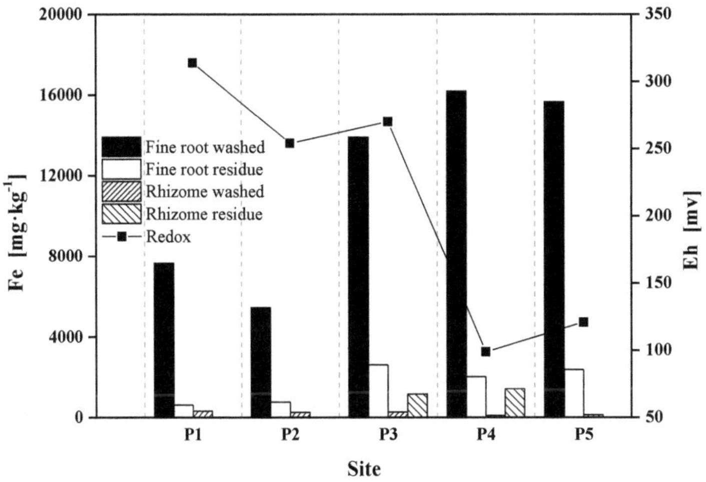

Fig.2. Spatial redox state variation and Fe plaque formation on root the system along the seepage water pathway from P1 to P5 (details compare see Fig.1)

It suggests that fine roots are the main tissue for Fe plaque formation. The data also show that the Fe plaque formation on fine root surfaces is significantly enhanced along the stream path increasing from a minimum of 5448 $mg \cdot kg^{-1}$ DW at P2 to a maximum of 16189 $mg \cdot kg^{-1}$ DW at P4. The average content of Fe plaque was calculated to 88.0±4% of total Fe in fine roots. The data indicate Fe plaque acting as the dominating form of storage of Fe^{2+} ions after oxidation on fine root surfaces instead of internal ad- and absorption. At the same time the redox potential (Eh) decreased along water path, transforming the root environment from an oxidizing to a reducing miliau. The reductive environment may play a positive role in Fe plaque formation. The ferric Fe to ferrous Fe transition process in the rhizosphere could be significantly enhanced by a highly reducing environment driven by decaying organic matter from plant litter, duckweed (Mkandawire & Dudel 2005) and filamentous algae (Dienemann et al. 2002) after annual summer mass development within the tailing pond. With the assistance of micro oxidation zones on reed root surfaces by aeration from aboveground plant parts via aerenchyma, the increased Fe^{2+} availability provides a major Fe source to support a relatively strong formation of Fe plaque by oxidizing it to Fe oxides, like it was observed in former studies (Greipsson and Crowder 1992; Taylor et al. 1984; Ye et al. 2001). We suggest that the change in redox state by increasing microbial organic matter decay (C as electron donator) in stagnant water bodies in (late) summer is the main driving force for Fe plaque formation and associated U removal.

U distribution in the root system

The redox state data from all sites correlate negatively with the U accumulation on the surface of the root system (Fig. 3). Comparing with the Fe plaque (R^2 =0.5704), the U (R^2 = 0.8707) was more sensitive to the variation of redox state and had a relatively higher accumulation strength under reduction environment.

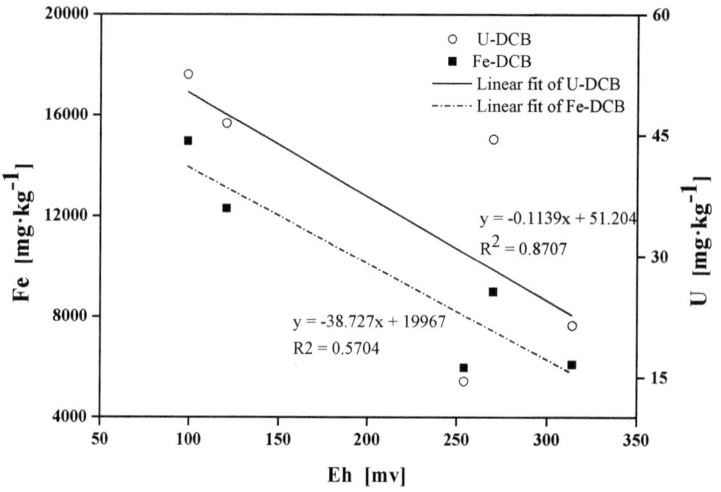

Fig.3. Redox state correlation with U and the Fe plaque in the root system along the seepage water pathway from P1 to P5 (details compare see Fig.1)

Most of theU has been found in fine roots while only trace amounts of this element were detectable in rhizomes. Abundant U was also detected in all selected DCB extractions with an average concentration of 27.1±16 mg·kg^{-1} DW. The data indicated that the rhizosphere U tends to be fixed to coatings on the root surface instead of internal tissues (Fig. 4).

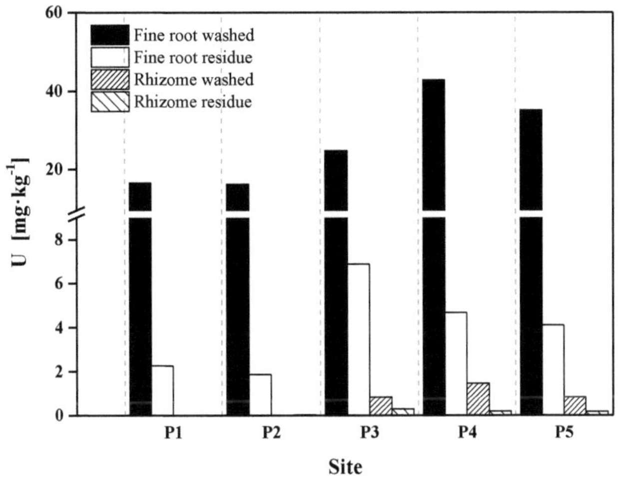

Fig.4. Spatial U distribution on fine root and rhizomes of common reed

However, the spatial distribution and accumulation of U on root surfaces is different despite a balanced in situ U content in rhizosphere interstitial water from P1 to P5 (average concentration in late summer/autumn: 47.5±5 $\mu g \cdot L^{-1}$). It suggests a similar exogenous U supply level along the stream path. In contrast, the output area showed a relatively high U accumulation capacity on root surfaces with up to 42.8 $mg \cdot kg^{-1}$ DW at P4 being almost 3 times higher compared to the input zone. When comparing this with the data from the spatial distribution of Fe plaque , a relation may be seen between increased Fe plaque on root surfaces and an increased content of U in DCB extractions reaching 87±5% of total root adsorption. The enrichment efficiency for U by Fe plaque was relatively higher compared to other elements. Former reports show that 75-89% of Zn could be found in DCB extracts associated with Fe plaque (Zhang et al., 1998). For Cu and Ni, only 30-40% could be detected (Ye et al. 2001). This may be caused by different physic-chemical characteristics of different element species in the respective water and related bonding forces to the Fe plaque. Altogether the results indicate that the Fe plaque had a positive effect on U accumulation on root surfaces. Furthermore the Fe plaque thickness may operate as a physicochemical diffusive barrier interfering the U adsorption inside root tissues instead of increasing the metal removal process as shown for other heavy metals (Ye et al., 2001). Although U has been observed mainly on the root surface, a discrepant distribution pattern between U and other elements like arsenic (As) also exists. The overall U in water remains at a low level while As in water maintained at a significantly higher content than U

(As - U concentration ratio was respectively 1.26, 1.58 and 2.04 from P1 to P3) (Fig.5). Hence the availability of As was higher than U.

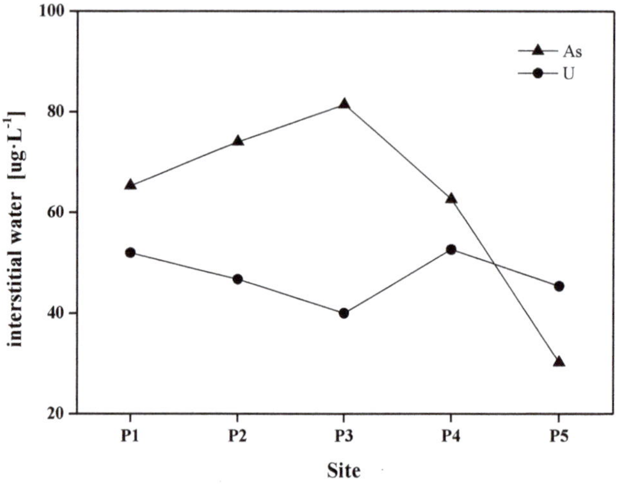

Fig.5. Spatial U and arsenic distribution in interstitial water surounding Fe plaques from sampling point P1 to P5 (compare Fig. 1)

U absorption strength was enhanced along with the decrease of As - U concentration ratio, which was only 1.19 and 0.66 near the outflow (P4 and P5). It suggests a significant negative impact of As on U adsorption on Fe plaque by high As to U ratio in water and related accumulation of U within the entire rhizosphere. The EDAX mapping analysis also intuitively showed that As was tightly bond with Fe on the root surface and the enrichment position was highly coincident with U (Fig.6).

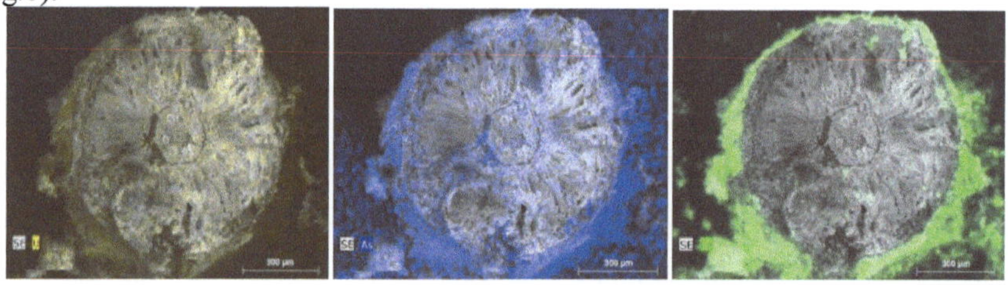

Fig.6. Distribution of U (yellow, left photo), As (blue, middle) and Fe (green, right) in and on fine roots of *Phragmites australis* in autumn 2012 (REM-EDAX mapping analysis).

This suggests a strong co-precipitation effect for both elements (Fig. 6 middle and right site). However, U was detected to be more irregular distributed around the root and partially between sites and shows a different distribution compared with Fe. The difference between those metals and arsenic indicate that the formation of Fe plaque may affect the element accumulation process in different ways due to their various physic-chemical and biochemical properties. As and U were both proven to bind fast and tight with Fe hydrated oxides like goethite (Fredrickson et al., 2000; Fritzsche et al., 2006). Former studies also show that As is a highly sensitive and attractive element to Fe plaque mainly composed by a series of Fe hydrated oxides (Fritzsche et al., 2011). Therefore, a relatively high content of As in interstitial water probably leads to a highly competitive adsorption process with U on similar binding sites of Fe plaque and hence directly impairs on U in coatings on the root surface.

Conclusions

The results demonstrated that the formation of Fe plaque widely exists in natural wetlands and is very much seen on fine root surfaces. U accumulation in the dimension of one order of magnitude compared with U concentrations of the water body was associated with the dimension of Fe plaque formation. The U ad- and absorption pattern is influenced by the change of Fe plaque formation at different reduction states (Eh) of interstitial waters in turn affected by the availability of degradable organic matter (microbial activity). The competitive adsorption by other highly bio-available elements like As, could also significantly interfere with the specific U distribution and accumulation on and in roots. The results suggest that Fe plaque act as both efficient medium increasing the U adsorption from the rhizosphere and may form a physicochemical barrier disturbing the transportation from interstitial water into the plant. Beneath dead organic matter in the rhizosphere root surfaces including plaques generate the primary U accumulation and storage zone and transfer in aboveground tissues was negligible.

References

Anderson et al. (1993) The potential role of rhizosphere microbial communities in bioremediation of agrochemical wastes. Abstracts of Papers American Chemical Society 205: 1-2.

Batty et al. (2000) The effect of pH and plaque on the uptake of Cu and Mn in Phragmites australis (Cav.) Trin ex. Steudel. Annals of Botany 86: 647-653.

Baumgartner et al. (2000) Plant uptake response to metals and nitrate in simulated uranium mill tailings contaminated groundwater. Water Air and Soil Pollution 118: 115-129.

Boileau et al. (1985) Uranium accmulation in the lichencladonis-rangiferina. 1. uptake of cationic, neutral, and anionic forms of the uranyl-ion. Canadian Journal of Botany-Revue Canadienne De Botanique 63: 384-389.

Carvalho et al. (2011) Radionuclides in plants growing on sludge and water from uranium mine water treatment. Ecological Engineering 37 : 1058-1063.

Carvalho et al. (2005) Environmental impact of uranium mining and ore processing in the Lagoa Real District, Bahia, Brazil. Environmental Science & Technology 39 : 8646-8652.

Christensen, K. K., and K. Sand-Jensen (1998) Precipitated iron and manganese plaques restrict root uptake of phosphorus in Lobelia dortmanna. Canadian Journal of Botany-

Revue Canadienne De Botanique 76 : 2158-2163.
Curl, E. A., and B. Truelove (1986) Advanced Series in Agricultural Sciences, 15. the Rhizosphere. X+288p. Springer-Verlag: Secaucus, N.J., USA; Berlin, West Germany. Illus
DIN-EN-ISO-17294-2 (2004) Wasserbeschaffenheit - Anwendung der induktiv gekoppelten Plasma-Massenspektrometrie (ICP-MS) - Teil 2. Bestimmung von 62 Elementen (ISO 17294-2:2003), Deutsche Fassung EN ISO 17294-2:2004, Berlin, Deutsches Institut für Normung, p. 24.
Dudel et al. (2001) Principles and limitations for natural attenuation of radionuclides in former uranium mining and milling sites. 8th international conference radioactive waste management and environmental remediation. New york p8.
Duff et al. (1999) Factors influencing uranium reduction and solubility in evaporation pond sediments. Biogeochemistry 45 : 95-114.
Finch, R., and T. Murakami (1999) Systematics and Paragenesis of Uranium Minerals. Reviews in Mineralogy 38 : 91-179.
Fredrickson et al. (2000) Reduction of U(VI) in goethite (alpha-FeOOH) suspensions by a dissimilatory metal-reducing bacterium. Geochimica Et Cosmochimica Acta 64 : 3085-3098.
Fritzsche et al. (2011) Arsenic strongly associates with ferrihydrite colloids formed in a soil effluent. Environmental Pollution 159: 1398-1405.
Greipsson, S., and A. A. Crowder (1992) Amelioration of copper and nickel toxicity by iron plaque of rice (oryza-sativa). Canadian Journal of Botany-Revue Canadienne De Botanique 70: 824-830.
Hinsinger et al. (2003) Origins of root-mediated pH changes in the rhizosphere and their responses to environmental constraints. A review: Plant and Soil 248: 43-59.
Hobbs, D. T., and D. G. Karraker (1996) Recent results on the solubility of uranium and plutonium in Savannah river site waste supernate. Nuclear Technology 114: 318-324.
Hsi, C. K. D., and D. Langmuir (1985) adsorpiton of uranyl onto ferric oxyhydroxides - application of the surface complexation site-binding model. Geochimica Et Cosmochimica Acta 49: 1931-1941.
Jackson et al. (2005) Characterization of colloidal and humic-bound Ni and U in the "dissolved" fraction of contaminated sediment extracts. Environmental Science & Technology 39: 2478-2485.
Johnson, D. B., and K. B. Hallberg (2005) Acid mine drainage remediation options: a review. Science of the Total Environment 38: 3-14.
Kalin et al. (2002) cological water treatment processes for underground uranium mine water. progress after three years of operating a constructed wetland. Uranium in the Aquatic Environment 587-596.
Lapham et al. (1989) Health implications of radiomnuclide levels in cattle raised near U mining and milling facilities in Ambrosia lake, Nex-mexico. Health Physics 56: 327-340.
Liu et al. (2004) Do iron plaque and genotypes affect arsenate uptake and translocation by rice seedlings (Oryza sativa L.) grown in solution culture?. Journal of Experimental Botany 55: 1707-1713.
Mkandawire, M. and Dudel, E.G. (2005): Accumulation of arsenic in Lemna gibba L.

(duckweed) in tailing waters of two abandoned U mining sites in Saxony, Germany. Science of the Total Environment 336(1-3): 81-89

Mkandawire et al. (2005) Resource manipulation in enhancing U and As attenuation with Lemna gibba L. (duckweed) in tailing waters, Journal of Water, Air and Soil Pollution 165 (1-4), 83-101

Moyes et al. (2000) Uranium uptake from aqueous solution by interaction with goethite, lepidocrocite, muscovite, and mackinawite: An X-ray absorption spectroscopy study. Environmental Science & Technology 34: 1062-1068.

Otte et al. (1989) Iron plaque on roots of aster-tripolium L-interarction with zinc uptake. New Phytologist 111: 309-317.

Pratas et al. (2014). Potential of aquatic plants for phytofiltration of uranium contaminated waters under laboratory conditions Ecological Engineering 69, 170-176.

Read et al. (1993) The migration of uranium into peat-rich soils at broubster, Caithness, Scotland, UK. Journal of Contaminant Hydrology 13: 291-308.

Taylor et al. (1984) Formation and Morphology of an iron plaqueon the roots of typha-latifolia L grown in solution culture. American Journal of Botany 71: 666-675.

Xu et al. (2009) Effect of Iron Plaque Formation on Phosphorus Accumulation and Availability in the Rhizosphere of Wetland Plants. Water Air and Soil Pollution 200: 79-87.

Ye et al. (2001) Copper uptake in Typha latifolia as affected by iron and manganese plaque on the root surface. Canadian Journal of Botany-Revue Canadienne De Botanique 79: 314-320.

Zafrir et al. (1992) Uranium in plants of southern Sinai. Journal of Arid Environments 22: 363-368.

Zhang et al. (1998) Effect of iron plaque outside roots on nutrient uptake by rice (Oryza sativa L.). Zinc uptake by Fe-deficient rice. Plant and Soil 202: 33-39.

Zimmermann et al. (2000) Nutrition-physiological responses of spruce to different levels of sulphur dioxide stress in the Erzgebirge Mountains and the Thuringian Forest, Germany. Forstw. Cbl. 119: 193-207.

I love you, my sweet ligands
And we love you, kind uranyl
Oh no! Is that acid?
We will miss you!
You were the best!

Temporal and Special Variation of Pore Water and Seepage Quality of an Abandoned Uranium Milling Tailings Impoundment

Lechang Xu[1], Xueli Zhang[1], Jie Niu[1], Hui Zhang[1]

[1]Beijing Research Institute of Chemical Engineering and Metallurgy, CNNC, P. O. Box 234, Beijing 101149, China

Abstract. A uranium milling tailings impoundment is enclosed by 9 artificial dams and covers 1.7km^2. The impoundment cease to receive uranium tailings in 1994 and the dams are reinforced in 2008. The impoundment will be covered and stabilized in a few years. A large number of tailings' water and seepage need to be treated and then discharged into the Xiangjiang river during covering. Pore water of 14 boreholes and seepage from 6 pump house were monitored to understand pore water and seepage quality. The water quality monitoring indicates that there is distinctly temporal and special variation of pore water and seepage qality in different position and time. The detected primary constituents of pore water includes U、^{226}Ra、Mn、SO_4^{2-}、Cl^-、NO_3^-、NH_3-N、F^- and other heavy metal.

Introduction

Uranium milling tailings impoundment stores a lot of tailings pore water and seepage containing radioactive and nonradioactive constituents. The tailings solution need to be dewatered and treated during tailings covering so that mechanical execution on the soft tailings can be carried out and impact on groundwater reduced. Understanding the solution characteristics is important to tailings water treatment including.

Site history

A uranium milling tailings impoundment is surrounded by 5 hills and 9 artificial dams and covers 1.7km^2. (Fig.1). It is close to Xiangjiang River, a large river, the nearest distance is only 2km. It also situated in densely populated area. Inlet of tailings solution to Xiangjiang River is located at the upstream of domestic water

B.J. Merkel and A. Arab (eds.), *Uranium – Past and Future Challenges*,
DOI 10.1007/978-3-319-11059-2_58

pumping station, which supplies drinking water for people in near city. The impoundment was built in 1963 and ceased to receive uranium milling tailings in 1994. After that the waste from titanium pigment plan was recharged into the impoundment and ceased to do in 2003. The dams are reinforced in 2008. The impoundment will be covered and stabilized in a few years.

At first, the tailings were recharged into impoundment at the starter dam and water in tailings was cleared and recycled at the first water dam. The tailings were recharged at the first water dam and the second water dam was used for clear and recycle of water in tailings from 1970 to 1979. The tailings were recharged into the second tailings pond from 1979 to 1982, and then the tailings recharged all round and the water evacuated in the centre.

There are 6 pump houses at close to relevant dams to collect seepage from the impoundment (Fig.1), Ximei pump house of which is located between the first and second water dam. The monitor boreholes were constructed around the represent dams to understand tailings pore water (Fig,1 and Table 1). Shallow (8.20-16.65m depth) and deeper (16.5-21.5m depth) monitoring boreholes were constructed at the adjacent monitoring point. Monitoring boreholes were set up at three different distances from the second water dam.

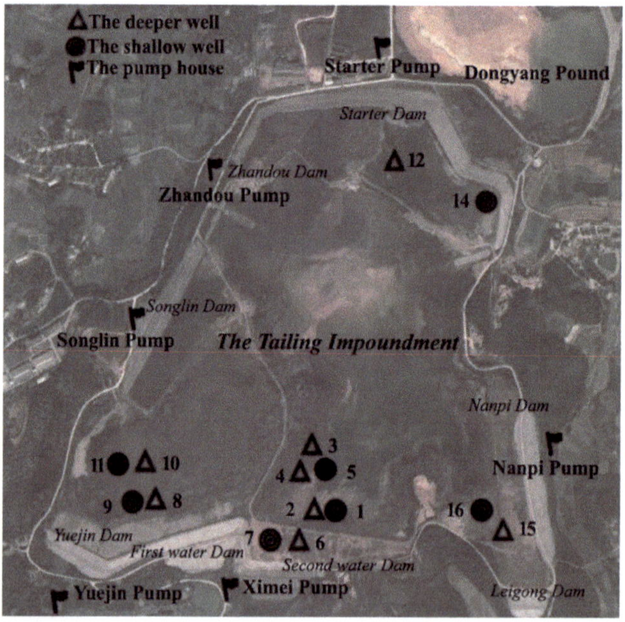

Fig.1. The impoundment and monitoring boreholes

Quality of tailings pore water and seepage

Table 1. Parameters of monitoring boreholes

Borehole No.	Elevation (m)	Borehole depth (m)	Water table Depth (m)	Water table Elevation (m)	Borehole classficatin	Location
ZK1	94.50	8.20	2.80	91.70	shanllow	Sencond water dam
ZK2	94.50	16.50	3.10	91.4	deeper	
ZK3	93.80	17.50	1.54	92.26	deeper	
ZK5	93.80	8.50	1.40	92.40	shallow	
ZK6	96.80	17.50	7.50	89.30	deeper	
ZK7	96.80	12.50	7.25	89.55	shallow	
ZK8	91.50	21.50	5.72	85.78	deeper	Yuejin dam
ZK9	91.50	16.65	5.65	85.85	shallow	
ZK10	91.50	19.38	1.27	90.23	deeper	Songlin dam
ZK11	91.50	13.00	1.35	90.15	shallow	
ZK12	97.20	17.50	13.00	84.20	deeper	Starter dam
ZK14	96.80	11.00	5.70	91.10	shallow	
ZK15	92.90	18.20	1.00	91.90	deeper	Nanpi and Leigong dam
ZK16	92.90	11.00	1.05	91.45	shallow	

Sampling and monitoring/analysis methods

Water was sampled as reference to technical specifications for environmental monitoring of groundwater (HJ/T 164 2004). Monitoring and analysis methods applied the national standards and ICP (Table 2). Pore water was usually sampled three times covering normal water period, wet and low water season.

Table 2. Monitoring and analysis methods applied for tailings pore water

Parameters	Methods	Parameters	Methods	Parameters	Methods
pH	GB 6920	NH_3-N	HJ 535-2009	Pb	ICP
U	GB/T 6768	Mn	GB 11906	Ni	ICP
^{226}Ra	GB 11214	Fe	ICP	Cu	ICP
F-	GB 7484	Cd	ICP	Zn	ICP
SO_4^{2-}	GB5750-85	Cr	ICP		

Water quality anlysis

The tailings' pore water and seepage quality is given in Table 3 and Table 4.

Table 3. Tailings pore water quality

No	U / μg·L^{-1}		^{226}Ra / Bq·L^{-1}		Cd	Cr	Pb	Ni
	Range	Mean	Range	Mean				
ZK7	51.7-380.8	144.2	0.206-0.246	0.151	<0.02	<0.02	<0.05	<0.05
ZK1	135.6-234.2	183.2	0.351-0.774	0.551	<0.02	<0.02	<0.05	<0.05
ZK5	3.1-10.8	7.0	0.278-0.471	0.375	<0.02	<0.02	<0.05	<0.05-0.24
ZK9	4.2-52.5	22.8	0.340-0.643	0.527	<0.02	<0.02	<0.05	<0.05-0.31
ZK11	19.0-26.3	22.7	0.236-0.338	0.287	<0.02	<0.02	<0.05	<0.05-0.32
ZK16	156.7-206.8	181.8	0.281-0.819	0.550	<0.02	<0.02	<0.05	<0.05
ZK14	103.3-578.3	362.1	0.319-1.21	0.633	<0.02	<0.02	<0.05	<0.05
ZK6	7.0-561.5	284.3	0.254-0.339	0.297	<0.02	<0.02	<0.05	<0.05-0.39
ZK2	58.6-526.7	346.2	0.573-0.136	0.977	<0.02	<0.02	<0.05	<0.05-0.25
ZK3	6.4-39.6	17.8	0.256-0.421	0.345	<0.02	<0.02	<0.05	<0.05-0.38
ZK8	27.6-146.2	76.2	0.331-0.626	0.441	<0.02	<0.02	<0.05	<0.05-0.46
ZK10	7.6-26.4	16.9	0.254-0.426	0.345	<0.02	<0.02	<0.05	<0.05
ZK15	342.7-593.1	484.4	0.212-0.986	0.570	<0.02	<0.02	<0.05	<0.05-0.35
ZK12	17.2-157.3	86.4	0.231-0.756	0.445	<0.02	<0.02	<0.05	<0.05
Disch.	300		1.1		0.1	1.5	1.0	1.0

Table 3 (cont.). Tailings pore water quality

No	NH_3-N		F^-		Cu	Zn	Mn	
	Range	Mean	Range	Mean			Range	Mean
ZK7	<0.05-1.07	0.36	7.18-8.19	5.12	<0.05	<0.1-0.11	1.07-1.52	1.30
ZK1	1.45-72.4	44.9	4.94-6.18	5.67	<0.05	0.12-1.08	0.58-5.64	3.21
ZK5	64.9-78.6	71.7	1.90-2.39	2.15	<0.05	<0.1-1.03	0.32-0.53	0.43
ZK9	2.71-53.7	35.5	1.98-12.1	5.58	<0.05	<0.1-3.19	0.76-2.71	1.49
ZK11	50.8-54.4	52.6	3.61-5.89	4.75	<0.05	<0.1-1.25	0.55-0.61	0.58
ZK16	45.3-60.8	53.1	7.63-8.35	7.99	<0.05	<0.1	2.06-3.44	2.75
ZK14	3.79-29.3	16.8	2.17-6.21	3.74	<0.05	<0.1-0.68	0.82-17.5	6.71
ZK6	3.88-22.6	13.2	5.72-9.10	7.41	<0.05	0.33-1.42	8.29-22.6	15.4
ZK2	0.57-84.2	46.9	4.39-6.28	5.595	<0.05	<0.1-2.84	1.24-6.26	3.84
ZK3	23.6-61.1	43.9	2.52-3.21	2.93	<0.05	0.15-2.21	0.81-0.96	0.90
ZK8	2.04-23.1	13.8	3.86-13.0	7.21	<0.05	<0.1-3.68	2.04-5.0	3.66
ZK10	60.8-77.6	68.9	5.13-7.61	6.29	<0.05	<0.1-0.98	0.69-0.85	0.76
ZK15	22.1-47.6	31.0	6.04-8.13	7.43	<0.05	<0.1-1.53	14.3-22.6	19.1
ZK12	5.16-59.6	36.4	3.77-6.45	5.42	<0.05	<0.1-2.34	1.43-3.42	2.29
Disch.	15		10		0.5	2.0	2.0	

Table 3 (cont). Tailings pore water quality

No.	pH (no dimension)	SO_4^{2-} (g/L) Range	SO_4^{2-} (g/L) Mean	Cl^- Range	Cl^- Mean	NO_3^- Range	NO_3^- Mean	Fe
ZK7	6.6-7.2	2.08-2.09	1.39	45-120	55	44.6-130	58.2	<0.1-0.62
ZK1	7.0-7.8	2.24-2.36	2.31	47-185	127	4.16-9.84	6.84	<0.1-0.19
ZK5	6.8-10.6	2.08-2.14	2.11	692-704	698	14.2-29.7	21.9	<0.1-0.54
ZK9	4.2-10.1	1.95-2.50	2.27	36-1383	862	11.3-146	86.2	<0.1-0.22
ZK11	5.4-9.2	0.92-1.02	0/97	328-423	376	24.1-33.7	28.9	<0.1-0.35
ZK16	7.5-7.9	2.33-2.65	2.49	498-607	553	50.6-69.9	60.2	<0.1
ZK14	6.4-7.5	1.46-2.05	1.78	11-80	53	32.1-652	246	<0.1-0.54
ZK6	5.7-6.8	2.28-2.29	2.28	45-125	85	11.1-22.8	17.0	<0.1-2.64
ZK2	7.3-8.4	1.34-3.35	2.56	33-101	75	7.08-56.3	23.8	<0.1-0.24
ZK3	9.3-10.2	1.75-2.43	2.06	214-643	485	46.8-68.9	58.4	<0.1-0.35
ZK8	4.0-7.3	2.11-2.93	2.38	30-151	97	5.02-17.1	9.30	<0.1-4.46
ZK10	9.1-9.8	2.45-2.84	2.58	826-1056	960	146-189	163	<0.1-0.18
ZK15	5.9-7.9	1.98-2.44	2.23	286-365	324	40.1-82.1	59,0	<0.1-0.24
ZK12	8.8-10.0	1.46-1.99	0.81	36-52	46	68.1-280	184	<0.1-0.21
Disch. limit	6-9	—		—		—		—

Table 4. Seepage quality （mg/L, except for U μg/L, μ^{226}Ra Bq/L, SO_4^{2-}g/L, pH no dimension）

No.	Yuejing pump Range	Yuejing pump Mean	Songlin pump Range	Songlin pump Mean	Ximei pump Range	Ximei pump Mean	Napi pimp Range	Napi pimp Mean	Zhandouu pump Range	Zhandouu pump Mean
pH	4.4-4.7		6.4-8.8		4.7-5.9		3.3-4.3		6.7-6.8	
U	26.1-39.4	34.7	239-1090	622	82.7-243	164	182-340	246.1	58.6-67.2	62.9
^{226}Ra	0.5-4.0	2.5	37.2-72.1	56.7	2.1-13.6	7.48	4.0-6.4	5.0	7.60-15.9	11.8
NH_3-N	2.23-4.18	3.41	33.9-79.6	54.82	12.0-15.0	13.41	7.49-13.1	11.0	0.32-3.52	1.92
F-	0.92-2.38	1.89	5.18-9.23	6.93	4.10-9.31	5.83	2.94-6.64	4.94	1.68-6.95	4.32
Mn	3.58-9.35	6.93	4.07-5.43	4.78	12.7-22.8	16.47	12.6-34.3	24.00	0.64-1.76	1.20
SO_4^{2-}	1.04-1.18	1.12	1.73-2.13	1.99	1.13-2.35	1.72	1.25-2.13	1.71	1.37-1.45	1.41
Cl-	24-52	38	68-141	94	123-171	144	50-141	92	27-48	38
NO_3^-	31.8-59.2	40.1	64.3-115	91.47	21.9-89.5	60.8	21.0-37.5	26.72	101-263	182
Zn	0.35-1.22	0.65	<0.1-2.68		0.30-1.47	0.78	0.69-2.79	1.96	0.13-0.65	0.39
Cd	<0.02	<0.02	<0.02-0.1		<0.02	<0.02	<0.02	<0.02	<0.05	<0.05
Pb	<0.05	<0.05	<0.05	<0.05	<0.05	<0.05	<0.05	<0.05	<0.05	<0.05
Cr	<0.02	<0.02	<0.02	<0.02	<0.02	<0.02	<0.02	<0.02	<0.02	<0.02
Cu	<0.05-0.08		<0.05-0.08		<0.05-0.45		<0.05-0.15		<0.05	<0.05
Ni	0.08-0.13	0.10	0.09-0.29	0.18	0.11-0.31	0.20	0.25-0.82	0.60	<0.05-0.06	

Fe	<0.1	<0.1	<0.1	<0.1	≤0.1	<0.1-0.21	<0.1	<0.1

Pore water and seepage quality is complicate. It contains higher U, ^{226}Ra, NH_3-N, F^-, Mn, SO_4^{2-}, NO_3^-. NH_3-N and Mn of pore water and U and Mn of seepage are more than discharge limit (GB 23727-2009 2009; GB 8978-1996 1996). Other heavy metal concentrations are low. pH of pore water is from 4.0 to 10.6. pH of seepage is from 3.3 to 8.8, less than pH of pore water quality. pH of some water samples are more than discharge limit. Some seepage samples of Songlin pump house are alkaline and the other seepage is acid..

Special Variation of Pore Water and Seepage Quality

For integrated pore water quality (Fig.2), NH_3-N and Cl^- concentrations of deeper pore water are slightly less than the ones of shallow pore water, and U、^{226}Ra、F^-、SO_4^{2-}、NO_3^- more than the shallow pore water. Mn concentrations of deeper pore water are more than the ones of shallow pore water 1.9 times.

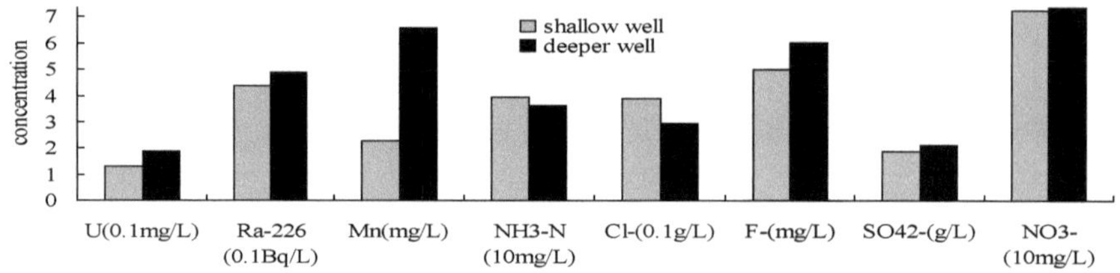

Fig.2. Integrated pore water quality of shallow and deeper boreholes

Three group of pore water monitoring boreholes (i.e., zk7-zk6, zk1-zk2, zk5-zk3) at different distance from the second water dam (Fig.1). For shallow pore water, NH_3-N and Cl^- concentrations increased with the increase of distance from the dam. For deeper pore water, NO_3^- concentrations increased with the increase of distance from the dam and Mn、F^- concentrations reduced.

Seepage volume from Zhandou dam is less and there is sometimes no seepage.

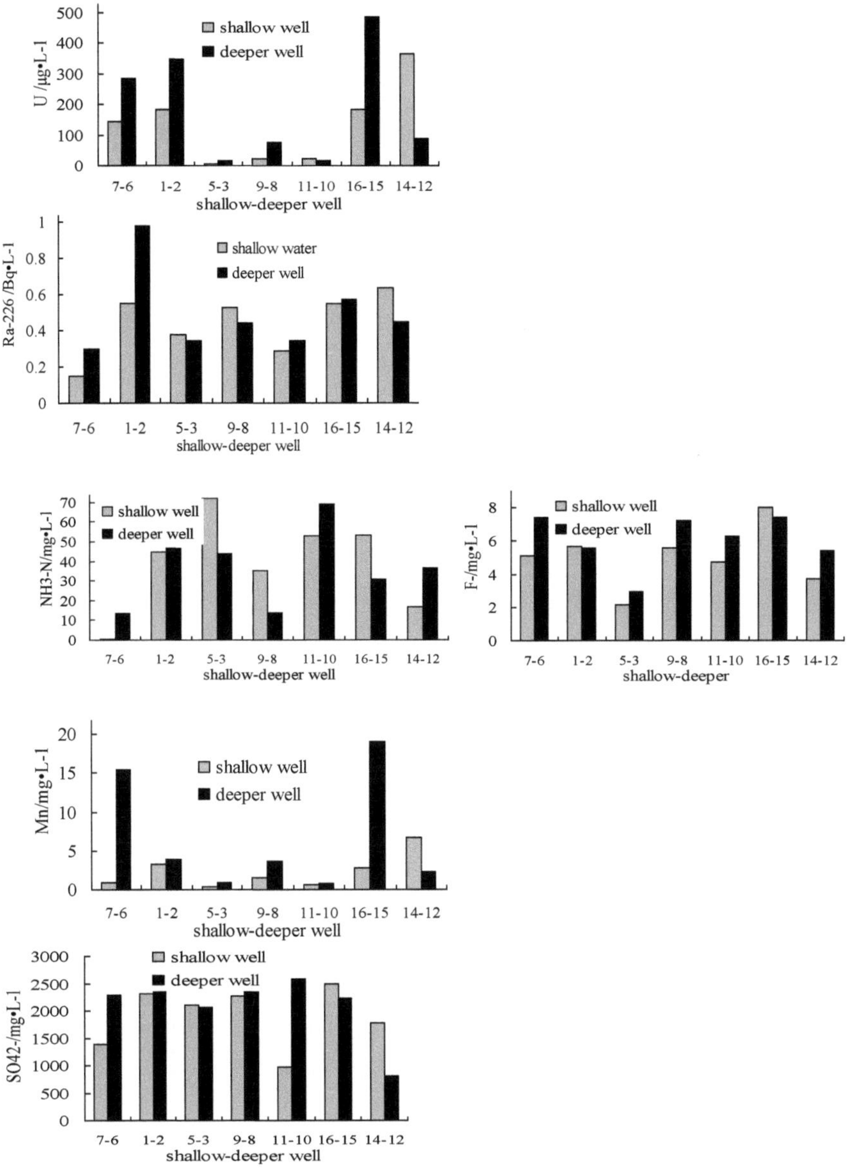
U /μg•L-1
shallow well
deeper well
500
400
300
200
100
0
7-6
1-2
5-3
9-8
11-10
16-15
14-12
shallow-deeper well
Ra-226 /Bq•L-1
shallow water
deeper well
1
0.8
0.6
0.4
0.2
0
shallow-deeper well
NH3-N/mg•L-1
shallow well
deeper well
70
60
50
40
30
20
10
shallow-deeper well
F-/mg•L-1
8
6
4
2
shallow-deeper
Mn/mg•L-1
20
15
10
5
shallow-deeper well
SO42-/mg•L-1
3000
2500
2000
1500
1000
500
shallow-deeper well

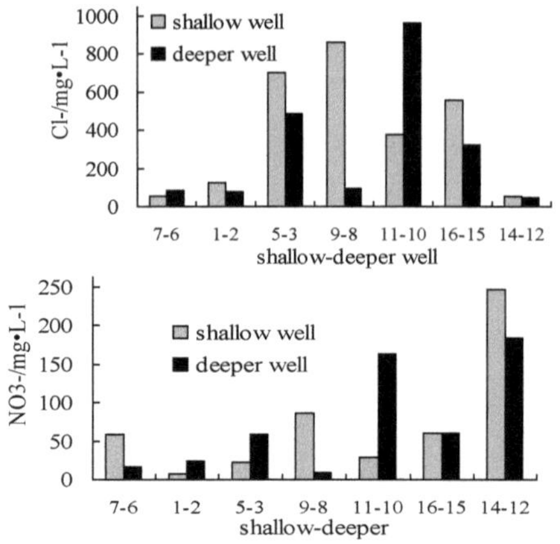

Fig.3. Pore water quality of different shallow and deeper boreholes

Water quality variation is obviously different from pore water and seepage sampling points from Table 3, Table 4 and Fig. 3.

Temporal Variation of Seepage Quality

In general, seepage constituents' concentration in wet water season is less than in low and normal water season, and U, ^{226}Ra and Mn concentrations from Songlin pump are higher than in low water season (Fig. 4)

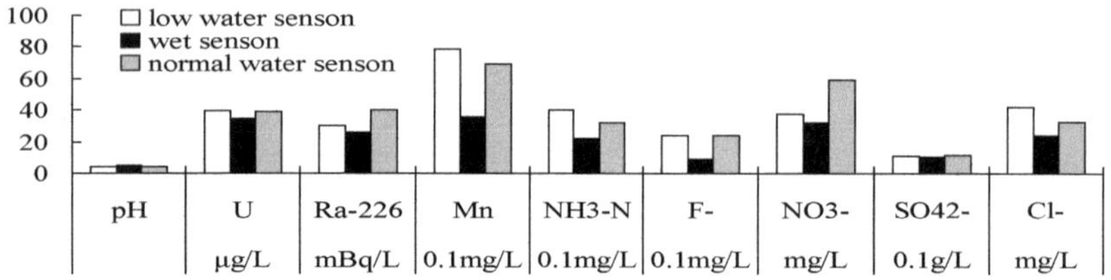

(a) seepage from Yuejin pump

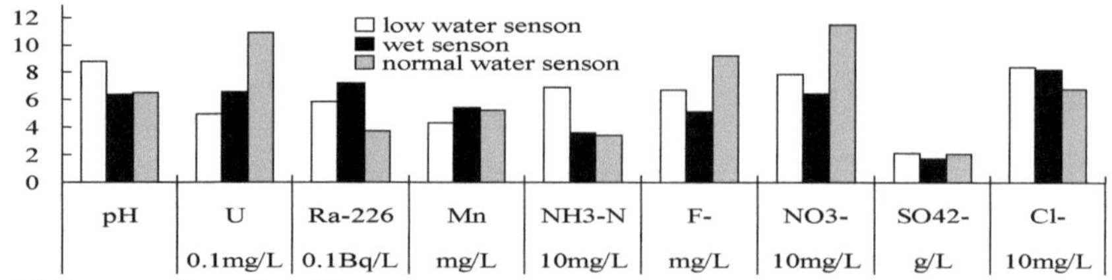

(b) Seepage from Songlin pump

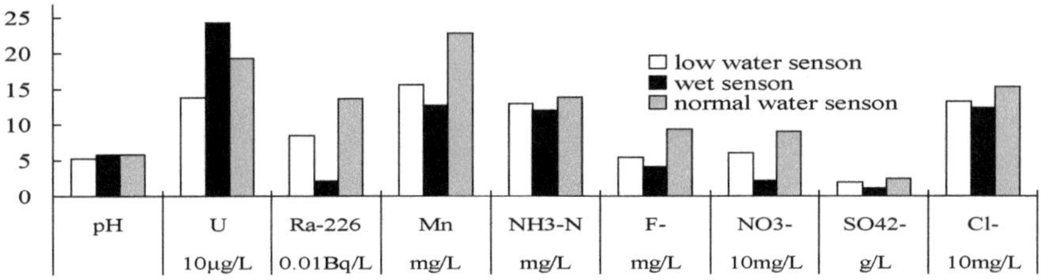

(c) Seepage from Ximei pump

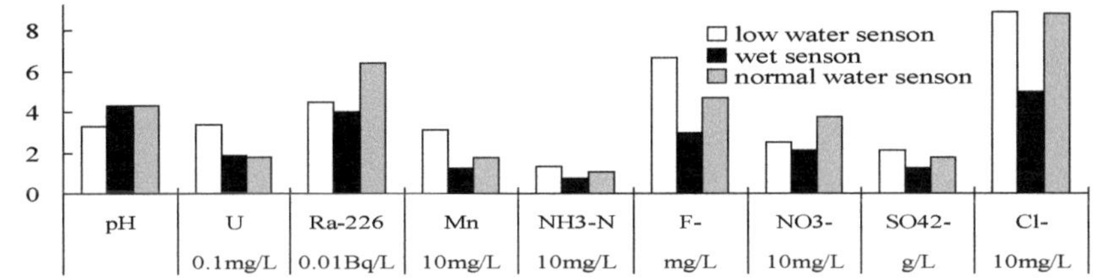

(d) Seepage from Nanpi pump

Fig.4. Seepage quality in different water seasons

Conclusion

Pore water and seepage quality are complicate. It contains higher U, ^{226}Ra, NH_3-N, F^-, Mn, SO_4^{2-}, NO_3^-, U, NH_3-N and Mn of which exceed discharge limit. quality variation of pore water and seepage is obviusly at different sampling poits. For integrated pore water quality, NH_3-N and Cl^- concentrations of deeper pore water are slightly less than the ones of shallow pore water, and U、^{226}Ra、F^-、SO_4^{2-}、NO_3^- more than the shallow pore water. In general, seepage constituents' concentration in wet water season is less than in low and normal water season.

References

HJ/T 164 (2004). Technical specifications for environmental monitoring of groundwater.
GB 23727—2009 (2009). Regulations for radiation and environment protection in uranium mining and milling.
GB 8978—1996 (1996). Integrated wastewater discharge standard

Today is a great day for work! All readings on my detector are correct and reproducible!
15.08
Oh, look ...
You forgot to plug it in.
-0.0002

Neotectonics influence of identified active geological structures on the safety of uranium tailings production Dniprodzerzhynsk industrial agglomeration (Ukraine)

Yuliia Yuskiv[1], **Valentin Verkhovtsev**[1]

[1]Institute of Environmental Geochemistry of National Academy of Sciences of Ukraine (Kyiv), Ukraine; yuskiv_yu@ukr.net

Abstract. The main results of large-scale (1:25 000) neotectonics mapping within Dniprodzerzhynsk industrial site in the Dnipropetrovsk region of Ukraine are given. Neotectonic mapping was made on the basis of morphosructural (morphotectonic) methods. Neogeotectonic study allows to determine the influence of recent activity geostructures and their activation impact on rock properties (physical, mechanical, filtration and other) on which tallings are located. The location area safety of tailings facilities and uranium production waste of the Dneprodzerzhinsk industrial site by neotectonic criteria is evaluated.

Introduction

The study area is located in the Dnipropetrovsk region of Ukraine within the industrial agglomeration Dniprodzerzhynsk. In geostructural relation research area is located in the junction zone of two major tectonic geostructures East European platform: the north-eastern slope of the Ukrainian Shield and south-western side of the Dnieper-Donets basin. The territory of Dnipropetrovsk is located in zone of Dneprodzerzhinsk fault which extends diagonally from northwest to southeast. It divides Kryvorizhskyy and Dnipropetrovskyi blocks of Srednepridneprovsky megablock. (Yuskiv 2014).

From 1949 to 1991 Industrial Enterprise «Pridneprovskiy Chemical Plant» provided the technological processing of uranium ore. As result of it working seven tailings («Dniprovske», «Zakhidne», «Central Yar», «Pivdenno-Skhidne», «Lanthanum fraction», «Sukhachivske» (Ist, IInd sections)), two storage of uranium waste («Baza S», «Domenna pich-6») were formed (Fig. 1). At present, about 42 million tons of waste with total activity $3{,}2\times10^{15}$ Bq (86,000 Ci) accumulated in tailings (Verkhovtsev et al. 2010; Kovalenko 2008).

B.J. Merkel and A. Arab (eds.), *Uranium – Past and Future Challenges*,
DOI 10.1007/978-3-319-11059-2_59

Impermeability of these objects is the one of the main requirements for their existence. It is necessary to find out possible ways of radionuclide migration in case of decapsulation. The most dangerous environmental consequence is possible radionuclide migration into underground aquifers and mines.

Neotectonic mapping allows to determine areas of surface high correlation and groundwater as a possible migration routes of highly toxic elements and their penetration into underground aquifers and mines.

Terminological aspects and methods of research

Neotectonics – tectonic movements, breach and changes in the landscape of the earth's surfaces that occurred during the Paleogene and Neogene periods and were continued during the Quaternary period.

Lineament – the most common form of elementary line change of geographic, geological and geophysical environments. They can be faults, zones of increased fracturing, structural mismatch, stratigraphic, lithological and petrographic boundaries, linear geophysical anomalies, relief elements and other components of the landscape, also including not related to geological processes. Preferably lineaments include: 1) a linear depth inhomogeneity, 2) fault (fault zone) in the hard shell of the Earth, and 3) linear landscape elements organized on the surface.

Lineament system – a system that created straight (or close to those) and conjugated perpendicular lineaments of two directions.

Ring structure –diverse (by the genesis, size, form of display on the surface, etc.) structural and geological objects which have central symmetry in the cross section of the earth's surface.

The methods of neotectonic mapping:

1) morphostructural (morphotectonic) analysis of topographic base includes the morphographic and morphometric features study of relief and comparing obtained results with a priori geological and geophysical data. Areal research of relief morphographic features is provided by the method of directional directions and anomalies. Research morphometric features of relief is passed using morphometric methods of finding geostructures, defined by Filosofov (Filosofov 1975; Verkhovtsev et al. 2010; Grohmann 2011); 2) structural and geological interpretation of images materials aerial survey.

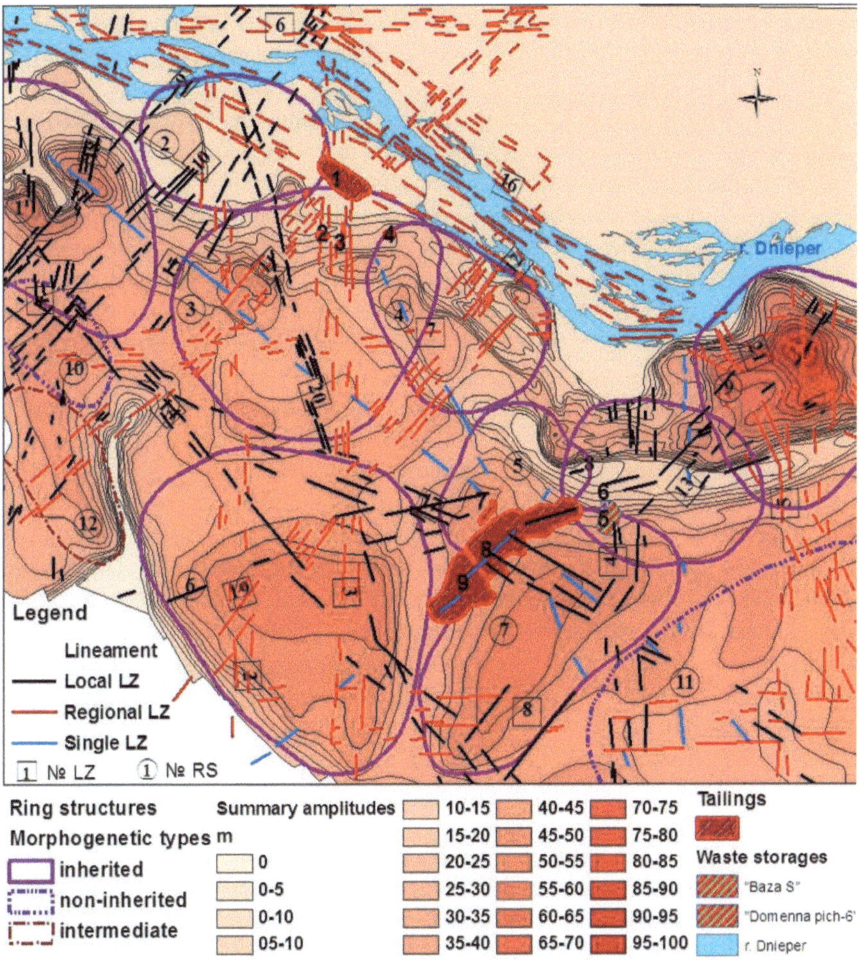

Fig.1. Neotectonic map of the territory of uranium production legacy site at Dniprodzerzhynsk. Composed by: Y. Yuskiv, V. Verkhovtsev. Scale 1:25 000 (reduced). Legend: Tailings and waste storages of uranium production: 1 – «Dniprovske», 2 – «Zakhidne», 3 – "Central Yar", 4 – «Pivdenno-Skhidne», 5 – «Baza S», 6 – «Domenna pich-6», 7 – «Lanthanum fraction», 8, 9 – «Sukhachivske» (Ist, IInd sections).

The main results of neotectonic studies.

We have constructed such maps of scale 1:25 000 as: streams erosion and watersheds; landscape indicators linear structures; base-level of 2 and 4 orders (of the streams erosion) and apical-level of 2 and 4 orders (of the watersheds), remainder map of difference between base-level of 4 order and apical-level of 4 order; linear structures; ring structures; summary amplitudes of Late Pliocene-Quaternary vertical movements of crustal movements (hereinafter - the summary amplitudes) and resulting map of neotectonics (Fig. 1).

Two types of objects (lineament zones (LZ) and ring structures (RS)) have been detected and studied as a result of morphotectonic mapping. A detailed study of these structures showed that first reflect mainly different order of discontinuous linear structure, and the second – volume-area discontinuity of tectonosphere.

21 lineament zones (LZ) were detected within the study area. These zones are ranked on: regional and local. Lineaments are forming two dominant systems – orthogonal (0°⊥90°, ±5°) and diagonal (40-45°⊥310-315°), two intermediate diagonal systems (25-30°⊥295-300°; 74-80°⊥345-350°) and one depressed area (15-20°) with single lineament zone (Table 1)

Ring geostructures. Depending on the nature, the identified RS are divided into several morphogenetic types as originally defined by Filosofov with partial changes and additions by Verkhovtsev (Filosofov 1975; Verkhovtsev et al. 2010): 1) inherited, 2) non-inherited 3) intermediate morphogenetic types (Table 2).

Inherited RS are fixed in terms of the central part position at Late Pliocene-Quaternary time. They are shown on all static morphometric maps and characterized by the coincidence contours which allocated to all built base-level and apical-level surfaces (the difference should not exceed 1/3 the size of a generalized circuit). In most cases they are clearly expressed in morphographic data (base-level and apical-level maps) and total amplitudes and confirmed by geological and geophysical data.

Non-inherited RS also reflected on all morphometric maps (base-level and apical-level maps) but the value of displacement contours to each other is more than 1/3 the size of a generalized circuit.

Intermediate RS occupy an intermediate position between inherited and non-inherited: they are fixed on both static morphometric maps junior orders and one of the older orders (base-level and apical-level maps of second and fourth orders).

12 ring structures (RS), which active on the latest stage (Late Pliocene–Quaternary), were allocated. There are: nine inherited, two non-inherited and one intermediate morphogenetic types. Their sizes are from 3936x2198 m to 10474x7678 m, and the estimated (calculated) depth – from 1158.4 to 5237 m.

Summary amplitudes of late Pliocene-Quaternary vertical crustal movements show that within the study area they have differentiated and highly significant intensity (maximum values exceeding +95 m, minimum – 0 m). Minimal summary amplitudes (from 0 to 10 m) are marked in the northern part of the area, and tend to the valleys of the Dnieper River and river Syha Syra. Maximum values are marked in the north-eastern corner of the study area and reaches +90 m. In general on the area dominate values from +25 to +40 m.

All identified structures were confirmed in geological mapping data and at hydrogeochemical modeling. The correlation between the identified local areas of high radon activity and active fault structures are visible with comparing the results of neotectonic studies on radioecological data.

Table 1. Basic information about the lineament zones of the territory of uranium production legacy site at Dniprodzerzhynsk

№	*Rank of LZ*	*Az⁰*	*L* / *W, m*	*N* / *$L_1 - L_2$, m*	*N_e* / *N_w*	*Matching with known fault*	*Estimated morpho-type*
			LZ dominant orthogonal system 0 ⊥ 90 (±5)				
1	local	356-358	15 000 / 216-890	29 / 160-1 544	18 / 14	–	reset-shift
2	regional	358-0	17 208 / 0,309-0,897	25 / 160-1 214	20 / 16	–	reset-shift
3	regional	359-1	20 441*2 / 295-976	43 / 220-1 626	22 / 27	+	shift-reset
4	local	357-3	13 319 / 656-1 200	32 / 138-1 547	15 / 19		shift
5	regional	358-1	15 / 948-1 648	31 / 119-1 618	15 / 22		shift- upthrust
6	regional	90-91	14 551*2 / 326-748	45 / 215-1 035	14 / 32	+	upthrust
7	regional	88-90	24 720*2 / 252-764	39 / 88-1 716	19 / 29	–	shift- upthrust
8	regional	88-90	17 185 / 855-1 625	27 / 230-2 268	12 / 18	–	shift
			LZ diagonal dominant system 40-45⁰ ⊥ 310-315⁰				
9	local	40-45	10 226*2 / 260-634	28 / 135-1106	16 / 14	–	shift
10	local	40-45	8 167*1 / 269-1047	32 / 102-1300	11 / 24	–	shift- upthrust
11	regional	40-45	19159*2	73	36	+	shift

			387-929	107-1732	43		
12	regional	40-45	16 579*2 / 852-1922	54 / 161-1860	26 / 32	+	shift
13	local	39-45	18 833 / 516-1500	20 / 286-2968	11 / 13	–	shift
14	local	310-315	22 973*2 / 338-948	30 / 219-2704	18 / 16	–	shift
			LZ intermediate diagonal system $25\text{-}30^0 \perp 295\text{-}300^0$				
15	local	25-30	16 790*1 / 418-899	33 / 123-1362	12 / 22	–	upthrust-shift
16	regional	295-300	26 602*2 / 2300- 4303	158 / 116-2062	100 / 66	Dniepro-dzerzhinsky fault	shift-reset
17	local	295-300	28 305*1 / 513-1420	37 / 144-2865	21 / 19		shift
			Lineament zone of the depressed direction $15\text{-}20^0$				
18	local	15-20	13 497*1 / 170-616	19 / 169-835	9 / 12	–	upthrust-shift
			LZ intermediate diagonal system $74\text{-}80 \perp 345\text{-}350^0$				
19	local	74-80	19 646 / 727-1645	21 / 121-2858	9 / 13	–	upthrust-shift
20	local	345-350	19 689 / 320-715	31 / 220-1094	14 / 15	+	shift
21	regional	334-328	13 620 / 1136-1352	26 / 160-1780	14 / 16	–	shift

Note: Az^0 – the general strike direction (in degrees); L - length; W – width of LZ; N – number of basic lineaments that are included in LZ; L_1 – minimum ; L_2 – maximum length of lineaments; N_e – the number of erosion; N_w – the number of watershed landscape indicators lineaments;+ – LZ completely coincides with known fault. *1 – LZ goes beyond the study area in one direction; *2 – too, but in both directions.

Table 2. Basic information about ring structures of the territory of uranium production legacy site at Dniprodzerzhynsk

№	*Size (diameter), m*	*base-level of 2 order*	*apical-level of 2 order*	*base-level of 4 order*	*apical-level of 4 order*	*Summary ampli-tudes, m*	*Morpho-graphic data*	*The cal-cula-tion depth, m*
				Inherited RS				
1*	8154x 5180	++	++	++	++	++↑ 40-65	+	4077x 2590
2	3558x 5045	++	++	++	++	+15-20	++	1779x 252265
3	6991x 6647	++	++	++	++	++↑ 35-55	++	3495,5x 3323,5
4	5468x 3707	++	++	++	++	+↑ 30-45	++	2734x 1853,5
5	4450x 2994	++	++	++	++	+↕ 35-50	++	2225x 1497
6	9177x 7040	++	++	++	++	++↑ 40-60	++	4588,5x 3520
7	8701x 4000	++	++	++	++	++↑ 45-60	++	4350,5x 2000
8	5519x 4202	++	++	++	++	+↕ 5-25	++	2759,5x 2101
9	6201x 5515	++	++	++	++	++↑ 10-100	++	3100,5x 2757,5
				Non-inherited RS				
10	3936x219 8	+	+	++	++	+↑30-45	+	1968x 1099
11*	10474x 7678	+	+	++	++	+↑35-45	+	5237x 3839
				Intermediate RS				
12*	5245x231 7	+	+	+	–	+↑35-50	+	2622,5x 1158,5

Note: * – RS goes beyond the region ++ – a complete, + – fragmentary (unclear) reflection of analyzed in parameters RS; - – the structure is not visible in relevant indicators; ↑ – coincidence RS circuit area with higher values of summary amplitudes, ↓ – coincidence RS circuit area with lower values of summary amplitudes (in meters).

Influence of identified active geological structures on the safety of uranium tailings production, prediction of places of active absorption of polluted water by neotectonic data

The most dangerous areas are allocated in combination of the neotectonic parameters:

1. the presence of LZ (active on the neotectonic stage of faults development). The most "favorable" those, among which stands out the highest number of basic lineaments is noted preference streams erosion indicators over watershed indicators.
2. the presence of local, usually positive RS (increased values of summary amplitudes). "Perspective" section is allocated in the peripheral parts of these structures, characterized by increased fracturing, or on slight distance from its outer contour.
3. the presence coincidence with LZ and positive RS local anomalies of summary amplitudes increased values, which contributes to hydrogeological structures reveal.

The main results of this analysis:

Tailing "Dniprovske" is located in a dangerous place by neotectonic criteria. It is confined to the node intersection of three LZ №3 and №11, №16. Especially dangerous zones are № 3 and № 11 because the migration of contamination is supposed by them (to the north and north-east). The RS № 2 and 3 affect directly at the tailing as it is located in their peripheral parts. At the same time, by vertical movements, it is a relatively quiet place and characterized by summary amplitudes of +5 m or less.

Tailings "Zakhidne" and *"Central Yar"* are located directly within the local LZ №3 and № 3 RS, and it may be the ways of active contamination migration. There are local anomalies of summary amplitude values +30-35 m on south of the tailings.

Tailing "Pivdenno-Skhidne" is located in the best neotectonic conditions.

The rise of radionuclides migration is assumed on the territory where the *tailings "Lanthanum fraction", "Sukhachivske» (Ist and IInd section), "Domenna pich-6", "Baza S".* It is caused by the influence identified LZ № 4, 17, 19 and erosion form that associated with of tailings. Tailing "Sukhachivske" is framing at RS № 7 also observed crossing at RS № 5, 8, 7. Average summary amplitude is +40-50 m, so they, presumable, have some minimal impact.

References

Grohmann C. H. C. Riccomini and M. A. C. Chamani. "Regional scale analysis of landform configuration with base-level (isobase) maps." Hydrology and Earth System Sciences 15, pp.1493-1504, 2011.

Filosofov V.P. Fundamentals of morphometric methods in search of tectonic structures. Saratov University Publishing House, Russia, Saratov, 1975. (in Russian)

Kovalenko G.D. Radioecology of Ukraine. Izdat. Dom "INŽĖK", 2008. (in Russian)

Verkhovtsev V.G., Lysychenko G.V., Yuskiv Y.V. Safety assessment of the radioactive waste tailing dumps area of Dniprodzerzhynskyi industrial region by neotectonic criteria. Scientific Papers «Geochemistry and Ecology», Ukraine, vol. 16, pp. 98-108, 2010. (in Ukrainian).

Yuskiv Y.V. Up-to-date geotectonics of location area of tailings facilities of uranium production within industrial agglomeration of Dneprodzerzhinsk (Ukraine). International scientific journal Science and world, Russia, № 2 (6), 2014, Vol. I (in Russian).

...sometimes life seems like a bully at school who always picks at you and doesn't leave you alone.

Re-Engineering Antibodies for Optimum Performance in Uranium Sensors

D. A. Blake[1], B. Ban[1,2], X. Li[1,3], R. C. Blake II[2], G. A. Jairo[1], Y. Sun[1]

[1]Department of Biochemistry and Molecular Biology, Tulane University School of Medicine, New Orleans LA 70112 USA;
[2]Division of Basic Pharmaceutical Sciences, College of Pharmacy, Xavier University of Louisiana, New Orleans, LA 70125 USA;
[3]Institute of Vegetables and Flowers, CAAS, Beijing, China

Abstract. Our laboratory has developed antibody-based sensors that provide on-site, near real-time information about uranium in environmental samples. The monoclonal antibody at the heart of this technology (clone 12F6) specifically recognizes U(VI) complexed to a chelator (2,-9-dicarboxyl-1-10-phenanthroline). Mutagenesis of a single amino acid in the 12F6 antibody light chain significantly increased its binding to chelated U(VI). Insertion of a gold-binding peptide sequence enhanced the antibody's ability to complex with the gold nanoparticles, a process required for the development of simple lateral flow strips for uranium monitoring.

Introduction

The detection and quantification of uranium is important both for monitoring remediation efforts at previously contaminated sites and for protecting personnel and the environment during current and future mining/processing operations. Most existing technologies for uranium analysis (ICM-MS and AAS) require extensive sample pretreatment and complex instrumentation at a centralized facility (Rozmaric et al., 2009). Simpler instrumentation like the Kinetic Phosphorescence Analyzer (KPA) is also relatively expensive and therefore used primarily in a laboratory setting; in addition, the KPA is useful only for the analysis of uranium and lanthanides and the reagents required are expensive and can be subject to interference from sample matrixes (Brina and Miller, 1993).

Our laboratory has previously described immunosensor-based measurements of uranium that have been used in the field to provide near real-time data about changes in U(VI) levels in groundwater during *in situ* bioremediation experiments (Melton et al., 2009). These assays were based on the ability of the antibody used in the assay (clone 12F6) to bind tightly and specifically to uranyl ion (UO_2^{2+})

B.J. Merkel and A. Arab (eds.), *Uranium – Past and Future Challenges*,
DOI 10.1007/978-3-319-11059-2_60

complexed to 2,9-dicarboxy-1,10-phenanthroline (UO_2^{2+}-DCP) (Blake et al., 2004). In this chapter, we describe the expression of this antibody as a recombinant protein and subsequent protein engineering to enhance its performance in uranium immunosensors.

Engineering of antibody 12F6

Molecular modeling and site-directed mutagenesis

Antibodies have been generated for millions of different antigens and experimental structural information (NMR data or crystal structure) is not available for every antibody of interest. However, because all antibodies share similar 3-dimensional characteristics, a procedure known as homology modeling can often provide the investigator with a reasonable approximation of the antibody binding site (Al-Lazikani et al., 1997; Morea et al., 1997). This modeling can help generate hypothesis-driven experiments through visualization of the molecular environment at the antigen binding site. Site-directed mutagenesis can then be used to investigate and identify amino acid residues of the protein that are involved in ligand binding (Delehanty et al., 2003). This approach was applied to the 12F6 antibody and the results are summarized in the space-filling model of the antibody binding site shown in Fig. 1. The antibody was expressed as a recombinant fragment as described in (Delehanty et al., 2003) and modeled using the Swiss-Model automated modeling server at ExPAsy (Peitsch, 1996). The specific animo acid residues that were predicted to interact with chelated $UO_2^{2}+$ were mutagenized using a Quick-change Site-directed Mutagenesis Kit (Stratagene, La Jolla, CA). All of the amino acid residues highlighted in Fig. 1 were mutated to alanine (with the exception of Y^{49} in the light chain, which for technical reasons was mutated to valine). The resulting mutated antibodies were expressed in culture and their binding activities were compared to the unmodified antibody using indirect ELISA. Those antibodies that retained activity were also analyzed by kinetic exclusion analysis (Blake et al., 2004).

The mutation most deleterious to antigen recognition was the conversion of Y^{49} in CDR2 of the light chain to valine (see Fig. 1). This mutation reduced the binding of the antibody to 6.4% of that observed in the wild-type. In contrast, alanine conversion of the neighboring lysine residue in CDR2 of the light chain, K^{50}, had little or no effect on binding to the immobilized metal-chelate-protein conjugate and actually increased binding to the soluble UO_2^{2+}-DCP complex. Alanine conversion of residue H^{34} in CDR1, hypothesized to interact either indirectly with UO_2^{2+} moiety of the metal-chelate complex or directly with the carboxylates of the phenanthroline moiety, decreased binding to ~30% of unmodified 12F6.

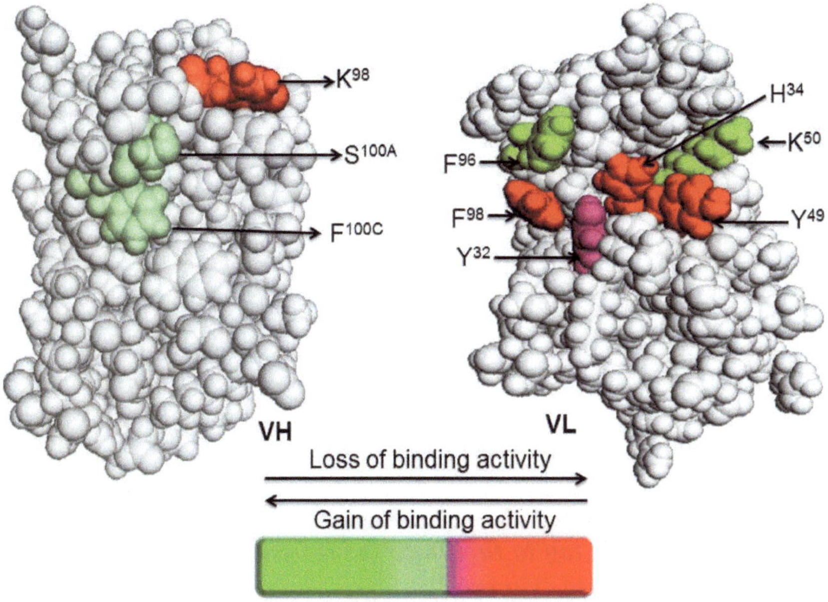

Fig.1. Molecular modeling and site-directed mutagenesis of the uranium-specific 12F6 antibody. VH, variable region of the antibody heavy chain; VL, variable region of the antibody light chain. The following single letter designations were used for mutated amino acids: F, phenylalanine; H, histidine; K, lysine; S, serine; Y, tyrosine

Tyrosine and serine residues occur with high frequency in antibody binding loops (Birtalan et al., 2008; Koide and Sidhu, 2009) and these residues have the potential to interact via both hydrogen bonds and hydrophobic stacking interactions with the UO_2^{2+}-DCP antigen. In addition, the phenylate moiety of tyrosine has been shown to coordinate directly with uranium in some protein crystal structures (Larsen et al., 1999; Pokkuluri et al., 2002). Alanine mutagenesis of Y^{32} in CDR1 of the light chain or Y^{98} and S^{100A} in CDR3 of the heavy chain had an intermediate effect in indirect ELISA, reducing binding to the immobilized UO_2^{2+}-DCP by less than 50%. Conversion of F^{100C} to alanine in the heavy chain similarly reduced binding by ~25%.

The only mutation that provided an antibody with a higher affinity for the UO_2^{2+}-DCP complex was the conversion of the lysine at position 50 in the heavy chain to an alanine ($K^{50} \rightarrow A$). This antibody was subsequently expressed as a single polypeptide fragment (scFv) by linking the heavy chain and light chain variable regions of from the parent 12F6 monoclonal antibody via a flexible linker arm. Each V_H-linker-V_L construct was inserted into a bacterial expression vector (POE-c-myc) that directed secretion of the recombinant antibody to the periplasm of *E. coli.* The recombinant antibodies were purified from periplasmic extracts and kinetic exclusion analysis was used to determine binding constants for both metal-free DCP and the UO_2^{2+}-loaded DCP complex. The values of binding constants (K_a) for selected parental and mutant scFv antibodies are summarized in Table 1. The K^{50}→A mutant had a small but significant (p=0.0052) increase in affinity for the UO_2^{2+}-DCP complex and a decreased affinity for metal-free DCP (p<0.001). Thus, this mutant not only bound more tightly to UO_2^{2+}-DCP but also was able to discriminate more efficiently between the metal-loaded and metal-free chelator, which is an important property for the sensor development.

Table 1. Binding constants[a] for 12F6 and the K^{50}→A Mutant

Antibody Construct	Ligand[b]	K_a, M^{-1}
12Fv scFv	UO_2^{2+}-DCP	$129.9 \pm 15.2 \times 10^6$
K^{50}→A scFv	UO_2^{2+}-DCP	$212.8 \pm 22.0 \times 10^6$
12F6 scFv	Metal-free DCP	$1.25 \pm 0.09 \times 10^6$
K^{50}→A scFv	Metal-free DCP	$0.80 \pm 0.06 \times 10^6$

[a] Equilibrium association constants (K_a) were determined as described in (Blake et al., 2004).
[b] DCP is an abbreviation for 2,9-dicarboxyl-1,10-phenanthrene, a specific chelator of UO_2^{2+}

Insertion of a peptide tag to facilitate the transduction of antibody binding into a detectable signal

The interaction of an antibody with an antigen is not readily detectable, and the function of a biosensor is to transform this binding event into a quantifiable signal. Previous work in our laboratory has utilized fluorescence-based transduction techniques in a kinetic exclusion format (Blake et al., 2001; Fisher et al., 2011; Melton et al., 2009; Yu et al., 2005). However, such sensors were not amenable to further miniaturization or multiplexing because the size of the pump required to move liquid required a good deal of power and the single observation cell made multiplexing beyond 2 analytes technically very difficult (Glass et al., 2004). Paper has re-emerged as a promising support material for sensing devices, because paper-based sensors represent cheap, portable and simple detection tools. They are small, easy to use, portable and low cost; they require minimal sample volume and the assay can be performed by individuals with very little scientific training (Posthuma-Trumpie et al., 2009). Lateral flow immunoassays using paper-based platforms can provide semi-quantitative analyses using even naked-eye detection of colorimetric changes, and multiplexing is as simple as running two strips at the same time.

The standard method for labeling the antibodies used in such lateral flow devices is the adsorption of the antibodies to gold nanoparticles (Thobhani et al., 2010). However, during previous studies with lateral flow immunoassays (Lopez et al., 2013a; Lopez et al., 2013b), we discovered that this adsorption process was inefficient and led to significant loss of antibody binding activity. We therefore initiated studies to insert specific peptide tags into our engineered antibodies, using a previously described peptide with gold-binding activity (GBP, Kacar et al., 2009). Two new constructs of the 12F6 scFv were created: 12F6 with 2 GBP repeats at the carboxyl-terminal of the scFv, and 12F6 with 2 GBP repeats replacing the flexible linker region between the light and heavy chain variable regions of the scFv. The new constructs, when expressed in bacterial cell cultures, retained their ability to bind to UO_2^{2+}-DCP complexes and exhibited the expected molecular sizes in western blots. The abilities of these 2 new constructs to interact with gold nanoparticles were compared to the unmodified scFv as shown in Fig 2, below. Clearly, insertion of a specific gold-binding sequence at the carboxyl terminal of the protein enhanced the ability of this construct to interact with the gold nanoparticles. In contrast, insertion of the identical sequence in the linker region of the scFv actually inhibited gold binding. The most likely explanation of this finding is that the linker region is involved in aligning the light and heavy chain variable regions, and any modifications in this region may have global effects on protein folding that inhibited adsorption to the gold nanoparticles.

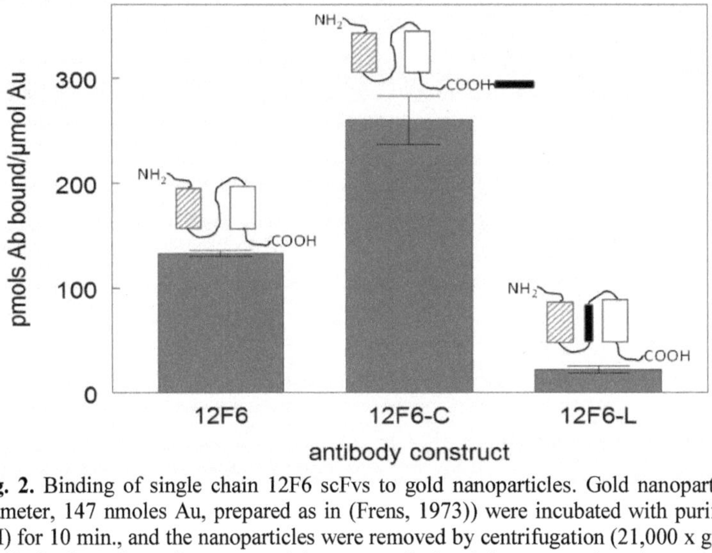

Fig. 2. Binding of single chain 12F6 scFvs to gold nanoparticles. Gold nanoparticles (20 nm diameter, 147 nmoles Au, prepared as in (Frens, 1973)) were incubated with purified scFv (90 nM) for 10 min., and the nanoparticles were removed by centrifugation (21,000 x g for 45 min.). Antibody bound to the nanoparticles was calculated by comparing the activity in these supernatants with supernatants prepared in the absence of nanoparticles. 12F6, unmodified scFv; 12F6-C, scFv with 2 gold-binding repeats at the carboxyl terminus; 12F6-L, scFv with 2 gold-binding repeats in the linker region. Hatched rectangle, heavy chain variable region; white rectangle, light chain variable region; black oblong, gold binding peptide, $(MHGKTQATSGTIQS)_2$

Discussion and conclusions

Antibody engineering techniques, which permit molecular manipulation of antibody structure and function, have the potential to revolutionize the way scientists generate binding proteins for specific applications. In this study we re-engineered existing monoclonal antibodies to chelated uranium so that the antibodies retained their binding to chelated uranium but decreased their binding to the metal-free chelator, 2,9-dicarboxyl-1,10-phenanthroline. This re-engineering solved a practical problem with the current assay (Melton et al., 2009). The 12F6 scFv bound approximately 104-fold more tightly to the UO_2^{2+}-DCP complex than to metal-free DCP (see Table 2). The re-engineered antibody binds ~2x more tightly to the UO_2^{2+}-DCP and ~1/3 less tightly to metal-free DCP. These differences in affinity more than doubled the "window" in affinities between the uranium-loaded and metal-free chelator to 266 fold.

The relatively small "window" in affinities meant that the DCP concentrations added in our previous field assays had to be relatively low (200 nM). In these field assays, the uranium in the groundwater was released from its natural complexants by acidification and then diluted into a buffer containing 200 nM DCP, a strategy that both neutralized the sample and allowed the DCP complex to form (Melton et al., 2009). The re-engineered antibody will tolerate at least twice as much DCP in the field assay and the use of higher DCP concentrations may permit this chelator to out-compete the other uranium complexants in the groundwater sample. Experiments are currently underway with field samples to determine if this higher chelator concentration will allow us to avoid our current acid pretreatment step, which requires transporting strong acid to the field site and checking the pH of both the acidified and diluted/neutralized sample before assay. New antibodies that are less sensitive to the concentration of DCP will also increase the precision of our environmental assays and make them easier for less experienced analysts to use. As we move forward to even simpler lateral flow immunoassay formats, our new constructs with higher gold binding ability will provide uranium test strips with brighter signals and enhanced ability to detect uranium in groundwater samples.

Acknowledgments

Supported by the U.S. National Science Foundation (OISE-1253272) and by a grant from the Office of Science, Department of Energy (DE-SC0004959). We also acknowledge funding from Tulane University School of Medicine.

References

Al-Lazikani B, Lesk AM, Chothia C (1997) Standard conformations for the canonical structures of immunoglobulins. J Mol Biol 273: 927-948

Birtalan S, Zhang Y, Fellouse FA, Shao L, Schaefer G, Sidhu SS (2008) The intrinsic contributions of tyrosine, serine, glycine and arginine to the affinity and specificity of antibodies. J Mol Biol 377: 1518-1528

Blake DA, Jones RM, Blake RC 2nd, Pavlov AR, Darwish IA, Yu H (2001) Antibody-based sensors for heavy metal ions. Biosensors & Bioelectronics 16: 799-809

Blake RC 2nd, Pavlov AR, Khosraviani M, Ensley HE, Kiefer GE, Yu H, Li X, Blake DA (2004) Novel monoclonal antibodies with specificity for chelated uranium(VI): isolation and binding properties. Bioconjugate Chemistry 15: 1125-1136

Brina R, Miller AG (1993) Determination of uranium and lanthanides in real-world samples by kinetic phosphorescence analysis. Spectroscopy 8: 25-28, 30-21

Delehanty JB, Jones RM, Bishop TC, Blake DA (2003) Identification of important residues in metal-chelate recognition by monoclonal antibodies. Biochemistry 42: 14173-14183

Fisher RA, Melton SJ, Blake DA (2011) A submersible immunosensor. Int J Environ Anal Chem 91: 123-137

Frens G (1973) Controlled nucleation for the regulation of the particle size in monodisperse gold suspensions. Nature (London), Phys Sci 241: 20-22

Glass TR, Saiki H, Blake DA, Blake RC 2nd, Lackie SJ, Ohmura N (2004) Use of excess solid-phase capacity in immunoassays: Advantages for semicontinuous, near-real-time measurements and for analysis of matrix effects. Analytical Chemistry 76: 767-772

Kacar T, Zin MT, So C, Wilson B, Ma H, Gul-Karaguler N, Jen AK, Sarikaya M, Tamerler C (2009) Directed self-immobilization of alkaline phosphatase on micro-patterned substrates via genetically fused metal-binding peptide. Biotechnol Bioeng 103: 696-705

Koide S, Sidhu SS (2009) The importance of being tyrosine: Lessons in molecular eecognition from minimalist synthetic binding proteins. ACS Chem Biol 4: 325-334

Larsen TM, Boehlein SK, Schuster SM, Richards NG J, Thoden JB, Holden HM, Rayment I (1999) Three-dimensional structure of *Escherichia coli* asparagine synthetase B: A short journey from substrate to product. Biochemistry 38: 16146-16157

Lopez MAM, Pons J, Blake DA, Merkoci A (2013a) All-integrated and highly sensitive paper based device with sample treatment platform for Cd^{2+} immunodetection in drinking/tap waters. Anal Chem 85: 3532-3538

Lopez MAM, Pons J, Blake DA, Merkoci A (2013b) High sensitive gold-nanoparticle based lateral flow Immunodevice for Cd^{2+} detection in drinking waters. Biosens Bioelectron 47: 190-198.

Melton SJ, Yu H, Williams KH, Morris SA, Long PE, Blake DA (2009) Field-based detection and monitoring of uranium in contaminated groundwater using two immunosensors. Environ Sci Technol 43: 6703-6709

Morca V, Tramontano A, Rustici M, Chothia C, Lesk AM (1997) Antibody structure, prediction and redesign. Biophys Chem 68: 9-16

Peitsch MC (1996) ProMod and Swiss-Model: Internet-based tools for automated comparative protein modelling. Biochem Soc Trans 24: 274-279

Pokkuluri PR, Gu M, Cai X, Raffen R, Stevens FJ, Schiffer M (2002) Factors contributing to decreased protein stability when aspartic acid residues are in β-sheet regions. Protein Sci 11: 1687-1694

Posthuma-Trumpie GA, Korf J, van Amerongen A (2009) Lateral flow (immuno)assay: Its strengths, weaknesses, opportunities and threats. A literature survey. Anal Bioanal Chem 393: 569-582

Rozmaric M, Ivsic AG, Grahek Z (2009) Determination of uranium and thorium in complex samples using chromatographic separation, ICP-MS and spectrophotometric detection. Talanta *80*, 352-362.

Thobhani S, Attree S, Boyd R, Kumarswami N, Noble J, Szymanski M, Porter RA (2010) Bioconjugation and characterisation of gold colloid-labelled proteins. J Immunol Methods 356: 60-69

Yu H, Jones RM, Blake DA (2005) An immunosensor for autonomous in-line detection of heavy metals: Validation for hexavalent uranium. Int J Environ Anal Chem 85: 817-830

Longevity Estimates for a Permeable Reactive Barrier System Remediating a ^{90}Sr Plume

Jutta Hoppe[1,3], David Lee[2], Sung-Wook Jeen[2,4], David Blowes[1]

[1]Department of Earth and Environ. Sci., University of Waterloo, ON. Canada
[2]Atomic Energy of Canada Ltd., Chalk River Laboratories, Chalk River, ON, Canada
[3]Present address: Gerencia de Sustentabilidad, Fundacion Chile, Santiago, Chile
[4]Present address: Department of Earth and Environ. Sci. & The Earth and Environ. Sci. System Research Center, Chonbuk National University, Jeonju-si, Jeollabuk-do, 561-756, Republic of Korea

Abstract. A permeable reactive barrier system, known as the Wall and Curtain, was installed at the Chalk River Laboratories, Chalk River, Ontario, in 1998, to intercept a ^{90}Sr plume. The system employs clinoptilolite, a zeolite, as a reactive material which adsorbs ^{90}Sr. Reactive transport simulations of the site were conducted using the numerical code HydroGeoSphere to provide longevity estimates for the system. The HydroGeoSphere simulations included three solutes, for which zoned distribution coefficients were specified. Longevity estimates derived from the simulation were between 70 and 100 years for the Wall and Curtain system.

Introduction

Strontium-90 (^{90}Sr) is a mobile radionuclide which decays through beta decay with a half-life of 28.8 years to yttrium-90 (^{90}Y). ^{90}Y, another pure beta emitter, decays with a half-life of 64.4 hours to stable zirconium-90 (^{90}Zr). ^{90}Sr is a by-product of fission reactions in groundwater flow systems, ^{90}Sr is a mobile contaminant that can remain an environmental concern in aquifers for up to 10 half-lives, i.e. 300 years (Wallace et al., 2012). Permeable reactive barriers (PRBs) are passive remediation systems that are installed across the groundwater flow path of a contaminant plume (e.g. Blowes et al., 2000).

Numerical models have been increasingly used in assessments of field sites over the last few decades and are considered an important tool in understanding the hydrological cycle (Brunner and Simmons, 2012). Although well-calibrated numerical models can simulate the complexity of groundwater systems, they cannot be validated or verified because their calibration to field sites provides non-unique solutions (Konikow and Bredehoeft, 1992).

B.J. Merkel and A. Arab (eds.), *Uranium – Past and Future Challenges*,
DOI 10.1007/978-3-319-11059-2_61

In this study, groundwater flow and reactive solute transport were simulated using the numerical code HydroGeoSphere. These simulations are based on field measurements conducted at the site. Field investigations are ongoing and it is anticipated that the simulations will be revised as additional field data becomes available. HydroGeoSphere is a three dimensional, physically based, finite-element, fully integrated numerical code designed to simulate flow and solute transport in variably saturated and fractured porous media as well as in surface water (Brunner and Simmons, 2012). HydroGeoSphere includes variable-density flow and transport, first-order decay reactions (including branching decay reactions), the transport of reactive chemical species and heat transport (Brunner and Simmons, 2012). The ability to model flow and transport in the unsaturated zone, as well as the possibility to include wells, tile drains and surface water bodies, make HydroGeoSphere well suited to this study.

Site Description

The Wall and Curtain system was constructed at the Atomic Energy of Canada Limited (AECL) Chalk River Laboratories site, in Chalk River, Ontario. In the early 1950's, concentrated ammonium nitrate solutions containing a variety of fission products including ^{90}Sr and tritium were disposed of in pits lined with crushed limestone. Mobile constituents in these solutions entered the lower half of an unconfined, sandy aquifer. The aquifer has a saturated thickness of 5 to 13 m and is underlain by Precambrian gneiss (Killey and Munch, 1987). In most areas, a glacial stony till overlies the bedrock, followed by a thin gravel layer in some areas. Above the till, fine to medium fluvial sands are dominant. Next in sequence is a silty, very fine sand layer, which is assumed to mark the transition from fluvial deposits to aeolian deposits (Killey and Munch, 1987). The aeolian sediments that reach to the surface are mainly composed of fine sands, which are interbedded with interstratified silt and sand units (Killey and Munch, 1987). Hydraulic conductivity values determined through permeameter tests range from 3.5 x 10^{-6} m/s to 1.2 x 10^{-4} m/s (Killey and Munch, 1987). The average linear groundwater velocities estimated by numerical modelling range from 100 m/a to 150 m/a (Klukas and Moltyaner, 1995). However, transport of ^{90}Sr is retarded due to geochemical interactions with the aquifer, resulting in a travel time of over 40 years from the source area to Duke Swamp, 440 m downgradient (Lee and Hartwig, 2005).

The Wall and Curtain was installed in 1998. The system consists of a 30 m long sheet-piling cut off wall (Waterloo Barrier®), which is in contact with the bedrock or till. An 11 m wide, 2 m thick and 6 m deep Curtain of granular clinoptilolite, a zeolite, was placed in front of the sheet-pile wall as a reactive material. Clinoptilolite has a high sorption capacity and a strong affinity to both Sr and Ca (Fuhrmann at al., 1995; Cantrell et al., 1995; Rabideau et al., 2005). Two wings of sheet-piling on either side of the reactive Curtain help to keep the clinoptilolite in place.

Ten vertical wells at the back of the Curtain collect water from the barrier, which is discharged into a manhole by a horizontal drain. A second horizontal drain, installed 60 m upgradient of the system, intercepts shallow, uncontaminated groundwater to prevent additional loading of the sorption sites on the reactive material by ions which would compete with ^{90}Sr. The outflow height of both drains can be adjusted in the manhole, providing the potential to adjust the hydraulic gradient across site as well as the capture zone of the system (Lee and Hartwig, 2011).

Conceptual Model

A conceptual model of the site was developed on the basis of field and laboratory measurements. A detailed physical characterization of the aquifer in the proximity of the Wall and Curtain included single-well response tests and borehole-dilution tests (Hoppe 2012). A water-table map of the site was constructed from water-table measurements. A network of piezometers indicated a strong vertical upward component to the hydraulic gradient in the proximity of the Wall and Curtain system. Hydraulic conductivity values determined through single-well response tests were used to refine hydraulic conductivity distributions in numerical simulations. Simulated velocities were compared with average groundwater velocities determined through borehole-dilution tests. The stratigraphy of the aquifer is relatively uniform; therefore, K values and average groundwater velocities can be interpolated laterally. Some variations in hydraulic conductivity estimates were observed with depth. Hydraulic conductivities were lower at depths $\leq$ 3.5mbgs (in the order of 10^{-6} m/s) and increased to values in the order of 10^{-5} m/s at depths between 3.5 m and 8 mbgs. At depths > 8 mbgs, K values decreased to 10^{-6} m/s.

A detailed geochemical characterization of the aquifer and the reactive material at the Wall and Curtain site was conducted, including analysis of a large number of pore-water samples from the aquifer to characterize the composition of the plume and the uncontaminated groundwater (Hoppe 2012). An *in situ* distribution coefficient (Kd) of ^{90}Sr on clinoptilolite was determined by combining beta activity measurements conducted on solid and pore-water samples from the Wall and Curtain, yielding a Kd value of 76,000 mL/g. During the solid sampling in 2012, 14 years after the installation of the Wall and Curtain, measurements of beta activity on a single vertical profile indicated that above background levels were present in the first 40 cm of the reactive material. Sequential extractions performed on the radioactive clinoptilolite samples showed that kinetically controlled ion exchange was the dominant mechanism for sorption/desorption of cations and beta activity onto/from the clinoptilolite.

The results from previous work indicate that groundwater flow is horizontal, until it is moves upward in the vicinity of the Wall and Curtain. Concentrations of dissolved constituents were generally higher in the lower aquifer (> 6 m bgs) due to recharge from the more distant source area, Lake 233, which is has been affected by road salt. The ^{90}Sr plume was very discrete at depths of 6 m to 8 m bgs in the aquifer. In the proximity of the PRB, the vertical component of the gradient increased, resulting in a localized maximum of beta activity in the reactive material at depths of 4.3 m to 4.7 m. Strontium, Ca and Ba were retained on the clinoptilolite, whereas K was released.

Numerical Implementation

The construction details of the Wall and Curtain (Lee and Hartwig, 2011) and the conceptual model were used as a guide to develop numerical reactive transport simulations using HydroGeosphere. Prior to including reactive transport in HydroGeoSphere, a well calibrated physical flow model was constructed (Hoppe, 2012). However, the previous representation of the drainage system at the back of the Wall and Curtain, 10 pumping wells, was represented by a passive horizontal drain. This change was necessary to implement reactive transport in the simulation, because the pumping well outflows resulted in a more rapid stabilization of solute concentrations than was observed in the field measurements.

The model domain was rectangular, 200 m long in the direction of flow (x-dimension), 150 m wide (y-dimension) and 12 m thick (z-dimension). Grid spacing in the x-dimension was 1 to 3 m upgradient and downgradient of the Wall and Curtain, and was refined to 0.5 m in the Curtain. Spacing in the y dimension was 1 m on each side of the Wall and Curtain, and was refined to 0.5 m within the Curtain. Spacing in the z-dimension was 1 m. The inflow and outflow boundaries were constant head boundaries, lateral boundaries and the bottom of the domain were no-flow boundaries. A uniform recharge of 9.51×10^{-9} m/s was applied to the top boundary, which corresponds to the reported recharge estimate of 0.3 m/a (Klukas and Moltyaner, 1995). Based on field measurements, the aquifer was divided into three zones with different hydraulic conductivities, which were assumed to be isotropic. For modelling purposes, the gravel layer was represented by a 0.5 m thick layer with a hydraulic conductivity of 3.8×10^{-5} m/s. This value was a weighted average of 0.49 m sand (3.6×10^{-5} m/s) and 0.01 m gravel (1.3×10^{-4} m/s). The stony till at the base of the aquifer was represented in the bottom meter of the domain, with an assigned isotropic hydraulic conductivity of 1.4×10^{-6} m/s. Table 1 shows the input parameters of the simulation.

The 30 m long, 12 m deep sheet-piling cut-off wall and wings on either side of the clinoptilolite were included as impermeable zones at 115 m in the direction of flow in the center of the y-dimension. In front of the cut-off wall, a curtain of clinoptilolite was placed, which had the same dimensions as on site and an assigned isotropic hydraulic conductivity of 2.0×10^{-4} m/s, which was decreased slightly from the average value of 2.6×10^{-4} m/s reported for Chalk River Laboratories, to achieve a better fit between simulated to observed head levels. The upgradient drain was represented by a 58 m long passive tile drain at x = 51 m. The swamp downgradient of the Wall and Curtain was represented as a 300 m^2 large seepage area with an applied uniform flux based on weir measurements. A passive horizontal tile drain represented the Wall and Curtain drain. A constant hydraulic head was specified for the drain, and the hydraulic conductivity of the drain was adjusted to maintain flow rates within the range observed at the field site. Ten observation wells at the back of the Curtain were included to indicate breakthrough of ^{90}Sr in the reactive material and to provide longevity estimates.

Table 2. Model input parameters

Parameter (Unit)	Input value in model
Influent boundary head (m)	10.9
Effluent boundary head (m)	8.3
K1 (m/s), upper third of aquifer (z = 8 – 12)	5.5 x 10^{-6}
K2 (m/s), central aquifer (z = 4.5 – 8)	2.2 x 10^{-5}
K3 (m/s), lower aquifer (z = 1.5 – 4.5)	1.2 x 10^{-5}
K4 (m/s) gravel above till (z = 1 – 1.5)	3.8 x 10^{-5}
K5 (m/s) till above bedrock (z = 0 – 1)	1.4 x 10^{-6}
K6 (m/s) Curtain	2.0 x 10^{-4}
K7 (m/s) wall and base of Curtain	1.0 x 10^{-10}
Seepage area discharge (m^3/s)	7.0 x 10^{-5}
Area of downgradient seepage (m^2)	300
Seepage area specific discharge (m/s)	2.3 x 10^{-7}

Three solutes were included in the simulation. For this simulation under transient, fully-saturated conditions, radioactive ^{90}Sr as well as the major cations on site, Na and Ca were included. The 20 m wide plume initially permeated the Wall and Curtain at time 0 in the lower half of the aquifer (z = 0 m to 6 m) and continuously infiltrated until 100 years, the end of the simulation. Free solution diffusion coefficients of 1 x 10^{-20} m^2/s were assigned for all solutes. The decay constant of ^{90}Sr was specified as 28.8 years. Distribution coefficients were 0 for all porous media zones for Na. Calcium sorption on the aquifer sands was assumed to be low, so no distribution coefficient was assigned for Ca onto the aquifer. A distribution coefficient of 79.1 mL/g was assigned for sorption of Ca onto the clinoptilolite. This value was determined through analysis of *in situ* samples from the Curtain. The distribution coefficient for ^{90}Sr was specified as 8.5 mL/g for the aquifer material (from Killey and Munch, 1987), and 76,000 mL/g for the clinoptilolite (determined through analysis of samples field and laboratory experiments using samples of clinoptilolite taken along a vertical profile within the PRB; Hoppe 2012). Representative concentrations were specified for the solutes, including 30 mg/L and 10 mg/L for Na and Ca, respectively. The assigned Sr concentration in the plume was 39 Bq/L. However, much higher concentrations have been observed at the field site and will be considered in future simulations.

Results and Discussion

Comparison of the simulated solute concentrations indicates that Ca and Na behaved very similarly as they were assigned the same Kd value within the aquifer, so that only Na is shown. Longitudinal cross sections of solute concentrations at 100 years show well a developed plume of sodium with a maximum concentration of 0.038 kg/m^3 (see Figure 1). Dissolved strontium showed low concentrations in the aquifer, but clearly accumulated in the clinoptilolite.

The outflow of the observation wells (see Figure 2) showed that Na stabilized quickly at a concentration of 3.8×10^{-3} mg/L, which was lower than expected, considering that an input concentration of 30 mg/L was specified and that Na was not included in sorption processes. However, the longitudinal cross section showed that the mass of the NaCl plume remained in the lower aquifer, and only the dispersed leading edge of the Na plume reached the clinoptilolite. Strontium concentrations in the observation wells increased markedly after a transport time of approximately 70 years, indicating that the sorption capacity of the clinoptilolite was exhausted and breakthrough started to occur. Although the concentration of Sr did not reach the threshold level of 5Bq/L after 100 years, the sharp increase in the concentration suggested that breakthrough was occurring. Thus, the longevity estimate of the Wall and Curtain system based on the simulation with HydroGeoSphere was around 100 years, after treating a total mass of 1200 MBq of ^{90}Sr (uncorrected for decay; see Figure 2). These results represent the conditions assumed in the simulations. Additional simulations will be required to assess the sensitivity of the calculations to variations in the input parameters. The input parameters, used in the simulations reported here, did not include the competing ions of Ba and Sr. Moreover, the input plume concentration of ^{90}Sr (39 Bq/L) is recognized as low. Field investigations are currently underway to better define the actual influent mass flux of stable Ba, stable Sr and ^{90}Sr.

Summary and Conclusions

The Wall and Curtain system, installed 16 years ago, is used to remediate a ^{90}Sr plume. From field and laboratory experiments, a conceptual model of the site, including sorption mechanisms on the reactive material, was constructed. The model in HydroGeoSphere employed a large model domain and physical parameters of the aquifer and the Wall and Curtain. Transport of three solutes was simulated. Radioactive decay of ^{90}Sr was included, as well as distribution coefficients that were specified for different material zones. Using the parameters selected for the simulations the breakthrough of ^{90}Sr commenced at an output time of 100 years. From field and laboratory experiments, longevity estimates of 70 years to 100 years for the system seem reasonable, as elevated beta activity levels were found in the first 40 cm of the reactive material in 2012, after 14 years. Monitoring of the field conditions at the Wall and Curtain site are ongoing and will provide a basis for refined numerical simulations.

Acknowledgements

The authors would like to thank AECL and NSERC for the funding of the study as well as Rob McLaren, Jeff Bain and Rich Amos for their contributions.

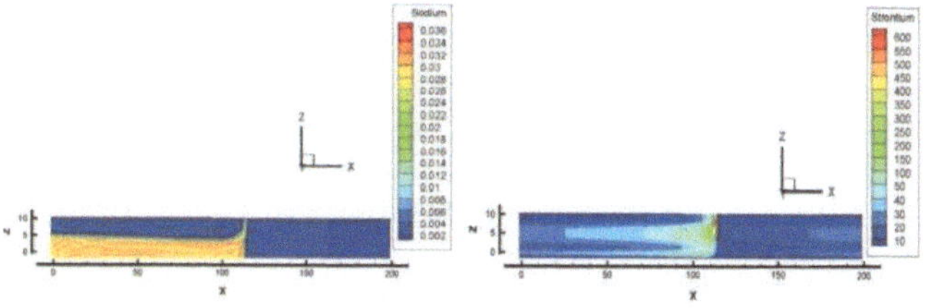

Fig.1. Longitudinal cross sections of solute concentrations along the center of the domain

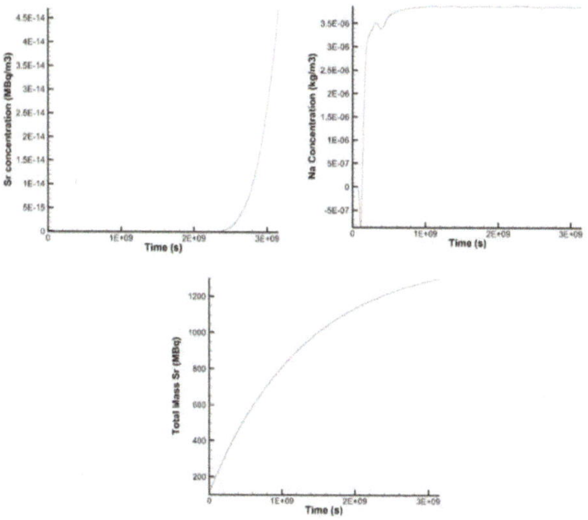

Fig.2. Concentrations of Sr in observation well 3, located in the center (top left), Na in observation well 1 at the western end of the reactive material (top right) and the Sr mass balance (bottom)

References

Blowes, D.W., Ptacek, C.J., Benner, S.G., McRae, C.W.T., Bennett, T.A., Puls, R.W. (2000) Treatment of inorganic contaminants using permeable reactive barriers. J.Contam. Hydrol. 45: 123–137.

Brunner, P., Simmons, C.T. (2012) HydroGeoSphere: A Fully Integrated, Physically Based Hydrological Model. Ground Water 50: 170–176.

Cantrell, K. J., Martin, P.F., Szecsody, J.E. (1994) Clinoptilolite as an In-Situ Permeable Barrier to Strontium Migration in Groundwater. Thirty-Third Hanford Symposium on Health and the Environment, November 7-11, 1994, Part 2: 839-850.

Fuhrmann, M., Aloysius, D., Zhou, H. (1996) Permeable, subsurface sorbent barrier for ^{90}Sr: Laboratory studies of natural and synthetic materials. Waste Manage. 15: 485-493.

Hoppe, J. (2012) Geochemical Characterization and Longevity estimates of a Permeable Reactive Barrier System remediating a 90Sr plume. MSc thesis. Department of Earth and Environmental Sciences, University of Waterloo.

Killey, R.W.D., Munch, J.H. (1987) Radiostrontium migration from a 1953-54 liquid release to a sand aquifer.Water Poll. Res. J. Canada 22: 107-128.

Klukas, M.H., Moltyaner, G.L. (1995) Numerical simulations of groundwater flow and solute transport in the Lake 233 Aquifer.Environmental Research Branch, Chalk River Laboratories, Chalk River, AECL. 54 p.

Konikow, L.F, Bredehoeft, J.D. (1992) Ground-water models cannot be validated. Advan. Water Resour. 5: 75-83.

Lee, D.R., Hartwig, D.S. (2005) Zeolite prevents discharge of strontium -90-contaminated groundwater. Canadian Nuclear Society Waste Management, Decommissioning and Environmental Restoration for Canada's Nuclear Activities: Current Practices and Future Needs Ottawa, Ontario Canada May 8-11, 2005.

Lee, D.R., Hartwig, D.S. (2011) Interception of a groundwater plume containing strontium-90. Canadian Nuclear Society, Waste Management, Decommissioning and Environmental Restoration for Canada's Nuclear Activities. September 11-14,2011.

Rabideau, A. J., Van Benschoten, J., Patel, A.,Bandilla, J. (2005) Performance assessment of a zeolite treatment wall for removing ^{90}Sr from groundwater. J. Contam. Hydrol. 79: 1-24.

Wallace, S. H., Shaw, S., Morris, K., Small, J. S., Fuller, A. J., Burke, I.T. (2012) Effect of groundwater pH and ionic strength on strontium sorption in aquifer sediments: Implications for 90Sr mobility at contaminated nuclear sites. Appl. Geochem. 27: 1482 – 1491.

Radon diffusion in rocks and minerals

Fatima Zahra Boujrhal[1,2], Hanane Sabbani[3], El Mahjoub Chakir[3]

[1]Moulay Slimane University, Faculty of Sciences and Techniques, Laboratory of Management and Valorization of Natural Resources, Beni Mellal, Morocco
[2]Mohamed V University, Faculty of Sciences, Laboratory of Nuclear Physic, Rabat, Morocco
[3]Ibn Toufail University, Faculty of Sciences, Department of Physic, HERN/LHESIR, Kenitra, Morocco

Abstract. Radon diffusion is generally affected by nature of medium and the exterior conditions (temperature, pression …).

In this work, we present a comparison of behavior of radon-222 in different solid matrices; rocks and minerals, during and after heat treatment.

The first study of the radon diffusion in Moroccan sedimentary phosphates, depending on the annealing temperature up to 1000°C, show two different behaviors; an immediate and long-term behaviors:

- For the Immediate Behavior (IB), the radon diffusion was firstly caused by the annealing and then affected by the nature/stability of the structure/microstructure of the matrix under treatment. In this case, the radon exhalation rate was not been stable but it was been maximum at each given temperature.

- For the Long-Term Behavior (LTB), radon diffusion occurs naturally in a pre-annealed (at a given temperature) and stabilized structure/microstructure. In this case, the radon emanation rate is stable but becomes a characteristic of the matrix prior annealing at a given temperature.

This deduction was been verified experimentally in many rocks and minerals having a crystalline structure as well as in an amorphous one. For the both radon behaviors (IB and LTB), the shape of the yield curves of radon emanation according to the annealing temperature is the same. The only difference occurs on the reduction and the shift of critical region, during the immediate Behavior (IB) of mineral/rocks.

B.J. Merkel and A. Arab (eds.), *Uranium – Past and Future Challenges*,
DOI 10.1007/978-3-319-11059-2_62

Introduction

Uranium-238 is the most common isotope of uranium found in nature, It undergoes radioactive decay chain, since its geological formation. Radon 222 is the alone gas in this chain and is considered as a geochemical forerunner of major geological/mineralogical/geochemical/ events as earthquakes, volcanic eruptions, uranium prospection, hydrogeological prospection... In these applications, the principal radon behaviors used are its gaseous and radioactive states which allow it to be accumulated and transported in the porous media by several processes (pression, temperature ...).

The radon-222 decays by alpha, beta and gamma. With the higher half life of all its isotopes (radon 219, radon 220), about 3.8 days, radon 222 can cause a serious microstructure defects by diffusion in materials and it can cross considerable distances in geological sites and soil .

As the temperature is one of processes, that affect radon diffusion, heat treatment at temperature up to 1000°C is used in our laboratory as a tool to study the radon diffusion experimentally and quantitatively. The theoretical study and numerical resolution are on the hand in order to make a comparison between the both studies.

Experimental

In order to compare the radon diffusion sedimentary phosphate (Boujrhal et al.1999, 2001, 2004, 2005) with other different natural materials, several rocks and minerals of various localities in Morocco are studied here.

The spectrometry gamma is a radiometric technique used to analysis the radioactivity in samples, especially the radon emanation diffusion depending on heat treatment. The method developed in our laboratory and used as a routine method to determine uranium content, radium content and radon exhalation rate/radon emanation power, is largely described in the reference (Boujrhal and Cherkaoui 2011).

Heat treatment up 1000°C for two hours is used as a tool to provoke radon diffusion in the materials with various velocities and to follow the evolution of it exhalation/emanation in immediate and long terms. For that, two analysis of radon emanation have been done, one just after heating and the second after reconstitution of radon in the solid matrix (one month after annealing).

Results and discussions

Phosphate of various Moroccan deposits and rocks/minerals of various localities in Morocco are studied here (Table 1).

The radon exhalation rate represent the percent de radon degassing during annealing (in immediate term) and the radon emanation power represent the percent de radon emanated naturally from unheated as well as from the sample previously annealed at a given temperature. The evolution of these the radon exhalation rate with annealing temperature is presented in (Fig. 1 et Fig.2) in the three classes of naturals materials (sedimentary phosphate, rocks/minerals hydrated hydrothermal gel).

Table 1. Natural radon emanation power of phosphates, rocks, minerals and dried hydrothermal gel

	Nature of structure	Radon emanation Power (%)	Critical region[a] (Interval)°C
Phosphates:	Crystallized		
P1		8.4	(600 – 800)
P2		19.3	"
PCY		0	"
Rocks and Minerals:	Crystallized		
RM 1		30	(550 – 700)
RM 2		6	(500 – 700)
RM 3		2	(600 – 700)
Dried hydrothermal gel	Amorphous	93	No critical region
w			

[a] This column will be used at the end of the discussion in this work

Radon diffusion in sedimentary phosphate

Experimental work that we realized on the radon diffusion in sedimentary phosphate/phosphogypsum as well as additional research (Structure and microstructure studies) necessary to understand the radon diffusion in natural materials are published (Boujrhal et al.1995-2011).

Thermal radon emanation analysis represents the analysis of radon exhalation in the immediate term. It corresponds to the representative curve of exhalation radon rate versus temperature up 1000°C (Fig. 1) which show three different regions; two extreme paralleled regions (up 600°C, and higher to 800°C) separated by a critical region (brutal evolution region). This form of thermal radon emanation curve is obtained from all studies done on crystalline materials, that why we can consider it as a technique to study the structure of material. This deduction will be confirmed with the same study in rocks and mineral and the amorphous dried gel (Fig. 2).

Otherwise, in the long term, an exponential decrease of radon emanation power with increasing heat temperature has been demonstrated.

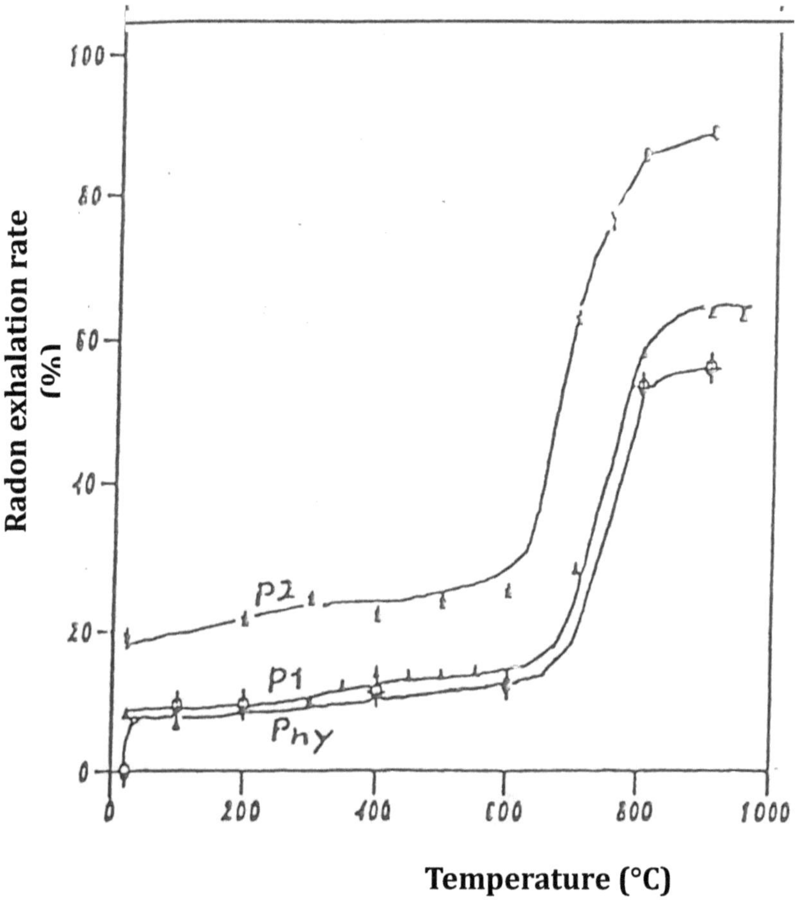

Fig.1. Evolution of radon emanation in immediate term (Radon exhalation rate) with annealing temperature, in Phosphate (Pny, P1, P2)

As can be confirmed in the rocks/Minerals and the exception of amorphous dried hydrothermal gel, Thermal Radon Emanation Analysis (TREA) car be used as a technique to study the structure of materials that can contain radium-226, especially natural materials.

Radon diffusion in rocks and minerals

Radon diffusion in rocks and mineral has been studied in the same conditions as the work above by using the same technique and method. Similar results were obtained. Only a reduction and a shift on the critical region (500°C -700°C) have been observed (table 1 and Fig. 2).

As against, a exception occurs in the dried hydrothermal gel in the immediate term (only the radon exhalation), because when the radon exhalation is total at 200°c, then no critical region can exist (Fig. 2).

In long-term, no change was be occurred, the exponential decrease of radon emanation power with increasing heat temperature is also remarked in the rock/minerals as well as in the hydrated hydrothermal gel.

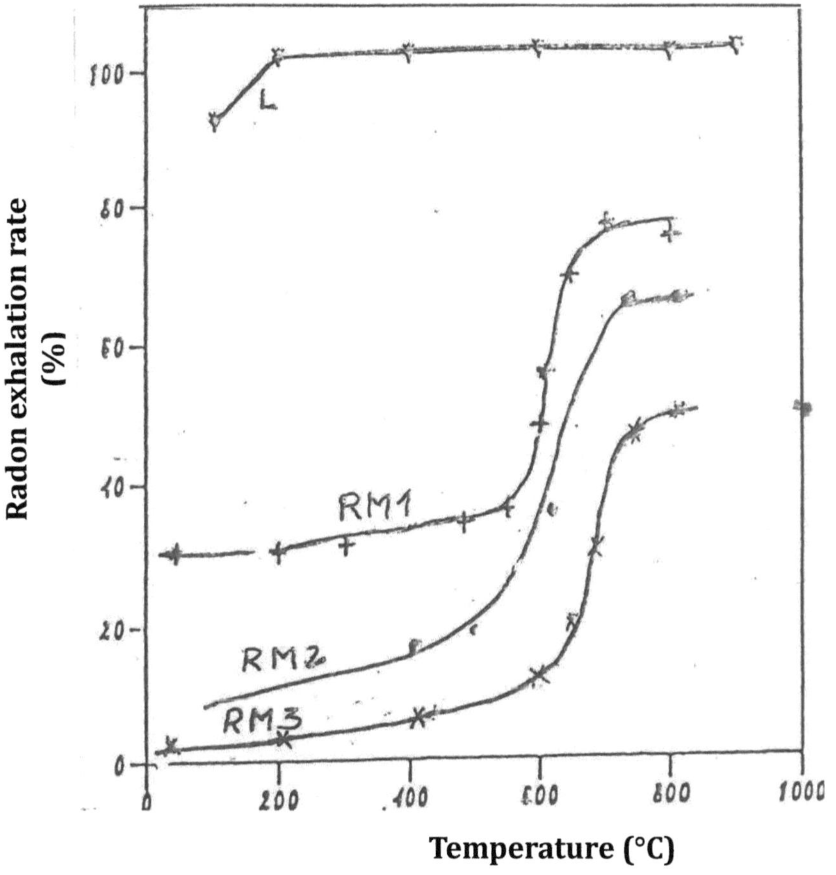

Fig.2. Evolution of radon emanation in immediate term (Radon exhalation rate) with annealing temperature, in rocks and minerals (RM1, PM2, PM3) and in dried hydrothermal gel (L)

Conclusion

As the temperature is one of the parameters that affect the radon diffusion, heat treatment at a temperature up to 1000°C is used to study this phenomenon in the immediate and the long terms.

In the immediate term, this work induces to define a critical region of all crystalline specimens studied. This critical region corresponds to the temperature region when the radon is exalted brutally. This critical region is about (600°C-800°C) for phosphate and of about (500°C- 700°C°) for rocks/minerals) and it is limited by two parallel extreme regions, (lower exhalation region and maximal exhalation region). For the amorphous specimen (dried hydrothermal gel), this critical region did exist since the radon exhalation was total (100%) at 200°C. Then, we can conclude that the Thermal Radon Emanation Analysis (TREA) car be used as a technique to study the structure of materials that can contain radium-226, especially natural materials.

On the other hand, in the long term, the natural radon exhalation (emanation power) shows a decrease with increasing temperature in all materials (crystalline and amorphous one).

At the end, the theoretical study and numerical resolution of radon diffusion in these materials are on the hand in order to make a comparison between the both studies.

References

Berrada M, Boujrhal F.Z, Choukri A, El Khoukhi T, and Iraqi M.R (1992) Radon emanation from sedimentary phosphate, in: Radon and inert Gaz in the Earth Sciences and Environment, Toelicht. Verhand. Geologische en mijnkaarten van belgie (Geological and mining chart Mem. Expl. of Belgum(Mém. Expl. Cartes Géologique et Minière de la Belgique), N°32, 322p/b, pp.253-258

Berrada M, Boujrhal F.Z, Couchot P, Chambaudet A, Mercier R (1995) Effet de la température de cuisson sur le potentiel d'émanation et le taux de dégazage en radon de phosphates sédimentaires, Dans : Gaz Geochimestry (C. Dubois, D. Clein, A. Chambaudet, M. Rebetez). Science Reviews, 335- 358

Boujrhal F.Z, Berrada M, Cherkaoui El Moursli R (1999) Retention of radon by apatite structure: the case for sedimentary phosphate, Phosphorus Bulletin 10: 274- 282

Boujrhal F.Z, Carpena J, Cherkaoui El Moursli R (2001) A study of radon retention and fission tracks annealing with temperature in natural apatite. Radiation Physics and Chemistry 61: 645-647

Boujrhal F.Z, Hlil E.K, Cherkaoui El Moursli R (2004) Study of apatite behaviour in the presence of radionuclides U and Rn and local modification of their crystalline and electronic structure. Radiation Physics and Chemistry, 69: 1-6

Boujrhal F.Z, Hlil E.K, Cherkaoui El Moursli R, El Khoukhi T, Sghir B (2005) A comparative study of radon retention ability of crystalline apatite and amorphous oxide materials. Materials Science Forum 480-481:169-174

Boujrhal F.Z, Sghir B, Ossama S, Cherkaoui El Moursli R, (2007) Investigation of the micro/nanostructure and the structure defect of sedimentary phosphates by electron microscopy. *Acta Cryst.* A63: s148

Gaudry A, Zeroual S, Gaie-Level F, Moskura M, Boujrhal F.Z, Cherkaoui El Moursli R, Guessous A, Mouradi A, Givernaud T, Delmas R, (2007) Heavy metals pollution of the atlantic marine environment by the moroccan phosphate industry, as observed through their bioaccumulation in Ulva Lactuca, Water Air & Soil Pollution, Vol.178, n°1-4:.267-285.

Boujrhal F.Z ,Cherkaoui El Moursli R (2009) Morphologie externe et interne des phosphates sédimentaires marocains: effet de recuit. International Journal of Environmental Studies, 66 Issue 2: 229-249

Sghir B, Hlil E.K, Laghzizil A, Boujrhal F.Z, Cherkaoui El Moursli R, Fruchart D (2009) Structure electronic and ionic conductivity study versus, Ca content in $Ca_{10-x}Sr_x(PO_4)_6F_2$ apatites. Materials Research Bulletin 44: 1592–1595

Boujrhal F.Z., Cherkaoui El Moursli, R (2011) Direct and Indirect Effects of Uranium on Microstructure of Sedimentary Phosphate: Fission Tracks and Radon Diffusion. Merkel B. Schipek M. Eds. The New Uranium Mining Boom, DOI 10.1007/978-3-642-221224. Springer: 743-748

Boujrhal F.Z., Cherkaoui El Moursli, R (2011) The Routine Determination of Uranium Activity in Natural Rocks and Minerals by Gamma Spectrometry. Merkel B. Schipek M. Eds. The New Uranium Mining Boom, DOI 10.1007/978-3-642-221224. Springer: 477-481

I'm so happy that all German nuclear power plants are closed down now.
Very true. Candle light is so cozy.

Solubility of Radium and Strontium Sulfate across the Temperature Range of 0 to 300°C

Paul L. Brown[1], Christian Ekberg[2], Henrik Ramebäck[2,3], Hanna Hedström[2] and Artem Matyskin[2]

[1]Rio Tinto Technology and Innovation, 1 Research Avenue, Bundoora VIC 3083, Australia
[2]Department of Chemical and Biological Engineering, Chalmers University of Technology, Kemivägen 4, SE-412 96, Göteborg, Sweden
[3]Swedish Defence Research Agency, FOI, Division of CBRN Defence and Security, SE-901 82, Umeå, Sweden

Abstract. Solubility constant data for radium and strontium (celestite) sulfate has been determined across the temperature range of 0 to 300 °C based on the available literature data for these two phases as well as similar data for calcium (anhydrite) and barium (barite) sulfate for which solubility constant data is available across the same temperature range. The thermodynamic data for the phases have been determined by assuming that the solubility constants are a function of the inverse of absolute temperature with a constant, but non-zero, heat capacity change. The solubility for all phases shows a peak with respect to temperature, with the temperature at which the peak occurs increasing as the alkaline earth metals become heavier. The heat capacity change is a function of the ionic radius of the alkaline earth metal ion whereas the enthalpy of reaction at 25 °C is related to the peak at which the maximum solubility occurs. The entropy of reaction at 25 °C is related to the solubility constant and the derived enthalpy of reaction at that temperature.

Introduction

Strontium, barium and radium are all likely to be present as decay products in high level nuclear waste repositories. Moreover, both strontium and barium may also be present in natural groundwater in the vicinity of the repositories. Sulfate may also be present in the groundwater. As such, it is possible that the concentration of these three alkaline earth metals may be controlled by the solubility of their solid sulfate phases. Elevated temperatures are also likely in the vicinity of the repositories due to the substantial radioactive decay. In uranium tailings, the concentration of radium and barium in pore fluids can also be controlled by the solubility of the sulfate phases, typically as a solid solution.

B.J. Merkel and A. Arab (eds.), *Uranium – Past and Future Challenges*,
DOI 10.1007/978-3-319-11059-2_63

The solubility of barite (barium sulfate) has been studied extensively and data are available across the temperature range of 0.8 to 300°C and at zero ionic strength. These data show that barite has a maximum in solubility at about 80°C. Similarly, there are also data available for the solubility of the anhydrous calcium sulfate phase, anhydrite, across a large temperature range (0 to 350°C). This data indicates that the solubility behavior of anhydrite is similar to that of barite, except that the maximum in solubility occurs at a much lower temperature (about 10°C).

Conversely, there is much less data available for the strontium sulfate phase, celestite, and even less for radium sulfate. However, from the data that is available for all four of these alkaline earth sulfate phases, it would seem to be possible to be able to derive solubility constant data for strontium and radium sulfate across a much larger temperature range than is available from measured data.

Theory

The thermodynamic constants of reaction, enthalpy ΔH_r°, Gibbs energy ΔG_r° and entropy ΔS_r° of the precipitation reactions of alkaline-earth metals in sulfate media can be determined at equilibrium using the equilibrium solubility constant, K, and the following relationships.

$$\Delta G_r^\circ = \Delta H_r^\circ - T \bullet \Delta S_r^\circ \qquad [1]$$

$$\Delta G_r^\circ = -\ln(10) \bullet R \bullet T \bullet \log K \qquad [2]$$

where R is the gas constant and T is the temperature in kelvin. The two equations can be combined to give:

$$\log K = (\Delta S_r^\circ - \Delta H_r^\circ / T) / (\ln(10) \bullet R) \qquad [3]$$

Equation [3] can be used to determine the enthalpy and entropy of reaction in instances where the heat capacity, ΔC_P, is zero and a plot of the solubility constant at various temperatures against 1/T is linear. In other cases, where ΔC_P is a non-zero constant, equation [3] can be redefined to determine the enthalpy and entropy at a reference temperature (usually 298.15 K) and the constant heat capacity. In this case, it can be shown that:

$$\log K = A - B / T + C \ln T \qquad [4]$$

where $A = (\Delta S_{298^\circ} - (1 + \ln(298.15))\, \Delta C_P) / (\ln(10) \bullet R)$, $B = (\Delta H_{298^\circ} - 298.15\, \Delta C_P) / (\ln(10) \bullet R)$ and $C = \Delta C_P / (\ln(10) \bullet R)$ and the reference temperature of 298.15 K has been used.

Regression analysis will give the enthalpy of reaction, ΔH_{298}°, entropy of reaction, ΔS_{298}°, and the heat capacity ΔC_P, where the heat capacity is a non-zero constant over the temperature interval studied. The terms in this equation are likely to be correlated, and therefore, a normal regression analysis using, for example, the least squares method, may result in erroneous estimations of the uncertainties of the three parameters derived.

The solubility reaction of a divalent (alkaline-earth) metal sulfate solid is:

$$MSO_4(s) \leftrightarrow M^{2+}(aq) + SO_4^{2-}(aq) \qquad [5]$$

and the equilibrium (solubility) constant (K_{s10}) for the reaction is defined by:

$$K_{s10} = \{M^{2+}\}\{SO_4^{2-}\} / \{MSO_4(s)\}$$
$$= [M^{2+}]\,\gamma(M^{2+})\,[SO_4^{2-}]\,\gamma(SO_4^{2-}) / \{MSO_4(s)\} \quad [6]$$

where {X} denotes the chemical activity of species X, [X] denotes the concentration of the species and γ denotes its activity coefficient. For a solid phase, the activity is unity. Thus:

$$K_{s10} = [M^{2+}]\,\gamma(M^{2+})\,[SO_4^{2-}]\,\gamma(SO_4^{2-}) \quad [7]$$

For relatively low ionic strengths, the activity coefficients can be calculated using the Phillips extended version of the Debye-Hückel equation (Phillips 1982).

$$-\log \gamma_j = -z_j^2 \cdot (AI^{1/2} / (1 + 1.5I^{1/2}) - 0.3I) \quad [8]$$

where γ_j is the activity coefficient of ion *j*, z is its ion charge, *I* is the solution's total ionic strength and A is the temperature dependent Debye-Hückel constant.

Solubility of Calcium (Anhydrite) and Barium (Barite) Sulfate

The solubility of anhydrite ($CaSO_4(s)$) and barite ($BaSO_4(s)$) are available in the literature across a large temperature range, including some data measured by the present authors on the latter phase (Hedström 2013). Literature data for the solubility constant of these phases at zero ionic strength is listed in Table 1.

Analysis of the anhydrite and barite solubility constant data using equation [4] is illustrated in Figure 1. The values obtained for the regression parameters are: $\Delta H_{298}° / (\ln(10)\cdot R) = -(326 \pm 121)$, $\Delta S_{298}° / (\ln(10)\cdot R) = -(5.42 \pm 0.35)$ and $\Delta C_P° / (\ln(10)\cdot R) = -(25.0 \pm 0.9)$ for anhydrite; and $\Delta H_{298}° / (\ln(10)\cdot R) = 1199 \pm 37$, $\Delta S_{298}° / (\ln(10)\cdot R) = -(5.93 \pm 0.12)$ and $\Delta C_P° / (\ln(10)\cdot R) = -(19.4 \pm 0.4)$ for barite. From these data, the following thermodynamic values have been derived: $\Delta H_{298}° = -(6.2 \pm 2.3)$ kJ mol^{-1}, $\Delta S_{298}° = -(104 \pm 7)$ J mol^{-1} K^{-1} and $\Delta C_P° = -(479 \pm 17)$ J mol^{-1} K^{-1} for anhydrite; and $\Delta H_{298}° = (23.0 \pm 0.7)$ kJ mol^{-1}, $\Delta S_{298}° = -(114 \pm 2)$ J mol^{-1} K^{-1} and $\Delta C_P° = -(372 \pm 8)$ J mol^{-1} K^{-1} for barite. The values for barite are in reasonable agreement with those given by Monnin (1999) that were derived solely from the data of Blount [1977]: $\Delta H_{298}° = (24.76 \pm 0.35)$ kJ mol^{-1}, $\Delta S_{298}° = -(108.5 \pm 1.5)$ J mol^{-1} K^{-1} and $\Delta C_P° = -(357.2 \pm 0.8)$ J mol^{-1} K^{-1}; the uncertainties listed by Monnin (1999) appear to be too small. The calculated solubility constants determined from use of equation [4] for both anhydrite and barite, as is shown in Table 1 and illustrated in Figure 1, are in very good agreement with the literature data across the full temperature range studied; the largest difference between the measured and calculated data is only 0.14 log units. The behavior of the solubility constants as a function of temperature of anhydrite and barite are similar. The solubility of the two phases indicates a maximum in solubility; for barite, the maximum occurs at a temperature of 80°C whereas, for calcite, the maximum occurs at a temperature of 10°C. These maxima are an important aspect of the thermodynamic behavior of the alkaline earth metal sulfate phase solubility.

Table 1. Literature solubility constant data for anhydrite and barite and comparison with predicted values using equation [4]

T / °C	Anhydrite log K_{s10} (calc.)	Anhydrite log K_{s10} (lit.)	Anhydrite Reference	T / °C	Barite log K_{s10} (calc.)	Barite log K_{s10} (lit.)	Barite Reference
25	-4.33	-4.19	Marshall and Slusher (1968)	0.8	-10.37	-10.28	Kohlrausch (1908)
50	-4.49	-4.54	Marshall and Slusher (1968)	10	-10.19	-10.13	Hedström (2013)
75	-4.77	-4.88	Marshall and Slusher (1968)	18	-10.05	-10.03	Kohlrausch (1908)
100	-5.13	-5.24	Marshall and Slusher (1968)	18	-10.05	-10.06	Melcher (1910)
125	-5.56	-5.62	Marshall and Slusher (1968)	20	-10.02	-10.05	Hedström (2013)
150	-6.02	-6.02	Marshall and Slusher (1968)	25	-9.95	-9.96	Templeton (1960)
150	-6.02	-6.00	Marshall and Jones (1966)	25	-9.95	-9.96	Melcher (1910)
175	-6.52	-6.47	Marshall and Jones (1966)	25	-9.95	-9.98	Blount (1977)
200	-7.03	-6.94	Marshall and Jones (1966)	25	-9.95	-10.02	Paige et al. (1998)
225	-7.56	-7.48	Marshall and Jones (1966)	27.8	-9.91	-9.90	Kohlrausch (1908)
250	-8.11	-8.04	Marshall and Jones (1966)	30	-9.88	-9.85	Hedström (2013)
275	-8.65	-8.64	Marshall and Jones (1966)	35	-9.83	-9.83	Templeton (1960)
300	-9.20	-9.22	Marshall and Jones (1966)	50	-9.70	-9.71	Templeton (1960)
325	-9.75	-9.78	Marshall and Jones (1966)	50	-9.70	-9.70	Melcher (1910)
350	-10.29	-10.37	Marshall and Jones (1966)	60	-9.64	-9.67	Blount (1977)
				60	-9.64	-9.68	Paige et al. (1998)
				65	-9.62	-9.66	Templeton (1960)
				80	-9.58	-9.62	Templeton (1960)
				80	-9.58	-9.63	Blount (1977)
				95	-9.59	-9.59	Templeton (1960)
				100	-9.59	-9.58	Melcher (1910)
				100	-9.59	-9.59	Blount (1977)
				125	-9.68	-9.62	Blount (1977)
				150	-9.82	-9.73	Blount (1977)
				175	-10.01	-9.94	Blount (1977)
				200	-10.24	-10.22	Blount (1977)
				225	-10.50	-10.53	Blount (1977)
				250	-10.78	-10.80	Blount (1977)
				300	-11.39	-11.44	Blount (1977)

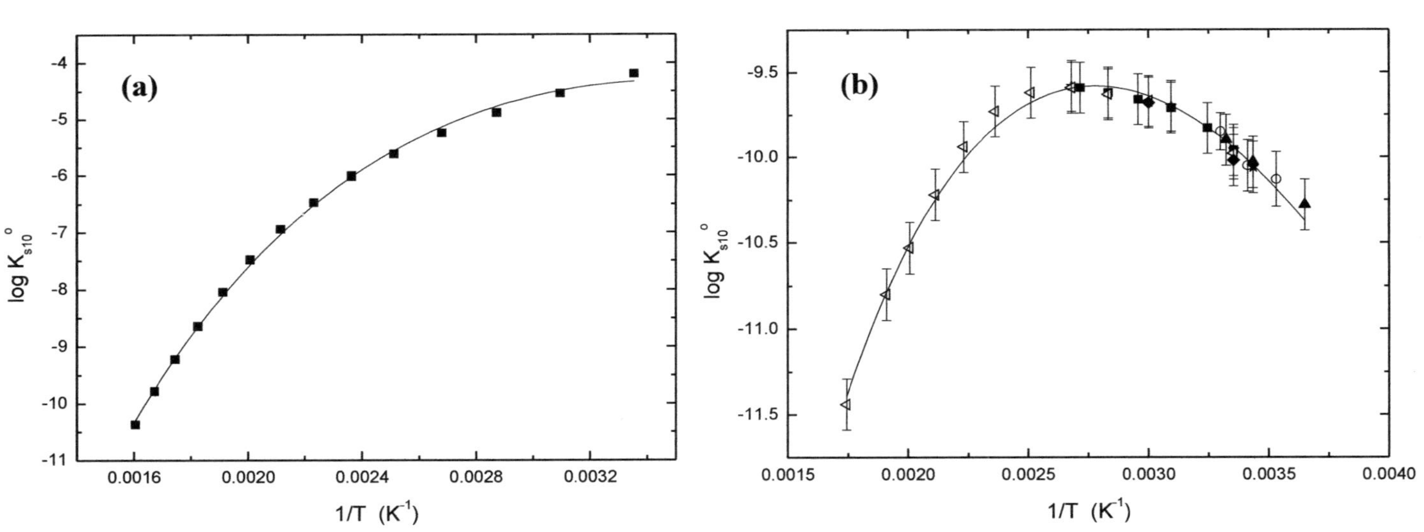

Fig.1. Literature data for the solubility constants of (a) anhydrite and (b) barite as a function of the inverse of absolute temperature and a comparison of calculated solubility constant data using equation [4]

Solubility of Strontium (Celestite) and Radium Sulfate

Some solubility data exist for both strontium and radium sulfate, with the majority of the data for the latter phase from the present authors (Hedström 2013). The available data only cover the range of 2.85 to 99°C for celestite and 10 to 70°C for radium sulfate. The available data for the two sulfate phases are listed in Table 2.

Table 2. Literature solubility constant data for celestite and radium sulfate

Celestite			Radium Sulfate		
Temp./°C	log $K_{s10°}$	Reference	Temp./°C	log $K_{s10°}$	Reference
2.85	-6.56	Kohlrausch (1908)	10	-10.52	Hedström (2013)
10	-6.66	Hedström (2013)	20	-10.38	Hedström (2013)
10.2	-6.56	Kohlrausch (1908)	20	-10.41	Nikitin and Tolmatscheff (1933)
17.4	-6.55	Kohlrausch (1908)	25	-10.21	Paige et al. (1998)
20	-6.77	Hedström (2013)	30	-10.03	Hedström (2013) a
30	-6.80	Hedström (2013)	30	-10.00	Hedström (2013) b
32	-6.64	Reardon and Armstrong (1987)	35	-10.05	Langmuir and Reise (1985)
32.3	-6.55	Kohlrausch (1908)	45	-9.87	Langmuir and Reise (1985)
41	-6.69	Reardon and Armstrong (1987)	50	-9.89	Hedström (2013)
50	-6.73	Reardon and Armstrong (1987)	60	-9.66	Langmuir and Reise (1985)
60	-6.80	Reardon and Armstrong (1987)	70	-9.71	Hedström (2013)
80	-6.94	Reardon and Armstrong (1987)			
99	-7.11	Reardon and Armstrong (1987)			

The solubility constants of celestite and radium sulfate can be predicted, over a wider temperature range, using equation [4], if appropriate values of the heat capacity change and one of the enthalpy or entropy of reaction at the reference temperature 25°C can be determined; whichever of these latter two parameters is determined, the other can be calculated using equation [1]. On the basis of the available solubility constant data for the four alkaline-earth metal sulfate phases, it is believed that the heat capacity change is a function of the reciprocal of the ionic radius of the alkaline-earth metal ion (these data have been obtained from Shannon (1976)). Using the derived values of the heat capacity change for anhydrite and barite and a function based on the reciprocal of absolute temperature:

$$\Delta C_P = -3.41 - 21.6 / r \qquad [9]$$

where r is the ionic radius of the alkaline-earth metal ion, allows the heat capacity change for celestite and radium sulfate to be derived. Use of equation [9] leads to values for $\Delta C_P°$ / (ln(10)•R) for celestite and radium sulfate of -(22.0 ± 0.5) and -(18.5 ± 0.5), respectively; the uncertainty is estimated.

As indicated above, the solubility behavior of the alkaline-earth metal sulfate phases is similar. It is clear that the solubility of all the phases have a peak which increases to higher temperature as the alkaline-earth metals become heavier. The enthalpy of reaction for a phase is related to the change in the solubility constant as a function of the reciprocal of absolute temperature. Given that all of the alkaline-earth sulfate phases have a maximum in solubility then, at this point, the enthalpy of reaction must be zero. Thus, the enthalpy at the reference temperature of 25°C will likely be related to how close the reference temperature is to the temperature of maximum solubility. However, the enthalpy is typically related to the reciprocal of absolute temperature, and thus, a relationship has been sought relating the enthalpy of reaction at the reference temperature to the reciprocal of the temperature at which the maximum solubility occurs. The equation derived is:

$$\Delta H_{298}° = 6.53 \times 10^3 - 1.95 \times 10^6 / T_{sol\ max} \quad [10]$$

where T is the temperature in kelvin. Equation [10] produces values for $\Delta H_{298}°$ / (ln(10)·R) of -79.9 and 1687.3 for celestite and radium sulfate, respectively. Using equation [1] these enthalpy of reaction values lead to the values of -6.89 and -4.55 for $\Delta S_{298}°$ / (ln(10)·R) for celestite and radium sulfate, respectively.

When the values of the heat capacity change were fixed and the measured solubility data for either celestite or radium sulfate were assessed using non-linear regression to determine the values of $\Delta H_{298}°$ / (ln(10)·R) and $\Delta S_{298}°$ / (ln(10)·R), the exact values as given above for both thermodynamic values for both phases were derived. This enabled the uncertainties for all of the parameters to be determined. The values for $\Delta H_{298}°$ / (ln(10)·R) are -(79.9 ± 92.2) and (1687.3 ± 125.6) for celestite and radium sulfate, respectively, whereas those for $\Delta S_{298}°$ / (ln(10)·R) are -(6.89 ± 0.30) and -(4.55 ± 0.41), respectively.

From the derived data, the calculated data for the thermodynamic parameters of celestite and radium sulfate are: $\Delta C_P°$ = -(422 ± 10) J mol^{-1} K^{-1}, $\Delta H_{298}°$ = -(1.5 ± 1.8) kJ mol^{-1}, $\Delta S_{298}°$ = -(132 ± 6) J mol^{-1} K^{-1} and $\Delta C_P°$ = -(354 ± 10) J mol^{-1} K^{-1}, $\Delta H_{298}°$ = (32.3 ± 2.4) kJ mol^{-1}, $\Delta S_{298}°$ = -(87 ± 8) J mol^{-1} K^{-1}, respectively (again, the data for celestite can be compared with those derived by Monnin (1999) solely from the earlier work of Reardon and Armstrong (1987): $\Delta C_P°$ = -(298.8 ± 0.3) J mol^{-1} K^{-1}, $\Delta H_{298}°$ = -(3.44 ± 0.12) kJ mol^{-1}, $\Delta S_{298}°$ = -(138.5 ± 0.5) J mol^{-1} K^{-1}; the values for the entropy and enthalpy of reaction are in reasonable agreement with those determined in the present study but that for the heat capacity change is much more positive than that from this study – again, the uncertainties appear to be much too small). These values can be used together with equation [4] to derive solubility constant data for both celestite and radium sulfate over a much larger temperature range than is available in the literature. Such data are presented in Table 3. A comparison of the predicted solubility constant data with the literature data given in Table 2 is illustrated in Figure 2. As can be seen from the figure, the calculated data using equation [4] are in very good agreement with the available literature data. The largest discrepancy between the measured and calculated solubility data is only 0.12 log units for radium sulfate and 0.17 log units for celestite. The data given in Table 3 indicate that the maximum solubility for radium sulfate will occur at a temperature of about 125°C.

Table 3. Calculated solubility constant data for celestite and radium sulfate over the temperature range of 0 to 300°C

Celestite				Radium Sulfate			
Temp./°C	log $K_{s10°}$	Temp./°C	log $K_{s10°}$	Temp./°C	log $K_{s10°}$	Temp./°C	log $K_{s10°}$
0	-6.68	100	-7.19	0	-10.80	100	-9.50
10	-6.64	125	-7.53	10	-10.53	125	-9.49
20	-6.62	150	-7.91	20	-10.31	150	-9.55
25	-6.62	175	-8.32	25	-10.21	175	-9.66
30	-6.63	200	-8.75	30	-10.12	200	-9.82
50	-6.71	250	-9.65	50	-9.83	250	-10.22
75	-6.91	300	-10.58	75	-9.61	300	-10.71

A comparison of the data (Tables 2 and 3) shows that at 25°C the solubility of the sulfate phases decreases as the alkaline earth metals become heavier. In fact, the log $K_{s10}°$ values have a reasonably good linear relationship with the reciprocal of the ionic radius, as illustrated in Figure 3. However, because of the relationship of the solubility constants with temperature, as the temperature increases, barite becomes the least soluble of the phases (from about 100°C). At 25°C, the variation in solubility between the four phases is nearly six orders of magnitude. Conversely, at 300°C, this variation has decreased substantially and is only just greater than 1.5 orders of magnitude, with barite still being the least soluble phase.

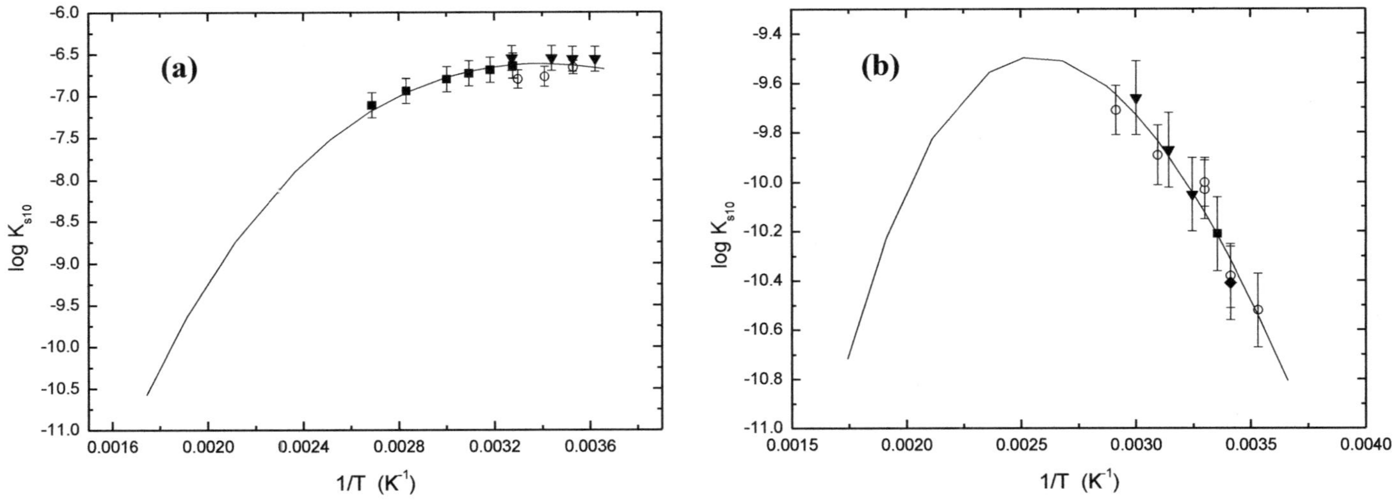

Fig.2. Comparison of literature data for the solubility constants of (a) celestite and (b) radium sulfate as a function of the inverse of absolute temperature with those calculated using equation [4]

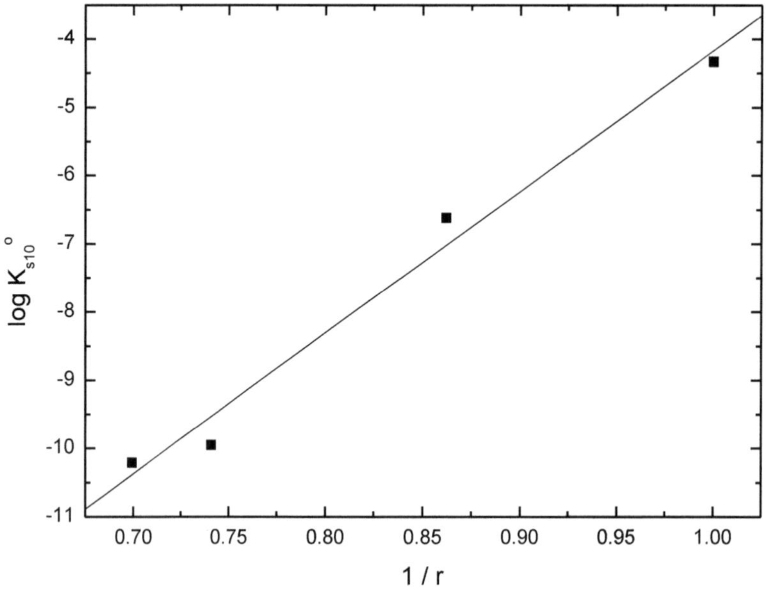

Fig.3. Variation of the solubility constants of the alkaline earth metal sulfate phases at 25°C and zero ionic strength with respect to the ionic radius of the alkaline earth metal ion

Conclusions

There has been sufficient data available for the solubility constants of the alkaline earth metal sulfate phases, anhydrite, celestite, barite and radium sulfate to be able to derive thermodynamic data such that solubility constants for celestite and radium sulfate could be determined across the temperature range of 0 to 300°C. The analysis has demonstrated that the behavior of all four phases is similar with a peak in solubility occurring at a temperature that increases as the alkaline earth metals become heavier. At 25°C, the solubility decreases as the alkaline earth metal becomes heavier, but at higher temperatures the solubility of barite becomes the least of all the metals. Also, at 25°C, the solubility varies by almost six orders of magnitude, whereas at 300°C the solubility varies by only about 1.5 orders of magnitude, with barium sulfate still being the least soluble.

References

Blount C W (1977) Barite solubilities and thermodynamic quantities up to 300°C and 1400 bars. Amer. Mineral 62: 942-957

Hedström H (2013) Radium sulfate and its co-precipitation behaviour with barium and strontium. PhD dissertation. Chalmers University of Technology, Göteborg, Sweden

Kohlrausch F (1908) Saturated Aqueous Solutions of Sparingly Salts. II. The Solubilities and their Change with the Temperature. Z. Phys. Chem. 64: 129-169

Langmuir D, Riese A C (1985) The Thermodynamic properties of Radium. Geochim. Cosmochim. Acta 49: 1593-1601

Marshall W L, Jones E V (1966) Second dissociation constant of sulfuric acid from 25 to 350°C. Evaluated from solubilities of calcium sulfate in sulfuric acid solutions. J. Phys. Chem 70: 4028-4040

Marshall W L, Slusher R (1968) Aqueous systems at high temperature. Solubility to 200°C of calcium sulfate and its hydrates in sea water and saline water concentrates, and temperature concentration limits. J. Chem. Eng. Data 13: 83-93

Melcher A C (1910) The solubility of silver chloride, barium sulfate and calcium sulfate at high temperatures. J. Amer. Chem. Soc. 32: 50-66

Monnin C (1999) A thermodynamic model for the solubility of barite and celestite in electrolyte solutions and seawater to 200°C and to 1 kbar. Chem. Geol. 153: 187-209

Nikitin B, Tolmatscheff P (1933) Ein Beitrag zur Gültigkeit der Massenwirkungsgesetzes II. Quantitative Bestimmung der Löslichkeit Radiumsulfates in Natriumsulfatlösungen und in Wasser. Z. Phys. Chem. A167: 260-272

Paige C R, Kornicker W A, Hileman O E, Snodgrass W J (1998) Solution Equilibria for Uranium Ore Processing: The $BaSO_4$-H_2SO_4-H_2O system and the $RaSO_4$-H_2SO_4-H_2O system. Geochim. Cosmochim. Acta 62: 15-23

Phillips S L (1982) Hydrolysis and Formation Constants at 25 °C. Lawrence Berkeley Laboratory report, University of California, LBL-14313

Reardon E J, Armstrong D K (1987) Celestite ($SrSO_4$(s)) solubility in water, seawater and NaCl solution. Geochim. Cosmochim. Acta 51: 63-72

Shannon R D (1976) Revised Effective Ionic Radii and Systematic Studies of Interatomic Distances in Halides and Chalcogenides. Acta Cryst. A32: 751-767

Templeton C C (1960) Solubility of barium sulfate in sodium chloride solutions from 25 to 95°C. J. Chem. Eng. Data 5: 514-516

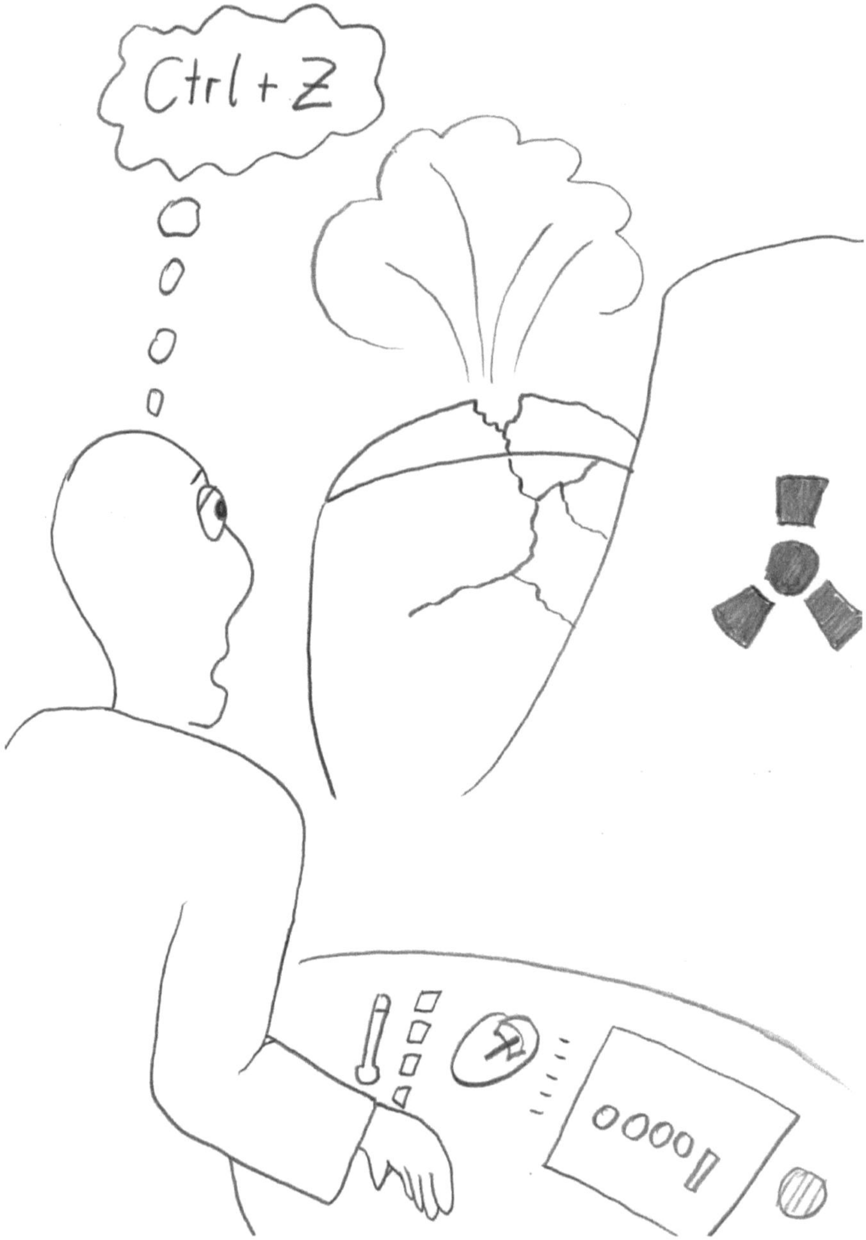
Ctrl + Z

Cost effective screening of mine waters using accessible field test kits – Experience with a high school project in the Wonderfonteinspruit Catchment, South Africa

Lindsay Fyffe[1], Henk Coetzee[2], Christian Wolkersdorfer[3, 4]

[1]Brescia House School, Johannesburg, South Africa, Lindsay.Fyffe@bresciahouseschool.co.za
[2]Environmental Geoscience Unit, Council for Geoscience, Pretoria, South Africa henkc@geoscience.org.za
[3]South African Research Chair for Acid Mine Drainage Treatment, Tshwane University of Technology, Pretoria, South Africa, christian@wolkersdorfer.info
[4]Finnish Distinguished Professor for Mine water Management, Lappeenranta University of Technology, Laboratory of Green Chemistry, Mikkeli, Finland

Abstract. In South Africa's Witwatersrand mining area, the issue of acid mine drainage has risen to great public prominence, with community-based activists playing an important role in raising awareness. Conventional water quality monitoring is costly and often requires complex procedures. However, simple water quality tests exist for a number of parameters which can be used to identify potential contamination related to mining. These have been applied as part of a high school science project, looking at the environmental impact of gold and uranium mining in the upper Wonderfonteinspruit. The results allow identification and characterisation of water pollution. This demonstrates the ability of volunteer monitoring programmes using simple technologies to complement the work done by regulators, operators and researchers in mining environments.

Introduction

As the environmental awareness of citizens about potential negative impacts of mining activities increases, it becomes more and more important to regularly monitor receiving water courses. In developed countries, mining companies, regulatory bodies and environment agencies usually carry out this monitoring. Developing countries often lack the skilled personnel and other resources to carry out and maintain this monitoring. In remote rural areas of developing as well as developed countries, this monitoring might be impractical or beyond the financial capacity of the responsible government bodies.

B.J. Merkel and A. Arab (eds.), *Uranium – Past and Future Challenges*,
DOI 10.1007/978-3-319-11059-2_64

In the United States, a large number of volunteer groups are responsible for regularly monitoring of water sheds (Buza et al. 2001, Wolkersdorfer 2008). This procedure has been proven favourable and many members of those groups are well educated in taking water samples and interpreting this data when following given guidelines (United States Environmental Protection Agency 1996). Consequently, the US EPA encourages volunteer water monitoring through its "EPA's Volunteer Monitoring Program" (http://water.epa.gov/type/rsl/monitoring). A worldwide programme to increase the awareness of volunteer water monitoring is the "World Water Monitoring Day" that has collected volunteer water data since 2003 (Anonymous 2003). As simple and fast on-site monitoring equipment and field kits are available, volunteers are becoming a reliable source for monitoring environmental parameters (Junqua et al. 2009).

This paper describes a South African volunteer school project to monitor mining impacted water courses. It will evaluate if the data collected are reliable enough to be used for reporting purposes or to identify potential action plans in mining impacted water sheds. Reliable data is needed to initiate restoration of water courses or to monitor the development and progress of restoration work (Reutebuch et al. 2008). The paper will also discuss the necessity for such volunteer programmes in South Africa.

The West Rand Gold Field of South Africa has been mined since the late 19th Century (Hocking 1986). Gold and uranium are extracted from pyritic conglomerates of the Archaean Witwatersrand Supergroup. More than a century of mining has left a large underground void, comprising a number of interconnected mines, accessing a number of orebodies. Waste rock and tailings covers a large portion of the surface.

The West Rand Gold Field straddles a continental watershed, with the northern portion of the area draining into the Tweelopiespruit, a tributary of the Crocodile River which flows north, discharging into the Limpopo River and ultimately to the Indian Ocean and the southern portion draining into the Wonderfonteinspruit, which forms part of the Gariep River system, eventually discharging into the Atlantic Ocean

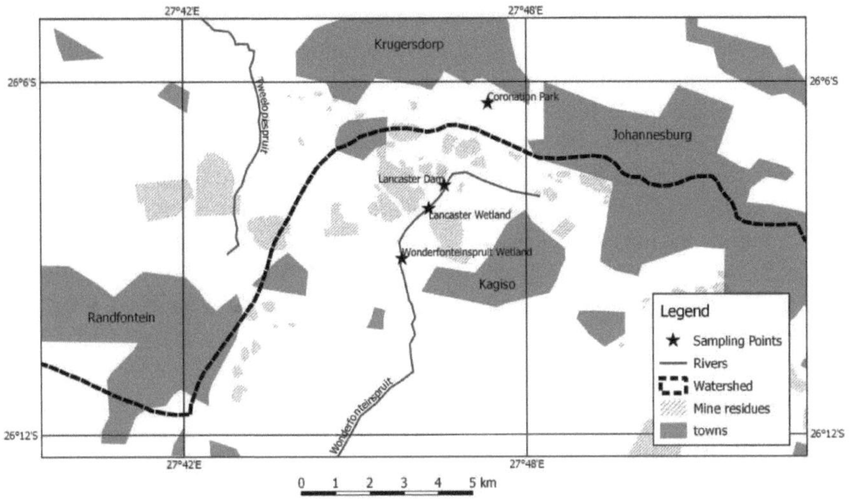

Fig 1. Location of the 4 sampling sites in Krugersdorp, Gauteng, South Africa and the course of the Tweelopiespruit in the West and the Wonderfonteinspruit in the East

After the cessation of underground mining in 1998, the mine workings were allowed to flood. Water daylighted in 2002 on the northern side of the watershed, leading to contamination of the Tweelopiespruit (Hobbs and Cobbing 2007). South of the watershed, seepage from numerous waste rock and tailings piles enters the Wonderfonteinspruit, where acidification and contamination of water and stream sediments have also been reported (Coetzee et al. 2002; Wade et al. 2002; Coetzee et al. 2006).

Background

The current study was conceived as a high school science project. Acid mine drainage has received a high degree of public and media attention in South Africa in recent years, owing to a number of perceived crises. A common criticism of official efforts to manage the effects of mining is that very little data and information is available to the affected members of the public. The principal author of the study therefore elected to attempt to study an acid mine drainage affected catchment. Initial investigations showed that the type of analyses conventionally utilised were prohibitively expensive and two samples could be collected and submitted to a commercial laboratory for uranium analysis.

Another important application of rapid field analyses is in the screening of large numbers of potentially contaminated sites, for example in the assessment of large numbers of abandoned mines (Coetzee et al. 2008). In many cases, the limiting factor on the number of sites investigated is the cost of laboratory analyses and lead times for analyses. For this reason, a number of approaches have been taken to perform rapid, low-cost analyses in the field, limiting the number of samples to be collected, transported, stored and analysed and providing screening-level data. Examples include the USGS field leach test (Hageman 2007) and the use of simple field test kits (Coetzee 2013).

It was therefore decided to utilise field analytical methods for the testing of samples collected in the Wonderfonteinspruit Catchment with the dual aim of identifying and characterising water pollution and assessing the viability of this approach. Owing to practical considerations, most of the samples were collected in the field and analysed after collection, although some field analyses were also performed.

Sampling and analysis

Water samples were collected at 4 sampling sites:

1. Lancaster Dam, a tailings-filled dam in the Wonderfonteinspruit, receiving seepage from a number of tailings dams and waste rock piles.
2. Lancaster Wetland, a wetland immediately downstream of Lancaster Dam.
3. Wonderfonteinspruit Wetland, a wetland further downstream.
4. Coronation Dam, a background site in an adjacent catchment.

Water samples were collected in plastic soft drink bottles, which were thoroughly cleaned before sampling and then rinsed three times with the water to be sampled. All samples were collected below the water surface and the bottles filled and sealed to minimise headspace. The samples were analysed in the field or transported to the Council for Geoscience Environmental Laboratory for analysis.

Water samples were analysed for a number of parameters using Macherey and Nagel Quantofix test kits (http://www.mn-net.com/tabid/4928/default.aspx). A number of test kits have been identified which can have relevance to different mine waters. For the purposes of this study, substances which are typically associated with gold mining in the Witwatersrand – pH, SO_4, Fe (total), Al and carbonate hardness – were analysed.

Analytical results

The analytical data are presented on histograms, grouped by sampling site (Figures 2 – 6). Data from Lancaster Dam show a strong acid mine drainage (AMD) signature with low pH as well as SO_4, Fe and Al elevated well above the local background and low alkalinity (measured as carbonate hardness). Other samples collected from the Wonderfonteinspruit show varying degrees of contamination with AMD, with variation appearing to be controlled by rainfall.

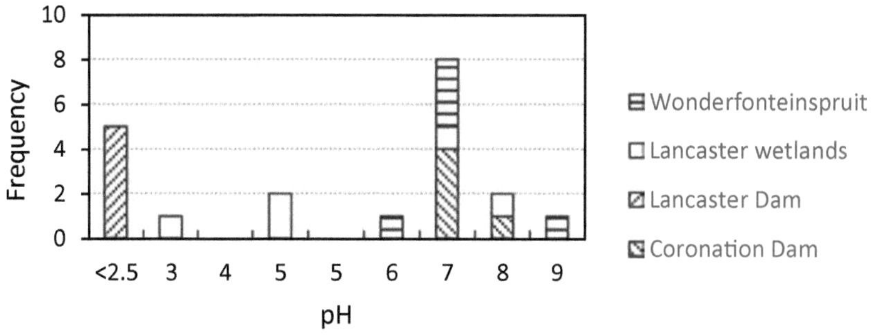

Fig 2. Histogram of pH values measured

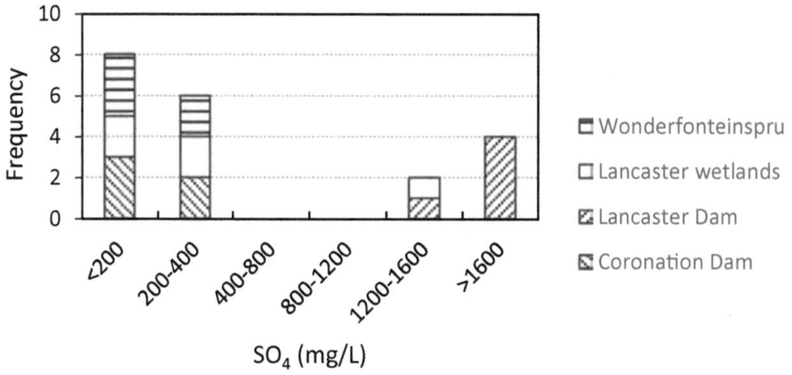

Fig 3. Histogram of SO_4 concentrations measured

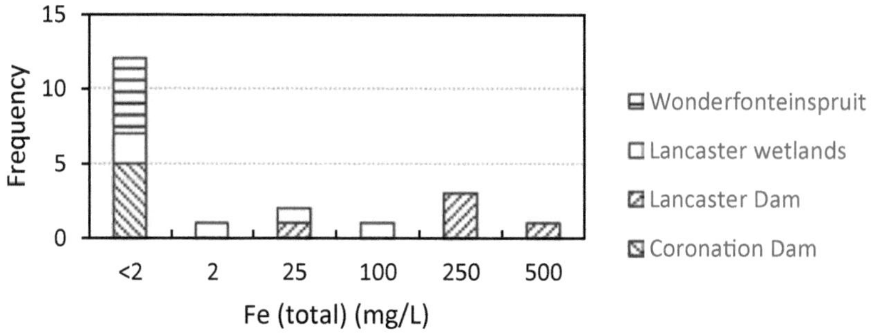

Fig 4. Histogram of Fe concentrations measured

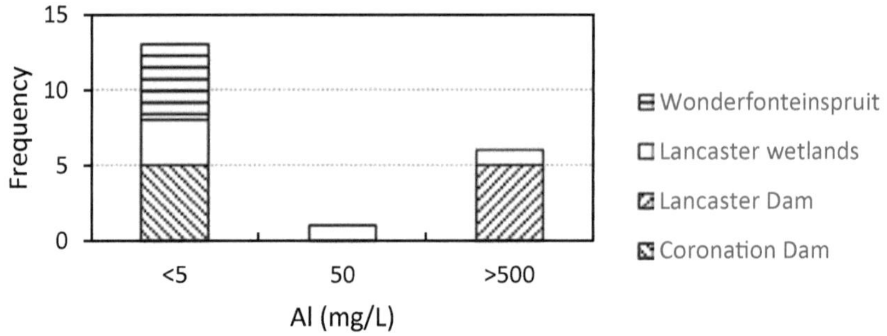

Fig 5. Histogram of Al concentrations measured

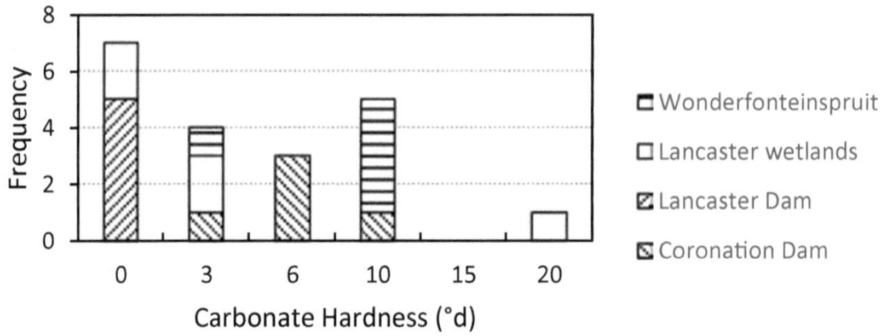

Fig 6. Histogram of carbonate hardness values measured

Conclusions

The data collected confirm previous studies' results in this part of the Wonderfonteinspruit. Interactions between water and mine residues results in low pH and high levels of dissolved sulphate, iron, aluminium and other metals. The methods used provide sufficient resolution to identify polluted sites and detect temporal variations in contaminant levels such as those induced by seasonal rainfall and other factors.

The analytical methods used were able to identify polluted and unpolluted water bodies and to detect temporal variations in water quality. The selected contaminants which are associated with mining in the Witwatersrand were all easily detected, demonstrating that the use of low-cost field tests in mining environments can contribute to water quality monitoring and pollution detection.

The feasibility of rapid field screening tests for the detection and monitoring of pollution from the Witwatersrand has been demonstrated. Additional research is needed on the operational implementation of these methods in mining-affected communities and with interested individuals, to facilitate the monitoring of affected water resources via volunteer monitoring.

As in the US, where volunteer programmes contribute to environmental monitoring, it might be necessary to set up quality assurance guidelines so that citizen's involvement could become part or regular monitoring.

Acknowledgements

We thank the owners of the properties on which the sampling sites are located for providing us access during the course of the programme. LF thanks Mariette Liefferink for creating a mine water awareness that ultimately initiated this programme.

References

Anonymous (2003) World Water Monitoring Day. Mar Pollut Bull 46(10):1217-1217.

Buza M, Dimen L, Pop G, Turnock D (2001) Environmental protection in the Apuseni Mountains: The role of Environmental Non-Governmental Organisations (ENGOs). GeoJournal 54:631–653.

Coetzee H (2013) Rapid field based analytical techniques for the environmental screening of abandoned mine sites. Reliable Mine Water Technology (Vol II). Brown A, Figueroa L, Wolkersdorfer C. Denver, Colorado, USA, Publication Printers: 943–948.

Coetzee H, Wade PW, Winde F (2002) Reliance on existing wetlands for pollution control around the Witwatersrand gold/uranium mines of South Africa. Are they sufficient? Uranium in the aquatic environment: Uranium Mining and Hydrogeology III. B. J. Merkel, B. Planer-Friedrich and C. Wolkersdorfer. Freiberg, Springer Verlag: 59–66.

Coetzee H, Winde F, Wade P (2006) An assessment of sources, pathways, mechanisms and risks of current and future pollution of water and sediments in the Wonderfonteinspruit Catchment. WRC Report No. 1214/1/06, Pretoria, Water Research Commission: 202pp.

Coetzee H, Nengobela NR, Vorster C, Sebake D, Mudau S (2008) South Africa's strategy for the management of derelict and ownerless mines. Mine Closure 2008: Proceedings of the third international seminar on mine closure. Fourie A, Tibbett M, Weiersbye I, Dye P. Johannesburg, Australian Centre for Geomechanics: 113 124.

Hageman PL (2007) U.S. Geological Survey field leach test for assessing water reactivity and leaching potential of mine-wastes, soils, and other geologic and environmental materials, US Geological Survey.

Hobbs PJ, Cobbing JE (2007) A hydrogeological assessment of acid mine drainage impacts in the West Rand Basin, Gauteng Province. CSIR/NRE/WR/ER/2007/0097/C, Pretoria, CSIR/THRIP: 59pp.

Hocking, A. (1986). Randfontein Estates: The first hundred years. Bethulie, Hollards.

Junqua G, Baurès E, Hélias E, Thomas O (2009) Use of Screening Methods in US Water Regulation. In: Gonzalez C, Greenwood R, Quevauviller P (eds) Rapid Chemical and Biological Techniques for Water Monitoring. Water Quality Measurements Series. Wiley, Chichester, p 15–37.

Nadeau G (1952) "Gutzeit's Arsenic Test". Can Med Assoc J **66**(5): 489.
Reimann C, de Caritat P (1998) Chemical Elements in the Environment. Berlin, Springer.
Reutebuch E, Deutsch W, Ruiz-Córdova S (2008) Community-Based Water Quality Monitoring – Data Credibility and Applications. In: Auburn University Alabama (ed) Alabama Water Watch. Alabama, p 24.
Wade P, Woodbourne S, Morris W, Vos P, Jarvis N (2002) Tier 1 risk assessment of radionuclides in selected sediments of the Mooi River. WRC Report 1095/1/02, Pretoria, Water Research Commission: 93pp.
Wolkersdorfer C (2008) Water Management at Abandoned Flooded Underground Mines – Fundamentals, Tracer Tests, Modelling, Water Treatment. Heidelberg, Springer.
United States Environmental Protection Agency (1996) The Volunteer Monitor's Guide to Quality Assurance Project Plans. vol EPA 841-B-96-003, p 59.

Dispersion Modelling of Natural Radionuclides ^{238}U, ^{232}Th and ^{40}K Released from Coal-fired Power Plants Operations

Maria de Lurdes Dinis[1,2], António Fiúza[1,2], Joaquim Góis[1,2], José Soeiro de Carvalho[1,2], Ana Cristina Meira Castro[1,2,3]

[1]Geo-Environment and Resources Research Centre (CIGAR) Engineering Faculty, Porto University (FEUP), Rua. Dr. Roberto Frias, 4200-465 Porto, Portugal
[2]Centre for Natural Resources and the Environment (CERENA), Instituto Superior Técnico - IST, Av. Rovisco Pais, 1049-001 Lisboa, Portugal
[3]School of Engineering Polytechnic of Porto (ISEP), Rua Dr. António Bernardino de Almeida, 431, 4200-07, Porto, Portugal

Abstract. The aim of this work was to simulate the radionuclides dispersion in the surrounding area of a coal-fired power plant, operational during the last 25 years. The dispersion of natural radionuclides (^{236}Ra, ^{232}Th and ^{40}K) was simulated by a Gaussian plume dispersion model with three different stability classes estimating the radionuclides concentration at ground level. Measurements of the environmental activity concentrations were carried out by γ-spectrometry and compared with results from the air dispersion and deposition model which showed that the stability class D causes the dispersion to longer distances up to 20 km from the stacks.

Introduction

Most environmental concerns are associated with uranium mining and milling sites, but the same concerns should be addressed to natural near surface occurrences of uranium as well as man-made sources such as technologically enhanced naturally occurring radioactive materials (TENORM). In particular, the energy production activities from coal-fired power plants are one of the major sources of increased exposure to man from enhanced naturally occurring materials.

Coal contains trace quantities of the naturally occurring radionuclides like uranium and thorium, as well as their radioactive decay products and ^{40}K. When coal is burned, minerals, including most of the radionuclides, do not burn and concentrate in the ash a few times more in comparison with their content in coal or surface soil. These materials may be released into the environment by continuous atmospheric emissions and dispersed by local meteorological conditions.

The radionuclides activities in fly ashes are extremely variable depending mostly on the origin and the type of coal. UNSCEAR (1982) reported ^{226}Ra con-

B.J. Merkel and A. Arab (eds.), *Uranium – Past and Future Challenges*,
DOI 10.1007/978-3-319-11059-2_65

centrations ranging from 4 to 15 Bq/kg in lignite (Italy) while in others types of coal the reported values were 0.52 Bq/kg (Wyoming, USA) and 100 Bq/kg (Brazil). In escaping fly-ash the average ^{226}Ra concentration was 15 Bq/kg in Wyoming (USA) and 560 Bq/kg in Hungary. In Greece the ^{226}Ra concentration in coal ranged from 44 to 236 Bq/kg and from 140 to 605 Bq/kg in the resulting fly-ashes (Manolopoulou and Papastefanou, 1992). The worldwide average ^{226}Ra concentration in escaping fly ash is 240 Bq/kg (Zeevaert et al., 2006).

The radionuclides concentration at a given place depends on many factors including the emission rate, the ash content of coal, the temperature of combustion, the efficiency of control devices, local atmospheric conditions (mostly wind speed and direction) and the distance to the receptor from the source. UNSCEAR (1982) reported low values for the annual effective dose at 1 km from a coal-fired power plant (Zeevaert et al., 2006) but in several countries, the reported values are relatively high such as those obtained in Greece (Papastefanou, 1996; Simopoulos and Angelopoulos, 1987), in Poland (Niewiadomski et al., 1986), in India (Mishra, 2004) and in several states of the USA (McBride et al., 1978).

The aim of this work was to study the radionuclides dispersion in the surrounding area of a Portuguese coal-fired power plant due to the continuous emissions over its period of operation. The dispersion of the natural radionuclides ^{238}U and ^{232}Th decay series and ^{40}K were simulated by a Gaussian plume dispersion model estimating the radionuclides ground level activity concentration up to a distance of 20 km from the coal plant. The results were transformed into doses using a simplified exposure pathway and a pre-defined critical group in order to perform a quantitative environmental risk assessment for the considered exposure scenario.

Methods and materials

Study area

The coal-fired power plant considered in this study is located in the southwest coastline of Portugal (at 6 km from the town of Sines) and it consists of four coal burning units, with an electrical energy output of 314 MW. This coal-fired power plant has been operational since 1985. It has two stacks, both with a height of 225 m and is fuelled by bituminous coal imported from many countries. Coal is stored at open air in 5 piles with approximately 1.3 x 10^6 tones.

Fifty sites were defined around the coal plant up to a distance of 20 km based on the possible TENORM contributions from the stacks, in addition to the expected atmospheric diffusion, height of the stacks, local meteorological conditions such as wind direction, in particular the prevailing wind-direction, velocity and frequency, human settlements distance and the accessibility of the locations.

Field measurements

The feed coal used in the selected coal-fired power plant is imported mainly from Colombia and South Africa and therefore the concentration of the radionuclides in coal, and consequently in ashes, is highly variable as shown in table 1. According to UNSCEAR (1982), the worldwide average concentration of ^{40}K, ^{232}Th and ^{238}U in coal is 50, 20, 20 Bq/kg, respectively.

Table 1. Average concentration of natural radionuclides in coal (Uslu and Gokmese, 2010)

Country	Calorific (kcal/kg)	Ash content (%)	Concentration in coal (Bq/kg)		
			^{226}Ra	^{232}Th	^{40}K
Colombia	7230	24	94-142	175-489	175-489
China	6390	21	17-160	7-122	16-220
India	4000	43	11-67	18-92	15-441
S. Africa	6430	24	151-248	125-204	148-204

In situ radiation measurements were carried out by high resolution gamma spectrometry with an HPGe detector at the previously defined sites; measuring time was set at 5000 seconds at each site. The radionuclides content of fly ash discharged through the stacks was also measured by gamma spectrometry.

All spectra were analyzed with Genie 2000 and ISOCS (In Situ Object Counting Systems) software from Canberra to identify and quantify the radionuclides detected in the fly ash which constitute the source term (Dinis et al., 2014). The ^{232}Th concentration was determined from the average concentrations of ^{228}Ac (911.07 and 968.09 keV), ^{212}Pb (238.63 and 300.09 keV) and ^{212}Bi (727.33 keV). The ^{226}Ra concentration was determined from the average concentrations of ^{214}Pb (295.21 and 351.92 keV) and ^{214}Bi (609.32 keV). The concentration of ^{40}K was determined directly by its gamma transition energy (1460.83 keV).

Air dispersion model

The average atmospheric transport, dispersion and deposition of the discharged radionuclides from the stacks, were calculated using the Gaussian plume dispersion model. The downwind radionuclides concentrations at ground-level for a given wind sector (defined by the wind direction) were calculated.

Eight wind sectors were adopted with a characteristic average wind speed and frequency in the following directions: N S E W NW NE SE SW.

The calculations were performed with three different stability classes: moderately unstable conditions (B), slightly unstable conditions (C) and neutral conditions (D) using site-specific meteorological data (wind speed and frequencies) of hourly surface observations from 2006 to 2012. Each of the three stability classes has a characteristic average wind speed that matches with the minimum, maximum and average wind speed verified for the considered period.

The annual radionuclides release rates available for the dispersion were calculated from the ash emission rate of the stacks (115 kg/h each one) and the measured radionuclides concentration in fly ash. The radionuclides concentration dispersion was calculated from the stacks location, at the mixing height, up to a distance of 20 km in each wind direction. The radionuclides concentrations at ground level, along each direction, were evaluated taking into account the respective average wind velocity and the frequency of the occurrence. The region where air concentrations and depositions were determined matches to the studied area with a point source in the center given by the two stacks. In particular, ground-level concentrations downwind (x direction) are more relevant, since radionuclides concentration will be highest along that direction. All mathematical calculations were set in MATLAB programming.

Hazards from the dispersion of atmospheric emissions

There are two possible scenarios resulting in the release of radionuclides to atmosphere from coal-fired power plants: releases of ash from the stack and release of ash from ash piles. Only radionuclides releases from the stacks were considered in this work.

Individual doses from atmospheric dispersion have been assessed for a hypothetical critical group which is assumed to have received the highest doses from the coal plant atmospheric emissions over time. In particular, the members of this critical group were considered to be constituted of self-sustaining farmers (Dinis et al., 2014).

Radionuclides emissions from a coal-fired power plant may result in public exposure from multiple pathways that include: i) external radiation from activity suspended in air or deposited on the ground and ii) internal exposure from the inhalation of airborne contaminants or ingestion of contaminated food products. The annual doses are presented only for inhalation that was considered to be the most significant exposure pathway taking into account the impact from exposure to the airborne effluents of the coal plant through the submersion in the contaminated plume and resuspension of the deposited activity (Dinis et al., 2014). The dose from inhalation of a radionuclide *i* in the plume ($D_{inp,i,}$) in Sv/y was calculated by the following equation:

$$D_{inp,i} = C_{air,i} . DC_{in,i} . T_{oe} . I_r . I_f \quad (1)$$

where $C_{air,i}$ (Bq/m^3) is the air average concentration of the radionuclide *i* for a continuous release from the stacks given as an output of the Gaussian plume model; $DC_{in,i}$ (Sv/Bq) is the effective dose coefficient for inhalation of radionuclide *i*; T_{oe} is the total outdoor exposure time (2922 h/y); I_r is the average inhalation rate of the individual exposed (0.83 m^3/h) and I_f is the inhalable fraction of the aerosol in the plume (assumed to be 1) (Papastefanou, 1996; Dinis et al., 2014).

Doses resulting from the inhalation of resuspended radionuclides depends on a resuspension factor (R_r, m^{-1}) given by the ratio between the concentration in air due to resuspension (Bq/m^3) and the concentration deposited in the superficial soil (Bq/m^2). This factor was calculated using the resuspension model developed by Garland assuming that the resuspended concentration varies over time (after deposition) as radionuclides become associated with the soil surface. The resuspension factor at time zero is assumed to be 1.2 x 10^{-6} m^{-1}; after 10 days is 10^{-7}; after 100 days is 10^{-8}; after 1000 days 10^{-9} and after 10000 days is 10^{-10}.

The dose from inhalation of a radionuclide *i* resuspended from the soil ($D_{inr,i}$) in Sv/y, was calculated by the following equation:

$$D_{inr,i} = C_{soil,i} . DC_{in,i} . S_r . \rho_s . D_{al} . I_f . T_{oe} . I_r \quad (2)$$

where, $C_{soil,i}$ is the radionuclides concentration in soil measured by gamma spectrometry in Bq/kg (Table 2); ρ_s is the average soil density (1600 kg/m^3); S_r is the soil resuspension factor (10^{-8} m^{-1}) and D_{al} is the depth of active soil layer (0.3 m) (Papastefanou, 1996; Dinis et al., 2014).

The total inhalation dose for each pathway was calculated considering the inhalation doses arising from the radionuclides considered in this study: ^{40}K, ^{226}Ra and ^{232}Th. The annual effective dose (Sv/y) was estimated by summing the total doses from these three radionuclides and both considered exposure pathways.

The total individual inhalation risk to the exposed receptor was calculated by:

$$R_{in} = \sum \left(D_{in,i} . RC_{in,i} \right) \quad (3)$$

where, for the considered exposure scenario, R_{in} is the total inhalation risk, $D_{in,i}$ is the inhalation dose resulting from radionuclide *i* through both pathways, expressed in Bq/y, and RC_{in} (risk/Bq) is the risk per unit of activity inhaled (EPA, 1995).

Results and discussion

The results obtained with the meteorological data from 2012 and for the considered stability classes (B, C and D) are presented in the following figures only for ^{226}Ra, although higher dose values were obtained for ^{232}Th. A 3D surface plot along with the contour map is presented to show the distribution of the ground-level concentration in the x-y plane and the effect of the stability class on that distribution (Fig.1, Fig.2 and Fig.3).

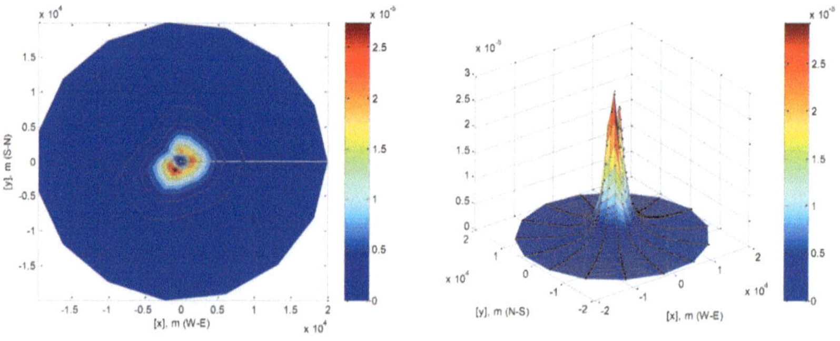

Fig.1. Downwind ground-level ^{226}Ra concentrations and 3D-surface plot, Bq/m^3 (2012, class B)

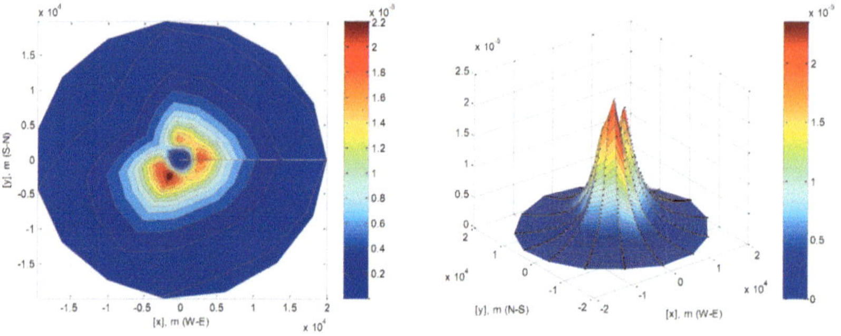

Fig.2. Downwind ground-level ^{226}Ra concentrations and 3D-surface plot, Bq/m^3 (2012, class C)

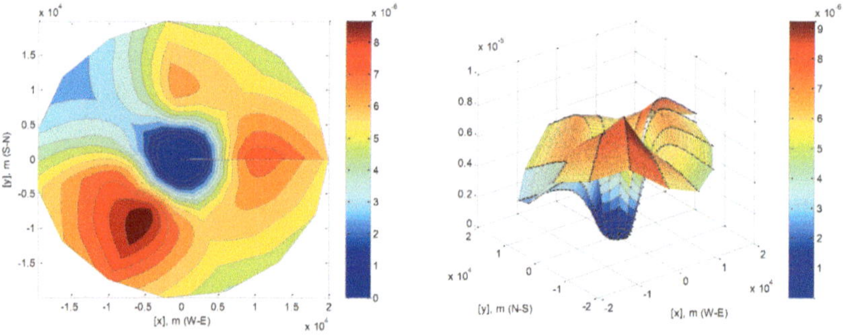

Fig.3. Downwind ground-level ^{226}Ra concentrations and 3D-surface plot, Bq/m^3 (2012, class D)

The average concentration for the radionuclides detected in the fly ash by gamma spectrometry was 901.44 Bq/kg for ^{40}K, 53.97 Bq/kg for ^{226}Ra and 40.13 Bq/kg for ^{232}Th. These values were used to calculate the emission rate for each one of the radionuclides and entered as an input in the dispersion model.

The radionuclides concentration in the prevailing wind direction showed an increase of 37% for ^{40}K, 44% for ^{226}Ra and 75% for ^{232}Th (SE) compared to average concentrations in other wind directions.

The results of the dispersion model showed that the stability class C originates the highest ground-level concentrations. Nevertheless, stability class D takes the dispersion to longer distances from the stacks; the dispersion occurs predominantly in the SW direction, matching with the Atlantic Ocean, followed by the NE-E and E-SE directions. Stability class B brings the plume to the ground very quickly resulting in higher concentration values near the stacks but a fair dispersion downwind. For stability classes C and D the dispersion reaches a distance of 20 km from the stacks although this effect is much more intense with class D. For locations near to the stacks (less than 2.5 km) it is possible to observe higher radionuclides concentration at ground level with stability classes B and C.

From the six years of site-specific meteorological data used to simulate the radionuclides dispersion and deposition on the ground the highest concentration values were obtained for 2008, 2011 and 2012. A comparison between the results obtained with meteorological data averaged from the period between 2006 and 2012 and each one of the considered stability classes is presented in Table 2.

Table 2. Radionuclides concentration, inhalation doses and effective doses

Nuclide	C_{air} (Bq/m^3)	C_{soil} (Bq/kg)	$D_{in,p}$ (Bq/y)	$D_{in,r}$ (Bq/y)	D_{in} (mSv/y)
Pasquill stability class (B) – average meteorological data from 2006-2012					
^{40}K	5.24×10^{-5}	361.8	1.30×10^{-1}	4.06	8.78×10^{-6}
^{226}Ra	3.10×10^{-6}	20.44	7.79×10^{-3}	2.29×10^{-1}	8.24×10^{-4}
^{232}Th	2.79×10^{-6}	23.59	7.79×10^{-3}	2.65×10^{-1}	2.98×10^{-2}
Pasquill stability class (C) – average meteorological data from 2006-2012					
^{40}K	8.39×10^{-5}	361.8	2.08×10^{-1}	4.06	8.94×10^{-6}
^{226}Ra	5.01×10^{-6}	20.44	1.26×10^{-2}	2.29×10^{-1}	8.44×10^{-4}
^{232}Th	4.38×10^{-6}	23.59	1.24×10^{-2}	2.65×10^{-1}	3.02×10^{-2}
Pasquill stability class (D) – average meteorological data from 2006-2012					
^{40}K	6.91×10^{-5}	361.8	1.73×10^{-1}	4.06	8.86×10^{-6}
^{226}Ra	4.16×10^{-6}	20.44	1.05×10^{-2}	2.29×10^{-1}	8.37×10^{-4}
^{232}Th	3.82×10^{-6}	23.59	1.05×10^{-2}	2.65×10^{-1}	3.01×10^{-2}

It is possible to observe that ^{40}K presents a higher ground-level concentration (C_{air}) than ^{226}Ra or ^{232}Th and that stability class C turn out higher average concentrations, followed by class D. It is assumed that C_{soil} remains constant for all stability classes.

The total annual effective dose was obtained by summing up the doses resulting from the considered radionuclides through both exposure pathways (D_{in}, mSv/y; 40K, ^{226}Ra and ^{232}Th). The resulting value for the annual effective dose was 0.031 mSv/y and this value was very similar for all the three stability classes used. It should be stressed that the highest contribution to the total effective dose comes from ^{232}Th individual inhalation dose. The resulting risk for the considered exposure scenario was 1.61×10^{-7}.

Conclusions

The aim of this work was to study the radionuclides dispersion in the surrounding area of a Portuguese coal-fired power plant. The effects of meteorological factors in the dispersion show that the atmospheric stability class affect the pattern of dispersion and the distance at which the radionuclides may be dispersed farther downwind; the stability class D causes the dispersion to longer distances, up to 20 km, and this matches with the results from the in situ measurements. Radionuclides downwind concentrations are mainly influenced by stability class, wind speed and source emission rate.

Average results for radionuclides concentration along the study area are similar for the three stability classes and this is reflected in the calculated doses and risk. The results showed that the total dose equivalent rate is approximately 0.031 mSv/y and the total annual risk is 1.61×10^{-7}. These values are lower than the effective dose recommended for the public (1 mSv/y) and EPA's risk value of 10^{-6} for the general public. Nevertheless, radionuclides concentration in the prevailing wind direction (SE) showed a significant differential enrichment (37% for ^{40}K, 44% for ^{226}Ra and 75% for ^{232}Th) and this evidently will reflect higher doses and higher risk in this particular direction.

References

Dinis ML, Fiúza A, Góis J, Carvalho JM, Castro ACM (2014), Modeling Radionuclides Dispersion and Deposition Downwind of a Coal-Fired Power Plant, Procedia Earth and Planetary Science 8; 59-63, doi:10.1016/j.proeps.2014.05.013.

EPA (1995), USEPA, Health Effects Assessment Summary Tables, Radionuclide Carcinogenicity Slope Factors: HEAST.

Manolopoulou M, Papastefanou C (1992), Behavior of natural radionuclides in lignites and fly ashes. J. Environ. Radioact. 16; 261-271.

McBride JP, Moore RE, Witherspoon JP, Blanco RE (1978). Radiological impact of airborne effluents of coal and nuclear plants. Science 202 (43-72); 1045-1050.

Mishra UC, 2004. Environmental impact of coal industry and thermal power plants in India. J. Environ. Radioact. 72 (1-2); 35-40.

Niewiadomski T, Jasinska M, Wasiolek P (1986), Enhancement of population doses due to production of electricity from brown coal in Poland. J. Environ. Radioact. 3; 273-292.

Papastefanou C (1996), Radiological Impact from Atmospheric Releases of ^{226}Ra from Coal Fired Power Plants. J. Environ. Radioact., 1-2; 105-114.

Simopoulos SE, Angelopoulos MG (1987), Natural radioactivity releases from lignite power plants in Greece. J. Environ. Radioact., 379-389.

UNSCEAR (1982), Ionizing Radiation: Sources and Biological effects, Report of the General Assembly with Scientific Annexes, Vol. 1, Annex E, New York.

Uslu I, Gökmeşe F (2010), Coal an Impure Fuel Source: Radiation Effects of Coal-fired Power Plants in Turkey. J Biol. & Chem., 38(4); 259-268.

Zeevaert T, Sweeck L, Vanmarcke H (2006), The radiological impact from airborne routine discharges of a modern coal-fired power plant, J. Environ. Radioact., 85; 1-22.

Sequential extraction of U and Th isotopes: study of their intrinsic distribution in phosphate and limestone sedimentary rock in comparison with black shale

Said Fakhi[1], Rabie Outayad[1], Elmehdi Fait[1], Zineb Faiz[1], C. Galindo[5], Abderrahim Bouih[3], Moncef Benmansour[3], Azzouz Benkdad[3], Ignacio Vioque[4], Marusia Rentaria[2], Abdelmjid Noureddine[5]

[1]University Hassan II Mohammedia -Casablanca, Faculté des Sciences Ben Sik, Morocco. E -mail: s.fakhi@univh2m.ac.ma.com
[2]Department de Recursos Naturales, Facultad de Ecologia Zootecnia of the Universidad Autonoma de Chihuahua Mexico
[3]National Centre for Energy, Sciences and Nuclear Techniques (CNESTN - Morocco), CEN Maamoura
[4]Department Fascia Palisade II, Secular Technical Superior de Arquitectura de Sevilla Universidad de Sevilla
[5]Institute Hubert Curien Multidisciplinary ULP/CNRS-In2p3 UMR 7178, 23 rue de Loess BP 28 F- 67037 Strasbourg Codex France.

Abstract. In this work, a study of the intrinsic behavior of U and Th isotopes in sedimentary rocks was performed by sequential extraction. The results for the phosphate layers and limestone interlayers of Oulad Abdoun Basin are compared to those obtained previously for Timahdit black Shale rocks. Characterization carried out by ICP-AES and SEM-EDX showed that apatite, calcite, dolomite, quartz and clays are the essential components of the inorganic matrix. The selective leached phases are treated radiochemically before measuring U and Th isotopes by alpha spectrometry. Among the results, it should be noted that the U isotopes were concentrated in an anaerobic environment. Humic acids that are the richest organic phase in U are responsible for the concentration of ^{234}U, ^{238}U in the carbonates and apatite. In the mineral phase, ^{232}Th, of terrigenous origin is partitioned between the silicate and pyrite minerals.

Introduction

Chemical reactions of metals in solid / water interfaces that contain many dissolved anions (carbonates, phosphates, silicates and humates) are dependent on the metals origin and distribution in the different phases of the solid compartment

B.J. Merkel and A. Arab (eds.), *Uranium – Past and Future Challenges*,
DOI 10.1007/978-3-319-11059-2_66

(Ivanovich 1991). In this context, the present work is focused, essentially, on the U and Th behavior in relation to the nature and intrinsic structure of the sedimentary rocks (McKelvery et al. 1956), (Roessler et al. 1980). The study was focused on the U and Th radioisotopes mode of the naturally occurrence in natural phosphate and limestone samples. The mineralogical composition of a limestone and phosphates rocks from Oulad Abdoun basin was determined by X-ray diffraction (XRD), scanning electron microscopy (SEM). A sequential leaching procedure (Blanco et al. 2005), (Schultz et al. 1998), based on the Tessier's method (Tessier et al. 1979) and (Quevauviller et al. 1993) was used to define the speciation of isotopes of U and Th. The leached phases are measured by alpha spectrometry after radiochemical treatment. The isotopic ratios such as $^{234}U/^{238}U$, $^{230}Th/^{238}U$ was used to determine the profile of the various phases that constitute the sedimentary rocks (Ivanovitch et al. 1992), (Fakhi et al. 2002). The results are compared with those obtained previously in black shale samples (Galindo et al. 2007)

Experimental section

Nature of sedimentary rocks

Two types of Oulad Abdoun basin samples were used in this investigation: phosphate layers and limestone interlayer samples. Sequential extraction for both type s of samples was compared to that of black shale (Rahall 1970) previously analyzed (Galindo et al. 2007).

Mineralogical characterization

Limestone and phosphate samples characterization was carried out by using a scanning electron microscope (SEM) coupled to a microprobe Energy Dispersive X-ray (EDX).The material used JEOL 6460LV is equipped with a beryllium window (ATW2). The digital image acquisition in both secondary (SEI) and backscatter (EIB) is performed by means of electronic imaging at resolutions equal to 3.5 nm The device is coupled to a microprobe energy dispersive X-ray (EDX). The device is characterized by a resolution that varies according to the measurement conditions from137 eV to 5.9 keV. The data provided by the system were analyzed using the Oxford-INCA software

U and Th isotope in leached phases

U and Th isotopes speciation in the leached phases was carried out by sequential extraction as described in Table 1 (Tissier et al. 1979) and (Schultz et al. 1998).

U and Th isotopes determination

Radiochemical separation

In general, the radiochemical separation of U and Th isotopes was conducted in the same conditions described previously (Galindo et al 2005).

Alpha spectrometry analysis

Different rooms of alpha spectrometry were used including, that described below:

Alpha spectra were recorded using 450 mm2 PIPS detectors mounted in vacuum chambers and coupled to the associated electronics Alpha Analyst (Canberra Electronique, Savigny-le-Temple France). The optimum source – detector distance was established at 6 mm. At this distance, the detection efficiency was about 25%. The acquisition times ranged from 1 to 14 days, depending on the ac-tivity level of the source. Background measurements were repeated all 10 counting for uranium determination and all 2 counting for thorium measurements. Spectra were analyzed with the Visual Alpha software (AMS).

Table 1. Sequential extraction procedure (RT means room temperature)

	Extracting agent	***Extracting conditions***	
		Shaking time	Temp.
Method 1			
F1. Water soluble	30 mL deionised water	2 h	RT
F2. Exchangeable	25 mL 1M MgCl2	24 h	RT
F3. Carbonates	35 mL 1M CH3COONa – 1M CH3COOH pH= 4.75	24 h	RT
F4. Manganese/iron hydroxides and oxides	35 mL 0.04M NH2OH.HCl - 25 % v/v CH3COOH pH=2 (adjusted with HNO3)	24 h	RT
F5. Organic matter + pyrite	25 mL 30% H2O2 – 15 mL 0.02M HNO3 (pH=2)	2h	85°C
	15 mL 30% H2O2 – 9 mL 0.02M HNO3 (pH=2)	3h	85°C
	35 mL 3.2M CH3COONa in 20%HNO3	0.5h	RT
F6. Residual	/	/	/
Method 2			
F1'.	15 mL 6M HCl	4 h	RT
	15 mL 48 % HF	4 h	RT
F2'. Residual	/	/	/

Results and interpretation

Mineralogical characterization of phosphate sample is illustrated in Fig. 1 and 2. In Fig. 1 is shown an image representing the morphology of a grain phosphate 6). This image shows clearly the existence of a crystalline apatite envelope characterized by a high porosity. Fig. 2 shows the main composition of the grain. The concentration of the main identified elements varies according to the sequence: O> Ca> P> F. This result is in good agreement with the stoichiometry of phosphate sedimentary terms (Renteria-Villalobos et al. 2010). Analysis by X-ray diffraction shows that the deposits are formed essentially of carbonate fluorapatite called francolite (Mc Connell 1938).

Fig.1. Typical morphology of phosphate pellets

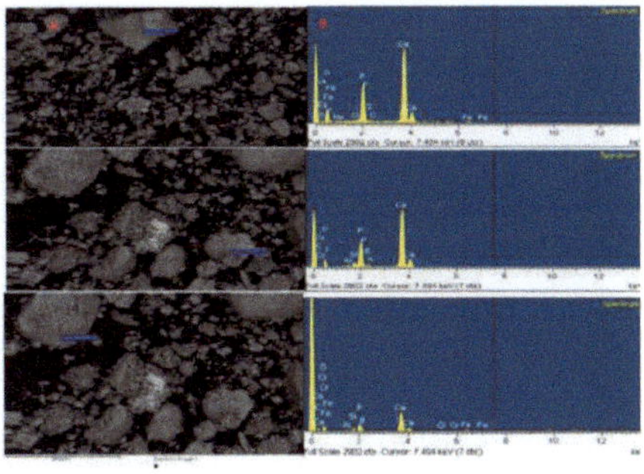

Fig.2. Morphology and composition of phosphate pellets

Radiochemical analysis

$^{230}Th/^{238}U$ isotopic ratio

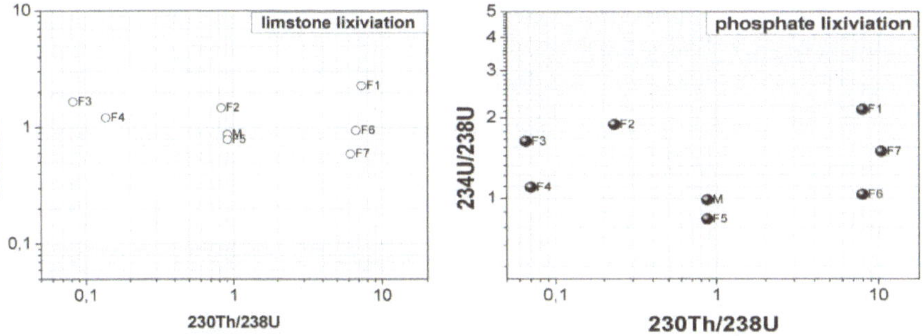

Fig.3. $^{230}Th/^{238}U$ and $^{234}U/^{238}U$ variation in different phases of samples

Values of isotopic ratios $^{230}Th/^{238}U$ (Fig. 3 and 4)) shows that the geochemistry of phosphate can represented by two models:

The first model, characterized by $^{230}Th/^{238}U > 1$

- For phosphate term (Fig.3) and limestone term (Fig.4), it covers the following sequence: F7 (residual phase), F6 (clay phase), F1 (water soluble). Radioactive disequilibrium marked by an excess of ^{230}Th is caused by the intrinsic exchange between the phases of sediment. U deficiency in the residual phase formed mainly by the kerogen and pyrite, in silicates can be explained by the ability of these phases to concentrate thorium. While U is accumulated in the adjacent phases such as humic acids.

- The second model is described by $^{230}Th/^{238}U < 1$. Since thorium is practically less soluble, excess of uranium is the result of an accumulation under reducing conditions of U (The model of U accumulation is observed in the F3 (carbonate), F4 (apatite) and F2 for the phosphate and limestone terms. U accumulation result of its reduction by organic matter intimately associated with these phases. In both facies, the raw rock (M) and organic phase (F5) have $^{230}Th/^{238}U$ slightly less than 1, despite the great power of U accumulation in this phases requires adsorption and/or complexation of U by organic matter associated with carbonates phosphate. While thorium is concentrated by pyrite coexisting with humic acids in F5.

234U/238U isotopic ratio

The values of $^{234}U/^{238}U$ found in analyzed samples (Fig.3 and Fig.4) can be classified into three patterns.

The first is described by the $^{234}U/^{238}U < 1$. It covers F5 (organic matter) for the phosphate term and F5 (organic matter) and F7 (residual phase) for limestone. For organic phase (F5), the radioactive disequilibrium is mainly caused by alpha recoil effect. Indeed ^{234}U, which has a greater mobility than that of ^{238}U, can be ejected from this phase or from its various interfaces by alpha recoil effect.

The second model is characterized by $^{234}U/^{238}U > 1$, described U intrinsic distribution.

- For the phosphate term, it is observed in: F1 (water soluble phase)> F2 (exchangeable)> F3 (carbonate)> F7 (residual phase).
- For the limestone facies, it is observed according to the sequence: F1 (water soluble phase)> F2 (exchangeable) = F3 (carbonate).

The second model showed the existence of an anomaly in the distribution of U isotopes in the various phases ($^{234}U/^{238}U > 1$). This anomaly is the result of the exchange reactions which may occur at the interfaces of the various phases. The second pattern is based on the adsorption of uranium taking place in a reducing environment at the interface of the phase/ water containing an excess of ^{234}U.

The third profile characterized by $^{234}U/^{238}U$ close to unity, indicating the secular equilibrium is found, for phosphates, in raw rock and clay part. In the limestone facies, it is also equal to 1 for the clay phase, but slightly less than unity for the overall rock.

To approach the intrinsic dynamic of phosphate and limestone sediments (Oulad Abdoun basin) we compared their profile with that previously established for black shale from Timahdit (Galindo et al.2007). This initiative is undertaken based on the work of Benalioulhaj (Benalioulhaj 1991) which concluded that the Timahdit oil shale series is an unstable facies lateral and synchronous at phosphates series of Basin Oulad Abdoun basin. He also concluded that the two basins have evolved in the same paleogeographic area within a marine golf.

The analysis of the curves representing the variation of the actinide amount in% of each phase: Fig. 4, Fig. 5 and Fig. 6, shows that the behaviour of actinides

in Oulad Abdoun Basin is similar to that previously observed in the sediments of black shale. Indeed, it has been shown that:
- The kerogen and minerals accessories such as pyrite have low capacity of uranium accumulation.
- The humic acid, precursor of the kerogen, is the phase most responsible for the U retention: 70% in black shale and around of 60% for Oulad Abdoun samples. Retention is controlled by adsorption or by complexes formation at OM / carbonate phosphates interfaces.
- The uranium is also accumulated in the carbonate and apatite terms.
- Soluble complexes with water and exchangeable phases contain negligible amounts of uranium (<1%). This result is of great environmental importance. It shows that the amount of uranium that water can be extracted from the rock is substantially negligible.
Fig. 4, 5 and 6 shows that terrigenous origin ^{232}Th and ^{230}Th radioactive descendant of ^{238}U seem accumulated in the organic fraction, in the silicates and pyrite. But during the testing of sequential extraction, we found that 49% of ^{232}Th are fixed in the final residue and 51% remains trapped in the kerogen which is resistant to acid attack (F7). The absence of ^{232}Th in apatite and in carbonates can be attributed to their diagenetic origin.

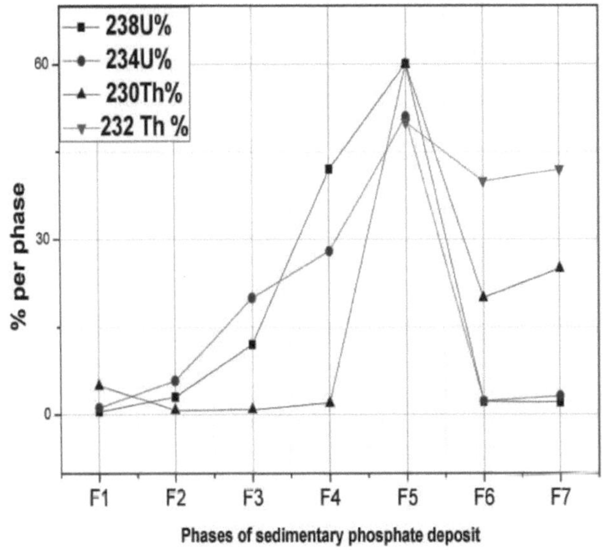

Fig.4. The amount isotopes of U and Th in % (phosphate)

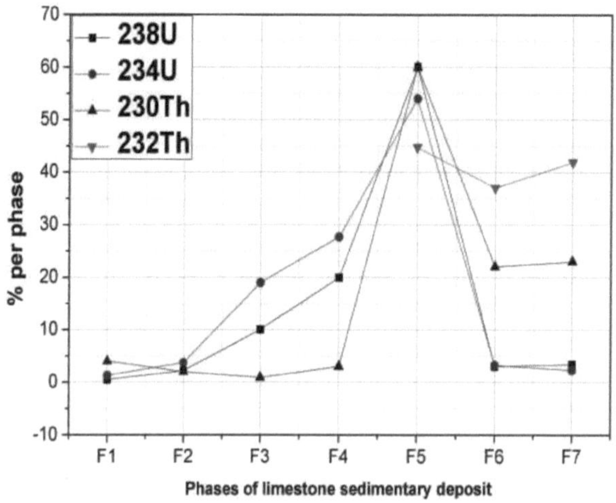

Fig.5. The amount isotopes of U and Th in % (limestone)

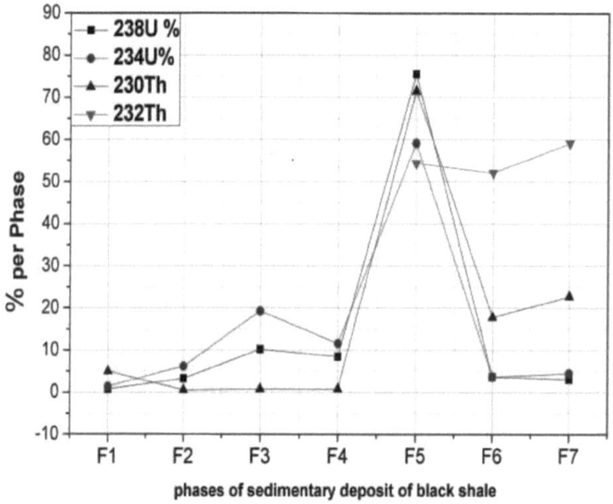

Fig.6. The amount isotopes of U and Th in % (black shale)

Conclusions

Deposits in sedimentary phosphate layers and interlayers essentially consist of limestone. U in the form of U(VI) is associated with humic acids. It has been deposited under anoxic reducing paleoenvironment. Carbonates and apatite contribute also to the retention of uranium. ^{232}Th is partitioned between silicate minerals and pyrite. The low activity ratios of $^{232}Th/^{238}U$ confers to the sediment a marine character. The absence of U in interstitial confirmed in the water leaching fraction of the procedure, means that the material has a high capacity of U retention. Therefore, naturally the material does not pose groundwater contamination problems.

References

Blanco P, Vera Tome F, Lozano J C (2005) Fractionation of natural radionuclides in soils from a uranium mineralized area in the south-west of Spain. Journal of Env. Radioactivity 79: 315-330

Fakhi S, Aadjour M, Fendan B, Andres Y, Abbe J C, Mokili M, El Morabet A (2002) Analyse géochimique par spectrométrie gamma et par fluorescence X (EXDRF) des dépôts mésozoïques des forages OYB-1 et NDK-2 (bassins de Doukkala et d'Essaouira, Maroc). Bulletin de l'Institut Scientifique, Rabat, section Sciences de la Terre, 2002, n° 24, 59-69

Galindo C, Mougin L, Fakhi S, Nourreddine A, Lamghari A, Hannache H (2007) Distribution of naturally occurring radionuclides (U, Th) in Timahdit black shale (Morocco). Journal of environmental Radioactivity 92 (1): 41-54

Galindo C, Mougin L, Nourreddine A N (2005) An improved radiochemical separation of uranium and thorium in environmental samples involving peroxide fusion, submitted to publication in Applied Radiations and Isotopes

Ivanovich M, Harmon R S, (1992) Uranium-series disequilibrium: applications to Earth, Marine, and Environmental Sciences, 2nd edition, Clarendon Press, Oxford, New-York

Ivanovich M (1991) Aspects of uranium/thorium series disequilibrium applications to radionuclide migration studies. Radiochim Acta 52–53, no. 1: 257–268

McKelvey, Carswell (1956) Uranium in the Phosphoria Formations. In: L. Page, H. Stocking and H. Smith (eds). Contribution to the geology of uranium and thorim by the United States Geological Survey and Atomic Energy Commission for the United Nations International Conference on the peaceful Uses of Atomic Energy, Geneva, Switzerland, 1955, US Geological Survey, prof. paper, 300: 483-494

McConnell D (1938) A structural investigation of the isomorphism of the apatite group. American Mineral 54: 1379-1391

Rahhali, I. 1970, Foraminifères benthoniques et pélagiques du synclinal d'El Koubbat (Moyen-Atlas, Maroc) Notes Serv. Géol. Maroc, 30, 225: 51-86

Schultz M K, Burnett W C, Inn K G W (1998) Evaluation of a sequential extraction method for determining actinide fractionation in soils and sediments. Journal of Environmental Radioactivity 40: 155-174

Rentería-Villalobos M, Vioque I, Mantero J, Manjon G (2010) Radiological, chemical and morphological characterizations of phosphate rock and phosphogypsum from phosphoric acid factories in SW Spain. Journal of Hazardous Materials 181: 193-203

Roessler C E, Kautz R, Bolch W E, Wethington J A (1980) The effects of mining and land reclamation on the radiological characteristics of the terrestrial environment of Florida's phosphate regions. The Natural Radiation Environment III, U.S. CONF - 780422.Vol 2: 1476-1493. Department of Energy, Washington

Tessier A, Campbell P G C, Bison M (1979) Sequential extraction procedure for the speciation of particulate trace metals. Anal. Chem 51: 844-851

Radium in Groundwater

Stephanie Hurst[1]

[1]Saxon State Ministry of the Environment and Agriculture, Archivstr. 1, 01097 Dresden, Germany

Abstract. The hydro-geochemical behavior of radium in groundwater may be not trivial, but is well known since many decades. Anyway comparatively few publications can be found on this topic. The isotope ratios of radium sampled during pumping tests can give information on ground water flow regimes, regarding the mixing behavior of groundwater. Radium-226 is from the radiological point of view the most important isotope in the uranium decay chain. Designing high level nuclear waste repositories this must be taken into account.

Introduction

The element radium was discovered in 1898 by Marie and Pierre Curie in the course of investigations on radiation. In the following three decades it was broadly used for different medical and other applications. Subsequent to new findings about the dangers of radiation radium disappeared nearly completely from public conscience. Due to the fact that there was no real practical use for radium or radium compounds the element was more or less buried in oblivion. It only came back into public conscience when enhanced concentrations were measured in some drinking waters and mineral waters in the 80ies of the last century.

By contrast the properties of uranium were extensively investigated, first for weapon and energy issues, later regarding the safe disposal of nuclear waste. The following delineation may illustrate the relevance of radium for aquifer diagnosis and also for safety cases in preparation of nuclear waste repositories. It is based mainly on investigations carried out between 1989 and 1992 and takes into account some recent publications e.g. by Krest and Harvey (2003) or Krishnaswami and Cochran (2008).

B.J. Merkel and A. Arab (eds.), *Uranium – Past and Future Challenges*,
DOI 10.1007/978-3-319-11059-2_67

Naturally occurring radium isotopes

Radium isotopes are occurring in all three natural decay chains, i.e. Uranium -238 (Ra-226), Uranium-235 (Ra-223) and Thorium-232 (Ra-228 and Ra-224). All four natural radium isotopes result from an alpha decay of thorium isotopes.
The half-lives of these isotopes vary between several days (Ra-223: 11,4 d, Ra-224: 3,6 d), several years (Ra-228: 5,7 y) and 1602 years (Ra-226).

Behavior of radium under different hydro-geochemical conditions

The behavior of radium in water is characterized by the physical properties (decay, α-recoil) as well as the geochemical and hydraulic aquifer properties (i.e. essentially porosity) and the chemical properties of the groundwater.

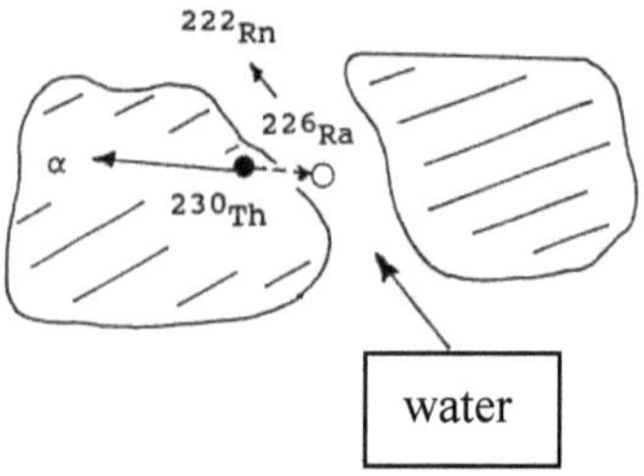

Fig.1. Alpha recoil: Following the radioactive decay of the parent isotope, the daughter atom recoils in the opposite direction from the path of particle ejection.

Depending on the concentration of the mother isotopes in the rock and the respective hydro-geochemical conditions the alkaline earth metal radium may occur in very small to extremely high concentrations in groundwater. Langmuir and Riese (1985) investigated the thermodynamic properties of radium on the basis of empiric data and extrapolation of thermodynamic properties of correspondent barium, strontium and calcium species. Bard et al. (1985) published thermodynamic data from several sources. There is a high probability that radium never occurs as pure phase in groundwater, but always as mixed phase with more common ions like Ba, as e.g. $Ba(Ra)SO_4$.

Interaction of radium with barium is independent from pH but highly dependent from the given sulfate concentration. Sulfate ions oppress desorption of Ra from rock surfaces. Radiolysis may – in an otherwise reducing environment – lead to generation of oxidants.

In the upper part of the aquifer organic colloids and clay minerals may have influence on radium behavior.

In general radium is relatively immobile in groundwater at/if

- enhanced pH- and Eh values
- enhanced sulfate ($SO_4^{2-)}$ concentrations are present or at sulfate saturation and at the same time absence of Ba-ions
- sulfide is oxidized to sulfate and radium is co-precipitated
- hydrocarbonate (HCO_3^-) is present in relevant concentration
- oxygen (O_2) is present and iron or manganese hydroxides are precipitated (co-precipitation)
- radium ions are fixed/bound in clay interlayer

On the other hand radium is relatively mobile at/if

- low pH- and Eh values
- relevant iron or manganese colloid concentrations are given (complexation)
- high mineralization of water connected with low dissociation ratio of water components and radium is migrating as loosely bound $Ba(Ra)SO_4$

This is of course a short simplified description that gives a rough idea but cannot make up for a sound water chemical case-by-case investigation.

Radium isotopes as tool for aquifer diagnosis

A comprehensive aquifer diagnosis is conducted on the base of geophysical, hydraulic, hydro-chemical and isotope hydrological investigations.
The method presented as follows may be used in addition to get more specific information on aquifer characters.
The change in radium isotope ratios can be interpreted to get better insight in the mixing behavior of groundwater from different aquifer areas or different aquifers.

It is based on the following facts:

a) groundwater systems are branded by a defined radium isotope ratio
b) radium isotopes have different half-lifes
c) different systems are mixed by hydraulic changes (e.g induced by pumping)

These facts have been published and rudimentary systemically interpreted e.g. by Cherdyntsev (1971) or Zukin et al. (1987).

They described that $^{224}Ra/^{228}Ra$ ratios below 1 may be the result of a mixture of different groundwater or a long distance/migration from an area with high ^{228}Ra to an area with low ^{228}Ra may explain the decrease of the above mentioned ratio.

These approaches were picked up and elaborated (see Hurst 2014). The following assumption was the starting point:
Is groundwater migrating from point A to point B

- in a representative elementary volume (REP)
- under constant hydraulic conditions
- under constant hydro- and geochemical conditions
- at homogeneous radionuclide concentration of bedrock
- in a clearly laid out time

the radium isotope concentration and the $^{224}Ra/^{228}Ra$ ratio remain unchanged.
If one of these assumptions does not apply, the radium concentration and the $^{224}Ra/^{228}Ra$ ratio will not remain unchanged. The initial $^{224}Ra/^{228}Ra$ ratio, anyway, may vary. From this assumption different models were developed. They can be applied to get information about

- the flow time of groundwater,
- the mixing of groundwaters,
- double porosity situations
- groundwater contamination

Krishnaswami and Cochran (2008) as well as Krest and Harvey (2003) additionally describe methods to get information of mixing between ocean water and ground water.

Flow time model

A change in hydraulic conditions (e.g. during a pumping test) leads to a decrease of the $^{224}Ra/^{228}Ra$ ratio due to the faster decay of ^{224}Ra.

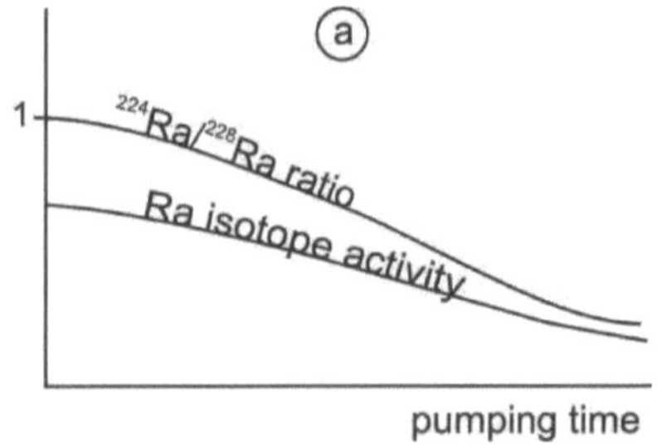

Fig.2a. Ra ratio and Ra activity in a homogeneous system decreasing with advanced pumping time.

Mixing models

The following figure (a) and (b) show possible increases in the Ra isotope activity during a pumping test due to different flow velocities, the figures (c) and (d) decreases of the activity due to migration from areas with higher activities (in the aquifer rock) to areas with lower activities.

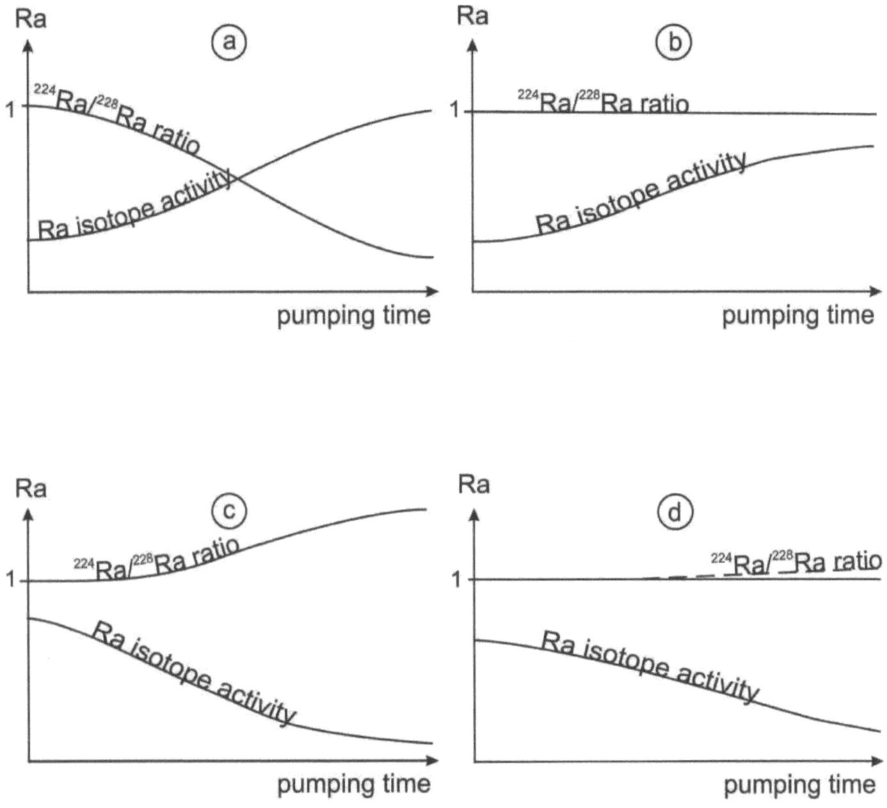

Fig.2b. see text above

Additionally radium isotope ratios can help to find out the influence of radium contamination in groundwater systems. Double porosity systems can also be identified by Ra isotope ratios (fig. 2 c).

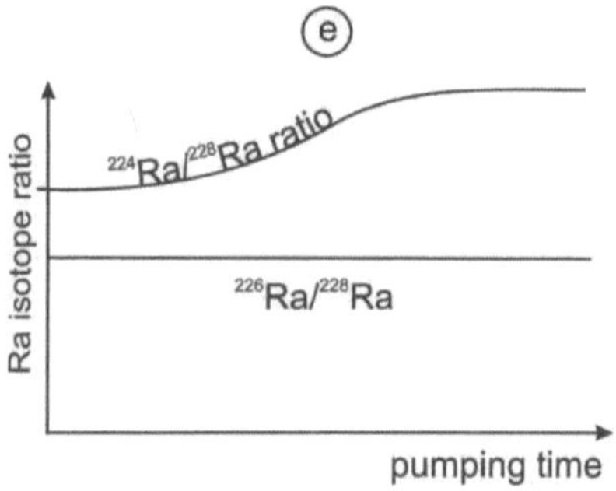

Fig.2c. This figure shows the typical development of the $^{224}Ra/^{228}Ra$ ratio and $^{226}Ra/^{228}Ra$ ratio in a double porosity system

Evaluating data from pumping tests

During several pumping tests in different aquifers and different depths in Northern Bavaria and the Czech Republic radium isotope ratios were determined.

Example 1

During two pumping tests in the preliminary drilling of the deep drilling project (KTB-VB; Northern Bavaria) the $^{226}Ra/^{228}Ra$ ratios changed as follows:

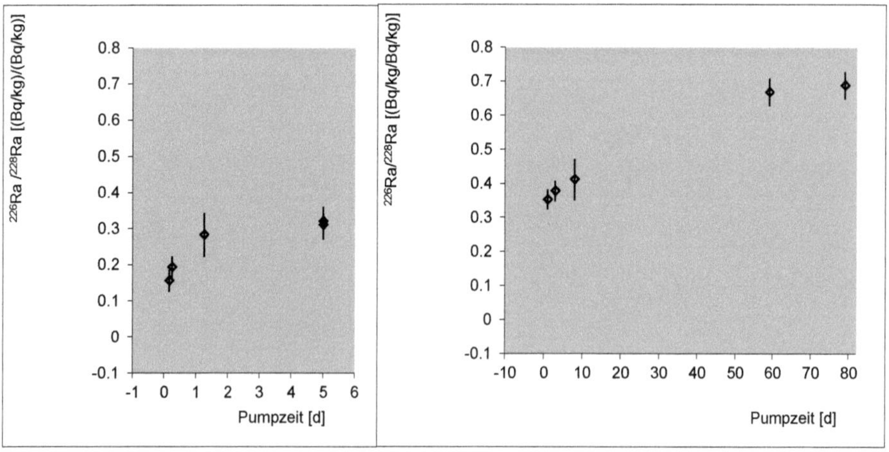

Fig.3. see text above, x-axis: pumping time

^{228}Ra was due to a fissure with high ^{232}Th content in the depth area of 4000 m very high. The change in the $^{226}Ra/^{228}Ra$-ratio showed that fluid from other areas (with relatively higher ^{226}Ra concentration, i.e. lower ^{228}Ra concentration) was increasingly contributing to the pumping water. It showed also that the thorium affected fissure was only a local phenomenon. Additionally the velocity of the rising pumping water could be estimated by the $^{224}Ra/^{228}Ra$ ratio in the pumping fluid.

Example 2

During a pumping test in an aquifer used for mineral water procurement the following results were gained (x-axis: pumping time):

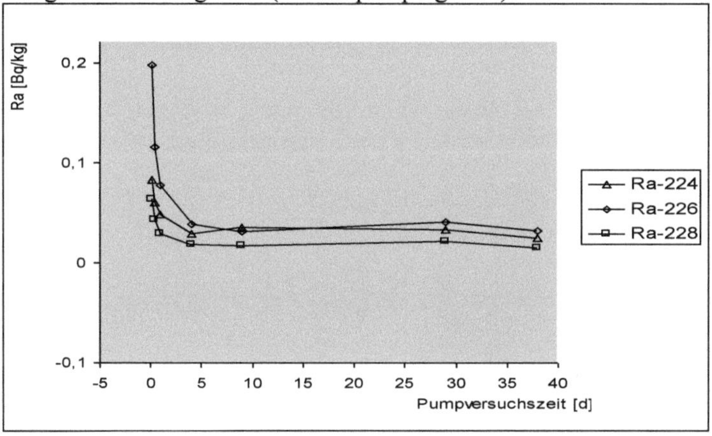

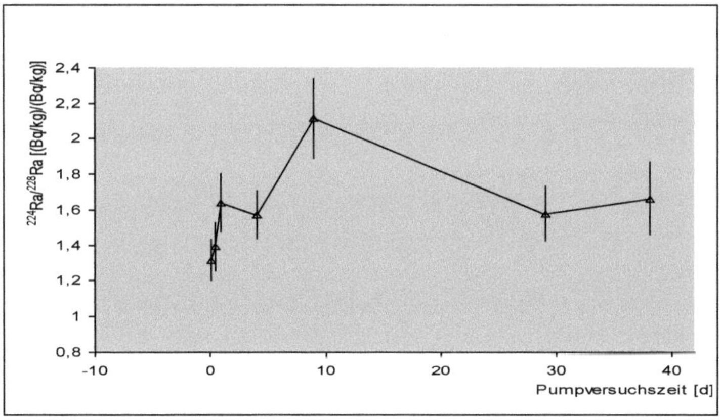

Fig.4. see text, x-axis: pumping time

It showed that after several days of pumping an area with relatively lower Ra activity was contributing to the system. The enhanced $^{224}Ra/^{228}Ra$ ratio after 10 pumping days is due to the fact that the water is migrating from a more distant area where the relatively immobile ^{228}Th is precipitated to the pore (i.e. fissure) system and produces (the relatively more mobile) ^{224}Ra into the aquifer. After 30 days this ratio is decreasing again, showing, that the water is transported from another more distant area. At the same time a (relatively late) quasi stationary state is gained. This circumstance and the changes of the activity ratios show that the pumped water comes from different source areas with relatively small porosity (no larger fissures or faults).

Designing HLW repositories

^{226}Ra from spent nuclear fuel is increasing over thousands of years relatively slowly until after 10000 years it becomes – also due to its decay products - a risk issue.

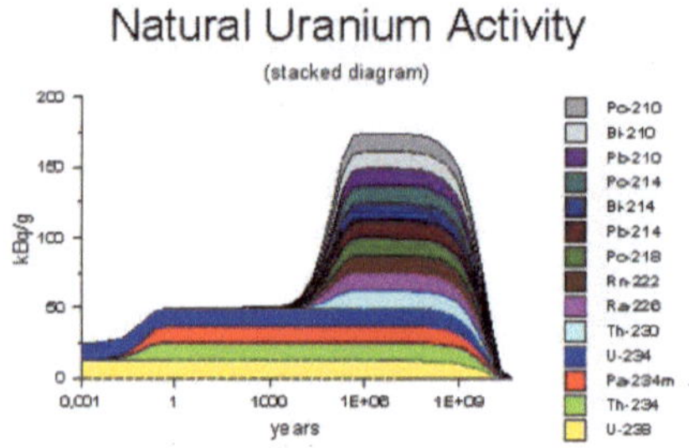

Fig.5. see text © Wise 2014

This has to be taken into consideration designing HLW repositories. During the exploration phase the natural radium isotopes – applying the models shown above - may help to identify the groundwater i.e. aquifer situation.

References

Bard A J, Parsons R, Jordan J (1985) Standard potentials in aqueous solution. Monographs in electroanalytical chemistry and electrochemistry. New York, 1985

Hurst S (2014) Radium- und Radon-Isotopenuntersuchungen als Hilfsmittel für die Aquiferdiagnose unter besonderer Berücksichtigung der geochemischen und hydrochemischen Verhältnisse im Grundwasserleiter. Unveröffentlichte Dissertationsschrift.

Krest J M, Harvey J W (2003) Using natural distributions of short lived radium isotopes to quantify groundwater discharge and recharge. Limnology and Oceanography 48 (1), 290

Krishnaswami S, Cochran J K (eds) (2008) Radioactivity in the environment, Vol 13 M.S. Baxter (series editor), Elsevier

Langmuir D, Riese C (1985) The thermodynamic properties of Radium, Geochimica et Cosmochimica Acta, Vol. 49, 1593 – 1601

Wise 2014 http://www.wise-uranium.org/rup.html

Speciation analysis based design of mine water treatment technologies

Andrea Kassahun[1], Corinne Lietsch[2], Nils Hoth[2], Michael Paul[1]

[1]Wismut GmbH, Jagdschänkenstr. 29, 09117 Chemnitz, Germany
[2]TU Bergakademie Freiberg, Institute of Mining and Special Civil Engineering, Gustav-Zeuner-Str. 1A, 09599 Freiberg, Germany

Abstract. The strong dependence of water treatment efficiency from pollutant speciation was demonstrated at the example of uranium removal from TMF seepage water by ion exchange. The pH dependence of carbonate equilibria was used to manipulate uranyl speciation in the investigated seepage water. Ion exchange experiments were carried out with TMF seepage feeds of the same origin but of different pH and thus of different uranyl speciation. Geochemical modeling and TRLFS measurements revealed calcium uranyl carbonate complexes to hamper sorption to ion exchange resins, whereas non-calcium uranyl carbonate complexes promote sorption. The manipulation of uranyl speciation by pH adjustment caused a rise in ion exchange efficiency from insignificant to technically feasible levels.

Introduction

Within Europe's largest environmental remediation program, WISMUT operates six water treatment plants to clean up effluents from flooded underground mines and seepage water from tailing management facilities (TMF) of the former East German uranium mining industry. A total annual throughput of ~ 20 Mm^3 water is treated for uranium, arsenic, 226-radium and heavy metal removal by modified lime precipitation. After removal of first flush pollution during the past decade(s), continuous water treatment is required. Pollutant concentrations in mine and seepage water stay well above regulatory limits for discharge in the long term. According to current estimates, water treatment is expected to continue until 2040.

Changes in water composition over time necessitate adjustments in water treatment methods. For example, ion exchange techniques may become superior to the presently used precipitation technology when contaminant spectrum is narrowing and contaminant concentrations are decreasing. The paper presents a feasibility test of TMF seepage treatment by ion exchange to remove uranium as sole contaminant at the example of the Culmitzsch mill tailings pond.

B.J. Merkel and A. Arab (eds.), *Uranium – Past and Future Challenges*,
DOI 10.1007/978-3-319-11059-2_68

TMF seepage water quality

Tab. 1 shows chemical data of seepage water from the Culmitzsch facility (Thuringia, Germany). Since contaminant retention by ion exchange strongly depends on solute speciation, seepage water speciation was modeled with the hydrogeochemical software package PhreeqC (Parkhurst and Appelo 1999). The implemented thermodynamic database Wateq4f was extended for ternary complexes of alkaline earth uranyl carbonates $M_xUO_2(CO_3)_3^{(2x-4)}$ (Lietsch et al. 2014). Calcium uranyl carbonate complexes were described earlier to dominate uranyl speciation in carbonate- and calcium-containing mine water from WISMUT sites at pH ~ 7 (Bernhard et al. 1998).

Table 1. Chemical data for Culmitzsch seepage water #E-394 (2013/08/08)

sample	pH [-]	EC[a] [µS/cm]	SO_4 [mg/l]	HCO_3 [mg/L]	Cl [mg/l]	Na [mg/l]	Ca [mg/l]	Mg [mg/l]	U [mg/l]
#E-394	7.3	8,100	3,620	657	496	1,350	292	301	3.1

[a] EC: electrical conductivity

The modeled distribution of uranyl species is shown in figure 1. The modeling indicates a dominance of alkaline earth uranyl carbonates. 90 % of the total dissolved UO_2 from calcium uranyl carbonate complexes, magnesium uranyl carbonates account for 5 % of total dissolved UO_2 and less than 0.5 % is assigned to non alkaline earth uranyl carbonates. Only half of the total dissolved UO_2 is associated in charged species.

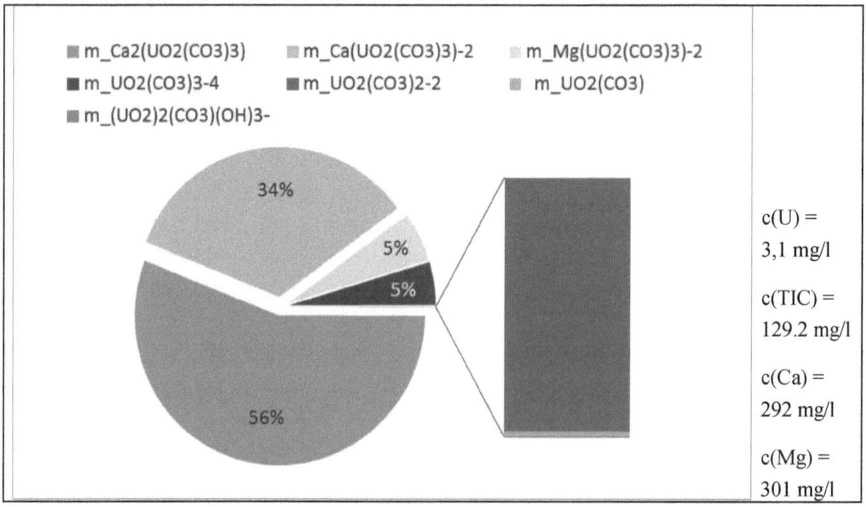

Fig. 1. Modeled distribution of UO_2-species (percent) in Culmitzsch seepage water #E-394 (chemical data see Tab. 1)

Shift in species distribution

As ion exchange is related to charges, a shift in species distribution was intended to increase the percentage of charged uranyl species. Formation of neutral and most stable $Ca_2UO_2(CO_3)_3$ complexes should be suppressed by minimization of carbonate and hydrogen carbonate ions in solution. Since carbonate equilibria are pH dependend, drop in pH drives carbonate speciation towards dissolved $CO_{2(aq)}$. Fig. 2 displays the modeled change in concentrations of HCO_3^-, $Ca(HCO_3)^+$ and $CO_{2(aq)}$ at dropping pH from 7.3 to 3 for Culmitzsch seepage water #E-394. When pH drops below 5.5, HCO_3^-- and $Ca(HCO_3)^+$-concentrations substantially decline and $CO_{2(aq)}$ starts to dominate TIC. Comparably, the drop in pH significantly changes uranyl speciation. As illustrated in fig. 3, calcium uranyl carbonate concentrations dominate total dissolved uranyl concentrations at pH $\geq$ 6 and fall below 25% of total UO_2 at pH = 5.5. At the same time, UO_2CO_3 concentrations rise and dominate total UO_2 at pH 5.5. With further drop in pH to pH=5, calcium uranyl carbonates disappear completely and uranyl sulfate complexes arise. Total UO_2 at pH=5 is formed by UO_2CO_3 and UO_2SO_4. When pH drops below 5, uranyl sulfate complexes become dominating and at pH<4.5 uranyl sulfates are the only complexes present.

The modeled species distribution indicates, that calcium uranyl carbonate complexes are suppressed at pH values of pH < 5.5 leading to rise in UO_2CO_3 (5<pH<5.5) and UO_2SO_4 concentrations (pH<5), respectively.

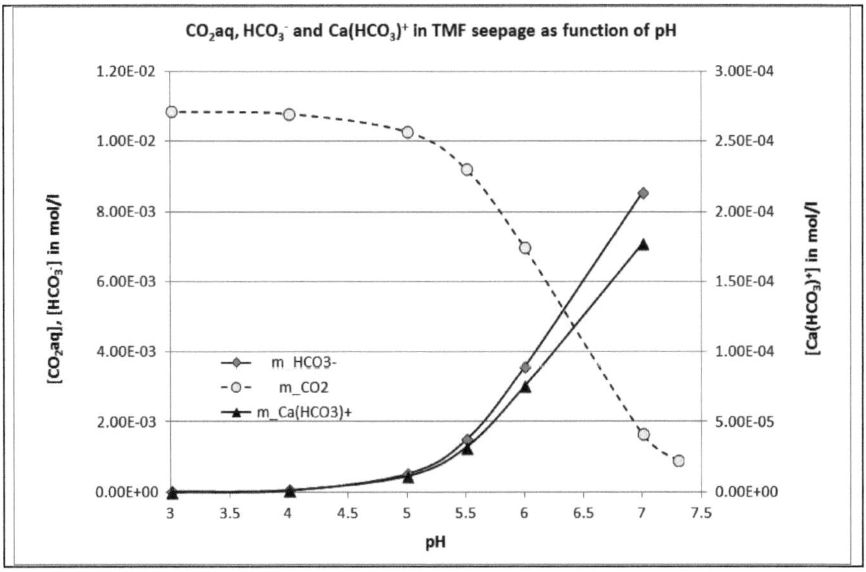

Fig. 2. Modeled pH dependence of TIC species distribution (concentrations in mol/l) in Culmitzsch seepage water #E-394 (chemical data see Tab. 1)

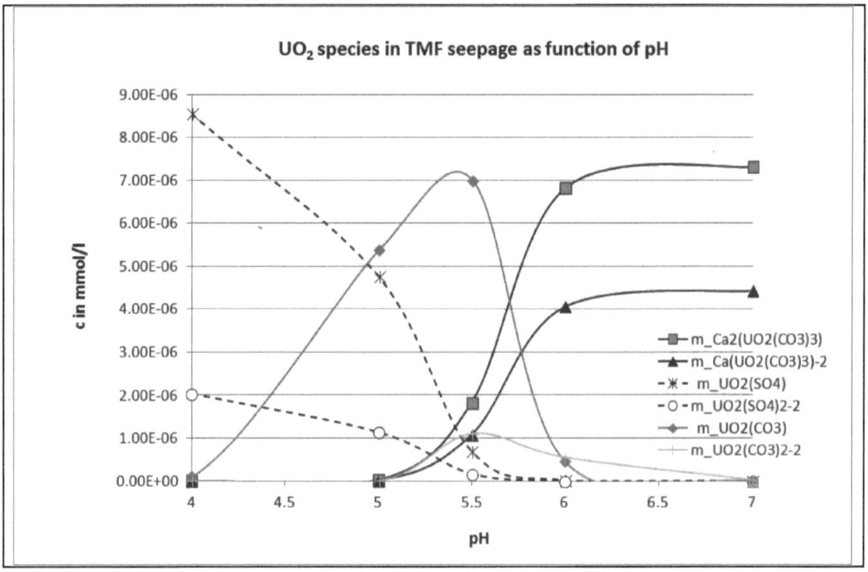

Fig. 3. Modeled pH dependence of UO_2 - species distribution (concentrations in mol/l) in Culmitzsch seepage water #E-394 (chemical data see Tab. 1)

Across the investigated pH-range of 4≤pH≤7 all dominant uranyl species are neutral. However, uranyl complexation chances significantly and may itself affect uranyl sorption to ion exchange resins.

Ion exchange tests

Experimental

A strongly basic anion exchange resin (Lewatit DW 630; Lanxess) and a weakly acidic cation exchange resin with chelating groups (Lewatit TP 207, Lanxess) were tested for removal of uranyl from TMF seepage # E-394 at pH-values of pH=7.3 (original), pH=6.5 (prevailing of calcium uranyl carbonate complexes) and pH=5.5 (prevailing of uranyl carbonate complexes). The resins (5 ml) were placed in glass columns and percolated with TMF seepage water at constant velocity of 50 ml/h (10 bed volumes / h). Prior to column percolation, the resins were pretreated by immersion in water for 48 h. TMF seepage water # E-394 was sampled from a seepage collector downstream Culmitzsch mill tailings pond B. Acidification was done using 0.1 molar hydrochloric acid. For preservation of carbonate equilibria, degassing of CO_2 from bulk water was prevented by use of gas tight bags as storage containers (Tesseraux). The bags were connected via 3-way valves and tubes to a peristaltic pump (Ismatec IPC8) in the feed of the columns. Column effluents were collected continuously and sampled twice a day for determination of pH, el. conductivity and uranium concentration. Additionally, feed water and column percolates were analyzed weekly for magnesium, calcium, sodium, chloride, sulfate, carbonate and hydrogen carbonate (WISMUT analytical laboratories).

Resin regeneration was preformed according to producer instructions using acidic sodium chloride solution (1 molar) and 1 molar sulfuric acid for Lewatit DW 630 anion exchange resin and for Lewatit TP 207 cation exchange resin, respectively. Percolation velocity at regeneration was 5 ml/h. Each bed volume was collected separately for pH measurement. For analysis of uranium and el. conductivity, always two successive bed volumes were combined.

Results

Fig. 4 and 5 show uranium breakthrough concentrations as functions of the exchanged bed volumes for the two different resins each with three different feed pH values.

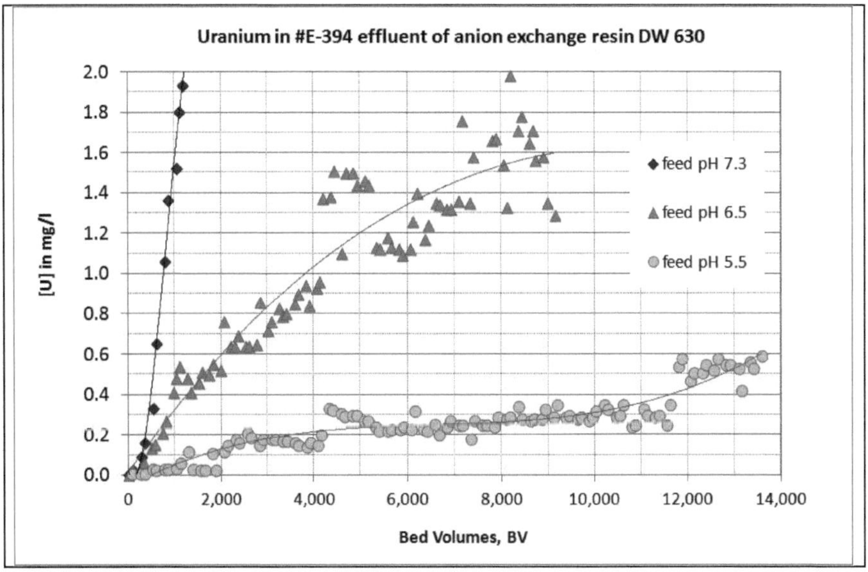

Fig. 4. Uranium breakthrough concentrations in column effluents of ion exchange experiments using anion exchange resin Lewatit DW 630 and three feed pH values (feed: Culmitzsch seepage water #E-394)

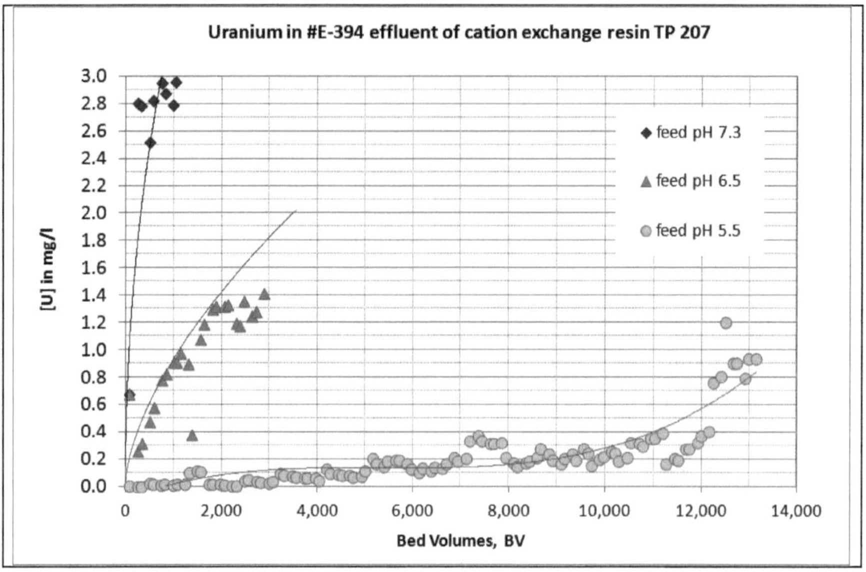

Fig. 4. Uranium breakthrough concentrations in column effluents of ion exchange experiments using cation exchange resin Lewatit TP 207 and three feed pH values (feed: Culmitzsch seepage water #E-394)

At the original pH of 7.3, uranium retention of both ion exchangers is negligible. Uranium retention increased with decreasing pH. Even though no substantial change in uranyl speciation was modeled when pH was lowered from 7.3 to 6.5, effluent concentrations stayed below present uranium discharge limits of 0.5 mg/l for 500 to 1,000 bed volumes. Uranium retention by ion exchange increased significantly at pH=5.5. Uranium concentrations < 0.5 mg/l in the effluents of both resins were measured until the exchange of ~ 12,000 bed volumes. The smaller the intended effluent concentrations, the more preferable proofed the use of the cation exchange resin with chelating groups (Lewatit TP 207) as compared to the anion exchange resin (Lewatit DW 630). For example, uranium effluent concentrations stayed below 0.2 mg/l for 5,000 BV at TP 207 but only for 2,400 BV at DW 630 percolation.

Resin regeneration was preformed to proof reversibility of the measured uranyl sorption and hence ion exchange as predominant retention reaction. Tab. 2 compares the uranium loads of the resins (feed pH=5.5) with the total amount of uranium eluted at regeneration.

Table 2. Resin uranium loads (feed pH=5.5) and total uranium desorption at resin regeneration

Resin	Lewatit DW 630	Lewatit TP 207
Uranium load	38.5 mg/ml	37.6 mg/ml
Total amount Uranium desorbed	40.4 mg/ml	47.6 mg/ml

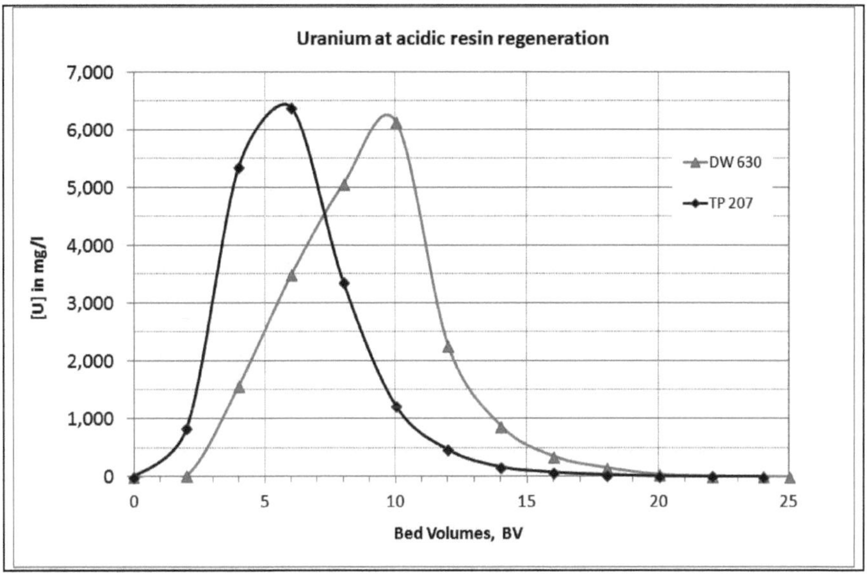

Fig. 5. Uranium breakthrough concentrations in column effluents at acidic ion exchange resin regeneration

The numbers indicate complete uranium desorption. Due to experimental errors at regeneration, calculated desorbed amounts exceed uranium loads by 5 – 25 % (small sample volumes as compared to tube dead volumes, sample merging and dilution prior to analysis). The differences in uranium loads between the resins result from different column runtimes. Fig. 5 shows the uranium breakthrough concentrations at regeneration indicating complete regeneration after 20 BV. Uranyl sorption by both resins was shown to be fully reversible and hence is assigned to ion exchange.

Verification of uranyl speciation

The best test results for uranyl retention from TMF seepage water #E-394 by ion exchange were obtained at feed pH 5.5 with neutral UO_2CO_3 – complexes dominating uranyl speciation. At higher pH, calcium uranyl carbonate complexes are prevailing with ~ 70 % neutral $Ca_2UO_2(CO_3)_3$ and ~ 30 % negatively charged $CaUO_2(CO_3)_3^{2-}$ - complexes. Despite the existence of charged species at pH>5.5, uranyl sorption at ion exchange resins was minor. If charge is considered to dominate ion exchange, modeled uranyl speciation and experimental results are contradictory. To check modeling results, uranyl speciation was measured in two TMF seepage water feeds (pH values 7.3 and 5.5). Analysis was performed at Helmholtz Zentrum Dresden Rossendorf by Dr. Robin Steudtner using time-resolved laser-induced fluorescence spectroscopy (TRLFS). TRFLS is sensitive to electron states caused by the ligand arrangements and thus provides direct evidence about uranium speciation (Baumann et al. 2012, see also for analytical details). Briefly, fluorescence of uranyl complexes in frozen samples (T= -120°C) was excited by a 266 nm laser beam with 1,000 µJ puls energy. TRLFS spectra were measured from 400 to 600 nm. The measured fluorescence bands and lifetimes are characteristic for uranyl speciation. Identification is performed by comparison to previously recorded reference substances. Fig. 6 shows the identified fluorescence bands (position of emission maxima; left) and lifetimes (exponential decay of fluorescence; right) of the feed solutions.

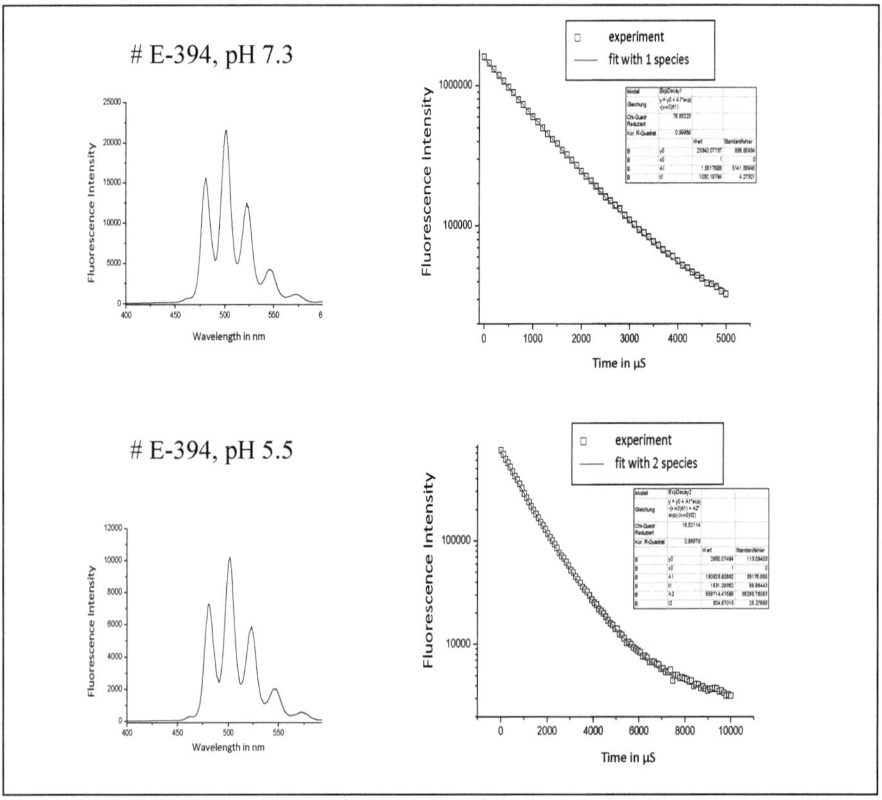

Fig. 6. TRLFS fluorescence bands (left) und lifetime curves (right) of # E-394 TMF seepage at pH 7.3 (above) and pH 5.5 (below); Data R. Steudtner, V. Brendler, HZDR

The fluorescence bands of the feed solutions (fig. 6 left) are typical finger spectra of U(VI) ions. Lifetime curve fits (fig. 6 right) indicate the presence of calcium uranyl carbonates at pH 7.3 (original TMF seepage water pH). Quantification of charged fraction was not possible because reference data of $CaUO_2(CO_3)_3^{2-}$ for cryo-TRLFS are lacking. In contrast, at pH 5.5 (pH adjusted TMF seepage water) a significant longer fluorescence lifetime was obtained. Lifetime curve fits revealed the presence of two uranyl species at pH 5.5: UO_2CO_3 and a species with very long lifetime but unknown reference data. The presence of the prevailing species modeled thus was confirmed by TRLFS measurements. Since species identification is restricted to cryo-TRLFS reference data availability, more detailed information on e.g. the $CaUO_2(CO_3)_3^{2-}$ fraction at pH 7.3 or the nature of the minor second species at pH 5.5 could not be obtained yet.

Although not obvious, neutral UO_2CO_3 complexes seem to undergo extensive sorption by ion exchange (pH 5.5 experiments), whereas calcium uranyl carbonate complexes hinder sorption to ion exchange resins (pH 7.3 and pH 6.5). This is explainable by complex size and stability and in analogy to biosorption behavior of uranyl carbonate complexes (Leavitt 2011).

Conclusion

Uranium removal efficiency from TMF seepage water by ion exchange strongly depends on pollutant speciation. The prevailing uranyl species are predictable by geochemical modeling using adequate data bases, including ternary complexes of alkaline earth uranyl carbonates. Modeling can be used to test geochemical conditions for optimization of retention reactions. In case of the investigated uranium retention from TMF seepage water, ion exchange is restrained by the dominance of calcium uranyl carbonate complexes, both neutral and negatively charged. A change in uranyl speciation can be achieved by pH adjustment. At pH decrease to pH = 5.5, calcium uranyl carbonates are suppressed and uranyl carbonates become the prevailing species. Unless their neutrality, UO_2CO_3 - complexes undergo extensive sorption to ion exchange resins. Thus, ion exchange efficiency rises at pH adjustment to pH 5.5 from insignificant to technically feasible levels. The performance comparison of an anionic and a cationic-chelating resin revealed the cationic-chelating resin superior in achieving uranium effluent concentrations < 0.2 mg/l.

Acknowledgments

This work is part of WISMUT TMF seepage water treatment technology development. Special thanks to Dr. Jan Laubrich for request and support, Birgit Zöbisch and Yvonne Schlesinger for experimental work, Helga Nürnberger and Katharina

Aranyos for chemical analysis and Dr. Vincent Brendler and Dr. Robin Steudtner for uranyl species measurements.

References

Baumann N., Arnold T., Lonschinski M. (2012) Speciation of Uranium in Seepage and Pore Waters of Heavy Metal-Contaminated Soil. In: Kothe, E., Varma, A. (eds.): Bio-Geo Interactions in Metal-Contaminated Soils, Springer Soil Biology Series 31

Bernhard G., Geipel G., Brendler V., Nitsche H. (1998) Uranium speciation in waters of different uranium mining areas. Journal of Alloys and Compounds, 271-273: 201-205

Leavitt J. (2011) The effects of biosorption on uranium transport in a bioremediated aquifer. Ph.D. Thesis University of New Mexico

Lietsch C., Hoth N., Kassahun A. (2014) Investigation of Phenomena in Uranium Mine Water using Hydrogeochemical Modeling – a case study. Proceedings UMH 2014, this volume

Parkhurst D.L., Appelo C.A.J. (1999) User's guide to PHREEQC (version 2)

225 years uranium and radioactivity cross-links around the Brocken – Klaproth, Elster and Geitel, Nazi research, Wismut prospection, and recent anomalies

Friedhart Knolle[1], Frank Jacobs[2], Ewald Schnug[3]

[1] GeoPark Harz . Braunschweiger Land . Ostfalen Network, Grummetwiese 16, D-38640 Goslar, Germany, E-Mail: fknolle@t-online.de
[2] Geoscience Advice & Support, Glockengiesserstr. 1, D-38640 Goslar, Germany
[3] Technical University Braunschweig - Faculty 2 Life Sciences, Pockelsstraße 14, D-38106 Braunschweig, Germany

Abstract. When Martin Heinrich Klaproth, born in Wernigerode, presented the discovery of “uranite”, he could not know which role this new element would play. In 1823, Alexander von Humboldt published the idea of the “Classic Geological Square Mile” later used for parts of the Northern Harz area because of their geodiversity. Between 1880 and 1920, the two physicists Julius Elster and Hans Friedrich Geitel worked in the Harz area – Geitel coined the term “atomic energy” in 1899. A cluster of industry and the Mining Academy Clausthal developed in this area because of the diversity of mineral deposits and mining. In the Nazi era, this cluster conducted military research and partly also worked on the possible use of uranium (U) for weapon production. The German U reserve was placed in Staßfurt. After WW II, Wismut carried out prospection work in the Harz region, but no U deposits were found worth mining. The U source of an anomaly in surface waters within the agricultural landscape Magdeburger Börde is probably to be found in the intensive and long-time use of mineral phosphate fertilisers.

Introduction

Due to the high geodiversity of the Harz Mountains and their forelands, mining of different ore types and subsequent processing and smelting played a key role in this northernmost middle mountain range of Germany. Mining started as early as in the bronze age more than 3000 years ago (Monna et al. 2000) and lasted until 2007. For decades, U was also prospected for, beginning after WW II.

B.J. Merkel and A. Arab (eds.), *Uranium – Past and Future Challenges*,
DOI 10.1007/978-3-319-11059-2_69

1789: Martin Heinrich Klaproth discovers U

The chemist Martin Heinrich Klaproth was born in the Harz city of Wernigerode in 1743. When he discovered "uranite" in his Berlin laboratory following studies on pitchblende (Klaproth 1789) and presented it to the scientific community, he could not know how important this new element, later renamed "uranium", would become. The house where Klaproth was born still exists and can be found at Liebfrauenkirchhof No. 5 in Wernigerode.

1823: Classic Geological Square Miles

It was Klaproth's later university patron Alexander von Humboldt who coined the idea of the "Classic Geological Square Miles" for parts of Central Germany in 1823 (Humboldt 1823). The acreage of the classic geological square mile corresponds with the Prussian square mile and represents an area rich in geological diversity. This might be the reason why the term was used more and more for parts of the Northern Harz area northwest of the Brocken (Geopark Harz . Braunschweiger Land . Ostfalen 2013).

Right through the city of Klaproth's birth and near Wernigerode Castle runs the Northern Harz Boundary Fault, a deep tectonic structure. Extending to the north, the sediments of the Subhercynian Cretaceous Basin stretch out over an area where Wismut geologists in later years were prospecting for another U deposit of the Königstein type.

South of this fault, the Harz basement block crops out, consisting of rocks from the Paleozoic. Slates and greywackes with partly embedded flinty slates and limestones emerged from the oceanic sedimentation of the Devonian and the Lower Carboniferous, accompanied by volcanic activities in different phases. At the turn Lower/Upper Carboniferous, the Variscan collision folded these sediments and led to a regional metamorphosis. By this plate collision, rocks were pressed down in the deeper earth's crust, melted and rose upwards as magma, until the intrusive bodies became stuck and cooled down. In the course of these processes the Brocken Granite and Gabbro complex emerged. Wismut geologists hoped that the granite contains mineable U minerals, but failed.

The Harz fold-belt has then been partly eroded in the Lower Permian (the Rotliegend Formation bearing U minerals) and, in the Upper Permian, for a time was flooded by the Zechstein ocean (the Kupferschiefer Formation also containing U minerals).

In the course of the Mesozoic, massive sediment layers developed on the folded Harz rocks – for the most part under the cover of the sea – such as clay, sandstone and limestone of the Triassic, Jurassic and Cretaceous. Since the Upper Cretaceous and further in the Tertiary, the half-horst of the Harz basement has risen for more than 5000 m in the wake of the Saxonian tectonics along the Northern Harz

Boundary Fault. At the same time, mesozoic sediments north of the Harz basement were dragged upwards and turned locally in an upright position so that their once horizontal stratification planes now stand vertically, causing the high geodiversity in the region described by many authors. Following the pre-Saxonian and Saxonian tectonics, ore mineralisations partly filled the tectonic structures, mostly under Zechstein coverage (Franzke and Schwab 2011).

Due to this high geodiversity, ore mining, processing and smelting and later also rock quarrying have played a key role in the Harz area, for more than 3000 years now. Regional examples include the mining of copper, silver-bearing lead and zinc minerals (the mines of Rammelsberg, Clausthal-Zellerfeld, Sankt Andreasberg and Mansfeld being the internationally best known examples), quarrying of anhydrite, gypsum and dolomite along its southern edge, and paleobasalt, gabbro, granite, reef limestone, roofing slate and greywacke throughout the region's Paleozoic area. Originally, human impact was only small-scale, but over the course of mining history, alongside increase in technological possibilities, mining and smelting began to have increasingly serious effects on the surrounding nature and landscape (Knolle et al. 2011). Since the 1950s, the whole eastern part of the Harz Mts. was screened repeatedly by Wismut prospectors and geologists for radioactive anomalies (see below).

1899: The term "Atomic Energy" was formed in the Harz area

Both the teachers and physicists Julius Elster (*1854 in Blankenburg/Harz, † 1920 in Bad Harzburg) and Hans Geitel (*1855 in Braunschweig, † 1923 in Wolfenbüttel) were born in the Harz foreland and spent their working lives there. They faced very limited technical facilities, but provided important basic research in the field of radioactivity, atmospheric electricity, photoelectric effects and gas discharge physics (Fricke 1992). Elster's and Geitel's most important scientific achievement is the discovery of the photo cell. With the knowledge that photoelectricity produced was proportional to light intensity, they provided basic data for modern photoelectric photometry. A major focus of Elster and Geitel lay on atmospheric electricity – their theory of the electronic charging in thunder clouds through the friction of water droplets can still be found in physics textbooks. For use in describing the distribution of charges, Elster and Geitel developed the model example of a spherical capacitor. They introduced a new era of research when they applied the theory of ions to the field of atmospheric electricity. For a long time they hypothesised that sunlight induces the release of ions before discovering that environmental radioactivity is causative. To prove this theory, they carried out underground measurements in the Baumann's and Iberg Dripstone Caves in the Harz. In the end, they provided evidence of a higher radioactivity inside the natural caves than in the air outside. In 1907, the two scientists concentrated on the relationship of radioactivity and geothermal energy. Elster and Geitel were the first to view radioactivity as a phenomenon of atomic disintegration processes and to postulate a radioactive decay-chain of elements. They also analysed cathode rays and had glass cells produced, in which the electrode had the form of a filament which could be heated electrically. With that, Elster and Geitel created hot-cathode tubes which are nowadays irreplaceable in electronics. In recognition of their scientific accomplishments they were repeatedly honored and even nominated seven times for the Nobel Prize for Physics (Fricke 1992).

Since 1933: Nazi atomic research and development, U storage

The Harz area is known for having had the largest underground weapon production site of its time – the KZ Mittelbau-Dora installations near Nordhausen at the southern Harz rim. Less known is the fact that the diversity of ore deposits in this area brought a cluster of mining and metallurgy related industry to the area. Some mining facilities cooperated intensively with the Bergakademie Clausthal (later Technical University Clausthal) and other universities. During the Nazi regime, they also carried out military research, and several companies and institutes worked on the possible use of U in weapon production in WW II.

Only fragments of this research work are known so far. Prof. Dr.-Ing. Max Paschke, who was arrested after WW II due to his Nazi implications, carried out experiments to produce graphite-free U carbide in his Institute for Ferrous Metallurgy, Foundry and Enamel Technology of the Bergakademie Clausthal (Halstead Exploiting Centre 1945). U carbide (UC) can be used as nuclear fuel or as a substitute for tungsten and molybdenum in steels. The experiments were suggested by Gebr. Borchers A.G. in Goslar – their CEO Dr. Friedrich Borchers was under house arrest after WW II because of Nazi collaboration (Knolle and Schyga 2012). This document from 1943 was later found by the Allies within the secret reports from 1939 - 1945 of German atomic research („Haigerloch Documents“). The Bergakademie Clausthal also cooperated with other firms in the region on projects involving German atomic research – alongside the Grona GmbH (high pressure and hollow charge research), the Phywe AG (ultra-zentrifuges) was also involved, both situated in Göttingen (Baranowski 2013).

Besides this, in Staßfurt in the northeastern Harz foreland U supplies were stored. It was the stock of approximately 1,100 t of U ore and U oxide (yellow cake) which the US military secured in April 1945 during a commando action in Staßfurt before the Red Army occupied the storage area at the Friedrichshall shaft. This stock represented the Belgian and French U reserves which had been transported to the Harz region and which – under surveillance of the SS – were part of the raw materials stock of the Reich.

1949 - 1973: Wismut prospection during the Cold War

After WW II, SAG, later SDAG, Wismut exploited the U deposits of the Erzgebirge, and also carried out intensive prospection work in the whole GDR, including the eastern Harz area. Here, they worked from 1949 to 1973. There were clear geochemical indications for possible larger U mineralisations in this area, like red bed sediments, copper shale and Königstein-type Cretaceous sediments at the rim of the Harz Mts. and a mineralisation of the Bi-Co-Ni formation at the Brocken granite complex near Wernigerode (Fig. 1). However, no U deposits were found worth mining. The largest concentration of U minerals was found in a small deposit bound to the Hornburg Anticline, northeast of Sangerhausen (Stedingk 2002; Runge 2006; see below).

Fig.1. Wismut shaft safekeeping near Drei Annen-Hohne, Wernigerode (photo: Dieter Schulz)

U rock and water anomalies in the Harz forelands

Within the inner Harz Mts., only in the Roter Bär learning mine (Sankt Andreasberg, Lower Saxony), a small uraninite mineralisation is known. The situation is different in the immediate Harz forelands. As mentioned above, alongside the eastern and south-eastern Harz rim a sedimentary complex of Permian rocks crops out in Lower Saxony, Thuringia and primarily in the Mansfeld and Sangerhausen area of Saxony-Anhalt. Especially the copper shales and the Rotliegend red sands contain U minerals. This results in regionally increased U and radioactivity values in runoff water of historic mine galleries and groundwater (Landesamt für Umweltschutz Sachsen-Anhalt 2007). Examples for mine run-off water are Erdeborner Stollen (0.41 Bq/l U-238) and Schlüsselstollen (0.37 Bq/l U-238). The highest groundwater concentrations were found in the regions Mansfeld (30 µg/l U), Hettstedt (100 µg/l U), Eisleben/Wimmelburg (90 µg/l U) and Hornburg (120 - 170 µg/l U).

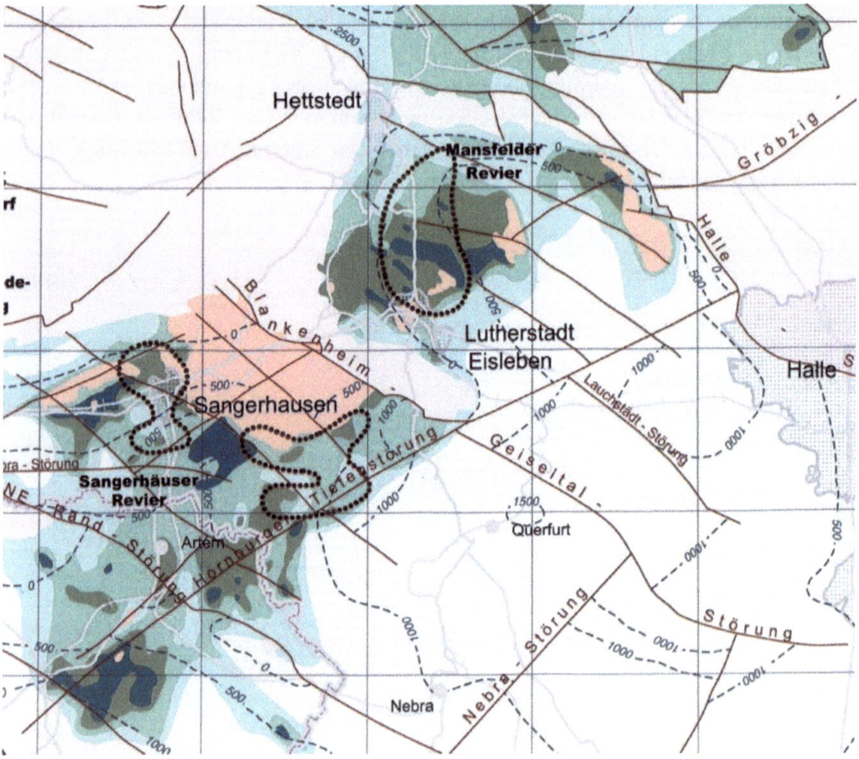

Fig.2. Overview of Mansfeld und Sangerhausen copper shale mining areas showing details of the orebody parameters like copper dissemination, oxidation zones („Rote Fäule"), tectonic structures and position of Zechstein basis; the dotted lines indicate natural radiogene anomalies (from Stedingk and Rentzsch 2003)

These groundwater anomalies are connected to natural radiogene anomalies in copper shales and red molasse sandstones of Permo-Carboniferous age in the Mansfeld and Sangerhausen mining areas (Fig. 2). In the case of the Hornburg orebody, Stedingk (2002) reports for the copper shale a concentration of 180 ppm U and for the molasse sandstone 170 - 260 ppm U resulting in a 400 t U deposit in the sandstone.

The mapping work for the Geochemical Atlas of Germany reveals a U anomaly in surface waters located in the agricultural landscape of the Magdeburger Börde north of the Harz Mts. with a peak value of 7.84 µg/l in the Beber River, north-west of Magdeburg (Fig. 3; Birke and Rauch 2008).

According to Birke and Rauch (2008) and Schnug and Haneklaus (2014), the source of this U is probably to be found in the intensive and long-term use of mineral phosphate fertilisers in the intensively used agricultural areas in the northern Brocken foreland – see also Birke et al. (2009).

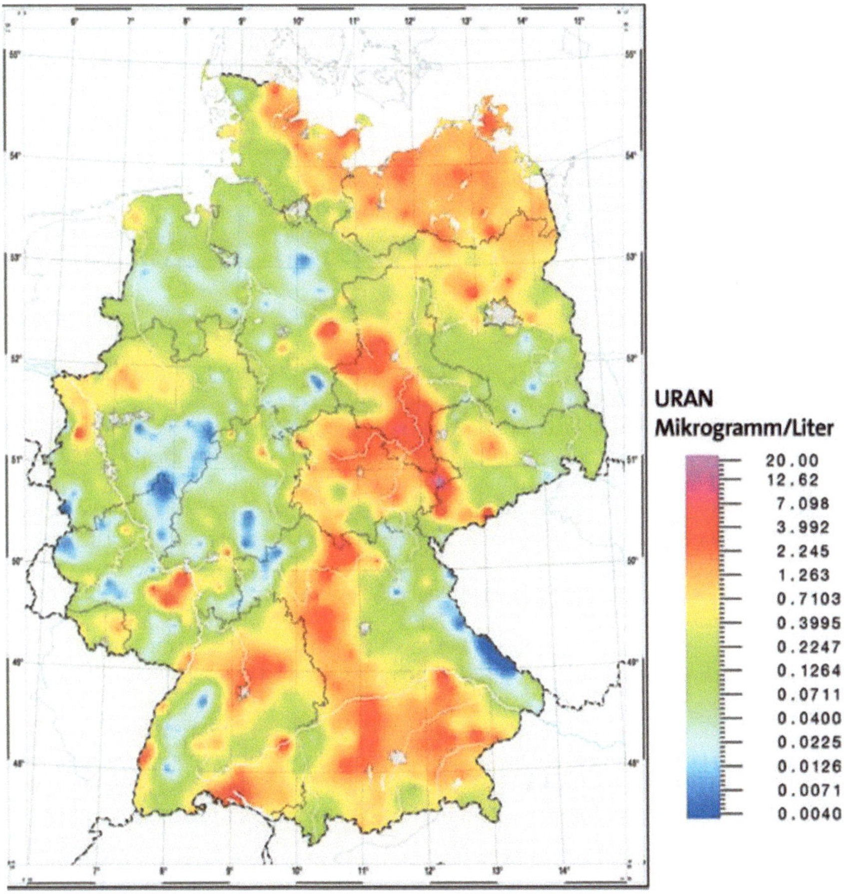

Fig.3. U content in surface waters of the Federal Republic of Germany (from Birke and Rauch 2008)

Acknowledgement

The authors wish to express their most sincere gratitude to Manfred Birke, Silvia Haneklaus, Marion Seitz and Klaus Stedingk for information, advice and support.

References

Baranowski F (2013) Rüstungsproduktion in der Mitte Deutschlands 1929 - 1945. Rockstuhl, Bad Langensalza

Birke M, Rauch U (2008) Uranium in Stream Water of Germany. In: De Kok L J, Schnug E: Loads and fate of fertilizer derived uranium. Backhuys, Leiden, The Netherlands, pp. 79-90

Birke M, Rauch U, Lorenz H (2009) Uranium in stream and mineral water of the Federal Republic of Germany. Environ Geochem Health 31: 693-706

Franzke HJ, Schwab M (2011) Harz, östlicher Teil mit Kyffhäuser Kristallin. Slg geol Führer 104, Borntraeger, Stuttgart

Fricke RGA (1992) J. Elster & H. Geitel – Jugendfreunde, Gymnasiallehrer, Wissenschaftler aus Passion. Döring Druck, Braunschweig

Geopark Harz . Braunschweiger Land . Ostfalen (2013) Geopark Harz . Braunschweiger Land . Ostfalen. The Classic Square Miles of Geology. Königslutter - Quedlinburg

Halstead Exploiting Centre (1945) Reports on Experiments on the Manufacture of a Graphite Free Uranium Carbide from Uranium Oxide and Carbon. Report Gr. 2/HEC/10402, Imperial War Museum, UK

Humboldt A von (1823): Essai géognostique sur le gisement des roches dans les deux hémisphères. Paris

Klaproth MH (1789) Chemische Untersuchung des Uranits, einer neuentdeckten metallische Substanz. Chem. Ann. Freunde Naturlehre (2): 387-403

Knolle F, Ernst WHO, Dierschke H, Becker T, Kison HU, Kratz S, Schnug E (2011) Schwermetallvegetation, Bergbau und Hüttenwesen im westlichen GeoPark Harz – eine ökotoxikologische Exkursion. Braunschweiger Naturkundliche Schriften 10 (1):1-44

Knolle F, Schyga P (2012) Gebr. Borchers/H.C. Starck in der NS-Zeit – Umweltgeschichte, Rüstungsproduktion und -forschung, Zwangsarbeit. Spuren Harzer Zeitgeschichte 5:1-85, Clausthal-Zellerfeld

Landesamt für Umweltschutz Sachsen-Anhalt (2007) Urankonzentrationen im Grundwasser von Sachsen-Anhalt. 26 pp.

Monna F, Hamer K, Lévêque J, Sauer M (2000) Pb isotopes as a reliable marker of early mining and smelting in the Northern Harz province (Lower Saxony, Germany). J Geochem Explor 68: 201-210

Runge W (2006) Chronik der Wismut. CD-ROM, Wismut GmbH, Chemnitz

Schnug E, Haneklaus S (2008) Dispersion of uranium in the environment by fertilization. In: Merkel B J, Hasche-Berger A: Uranium, Mining and Hydrogeology, Springer, Berlin

Schnug E, Haneklaus N (2014) Uranium in phosphate fertilizers – review and outlook. This volume

Stedingk K (2002) Potenziale der Erze und Spate in Sachsen-Anhalt. In: Rohstoffbericht 2002 – Verbreitung, Gewinnung und Sicherung mineralischer Rohstoffe in Sachsen-Anhalt. Mitt Geologie Sachsen-Anhalt, Beih 5: 75 - 132, Ed Landesamt für Geologie und Bergwesen Sachsen-Anhalt, Halle (Saale)

Stedingk K, Rentzsch J (2003) Übersichtskarte Tiefliegende Rohstoffe und Energierohstoffe in Sachsen-Anhalt 1 : 400 000, Blatt II: Potenziale der Erze und Spate. Ed Landesamt für Geologie und Bergwesen Sachsen-Anhalt, KTR 400, Bl. II, Halle (Saale)

Contamination of Water Bodies Affected by Post-Mining Activities in the Light of the European Water Framework Directive

Elke Kreyßig[1], Jana Götze[1]

[1]Wismut GmbH, Jagdschänkenstr. 29, D-09117 Chemnitz

Abstract. According to the EU Water Framework Directive (EU-WFD), all European water bodies should achieve the so called “good water status” by 2027 at the latest. In contrast, the German Wismut remediation project, focusing on the elimination of the consequences of a long-lasting uranium industry from the cold war era, will continue beyond that deadline. In spite of intensive and costly water management and treatment measures an impact on water bodies surrounding the sites under remediation cannot be totally avoided. The presentation focuses on the quality levels reached for selected water bodies in comparison to the benchmarks of the EU-WFD and on ways how to deal with the latter.

Introduction

Germany has set the objective to achieve the so called 'good water status' of its water bodies by 2015, at the latest by 2027. This aim is based on the EU Water Framework Directive (WFD) adopted in 2000. To this directive all EU member states feel committed. Simply spoken, the "good water status" means that nearly natural biological conditions can be achieved within the water bodies. As a fundamental requirement a number of well-defined environmental quality standards have to be met, inter alia with regard to volumes and water quality. In this respect, the target for the given standards is the habitat's faunal biodiversity rather than the human being. Besides fish species the base line for rivers is defined by benthic invertebrates. Most of the quality standards are internationally defined, partly supplemented with national or even local standards. Internationally and nationally applicable criteria are implemented in the German Federal Water Act and related regulations, e.g. Surface Water Ordinance (OGewV). Some relevant benchmarks which Wismut has to deal with are shown in Table 1.

B.J. Merkel and A. Arab (eds.), *Uranium – Past and Future Challenges*,
DOI 10.1007/978-3-319-11059-2_70

Table 1. Selected quality standards of OGewV; mean values 2013

Parameter	Reference	Material Phase	Standard	Unit
As	Annex 5	Suspended	40	[mg/kg]
Cu	Annex 5	Solids	160	[mg/kg]
Cr	Annex 5	or	640	[mg/kg]
Zn	Annex 5	Sediments	800	[mg/kg]
Cd	Annex 7	Dissolved	0.08… 0.25 *	[µg/l]
Ni	Annex 7	Dissolved	20	[µg/l]
Pb	Annex 7	Dissolved	7.2	[µg/l]
Se	Annex 7	Dissolved	3	[µg/l]
TI	Annex 7	Dissolved	0.2	[µg/l]

* depending on water hardness

Another target of the Federal Government focuses on the elimination of the consequences of a long-lasting uranium industry from the cold war era: In the areas of Saxony and Thuringia the fourth largest uranium exploitation of the world took place with a production of 231,000 t of uranium in total. The responsibility for the remediation process which began in 1991 was handed over to Wismut GmbH. It is a giant long term project, and especially the included water treatment process will take much longer than 2027. Thereby surface and groundwater in the catchment area of remediation sites get unavoidably affected - be it due to rising groundwater levels from formerly dewatered mines, due to seepage water or deliberate discharges which emerge as a result of the remediation process.

In the following the main impacts in terms of type and quantity, their effects in surface water bodies as well as a first evaluation with respect to the given quality standards will be presented. Furthermore, a summary will be given on the temporal connection between the two simultaneous lines of development, namely the ongoing remediation process and the parallel development of benchmarks for aquatic quality standards. Finally, a first prediction of complying with the standards will be made.

Main tasks of remediation, affected receiving streams

The legacy of a long term Uranium mining history consists primarily of dewatered aquifers, piled up waste rocks and large tailing ponds, which store the remaining materials from the milling process. Besides them there were a lot of industrial areas and facilities to be remediated, but most of them have lost their relevance after being demolished. With a view to the remaining tasks, the following sources of contamination have to be dealt with:

1. During the controlled flooding of dewatered mines the groundwater levels rise steadily. In this context, the formerly aerated rock layers are submerged and uranium is washed out together with associated heavy metals. In these cases, the sources of contaminants are "newly" generated by the remedial action. Their characteristics, however, are already and unavoidably determined in the history of the mining process. After reaching the natural groundwater level these contaminants would discharge into the neighboring rivers or, in the case of groundwater, would enter the natural aquifer.

2. Precipitation on piles and tailings causes a steady effluence of seepage and contaminated ground water. With them an outflow of natural radionuclides, heavy metals and salinity can be observed, which are contained inside the stored material.

With a view to water quality the main activities include

- to catch the main flow of pollutants and pipe them to a water treatment process, whenever possible;
- to discharge small contaminated fluxes of water in a tamper-proof way;
- to minimize the contamination loads by compaction, encapsulation and covering for long term disposal.

In addition it should be pointed out that the named targets are overlain with more problems, e.g. Radon exhalation, seismic or geo-mechanical issues. Partly it competes against solutions for water and it requires an individual optimization for each remediation solution (environmental evaluation).

In total, the following rivers are affected by remediation: in Saxony the Elbe river between Schmilka and Dresden and the Zwickauer Mulde river between Aue and Crossen, in Thuringia the Weiße Elster river between Greiz and Gera (Fig. 1).

Fig.1. Elbe river basin and the two areas affected by remediation activities of Wismut GmbH (Modified form FGG Elbe)

Contamination source strengths, results of water treatment

The types of mobile contaminants are varying from remediation site to remediation site, sometimes also over time. Table 2 gives a summarized overview by location, type and concentration. Therefore, the principal contaminants to deal with are uranium, Radium-226, selected heavy metals and salinity parameters.

Table 2. Mine and tailings water qualities in remediation objects of Wismut GmbH (2013, mean values; bold numbers show the main pollutants)

		mine water					tailings water	
Contaminant		König-stein	Gitter-see	Schlema	Pöhla	Ronne-burg	Crossen	Seeling-städt
pH		**2.9**	6.8	7.2	7	6.3	8.9	8.1
GH	[°dH]	40	42	50	10	**170**	11	**110**
$SO4^{2-}$	[mg/l]	**1030**	**950**	580	**1400**	**3000**	**1400**	**3600**
Fe	[mg/l]	**60**	17	5	0.8	**200**	0,8	0,4
Mn	[mg/l]	**9**	2	2	0.2	**7**	0,08	2
U	[mg/l]	**10**	0.07	2	0.02	0.8	**4.3**	1.3
Ra-226	[Bq/l]	**6.2**	0.024	**2.0**	**4.0**	0.37	0.18	0..13
As	[µg/l]	65	24	**1400**	**1900**	**270**	**370**	32
Cu	[µg/l]	**50**	<20	<5	-	<20	<20	<20
Cr	[µg/l]	**40**	<5	<5	-	<20	-	<20
Zn	[µg/l]	**6600**	20	6	15	**480**	13	19
Cd	[µg/l]	**80**	<1	<1	-	2	-	<1
Ni	[µg/l]	**460**	<10	8	-	**800**	<20	<20
Pb	[µg/l]	**540**	<0,5	<5	-	<20	<20	<20
Se[a]	[µg/l]	<1	<1	<1	<1	3	1	2
Tl[a]	[µg/l]	**16**	<0.2	<0.2	<0.2	3	<0.5	<0.5

[a] environmental quality standards, extended in 2011 according to OGewV; preliminary results from a special screening program

Meanwhile seven water treatment plants have been established for a significant depletion of contaminated water. The treatment method is a combined physical-chemical, in essence a modified lime precipitation process. It eliminates Uranium, Radium-226 and a great number of different heavy metals in a very reliable way. In contrast, a separation process for salinity is not included. In this way an amount of roughly 18 Mm^3, i.e. 2,000 m^3/h, is being treated every year. After the treatment, contaminant levels are down to the discharge concentrations as shown in Table 3.

The results of this costly and intensive treatment process show that most of the effluent concentrations are already in compliance with the levels set by the OGewVO quality standards (Table 1).

Table 3. Discharge concentrations after water treatment (2013, mean values)

Contaminant		König-stein	Gitter-see[b]	Schlema	Pöhla	Ronne-burg	Crossen	Seeling-städt
pH	-	7.6	7.8	7.1	8.1	8.1	7.4	7.8
GH	[°dH]	54	-	50	-	170	-	114
$SO4^{2-}$	[mg/l]	930	940	560	<5	3000	1400	3500
Fe	[mg/l]	0.06	1.3	0.4	0.03	0.13	0.14	0.14
Mn	[mg/l]	3	-	1.6	0.007	0.07	-	0.29
U	[mg/l]	0.10	0.071	0.16	0.01	0.03	0.16	0.19
Ra-226	[Bq/l]	0.01	0.02	0.03	0.01	0.02	0.01	0.07
As	[µg/l]	1	-	64	42	0.7	5	2.3
Cu	[µg/l]	5.2	-	<5	<5	20	5	3.4
Cr	[µg/l]	<5	-	<5	<5	0.7	<1	1.24
Zn	[µg/l]	60	-	8	-	8	-	10
Cd	[µg/l]	3	<0.5	<1	<1	0.7	0.4	0.4
Ni	[µg/l]	39	<1	8	<5	15	9	4
Pb	[µg/l]	2	<2	<5	<5	14	2	2
Se[a]	[µg/l]	<1	<1	<1	<1	<1	11	2
Tl[a]	[µg/l]	8	<0.2	<0.2	<0.2	0.8	<0.5	<0.5

[a] environmental quality standards, extended in 2011 according to OGewV; preliminary results from a special screening program
[b] discharge will end in 2015 after completion of a drainage adit

It has to be pointed out that the amount of residues which is inevitably produced during this process is enormous. They have to be stored in an engineered facility for long-term safe disposal.

Impacts to receiving streams, evaluation by WFD standards

In addition to contamination sources and discharges also rivers, crossing the remediation area, are intensively monitored over a long term. As a result, the concentration loads shown in Table 4 have been recorded. Comparison of the measured concentrations inside the receiving water bodies with the benchmarks of OGewVO for surface water bodies (Table 1) shows that there are only a few cases where the standards were exceeded (compare Table 3, discharge concentrations). It has to be pointed out that in the case of the Mulde River the main impacts from As and Zn are not Wismut-related (Meyer, J. et al., 2008).

In summary it can be said that quality standards in case of the big rivers Elbe, Zwickauer Mulde and Weiße Elster (1st order watercourses) are not out of compliance by Wismut discharges. In addition, the discharged loads will be declining over time as the huge contaminant potential of the sites under remediation is, after all, limited. Only in the case of small water bodies, sometimes only 5 km long (2nd order watercourses, e.g. Wipse and Culmitzsch in Thuringia), for which discrete benchmarks (Uranium, salinity) have been announced, the good water status will not be achieved. It is not to be expected that this objective might be achieved by 2027 within reasonable spending limits. Among the main causes for non-compliance with the 'good water status' are the salinity loads which pass through the treatment unhindered.

Table 4. Water qualities in surface water bodies, influenced by remediation objects of Wismut GmbH (2013, mean values); bold characters characterize a non-compliance with the requirements for the 'good water status' according to the current OGewVO

Contaminant	state	Saxony			Thuringia				
	Site	Königstein	Schlema	Crossen	Ronneburg		Seelingstädt		
	river 1st Order	Elbe	Zwickauer Mulde		Weisse Elster				after impact
	stream 2nd Order				Gessenbach	Wipse	Fuchsbach	Culmitzsch	
pH	-	8.3	7.5	7.5	8.0	7.7	8.2	7.9	7.9
GH	[°dH]	10	4.5	6.2	23	106	18	63	13
$SO4^{2-}$	[mg/l]	71	55	67	245	1900	117	1500	180
U	[mg/l]	0.001	0.007	0.007	0.018	0.014	0.036	0.094	0.004
Ra-226	[Bq/l]	<0.01	0.010	0.015	-	-	0.012	0.019	0.010
As	[mg/kg]	-	**191**[b]	**104**[b]	30	30	22	30	-
Cu	[mg/kg]	-	195	-	**800**	170	100	50	-
Cr	[mg/kg]	-	51	-	30	40	150	30	-
Zn	[mg/kg]	-	**930**[b]	-	400	190	560	230	-
Cd	[µg/l]	<0.25	<1	-	0.2	**< 0.3**	-	0.2	0.1
Ni	[µg/l]	3.2	8	5.9	20	10	1.6	6.3	4.6
Pb	[µg/l]	-	<5	<1	<1	<2	<1	<2	-
Se[a]	[µg/l]	<1	<1	<1	1	1	<1	<1	<1
Tl[a]	[µg/l]	<0.2	<0.2	<0.2	<0.2	**0.3**	<0.2	0.2	<0.2

[a] environmental quality standards, extended in 2011 according to OGewV; preliminary results from a special screening program

[b] not significantly caused by the Wismut Uranium remediation project

Moreover, a tightening/extension process for the rating spectrum by EU and German authorities is to observe. For instance, Selenium and Thallium have been added in 2011 only. The introduction of Uranium as a benchmark is being repeatedly discussed, others might follow. If such parameters are significantly contained inside the source-term, a significant adjustment of water treatment methods could become necessary. Furthermore this leads to new classification results or downgrading of quality levels, respectively.

Whatever benchmarks for a 'good water status' will be developed, it needs to be noted, that the approved Wismut remediation strategies, generated in the 1990ies, cannot be fundamentally changed in retrospect. This is not possible, due to the dimension of remediation and amounts of water to be treated. Accordingly, it is to be acknowledged that most mining activities have been done in an area of elevated background levels, whereby, as a rule, the pre-mining state can never be restored because of intensive mining activities.

Nevertheless Wismut is obliged to improve its treatment efficiency as far as possible, i.e. as far as compatible with reasonable spending limits. It was done in the past and will continue to be done in the future as well. By way of example, Fig. 2 illustrates that rising amounts of water did not lead to similar emission loads of Uranium into the rivers.

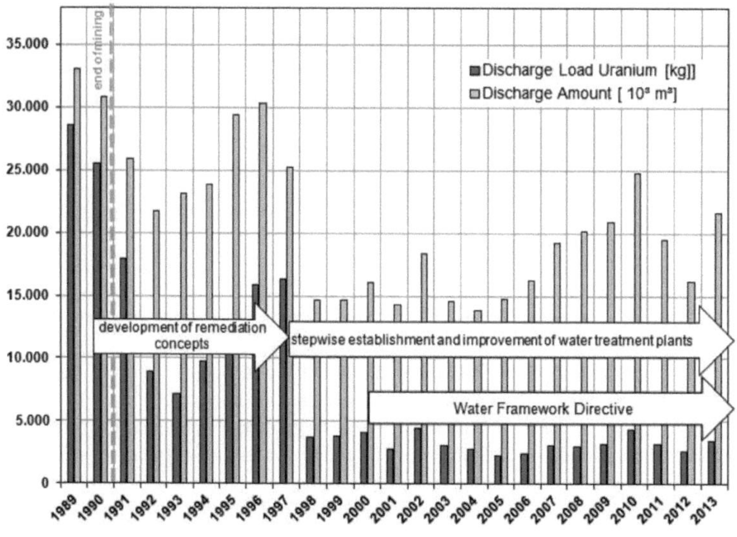

Fig.2. discharged amounts and loads of Uranium temporally related to the fundamental remediation process as well as to WFD

Furthermore the WFD-implemented exemption clauses offer enough flexibility to deal with all the inherited problems. This includes the possibility for:

- substitution of quality standards by local background levels;
- extending deadline up to 2027;
- less stringent environmental objectives;
- possibility of temporary deterioration, applying in case of rising water levels in compliance with strict exemption rules.

Conclusion

The Wismut remediation program causes discharges, which require an immense and ongoing effort in treatment. With regard to the big rivers Elbe, Zwickauer Mulde and Weiße Elster, discharges of water after treatment are not out of compliance with current quality standards by WFD as well as by OGewVO. As the pollution potential at the sites under remediation is huge but limited, these discharges have furthermore a declining character.

In the case of small water bodies to which additional local criteria will be applicated (e.g. Wipse und Culmitzsch), the 'good water status' cannot be achieved by 2027 within reasonable spending limits. The same applies to most of the affected groundwater bodies.

The ongoing initiatives at tightening the current quality standards and to extend them, makes compliance more and more difficult. It complicates the process by ongoing new-classification and downgrading of water bodies. Reaching zero emissions is not possible in the case of the remaining legacies. Such an approach would be related to disproportionate effort (costs, energy) and significantly increasing residues.

Both authorities and Wismut GmbH search for a realistic measure of technical, operational and economic feasibilities by using exemption clauses, defined in the Federal Water Act and in the WFD, respectively. For both partners the learning process has only just begun.

References

European Parliament & Council (2000): Directive 2000/60/EC, Water Framework Directive;

German Federal Government (2009): Federal Water Act (FWA);

German Federal Government of BRD (2011): Surface Water Ordinance (OGewV);

Meyer, J., Paul, M., Jenk, U. (2008): Mine Water Hydrology of the Schneeberg Mine (Saxony) Fifty Years after Flooding; 10th IMWA Congress, conference proceedings: 163ff

Changes of water composition in filtration processes due to natural geological formations obtained from opencast mines

Adam Marek[1], Justyna Sobolczyk[1], Waldemar Bicz[1]

[1]Poltegor- Institute Institute of Open Cast Mining, ul. Parkowa 25, 51-616 Wrocław

Abstract. In the article research results of the impact of natural geological formations, obtained from opencast mines in Poland, on changes in water chemical and microbiological composition in filtration processes at lysimeters stands were presented. Deposits, applied in the studies, improved the quality of water, particularly in terms of reducing the manganese and iron and enriching in minerals, such as calcium and magnesium. The microbiological quality of the filtered water improved in relation to the raw water.

Introduction

Opencast mines of brown coal and rock raw materials are the source of obtaining natural geological formations which can be used to build filtration deposits in order to improve the water content. So far, rock raw materials have been used to a small extent due to the necessity of conducting chemical and microbiological studies and developing directives of packaging obtained mineral raw materials. Nevertheless, the problem of the influence of geological formations properties in mines on the changes of water quality are the crucial issue, especially when controlling the process of water inflow to post-exploitation excavations in opencast mines. The analysis of the domestic filtration and sorptive market proved that numerous mineral, natural and thermally transformed products exist in the trade and are useful when building filtration deposits to purify the drinking water (Table 1) (Schmidt 2011).

B.J. Merkel and A. Arab (eds.), *Uranium – Past and Future Challenges*,
DOI 10.1007/978-3-319-11059-2_71

Table 1. Obtained domestic mineral raw materials

Item	Name	Chemical composition	Output venue	Voivodeship	Notes
1.	Jurassic limestone	$CaCO_3$9195%; Mg CO_3 0,9-2,5%; SiO_2 3,4- 4,0 %; Fe_2O_3 0,2-0,9%; Al_2O_3- 0,4-0,5%	Bukowa/ Małogoszeny	Silesian	
2.	Cambrian limestone	$CaCO_3$ min. 91%; $MgCO_3$ max. 5,9 %, NR+ SiO_2 max 3,0%.	Wojcieszów near Jelenia Góra	Lower Silesian	Mined mineral aggregate from carbonate rocks.
3.	Limestone deposit Górażdże	$CaCO_3$, Mg CO_3 and Cd, Cu, Ni, Pb, Zn, Ba, which are released in the working process	Wojcieszów near Jelenia Góra	Opole	Mined mineral aggregate from siliceous limestone rocks.
4.	Limestone from Tarnowo Opolskie	Mineral aggregate from the sedimentary carbonate rock.	Tarnów Opolski	Opole	
5.	Dolomite from Romanów	$CaMg(CO_3)_2$ more than 95%	Romanów	Lower Silesian	
6.	Triassic Dolomite	30% CaO, MgO roughly 20% in weight	Siewierz	Silesian	
7.	Dolomite from Rędziny	30% CaO, MgO roughly 20% in weight	Rędziny	Lower Silesian	
8.	Magnesite from Grochów	Contains over 45% MgO, which after roasting equals: MgO- 80-86%; CaO 1- 1,5%; SiO_2 4,0- 8,0 %; Al_2O_3 0,3-1,0%, Fe_2O_3 0,8-4,0%.	Grochów	Lower Silesian	
9.	Filtration grit from Brzezia	85-99% $Si0_2$	Brzezia near Opole	Opole	Mined in accumulative old river bed Triassic

The chemical composition of the aforementioned mineral deposits can be used for technological processes of water purification.

Given various types of pollution which affect water, its quality and resources have been recently deteriorating. What caused such a phenomenon was the environment pollution resulting from the impact of industry, agriculture and lack of a sewage system in certain locations. Another cause is related to an increased life standards and climate change which limit the amount of available water. In order to point the most frequent water pollutions, one shall mention biological and chemical contaminations, including organic and non-organic, as well as medicaments and hormones (Bożek and Konieczny).

Microbiological pollutions apply not only to microorganisms' cells but also to their deposits and EPS (Extracellular Polymetric Substances), which is referred to as biofilm that emerges on the brink of water and surface environments. They hinder the water purification and distribution systems, as well as pose a threat to public health. Functioning of single microorganisms' cells and their deposits is the consequence of chemical contaminations, accumulation of biodegradable factors microorganisms live on (Lakretz et al. 2014). The major cause of deterioration of water quality is the natural organic matter and nitrogen (Fan X et al. 2014). As a result, water resources as a raw material decrease but it is renewable, which is however a slow process due to anthropogenic factors (Bożek and Konieczny). The quest of new methods of water treatment in order to improve its quality is highly reasonable especially that, according to WHO, "Drinking water suppliers are always responsible for the quality and security of water they produce" (Marek et al. 2012).

In literature one can distinguish diverse methods of water purification, among which filtration is the most common way as it removes most of pollution both with regard to surface water and underground water. In this process molecules are fixed in the filtration material, and these previously held serve as an additional place (surface) to eliminate pollution (Lakretz et al. 2014). An important aspect of water filtration process is its speed, whose fluctuations may rapidly and finally reduce the filtration effectiveness by means of keeping molecules in filtration medias (No-Suk et al. 2010). The most popular filtration techniques encompass coagulation, fast filtration, where sedimentation is omitted. Coagulation and fast filtration are commonly applied since this way of water purification management is effective for the wide range of pollution (Logsdon 2000; Lakretz; et al. 2014; Al.-Gabr et al. 2014; Fan et al. 2014). Another more specialist form of traditional filtration is a membrane filtration as a substitute for conventional methods (Targo O 1996).

Similar to the studied object, most communal water intakes are most commonly located in river valleys (Marek et al. 2012), where early filtration processes take place. When water goes through rock masses, transporting pollutions, in the meantime it is subject to filtration processes which are defined by means of environment carriers' properties (Derx et al. 2013). What turns out to be a common problem of such intakes is the existence of iron and manganese ions, with low level of oxidization. The presence of these compounds leads to deterioration of organoleptic properties and precipitation of deposits in pipelines and fittings. The methods of manganese elimination, similar to iron, are mainly concerned with oxidization and precipitation in the form of the deposit. It is possible in filtration processes, which jointly with pathogenic micro-organisms elimination was the subject of the presented work.

Materials and Methods

Research work

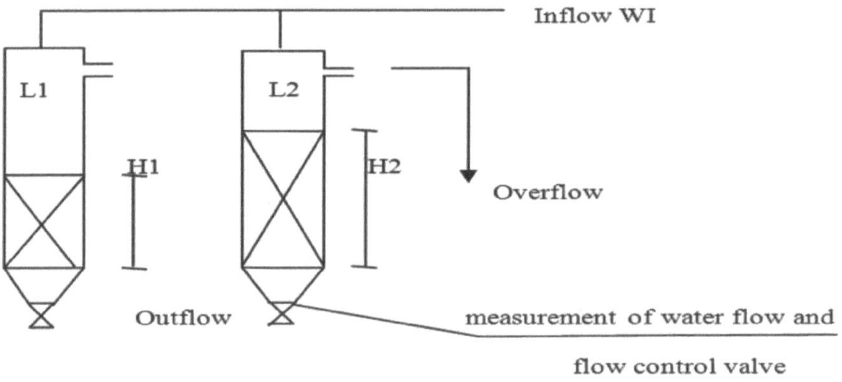

Fig.1. Testing station

Table. 2. Lysimeter filling specification

Marking	Sorbent material	Filling height [m]	Granule size, [mm]
L1	Wojcieszów Limestone	0,5	2,0-8,0
L2	Hydrocleanit	0,25	2,0-4,4
	Romanowo Dolomite	0,25	1,0-4,0
	Mietków Sand	0,22	0,5-1,6

In the case of all lysimeters, a sinter replacing the supporting layer was provided below the researched filling. In order to limit aeration of flowing water, a lysimeter was provided with the overflow which serves as "a vent valve", which prevents from negative pressure (Fig.1). The table 2 presents the filling specification.

The speed of water flow through deposits was up to 0,5 m/h, and every water conduit was provided with flow-meters. A testing station was powered by the pump system in 6 aquifer recharge wells located along the bank of the Nysa Kłodzka at the distance of 70-100 m from the river. The experiment was conducted twice per variant.

Classification of water quality from the well (WI) and from lisymeters (L1, L2) was conducted on the basis of the Health Minister's Directive dated 29 March 2007 on the quality of water devoted to consumption (Journal of Laws, No. 61, item 417), with further amendments dated 20 April 2010 (Journal of Laws, No. 72, item 466). In the given thesis basic physiochemical analyses were conducted, yet only selected ones were presented.

Microbiological studies

In order to define the quality of surface waters, microbiological analyses of the number of *E. coli* bacteria, bacteria from the group of *coli* and bacteria from the group of *coli* (faecal type) were carried out in accordance with the PN-EN ISO 9308-1 (2000) norm. The number of *Enteroccocus faecalis* were analysed in line with the PN-EN ISO 7899-2 standard.

The total number of colonies at the temperatures 22°C (TNM 22°C) and 36°C (TNM 22°C)were conducted according to the PN-EN ISO 6222 standard.

Chemical studies

Taking water samples, their preparation to marking and physiochemical studies were conducted on the basis of the applicable norms and regulations, and developed on their basis quality system procedures are applied in accredited by PCA laboratory NU – 2 of Chemical Studies of Water Poltegor – Institute.

Markings of particular constituents were performed by means of the following methods:

- calcium and magnesium – complexometric with disodium versenate in line with PN-ISO 6058:1999 and PN-ISO 6059:1999,

- manganese and iron – by means of atomic absorption with spectrometer Perkin-Elmer 3100 type in accordance with PB-12.19, 12.15, 12.20 i 12.14, respectively,

Research samples were taken in accordance with PB-12.1 and PB-12.2.

For particular markings, water samples were consolidated in accord with PB – 12.1.

- calcium and magnesium via addition of 2-2,5 cm^3 of concentrated HNO_3 per 1dm^3 of water to pH<2,

Other markings were made with regard to not consolidated samples.

Results and discussion

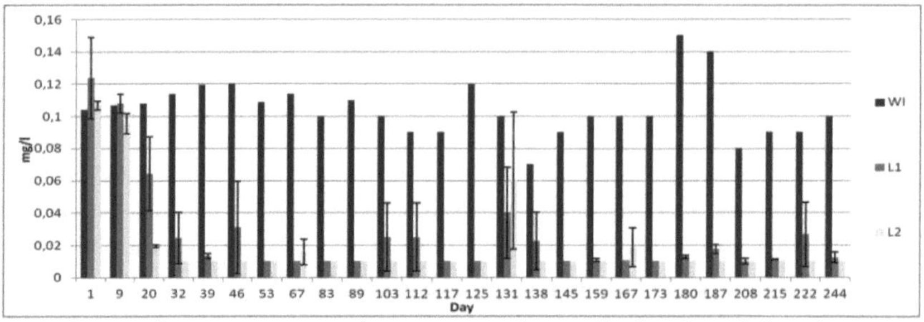

Fig.2. Change of manganese content in raw water and water flowing through deposits

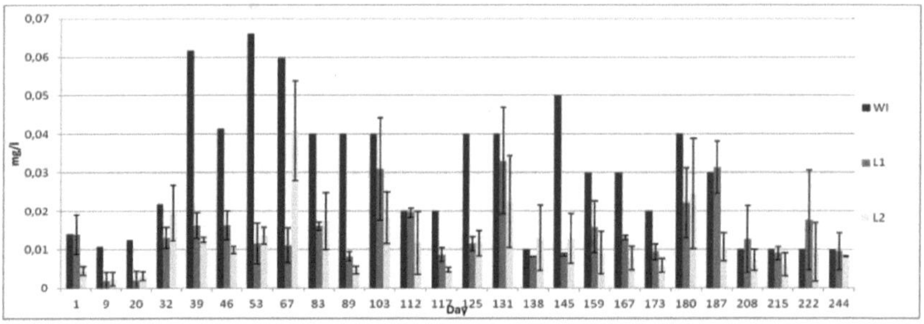

Fig.3. Change of iron content in raw water and water flowing through deposits

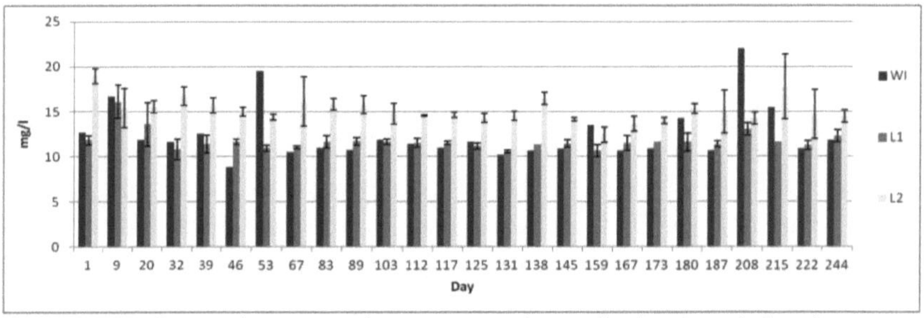

Fig.4. Change of magnesium content in raw water and water flowing through deposits

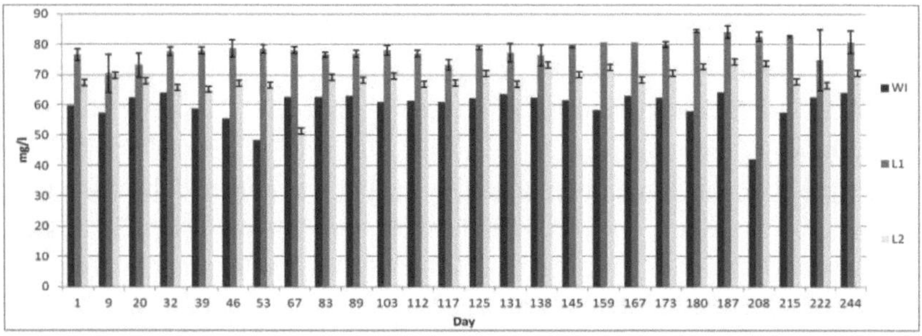

Fig.5. Change of calcium content in raw water and water flowing through deposits

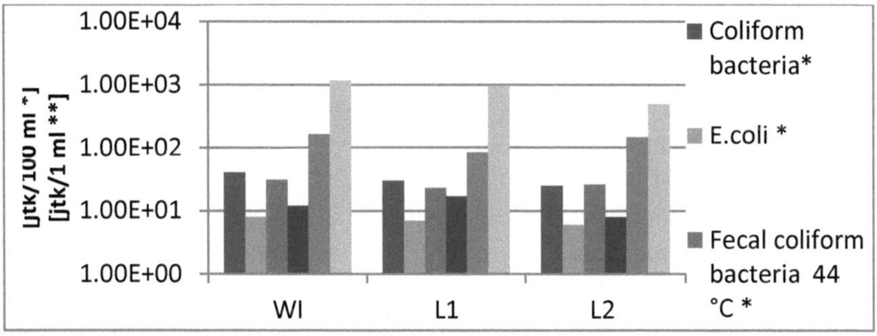

Fig.6. Changes of microbiological indices in raw water and water flowing through deposits

The control of the basic physiochemical studies when taking water samples indicates that raw water proved to be close to neutral reaction ~ 6,80 pH in the entire measurement period. In turn, waters taken from lisymeters proved to have higher reaction, yet within norms of water quality acceptable for consumption, i.e. in the range pH 7-7,6 (Data not show)

Measurements of water flow through lisymeters showed that lisymeter 1 had a three times higher water flow rate when compared to lisymeters L2. A cause of lower lisymeter flow rate was concerned with both differences of deposit height by roughly 40%, as well as lower filling granulation by about 50%. In lisymeter L2, one also observed an increased colmatage of deposit pores and successive increase of resistance to water flow (Data not show). The speed rate of flow through deposits is of paramount importance with respect to filtration efficiency, which is also emphasized by Logsdon 2000 in his studies. He presents that inappropriate management of filtration level increase may significantly worsen the filtration efficiency, particularly when eliminating protozoon from water (Logsdon 2000).

Analyzing conducted studies to see chemical efficiency of burden elimination, as well as enriching water in mineral materials, one can notice changes in these respects.

In raw water exceeding quality of drinking water in chemical terms applied to manganese. Elimination of this compound occurred in all lisymeters, yet with various effects. What was particularly effective was lisymeter L2 where elimination of manganese and iron was considerable, thanks to which water met drinkable water criteria. Maximum burden of manganese amount was obtained thanks to lisymeter 2 on 180th day where out of 0,14mg/l we obtained 0,01mg/l (Fig. 2). Although the amount of iron was as expected, one recorded its reduction roughly from 0,029 mg/l to 0,011 mg/l, with highest efficiency for the lisymeter L2 (Fig. 3). It indicates a better work. A good work efficiency of a two-constituent sand deposit and hard coal deposit in relation to the very sand is underscored by No-Suk et al. (2010). This efficiency mainly included reduction of the number of molecules as a whole and turbidity. In previous own studies conducted with regard to Cleanmag deposit where flowing water was the Poltegor- Institute well water, one also indicated reduction of manganese and iron with the presence of unbeneficial increase of pH (Marek et al. 2012).

The conducted studies proved also the existence of water enrichment in magnesium and calcium. In the case of lisymeter L2, an increase of the amount of magnesium was recorded as early as during the first 24 hours where 12,66 mg/l turned into 18,22 mg/l of magnesium. Similarly, on 32nd and 138th days one obtained more contents in relation to water of 5,11 mg/l and 5,80 mg/l, respectively (Fig. 4). Also, a considerable enrichment in calcium was noted in lisymeter L1 by 18,42 mg/l on average (Fig. 5). In previous studies conducted on the basis of Dolomit Romanowo deposit, an increase of magnesium and calcium content occurred (Marek et al. 2012).

When it comes to microbiological studies, raw water (WI) was characterized by microbiological pollutions, especially a large number of coli group bacteria (Fig. 6). As for other criteria, one also noticed exceeding which excluded the tested water from the group of drinkable waters. The water flowing through lisymeters was of better quality when compared to raw water. The best parameters of water were obtained for the water from lisymeter L2, that is with mixed filtration and sorptive deposits, which contained Mietków sand, Hydrocleanit and Romanowo dolomite. Eventually, water did not apply to drinkable water for bacteriologic reasons, but the reduced number of bacteria may contribute to limitation of application of disinfection means. Similar to studies conducted by Al-Gabr et al. (2014), who used filtration deposits with application of coagulation compounds, they tested the efficiency of eliminating filamentous fungi (*Aspergillus flavus)*. Their studies show a partial elimination of fungi; yet, in order to eliminate all of them, disinfection was indispensable. Among three they used, there were ceramic granules, active coal and sand, and the last two were more effective in the process of elimination of A. flavus.

Half-year lisymetric studies of infiltration Nyska Kłodzka river indicate that the best filtration and sorptive deposit was a three-layer deposit of Mietków sand, Hydrocleanit and Romanowo dolomite. Considering the above, selected sorbents can be applied to build a horizontal well in intake areas of Nysa Kłodzka's infiltration waters.

Conclusion

Opencast mines are the major supplier of mineral resources for production of preparations applied in processes of water treatment.

The conducted works of physiochemical properties of mined deposits of calcium and magnesium and quartz rocks indicate possibilities of applying produced natural aggregates and thermally transformed aggregates to build innovative filtration deposits in accord with the project Poltegor-Institute.

In order to carry out studies of beneficial changes in water content, one selected mineral resources which have an impact on the following: pH regulation, iron removal, manganese removal, enrichment in manganese and calcium. An average elimination of manganese was at the level of 75% and 81% respectively in lisymeters L1 and L2, and iron by 53% -L1 and 62% - L2. Average enrichment in mineral compounds was as follows: magnesium by 22% in lisymeter L2, and calcium by 30% in L1.

Performed works proved that in order to build innovative filtration deposits of drinking water, requiring large amounts of sorptive materials, the deposit of Mietków sand, Hydrocleanit and Romanowo dolomite turned out to be the most beneficial due to the technological universality of work.

References

Al-Gabr H M, Zheng T, Yu X (2014) Efficacy of two chemical coagulants and three different filtration media on removal of *Aspergillus flavus* from surface water. J Environ Sci 26: 274-280

Bożek M, Konieczny K, Technologie membranowe w uzdatnianiu wody do picia Wa Sci Technol

Derx J, Blaschke P, Farnleitner A H, Pang L, Blöschl G, Schijven J F (2013) Effects of fluctuations in river water level on virus removal by bank filtration and aquifer passage — A scenario analysis. J Contam Hydrol 147: 34-44

Fan X, Tao Y, Wang L, Zhang X, Lei Y, Wang Z, Noguchi H (2014) Performance of an integrated process combining ozonation with ceramic membrane ultra-filtration for advanced treatment of drinking water. Desalination 335:47- 54

Lakretz A, Elifantz H, Kviatkovski I, Eshel G, Mamane H (2014) Automatic microfiber filtration (AMF) of surface water: Impact on water quality and biofouling evolution. Water Res 48: 592-604

Logsdon G S (2000) Effective Managament And operation of coagulation and filtration. Water Air Soil Poll 123: 159–166

Marek A, Sobolczyk J, Bicz W (2012) Wstępne badania laboratoryjne zmian parametrów składu wód w procesach filtracji przez złoża sorbentowe. Gór Odkryw 5-6: 21-30

No-Suk P, Seong-Su K, Dong-Hark P, Sangyoung P, Seung-Il C (2010) The optimal composition of the filter-media for coping with daily flow-rate fluctuation. Korean J Chem Eng 27(5) 1492-1496

Schmidt Z (2011) Rozpoznanie i dobór materiałów filtracyjno- sorbcyjnych do budowy złóż dla oczyszczania wody Praca archiwaln IGO Wrocław, nr arch. 6388/IGO/1

Taro O, Hiroyuki K, Hiroyuki H, Katsumi O (1996) Performance of membrane filtration system used for water treatment. Desalination 106:107-113

The Regulations of the Ministry of Health regarding quality of water meant for human consumption [Journal of Laws no. 61 (2007), item 417] as amended [Journal of Laws no. 72 (2010), item 466]

The activities have been performed in the framework of the project no. NR09-0036-10/2011 co- financed by The National Centre for Research and Development

Distribution of uranium and thorium isotopes in colloidal and dissolved fraction of water from San Marcos Dam, Chihuahua, Mexico

Z.K. Ortiz-Caballero[1], A. Covarrubias-Muñoz[1], M.E. Montero-Cabrera[2], Rentería-Villalobos M.[1]

[1]Universidad Autónoma de Chihuahua, Periférico R. Almada km 1, 33820, Chihuahua, Chih., Mexico.
[2]Advanced Material Research Center, CIMAV, Miguel de Cervantes 120, 31109, Chihuahua, Chihuahua, Mexico

Abstract. In the NW of Chihuahua City, Mexico, San Marcos range formation is located. San Marcos area includes two uranium mineral outcrops, a river and its dam, which are the main water supply for the nearest agricultural lands. Assessing fractionation and distribution of U and Th isotopes contained in colloidal and dissolved fractions of surface water is the aim of this work. U and Th isotopes' activities were obtained by alpha spectrometry. ^{238}U and ^{232}Th contents in colloids ranged from 5 to 16, and from 0.3 to 0.5 mBq•g^{-1}, respectively, with activity ratio (AR, $^{238}U/^{234}U$) of up 6. The U distribution coefficient (k_{dU}) ranged from $1 \cdot 10^3$ to $7 \cdot 10^3$ L•g^{-1}. The results have shown a possible lixiviation of uranium from geological substrate into the surface water and an important fractionation of uranium and thorium.

Introduction

To assess the radiological contamination from natural sources, it is important to understand the behavior of radioisotopes released to the environment (Vera Tome et al., 2003). The concentration of radioisotopes in water is influenced by the lithology of the region (Chabaux F. et al., 2008), and in addition, they are strongly affected by interaction with suspended particulate matter (inorganic and organic), and by sedimentation. Sedimentation and resuspension are important processes in the migration of radioisotopes from water to sediment and vice versa. Thus, the main physical processes that have the control on the migration of radioisotopes in the different water bodies are diffusion and dispersion in water transport, ex-

B.J. Merkel and A. Arab (eds.), *Uranium – Past and Future Challenges*,
DOI 10.1007/978-3-319-11059-2_72

change phases solid/solution, deposition and remobilization in sediments, both suspended and precipitates (International Atomic Energy Agency (IAEA), 2010). In most rivers, the suspended and dissolved materials are an excellent source of information about the history of the material from its source to its transport through water (Tosiani et al., 2004).

According to the chemical properties of the various radioisotopes, their behavior in natural waters depends on three main factors: a) the chemical speciation of radioisotopes in solution, b) interactions with mineral or organic solids and c) reaction with colloidal material. The chemical speciation of uranium in solution has been the most extensively studied, where the pH and redox potential (Eh) are the main parameters that affect their chemical form. Thus, uranium under reducing conditions has valence of +4 and it is insoluble, but under oxidation conditions it takes valence of +6 which is its soluble form, being the most common the uranyl ion UO_2^{+2}. The solubility of uranium depends on its ability to form compounds; for example, compounds with carbonates inhibit the adsorption of uranyl, especially in alkaline solutions. In oxidized waters, uranyl compounds may predominate with hydroxides, carbonates, fluorides, sulfates or phosphates (Almeida et al., 2004). Otherwise, the actinides have high reactivity with mineral surface and with organic compounds, and it results in the removal of these from the solution (Chabaux et al., 2008). Studies on thorium (Th) in sediments are useful for characterizing water environments, as well as to provide information on the weathering processes. Detrital Th remains in its particulate form, whereas authigenic Th is scavenged and removed from the water to be mixed with the bed sediments. Moreover, the isotope ^{232}Th is a primordial radionuclide with a half-life comparable to the age of earth and its decay products exist in significant quantities in the environment (Marmolejo-Rodríguez et al., 2008). Therefore, more efforts are required to determine the distribution of this element.

In the natural environment, colloids affect the mobility of radioisotopes in water and soils, and thereby influence the division of these within surface environments (Degueldre, 2006). It is known that water carries a diversity of colloids with different properties, both physical and chemical, which can also modify the metal ion sorption onto mineral surfaces (Chabaux F. et al., 2008). Actinides as U and Th show high affinity to be adsorbed on organic and inorganic colloids.

Several recent investigations of colloidal radionuclides in natural waters have been carried out using ultrafiltration, where association of those with colloidal macromolecules and nanoparticles remain poorly quantified. The variability in the partitioning of radionuclides to the colloidal fraction may depend on specific water chemistry parameters, radionuclides concentrations, composition and/or concentration of organic matter, as well as the rejection characteristics of the ultrafiltration membranes (Guo et al., 2007).

San Marcos zone, at Chihuahua basin, Mexico, has been subjected to several studies since 2005 (Burillo Montufar et al., 2012; Rentería-Villalobos et al., 2013; Rentería Villalobos et al., 2007; Rentería Villalobos et al., 2005; Reyes-Cortés et al., 2010; Reyes-Cortes et al., 2007). Indeed, San Marcos vadose and ground waters behaves very similar to those from Peña Blanca uranium ore, and the zone may be considered as a natural analog for studying uranium-series isotopes transport in repositories of nuclear wastes (Burillo Montufar et al., 2012). San Marcos dam waters and sediments are being subjected of further studies, like in the present work.

It has been found that in water from San Marcos dam the activity concentration of ^{238}U has been as high as 7.7 Bq/L (Rentería Villalobos, 2007), and in coarse suspended solids (particles size >25 μm) from that water, the values of uranium ranged from 0.022 to 1.9 Bq/g (Rentería et al., 2009). In general surface water was filtered at 0.2 or 0.45 microns to separate nominally the particulate phase, which represents the products of physical and chemical weathering. The "dissolved phase" (<0.2 or 0.45 μm) of surface water was shown to be enriched with solutes derived from weathering and may also contain colloidal material (Dosseto et al., 2006). Therefore, determination of the distribution and fractionation of uranium and thorium in suspended particulate matter in San Marcos Dam is the main objective of this work.

Materials and methods

Study area

The area of study is shown in Fig. 1. San Marcos area is located at northwest of Chihuahua city, Mexico. It is a rhyolitic volcanic system, showing mainly rhyolitic tuffs and some Upper Cenozoic intermediate volcanic sequences. Its uranium mineralogical characterization showed the following radioactive species: uranophane, metatyuyamunite, uraninite, bequerelite and masuyite (Reyes-Cortés et al., 2010). San Marcos range formation includes San Marcos River and dam, and at least two uranium minerals outcrops. These outcrops are named Victorino and San Marcos I, which are of hydrothermal origin (Reyes-Cortes et al., 2007). The water of San Marcos River passes close to the outcrops and reaches to the dam, being the main storage of that water.

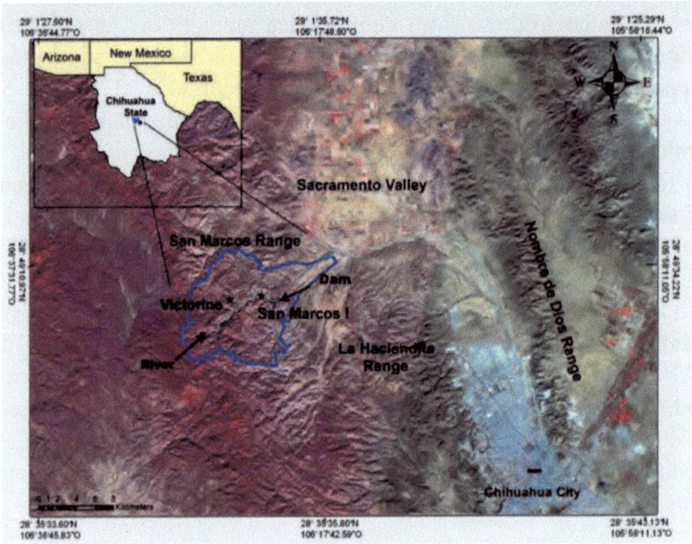

Fig.1. Location of San Marcos region. Sampling points and flux water in San Marcos dam

Sampling and experimental

Sampling was carried out in May 2012; five sampling points were taken at the surface level along the dam (5 kilometers of dam length), according with the water flux. The nearest point to the dam wall (3S) was sampled at the deepest place (19 m) of dam. There, other two sample points were taken in depth: first to 8 meters (3M) and the second to 15 meter (3P).

Water was collected in polyethylene 25 L containers, taking 2 containers for each sample. Geographic coordinates, temperature (T), total dissolved solids (TDS), redox potential (Eh), and pH parameters were measured in situ.

Water samples were filtered to remove the coarse size particle > 1 μm from suspended matter. Then, every water sample was filtered to obtain the colloidal fraction, using ultrafiltration equipment with a tubular membrane of molecular weight cut-off of 100 kDa. In the reject water from ultrafiltration process, containing colloidal fraction, U and Th isotopes were determined. Concentrations of these natural isotopes were measured using alpha spectrometry. Samples were spiked with tracers of ^{232}U and ^{229}Th, and put under radioanalytical analysis procedures. Total sample dissolution was performed by atmospheric acid digestion using HNO_3 and H_2O_2. UTEVA ion-exchange resins were used to isolate U and Th (Casacuberta et al., 2012; Michel et al., 2008). Then, those eluted fractions were electrodeposited on stainless steel planchets (Hallstadius, 1984). An alpha-spectrometry chain Alpha Analyst (PIPS, Mod. A 450–18 AM, CANBERRA) was used for alpha activity measurements. Radiochemical yield was determined by ^{232}U and ^{229}Th counting rates.

Results and discussion

TDS, pH, T data as well as geographic coordinates of surface water samples are shown in Table 1. In Table 2 are shown the activity concentrations of U and Th isotopes, as well as their activity ratios (AR), in colloidal fraction of surface water from San Marcos dam. In order to assess the distribution of uranium, a distribution coefficient k_d was calculated by the total uranium activity concentration ratio between colloidal and dissolved fractions, which is also showed in Table 2.

Table 1. Sampling points and physicochemical parameters of the surface water

Sample	pH Month	T (°C)	Eh (mV)	SDT (ppm)	Latitude	Length	Depth (m)
1S	8.6	30	193	60	28°47'23.78"	106°21'57.12"	0.4
2S	7.3	26	175	60	28°47'23.1"	106°21'35.9"	0.4
3S	6.7	28	178	60	28°47'30.8"	106°21'21.2"	0.4
3M	6.7	26	212	60	28°47'30.8"	106°21'21.2"	8
3P	6.7	26	206	60	28°47'30.8"	106°21'21.2"	15

Table 2. Specific activity, uncertainty and activity ratios of U and Th isotopes, as well as the U distribution coefficient (k_{dU}; $L \cdot g^{-1}$) in colloidal and dissolved fractions from surface water

Sample	^{238}U	^{234}U	^{232}Th	^{230}Th	$^{234}U/^{238}U$[a]	$^{230}Th/^{232}Th$[a]	$^{230}Th/^{238}U$[a]	k_{dU}[a]
Coloidal Fraction (mBq/g)								
1S	5.6±0.3	30.2±1.7	0.5±0.1	4.7±0.4	5.4	9.4	0.9	$6.8 \cdot 10^2$
2S	16.9±1.2	77.0±4.6	0.30±0.07	1.5±0.2	4.5	5.0	0.1	$5.2 \cdot 10^3$
3S	8.5±0.5	39.5±2.2	0.4±0.1	1.0±0.2	4.7	2.3	0.1	$3.7 \cdot 10^3$
3M	4.1±0.1	12.9±0.2	0.70±0.1	2.5±0.1	3.1	3.3	0.6	$1.0 \cdot 10^3$
3P	2.6±0.3	10.1±0.7	3.2±0.8	11.0±1.6	3.8	3.4	4.2	$9.8 \cdot 10^2$
Dissolved fraction (mBq/L)[a]								
1S	12±1	41±5			3.4			
2S	5±0.1	13±1			2.6			
3S	3±0.1	10±1			3.3			
3M	5±0.1	12±2			2.4			
3P	6±0.1	7±1			1.2			

[a] U specific activities in dissolved fraction are published in (Rentería-Villalobos et al., 2013).

From results showed in Table 1, the pH values decreased from 8.6 to 6.7 along the flux water. Nevertheless, this parameter was found slightly acid and remained constant with depth where these values are typically found in surface water. TDS, T, and Eh values were constant in both surface and depth sampling points. In colloidal matter, the U concentrations were higher in surface samples than in deeper ones; on the contrary, Th concentrations tend to be high in deep samples. The highest activity concentrations of uranium and thorium were at 2S and 3P sampling point, respectively.

In surface points, ^{238}U and ^{234}U activity concentrations were the highest one at middle point of dam, whereas in depth, concentrations of both uranium isotopes decreased. The high uranium contents in 3S sampling point can be attributed to the sum of concentrations from both San Marcos River and a spring located near to that point.

Uranium isotopes in dissolved fraction were also determined in these samples and published elsewhere (Rentería-Villalobos et al., 2013). From results, in dissolved fractions ^{238}U decreased along the dam and slightly increased with depth, while ^{234}U remained constant. Taking into account these results, it was possible to obtain the coefficient distribution k_{dU} for uranium. The distribution coefficient k_{dU} shows that most uranium was in the colloidal fraction of surface water (Table 2). In both colloidal and dissolved fractions, the ^{234}U activity concentrations are higher than those for ^{238}U.

Unlike, at surface level, the ^{232}Th contents are constant along the flux of water, whereas at depth its concentration increases. Th tends to be desorbed at low pH, below 5.0 (Anirudhan et al., 2013). In this case, samples showed pH from slightly acid to neutral (Table 1), allowing that ^{232}Th concentration remains constant along the surface level; it can be related with its detrital origin. In addition, the higher concentrations of thorium at deepest zone indicate that it is from colloidal fraction of bed sediments which are in solution by resuspention. Furthermore, ^{230}Th concentration trends to decrease along of surface water flux, and increase with depth. This can be explained by the fact of ^{230}Th is a daughter of ^{238}U decay chain. Uranyl cation is highly adsorbed in colloidal material of both origins inorganic and organic (Chabaux F. et al., 2008). Thus, it can be suggested that high ^{230}Th contents in colloidal fraction at input water point (1S) is due to concentration of uranium from San Marcos River, which its water runoff passes close to the outcrops.

On the other hand, the AR ($^{234}U/^{238}U$) was almost constant along the surface flux from input water to the dam wall, reaching 5.4, while this AR showed a slightly decreasing at depth. AR in the point 1S suggests that colloidal fraction is composed of sediments from the river, whereas the ^{238}U content in point 3P can be attributed to sediments resuspended from the dam bottom. The AR ($^{230}Th/^{232}Th$) in colloidal fraction showed values from 2.3 to 10.2. Results show that ^{230}Th is in higher concentration than ^{232}Th in colloidal fraction. This can be attributed to a preference of these colloids to adsorb uranium.

On the other hand, studies about preference to sorb actinides on colloids have been published. Indeed, sediments' hematite (α-Fe_2O_3), in colloids presented as a (α-$Fe_2O_3 \cdot H_2O$) or goethite (FeOOH), has been long recognized as uranium adsorbent at pH between 5 and 8.5 (Hsi and Langmuir, 1985). Therefore, the adsorption uranium capacity may be explained from the presence of Fe (III) as amorphous goethite resulting from hematite suspension coming from the sediments. Thus, in sediments taken from the bottom of San Marcos Dam, montmorillonite, orthoclase, kaolinite, quartz, muscovite, albite and hematite have been recognized by x-ray diffraction phase analysis.

Conclusions

For all the aforementioned, it is suggested that uranium radioisotopes are lixiviated from geological substrate into the surface water, whereas thorium is present in colloids due to its detrital origin. Thus, the fractionation of uranium and thorium in suspended particulate matter has in consequence that those are distributed non homogeneously along San Marcos dam.

References

Almeida , R.M.R., Lauria, D.C., Ferreira, A.C., Sracek, O., 2004. Groundwater radon, radium and uranium concentrations in Região dos Lagos, Rio de Janeiro State, Brazil. J. Environ. Radioact. 73: 323-334.

Anirudhan, T.S., Sreekumari, S.S., Jalajamony, S., 2013. An investigation into the adsorption of thorium(IV) from aqueous solutions by a carboxylate-functionalised graft copolymer derived from titanium dioxide-densified cellulose. J. Environ. Radioact. 116: 141-147.

Burillo Montufar, J.C., Reyes Cortés, M., Reyes Cortés, I.A., Espino Valdez, M.S., Hinojosa de la Garza, O.R., Nevárez Ronquillo, D.P., Herrera Peraza, E., Rentería Villalobos, M., Montero Cabrera, M.E., 2012. Uranium-series isotopes transport in surface, vadose and ground waters at San Marcos uranium bearing basin, Chihuahua, Mexico. Appl. Geochem. 27: 1111-1122.

Casacuberta, N., Lehritani, M., Mantero, J., Masqué, P., Garcia-Orellana, J., Garcia-Tenorio, R., 2012. Determination of U and Th α-emitters in NORM samples through extraction chromatography by using new and recycled UTEVA resins. Appl. Radiat. Isot. 70: 568-573.

Chabaux, F., Bourdon, B., Riotte, J., 2008. Chapter 3. U-Series Geochemistry in Weathering Profiles, River Waters and Lakes. Radioactivity in the Environment 13: 49-104.

Degueldre, C., 2006. Identification and Speciation of Actinides in the Environment, The Chemistry of the Actinide and Transactinide Elements, 3013-3085.

Dosseto, A., Turner, S.P., Douglas, G.B., 2006. Uranium-series isotopes in colloids and suspended sediments: Timescale for sediment production and transport in the Murray–Darling River system. Earth Planet. Sci. Lett. 246: 418-431.

Guo, L., Warnken, K.W., Santschi, P.H., 2007. Retention behavior of dissolved uranium during ultrafiltration: Implications for colloidal U in surface waters. Mar. Chem. 107: 156-166.

Hallstadius, L., 1984. A method for the electrodeposition of actinides. Nucl. Instrum. Methods Phys. Res. 223, 266-267.

Hsi, C.-k.D., Langmuir, D., 1985. Adsorption of uranyl onto ferric oxyhydroxides: Application of the surface complexation site-binding model. Geochim. Cosmochim. Acta 49, 1931-1941.

International Atomic Energy Agency (IAEA), 2010. Radionuclide transfers in freshwater ecosystems. Handbook of Parameter Values for the Prediction of Radionuclide Transfer in Terrestrial and Freshwater Environments Technical Reports No. 472, 117-131.

Marmolejo-Rodríguez, A.J., Caetano, M., Prego, R., Vale, C., 2008. Thorium accumulation in the sedimentary environment of the Vigo Ria (NW Iberian Peninsula). J. Environ. Radioact. 99, 1631-1635.

Michel, H., Levent, D., Barci, V., Barci-Funel, G., Hurel, C., 2008. Soil and sediment sample analysis for the sequential determination of natural and anthropogenic radionuclides. Talanta 74, 1527-1533.

Rentería-Villalobos, M., Ortiz-Caballero, Z.K., Velez-Sanchez-Verin, C., Montero-Cabrera, M.E., 2013. Distribution and fractionation of uranium isotopes in water from San Marcos Dam, Chihuahua, Mexico, in: VII, N. (Ed.), 7th International Symposium on Naturally Occurring Radioactive Material, Beijing, China.

Rentería, M., Silva, M., Reyes, M., Méndez, C.G., Burciaga, D., Herrera, E.F., Montero, M.E., 2009. Radionuclides present in surface water at the San Marcos Range, Chihuahua, Mexico, in: Lombi, E., Pierzynski, G., Adriano, D., Selim, M., Lepp, N., Alarcón Herrera, M.T. (Eds.), 10th International Conference on the Biogeochemistry of Trace Elements, Chihuahua, Chih., Mexico.

Rentería Villalobos, M., 2007. Modelo conceptual de las concentraciones de uranio en agua superficial y subterránea en la zona de San Marcos-Sacramento Ciencia y Tecnología Ambiental. Centro de Investigación en Materiales Avanzados, CIMAV, Chihuahua, México.

Rentería Villalobos, M., Montero Cabrera, M.E., Reyes Cortés, M., Herrera Peraza, E.F., Rodríguez Pineda, A., Manjón Collado, G., García Tenorio, R., Crespo, T., Valenzuela Hernández, M., 2007. Characterization of source rocks and groundwater radioactivity at the Chihuahua valley. RMxF S53, 16-22.

Rentería Villalobos, M., Montero Cabrera, M.E., Rodríguez Pineda, A., Reyes Cortés, M., Herrera Peraza, E.F., Valenzuela Hernández, M., 2005. Uranium series specific activities of rocks forming ranges around Chihuahua City, Mexico, 2nd International Conference on Radioactivity in the Environment. International Union of Radioecology, Nize, France.

Reyes-Cortés, M., Fuentes-Cobas, L., Torres-Moye, E., Esparza-Ponce, H., Montero-Cabrera, M., 2010. Uranium minerals from the San Marcos District, Chihuahua, Mexico. MinPe 99, 121-132.

Reyes-Cortes, M., Montero-Cabrera, M.E., Renteria-Villalobos, M., Fuentes-Montero, L., Fuentes-Cobas, L., Herrera-Peraza, E.F., Esparza-Ponce, H., Rodriguez-Pineda, A., 2007. Radioactive mineral samples from the northwest of Chihuahua City, Mexico. RMxF S53, 23-28.

Tosiani, T., Loubet, M., Viers, J., Valladon, M., Tapia, J., Marrero, S., Yanes, C., Ramirez, A., Dupre, B., 2004. Major and trace elements in river-borne materials from the Cuyuni basin (southern Venezuela): evidence for organo-colloidal control on the dissolved load and element redistribution between the suspended and dissolved load. Chem. Geol. 211, 305-334.

Vera Tome, F., Blanco Rodríguez, M.P., Lozano, J.C., 2003. Soil-to-plant transfer factors for natural radionuclides and stable elements in a Mediterranean area. J. Environ. Radioact. 65, 161-175.

Rn-222 - a potential health risk for thermal spas workers in Poland

Jakub Nowak[1], Chau Nguyen Dinh[1], Paweł Jodłowski[1]

[1]AGH University of Science and Technology, Faculty of Physics and Applied Computer Science, al. Mickiewicza 30, 30-059 Kraków, Poland; jakub.nowak@fis.agh.edu.pl

Abstract. This paper presents the assessment of radiation exposure due to inhalation of Rn-222 and its progeny for workers and clients of selected thermal water spas. The evaluation of the radon risk is based on the activity concentration of Rn-222 in the investigated thermal waters. For this purpose, a radon transfer coefficient which describes a fraction of radon transferred from swimming-pool water to the indoor swimming-pool air was calculated. The evaluated annual effective doses for the model clients and workers range in the intervals of 0.40 μSv to 194 μSv and 0.022 mSv to 10.8 mSv, respectively.

Introduction

Thermal water is defined as groundwater which temperature is higher than the annual average temperature of the air in the region. In Poland groundwater is regarded as thermal if its temperature exceeds 20 °C. Over the last two decades Poland has been observing a rapid growth of the utilization of thermal waters. Thermal waters are used mainly for recreational bathing and heating purposes (Kępińska 2006). The growth occurs chiefly in Podhale region (part of the Polish Inner Carpathians) where five bathing facilities and one heating plant were constructed (Górecki et al. 2010).

Rn-222 is a decay product of Ra-226 with a half-life of 3.823 days. As an inert gas, radon shows high mobility. Activity concentration of Rn-222 in groundwater may vary in a wide range from a fraction of Bq L^{-1} to hundreds of kBq L^{-1} (Böhm 2002; Nguyen et al. 2011). Radon released from thermal water into the ambient atmosphere is a source of health hazard to the workers and clients of thermal spas.

In accordance with the Polish law, spas are obligated to assess the risk of exposure to radiation from natural sources for their employees and members of the public (Atomic Law Act 2000). The aim of this work was the assessment of the effective doses resulted from the use of thermal waters for recreational bathing. The effective dose was evaluated for both workers and clients of thermal spas. The

B.J. Merkel and A. Arab (eds.), *Uranium – Past and Future Challenges*,
DOI 10.1007/978-3-319-11059-2_73

evaluation was carried out for 17 thermal waters both used and not used for recreational bathing: Bańska Niżna IG-1, Bańska Niżna PGP-1, Bukowina-T. PIG/PGNiG 1, Cieplice C-1, Duszniki-Zdrój GT-1, Jaszczurówka, Mszczonów IG-1, Lubatówka 12, Mszczonów IG-1, Poddębice GT-1, Rabka IG-2, Uniejów PIG/AGH-2, Ustroń U3, Ustroń U3A, Szymoszkowa GT-1, Zakopane 2, Zakopane IG-1.

Materials and Methods

The water samples were collected from three regions: The Polish Carpathians (11 samples), The Sudety Mountains (3 samples) and the central part of Poland (3 samples).

The Polish Carpathians belong to the hydrological Carpathian Province and are localized in the south of Poland. The region is divided into inner and outer parts, whose boundary is the Pieniny Klippen Belt. The Inner Carpathians are built of crystalline rocks and Mesozoic sediments, folded during the late Cretaceous. In the Outer Carpathians composed of the Cretaceous and Paleogene flysch formations folded during the Neogene, dominant are clastic and clay rocks: sandstones, mudstones, shales, claystones and conglomerates (Porowski 2006). The Podhale Trough, which belongs to Polish Inner Carpathians, is one of the most important geothermal water system in Poland. Thermal water has been reported in over a dozen boreholes, nine of them are exploited. The water belong to four hydrochemical types: HCO_3-Ca-Mg, SO_4-HCO_3-Cl-Na-Ca, SO_4-Ca-Na and SO_4-Cl-Ca-Na.

The Sudety Mountains is the region where the high concentration of Rn-222 in groundwater is the most likely in Poland. Groundwaters containing radon concentration of several hundred of Bq L^{-1} are quite common (Przylibski et al. 2004). This is due to the geological structure of the Sudetes, which are composed mainly of a complex mosaic of igneous and metamorphic rock types with increased uranium content (Ciężkowski et al. 2010).

The central part of Poland belongs to the hydrological province of the Paleozoic Platform. The sandstone formations are major aquifers in the province and the Cl-Na is the common hydrochemical type of the thermal waters in the region (Paczyński and Płochniewski 1996).

The samples of the thermal water for the determination of the Rn-222 activity concentration were collected "under the cap" at the laminar water flow (these condition assure that the radon would not escape from the sample) and kept into glass bottles. The samples were collected directly from the wellheads and transported to the laboratory as fast as possible. To make a correctionfor radon decay ($T_{1/2}$ = 3.82 days), the sample collection times were noted and a decay correcting factor was introduced. The radon activity was measured in a mixture of 10 mL of the collected water sample with 10 mL of a PerkinElmer™ mineral oil liquid scintillator placed in the glass vial using a Wallac™ Guardian Liquid Scintillation counter.

Results and discussion

The Rn-222 activity concentration

The obtained results showed that the activity concentration of Rn-222 in thermal waters vary from 0.3 Bq L^{-1} to 145 Bq L^{-1}. The activity concentrations for investigated waters from the Polish Carpathians and from the central part of Poland vary from 0.3 Bq L^{-1} to 36.3 Bq L^{-1}. The highest radon concentration was found in the water from Ustroń U3A (the Polish Outer Carpathians). Such low levels are characteristic of the limestone aquifer and waters with high temperature. The activity concentrations of radon-222 for 3 selected waters from the Sudety Mountains change in the range from 3.4 Bq L^{-1} to 145 Bq L^{-1}. The highest activity concentration of Rn-222 was found in the water sample from Lądek-Zdrój L-2 (the Sudety Mountains); such a high radon concentration is related to the presence of igneous rocks in the region.

The radon transfer coefficient (T_{Rn})

The radon transfer coefficient T_{Rn} is defined as a fraction of radon released from the thermal water in swimming pool into the air of swimming pool house. The T_{Rn} value was calculated based on data of Rn-222 concentration in swimming-pool water and Rn-222 concentration in swimming-pool air gathered from published articles (Vogiannis et al 2004; Song et al 2011). The obtained T_{Rn} coefficient value is equal to 10.3 Bq$\odot$m^{-3}/Bq$\odot$L^{-1}.

In next step, the verification of the calculated T_{Rn} coefficient was done based on direct measurements of indoor air Rn-222 concentration in Lądek-Długopole Spa (Lądek-Zdrój) which were made by Olszewski et al. (2006). The indoor air Rn-222 concentration in Lądek-Długopole Spa vary from 50 Bq m^{-3} to 2180 Bq m^{-3}. The mean value is 903 Bq m^{-3}. The indoor air Rn-222 concentration in Lądek-Długopole Spa obtained by using the radon transfer coefficient and the activity concentration of Rn-222 in water from Lądek-Zdrój L-2 is 1495 Bq m^{-3}. This value is in the range of the mentioned measurement of indoor Rn-222 concentrations. Therefore, the estimated value of T_{Rn} coefficient can be used for the assessment of Rn-222 concentrations in the indoor swimming-pool air, knowing the Rn-222 concentration in swimming pool water.

Effective dose from inhaled radon and its progeny

To evaluate the effective dose from radon and its progeny inhalation the approach of the dose conversion was used. The method is described in UNSCEAR reports (1993, 2000) as an alternative to a dosimetric approach. In accordance with these reports the annual effective dose (D_{Rn}) can be calculated by the following formula:

$$D_{Rn} = C_{airRn} \cdot F_{eq} \cdot K_{Rn} \cdot t,$$

where C_{airRn} is the indoor swimming-pool air Rn-222 concentration obtained using the T_{Rn} coefficient, F_{eq} is equilibrium factor for Rn-222 indoor, equals to 0.4 (UNSCEAR 1993), K_{Rn} is the dose conversion factor, equals to 9 nSv (Bq h $m^{-3})^{-1}$ (UNSCEAR 1993, 2000), t is the annual presence time at a swimming-pool.

The effective dose was calculated for the model clients and workers of thermal spas. The annual presence time at a swimming-pool for the clients and workers was assumed as 36 h (a three-hour visit to thermal spa per a month) and 2000 h (employees spend the whole working time at a swimming-pool), respectively. The evaluated annual effective doses in most spas for the model clients and workers range in the intervals of 0.40 μSv to 48,5 μSv and 0.022 mSv to 2.69 mSv, respectively. Significantly higher doses were obtained for water from Lądek-Zdrój L-2 both for the clients (194 μSv) and workers (10,8 mSv).

For most investigated waters the evaluated effective doses for the model clients of thermal spas are much lower than the annual radiation dose limit for members of the public,(ICRP 1991). Generally the doses do not exceed 5 per cent of the annual dose limit.

In the case of workers, the situation is a little bit complicated. For six thermal waters (in which content of Rn-222 is higher than 18 Bq L^{-1}: Zakopane 2, Szymoszkowa GT-1, Jaszczurówka, Ustroń U3, Ustoń U3A and Lądek-Zdrój L-2) the calculated effective doses exceed the annual radiation dose limit for incidentally exposed workers, which is 1 mSv (Regulation of the Council of Ministers 2005). However, the annual effective dose limit (20 mSv) for occupationally exposed workers wasn't exceeded. Nevertheless, for the water from Lądek-Zdrój L-2 the effective dose is higher than the dose limit (6 mSv), above this limit a personal dosimetry monitoring is obligated (Atomic Law Act 2000).

Conclusions

From radiation protection point of view, using investigated thermal waters for recreational bathing dose not cause the health risk for clients of thermal spas.

In the case of workers, calculations showed that, there should be concern when a thermal water with the Rn-222 content above 18 Bq L^{-1} is used for recreational bathing. For such activity concentration of Rn-222 in thermal waters radon would reach the level in the indoor swimming-pool air for which the effective dose from inhaled radon and its progeny exceeds the radiation dose limit for incidentally exposed workers. Additionally, for the thermal water with the content of Rn-222 higher than 100 Bq L^{-1} the effective dose exceed the limit 6 mSv. Therefore the classification of thermal spa's workers for a group of occupationally exposed workers should be considered. Basically, further research including in situ Rn-222 measurements and the survey of the working time is needed.

Acknowledgments: This work was supported by the "Doctus - Małopolska Scholarship for PhD" co-funded by the European Union (Project No. MCP.ZS.4110-29.1/2009) and the statutory research of the AGH University of Science and Technology (Grants No. 11.11.220.01 and 11.11.140.021).

References

Atomic Law Act of Parliament of Republic of Poland of 29 November 2000. Polish Official Journal 161 poz.1689 (in Polish)

Böhm C. (2002) Radon inWasser - Uberblick für den Kanton Graubünden. Jahresbericht Naturforschende GesellschaftGraubünden 111 (49)

Ciężkowski W., Chowaniec J., Górecki W., Krawiec A., Rajchel L., Zuber A. (2010) Mineral and thermal waters of Poland. Polish Geolgical Review 58(9/1): 762-774

Górecki W., Kępińska B., Sowiżdżał A. (2010) Wykorzystanie wód oraz energii geotermalnej w Polsce. In: Górecki W. (Eds.) Atlas zasobów wód i energii geotermalnej Karpatach Zachodnich. Akademia Górniczo-Hutnicza im. S. Staszica w Krakowie Wydział Geologii, Geofizyki i Ochrony Środowiska Zakład Surowców Energetycznych: 41-46 (in Polish)

ICRP (1991) 1990 Recommendations of the International Commission on Radiological Protection. ICRP Publication 60 . Ann. ICRP 21 (1-3)

Kępińska B. (2006) Geothermal energy - utilization in the World and Europe. Polityka Eergetyczna 9: 545-556 (in Polish)

Nguyen C.D., Duliński M., Jodłowski P., Nowak J., RóŜański K., Śleziak M., Wachniew P. (2011) Natural radioactivity in groundwater - a review. Isotopes in Environmental and Health Studies 47(4): 415-437

Olszewski J., Chodak M., Jamkowski J. (2008) An assay of the current exposure to radon of spa workers in Poland. Medycyna Pracy 59(1): 35-38

Paczyński B., Płochniewski Z. (1996): The Polish mineral and medicine waters. PIG, Warsaw: 1-108 (in Polish)

Porowski, A. (2006) Origin of mineralized waters in the Central Carpathian Synclinorium SE Poland. Studia Geologica Polonica 125: 1-67

Przylibski T. A., Mamont-Cieśla K., Kusyk M., Dorda J., Kozłowska B. (2004) Radon concentrations in groundwaters of the Polish part of the Sudety Mountains (SW Poland). Journal of Environmental Radioactivity 75: 193-209

Regulation of the Council of Ministers of 18 January 2005 on the radiation dose limits. Polish Official Journal 20 poz.168 (in Polish)

Song G., Wang X., Chen D., Chen Y. (2011) Contribution of ^{222}Rn-bearing water to indoor radon and indoor air quality assessment in hot spring hotels of Guangdong, China. Journal of Environmental Radioactivity 102: 400-406

Vogiannis E., Nikolopoulos D., Louizi A., Halvadakis C.P. (2004) Radon variations during treatment in thermal spas of Lesvos Island (Greece). Journal of Environmental Radioactivity 76: 283-294

UNSCEAR (1993) Sources and Effects of Ionizing Radiation. Report to the General Assembly,with scientific annexes

UNSCEAR (2000) Sources and Effects of Ionizing Radiation. Report to the General Assembly,with scientific annexes

Measurement of indoor radon, thoron and their progeny concentrations in the dwellings of district Hamirpur, Himachal Pradesh, India

Parminder Singh[1], Prabhjot Singh[1], B.S. Bajwa[1], Surinder Singh[1], B.K. Sahoo[2]

[1]Department of Physics, Guru Nanak Dev University, Amritsar, Punjab-143005, India
[2]Radiological Physics and Advisory Division, Bhabha Atomic Research Centre, Mumbai-400085, India

Abstract. In the present investigation indoor radon (^{222}Rn), thoron (^{220}Rn) and their progeny concentrations have been measured in the wide range of dwellings from 12 different villages of Hamirpur district, Himachal Pradesh, India by using LR-115 type-II based Pin-hole Radon-Thoron discriminating twin cup dosimeters, direct radon and thoron progeny sensors (DRPS/DTPS). As inhalation doses are predominantly due to daughter products of radon and thoron and not due to gases, it is important to measure the decay products directly for health risk assessments. In the study region different types of houses were selected randomly according to methodologies described by Radiological Physics and Advisory Division (RPAD), Bhaba Atomic Research Centre (BARC), Mumbai. The indoor radon concentrations in these villages have been found vary from 36.1 to 272 Bq/m^3, with average value of 125.54 Bq/m^3 and for thoron vary from 38.5 to 331 Bq/m^3, with average value of 140 Bq/m^3. The progeny concentrations of radon and thoron are found within the limits of 10.5 to 106.9 Bq/m^3 and 0.9 to 6 Bq/m^3 respectively, with average values 43.7 Bq/m^3 and 3.34 Bq/m^3 respectively. The average concentration values observed in these dwellings have been found within the safe limits as recommended by International Commission on Radiological Protection (ICRP) and United Nation Scientific Committee on the Effect of Atomic Radiation (UNSCEAR).

Keywords: Radon, Thoron, Progeny concentration, Dosimeter

B.J. Merkel and A. Arab (eds.), *Uranium – Past and Future Challenges*,
DOI 10.1007/978-3-319-11059-2_74

Introduction

Measurement of indoor radon, thoron and their daughter products is an important aspect as they contribute more than the 50% of total radiation dose inhaled by human beings from natural radioactive sources (UNSCEAR 1988). After smoking radon is the second main cause of lung cancer in human population (BEIR 1990). Radon (^{222}Rn) and thoron (^{220}Rn) with half life 3.82 days and 55.2 sec. are decay products of naturally occurring radioactive elements ^{238}U and ^{235}Th respectively. The contribution of radiation dose by thoron is negligible as compared to radon because of short half life but decay products of radon and thoron are more harmful and then they cannot be neglected (Ramola et al. 2010; BEIR 1990). The alpha emitting, short lived daughter products of radon (^{218}Po, ^{214}Pb and ^{214}Bi) and thoron (^{212}Pb and ^{212}Bi) are present everywhere in environment in attached or unattached fractions of air particles. During inhalation main fraction of these daughter products stay in lungs and irradiate the bronchial target cells by emission of alpha particles, which results in bronchial carcinomas (ICRP 1981; UNSCEAR 2000) and sometimes causes lung cancer (Ramachandran and sahoo 2009; Ramola et al. 2008).

The main sources of indoor radon, thoron and their decay products are the soil-gas, minerals and rocks in earth's crust, building material and ground water (Bajwa et al. 2003). The construction type of houses, type of soil, ventilation rate, atmospheric pressure and temperature also have a significant role in indoor gas measurements (Singh et al. 2002; Abu-Jarad and Fremlin 1983). In the present study latest passive techniques have been used in the dwellings of Hamirpur district Himachal Pradesh for indoor radon, thoron and their progeny concentration measurements, using LR-115 (type II) plastic track detectors and Direct Progeny Sensors (DRPS/DTPS) as developed and recommended by Radiation Protection Advisory Division (RPAD) (Sahoo et al. 2013; Mishra et al. 2010), BARC, Mumbai.

Geology of the area

The area of study region is Hamirpur district in Himachal Pradesh (HP) extends from31.27° to 32.10° North latitudes and from 75.80° to 76.52° East longitudes. HP is bounded from west and north by Jammu and Kashmir, east by Uttrakhand and in the south by Punjab. It is highly hilly area and population is very low. Soil contains large number of minerals and rocks, which contribute to radon, thoron and other radioactive gases.

Experimental technique

The measurement of radon, thoron and their progeny concentrations have been carried out by latest passive techniques developed by RPAD, BARC, Mumbai, using LR-115 type II plastic track detectors. Indoor radon and thoron concentrations are calculated using pin hole based twin cup dosimeters suspended inside the dwellings such that they are 30 cm away from any surface. Progeny concentrations are also calculated by suspending DRPS/DTPS in the indoor environment away from any door and window. The Pelliculable LR-115 of size 3 cm x 3 cm have been fixed on holders provided in dosimeters. After exposure for three months, the detectors were retrieved from dosimeters and etched in 2.5 N NaOH solution for 90 minutes at 60°C in a water bath without stirring. After etching and drying, the detector should be peeled off from its cellulose acetate base and the track counting can be done under spark counter Model PSI-SC 1. The operating as well as the pre-sparking voltage of the spark counter has been established prior to these measurements.

Pin hole based twin cup dosimeters

The newly designed this dosimeter has two compartments separated by a central pin-holes disc, acting as thoron discriminator. Four pin holes each with dimension of 2 mm length and 1 mm diameter are made in this circular disc. The schematic diagram of the dosimeter is shown in Fig.1. The dosimeter has a single entry through which gas enters the first chamber namely "radon+thoron" chamber through a glass fiber filter paper (thickness 0.56 μm) and subsequently diffuses to second chamber namely "radon" chamber through pin-holes cutting off the entry of thoron into this chamber due to its short half life. Each chamber is cylindrical having a length of 4.1 cm and radius 3.1cm. Chambers are internally coated with metallic powders to have zero electric field inside the chamber volume, so that the deposition of progenies formed from gases will be uniform throughout the volume.

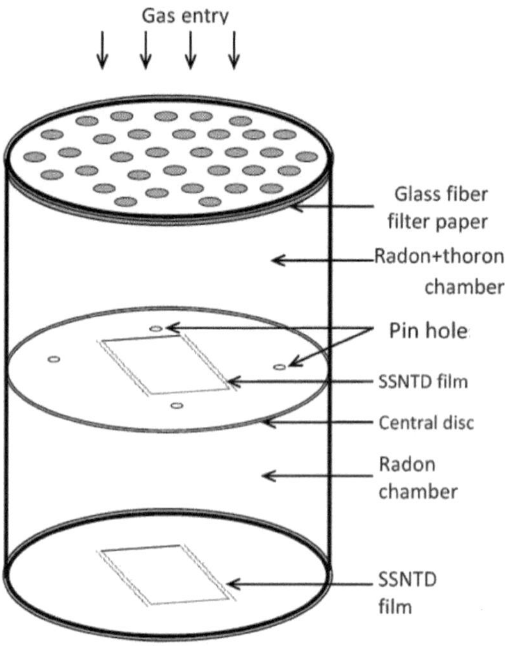

Fig.1. Pin hole based twin cup radon/thoron discriminating dosimeter

The radon (C_R) and thoron (C_T) concentrations are calculated by following relations (Sahoo et al. 2013):

$$C_R\,(Bq/m^3) = T_1/\,(d.k_r)$$
$$C_T\,(Bq/m^3) = (T_2 - d.C_r.k_r)/\,(d.k_t)$$

where, d is the number of days of exposure, T_1 and T_2 are the track densities (in tracks/cm^2) observed in 'radon' and 'radon+thoron' compartment respectively, k_r and k_t are the calibration factors for radon in 'radon' compartment and thoron in 'radon+thoron' compartment respectively. The values of these factors are as follows:

$k_r = 0.019 \pm 0.003$ tracks. Cm^{-2}/Bq.d.m^{-3}
$k_t = 0.016 \pm 0.005$ tracks. Cm^{-2}/Bq.d.m^{-3}

Direct radon/thoron progeny sensors

Deposition based Direct Radon and Thoron progeny sensors (DRPS and DTPS) have been developed for measurement of radon and thoron progeny concentration in the environment. These are made of passive nuclear track detectors (LR-115) mounted with absorbers of appropriate thickness. For thoron progeny, the absorber is 50 μm aluminized mylar which selectively detects only 8.78 MeV α- particles emitted from ^{212}Po; while for radon progeny, the absorber is a combination of aluminized mylar and cellulose nitrate of effective thickness 37 μm to detect mainly 7.69 MeV α-particles emitted from ^{214}Po. The tracks recorded in the exposed LR-115 film are related to Equilibrium Equivalent Progeny Concentration (EEC) using the sensitivity factor. The number of tracks per unit area per unit time (T) can be correlated to the Equilibrium Equivalent Progeny Concentration (EEC) (Mishra et al. 2008, 2010) in air using the Sensitivity factor (S) as:

$$EEC\ (Bq.m^{-3}) = T\ (Tracks.cm^{-2}.d^{-1})\ /\ S\ (Tracks.cm^{-2}.d^{-1})/EEC\ (Bq.m^{-3})$$

Where sensitivity factor (S) for DTPS is 0.94 ($Tracks.cm^{-2}.d^{-1}$)/EEC ($Bq.m^{-3}$) to measure thoron progeny and to find radon progeny concentration the sensitivity factor for DRPS is 0.09 ($Tracks.cm^{-2}.d^{-1}$)/EEC ($Bq.m^{-3}$).

Results and Discussion

The indoor concentration of radon, thoron and their daughter products in the winter season of year 2013, at different locations of Hamirpur district Himachal Pradesh are calculated by using latest developed pin hole based twin cup dosimeters and Direct radon/thoron progeny sensors (DRPS/DTPS). In every village 5-6 different types of houses were selected according to their construction material and ventilation conditions. The observed concentration values of each village of Hamirpur district are represented in Table 1. The indoor radon concentrations in these dwellings have been found vary from 36.1 to 272 Bq/m^3, with average value of 125.54 Bq/m^3 and thoron concentration vary from 38.5 to 331 Bq/m^3, with average value of 140 Bq/m^3. At few dwellings in villages Chakrauli, Romera and Galote show slightly high values of radon and thoron concentration, this may be due to the difference in construction material of houses and ventilation conditions. The progeny concentrations of radon and thoron are found within the limits of 10.5 to 106.9 Bq/m^3 and 0.9 to 6 Bq/m^3 respectively, with average values 43.7 Bq/m^3 and 3.34 Bq/m^3 respectively. It can be observed from Table 1, that the average values of indoor radon, thoron and their progeny concentrations are lies within the safe limits as recommended by ICRP i.e. 200-600 Bq/m^3 (ICRP 1993). The Variation in concentration values at different locations may be due to uranium concentrations in the soil of the area (Virk et al. 1999), building materials used during construction of houses, ventilation condition of the dwellings, temperature and pressure.

Table 1. Average Indoor radon, thoron and their progeny concentrations at different locations in Hamirpur district of Himachal Pradesh

Sr. no.	Location	Radon conc. (Bq/m^3)	Thoron conc. (Bq/m^3)	Radon Progeny conc. (Bq/m^3)	Thoron Progeny conc. (Bq/m^3)
1.	Chakrauli	176.1	316.1	72.8	4.4
2.	Romera	272.0	201.1	51.6	3.7
3.	Khaya	99.8	119.4	39.3	2.5
4.	Loharian	147.1	67.9	62.8	4.4
5.	Hamirpur	51.7	38.5	10.5	0.9
6.	Tikker	103.1	97.9	39.4	5.2
7.	Galore Khas	36.1	53.3	10.8	1.0
8.	Badsar	40.3	54.5	13.7	1.2
9.	Galote	214.8	331.1	106.9	6.0
10.	Chakmoh	71.3	117.3	27.9	3.0
11.	Baroha	116.0	78.8	26.1	3.7
12.	Asthota	133.4	144.9	39.4	4.0
	Average	125.54	140	43.7	3.34

According to the type of construction material and ventilation rate, we have divided the houses in two categories. The mud houses, earthen flooring and poor ventilation conditions considered in type A while concrete/brick houses, concrete floor and roof and good ventilated dwellings consider in type B. The average concentrations of radon, thoron and daughter products in both types of dwellings are represented in Table 2. The average concentration values in A type houses are higher as compared to B type. The higher values in A type dwellings is due to ground floor directly constructed on top soil allows more radon/thoron to diffuse inside the house because of higher porosity of mud floor and walls. Also poor ventilation condition accumulate higher levels of radon/thoron inside the dwellings, results in high values of radon, thoron and their progeny concentrations as compared to B type dwellings.

Table 2. Concentration of radon, thoron and their progenies in A type and B type houses

S. No.	Types of dwellings	No. of dwellings	Avg. radon conc. (Bq/m^3)	Avg. thoron conc. (Bq/m^3)	Avg. radon progeny conc. (Bq/m^3)	Avg. thoron progeny conc. (Bq/m^3)
A	Mud houses/ poor ventilation	25	136. 5	145.7	53.6	6.2
B	Concrete houses/ good ventilation	32	83.2	103.1	31.2	3.2

Conclusion

The indoor radon, thoron and their progeny concentrations have been determined from selected locations in Hamirpur district Himachal Pradesh and found that average values of radon and thoron are 125.54 and 140 (Bq/m^3) respectively and radon and thoron progeny levels are 43.7 and 3.34 (Bq/m^3) respectively, which lies within the safe limits as recommended by International Commission on Radiological Protection (ICRP) and UNSCEAR. Also found that concentration levels in mud houses or in poor ventilation conditions is higher as compared to concrete houses or well ventilated.

Acknowledgement

We are thankful to the Board of Research in Nuclear Sciences (BRNS), Department of Atomic Energy (DAE), Government of India, for providing financial assistance and also grateful to the residents of Himachal Pradesh for their cooperation and help for placing dosimeters in respective dwellings.

References

Abu-Jarad F, Fremlin J H (1983) Effect of internal wall covers on radon emanation inside houses. Health Phys 44: 243-248.

Bajwa B S, Virk H S, Singh S (2003) A comparative study of indoor radon level measurements in the dwellings of Punjab and Himachal Pradesh, India. Radiat Meas 36: 457-460.

BEIR (1990) Committee on the Biological Effect of Ionizing Radiations (BEIR-V), National Research Council National Academy Press, Washington DC.

BEIR (1999) Committee on the Biological Effect of Ionizing Radiations (BEIR-VI), National Research Council National Academy Press, Washington DC.

ICRP (1981) International Commission on Radiological Protection, ICRP Publication-32, Pergamon Press, Oxford.

ICRP, (1993) International Commission on Radiological Protection, ICRP Publication-65, Pergamon Press, Oxford.

Mishra R, Mayya Y S (2008) Study of a deposition-based direct thoron progeny sensor (DTPS) technique for estimating equilibrium equivalent thoron concentration (EETC) in indoor environment. Radiat Meas 43: 1408-1416.

Mishra R, Prajith R, Sapra B K, Mayya Y S (2010) Response of direct thoron progeny sensors (DTPS) to various aerosol concentrations and ventilation rates. Nucl Instrum Meth B 268: 671-675.

Ramachandran T V, Sahoo B K (2009) Thoron in the indoor environment and work places. Indian J Phys 83(8): 1079-1098.

Ramola R C, Choubey V M, Negi M S, Prasad Y, Prasad G (2008) Radon occurrence in soil-gas and ground water around an active landslide. Radiat Meas 43: 98-101.

Ramola R C, Prasad G, Gusain G S, Rautela B S, Choubey V M, Sagar D V, Tokonami S, Sorimachi A, Sahoo S K, Janik M, Ishikawa T (2010) Preliminary indoor thoron measurements in high radiation background area of southeastern coastal Orissa, India. Radiat Prot Dosim, 141(4): 379-382.

Sahoo B K, Sapra B K, Kanse S D, Gaware J J, Mayya Y S, (2013) A new pin hole discriminated $^{222}Rn/^{220}Rn$ passive measurement device with single entry face. Radiat Meas 58: 52-60

Singh S, Kumar A, Singh B (2002) Radon level in dwellings and its correlation with uranium and radium content in some areas of Himachal Pradesh, India. Environ. Int. 28: 97-101.

UNSCEAR, United Nation Scientific Committee on the Effect of Atomic Radiation (1988) Sources, effects and risks of ionizing radiation. Report to the General Assembly, United Nations, New York.

UNSCEAR, United Nation Scientific Committee on the Effect of Atomic Radiation (2000), Sources and Biological effects of ionizing radiation. United Nations, New York.

Virk H S, Sharma N, Bajwa B S (1999) Environmental radioactivity: a case study in Himachal Pradesh, India. J. Environ. Radioact 45: 119-127.

Soils and ground water's radioactive contamination into the local zone of the "Shelter" object and industrial site of Chernobyl NPP

M. I. Panasyuk[1], I. A. Lytvyn[2], E. P. Liushnya, A. M. Alfyoroff, G. V. Levin, V. M. Shestopalov

[1]mipanasyuk@bigmir.net
[2]livan@ukr.net

Abstract. The purpose of radio-hydro-ecological monitoring of the "Shelter" object includes: evaluation of soil and groundwater radioactive contamination level and migration or gradual penetration of radio-nuclides to the environment around the safety block.

The research objects are: soils of aeration zone, groundwater and the sources of their contamination - surface and block waters.

Observation network includes about 50 wells located around the "Shelter" object and at the industrial site of the Chernobyl NPP.

The following activities are determined in the following groundwater samples: ^{90}Sr, ^{137}Cs, ^{3}H, $^{238, 239, 240}$Pu, ^{241}Am, U along with complete chemical analysis.

The set of radio-hydro-ecological monitoring works also includes the control of the following groundwater radioactive contamination sources:

1. Water clusters of high activity are located inside the "Shelter" object (4th block of Chernobyl NPP, the one destroyed by accident);
2. Water clusters in communications;
3. Infiltration waters, that deliver radio-nuclides from aeration zone during the filtration process of precipitation into the soil.

Everywhere on the industrial site of ChNPP unconfined aquifer flows in alluvial sands with sandy loam in its upper layer.

The depth of the groundwater level (that is aeration zone power) varies from 5 to 14 meters depending on the surface hypsometry. Power of unconfined aquifer is about 30 m. Groundwater movement is directed from south to north with a slope of 0.001 - 0.002 and actual speed 20-30 m/h. Unloading of unconfined aquifer is carried to the Pripyat river. Movement of groundwater under the "Shelter" object comes from turbine building in the north direction to the block of the reactor compartment space support systems and to the Cascade wall.

B.J. Merkel and A. Arab (eds.), *Uranium – Past and Future Challenges*,
DOI 10.1007/978-3-319-11059-2_75

^{90}Sr volumetric activities in the groundwater of the "Shelter" object local zone and adjacent area fluctuated within the range of 0,12-1700 Bq/l for the last 5 years.

Highest concentration of ^{137}Cs in the groundwater's samples are fixed in the range of 170-210 Bq/l.

Maximum concentration of tritium in the groundwaters of the "Shelter" object local zone and adjacent industrial site of ChNPP was up to 8000 Bq/l.

Content of each uranium isotopes: ^{234}U, ^{235}U, ^{236}U and ^{238}U in groundwater do not exceed the value 0.2 Bq/l.

Concentration of transuranium into the groundwater varies from 0.0005 to 0.054 Bq/l. The largest values are typical for $^{239+240}$Pu.

Thus, ^{90}Sr, ^{137}Cs, and ^{3}H have the most radiologically significant concentrations in the groundwaters.

During the last 9 years a substantial rising tendency of volumetric activities of ^{90}Sr in the samples of groundwater from the certain wells is observed (Fig.1). As Fig.1 shows, areas with ^{90}Sr concentration more than 50 Bq/l increased almost tenfold between 2003 and 2013.

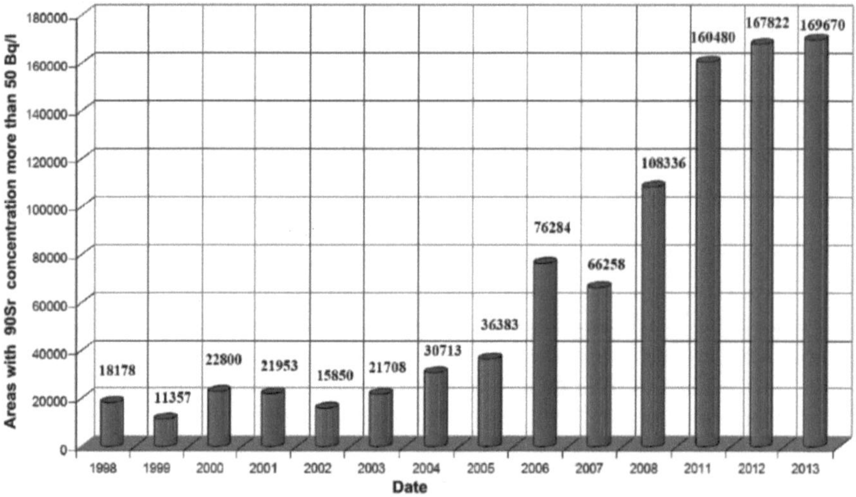

Fig.1. Distribution of areas with ^{90}Sr volumetric activity above 50 Bq/l on the radio-hydro-ecology monitoring site by years.

Fig.2 shows a map of ^{90}Sr volumetric activity distribution in the groundwaters within the area of "Shelter" object in 2013.

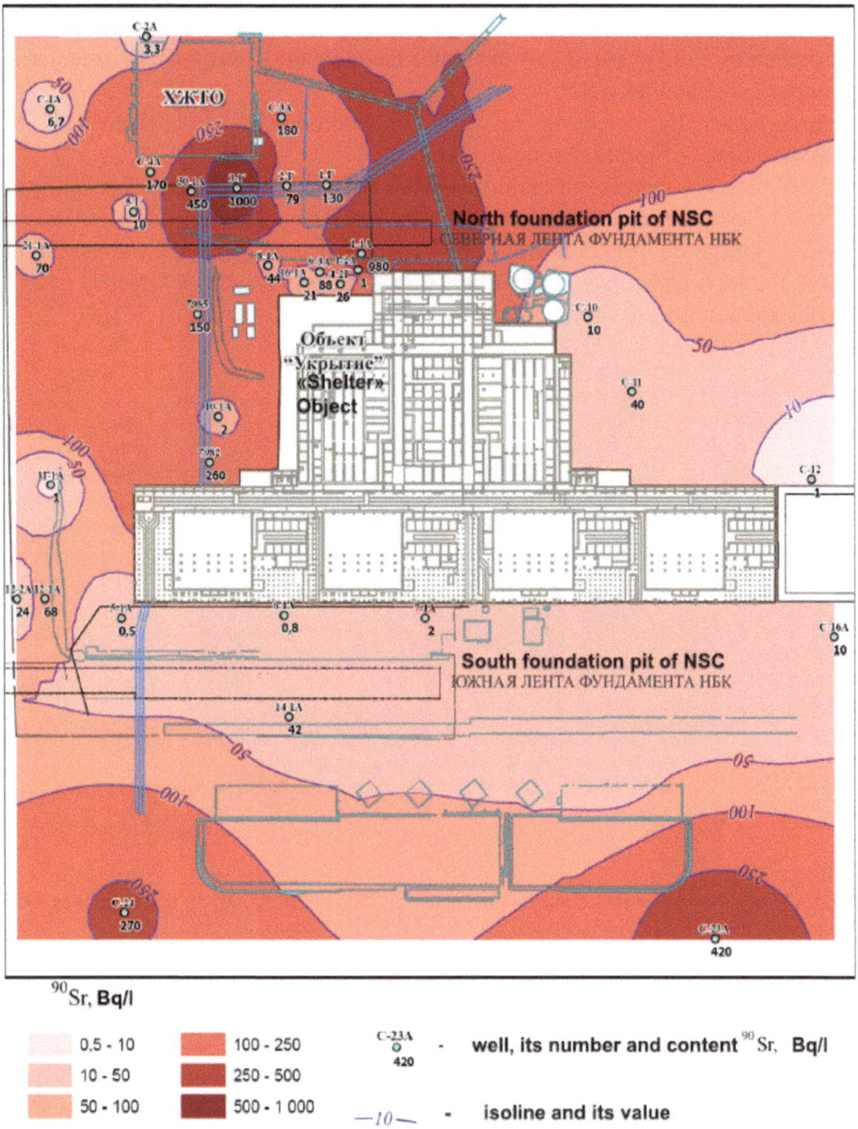

Fig.2. Map of ^{90}Sr volumetric activity distribution in groundwater within the area of "Shelter" object in 2013.

Increase of ^{90}Sr volumetric activities dynamics depends on the concentration in water of Ca^{2+}, K^{+}, Na^{+}, HCO_3^{2-} (Fig. 3) and also on pH index (Fig. 4).

Fig.3 contains a diagram that illustrates the relationship of major ions' concentrations and ^{90}Sr volumetric activity in the groundwater samples from C-24 well.

High values of ^{90}Sr volumetric activities in samples of groundwater from C-24 well can be caused by sorption processes between the solid and the liquid phases of waterbearing aquifer soil.

In these processes calcium ions most strongly reduce the absorption of radiostrontium by solid phase of soil and hence facilitate the growth of ^{90}Sr volumetric activities in the aquifer water.

In particular, by increasing the degree of desorption of earlier deposited ^{90}Sr from the surface of soil skeleton particles.

In its turn, the increase of major ions concentration in the water of unconfined aquifer is related most probably to the dissolution or alkali leach from concrete flooded with groundwaters from Cabling channel that connects the turbine building of 4th block with Open distribution system 750 and located 4-5 meters east of the observation well C-24.

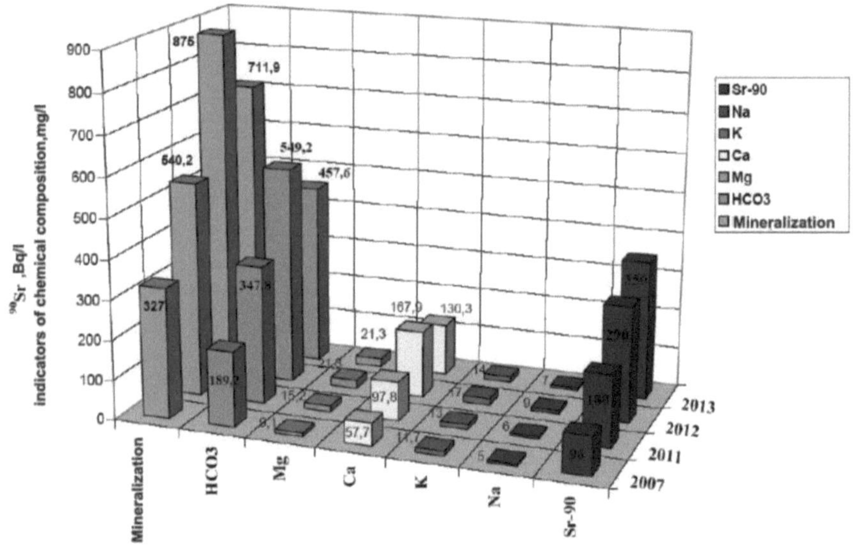

Fig.3. Changing of major ions concentration and ^{90}Sr volumetric activity in water samples from C-24 well.

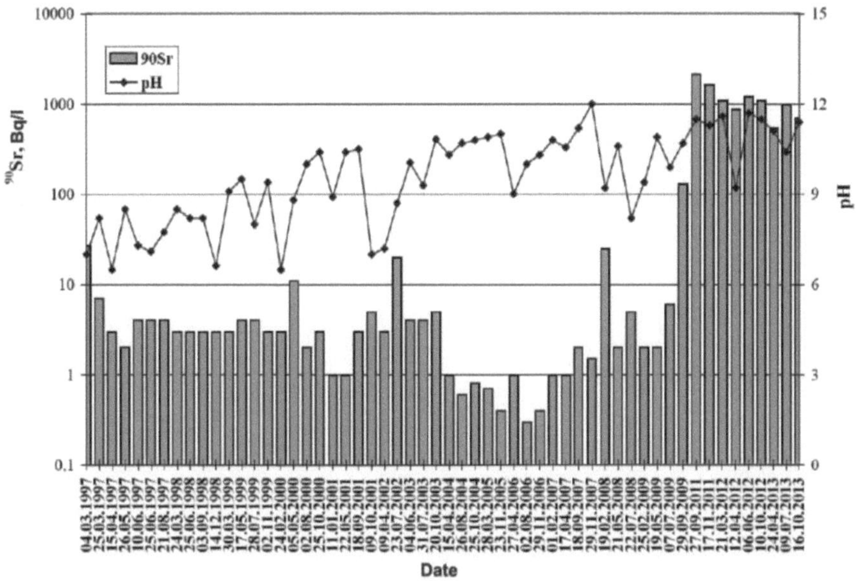

Fig.4. Dynamics of pH and ^{90}Sr volumetric activity in the groundwater samples from the 1-1A well

High concentrations of ^{90}Sr (130 -2100 Bq/l) in groundwaters' samples from the 1-1A well are preceded by increase in concentrations of ions (HCO_3^{2-} + CO_3^{2-}) and -K+, and also pH values (10-12).

High pH values, up to 12, probably are influenced by multiple factors, acting in a direction of ascending pH value:

- concentrations increase $(HCO_3 + CO_3)^{2-}$ as well as the alkali metals K+ and Na+;
- partial pressure reduction and consequently the concentration of CO_2;
- influence of groundwater radiolysis products generated by impact of ionizing radiation from radionuclides in high-active block-waters as they enter or migrate to the groundwaters.

It is known that when pH is 7-8, ^{90}Sr migration occurs mostly in ion-dispersible form. If pH increases from 8 to 11, migration capacity of ^{90}Sr reduces. Thus, considering also Ca^{2+} reduction in water, we can assume that in this period Ca^{2+} and ^{90}Sr settle in the sediment due to oversaturation with calcite.

For pH above 11, probably, ^{90}Sr is present in groundwater in the form of colloids or complex compounds, and migration of ^{90}Sr in unconfined aquifer from the place where block waters come is in the form of weakly exposed skeleton soil sorption.

Highest concentration of ^{137}Cs in the groundwater's samples are fixed in the range of 170-210 Bq/l. ^{137}Cs concentrations are strongly related to potassium and sodium in the groundwater.

Maximum ^3H concentrations correspond to the wells located in the immediate proximity of the places where block waters exit from the “Shelter” object premises (wells 4-3A, 16-2A, 1-3A, 9-3A, C-3Б) (Fig.5).

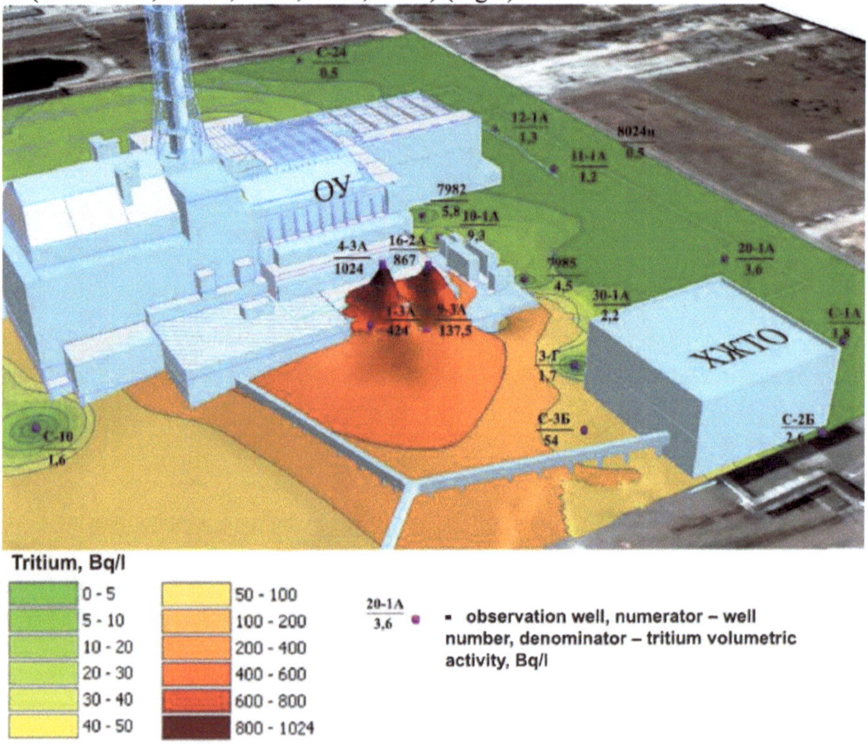

Fig.5. Map of tritium volumetric activity distribution in groundwater within the area of “Shelter” object in 2013.

To detect the places of block waters exit outside the "Shelter" object into the groundwater tracer method was applied. In December 2008, in block water of the Unit 001/3, that contained about 300 m^3 of high activity water, sodium bromide was added as a tracer to confirm the output of block water to the environment. Detection of bromine in wells, located downstream of groundwater, indicates the ways of block waters’ output. Inside the well 9-3A increase in the concentrations of bromine is observed, started from the middle of 2009 (Fig.6).

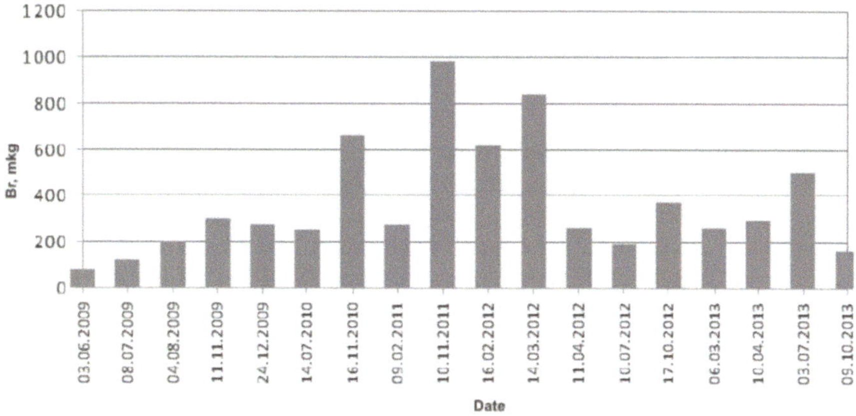

Fig.6. Dynamics of bromine concentration in the groundwater samples from the well 9-3A.

Radioactive contamination of soils is studied by means of gamma-logging and spectrometric gamma-logging of the wells and by the laboratory analysis of wells' core samples.

Maximum values of gamma-field (up to 10 R/h), according to the gamma - logging of wells, correspond to the interval, the so-called active layer at a depth of 1.7 to 10.1 m under the post-accident techno-genic and related to the pre-damage on the earth surface which had been exposed to radioactive materials during an accident.

Fig. 7 shows a map of the distribution of the specific ^{137}Cs activities in the soil of active layer at the site of excavation for the foundation pit of NSC following the gamma-logging results of the wells that were drilled for driving sheet piling wall recess.

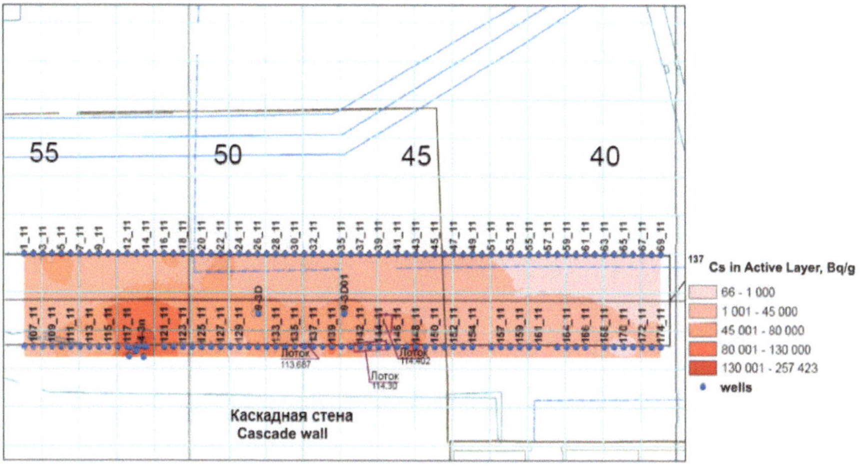

Fig.7. Forecast map of the ^{137}Cs specific activity distribution in the area 11 active layer.

As seen in Figure 7, fixed maximum specific activity of ^{137}Cs in the soil according to the gamma-ray logging reach values of $2.6*10^5$ kBq/kg.

This layer is characterized by a ground relating to high-level Radioactive Waste (RAW), with all the attendant problems in handling of the development of pits for the foundations of the New Safe Confinement (NSC).

As a means to facilitate a crunching it was designed a GIS project: "Radioactive contamination of soils and groundwater of ChNPP industrial site" that allowed of storing, processing and visualizing the obtained data.

For the purpose of forecasting of radiation situation changes into the soil and groundwater's implementation a specially developed three-dimensional model of radiation and hydroecological conditions in format MODFLOW is used, that allows of estimates of radionuclides spreading with the underground waters from the "Shelter" Object to places of unloading - into the Pripyat' riverbed. In addition, by means of this model, it was implemented an impact assessment on changes of levels and direction of groundwater flow from the construction of the pile foundation of NSC or in case of decommissioning of the pond - the cooler.

Investigation into the Transport of ^{238}U-series Radionuclides in Soils to Plants

Danyl Pérez-Sánchez[1], **Mike Thorne**[2]

[1]Radiological Protection of the Public and Environment, CIEMAT, Spain
[2]Mike Thorne and Associates Limited, County Durham, UK

Abstract. The migration of radionuclides in the ^{238}U-series in soils and their uptake by plants is of interest in various contexts, including the geological disposal of radioactive waste and the remediation of former sites of uranium mining and milling. In order to investigate the likely patterns of behaviour of U-series radionuclides being transported upward through the soil column, a detailed soil-plant model originally developed for studying the behaviour of radionuclides in soil-plant systems has been adapted to make it applicable to the U-series. In this paper, a brief account is given of the mathematical basis of the adapted model and key parameters are identified and then estimated. Hydrological data appropriate to a situation in which upwelling groundwater interacts with meteoric water in the soil zone are also selected, using meteorological data from a former uranium mining site in Spain as a key input. On this basis, a reference calculation are defined and run for a unit input rate of either U or Ra at the base of the soil column and for a simulation period of 5,000 years. Results for this case are then used to comment on potential long-term patterns of behaviour of U-series radionuclides in soils and their uptake by plants.

Introduction

The investigation of migration of radionuclides in the U decay series in soils and their uptake by plants is of interest in various contexts, including the disposal of radioactive waste and the remediation of former sites of uranium mining and milling and NORM wastes situations. However, there is only a limited literature on such upward transport. Furthermore, information on the sorption of these radionuclides to soils and their availability for plant uptake has recently been reviewed (Mitchell et al., 2013), providing an updated source of information relevant to modelling the transport and uptake of U-series radionuclides. Also in a previous paper (Pérez-Sánchez and Thorne, 2014); a mathematical model for the behaviour of ^{238}U-series in soils and plants was described for dose assessment.

Here, we bring together those two strands of work to describe briefly a new mathematical model to investigate the upward behaviour of ^{238}U-series radionu-

B.J. Merkel and A. Arab (eds.), *Uranium – Past and Future Challenges*,
DOI 10.1007/978-3-319-11059-2_76

clides in soils and their uptake by plants. Illustrative calculations undertaken using this new model are also described. Currently, the model is being applied in studies relating to the near-surface disposal of low-level radioactive wastes, taking into account both the transport of ^{238}U-series radionuclides in groundwater and potential human intrusion resulting in the contamination of surface soils.

Mathematical Structure and Parameters of the Model

The mathematical structure of the model is essentially identical to that described previously in Pérez-Sánchez and Thorne, 2014. That model was implemented in the AMBER 5.4 simulation system (Quintessa, 2012) that provides a simple method for encoding decay chains of any length. This meant that extending from the single-member decay to the full decay chain for U. Shorter-lived progeny of these radionuclides, notably the several generations of short-lived progeny of Rn, are considered to be present in secular equilibrium with their most immediate ancestor in the modelled chain.

The quantities that were generalized to be contaminant dependent were; Distribution coefficients (Kd values); Volatilisation rates from soil; Plant uptake rates from soil; Volatilisation rates from plants and Mineralisation rates of organic matter in soil.

Distribution coefficients are defined in terms of the maximum and minimum Kd values applicable to each contaminant and interpolated between them using the water content of each layer. The only restriction is that the Kd behaviour occurs over the same vertical extent for all members of the decay series. The hydrology model calculates Kd values on a normalised range of [0, 1] based on the water content of the soil. The actual Kd values are estimated in the transport model by scaling these normalised values to the specified Kd for each radionuclide in the decay chain.

For volatilisation from soil, there are three volatilisation rates (above, within and below the capillary fringe) with linear interpolation between those regions (Pérez-Sánchez and Thorne, 2014). The three rates have been made contaminant dependent (they can, therefore, all be set high for Rn to simulate its loss from soil) and are set to zero for all U-series radionuclides except for Rn, because a rapid losses from the unsaturated zone are expected, due to diffusion and pressure pumping, but little loss is expected from the saturated zone.

The volatilisation rate from plants has been generalised to a single value for each contaminant. Similarly, the mineralisation rate of organic matter in soil (or more strictly the rate of release of a contaminant from organic matter back to soil solution) has been made contaminant dependent. The volatilisation rates from plants are taken as zero for all members of U chain, except for Rn to ensure that ^{222}Rn produced in plant tissues is lost from them on a timescale of a few hours. This rate is not well known, but the value adopted is thought likely to overestimate ^{222}Rn retention in plant tissues, bearing in mind the rapidity of loss observed from mammalian soft tissues (ICRP, 1979).

The meteorological, chemical, physical and radiological data used are those to the Los Ratones former uranium mine, which is located in Extremadura region in the south-west of Spain. The mine was in use from 1960 to 1974, and restoration work on the tailings pile was undertaken in 1998 and 1999. The site has been subject to environmental sampling and characterisation during and after completion of the site remediation process and the data relating to the site have been obtained over a series of discrete sampling campaigns since the site remediation was undertaken (Vera Tomé et al. 2002, 2003). The mean monthly meteorological data values required for the model was adopted as standard annual cycle repeated for each year of the simulations and the effects of inter-annual variation were not taken into account.

Plants were taken to have a growth period starting at the beginning of April and finishing in September. The standing biomass, biomass production rates and removal rates adopted are listed in (Pérez-Sánchez and Thorne, 2014). The harvesting occurs at the beginning of October and is set to match annual production and the standing biomass after harvest represents roots and stubble, degrade to soil organic matter over the winter months.

Results and Discussions

Two alternative source terms were used. These were a flux of 1 mol/m^{2}y of ^{238}U to the base of the soil column or 1 mol/ m^{2}y of ^{226}Ra to the base of the soil column. It is emphasised that the inputs are treated by the model as tracers that do not perturb the system, so the results scale proportionately with the input flux. A value of 1 mol/m^{2}y was selected solely for convenience of normalisation. The ^{238}U source was primarily of interest in the context of uranium migration through the soil column, as there is little in-growth of long-lived progeny of ^{238}U within the 5,000 year simulation period. In contrast, the ^{226}Ra case is associated both with substantial decay of the ^{226}Ra and substantial in-growth of its progeny.

For the case of ^{238}U calculations, there is little in-growth of progeny, so attention is appropriately focused on the behaviour of the ^{238}U. The upward migration of the ^{238}U through the soil column is illustrated in Fig. 1. As expected, the concentration in soil layer 10 increases most rapidly and reaches the highest concentration. Further up the column, the concentration increases more slowly and the maximum concentration that is reached is progressively lower. However, even after 5000 years in a soil system with a strong advective upward flow of groundwater, there remains a strong concentration gradient between the uppermost soil layer and the lowermost, with ^{238}U concentrations in those two layers differing by almost four orders of magnitude. This, in itself, is sufficient to demonstrate the difficulty of using one or two compartments to represent deep soil systems, as is often done in assessment models.

The long timescale for migration is readily explained by noting that the upward flux of groundwater is typically 0.1 m/y and that most of the migration takes place through a zone that remains fully saturated throughout the year. As the water-filled porosity in saturated conditions is 0.5, the upward velocity of the groundwater is 0.2 m/y. However, in the fully saturated zone, the ^{238}U is taken to have a Kd value of 1.0 m^3/kg. Hence the vertical velocity of ^{238}U is 7.5 10^{-5} m/y and the characteristic time to cross a layer of 0.2 m thickness is 2651 years. Thus, as expected, most of the ^{238}U is localised in the lowermost three soil layers after 5000 years of simulation. To some degree, migration further up the soil column reflects the numerical dispersion that arises from discretising the soil column into layers of finite thickness rather than treating it as a continuous system.

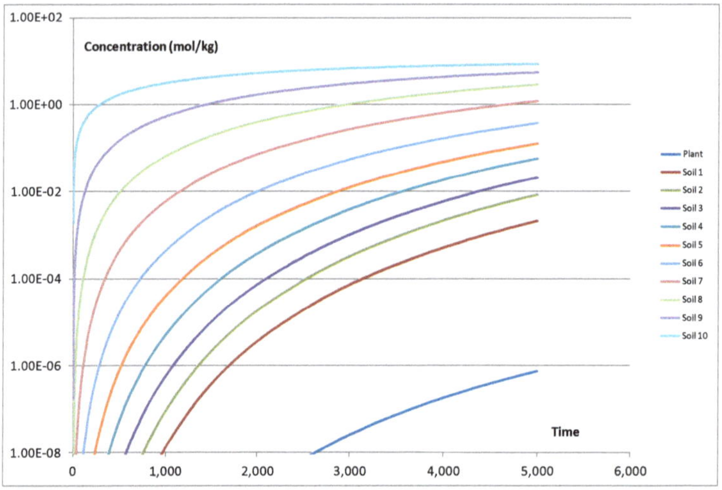

Fig.1. Concentrations of ^{238}U in Soils and Plants in the Reference Spanish Case

Over the 5000 year simulation period, the total amount of ^{238}U entering the soil column is 5000 moles. As ^{238}U is very long lived, radioactive decay can be neglected, so the total amount present in the soil column plus plants at 5000 years should also be 5000 moles. Based on the concentrations shown in Fig. 1 and using a dry bulk density of soil of 1.325 kg/ m^{-3} together with layer thicknesses of 0.2 m, the total amount present is determined to be 5000.49 moles. This confirms the accuracy of the numerical solution method over the extended simulation period adopted.

Although relatively little ^{238}U reaches the uppermost soil layer, there is sufficient to compute the plant:soil concentration ratio (i.e. the ratio of the plant concentration expressed on a dry mass basis to the concentration in the top 0.2 m of soil expressed on a dry mass basis). Results are calculated at the end of June, i.e.

part way through the growing season. Variations in the concentration ratio during the growing season are discussed elsewhere using a case in which the contamination is by irrigation (Pérez-Sánchez and Thorne, 2014).

The lower concentration ratios recorded in the first few years of the simulation reflect the fact that the plant concentration lags the soil concentration (being determined by uptake throughout the preceding part of the growth period). As the soil concentration is increasing with time, this means that the plant 'sees' a somewhat lower soil concentration than that existing at the time the concentration ratio is calculated. Beyond about ten years, this lag effect becomes negligible, as the fractional rate at which the soil concentration changes decreases. Thus, a constant ratio of 3.6 10^{-4} is established. This is about a factor of fifty higher than the recorded value. This arises because the soil is saturated to the surface in the early part of the growing period and this strongly suppresses uranium uptake by the roots. For comparison, in year 90 of the simulation, the concentration ratio increases from 3.6 10^{-4} at the end of June to 1.46 10^{-2} at the end of September, as the soil dries from a moisture content of 0.5 to a moisture content of 0.1.

In the case of ^{226}R, as was found with ^{238}U, concentrations of ^{226}Ra increase rapidly in the lowermost soil layer, but increase successively more slowly in the higher soil layers (Fig.2).

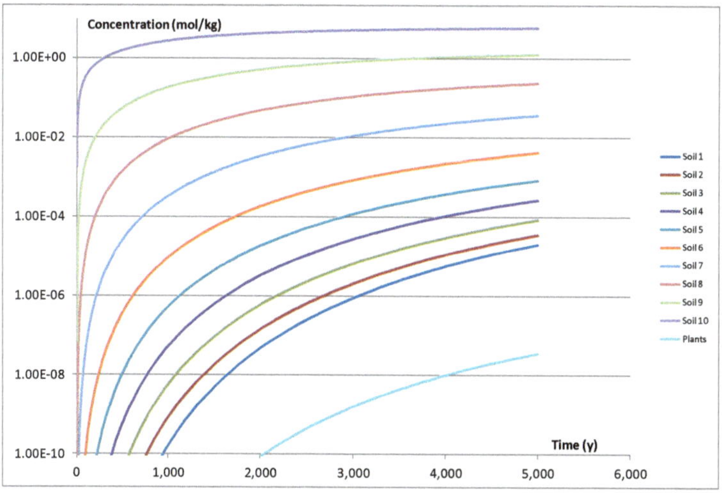

Fig.2. Concentrations of Ra in Soils and Plants in the Reference Spanish Case

Also, by the end of the simulation period the concentration of ^{226}Ra in the uppermost soil layer is more than four orders of magnitude lower than the concentration in the lowermost soil layer after 5000 years of simulation, the total inventory of ^{226}Ra present in the soil is 2040 moles. The total input is 5000 moles, but ^{226}Ra has a radioactive half life of 1600 years, so the decay corrected inventory is 2043 moles, in good agreement with the simulation.

For ^{226}Ra, the build-up of its shorter-lived progeny is of particular interest. It is immediately clear that the build up of the progeny follows similar kinetics to the build-up of ^{226}Ra. This can be brought out further by plotting the mass ratios of the four radionuclides in Fig. 3.

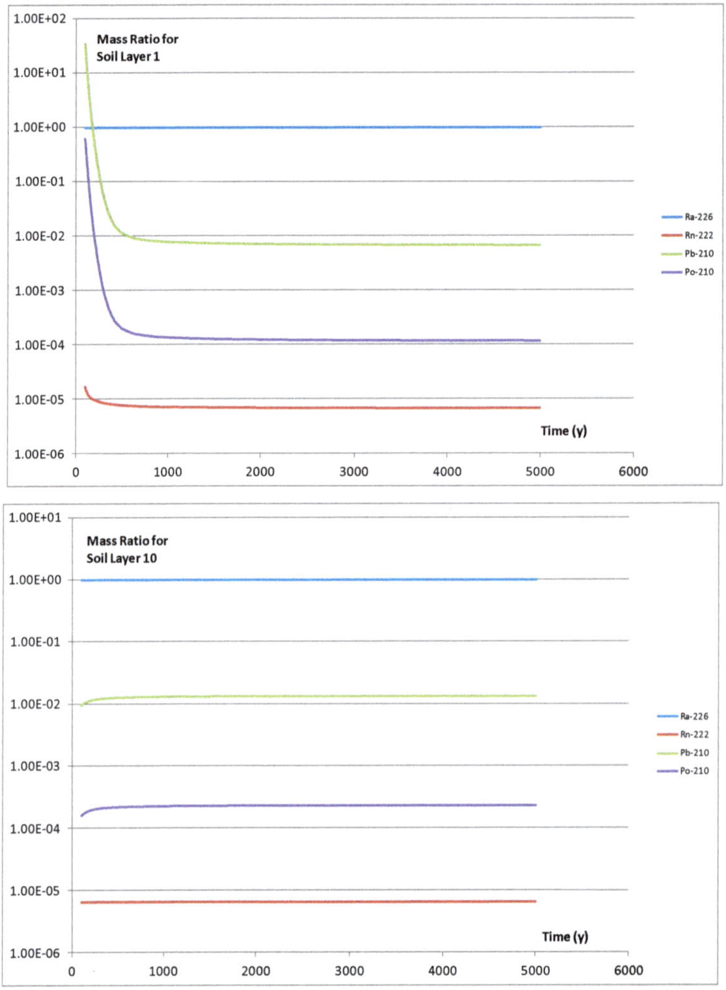

Fig.3. Mass Ratios in two Soil Layers for the Reference Case in which 226Ra enters the Base of the Soil

It will be observed that the mass ratios tend to asymptotic constant values at long times. However, in the cases of soil layers 1, these constant values are approached from above, whereas for soil layer 1, the constant values are approached much more rapidly and from below. The reason for this difference is that Rn is transported rapidly up the soil column giving rise to ^{210}Pb and ^{210}Po in the upper soil layers. Conversely, the ^{222}Rn concentration is somewhat decreased in the lowermost soil layer due to this rapid upward migration. This effect is of greatest importance at early times when little ^{226}Ra has been transported into the upper soil layers from the lowermost layer, i.e. the ^{222}Rn, ^{210}Pb and ^{210}Po in the upper soil layers are unsupported. At late times, ^{226}Ra migrates up the soil column and the ^{226}Ra in each soil layer becomes the predominant source of Rn in that layer.

As expected, secular equilibrium is closely approached in soil layer 10, though the values remain slightly low because of the upward migration of ^{222}Rn. In soil layer 1, the values are depressed below secular equilibrium, due to loss of ^{222}Rn generated in that layer to the overlying atmosphere. Note that ^{222}Rn transport in the saturated zone is solely in groundwater, whereas in the unsaturated zone it is mainly transported in the gas phase, as parameterised in the volatilisation model.

In the case of ^{226}Ra and progeny the concentration rates rapidly reach a long-term plateau. The concentration ratios at 5000 years are found to be 1.81 10^{-3}, 1.08 10^{-4}, 5.38 10^{-4} and 2.52 10^{-4} for ^{226}Ra, ^{222}Rn, ^{210}Pb and ^{210}Po, respectively. The non-zero value for ^{222}Rn arises due to production in the plant from ^{226}Ra present in plant tissues. Although this ^{222}Rn is rapidly lost by volatilisation, the process is not assumed to be instantaneous. The concentration ratios are lower than the ratio values because root uptake is suppressed due to the waterlogged condition of the soil early in the growing season, as discussed for ^{238}U.

Conclusions

The simulations reported in this study illustrate the wide variety of behaviour that can be exhibited by radionuclides in the ^{238}U decay chain in soils, even when the source term is limited to being a constant flux of either ^{238}U or ^{226}Ra. Hydrological conditions are a primary factor, both in respect of the overall advective flow deeper in the soil, which controls the rate of upward migration, and in the influence of seasonally changing flow directions closer to the soil surface, which can result in the accumulation of radionuclides at specific depths irrespective of changes in sorption between the oxic and anoxic regions of the soil.

However, such changes in sorption can also be significant in controlling the degree of accumulation that occurs, so there is a need to take both factors into account simultaneously when evaluating radionuclide transport in soils. This importance of seasonally varying factors in controlling radionuclide transport in soils even in very long-term simulations is a strong argument against the use of annually averaged parameters in long-term assessment models, unless those annually averaged parameters have been suitably calibrated against a more detailed model in which seasonal variations are taken into account.

The simulations reported here used a single annual hydrological cycle that was repeated throughout the simulation period. In practice, the hydrological cycle will vary somewhat from year-to-year. In particular, the depth of the water table will vary both seasonally and inter-annually. The effect of inter-annual variation may be to smear out to some degree the sharp accumulation fronts simulated in this study. The effects of such inter-annual variations should be a priority for further investigation.

With a water table that fluctuates from a substantial depth in soil to the surface soil layer, the seasonality of such variations in relation to the period of plant growth has a major impact on the degree of uptake of radionuclides by plant roots. In this study, it has been shown that high water content in the rooting zone early in the growth season can strongly suppress plant uptake of radionuclides, with the uptake occurring mainly later in the growth season as the soil dries out. It is emphasised that this result arises as a result of the particular way in which the model characterises the effects of soil water content on plant root activity. In practice, the effects of water content on root uptake are likely to be affected by the type of plant under consideration. Again, this is a topic that should be a priority for further investigation.

In long-term safety assessment studies it has sometimes been the practice to model the transport of ^{226}Ra in soil, but to assume that both ^{210}Pb and ^{210}Po can be treated as being present in secular equilibrium with the ^{226}Ra. This study illustrates that this simplification is not always appropriate. Where geochemical conditions are such that the ^{226}Ra migrates upward in the soil column faster than ^{210}Pb and ^{210}Po, disequilibrium is not a significant issue, as the ^{226}Ra supports ^{210}Pb and ^{210}Po at concentrations somewhat below those estimated on the basis of assumed secular equilibrium. This means that models based on the assumption of secular equilibrium should not be employed without a careful consideration of the hydrological and hydrochemical situation of interest.

References

ICRP (1979). ICRP Publication 30, Limits for Intakes of Radionuclides by Workers – Part 1, Annals of the ICRP, 2(3/4).

Mitchell N, Pérez-Sánchez D and Thorne M. C (2013). A review of the behaviour of U-238 series radionuclides in soils and plants, Journal of Radiological Protection, 33, R17-R48.

Pérez-Sánchez D and Thorne M. C (2014). Modelling the behaviour of uranium-series radionuclides in soils and plants taking into account seasonal variations in soil hydrology. Journal of Environmental Radioactivity, 131, 19-30.

Quintessa (2010). AMBER 5.4 Reference Guide,Version 5.4, QE-AMBER-1, version 5.4, December 2010. Quintessa Limited.

Vera Tomé F, Blanco Rodríguez P and Lozano J. C (2002). Distribution and mobilization of U, Th and Ra in the plant–soil compartments of a mineralised uranium area in south-west Spain. Journal of Environmental Radioactivity, 59, 41-60.

Vera Tomé F, Blanco Rodríguez P and Lozano J. C (2003). Soil-to-plant transfer factors for natural radionuclides and stable elements in a Mediterranean area. Journal of Environmental Radioactivity, 65, 161-175.

Development of a Biochemical Sensor for the Determination of Uranium in Aqueous Solutions

Thomas Streil[1], Broder J. Merkel[2], Corina Unger[3], Bianca Störr[4]

[1,3]SARAD GmbH, Wiesbadener Str. 10, 01159 Dresden, Germany
[2,4]Technische Universität Bergakademie Freiberg, Faculty of Geosciences, Geoengineering and Mining, Institute of Geology, Gustav Zeuner Str.12, 09599 Freiberg, Germany

Abstract. Currently there are only very expensive and time-consuming analytical methods for the determination of uranium in water such as the ICP-MS, which is solely useable in laboratories. Novel measurement systems based on biochemical reactions, which are made visible with the help of gold nanoparticles by color changing or fluorescence, shall represent an alternative. Within a project founded by the German Federal Ministry of Economics and Technology the SARAD GmbH worked together with the Technical University Bergakademie Freiberg to develop such systems.

Introduction

The uranyl ion (UO_2^{2+}) is the most stable chemical form of uranium in water. We worked and are still working on two different measurement systems to determine the uranyl concentration in aqueous solutions: on the one hand on a colorimetric system based on DNA and gold nanoparticles (AuNPs), on the other hand on a fluorescent measurement system based on DNA and fluorophores. It is possible to quantitatively detect uranyl contents from 1nM on with them. These systems consist of two parts: One is the biochemical sensor, which carries out a biochemical reaction in the presence of uranyl. The other is a device, which can interpret this reaction to the extent that the uranyl concentration of water samples can be determined without the usage of highly expensive and non-portable instruments.

B.J. Merkel and A. Arab (eds.), *Uranium – Past and Future Challenges*,
DOI 10.1007/978-3-319-11059-2_77

The sensors are based on irreversible reactions; therefore each sensor solution can only be used once.

Colorimetric Sensor

In the past years the aggregation of gold nanoparticles through DNA received more and more attention, mainly because DNA molecules can easily be synthesized and modified chemically. On a colorimetric system based on so called DNAzymes and gold nanoparticles has been worked on for several years (Lee et al. 2008).

The principle is that so called DNAzymes hybridized with gold nanoparticles are splitted in the presence of uranyl, which results in a disassembly of those gold nanoparticles and a color change of the sensor. The measure of the change is supposed to be proportional to the uranyl concentration. Desoxiribozyme or DNA enzymes (DNAzyme) are biocatalysators based on DNA. Biochemical studies show that for a efficient catalyses the value of the metal ions has to be two. That implies that nucleic-acid enzymes create specific sites for metal bonds. Uranyl is such a two-value metal ion. Biocatalysators accelerate biochemical reactions mainly by decreasing the activation energy of the reaction. During an enzymatic reaction the substrate is converted not the enzyme. A DNAzyme specific for uranyl represents a double-stranded DNA consisting of an uranyl-specific enzyme strand and the corresponding substrate strand. The substrate strand is elongated by two small oligonucleotides to which ends a gold nanoparticle is attached by a sulfhydryl group. In this aggregated state the sensor's color is violet. It absorbs primarily green and red light, as is seen in Fig.1. The peaks' maxima of its extinction curve are at 525 nm and 665 nm.

In the presence of uranyl the substrate strand will be cleaved at its ribonucleotide adenosine position in the middle. This position was inserted because of its higher susceptibility to hydrolytic cleavages. The cleavage results in a disassembly of the gold nanoparticles because of the substrate strand's separation. Depending on the amount of uranyl and therefore disassembly of the gold nanoparticles the sensor's color changes into a red color range. The extinction curve of the dissembled gold nanoparticles shows clearly one increased peak with its maximum at 525 nm and a decreased extinction at 665 nm (Fig. 1). That means a great amount of green light is absorbed, red light gets reflected. In conclusion the uranyl concentration within the sensor can be determined through the ratio of the extinction at 525 nm to the extinction at 665 nm. The higher this ratio the higher the uranyl concentration.

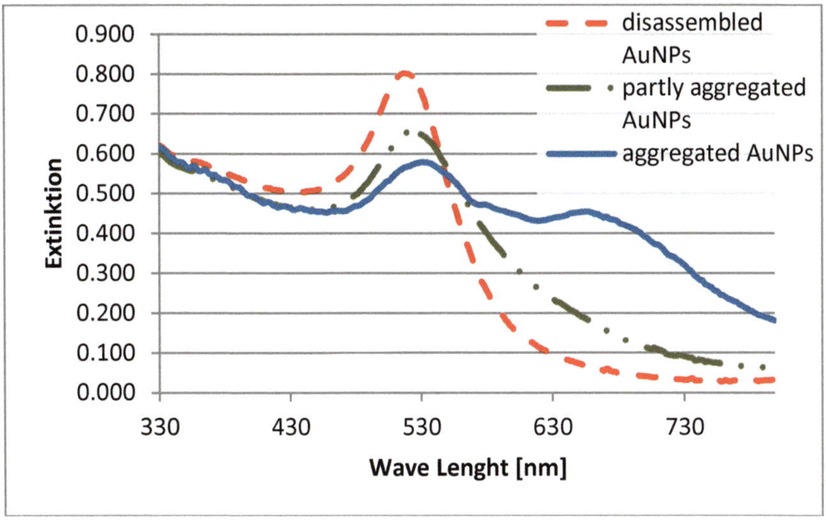

Fig.1. Extinction spectra of disassembled, partly aggregated and aggregates gold nanoparticles

The detection range of this sensor should have been between 50 nM and 2 μM (Lee et al. 2006). Unfortunately the tests performed to this day showed that no reproducible results can be achieved with this sensor. Measurements with the same AuNPs-DNA-aggregates showed very different results. Different angles like the addition of surfactants to attempt to control the aggregation of the gold nanoparticles were tried to improve the colorimetric sensor but to no vain.

Nevertheless, parallel to the work on the sensor a low cost, simple and portable system for its analysis was developed (Fig. 2). The main components are two LEDs which emit primary light within the relevant wavelengths (525 nm and 665 nm). Photo resistors are used as light detectors with which the extinction of the LEDs' light was measured. The LED's focused light rays are send through a sample one by one. This sample is placed inside a cuvette. As blank value the sample to be examined without the senor solution is used. This ensures that distortions of the measurement value by color impurities of the sample can be ruled out.

Test carried out with a prototype of the analysis system and first sensor solutions showed its suitability for the measurement purpose. Color changes of the sensor could be detected.

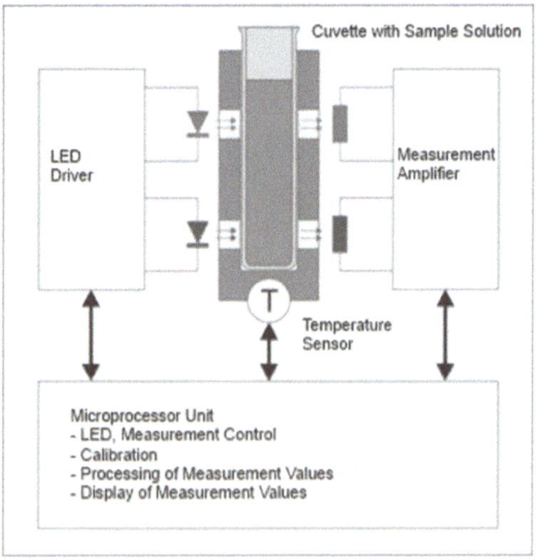

Fig.2. Schematic structure of the analysis systems for the colorimetric sensor

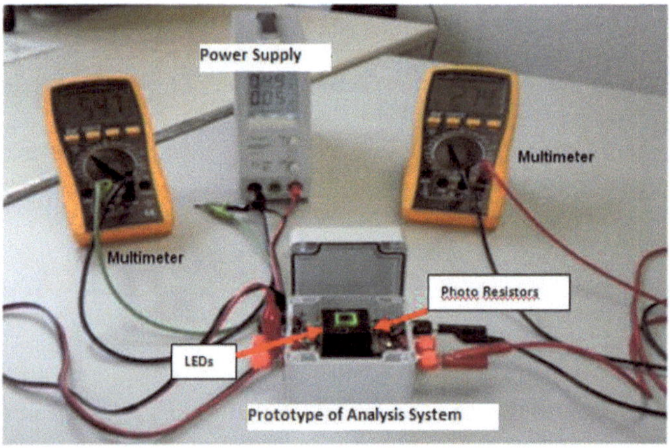

Fig.3. Measurement arrangement with prototype for first tests with the colorimetric sensor

Fluorescent Sensor

The fluorescent sensor was created on the basis of a publication by Liu et al. 2006. This sensor is also utilizes the reaction of DNAzymes in the presence of uranyl. But in this case the DNAzymes, which react under the presence of uranyl, are hybridized with fluorophores and quenchers. Fluorophores can be excited by light of a certain wavelength, which in turn will cause the emission of light (fluorescence) of a slightly higher wavelength. On one end of the enzyme strand and the opposite end of the substrate strand quenchers are hybridized. A fluorophore is attached to the other end of the substrate strand. In analogy to the colorimetric sensor, in the absence of uranyl the fluorescence is quenched because of the close proximity of fluorophore and quencher. In the presence of uranyl the substrate strand will be cleaved at the ribonucleotide adenosine position, which causes a release of the part of the substrate strand with the fluorophore. The fluorescence increases because it will no longer be quenched. The increased intensity of the emitted fluorescence is supposed to be proportional to the amount of uranyl concentration inside the sensor solution. To determine this concentration the ratio of the fluorescence intensity with uranyl to the intensity without uranyl (F/F_0) will be calculated.

The developed evaluation device for this sensor is very similar to the one for the colorimetric sensor. To excite the fluorescent solution light in the wavelength range of 490 nm is needed. For this we also used a LED. To prevent an overlap of light of the LED and the fluorescent signal, the latter is measured perpendicular to the input direction of the LED light (Fig. 4). A photo resistor serves as light detector once again.

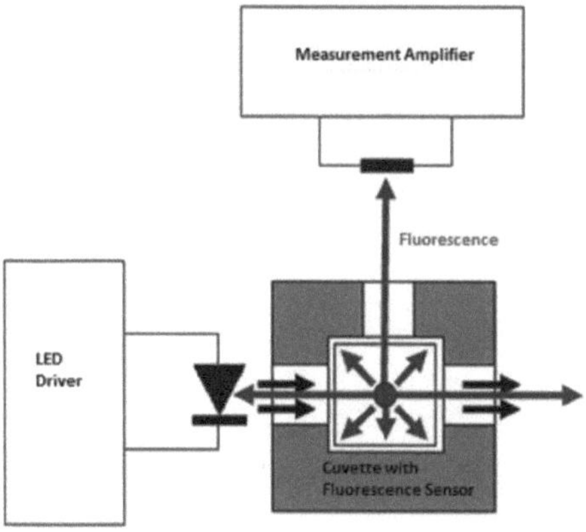

Fig.4. Schematic structure of the analysis systems for the fluorescent sensor

To minimize scattering light a narrow-band interference filter can be placed in front of the detector. The tests already performed with the prototype (Fig. 5) and the fluorescent sensor show the suitability of the developed analysis system for the measurement purpose. The fluorescence in different intensities generated within the sensor can be determined with it. In this case the blank value is the fluorescence of the sensor itself without any other solutions. Not all the fluorophore's fluorescence can be eliminated by the quencher. The sensor will always have an offset.

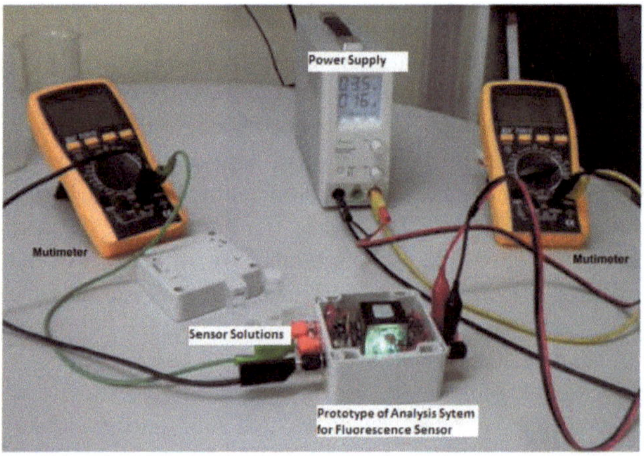

Fig.5. Measurement arrangement with prototype for first tests with the fluorescent sensor

The detection range of this sensor is supposed to be between 45 pM and 400 nM (Liu et al. 2006). However our first tests indicate that no uranyl determination under 5 nM (1,4 µg/l) will be possible. These first tests also point to reproducible results. A reaction time of 8 minutes in sufficient for this sensor as shown in Fig. 6.

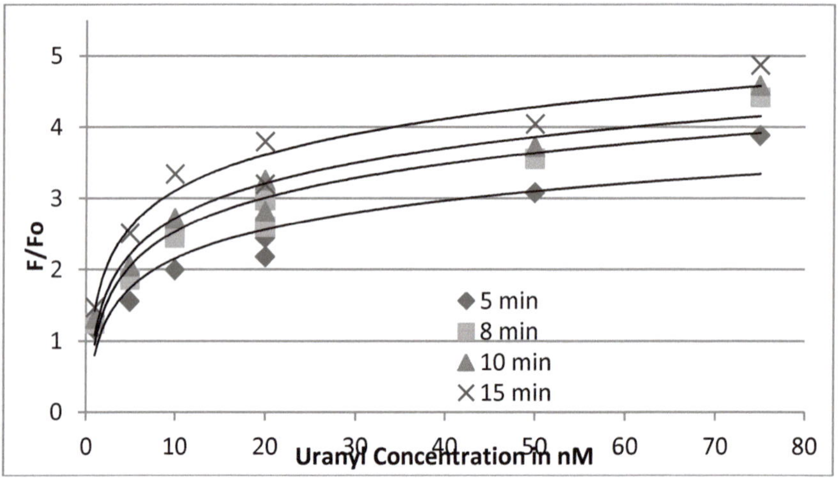

Fig.6. Schematic structure of the analysis systems for the fluorescent sensor

Conclusions and Outlook

For the time being, the focus is put on the fluorescent sensor. It including its analysis system is still in the first test phase. The results of first measurements were very promising but there is still the need to verify they suitability for the measurement purpose and stability completely by long term tests. Other parameters like the influence of temperature fluctuations, interfering ions or metal ions have to be studied. The sensor has to be tested under real conditions. Its life time has to be determined.

After that the possibility to immobilize the fluorescent sensor on solid surfaces like glass or cellulose by gold nanoparticles will be studied. For that the quencher has to be replaced by gold nanoparticles that are also able to annihilate the fluorescence.

The work on the colorimetric sensor is not finished. There will be attempts to solve the problems which occurred in the development process, because the detection principle seems to be very promising.

References

Jung Heon Lee, Zidong Wang, Juewen Liu and Yi Lu (2008): Highly Sensitive and Selective Colorimetric Sensors for Uranyl (UO22+): Development and Comparison of Labeled and Label-Free DNAzyme-Gold Nanoparticle Systems. Journal of the American Chemical Society 130 (43): 14217-14226

Juewen Liu, Andrea K. Brown, Xiangli Meng, Donald M. Cropek, Jonathan D. Istok, David B. Watson and Yi Lu (2006) A catalytic beacon sensor for uranium with parts-per-trillion sensitivity and millionfold selectivity. Proceedings of the National Academy of Sciences 104 (7): 2056-2061

I found out that the concentrations in both samples are the same.
$c_A = c_B$
Wait...
This way it looks more scientific.
$\int c_A^{\varsigma} = \int \left(\sum_{i=1}^{n} c_{B,i} \right) \cdot \cosh(0)$
$\varsigma = 2 \cdot \frac{\sinh(5)}{e^5 - e^{-5}}$
You're truly knowledgable!

Modelling of U series nuclides disequilibria – presentation a modelling tool

Juhani Suksi[1]

[1]University of Helsinki, Department of Chemistry, Laboratory of Radiochemistry

Abstract. U series radioactive disequilibrium has been used to obtain time constrains for geological and geochemical processes and to assess conditions under which disequilibrium formed. Recently developed DECSERVIS-2 simulation model offers an opportunity to study the behaviour of natural radioactive decay chains as a function of time under variable nuclide mass flows into and out of the system which greatly improves interpretation radioactive disequilibria and liable processes. In the present paper DECSERVIS-2 simulation model is introduced and its use as an interpretation tool demonstrated with selected simulation examples.

Introduction

U series disequilibrium always reflects water-rock interaction that has caused mass flow of U. Traditionally U series disequilibrium refers to a situation in the ^{238}U decay chain when its longest-lived nuclides ^{238}U, ^{234}U and ^{230}Th (Fig. 1) have different activity, i.e. activity ratio of the nuclides deviates from unity. With equal activities decay series is said to be in radioactive equilibrium. Complete equilibrium occurs when no mass flow of nuclides has occurred for two million years. Main reason for disequilibrium is different chemical properties of the nuclides which give them different mobility (Gascoyne 1992).

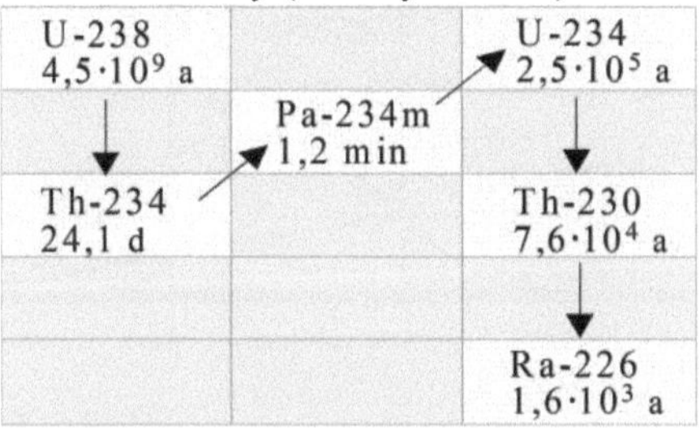

Fig. 1. Beginning of the ^{238}U decay chain with the longest-lived nuclides ^{238}U, ^{234}U and ^{230}Th. The half-lives are ideal to study U mass flow events occurred within the last glaciation cycle. ^{234}U is radioactive decay product and is therefore more mobile than ^{238}U

B.J. Merkel and A. Arab (eds.), *Uranium – Past and Future Challenges*,
DOI 10.1007/978-3-319-11059-2_78

U mass flow and time modify nuclides' activity ratios. Activity ratio can be above or below equilibrium value unity depending on U mass flow direction and time elapsed after and duration of mass flow. When the $^{230}Th/^{238}U$ and $^{230}Th/^{234}U$ activity ratios are unity, there has not been U mass flow for 450000 years which is the time limit to detect meaningful disequilibria. In geological dating and studying radionuclide transport activity ratios are analysed. Above activity ratios reveal immediately whether U has mobilised or accumulated in the system which allows development of realistic conceptual models for the more detailed study. With the DECSERVIS-2 simulation model (Azzam et al. 2009) decay chain evolutionary pathways to analysed activity ratios can be constructed and irrational pathways excluded. Decay chain evolution can be simulated under sudden, continuous and successive mass flow events into and out from the system and freely determined time intervals. In this paper the capacity of DECSERVIS-2 to study U decay series disequilibria is demonstrated.

On the basics of interpretation

Uranium is redox-sensitive element and can change oxidation state if conditions change which affects its mobility. Uranium (U) has two geochemically relevant oxidation states U(IV) and U(VI), the former being low solubility form of U and the latter the mobile form (Langmuir 1978; 1997). Thorium (Th) is practically immobile (Langmuir and Herman 1980) and can thus serve as a static marker for mobile U isotopes establishing fundamental basis for interpretation. Osmond et al. (1983) constructed conceptual model which explains formation of disequilibrium. The model describes evolution of ^{238}U, ^{234}U and ^{230}Th activities in redox-frontal system under steady-state flow field when water passes redox-front (Fig. 2).

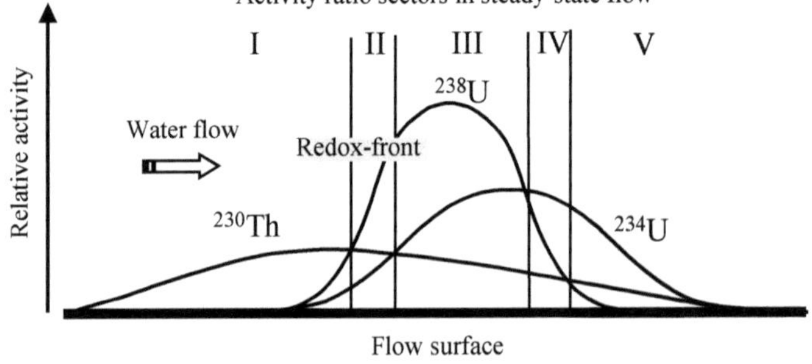

Fig. 2. Distribution of ^{238}U, ^{234}U and ^{230}Th activities on flow surface when water passes redox-front (Osmond et al. 1983). U isotopes are leached before redox-front leaving immobile ^{230}Th behind. Due to enhanced mobility of ^{234}U it moves ahead of ^{238}U. Deposition of U occurs in downstream of redox-front. Activity ratios of the nuclides define sectors from upstream (I) to downstream (V)

Nuclide activity curves formed activity ratios are seen to divide into five sectors. Sector I shows activity ratios formed in upstream side of the redox-front where U is dissolved leaving immobile ^{230}Th behind (^{230}Th/^{234}U>1 and ^{234}U/^{238}U<1). Due to the preceding radioactive chain decay (see Fig. 1) mobility of ^{234}U is enhanced which shifts its activity curve to downstream direction. The decay of ^{234}U to ^{230}Th causes stretching of the ^{230}Th peak as the ^{238}U and ^{234}U peaks advance. Of course, displacements and shapes of the ^{238}U and ^{234}U peaks and their displacement to the ^{230}Th peak are controlled by water flow rate. To proceed in interpretation activity ratios are displayed on diagram form where the ^{234}U/^{238}U ratio is plotted against the ^{230}Th/^{238}U ratio (Fig. 3). Horizontal line at ^{234}U/^{238}U=1.0 separates upstream activity ratios from downstream activity ratios. At the same time the diagonal line ^{230}Th = ^{234}U separates all activity ratios into upstream (lower right) and downstream (upper left) sets. The vertical line at ^{230}Th/^{238}U=1 allows more detailed separation: at the top of the diagram activity ratios plotting to the right (sector V) are further downstream while at the bottom activity ratios plotting to the right (sector I) are further upstream. Isochrones show evolution in closed system after a sudden deposition of U with different ^{234}U/^{238}U activity ratios (Thiel et al. 1983).

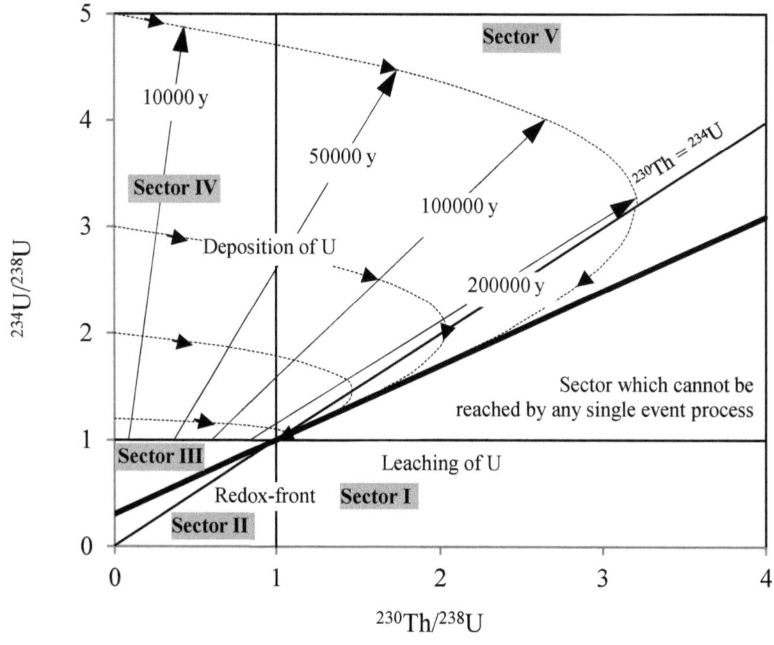

Fig. 3. Interpretation of activity ratios. Isochrones in sectors IV and V describe U series evolution after a sudden deposition of U. Activity ratios in a sector between sectors V and I cannot be reached by single event process. 2. Black diagonal constrains single event process evolution

All isochrones stop by below the diagonal $^{230}Th=^{234}U$ on the way to radioactive equilibrium. Activity ratios displaying in sector between sector I and the tangent of isochrones cannot be reached by a single U mass flow event. Access to this "sixth or forbidden" sector requires leaching of deposited U.

DECSERVIS-2 simulation model

The major work on the DECSERVIS-2 code was done during the development of DECSERVIS (Azzam and Suksi 2006). Fundamental radioactive decay equations were used to build up necessary decay chain equations (Bateman 1910, Ivanovich and Harmon 1992).

Decservis-2 simulation model was constructed to study decay chain evolution pathways and time that could explain measured activity ratios and their geography in the sector graph in Fig. 3.

Decservis-2 allows simulation of decay chain behaviour and evolution under different mass flow events (single or multiple, sudden or continuous) into or out from the system in freely determined time intervals. The benefit of studying alternative scenarios is that irrational ones can be excluded in order to find the most probable scenario. Paleohydrological information of the study site is necessary for constructing reasonable scenarios and to obtain realistic inputs.

Simulation examples

Carefully designed user interface allows easy operation of DECSERVIS-2 (Fig. 4). Input data are seen on the left and output data on the right both data sets clearly displayed. The nuclide input form has been supplemented with a time interval window for the start and end times of mass flow events. There is no limit for the number of input lines (specifying nuclide, time interval, and amount). In the case of more than one input line different entries are clearly separated by dotted lines which are automatically created when the entry has been accepted. The simulation time is inserted in a respective field below. Simulation output for the entire decay chain (nuclide activity, mass, number of nuclides, nuclide ratios) can be presented as a function of time with various graphical presentations such as solid curve and column diagrams or animation. The input and output data can be saved as a MS Excel file.

The use of interface is guided with an example in Fig. 4 where interface screen shot is presented and respective graphical output is presented in Fig. 5. Examples of other graphical outputs are presented in Figs. 6-8. Simulation input details are given and explained in figure captions.

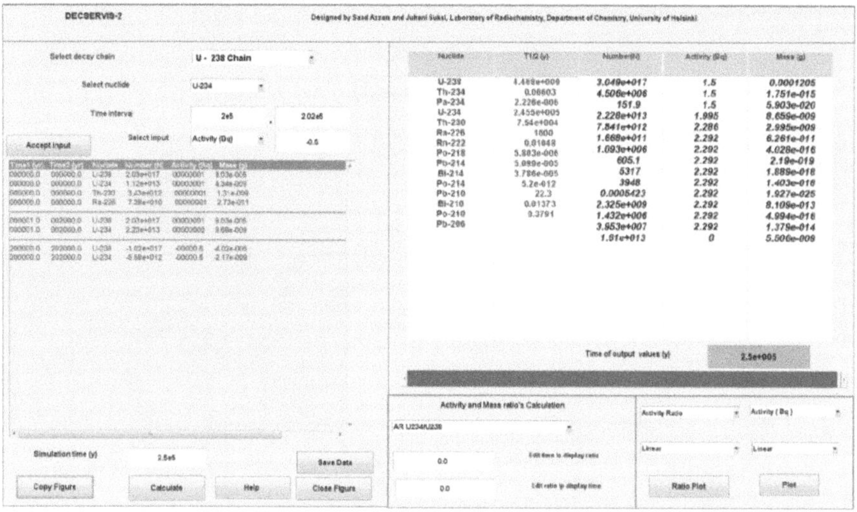

Fig. 4. Screen shot of the user interface with input data on the left and output data on the right. Simulation describes a case where 1 Bq of U with the $^{234}U/^{238}U$ activity ratio 2 accumulated for 2000 years on 1 Bq of U in radioactive equilibrium (in the first input nuclides' activity is 1 Bq). Then 188000 years later conditions changed and 0.5 Bq of U with the $^{234}U/^{238}U$ activity ratio 1 was leached for 2000 years. Simulation time is 250000 years

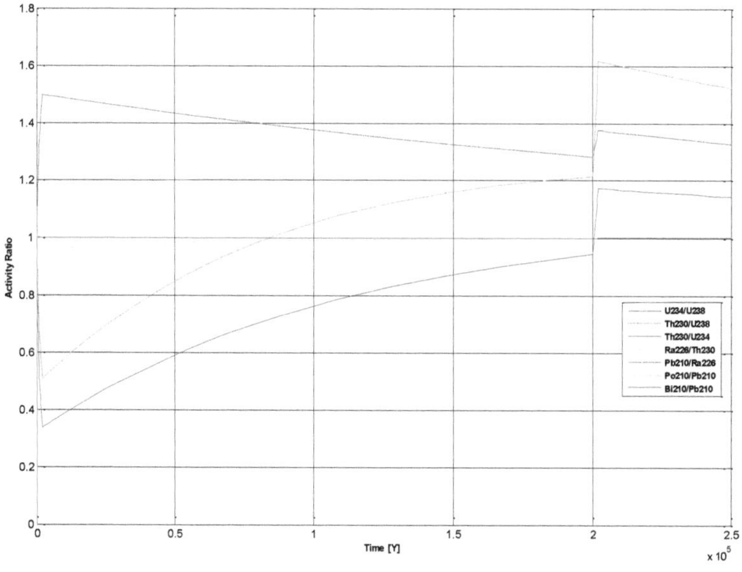

Fig. 5. An example of graphical output from a simulation example in Fig. 4. Output describes development of activity ratios after U accumulation 250000 years ago. U accumulation lasted for 2000 years. 188000 years later a change in conditions triggered U leaching lasted for 2000 years

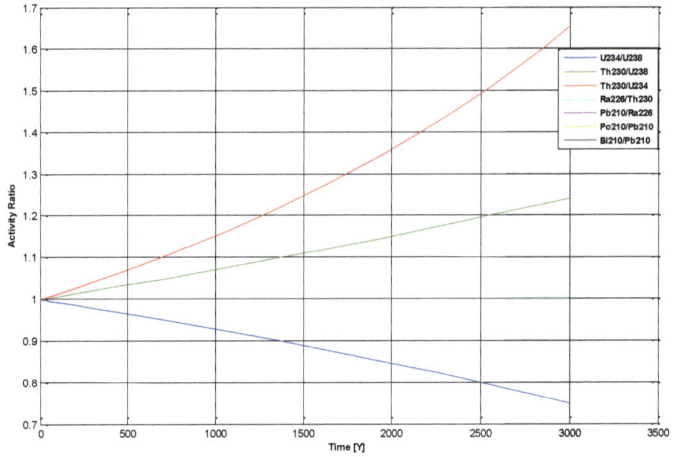

Fig. 6. Simulation of U leaching occurred for 3000 years. The amount of leached U is 20 % from the original amount. U series equilibrium was assumed before leaching. Leacing removes ^{234}U twice as much as ^{238}U. Activity ratios evolve to values found only in sector I (cf. Fig. 3)

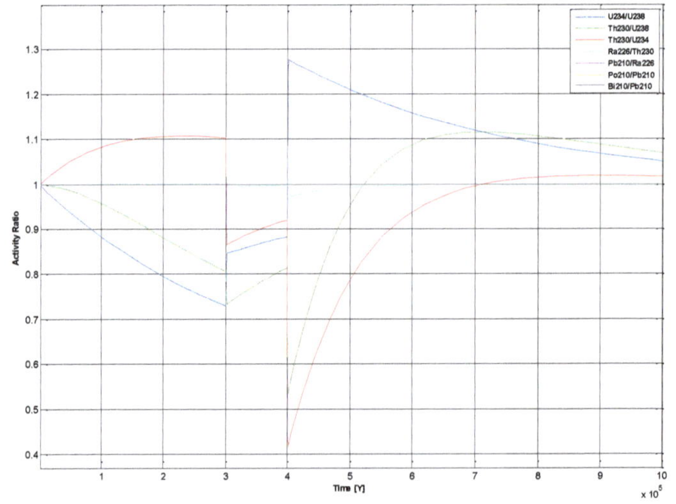

Fig. 7. Simulation starts with leaching of 0.001 % ^{238}U and 40 % ^{234}U from U in radioactive equilibrium. Leaching time is 300000 years. During leaching activity ratios remain in sector II (cf. Fig. 3). Then suddenly 10 % of U ($^{234}U/^{238}U$ =2) starts to accumulate for 2000 years driving evolution to sector III after which system remains closed for 100000 years (from 300000 y to 400000 y) keeping activity ratios in sector III. Then 60 % more U ($^{234}U/^{238}U$ =2) accumulates which drives evolution to sector IV. After 320000 years decay activity ratios attain sector V where decay continues for 280000 years. Simulation time is 1 million years

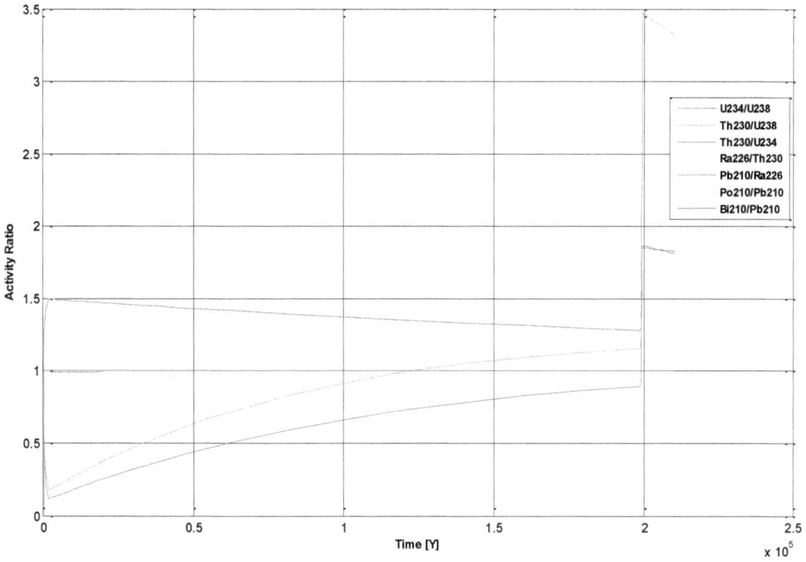

Fig. 8. Simulation example of evolution to the "sixth or forbidden" sector (cf. Fig. 3). The sector is reached if sufficient old U deposition is leached. In the example U deposited 200000 years ago is leached. After U deposition and before leaching there is no U mass flow for ~200000 years.

Conclusions

The main objective of developing DECSERVIS-2 simulation tool was to improve interpretation of natural decay series distribution in different geologic materials. The unique capacity of the tool was demonstrated by simulating the evolution of the ^{238}U decay chain with more than one and opposite U mass flow events. Equal complicated scenarios can be also calculated for other decay chains (the ^{235}U and ^{232}Th decay chains). There are no limitations for DECSERVIS-2 to calculate simulations.

DECSERVIS-2 has been built on a strong mathematical base and all details have been carefully tested. Equally comprehensive tools have not been found in the general literature. The tool was constructed for easy operation with an eye also on teaching and self-learning of natural decay series applications and problems.

DECSERVIS-2 has been developed for free use and it will be downloadable on request.

References

Miller X (1999) How to format conference proceedings. J Anal Chem 20: 721-732

Azzam S, Suksi J (2006) DECSERVIS: a tool for radioactive decay series visualisation. Central European Journal of Chemistry 4 (4): 822–834Miller X, Meier Y (2000) Further instructions. Sci de la Terre 62: 110-120

Azzam S, Suksi J and Ammann M (2009) DECSRVIS-2: A tool for natural decay series mass flow simulation. Applied Radiation and Isotopes 67: 1992–1997

Bateman H (1910) The solution of a system of differential equations occurring in the theory of radioa-active transformations. Proceedings of the Cambridge Philosophical Society 15: 423–427

Gascoyne M (1992) Geochemistry of the actinides and their daughters. *In:* Uranium Series Disequilibrium: Applications to Earth, Marine, and Environmental problems, (M. Ivanovich and R.S. Harmon, eds.): 33-55

Ivanovich M, Harmon R.S (1992) Uranium-series Disequilibrium, Applications to Earth, Marine, and Environmental Sciences, second ed. Clarendon Press, Oxford.

Langmuir D (1978) Uranium solution-mineral equilibria at low temperatures with applications to sedimentary ore deposit. Geochim. Cosmochim. Acta 42: 547-569

Langmuir D. and Herman J.S (1980) The mobility of thorium in natural waters at low temperatures. Geochim. Cosmochim. Acta 44: 1753-1766

Langmuir D (1997) Aqueous environmental geochemistry. Prentice-Hall, Inc., 600 p.

Thiel K, Vorwerk R, Saager R, Stupp HD (1983) ^{235}U fission tracks and ^{238}U-series disequilibria as a means to study recent mobilisation of uranium in Archaean pyritic conglomerates. Earth Planet Sci. Lett. 65: 249-262

Comparison of approaches in Slovenia and Kazakhstan in managing exposure to radon

Ivan Kobal[1], Janja Vaupotič[1], Asta Gregorič[1], Bolat Uralbekov[2]

[1]Jožef Stefan Institute, Ljubljana, Slovenia
[2]al-Farabi Kazakh National University, Almaty, Kazakhstan

Abstract. Increasing attention has been paid to radon in both Slovenia and Kazakhstan. The paper reviews their activities in radon measurements and dose estimates for radon exposure mitigation. Slovenia succeeded to accomplish the main goal of managing exposure to radon in homes and at workplaces at acceptably low levels, while in Kazakhstan there are still quite a number of dwellings with potential radon risk needing further monitoring and mitigation measures.

Introduction

It is now well known that on the world average, radon, thoron and their decay products contribute more than half to the effective dose that a member of the general public receives from all natural radioactivity (UNSCEAR 2000), and are a major cause of lung cancer, second only to cigarette smoking (Darby et al. 2005). In this issue, thoron may contribute significantly and therefore its inclusion in radon survey is highly recommended (Akiba et al. 2010; Tokonami 2010).

In this paper, approaches of managing exposure to radon in dwellings and at workplaces in Slovenia (area 20 thousand km^2, population 2 million) and Kazakhstan (area about 2,725 thousand km^2, population about 17 million) will be reviewed and compared.

Radon monitoring and safety standards

The national radon survey in Slovenia started in 1990, soon after the US Environmental Protection Agency (USEPA) issued an initiative, recommending that schools, with the population most vulnerable to radiation, be surveyed for radon (USEPA 1989; USEPA 1993). It was carried out in all the Slovene kindergartens and schools and in about 1400 dwellings. This was followed by checking a number of workplaces, and additional analyses were performed in

B.J. Merkel and A. Arab (eds.), *Uranium – Past and Future Challenges*,
DOI 10.1007/978-3-319-11059-2_79

buildings with elevated radon levels, in order to design mitigation measures if necessary to be undertaken. Relationship between radon levels and geology was sought.

In Slovenia, radon is regulated within two administrative acts, prepared according to the international recommendations (UNSCEAR 2006; ICRP 1994; BSS 2000), and also on the basis of radon research performed in living and working environment in Slovenia.

According to the first act "Decree on dose limits, radioactive contamination and intervention levels" (ULRS 2004), indoor radon levels should not exceed 400 Bq m^{-3} in homes and 1000 Bq m^{-3} at workplaces.

Second act "Rules on the requirements and methodology of dose assessment for the radiation protection of the population and exposed workers" (ULRS 2003) defines the limit of the effective dose of 6 mSv per year, due to exposure to radon and thoron decay products, for members of general public and workers, who are not additionally exposed to other sources of ionising radiation during their work process.

Radon survey in Kazakhstan started by identifying radon-prone areas in the country based on geological characteristics, including radium content in soil. The Rn measurements have been performed mainly by the State Sanitary stations and the Volkovgeologia Institute within the Radiological Survey Programme under of Ministry of Mineral Resources. In addition, research organizations Solo and EcoService carried out radon measurements in dwellings and workplaces in various regions and cities of Kazakhstan. Indoor radon levels around the former Kazakhstan uranium mines have also been obtained within several international projects (NATO Science for Peace (SfP) RESCA project, the Joint collaboration between Norway, Kazakhstan, Kyrgyzstan, Tajikistan (JNKKT) project) (Salbu et al. 2011).

In Kazakhstan, several standards regulating radon levels in indoor air were inherited from the period of the Union of Soviet Socialist Republics (USSR), the major of them being in NRB-96 and NRB-99. Recently, the Kazakhstan government issued new document titled "Sanitary requirements for radiation safety security" (RS 2012). In this document, the limit for Rn level expressed by $C_{RnDP}+4.6C_{TnDP}$ is 100 Bq m^{-3} for new and 200 Bq m^{-3} for old dwellings. $C_{RnDP}<1200$ Bq m^{-3} is recommended for the personnel of A category (corresponds to the occupational dose limit of an effective dose of 20 mSv per year). The guidance level for Rn concentration in drinking water is 60 kBq m^{-3}.

Radon measuring equipment and dose calculations

In both countries several devices have been used to measure activity concentration of Rn (C_{Rn}), equilibrium equivalent activity concentration of RnDP (C_{RnDP}), equilibrium factor between Rn and RnDP (F_{Rn}) and in some cases, also the fraction of unattached RnDP (f_{un}).

Slovenia

Several complementary measuring devices have been used in Slovenia. Radon scintillation cells (Kristan and Kobal 1974; Vaupotič et al. 1992; Quindós-Poncela et al. 2003) have been used to obtain instantaneous radon concentrations. Average radon concentration has been measured by exposing electret ionization chambers (Rad Elec Inc., USA) or solid state nuclear track detectors (SSNTD) of various manufacturers (e. g., German Nuclear Centre Karlsruhe, Japanese National Institute of Radiological Sciences, Hungarian Radosys, Swedish Landauer Nordic, and others). After exposure, electrets were read in the laboratory while track detectors were sent back to the manufacturer for etching and data evaluation. AlphaGuard radon monitors (Saphymo, Germany) and Radon Scout devices (Sarad, Germany) were used to follow diurnal variations of Rn concentration only, while various EQF radon monitors (Sarad, Germany), to follow also diurnal variations of equilibrium equivalent activity concentration of RnDP, equilibrium factor between Rn and RnDP, and unattached fraction of RnDP. All these devices have been checked at intercomparison experiments organised annually by the Slovenian Nuclear Safety Administration (Križman 2001) and were calibrated in the Radon Chamber at the Henryk Niewodniczański Institute of Nuclear Physics, Polish Academy of Sciences, Kraków, Poland (Kozak et al. 2009).

Kazakhstan

Active and passive devices have been used for radon measurements within various radon studies and surveys. All devices manufactured in Countries of Independent states determine RnDP but not Rn levels, as the regulatory documents are based the RnDP limits for dwellings and workplaces. The most widely used devices for instantaneous screening of RnDP levels are the Ramon spectrometric systems (SOLO) and Russian devises produced by NPP «Doza» (RRA and others). At sites with high Rn levels, SSNTD of different manufacturers (e. g., Japanese National Institute of Radiological Sciences, Hungarian Radosys, and others) have been used to obtain the long-term radon exposure assessment (from several months up to a year). Detectors were exposed in various rooms of public and private buildings (dwellings). After exposure, they were collected and sent back to the manufacturers for reading and evaluation. All the Rn measuring devices were included in the intercomparison experiments performed within several multinational projects, such as RESCA and JNKKP (Salbu et al. 2011).

Dose calculations

For a person spending or working for t hours in a room with equilibrium equivalent activity concentration C_{RnDP}, the effective dose E_{eff} received was calculated using the equation (James 1988):

$$E_{\mathrm{eff}} = C_{\mathrm{RnDP}} \times t \times f_{\mathrm{DC}} \tag{1}$$

with f_{DC} as the dose conversion factor for RnDP, being 9 nSv (Bq m^{-3} h)$^{-1}$ as recommended by UNSCEAR (2000) for both home and workplace, or 1.1 mSv (Bq m^{-3} h)$^{-1}$= 4 mSv WLM^{-1}and 1.4 mSv (Bq m^{-3} h)$^{-1}$= 5mSv WLM^{-1} as recommended by the International Commission of Radiological Protection (ICRP 1994) for home and workplace, respectively. When only C_{Rn} was measured, C_{RnDP} in Eq. 1 was substituted by $C_{Rn} \times F_{Rn}$, with F_{Rn}= 0.40 for indoor air (ICRP 1994).

Radon in dwellings

Slovenia

In 1994, solid state nuclear track detectors were exposed for three winter month in living rooms of about thousand randomly selected dwellings (Križman et al. 1996). Radon concentration ranged from 7 to 1890 Bq m^{-3}, with a median of 54 Bq m^{-3}. The national limit of 400 Bq m^{-3} (ULRS 2004) was exceeded in 4.5 % of dwellings. In the period 2011–2012 annual concentrations of radon were measured in 400 dwellings, one in each 7 km × 7 km square within a regular grid, using SSNTD, in cold and warm periods. Radon concentrations ranged from 21 to 5798 Bq m^{-3} with a median of 99 Bq m^{-3} and the national limit was exceeded in 14.6 % of dwellings (Vaupotič et al. 2013). The owners of these dwellings were advised to ask for additional measurements to check whether mitigation is needed. Several did so and had their buildings remediated.

Kazakhstan

The RnDP levels in air of some dwellings were measured in almost all provinces of Kazakhstan and were mostly found to be below the limit of 200 Bq m^{-3}. However, several 'hot places' with extremely high levels in dwellings were identified at the Wolfram mining and milling enterprises in the vicinity of Akshatau villages in Dzeakazgan region and Zhezkazgan, Aktorgay and, Aksai in West Kazakhstan oblast and Gorniy village in North Kazakhstan (Sevostiyanov 2005), and many villages close to the former uranium mining facilities (Kurday site, Dzhambul region and others). Sevostianov also reported that RnDP activity concentration in 70 % of the surveyed buildings of Zhezkazgan, Makinsk, Shuchinsk, Akshatau, Aktogay, Aryl-Balyk and Balkashino exceeded limit level of 200 Bq m^{-3}.

Soroka and Molchanov (1998) reported that RnDP concentrations in 44 % of the studied houses of Achatau village exceeded the national limit of 200 Bq m^{-3} with a peak concentration at 37,650 Bq m^{-3}. They concluded that high concentration of RnDP levels in dwellings of Akshatau were caused by high Rn exhalation rate from ground. Salbu et al. (2011) published results on Rn at the Kurday site, where indoor air RnDP concentrations varied from 130 to 1200 Bq m^{-3}.

Radon in workplaces

Slovenia

In Slovenia a great attention has been paid to radon monitoring in kindergartens and schools. In each of all 730 kindergartens and each of all 890 schools, radon concentration was measured with alpha scintillation cells in a selected room, closed overnight prior to air sampling in wintertime (Vaupotič 2003). The majority of elevated radon levels were found on carbonate rocks and crossed by several tectonic faults. In 46 kindergartens and 77 schools the national radon limit for dwellings of 400 Bq m^{-3} (ULRS 2004) was exceeded. In all the rooms of these buildings radon was additionally checked and curative measures have been undertaken.

In 64 rooms of 24 public buildings, such as health care centres, community, post, customs and police offices, bus and railway stations and university premises (Žvab et al. 2006), radon exceeded 1000 Bq m^{-3} in 6 rooms. For 15 persons (2.5 %), of the total 528 included into dose calculation, the annual effective dose was more than 6 mSv. Managements of 6 rooms with highest radon levels were recommended to undertake curative measures.

Within 201 rooms of 26 major hospitals (Vaupotič 2003), radon concentration was higher than 1000 Bq m^{-3} in 5 rooms. Of total 1025 persons included into dose calculations only four received the annual effective doses higher than 6 mSv. In 5 rooms with highest radon levels radon was reduced below the national level (ULRS 2004).

In 34 underground rooms of 10 major water supply plants, radon concentrations were in the range 37–2600 Bq m^{-3}; only in one room it was higher than 1000 Bq m^{-3}. Doses are low because of relatively short attendance times at these places (Vaupotič 2003).

Spas, using radium rich water, may also be workplaces with elevated radon levels. Therefore, in total 47 workplaces in five biggest spas were surveyed (Vaupotič 2003). Due to efficient ventilation based on air conditioning, radon concentration was below 200 Bq m^{-3} at all places, even by the swimming pool with radon concentration in water of 1400 Bq m^{-3} and in bathrooms with radon concentration in water of 3300 Bq m^{-3}. Thus, elevated exposure of personnel to radon is not of concern under present working regime.

In wine production, some premises are underground and elevated indoor air radon levels are thus expected. At 20 underground workplaces in eight biggest wineries, radon concentration in the range 33–456 Bq m^{-3} has been found using alpha scintillation cells and nuclear track detectors; only at one place it was higher than 1000 Bq m^{-3}, and only for a single place, annual effective dose of more than 6 mSv was estimated (Vaupotič 2008).

Radon has been monitored also in about 50 caves and with only a few exceptions, elevated radon levels have been observed (Vaupotič 2010). In 1995 a regular radon monitoring was introduced in the Postojna Cave (Vaupotič et al. 2001), the biggest show cave in Slovenia and one of the biggest in the world.

Kazakhstan

In the period from 1990 to 93, Volkovgeologia carried out preliminary measurements of RnPD at workplaces (public buildings and kindergartens) in Almaty. Several places with C_{RnDP} between 380 and 532 Bq m^{-3} were identified (Kaukov et al. 2008). In the period 1998–2001, at about 200 workplaces including schools and kindergartens Rn was monitored by SOLO company (Sevostianov 2005). This study showed that the geometric mean of Rn concentration in air was 204 Bq m^{-3}, with 9 workplaces with elevated Rn levels higher than 300 Bq m^{-3}. They attributed the elevated levels to the presence of tectonic faults.

Screening of Rn concentrations at workplaces performed by Volkovgeologia in the Karagandy region villages (Kaukov et al. 2008) showed the following ranges: 386–908 Bq m^{-3} in Aksu-Auly school, 226–351 Bq m^{-3} in Akshatau school, 254–302 Bq m^{-3} in Kyzyltoo school, with values reaching 23,018 Bq m^{-3} in the rooms of several private houses in the Akshatau settlement.

In 2004 and 2005, EcoService company studied about 120 workplaces in the Ust-Kamegorsk city in East Kazakhstan oblast. At 18 workplaces the national level of 200 Bq m^{-3} was exceeded (Kayukov et al. 2008). 17 settlements (with high potential radon risk) in the region were proposed to be further investigated.

Data concerning the radon concentrations outdoors and indoors in various workplaces located close the U mining sites in all Kazakhstan provinces and cities are still lacking, in spite of the fact that radon daughters are the largest contributors to the effective dose from natural exposure.

Conclusions

National radon programme in Slovenia started with measurements in all the kindergartens and schools, soon followed by a thousand randomly selected dwellings, without taking into account geology of the country. In contrast, in Kazakhstan geology was considered primarily and then, based in uranium and radium content in soil, regions of potentially elevated radon levels were identified as targets on which to start and concentrate radon survey in dwellings and at workplaces. These regions mostly coincided with uranium ore deposits, tectonic faults and abandoned uranium mining and milling sites. In Slovenia, elevated indoor radon levels were mostly found in south-west, the area covered by carbonates and crossed by several tectonic faults while no significant impact of the past uranium mining was observed. The programme, comprising also additional measurements in public buildings exceeding the national radon limit, was in Slovenia coordinated and funded by Radiation Protection Administration at the Ministry of Health and in Kazakhstan by the Ministry of Environmental Protection. Additional measurements in private buildings in Slovenia were paid by the owners.

For radon mitigation purposes in Slovenia, the Radon Center was established, in which radon experts performed thorough radon measurements providing data needed by the civil engineers to design appropriate mitigation measures to be undertaken by small private construction companies. In the majority of cases the main radon source was a sub-floor channel. After the floor was renewed and a fan was installed to aerate the channel, indoor radon concentration was kept below the national limit. In total, radon levels were successfully reduced in about 50 buildings.

In a Slovenian school with radon activity concentration reaching 7000 Bq m^{-3} in wintertime, 85 pupils receiving effective doses of 7–11 mSv y^{-1} were subjected to medical examination, comprising mutagenic tests for structural chromosomal aberrations and the micronucleus test. Although a statistically significant difference in cytogenetic change between the exposed and control group was observed, the results were understood only as a preliminary basis for further studies. In Kazakhstan, 28 children living in dwellings at a median indoor radon concentration 123 Bq m^{-3} and affected by childhood leukaemia were examined. The results suggest that they received similar doses as the control group living in other areas with enhanced natural radioactivity, and therefore continuation of the study did not seem to be justified.

National radon surveys in both countries markedly increased general awareness on the exposure to radon in living and working environment, as achieved by interviews and popular contributions in mass media.

Although different approaches in Slovenia and Kazakhstan, radon surveys in both countries have shown radon levels in homes and at workplaces pointing out buildings and regions of elevated values. Also, when necessary, mitigation measures have been implemented reducing exposure to radon sufficiently. Nonetheless, measurements have been carried out to a limited extent in a relatively low number of selected buildings only, and therefore the picture so obtained may not be generalised, and radon concentrations should be measured at least if elevated levels may be expected.

Acknowledgements

Radon research, presented in this overview, was financially supported by the Slovenian Research Agency, the Slovenian Radiation Protection Administration and the Slovenian Nuclear Safety Administration.

References

Akiba S, Tokonami S, Bochicchio F, McLaughlin J, Tommasino L, Harley N (2010) Thoron: its metrology, health effects and implications for radon epidemiology: a summary of roundtable discussions. Radiat Prot Dosim 141: 477–481

BSS (*Basic Safety Standards*). International basic safety standards for protection against ionizing radiation and for the safety of radiation sources.1–329 (International Atomic Energy Agency, Vienna, 2000).

Darby S, Hill D, Auvinen A, Barrios-Dios JM, Baysson H, Bochicchio F, Deo H, Falk R, Forastiere F, Hakama M, Heid I, Kreienbrock L, Kreuzer M, Lagarde F, Makelainen I, Muirhead C, Oberaigner W, Pershagen G, Ruano-Ravina A, Ruosteenoja E, Rosario AS, Tirmarche M, Tomasek L, Whitley E, Wichmann HE, Doll R (2005) Radon in homes and risk of lung cancer: collaborative analysis of individual data from 13 European case-control studies. Br Med J 330: 223–226

ICRP (*International Commission on Radiological Protection*). Protection against radon–222 at home and at work. Publication 65. 1–44 (Pergamon Press, Oxford, 1994).

James AC (1998) The basis for health concern. In: Nazaroff, W. W.; Nero, Jr. A. V. (eds.) *Radon and its decay products in indoor air*. 259–309 (John Wiley & Sons, New York, 1988)

Kayukov PG, Efremov GF, Berikbolov BR (2008) Report on measures. Study on radiological situation of territory of the Republic of Kazakhstan. In the frame of budget programme 011, Radiation safety securing. Results for 2004–2008 period. In 16 volumes

Kozak K, Mazur J, Vaupotič J, Kobal I, Janik M, Kochowska E. Calibration of the IJS-CRn and IFJ-PAN radon measuring devices in the IFJ-KR-600 radon chamber. (Jožef Stefan Institute Report, IJS-DP-10103, Ljubljana, 2009).

Kristan J, Kobal I (1974) A modified alpha scintillation cell for the determination of radon in uranium mine atmosphere. Health Phys 24: 103–104

Križman M, Ilić R, Skvarč J, Jeran Z. A national survey of indoor radon concentrations in dwellings in Slovenia. In: Glavič-Cindro, D. (ed.) *Proceeding of the Symposium on Radiation Protection in Neighbouring Countries in Central Europe*. 66–70 (Jožef Stefan Institute, Portorož, Slovenia, 1996).

Križman M (2001) Report on the intercomparison experiment on radon and progeny in air. (Slovenian Nuclear Safety Administration Report, URSJV RP 47/2001, Ljubljana, 2001).

RS (*Regulatory Standards*). Sanitary requirements for radiation safety security. Approved by the Government of Republic of Kazakhstan. No. 2012 (February 3, 2012).

Quindós-Poncela LS, Fernandez PL, Sainz C, Arteche J, Arozamena JG, George AC (2003) An improved scintillation cell for radon measurements. Nuc Instrum Meth A 512: 606–609

Salbu B, Stegnar P, Strømman G, Skipperud L, Rosseland B, Heier L, Lind O, Oughton D, Lespukh E, Uralbekov B, Kayukov P (2011) Legacy of uranium mining activities in central Asia - contamination, impact and risks.UMB Report. Draft project report of results obtained within the NATO RESCA project and the joint project between Norway, Kazakhstan, Kyrgyzstan and Tajikistan, 0805–7214. Vol. 164 (Norwegian University of Life Science, Åas, 2011) (restricted distribution).

Sevostianov VN (2005) Some aspects of the radon problem in Kazakhstan. *Radioactivity in the Environment, Vol. 7 (C)*. 409–419 (Elsevier, 2005).

Soroka Y, Molchanov A (1998) Radiation and radon survey of Akchatau (Kazakhstan) and experience with radon remedial measures. Radiat Prot Dosim 78: 231–236

Tokonami S (2014) Why is ^{220}Rn (thoron) measurement important? Radiat Protect Dosim 141: 335–339

ULRS (*Uradni list Republike Slovenije* - Slovenia's Gazette). Rules on the requirements and methodology of dose assessment for the radiation protection of the population and exposed workers. (UL RS 115/2003).

ULRS (*Uradni list Republike Slovenije* - Slovenia's Gazette). Decree on dose limits, radioactive contamination and intervention levels. (UL RS 49/2004).

UNSCEAR (*United Nations Scientific Committee on the Effects of Atomic Radiation*). Sources and effects of ionizing radiation. UNSCEAR 2000 Report to the General Assembly, with scientific annexes. Vol. 1 (United Nations, New York, 2000).

UNSCEAR, (*United Nations Scientific Committee on Effects of Atomic Radiation*). Sources to effects assessment for radon in homes and workplaces. UNSCEAR 2006 Report to the General Assembly, with scientific annexes.Vol. 2 (United Nations, New York, 2008).

USEPA (*US Environmental Protection Agency*). Radon measurements in schools - an interim report. (EPA 520/1-89-010, March 1989).

USEPA (*US Environmental Protection Agency*). Radon measurements in schools - revised edition. (EPA 402-R-92-014, July 1993).

Vaupotič J, Ančik M, Škofljanec M, Kobal, I (1992) Alpha scintillation cell for direct measurement of indoor radon. J Environ Sci Heal *A* A27: 1535–1540

Vaupotič J, Csige I, Radolić V, Hunyadi I, Planinić J, Kobal I (2001) Methodology of radon monitoring and dose estimates in Postojna Cave, Slovenia. Health Phys 80: 142–147

Vaupotič J (2003) Indoor radon in Slovenia. Nucl Technol Radiat Prot 18: 36–43

Vaupotič J (2008) Comparison of various methods of estimating radon dose at underground workplaces in wineries. Radiat Environ Biophys 47: 527–534

Vaupotič J (2010) Radon levels in Karst caves in Slovenia. Acta Carsologica 39: 503–512

Vaupotič J, Gregorič A, Leban M, Bezek M, Žvab Rožič P, Zmazek B, Kobal I (2013) Radon survey within a regular grid in homes in Slovenia. In: VII. Hungarian Radon Forum and Radon and Environment Satellite Workshop: Veszprém, Pannonian University Press, 195–200

Žvab P, Vaupotič J, Dolenec T (2006) Reasons for elevated radon levels inside the building in Divača. *Geologija* (Ljubljana) 49: 409–415

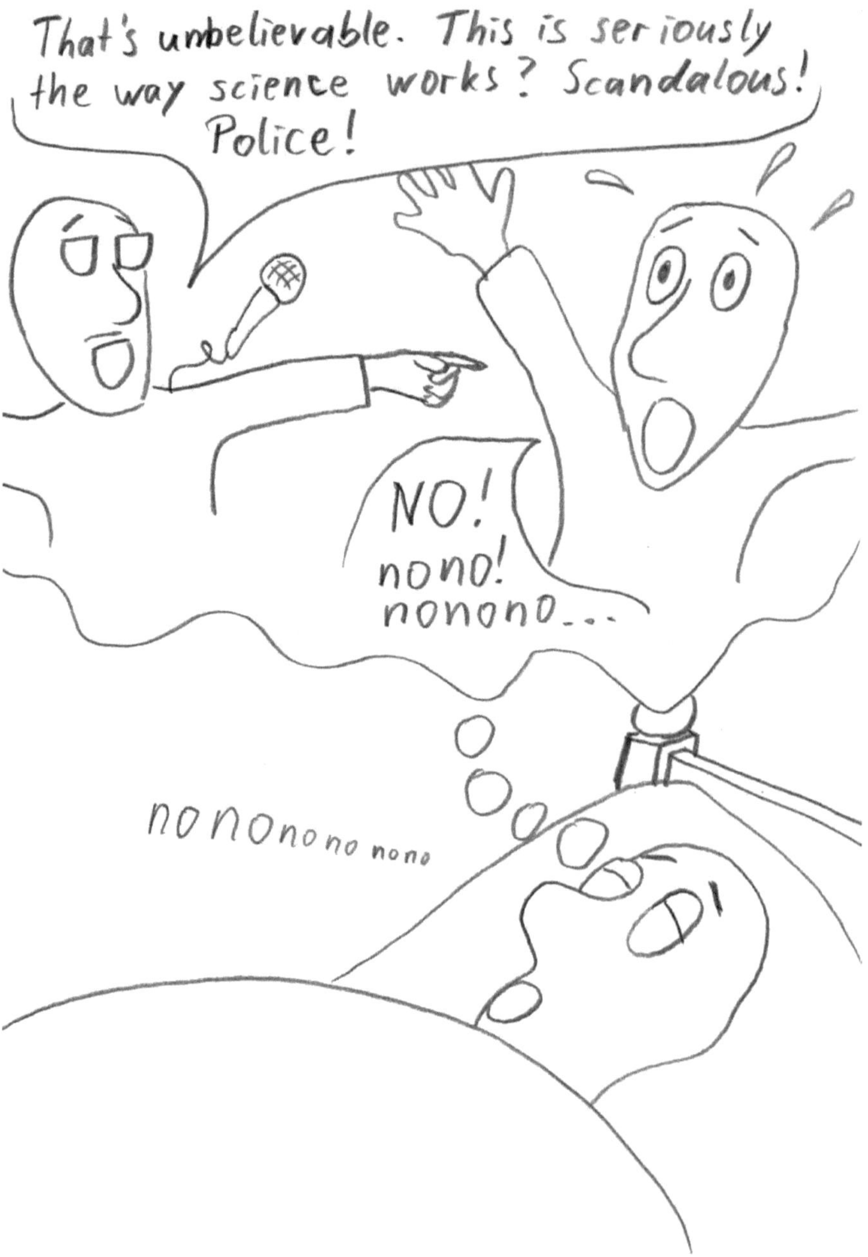
That's unbelievable. This is seriously the way science works? Scandalous! Police!
NO! nono! nonono...
nononononono nono

Reactive transport simulation applied on uranium ISR: effect of the density-driven flow

E. Bonnaud[1], V. Lagneau[1], O. Regnault[2], N. Fiet[2]

[1]Mines ParisTech, Centre de Géosciences, 35 rue Saint Honoré 77305 FONTAINEBLEAU CEDEX, France
[2]Areva Mines, Tour AREVA, 1 place Jean Millier, 92084 PARIS LA DEFENSE CEDEX, France

Abstract. The *in situ* recovery operation of a perched uranium mineralization within a thick, permeable aquifer can become a sensitive issue. Indeed, density difference between the injected high-density acidic solutions and fresh groundwater are liable to trigger density-driven migration of solution towards the bottom of the aquifer. The goals of this study are (1) to represent with a reactive transport model, the hydrogeological behaviour of such an ISL operation for quantifying the lost part of uranium and acid linked to the density driven flow, (2) to suggest some possible solutions to optimize the production.

Introduction

The *in situ* recovery (ISR) operation of a uranium mineralization requires some hydrodynamic conditions.

- The geological formation must be both porous and permeable enough to allow the leaching solutions to flow through the mineralization.
- The mineralization must be located within a confined aquifer so that dispersion of fluids out of the mineralization is limited, effectively limiting loss of uranium and contamination of groundwaters.

What about the exploitation of a perched uranium mineralization, within a thick and permeable aquifer? The density difference between the injected high-density acidic solutions and fresh groundwater can trigger density-driven migration of solutions towards the bottom of the aquifer. This phenomenon has to be taken into account from an operational point of view – overconsumption of reactive fluid, loss of dissolved uranium – and from an environmental point of view – evaluation of the aquifer condition at the end of exploitation, remediation and/or natural attenuation.

The main topics of this study are: (1) to represent with a reactive transport model, the hydrogeological behaviour of such an ISR operation for quantifying the

B.J. Merkel and A. Arab (eds.), *Uranium – Past and Future Challenges*,
DOI 10.1007/978-3-319-11059-2_80

loss of uranium and acid due to density-driven flow, (2) to test the sensitivity of operational parameters with a view to optimizing the uranium recovery and acid consumption. Simulations of this study were performed with the reactive transport code HYTEC (van der Lee et al. 2003, Lagneau and van der Lee 2010).

Description of reactive transport model

Dimensioning of the model

The study of the uranium *in-situ* recovery was performed in 3D, based on the geometry of an exploitation polygon with two adjacent hexagonal cells. The block consists of 10 injector wells and 2 producer wells (Fig.1). The radius of both cells is about 15 m.

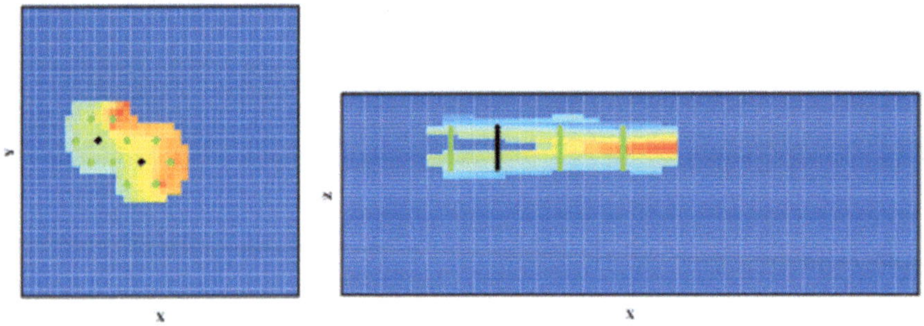

Fig.1. Dimensioning of the model. Injectors (green), producer (black) and the mineralization (only documented inside and close to the two cells).

Geochemical model

The mineralogical composition of the aquifer is defined by three geochemical units typical of a roll-front deposit (oxidized-, mineralized-, and reduced-facies). The mineralogical model, able to represent with precision the uranium recovery and the acid consumption, is composed of a collection of minerals and dissolved species, notably:

- A carrier phase of uranium : uraninite
- A carrier phase of Fe(III) within the oxidized area: goethite
- A « fast » acid consumer phase within the mineralized area: calcite
- A « slow » acid consumer phase: clay

In the model, uranium contents are only mentioned in the close vicinity of the two cells, which is enough to give a correct assessment of the uranium recovery processes: with strong (and balanced) injection/production rates, the flow stays mostly between the wells. To minimize the effects of boundary conditions on the hydrodynamic behavior, the dimensioning of the model is much larger than production block (+ 30 m minimum); however, no effective uranium dissolution is expected outside the close vicinity of the block.

The injection solutions are represented by a sulfuric acid solution rich in Fe^{3+} with concentrations evolving through time following production procedures: high acidification at the beginning, then decreasing until the end of the simulations.

Hydrodynamic parameters

The hydrodynamic characteristics are homogeneous within the aquifer. The porosity is 15 % and the (isotropic) permeability is 3.10^{-04} m/s. The regional water flow in the aquifer is ignored, compared to very high velocity field between the wells.

Concerning the hydrodynamic characteristics of the well field, the flux boundary conditions were imposed along the well-screen throughout the simulation: constant injection rate: 1.25 m^3/h/well and 2.5 m^3/h/well for the two wells shared by the two production cells; constant production rate: 7.5 m^3/h at each production well. Each cell is at hydrodynamic balance: total injection rate is equal to the total production rate. Well-screens are centered on the mineralization.

The flow within the aquifer was simulated using the density-flow module developed in HYTEC. The diffusivity equation and Darcy's law in porous media, including density effects, are:

$$\begin{cases} S_s \dfrac{\partial p}{\partial t} + \rho g q = \text{div}\left[\dfrac{\rho}{\rho_0} K_0 \left(\overrightarrow{\text{grad}p} + \rho g \overrightarrow{\text{grad}z}\right)\right] \\ \vec{U} = -\dfrac{K_0}{\rho_0 g}\left(\overrightarrow{\text{grad}p} + \rho g \overrightarrow{\text{grad}z}\right) \end{cases}$$

where p is the pressure (Pa), ρ is the density of water (kg.m^{-3}), K is the permeability (m.s^{-1}) and S is the storage coefficient (m^{-1}). The density is calculated locally, using concentrations from the chemical description of the system:
$\rho = 1.0 + 0.1\ x\ tds$ (total dissolved solids, mol/L).

The spatial distribution of a tracer introduced in the porous media by injectors is illustrated in the figure 2 after 20 days of simulation. Most of the tracer injected is captured back by the producer wells. However, some dispersion occurs around the cell, horizontally and vertically. When the density-driven flow is not taken into account in the model (Model WD), the tracer remains concentrated (vertically and horizontally) between the well-screens. However, when the density driven flow module is activated, the tracer plume is deformed downwards due to the higher density of the injected solution compared to the aquifer pore water. As a result, part of the injected solution migrates toward the bottom of the aquifer.

The density effects are obviously also at stake for *reactive* species, so that *e.g.* the pH, sulfate and uranium plume also display downwards migration.

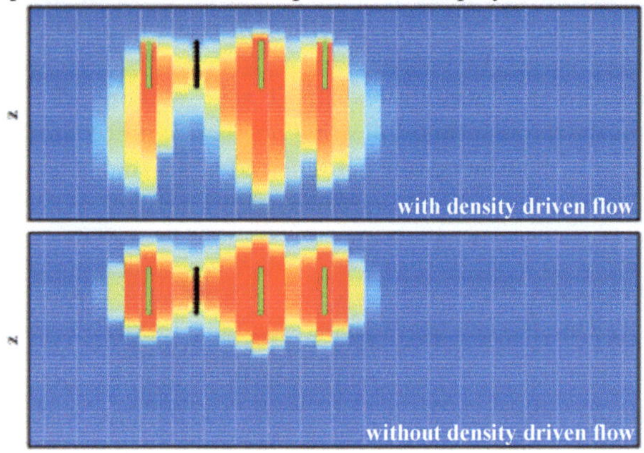

Fig.2. Vertical cross-section. Spatial distribution of a tracer introduced in porous media through injectors after 20 days of simulation with/without the density driven flow module (top/bottom).

Operational simulation

Impact of density driven flow on the production

In order to quantify the loss of solutions related to the density driven flow, two simulations were performed (Fig.3):

- a model without density driven flow (model WD),
- a model that takes the density driven flow into account (model D)

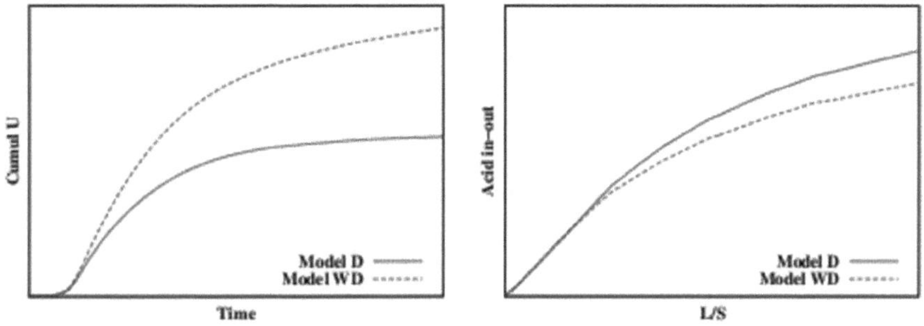

Fig.3. Comparison of the uranium recovery (left) and acid consumption (right) between the model D and the model WD.

With the model D, only 60 % of uranium is recovered at the end of the simulation whereas uranium is totally produced with the model WD. Likewise, acid consumption (difference between acid injected and produced back at the producer wells), it is greater (13 %) with the model D.

These results illustrate the importance to take density effects into account when performing (reactive transport) simulation ISR for a perched mineralization.

Sensitivity of the permeability on the production

Since the permeability controls the rate of solutions migration toward the bottom of the aquifer, its impact on the solutions loss has been studied. Two simulations were performed: case A (K: 5.10^{-05} m.s^{-1}, × 0.5) and case B (K: 2.10^{-04} m.s^{-1}, × 2).

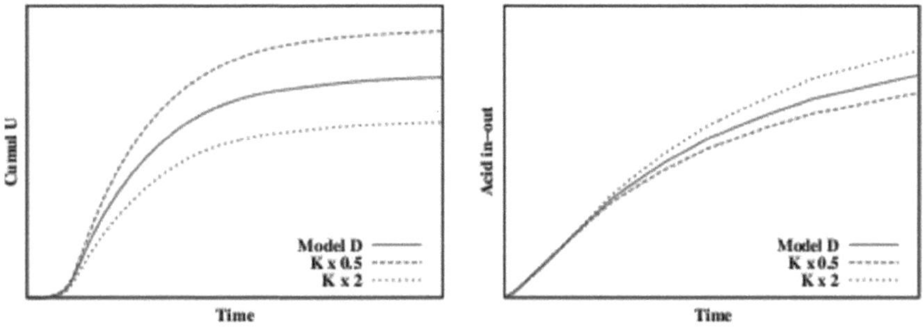

Fig.4. Impact of the permeability on the uranium recovery (left) and the acid consumption (right).

Lower permeability (-50 %) decreases the plume migration under the block, As a result, uranium recovery is increased (+21%) compared to reference model D and acid consumption decreases (-8%) (Fig.4). Conversely, with permeability two times higher, total recovery decreases (+20%) and the acid consumption increases (+11%).

In the case of perched mineralization, the permeability has a huge impact on uranium recovery and acid consumption. It is essential to acquire accurate data for this parameter in order to be able to best represent exploitation of such mineralization.

Additionally, we note that density-driven flow effect on the solutions loss could be reduced for anisotropic permeability (vertical permeability < horizontal permeability). Indeed, the horizontal flow would therefore be increased, thus reducing solutions losses toward the bottom of the aquifer.

Optimization of exploitation flow rates

The flow of acidic solutions through the mineralization is controlled by the exploitation flow rate and the density contrast. The competition between these two parameters is first of all studied with a 2D schematic model with an injector-producer dipole (flow rate in=out). The injected tracer is denser than the fresh water within the aquifer. Two simulations have been compared: flow rate in case 1 is ten times higher than in case 2 (for the quantity of tracer to be similar in both cases, the duration of the simulation of the case 2 needs to be ten times higher).

When the flow rate of wells is high (case 1), the flow of tracer takes place predominantly horizontally from the injector to the producer: the flow induced by wells is thus controlling compared to the density-driven flow. Conversely, when the flow rate of injection is ten times less (case 2), the tracer migrate progressively toward the bottom of the aquifer without reaching the producer: the density driven flow largely dominate.

Thus, the results of simulation show that the density-driven flow becomes preponderant when exploitation flow rates decrease. To enhance the solutions recovery, it would be appropriate to apply exploitation flow rates the highest possible (as far as allowed by exploitation constraints).

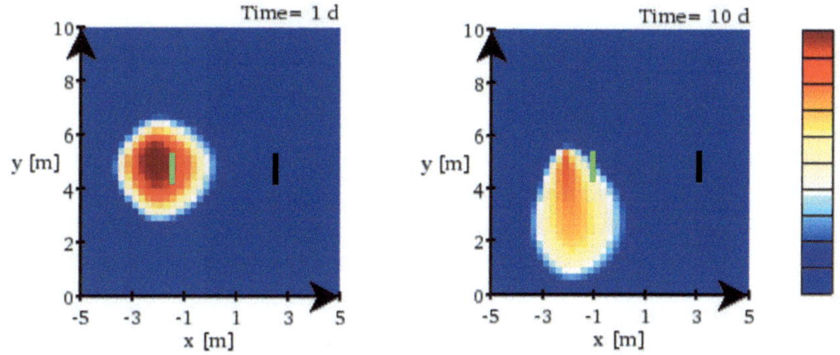

Fig.5. Effect of exploitation flow-rate on the flow of an injected solution, denser than water within the aquifer. Injector (green) – producer (black) dipole. Flow rate are ten times higher in the case 1 (left). For the quantity of tracer to be similar in both cases, the duration of simulation of the case 2 needs to be ten times longer.

Then, two simulations with different flow rates were performed based on model D: Case A (flow rate: 1.00 m^3/h/inj, × 0.8), and case B (flow rate: 1.75 m^3/h/inj, × 1.4). Results are presented as a function of ratio L/S (injected liquid to facilitate the comparison between models. The uranium loss caused by density driven flow decreases when exploitation flow rates increase (Fig.6). *E.g.* for an exploitation flow rate of 1.75 m^3/h/well for example, uranium recovery is improved by 12 % compared to the model D and the acid consumption is reduced to 7 %.

According to the results of models, the losses caused by density driven flow could be significantly limited with the increase in exploitation flow rate. Alternatively, the decrease of the size of exploitation cells has the similar effect.

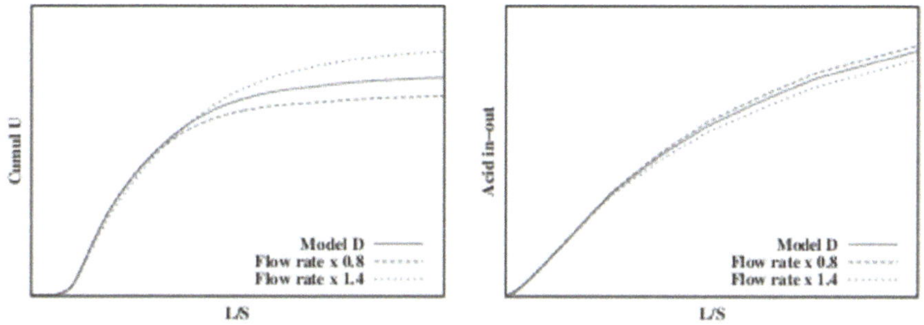

Fig.6. Effect of exploitation flow rates on the recovery of uranium (left) and acid consumption (right).

Conclusion

Reactive transport models highlight the large influence of the density effects on the migration of leached solutions recovery in the case of perched mineralization. From this study, this mechanism is responsible of the loss of up to 38 % of uranium and 13 % of sulfuric acid overconsumption.

The migration rate of the acid plume toward the bottom of the aquifer is controlled by the permeability of the porous media. The sensitivity study showed that increased exploitation flow rate could reduce uranium loss and limit acid consumption.

These results provide new insights for the optimization of the production process of uranium by ISR in the case of a perched mineralization.

References

Lagneau V., van der Lee, J., 2010. Operator-splitting-based reactive transport models in strong feedback of porosity change: the contribution of analytical solutions for accuracy validation and estimator improvement. Journal of Contaminant Hydrology, 112, 118-129.

van der Lee J., De Windt L., Lagneau V., Goblet P., 2003. Module-oriented modelling of reactive transport with HYTEC. Comput. Geosci. 29, 265-275.

Uranium contamination of soil and groundwater by phosphate fertilizer application

Mandy Hoyer[1]

[1]Institute of Geology, TU Bergakademie Freiberg, Gustav-Zeuner-Str.12, 09599 Freiberg, Germany

Abstract. Besides P, the application of phosphate fertilizers also introduces U to agriculturally used land. Still today there are many unknowns regarding the behavior of U in the subsurface. Therefore, in this study two different sites of agriculturally used land in the federal state of Saxony, Germany, were studied in detail by soil and water sampling with subsequent analysis of the samples' composition and properties in the laboratory. Additionally, long-term land use-, weather- and water and soil analytical data from state authorities was considered. It was shown that the fate of U is influenced by soil properties (cation-exchange capacity (CEC), humic and organic matter content, the fraction of cohesive material (clay and silt)), fertilization (type, amount and frequency), ploughing depth, precipitation, and the factor time.

Introduction

Phosphate fertilizers are used in agriculture to satisfy the plants' need for the macro-nutrient P. Depending on the type and origin, such fertilizers contain between below 1 and more than 100 mg U/ kg (Kratz et al. 2008a; Taylor 2013) which together with P will consequently be introduced to the soil being fertilized. After P fertilization, U will be more or less effectively sorbed in the soil zone or washed out into the groundwater zone depending on the boundary conditions and the properties and composition of soil, groundwater and the aquifer. Less than 0.025 mg U/(kg dry matter) will be taken up by plants (Kratz et al. 2008a, based on yield data from Kerschberger et al. 1997).

Study sites

Fig.1 shows the location of the study sites in the federal state of Saxony, Germany. The lysimeter station in Brandis is located east of Leipzig and the Hilbersdorf

B.J. Merkel and A. Arab (eds.), *Uranium – Past and Future Challenges*,
DOI 10.1007/978-3-319-11059-2_81

intensive test site is located east of Freiberg at an altitude of 128 and 425 m.a.s.l., respectively. They are managed by state authorities, namely the Saxon State Ministry for the Environment and Agriculture (SMUL) and the Saxon State Office for Environment, Agriculture and Geology (LfULG), respectively.

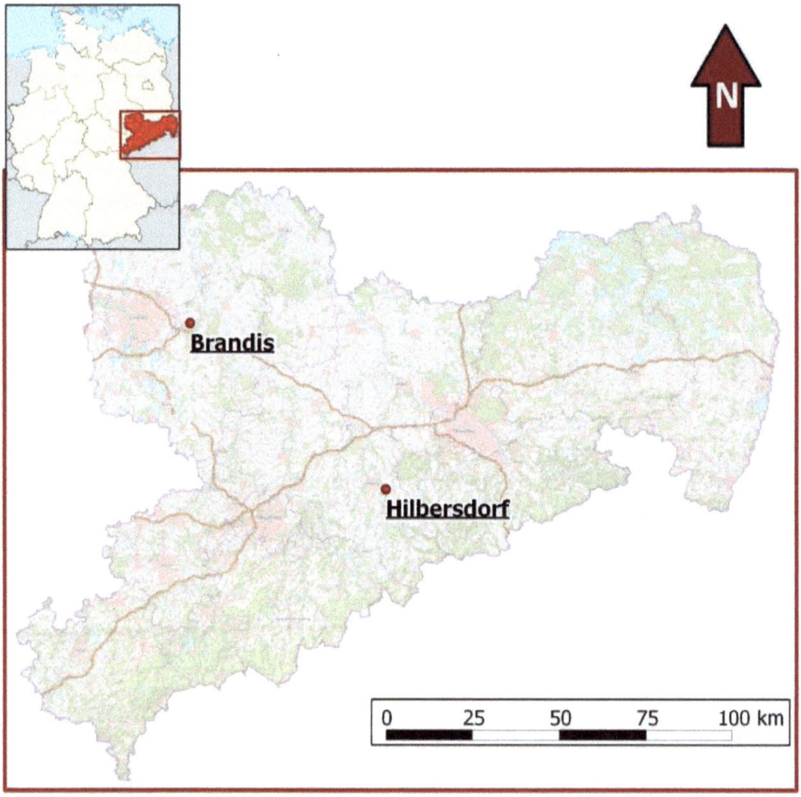

Fig.1. Location of the lysimeter station in Brandis and the intensive test site in Hilbersdorf in Saxony, Germany (based on map data by Roethig 2014; compiled with Quantum GIS 1.8.0. Lisboa).

Both sites have been used for agriculture for several decades already. Fertilization records for Brandis do exist since 1980 and for Hilbersdorf since 1995. The average annual temperature is 8 and 9°C, and the average annual precipitation is 655 and 628 mm, respectively. Consequently both sites belong to the temperate zone. The upper soil zone of both locations is made of highly silty sand forming oxidized, temporarily waterlogged soils with a pH of around 6.

Methodology

At both sites soil samples were taken from about 12 and 26 cm depth for subsequent analysis for their grain size distribution, CEC, carbonate content and total organic carbon (TOC). All details can be found in Hoyer (2013).

Seepage water samples were taken from 3 m depth at the lysimeter station in Brandis and from 80 cm depth at the intensive test site in Hilbersdorf. The in-situ parameters temperature, redox potential, pH and electrical conductivity were measured and the samples were stabilized as needed. Then the samples were analyzed for major cat- and anions, trace elements, TIC and DOC (see Hoyer (2013)). And finally Spearman rank correlation analyses (level of significance < 0.05) was done to identify factors influencing U's distribution and availability.

In addition to own analyses and experiments data from the state authorities LfULG, SMUL, and the German Weather Service (DWD) was used. This data comprised results from soil-, percolation water and groundwater analyses, information on ploughing, fertilization, and precipitation. The water analytical results were post-processed by calculating monthly or annual means. In Hilbersdorf for percolation water sampling there are four suction cups per depth. Their results were taken into account by calculating means. Throughout this work, in case concentrations were below the detection limit of the respective element, they were replaced by 0.3-times the detection limit of the respective element.

As no samples of the fertilizers applied at the study sites could be obtained, typical minimum, mean and maximum U contents of the respective fertilizer types compiled by Taylor (2013) were used for U input calculations. For more information on all methods used see Hoyer (2013).

Results and discussion

Change in U contents in soil and water due to fertilization

Fig.2 and Fig.3 show the relationship between U input from phosphate fertilization and U concentration in soil, seepage water and groundwater in Brandis and Hilbersdorf, respectively.

The arable land in Brandis has been fertilized regularly in the 1980s. Since then, no more P fertilizer has been applied except for some applications in 2000 and 2008. Even if there is no groundwater data before 1993, it seems like U concentrations in at least some monitoring wells (MWs) significantly decreased. At MW1 and MW2 within about 10 years it dropped from about 15 to 8 μg/L and 9 to 3 μg/L, respectively. That means that about 10 years ago, U contents in the groundwater of these wells was close to or even exceeded the MCL (10 μg/L) of the German Drinking Water Directive even if no more P fertilizer had been applied for 10 years already. That underlines the hazard potential for groundwater pollution with U at this site.

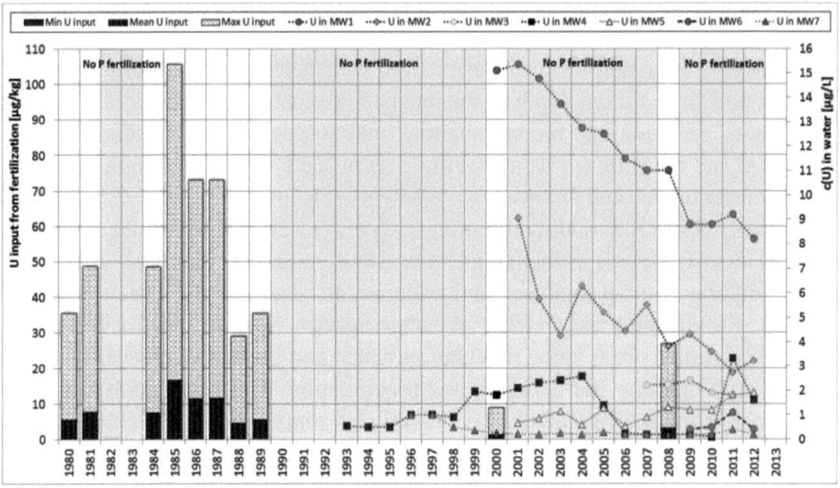

Fig.2. Relation between U input from fertilization (based on typical U contents in commercial P fertilizers according to Taylor (2013)) and U concentration in groundwater from monitoring wells in the surrounding of the lysimeter station in Brandis (data from SMUL; detection limit is 0.1 µg U/L).

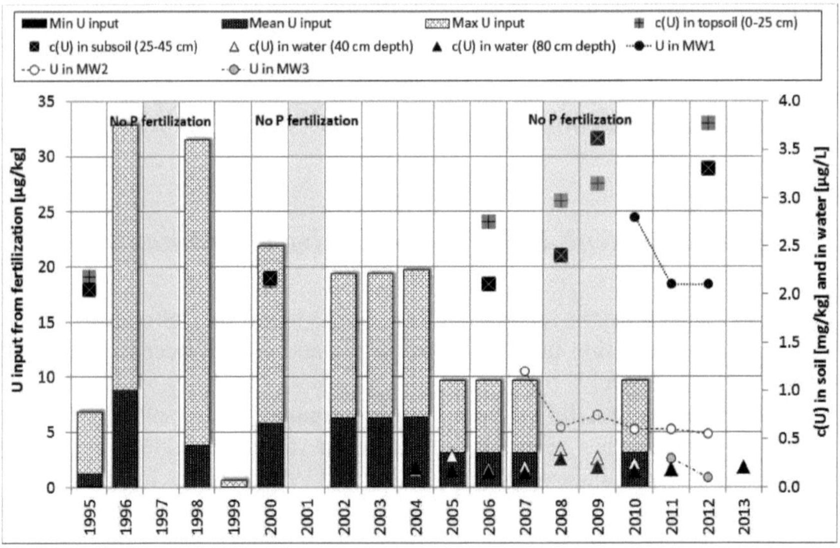

Fig.3. Relation between U input from fertilization (based on typical U contents in commercial P fertilizers according to Taylor (2013)) and U contents in soil, percolation water and groundwater from monitoring wells (MW) in the surrounding of the intensive test site in Hilbersdorf (data from the LfULG; detection limits are 0.5 µg U/L and 0.03 mg PO_4/L).

In Hilbersdorf (Fig.3), data on P fertilization and U contents in soil is available since 1995. Water analytical results recording, however, did not start before 2004 for percolation water samples at 40 and 80 cm depth at the intensive test site itself and in 2007 at surrounding monitoring wells. As opposed to most of the monitoring wells in Brandis, U concentrations in Hilbersdorf groundwater show a constant or slightly decreasing trend of (far) below 3 µg/L. Percolation water of the site itself shows rather constantly low U contents, too. However, in Hilbersdorf the annual U input from fertilization was mostly only about half that of Brandis. The applied U was retained rather effectively by the soil in Hilbersdorf. Consequently the U content in soil nearly doubled from about 2 to about 4 mg/kg within 17 years.

In Brandis the U content in soil was determined only once, in 2013 within this study. At that time, U contents were at about 2 mg/kg, which is in the typical range of such types of soil, i.e. not elevated (Utermann and Fuchs 2008).

The fate of U in the subsurface

Now the question arises why U is rather effectively prevented from reaching the groundwater zone in Hilbersdorf but in Brandis not.

For sure, the difference in the **amount of the fertilizer applied** plays a role. If it is too high, the soil's capacity to retain U will no longer be sufficient. This capacity depends on the **fraction of cohesive material**, i.e. clay and silt (32-43 % in Brandis; 46-51 % in Hilbersdorf), the **humic matter content** (very low to low (2 vol%) in Brandis, and high (4-5 vol%) at the surface to low (0.5 vol%) in the subsurface in Hilbersdorf) and the **organic carbon content** (low in Brandis (about 1 mass% C), and medium in the topsoil (about 2.5 mass% C) and low (about 0.5 mass% C) in the subsoil of Hilbersdorf) as correlation analysis showed. The higher the values of these properties are the better U is retained by the soil. If the **pH** of the respective soil is low, U will mainly be present as positively charged $(UO_2)^{2+}$ which is will be present in mobile complexes in the first place but can also be bound by **clay minerals** and humic matter (Merkel 2009). The capacities of the studied soils to retain cations, their **CEC**, are medium (8 and 12 meq/kg) for the sub- and topsoil in Hilbersdorf as compared to a low CEC of 6 and 7 meq/kg in sub- and topsoil in Brandis. That means the CEC of the sub- and topsoil in Hilbersdorf is higher than in Brandis by 70 and 30 %, respectively.

In addition to the soil properties and the amount of fertilizer applied, the **type of fertilizer** used is important. That is because the fertilizers do not only differ in their P and U contents but also in their solubility and plant availability (Kratz and

Schnug 2008b). P and U can form soluble and insoluble complexes depending on pH, U-, P- and other ions' concentrations, and boundary conditions like **moisture content and temperature**. These complexes can sorb on **Fe- and Al (hydr)oxides** as well as on organic matter (Merkel 2009; Yamaguchi et al. 2009). As time passes, binding strength and consequently immobilization of U and P increases due to diffusion of these components into for example iron-containing particles (Taylor and Kim 2008; Cheng et al. 2004).

The impact of precipitation and ploughing

Fig.4 shows a detailed view of the year 2008 in Hilbersdorf with monthly precipitation, fertilization, ploughing, and monthly means of U and P concentrations in seepage water.

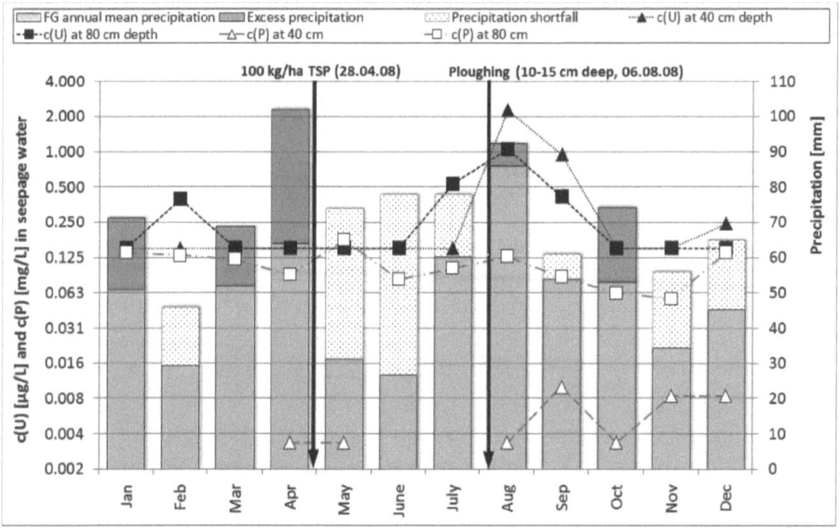

Fig.4. Impact of P fertilization, precipitation and ploughing on U and P contents in seepage water (c(U) and c(P)) at the intensive test site in Hilbersdorf (The long-term mean precipitation data from the nearest DWD station in Freiberg was used as a reference for comparing the precipitation measured at the intensive test site in Hilbersdorf in 2008 with. All data stems from the LfULG; only the precipitation data for Freiberg stems from the DWD.).

U concentrations throughout the year are mostly below or close to the detection limit (0.5 µg U/L). Triple superphosphate had been applied in the end of April. After a very dry beginning of the summer with monthly precipitation up to 50 % below average, in July and August U contents started to increase. In August, after the field had been ploughed and precipitation had increased U concentrations were highest (1-2 μg U/L) and slowly decreased again within the following months. So, precipitation and ploughing seem to impact U mobilization after fertilization.

Inorganic phosphate concentrations in seepage water (here expressed as P), however, did not vary considerably. What is decisive, however, is the great difference (factor 8 to 45) between P contents determined for percolation water sampled at 40 and 80 cm, respectively. Furthermore, it is surprising that the contents at deeper sampling depth are higher than those at shallower depth. As the main input of P probably occurs by P fertilization, one would expect higher P contents at shallower than at deeper depth. However, when examining the P content in soil, this expected distribution can actually be observed. In 2006, the year when the soil was analyzed for its P content for the last time, total P in topsoil was twice as high as in subsoil (0.2 % PO_4 at 0-25 cm depth as compared to 0.1 % PO_4 at 25-45 cm depth). So, the upper soil horizon is an effective sink for P. After fertilizer application P is immediately bound as Ca-, Al- and Fe phosphates. Because these minerals have very low solubility products, only small amounts of the respective minerals dissolves and so P is available for plant growth for several years after a P fertilization measure. Plants extrude organic acids that mobilize P by forming soluble chelate complexes. These complexes are taken up by the plants and incorporated in their cells (e.g. as phytic acid). After harvest microorganisms decompose the remaining dead plant residues and release the contained P while organic carbon is metabolized. The concentration distribution of total organic carbon (TOC), humic matter content and P support this hypothesis. The TOC content decreases with depth (1.5 %C in topsoil vs. 0.7 %C in the deeper soil); the content of humic matter, a product of TOC decomposition, decreases as well (4-5 vs. 0.5 vol%) and P content increases as mentioned before.

Conclusions

In this work, the fate of U was studied at two different agricultural soils. It was shown that important factors ruling the behavior of U in the subsurface are the properties and composition of the soils, fertilization and ploughing practices, precipitation and time.

Although the studied soils have a certain capacity to retain U, this capacity will be exceeded and U will reach the groundwater zone if U input is too high and/or occurs over long periods of time as shown for the study site in Brandis.

In Germany, about 70 % of drinking water is produced from groundwater. According to the German Drinking Water Directive, the MCL for U in drinking water is 10 μg U/L. If fertilization practices are not changed, this limit will be exceeded in many agriculturally used areas in Germany within the following decades and expensive water treatment will be needed.

Until now, unlike for Cd there is no obligation to label the U contents in fertilizers. To enable farmers to wisely select P fertilizers low in U, labeling has to be commanded by law.

There is a great demand for future studies on the fate of U in the subsurface as there are still many unknowns. For example, the mechanisms of U and P release from different types of fertilizers are not yet understood. Additionally, microorganisms and organic acids seem to strongly impact the mobility of U – especially in micropores. Besides that the influence of different types of organic matter and clay minerals still needs further studies. These studies should comprise two parts: laboratory experiments (flow-through and batch experiments) and corresponding long-term field monitoring. Furthermore P analytics has to distinguish between inorganically and organically bound P to better understand the ruling processes. For proper interpretation it is crucial that the partners involved communicate to each other, which studies are intended and which data is needed for that purpose and at which spatial and temporal resolution.

References

Cheng T, Barnett M O, Roden E E, Zhuang J (2004) Effects of phosphate on uranium(VI) adsorption to goethite-coated sand. Environmental Science & Technology. Environ. Sci. Technol 38 (22): 6059–6065 (http://pubs.acs.org/doi/pdf/10.1021/es040388o)

Hoyer, M. (2013): Uranium contamination of soil and groundwater by phosphate fertilizer application; FOG – Freiberger Online Geoscience, Vol. 35 (http://tu-freiberg.de/geo/fog)

Kerschberger, M., Hege, U., Jungk, A., 1997. Phosphordüngung nach Bodenuntersuchung und Pflanzenbedarf. Verband Deutscher Landwirtschaftlicher Untersuchungs- und Forschungsanstalten (http://www.vdlufa.de/joomla/Dokumente/Standpunkte/0-4-phosphor.pdf. Accessed 14 October 2012)

Kratz, S., Knappe, F., Rogasik, J., Schnug, E., 2008a. Uranium balances in agroecosystems, in: Kok, L.J. de, Schnug, E. (Eds.), Loads and fate of fertilizer-derived uranium. Backhuys Publishers; Margraf Publishers, Leiden, Weikersheim, Germany, pp. 179–190.

Kratz, S., Schnug, E., 2008b. Agronomische Bewertung von Phosphat-Düngern: Ressourcen schonender Einsatz von Phosphor in der Landwirtschaft. JKI - Julius-Kühn-Institut, Bundesforschungsinstitut für Kulturpflanzen (http://www.jki.bund.de/fileadmin/dam_uploads/_koordinierend/bs_naehrstofftage/phosphor_landwirtschaft/15_Kratz.pdf. Accessed 19 April 2013)

Merkel, B.J., 2009. Uran im Trink- und Mineralwasser: Presentation in the "Centrum für Umwelt und Technologie" (http://www.cut-os.de/downloads/Vortrag_Merkel_internet.pdf. Accessed October 17th, 2012)

Roethig, S., 2014. Serviceanmeldung. Staatsbetrieb Geobasisinformation und Vermessung Sachsen (http://www.landesvermessung.sachsen.de/ias/basiskarte4/service/register. Accessed 13 June 2014)

Taylor, M., Kim, N., 2008. The fate of uranium contaminants of phosphate fertilizer, in: Kok, L.J. de, Schnug, E. (Eds.), Loads and fate of fertilizer-derived uranium. Backhuys Publishers; Margraf Publishers, Leiden, Weikersheim, Germany, pp. 147–155

Taylor, M., 2013. Trace Element Contaminants and Radioactivity from Phosphate Fertiliser: Waikato Regional Council document 2015563
Utermann, J., Fuchs, M., 2008. Uranium in German soils, in: Kok, L.J. de, Schnug, E. (Eds.), Loads and fate of fertilizer-derived uranium. Backhuys Publishers; Margraf Publishers, Leiden, Weikersheim, Germany, pp. 33–46
Yamaguchi, N., Kawasaki, A., Iiyama, I., 2009. Distribution of uranium in soil components of agricultural fields after long-term application of phosphate fertilizers. Science of The Total Environment 407 (4): 1383–1390 (http://www.sciencedirect.com/science/article/pii/S0048969708010395)

This is what you'll get when you order a huge lab on the internet

Investigation of Phenomena in Uranium Mine Waters using Hydrogeochemical Modeling – a case study

Corinne Lietsch[1], Nils Hoth[1], Andrea Kassahun[2]

[1]TU Bergakademie Freiberg, Institute of Mining and Special Civil Engineering, Gustav-Zeuner-Str. 1A, 09599 Freiberg/Sachs
[2]Wismut GmbH, Jagdschänkenstr. 29, 09117 Chemnitz

Abstract. One of the main aspects in the remediation of closed uranium mine sites is the treatment of mine water. In order to understand speciation and processes in uranium mine waters, hydrogeochemical modeling is carried out. The importance of modeling in mine water treatment is shown by the example of two case studies. The first deals with precipitation processes in a stripping column. In the second case uranium speciation of different mine waters is characterized. By means of modeling results, treatment technologies can be adjusted and improved.

Introduction

Since the closure of uranium mines in the early 1990s, their remediation is an important task in the German federal states of Saxony and Thuringia. It is carried out by Wismut GmbH within one of the largest environmental remediation programs in Europe. A main aspect of the remediation is the treatment of mine waters, like pit water, drainage water, pore water and dump leachate. As the mine waters contain contaminants, such as uranium, radium, arsenic and heavy metals, they have to be treated in order to protect ground water and surface water around old uranium mine sites from pollution.

Mine water treatment is a challenging task, which is regulated by limiting concentrations. The water treatment processes need to guarantee a sufficient removal of pollutants, so that the receiving waters are not put at risk. Therefore, reaction processes during the treatment have to be understood as thoroughly as possible. One tool for the comprehension of conditions and processes in mine water before or during the treatment is hydrogeochemical modeling. Based on standard analyses (environmental parameters, main element concentrations etc.) of water samples, conclusions about speciation of elements or saturation states of mineral phases can be made. Furthermore, reaction processes in treatment steps can be reproduced by hydrogeochemical modeling.

B.J. Merkel and A. Arab (eds.), *Uranium – Past and Future Challenges*,
DOI 10.1007/978-3-319-11059-2_82

This article deals with two case studies, in which hydrogeochemical modeling is used to investigate phenomena in mine waters of Wismut's mine sites. Modeling is carried out with the hydrogeochemical software PhreeqC (Parkhurst and Appelo 1999).

Subject of the first case study is the hydrogeochemical understanding of precipitation processes in a CO_2 stripping column. The column is one of several pretreatment steps of a mine water treatment plant in the Saxonian Ore Mountains. During the stripping, solids precipitate on the synthetic filling material of the column. In order to prevent this process, hydrogeochemical modeling is used to identify oversaturated mineral phases inside the stripping column. Modeling results provide information about possible prevention measures.

The second case study deals with uranium complexation in Wismut's mine water. In most cases dissolved uranium occurs as uranyl (UO_2^{2+}) species. Predominant uranyl species influence the sorption and precipitation of uranium due to their different charge states. By performing hydrogeochemical modeling, the typical species distribution in waters of different mine sites is characterized.

Reproduction of chemical reaction processes in a CO_2 stripping column by means of hydrogeochemical modeling

Description of the task

Most of the mine water treatment plants operated by Wismut use precipitation processes for the effective removal of uranium and other contaminants. During a pretreatment step relatively stable uranyl carbonate complexes are destroyed. This is done by decreasing the pH value significantly. Afterwards, the pH value needs to be raised again by stripping CO_2 out of the solution.

Inside the stripping column solids precipitate on the synthetic filling material of the column, which is an unintentional process. Hence, the filling material has to be exchanged frequently, which requires a high maintenance and leads to additional costs.

By means of hydrogeochemical modeling, chemical reaction processes inside the stripping column are investigated with a focus on the influence of pH value.

Hydrogeochemical investigations

Hydrogeochemical modeling with the software PhreeqC is carried out based on analytical results of samples from the mine water treatment plant. Main objective is the characterization of the raw water, as well as of influent and effluent of the stripping column. Table 1 provides the measurement results of environmental parameters, which illustrate the change of the hydrochemical environment between the three sampling points (raw water, influent and effluent). This is due to the addition of acid and the subsequent stripping. One parameter is the redox potential (E_h), which increases significantly. Decrease in DIC is due to the removal of CO_2 during the stripping process. Consequently, a change in precipitation tendency of mineral phases can also be expected.

Table 1. Collection of environmental parameters, field measurements

Sample	pH [-]	T [°C]	EC[a] [μS/cm]	E_h [mV]	DIC[b] [mmol/L]
Raw Water	7.05	25.8	2010	45	11.3
Influent	3.80	25.8	2400	355	4.95
Effluent	3.76	27.2	2410	608	0.60

[a] EC: electrical conductivity

[b] DIC: dissolved inorganic carbon (concentration)

The evaluation of analysis data provides information about saturation indices (SIs) of relevant mineral phases. Saturation indices are calculated with the hydrogeochemical software PhreeqC. Positive SI represents oversaturation, negative SI shows undersaturation and SI≈0 stands for equilibrium conditions of the solution regarding the respective phase.

Relevant mineral phases are plotted depending on the respective sampling point (cf. Fig. 1-2). As the raw water contains 4-5 mg/L Fe, iron species are generally of major interest, e.g. ferrihydrite (Fig. 1(A)). In contrast, carbonate species become undersaturated in correlation with CO_2 removal. The saturation of gypsum is not influenced by the pretreatment processes. Fig. 1(B) shows the characteristic behavior of jarosite phases when passing the stripping column. In the raw water jarosites are undersaturated. The addition of acid causes a decrease of saturation indices, however, stripping increases the SI significantly. As it is visible in Fig. 1(B), there is oversaturation of jarosites inside and after the stripping column, which is presumed to be a major reason for precipitation.

Fig. 2 shows the results of two different approaches. The continuous line in each diagram describes the development of SIs, based on real analyses from the three sampling points (cf. Fig. 1). Beyond that, saturation conditions in case of further pH decrease are modeled. Therefore, the raw water (based on real analysis) is manipulated in such a way as to model the reaction processes pH decrease and stripping. First, the pH value is decreased on pH 3.8, 3.5, 3.0, 2.5, 2.0 and 1.5, respectively. Then the solutions are equilibrated with air. The modeling results are illustrated by the non-continuous lines. Aim of the modeling is the prognosis of a starting pH value, which prevents or reduces precipitation inside the stripping column.

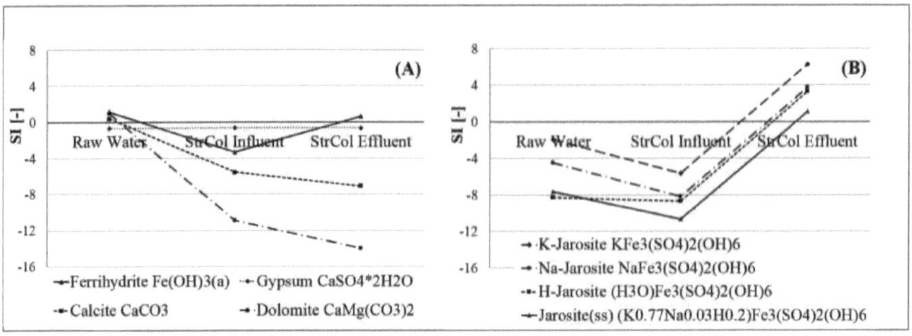

Fig. 1. Saturation indices of relevant mineral phases at the three sampling points; calculated with PhreeqC (Parkhurst and Appelo 1999). (A) SIs of Ferrihydrite, Gypsum, Calcite, Dolomite, which have little influence on precipitation. (B) SIs of different Jarosites, which are related to precipitation in the stripping column.

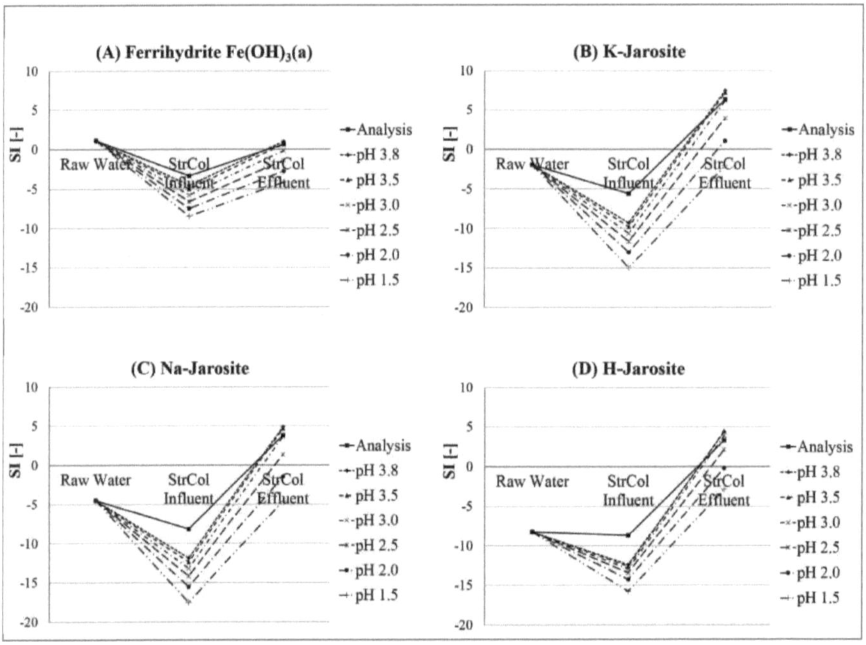

Fig.2. Saturation indices of (A) Ferrihydrite, (B) K-Jarosite, (C) Na-Jarosite and (D) H-Jarosite. Comparison of evaluated analysis and modeled SIs for decreasing pH value.

Results of hydrogeochemical modeling

According to Fig. 2, a decrease of pH value causes a simultaneous decrease of saturation indices. In Table 2 the necessary pH decrease for undersaturation of a respective mineral phase inside the stripping column is summarized.

Table 2. Results of modeling: prognosis of necessary pH values in order to prevent oversaturation inside the stripping column. Evaluation of the mineral phases described in Fig. 2

Mineral Phase	Necessary pH value
Ferrihydrite	3.0 (-3.5)
K-Jarosite	1.5 (-2.0)
Na-Jarosite	2.0 (-2.5)
H-Jarosite	2.0

Concluding from the modeling results, the precipitation inside the stripping column is caused by oversaturated conditions of iron mineral phases, such as ferrihydrite, K-jarosite, Na-jarosite and H-jarosite. By means of a further decrease of the pH value in front of the stripping column, oversaturation – or in other words precipitation – can be prevented.

The recommended pH values (Table 2) are considerably lower than the pH values used at the mine water treatment plant of Wismut GmbH. Further pH decrease causes high additional costs for hydrochloric acid. Furthermore, the chloride load in the mine water increases. In practice, these disadvantages need to be considered compared to the improvements in maintenance due to the prevented precipitation. These aspects have already been tested at the mine water treatment plant, where a lower pH value shows significant improvements.

Determination of uranium species distributions in mine waters by means of hydrogeochemical modeling

Uranyl complexes in hydrogeochemical modeling

Uranium speciation, which is an important characteristic of mine waters, influences uranium sorption and precipitation during the water treatment process. Dissolved uranium usually occurs as uranyl (UO_2^{2+}) which builds complexes with carbonate, sulfate, phosphate, hydroxide, etc. Uranyl carbonate complexes are of particular importance at neutral pH value due to their stability and dominance.

In order to identify typical uranium species and complexes in mine waters, hydrogeochemical modeling is performed. For this purpose, the hydrogeochemical software PhreeqC is used with a modified thermodynamic database, as standard databases in PhreeqC do not contain ternary complexes of alkaline earth uranyl carbonates ($M_xUO_2(CO_3)_3^{(2x-4)}$-complexes).

Since the formation of $Ca_2UO_2(CO_3)_3^0$ was introduced and validated (Bernhard et al. 1996, 2001), ternary complexes of alkaline earth uranyl carbonates have been described by different references. Thermodynamic parameters were measured and calculated not only for calcium and magnesium, but also for strontium and barium (Dong and Brooks 2006). Nevertheless, thermodynamic parameters of $M_xUO_2(CO_3)_3^{(2x-4)}$-complexes have not been included in PhreeqC databases yet. Therefore, reaction equations and log-K values need to be added manually. In detail, anionic and zero-valent species of calcium, magnesium, strontium and barium are considered. The selection of particular complexes is based on preliminary modeling tests. The formation of strontium and barium uranyl carbonates is of minor importance and therefore disregarded. The following reaction equations and log-K values are added to the PhreeqC Input File (Dong and Brooks 2006):

$$2Ca^{2+} + UO_2^{2+} + 3CO_3^{2-} \Leftrightarrow Ca_2UO_2(CO_3)_3^0 \qquad \log K = 30.70$$
$$Ca^{2+} + UO_2^{2+} + 3CO_3^{2-} \Leftrightarrow CaUO_2(CO_3)_3^{2-} \qquad \log K = 27.18$$
$$Mg^{2+} + UO_2^{2+} + 3CO_3^{2-} \Leftrightarrow MgUO_2(CO_3)_3^{2-} \qquad \log K = 26.11$$

Results of hydrogeochemical modeling

The addition of the selected $M_xUO_2(CO_3)_3^{(2x-4)}$-complexes causes a dominance of those species at neutral and alkaline pH values. Fig. 3 shows the results of hydrogeochemical modeling for three mine water samples of Wismut's mine sites. The water samples are characterized by different pH values. Fig. 3(A), (B) and (C) show the uranium species distribution in mine water with slightly alkaline, neutral and acidic pH value, respectively. At alkaline (Fig. 3(A)) and neutral (Fig. 3(B)) pH values calcium uranyl carbonates dominate in these mine waters. In contrast, at acidic pH values (Fig. 3(C)) uranyl sulfates hold the largest fraction in the speciation.

Without the consideration of the $M_xUO_2(CO_3)_3^{(2x-4)}$-complexes, $UO_2(CO_3)_3^{4-}$ and $UO_2(CO_3)_2^{2-}$ are of great importance for the investigated mine waters at neutral and alkaline pH value. Hence, with the standard databases of PhreeqC no good prediction of the uranium speciation for waters at these pH values is possible. However, the speciation of the water in Fig. 3(C) does not depend on $M_xUO_2(CO_3)_3^{(2x-4)}$-complexes. Calcium uranyl carbonates have little influence on the species distribution of waters with low pH value, as they occur only at neutral and alkaline pH value (Nair et al. 2013). At around pH 8 and higher they are the dominating species and other uranyl species can be disregarded.

The dominating uranium species are essential for characterizing the mine waters. Treatment processes, such as precipitation and ion exchange, depend on the ion charge of the uranyl species. However, the distinction of uranium species is a challenging task. Thermodynamic databases used for the hydrogeochemical modeling influence the species distribution to a great extent. Therefore, different databases and additional thermodynamic data can result in different species distributions. Especially the distinction of $Ca_3UO_2(CO_3)_3^0$ and $CaUO_2(CO_3)_3^{2-}$ is still under investigation. As it is very challenging to analyze uranium species, the thermodynamic parameters of those species have to be further investigated, so that the databases could be improved. That would be of great value for the evaluation of treatment processes in uranium mine waters. Therefore, future work on that topic together with Dr. Brendler of Helmholtz-Zentrum Dresden-Rossendorf is planned.

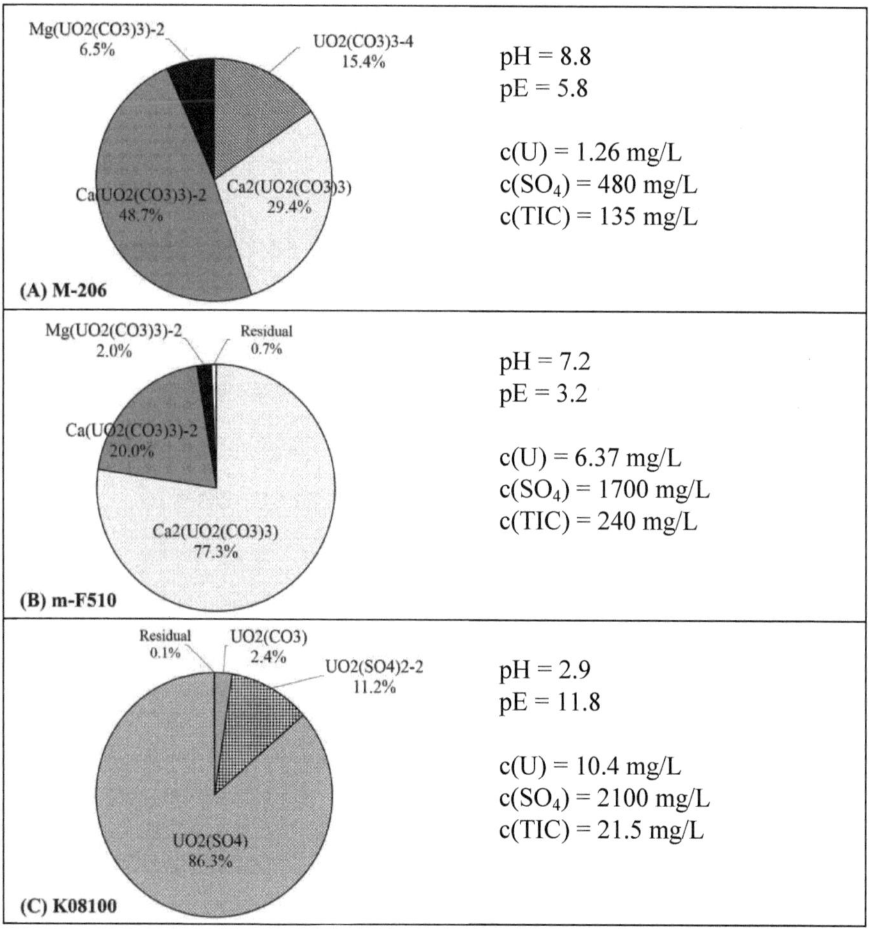

Fig.3. Results of the hydrogeochemical modeling of uranium species distribution for three Wismut mine waters; next to the pie charts: some milieu parameters and concentrations for the characterization of the respective mine water (TIC = total inorganic carbon)

Conclusion

Hydrogeochemical modeling has an important role in understanding the hydrochemical reaction processes in mine waters. It is a tool to predict saturation indices, speciation and simple reactions by means of the evaluation of standard analyses. Thus, it is possible to find solutions to issues of less complexity, like precipitation processes. Prognoses of hydrogeochemical modeling should always be validated with experiments and analyses.

More complex topics, such as the uranium speciation, need further investigations. Thermodynamic data has to be compared between different references and – if possible – validated by experiments.

References

Bernhard G., Geipel G., Reich T., Brendler V., Nitsche H. (1996) Speciation of uranium in seepage waters of a mine tailing pile studied by time-resolved laser-induced fluorescence spectroscopy (TRLFS). Radiochim Acta 74: 87-91

Bernhard G., Geipel G., Reich T., Brendler V., Amayri S., Nitsche H. (2001) Uranyl(VI) carbonate complex formation: Validation of the $Ca_2UO_2(CO_3)_3$(aq.) species. Radiochim Acta 89: 511-518

Dong W., Brooks S.C. (2006) Determination of the Formation Constants of Ternary Complexes of Uranyl and Carbonate with Alkaline Earth Metals (Mg^{2+}, Ca^{2+}, Sr^{2+} and Ba^{2+}) Using Anion Exchange Method. Environ Sci Techno. 40: 4689-4695

Nair S., Karimzadeh L., Merkel B.J. (2013): Surface complexation modeling of Uranium(VI) sorption on quartz in the presence and absence of alkaline earth metals. Environ. Earth Sci. 71: 1737-1745

Parkhurst D.L., Appelo C.A.J. (1999) User's guide to PHREEQC (version 2)

3D Reactive Transport simulations of Uranium In Situ Leaching: Forecast and Process Optimization

Olivier Regnault[1], Vincent Lagneau[2], Nicolas Fiet[1]

[1]AREVA Mines, Tour Areva, 1 place Jean Millier, 92084 Paris La Défense, France
[2]MINES ParisTech, PSL Research University, Centre de géosciences, 35 rue St Honoré 77300 Fontainebleau, France

Abstract. 3D reactive transport model is an efficient tool to simulate uranium ISR operation: indeed the whole process is based on reacting fluid circulation in porous media. The simulation can help quantify the processes at stake and optimize the ore dissolution rate and the use of reagents. This study details the application of a 3D reactive transport approach using the code HYTEC at operation scale in Kazakhstan.

Introduction

In situ Recovery (ISR), or solution mining, is an alternative technique for recovering (usually) deep, poor grade ores. The process involves a series of injection and producer wells, and the circulation of a leaching solution (acidic, basic and/or oxidizing). The leaching solution interacts with the ore body, and solubilizes the desired element. *E.g.* since 2009, ISR has become the leading technique for uranium exploitation.

Reactive transport modeling is an efficient tool to investigate the behavior of ISR exploitation: indeed, the whole process is based on reacting fluid circulation in porous media.

The simulation can help understand the processes at stake, optimize the ore dissolution rate, minimize the residual ore after production, and minimize the use of reagents. It can also help understand and predict the long-term impact of the exploitation.

A first stage of the simulation is to understand and calibrate the phenomena at stake: pH buffering by several mineral phases, dissolution of *in situ* oxidized iron bearing phases and oxidative dissolution of uranium, kinetic controls. The simulation of ISR operations faces several challenges: strong reactivity of the fluids and

B.J. Merkel and A. Arab (eds.), *Uranium – Past and Future Challenges*,
DOI 10.1007/978-3-319-11059-2_83

fast fluid flow are numerically challenging. Moreover, the large number of wells imposes to cover large domains. Finally, a correct description of the heterogeneity of the medium is key for the simulation to be predictive: 1D or 2D, or 3D with homogenized sections of space fail to reproduce production data. As a result, large simulations, 3D with over 10^5+ cells, have to be undertaken.

Material and Methods

Code HYTEC

Hytec is a reactive transport code developed at MINES ParisTech (van der Lee et al. 2003, Lagneau and van der Lee 2010). The code solves the chemical equations for most geochemical reactions, including aqueous complexation, redox, dissolution/precipitation, sorption (using several formalisms), each at equilibrium or under kinetic control. The hydrodynamic module copes with flow (saturated, unsaturated or two-phase, at steady- or transient-state), transport (water and gas phases), and heat transfer. The resolution is performed using a finite volume discretization over a Voronoi mesh (nearest neighbor) and a semi-implicit adaptive time-discretization. The coupling between chemistry and transport is performed using a sequential iterative approach. The coupling with flow and heat transfer is explicit.

Geological model

The ore body lies within a heterogeneous aquifer formation. The underlying heterogeneity is two-fold and derives from the genesis of the ore (Dahlkamp 1991) (Hobday and Galloway 1999). In a first time, the sedimentation under continental conditions builds a reduced, channelized aquifer. Second, an oxidizing front moves through the aquifer; mobile U(VI) in the oxidized part of the aquifer is then precipitated into U(IV) low solubility minerals such as uraninite. These genetic processes lead to two superimposed heterogeneities: lithofacies coarser or finer sands down to clays, chemical facies (oxidized, mineralized and reduced), and finally uranium grade within the mineralized facies.

The model is based on block-models of the ore deposit, using geological data (drill core observations and geophysical logging) and geostatistical reconstruction using stochastic simulation methods (Petit et al. 2012).

Finally, a pattern of injection and production wells is set over the geometry (Fig. 1).

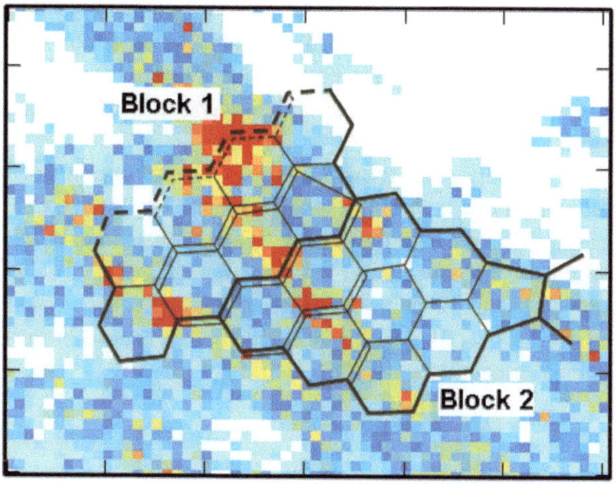

Fig.1. Example of uranium distribution in the system: accumulation (*i.e.* integral concentration over the height of the aquifer) on the xy axis for 2 technological blocks. The hexagonal pattern of the well field is figured: injectors at the hexagon vertices, producers at the center.

Geochemical model

A geochemical model is devised to represent all three geochemical facies (oxidized, mineralized and reduced) according to laboratory and field observations (production data). The main mineral interactions with sulfuric acid leaching solution taken into account are the oxidative dissolution of U(IV)-bearing phase with Fe(III), acid-base dissolution/precipitation of and carbonates, clays and ferric minerals.

Results and discussion

Simulation of historical data

The first set of simulation is dedicated to reproduce ISR behavior both in terms of uranium recovery and acid consumption at technological blocks scales and to test the robustness and the prediction quality of the method on different technological blocks.

The simulations were run over a 2-year period. The simulation follows the exploitation conditions: wells position, injection and production rate history (including closing wells, or newly developed injectors), and concentration of the injection fluids (acidity, majors and Fe(III) content).

First of all, the results are correct in terms of processes: progression of a pH front then oxidative dissolution of the uranium, and increased Fe(III) due to acidic dissolution of Fe(III) bearing phases. Then, the comparison with field data fits correctly, both in terms of uranium produced and acid consumption (Fig. 2). This result is all the more satisfying that the number of fitting parameters is very small: concentration of carbonates and Fe(III) bearing mineral (homogeneous in the mineralized and oxidized facies respectively) and kinetic rate of the clay.

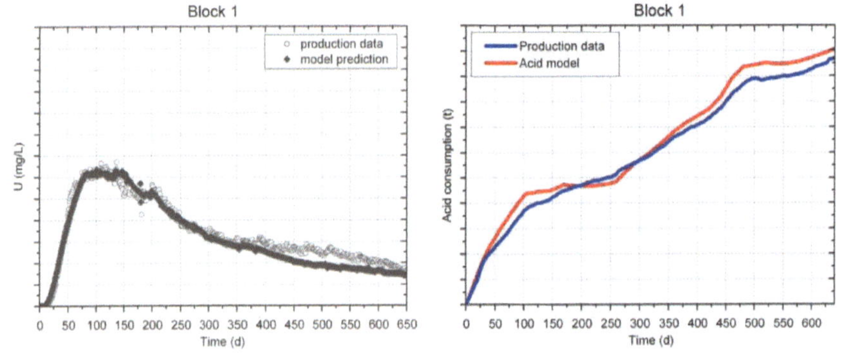

Fig.2. Comparison of simulation results and field data for block 1: uranium concentration produced by the field mixing of all the producer wells (left), acid consumption (right).

The robustness of the method was tested by applying the calibration of the model obtained with block 1 as is on different blocks from the same area of the deposit: different well field designs and operational histories (acidification, flow rates), geology similar based on specific block-model for the second zone. The comparison between production data and the prediction of the model shows a very nice match and illustrates the potential of the method to take into account the impact of both geological media heterogeneity and operational histories on uranium recovery behaviour at field scale (Fig 3.).

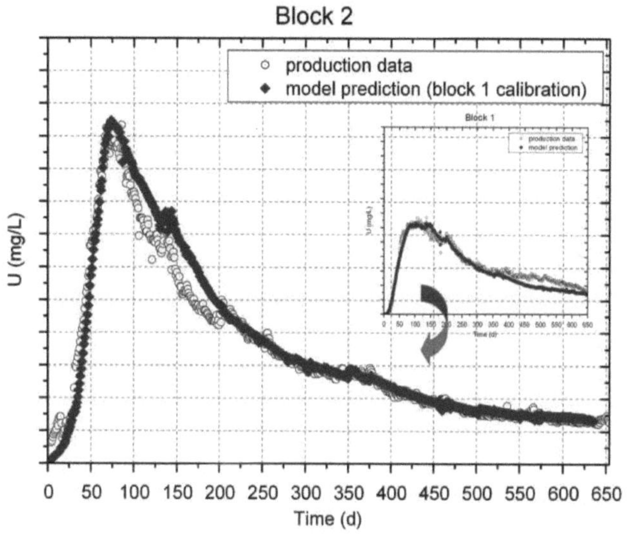

Fig.3. Comparison of simulation results and field data for uranium concentration produced by the block 2: good quality of prediction on block 2 using calibration obtained on block 1.

Simulation of optimization scenario

The modeling approach gives also the user the ability to test a large choice of production scenario. Indeed, the main operational levers are described explicitly within the 3D reactive transport tool Hytec: from acid addition and flow rates management to well field design evolution over the time.

One example of wells addition is illustrated in this paper (Fig. 4). The scenario aims to test the addition of 1 production wells along with 3 injection wells within a high concentrated uranium area. When comparing with the reference model (production history), the results of the “well addition” scenario shows, at block scale, a large and positive impact in terms of uranium recovery rate from the beginning of the acidification.

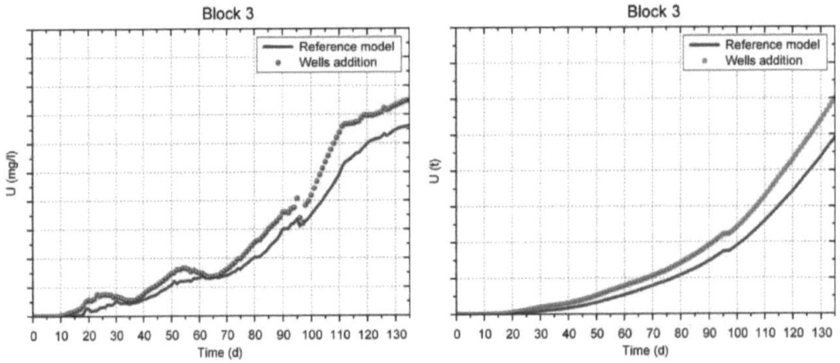

Fig.4. Comparison of 2 simulation results (block 3) for uranium concentration (left) and cumulated uranium production (right): the green curve shows the result for the reference model (operation history), the red dots show the well addition scenario (1 producer + 3 injectors).

Conclusions

The simulation of ISR exploitation gives very encouraging results, in that simulations can be performed using a limited number of fitting parameters. These simulations open the way to predictive simulations or a tool to test new production ideas. However, the simulation can only perform accurately if the geometrical description is sufficiently accurate. This leads to massive simulations (10^5+ cells), including numerous computationally challenging reacting fronts. These simulations were made possible by a combination of improved management of the code parallelization and performing computers.

Acknowledgements The authors would like to thank KATCO for supporting this work and providing all valuable data and constructive discussions.

References

Dahlkamp, F J (1991). Uranium ore deposits. Springer, Berlin Heidelberg New York.

Hobday D K, Galloway W E (1999). Groundwater processes and sedimentary uranium deposits. Hydrology journal, 7:127-138.

Lagneau V, Van der Lee J (2010). HYTEC Results of the MoMas Reactive transport benchmark. Computational Geosciences, 14:435–449.

Petit G, De Boissezon H, Langlais V, Rumbach G, Khairuldin A, Oppeneau T, Fiet N (2012). Application of stochastic simulations and quantifying uncertainities in the drilling of roll front deposit uranium. 9 International Geostatistics Congress, Oslo.

van der Lee J, de Windt L, Lagneau V, Goblet P (2003). Module-oriented modeling of reactive transport with HYTEC. Computers and Geosciences, 29:265-275.

Where has all the uranium gone? Or what feeds Dimona – circumstantial evidence for an illicit fate of uranium from rock phosphate processing

Ewald Schnug[1]

[1]Technical University Braunschweig - Faculty 2 Life Sciences, Pockelsstrasse 14, D-38106 Braunschweig, Germany

- in memoriam Inge Lindemann (May 3, 1958 – August 4, 2011) -
(Wippel et al. 2012)

Abstract. Sedimentary rock phosphates contain up to several hundred mg/kg of uranium (U), which seems low when compared to the U concentration in typical mined U minerals, but is significant in terms of U masses. U in rock phosphate is easy to recover, and sufficiently high stock market prices for U provided, will supply nuclear fuel cycles as a secondary U source. The paper presents a circumstantial case study on the recovery, use and fate of U derived from phosphate mining in the Negev desert and supports evidence that Israel's nuclear facility at Dimona might be fed by U derived from refining rock phosphates at the Rotem Amfert mine for food and fertilizers. Evidence for U enrichment provides an archived compound fertilizer which contains depleted U.

Introduction

U is a naturally accompanying element in sedimentary rock phosphate deposits where it has been precipitated mainly in phosphates of fossil apatites from the seawater during sedimentation processes. All U isotopes (^{233}U, ^{234}U, ^{235}U, ^{236}U, ^{237}, ^{238}U) are radioactive. The most abundant U isotopes are ^{238}U, which accounts for around 99.3% of U found in nature and ^{235}U being liable for 0.7%, respectively. ^{235}U is the only fissile isotope in nature, making U to the most concentrate source of energy on earth: one kg of U with the natural concentration of ^{235}U contains 50.000 times more energy than one kg of firewood and if technically enriched to 3.5% ^{235}U then almost 3000.000 times more (Schnug and Haneklaus 2012).

It is human nature that everything man has laid his hand on sooner or later was used to attack and kill his own fellows: a million years ago man made his fire from

B.J. Merkel and A. Arab (eds.), *Uranium – Past and Future Challenges*,
DOI 10.1007/978-3-319-11059-2_84

wood (Bernaa 2011), but also bows and arrows. A million of years later man made heat from U, but also nuclear weapons. But this time the first friendly use of nature's present (1954 the first nuclear power station became critical in Obninsk, Russia) emerged as late as nine years after the first nuclear bomb was dropped by the United States on Hiroshima, Japan in 1945 (Anonymus, 2013a).

Ever since an advantage in military tactics was keeping your enemy in the dark about what weapons you possess (Giles 1910; Frisbee 2013). This is of no interest if a weapon is a commonplace commodity like a machine gun, but for sure if it comes to nuclear weapons. U is commonly retrieved from minerals with high U content like for instance uraninite (pitchblende, $UO_{2)}$). The safeguards of EURATOM and IAEA supervise meticulously mining, processing and use of U from mined sources. But there are secondary U sources with much lower, however, noteworthy U concentrations such as sedimentary rock phosphate deposits, which escape the watch of the safeguards (Astley and Stana 2013). These are commonly mined for phosphates that are mainly used as fertilizers in agriculture (Figure 1).

Fig.1. IMC Agrico - New Wales Plant, Polk County, Florida (Connett 2001).

Out of the 1630 phosphate mines known worldwide in 2002 6% show significant concentrations of U in the minerals. These are located in countries like Afghanistan, Angola, Australia, Belgium, Brazil, Canada, Central-Africa, Ecuador, Finland, Greenland, Hungary, India, Israel, Jordan, Mauretania, Morocco, Mozambique, New Zealand, Saudi Arabia, Senegal, South Africa, Spain, Sweden,

Syria, Togo, Turkey, United States (Orris and Chernoff 2002). However, no data exist which reveal how much U has been or is still retrieved from these operations.

Circumstantial evidence for the fate of uranium from rock phosphate processing

Comparing the U/P_2O_5 relationship in manufactured phosphate fertilizers reveals a slightly but distinctively smaller index in older samples indicating a 13.6% to 70.2% lower U content (Figure 2). This can be interpreted as circumstantial evidence that during the time given (1975-1985) U extraction was common at least in some of the plants contributing phosphate fertilizers to the international fertilizer market.

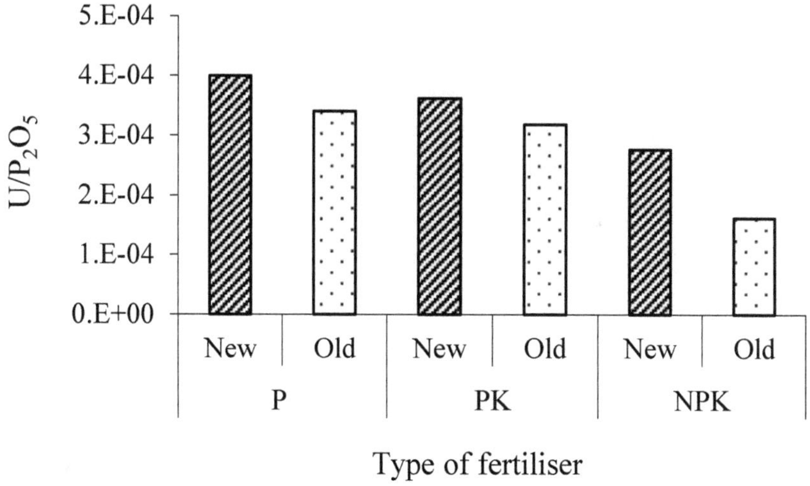

Fig.2. U/P_2O_5 in old (manufactured 1975-1985) and new (manufactured 1995-2005) collections of P-containing fertilizer samples (Sattouf 2007).

Common opinion is that a new, third wave of U extraction is soon to come, but that there are no significant plants in operation at the moment (Astley and Stana 2013). There are rumors about such plants, like the one in Homs, Syria (Reuters 2011), Al quaim in Iraq (Anonymus 2013b) or Rotem Amfert in Israel (Lindemann 2007; NTI 2011), but no proof is available.

Shrouded in legends is especially the phosphate mine at Rotem Amfert in Israel's Negev desert, because its only 8 km away from the Negev Nuclear Research Center close to the town Dimona (Figure 3). The Dimona reactor became critical between 1962-1964 employing U from obscure source and origin (Burr and Cohen 2013). The facts have inspired many fiction stories (Davenport et al. 1978), the most popular one "Triple" written by Ken Follet in 1979.

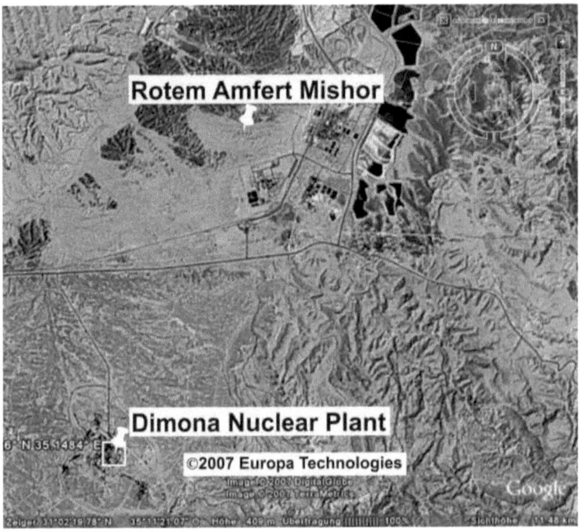

Fig.3. Map of the Rotem Amfert phosphate mine facilities and Negev Nuclear Research Center in the Negev desert of Israel.

Uranium removal during phosphate fertilizer manufacturing

All fertilizer products that contain mineral phosphates unfold U in varying concentrations. The U content is consistently correlated to the phosphate content. Rock phosphates from Israel and Morocco contain the highest U concentrations (Lindemann 2007; Sattouf 2007). As a rule of thumb phosphate fertilizers contain between 200 and 300 mg/kg U per kg P_2O_5 which contaminates agricultural soils (Schnug and Haneklaus 2014).

The technology for extracting U during manufacturing of phosphate fertilizers is a well-established procedure that is known since long (Astley and Stana 2013; IMPHOS 2009; Ragheb 2010). First plants for retrieving U from phosphate fertilizer production for military stockpiling went operational already in the 1950s and run for about 10 years. The U extraction business halted for about 10 years and commenced in the 1970s for another 20 years, mainly for supplying fuel for nuclear power plants. Plants known for U recovery from rock phosphates were operated in Belgium, Canada, China, Iran, Iraq, Israel, Taiwan and the United States of America. After another 20 years of inactivity U extraction from phosphates now sees its renaissance in the light of resource conservation and sustainability, carbon dioxide mitigation, and soil protection (Astley and Stana 2013; Schnug and Haneklaus 2014). The most recent process employs the extraction of U from acid by solvent extraction. The recovery rates range from 0.1-7 g U/kg P_2O_5 (Astley and Stana 2013). In total, approximately 17,000 tons U have been recovered from rock phosphate processing in the United States of America (Astley and Stana 2013).

After the mysterious incident of gathering the U for getting the Dimona reactor started no further reports on U deliveries to this site exist. Already in 1953 the Science Corps C, a special unit of the Israel Defense Force's Science Corps perfected a process for extracting U from phosphate deposits (Kahaha 2006), but domestic production is supposed to amount only to 10 tons per year (NTI 2011).

Approximately since 2005 "white" phosphoric acid is produced at Rotem Amfert to supply the feed and food industry (Imas 2007). This "white" phosphoric acid is virtually free of heavy metals, especially U. Own analyses of raw material from the site in 2007 reveal U concentrations of 120 mg/kg U. The "green" phosphoric acid, which is the first intermediate product of the extraction with sulfuric acid, contains 187 mg/kg U. At a specific weight of 1.38 for the "green" phosphoric acid this means that nearly all the U contained in the rock phosphate has been transferred into the acid. The "standard" PK type fertilizers from the plant contain about 59 mg/kg U, which is typical for these products, but a "special" NPK product deriving from the "white" phosphoric acid production line turns up with only 4.5 mg/kg U. This is clear evidence that U is removed during the production of phosphate fertilizers at Rotem Amfert. Considering an annual amount of 3,290,000 tons of mined rock phosphates and 2,724,000 tons processed at the side (Imas 2007) the amount of U which can be potentially retrieved is 326 tons, equaling 2.3 tons of ^{235}U. Based on the real output of the "white" phosphoric line production the realistic amount of U retrieved is about 18 tons per year. These amounts are sufficient to keep the Dimona reactor running, or to assemble 5-15 nuclear warheads.

Hiding the tracks

In 2007 Sattouf made a rather strange observation while screening the isotope pattern of one of the world's largest and most versatile collection of phosphate fertilizers maintained at the then Institute for Plant Nutrition and Soil Science of the Federal Research Station at Braunschweig, Germany by TIMS (Thermal Induced Mass Spectrometry): in one of the NPK type fertilizers which originated from the Anglo-Continentale (formerly Ohlendorff′sche) Guano-Werke in Hamburg, Germany he found a completely atypical relation of the isotopes $^{235}U/^{238}U$ indicating a depletion of the isotope ^{235}U (Figure 4) .

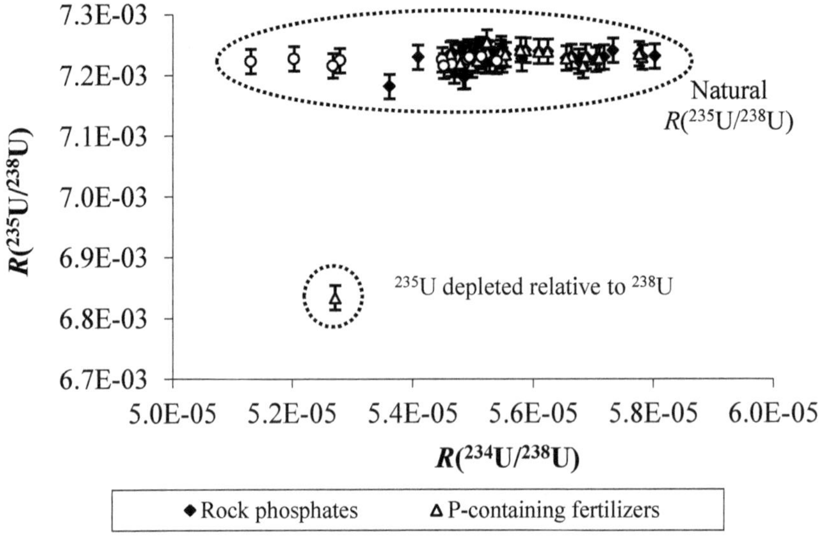

Fig.4. $R(^{235}U/^{238}U)$ and $R(^{234}U/^{238}U)$ in different phosphate containing fertilizers (Sattouf 2007).

During the enrichment process the conversion of the gaseous UF_6 to the solid U_3O_8 is costly and therefore the remaining "depleted" waste stored as gas until further disposal (Long 2002). The only reason to solidify "depleted" UF_6 is when there is a need for metallic depleted U, which is in most cases the production of Depleted Uranium weapons. DU in a fertilizer is clearly a hint that somewhere during manufacturing U has been removed, depleted of ^{235}U and the "waste" of the enrichment process afterwards returned into the fertilizer. The NPK fertilizer with DU found by Sattouf (2007) had exactly the same total U concentration like other specimens of the same fertilizer type. A hint from where the phosphate component was coming was found in the relation of strontium isotopes: A value for $R(^{87}Sr/^{86}Sr) > 0{,}70883$ indicates a Florida phosphate from the United States of America (Sattouf 2007).

Hiding a nuclear program? *Honi soit qui mal y pense!* (Edward III. Of England, 1312–1377).

References

Anonymus (2013a) A Brief History of Nuclear Weapons States http://asiasociety.org/asia101/brief-history-nuclear-weapons-states (visited on June 26, 2014)

Anonymus (2013b) Al-Qaim Phosphate plant http://www.iraqwatch.org/entities/images/qaim.htm

Astley V, Stana R (2013) There and Back Again 2.5. Who did what in solvent extraction a demonstrated and proven technology for uranium recovery from phosphoric acid. Symphos 2013, Morocco 6th to 10th May 2013, http://ebookbrowsee.net/t-s-14-symphos-2013-vaughn-astley-pdf-d528214276, (visited on June 26, 2014)

Bernaa F, Goldberga P, Horwitzc LK, Brink J, Holdt S, Bamford M, Chazan M (2011) Microstratigraphic evidence of in situ fire in the Acheulean strata of Wonderwerk Cave, Northern Cape province, South Africa. http://www.pnas.org/content/109/20/E1215, (visited on June 26, 2014)

Burr W, Cohen A (2013) Israel´s secret uranium buy. http://www.foreignpolicy.com/articles/2013/07/01/israels_secret_uranium_buy

Connett M (2001) Photographs of the Florida Phosphate Industry. http://fluoridealert.org/content/phosphate-photos/

Davenport E, Eddy P, Gillman P (1978) The Plumbat affair, Philadelphia, PA (Lippincott)

Frisbee jr. WS (2013) Tactics. http://www.military-sf.com/Tactics.htm

Follett K (1979) Triple. Fine blend N.V.

Giles L (1910) Sun Tzu on the arts of war. http://www.au.af.mil/au/awc/awcgate/artofwar.htm (visited on June 26, 2014)

Imas P (2007) personal communication at Rotem-Amfert 14.03.20017

IMPHOS (2009) Newsletter Uranium recovery from phosphoric acid. Phosphate Newsletter 26, 11 http://www.imphos.org/ download/imphos_news_special26_web.pdf

Kahaha E (2006) Historical Dictionary of Israeli Intelligence . Scarecrow Press, Inc., p. 216

Lindemann I (2007) Futter für Dimona. Strahlentelex 496-497, 6-10

Long ME (2002) Half life: The lethal legacy of America's nuclear waste. National Geographics 202: 1-33

NTI (2011) Rotem Amfert Negev Ltd.. http://www.nti.org/facilities/336/; http://www.globalsecurity.org/wmd/world/israel/mishor_rotem.htm

Orris GJ, Chernoff CB (2002) Data set of world phosphate mines, deposits, and occurrences - Part B. Location and mineral Economic data: U.S. Geological Survey Open-File Report 02-156A, 328 p. http://geopubs.wr.usgs.gov/open-file/of02-156/ OF02-156B.PDFhttp://geopubs.wr.usgs.gov/open-file/of02-156/OF02-156B.PDF

Ragheb M (2010) Uranium resources in phosphate rocks. https://netfiles.uiuc.edu/mragheb/www/NPRE%20402%20ME%20405%20Nuclear%20Power%20Engineering/Uranium%20Resources%20in%20Phosphate%20Rocks.pdf

Reuters (2011) U.N. nuclear agency inspects Syria's Homs site. http://www.reuters.com/article/2011/04/01/us-syria-iaea-nuclear-idUSTRE7302LT20110401, visited on 27.06.2014

Sattouf M (2007) Identifying the origin of rock phosphates and phosphorus fertilisers using isotope ratio techniques and heavy metal patterns. Diss. TU-Braunschweig: http://rzbl04.bib lio.etc.tu-bs.de:8080/docportal/ servlets/MCRFileNodeServlet/DocPortal_derivate_00004401/sattouf.pdf?hosts=local

Schnug E, Haneklaus N (2014) Uranium in phosphate fertilizers – review and outlook. This volume

Wippel G, Schnug E, Schmidt M, Lange F, Haneklaus S, Dersee T (2011) Nachruf Inge Lindemann. Strahlentelex 592-593/2011, http://www.strahlentelex.de/Stx_11_592_S12-13.pdf

Planning of reactive barriers – an integrated, comprehensive but easy to understand modeling approach

Markus Zingelmann[1], Mandy Schipek[1], Arnold Bittner[2]

[1] Beak Consultants GmbH, Am St. Niclas Schacht 13, 09599 Freiberg, (markus.zingelmann@beak.de or mandy.schipek@beak.de)
[2] SLR Environmental Consulting (Namibia) (Pty) Ltd, Windhoek, (abitt-ner@slrconsulting.com)

Abstract. Permeable reactive barriers (PRB's) are of particular importance for uranium mining due to a cost-effective and long-term protection standard unlike other pump-and-treat methods. As a common approach, PhreeqC models are used to determine behaviour, effectivity and dimensioning of PRB's. However, calculations and simulation of PRB's frequently fail as comprehensive and reasonable planning and prediction tool, because requirements on PhreeqC models differ between users and developer on the one hand and the mining industry on the other hand. This paper demonstrates how reasonable planning but easy to understand solutions of in situ PRB's can be managed by a combined approach of PhreeqC and numerical groundwater flow modeling results.

Introduction

Reliable strategies for containment of mined metals are compulsory for most mine closure and environmental management plans. Due to a rising demand for uranium, mining is booming and has caused a uranium rush in Namibia (Merkel and Schipek 2011).

Since groundwater quality and quantity impacts of mining projects are affected by a progressive rise of public awareness, closure plans and environmental planning has become more and more important (Bittner and Zingelmann 2013). Most recently developed mining projects within the Erongo Region, Namibia, are carnotite-based deposits. In all cases, these deposits are connected to tertiary paleochannel systems which simultaneously represent aquifers with confined conditions. Fig.1 gives an overview of current uranium mining and exploration operations in Namibia, both paleochannel and hardrock deposit hosted.

B.J. Merkel and A. Arab (eds.), *Uranium – Past and Future Challenges*,
DOI 10.1007/978-3-319-11059-2_85

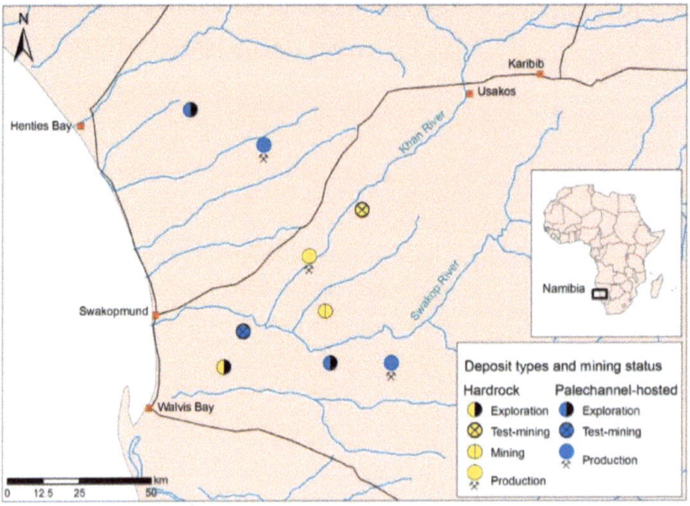

Fig.1. Overview of exploration and mining operations with regard to host-type, Erongo Region, Namibia

Besides, alluvial aquifers and seasonal perched water tables, groundwater recharge by floods in ephemeral rivers are widely distributed within the area. The paleochannels are mainly dominated by natural groundwater conditions with total dissolved solids (TDS) exceeding 10,000 mg/l on average and neutral pH varying between 6.5 and 7.5, whereas alluvial and perched aquifers show TDS below 5,000 mg/l and neutral pH conditions. Although groundwater conditions fulfil only in some cases class A-B (recommended for human consumption) according to Namibian drinking water standards (see MAWF 1956), it is widely used as source for irrigation and crop cultivation.

For this reason and because water is a scarce commodity in semi-arid Namibia anyway, measures for groundwater protection and remediation planning play a key role as part of most management plans. In this regard, PRB's are of particular importance for uranium mining due to a cost-effective and long-term protection standard unlike other pump-and-treat methods (Naftz et al. 1999). This paper demonstrates how reasonable planning but easy to understand solutions of in situ PRB's can be managed by a combined approach of PhreeqC and numerical groundwater flow modeling results. A showcase for treating groundwater and dimensioning of PRB's for efficient contaminant removal has been set up for paleochannel hosted deposits.

Methods

One of the main issues for the dimensioning of PRB's within the region was to use on-site materials. Against this background, Külls and Bittner (2013) have shown that natural on-site materials such as quartz sand and iron hydroxide coated clays have a significant retention potential and are adequate for the use in PRB's, therefore. Prior to detailed investigations of on-site material, zero valent iron (ZVI) shall be taken as reactive material, to show how integrated flow modeling and transport model solutions provide a user friendly estimation of dimensioning for PRB's.

Uranium and vanadium species and solubility depends on pH, redox conditions and the activity of other complexing species in the saturated zone (Langford 1974). Other parameters such as salinity and alkalinity are affecting transport and mobility of uranium and vanadium as well, as previous studies have proven (Mann and Deutscher 1978). Uranium species are mobile at pH < 5.5 and at >8.5 and at positive redox conditions, whereas mobility decreases in neutral and reducing conditions. In contrast, vanadium tends to form mobile species with increasing pH and with an increasing activity of negatively charged vanadium.

Above all, dispersion of uranium and vanadium is mainly dependent on groundwater movement and a general flow direction pattern. Previous investigation (Külls and Bittner 2013) in the area has shown that groundwater flow from the basement towards the alluvial aquifer system is a matter of fact. Flow velocities in the basement rocks can be specified with 10^{-9} to 10^{-7} m/s, whereas velocities in the confined paleo-channel are 10^{-7} to 10^{-6} m/s. Groundwater velocities within the alluvial channels could be determined with 10^{-5} to 10^{-4} and differs from velocities within the ephemeral river channels (10^{0} to 10^{-1} m/s). Hence, taking groundwater velocities and direction into consideration, suitable locations for PRB's as prevention measure for containment of mined metals should be downstream a proposed mine pit, cross-cutting the alluvial aquifer system as highlighted in Fig.2 below.

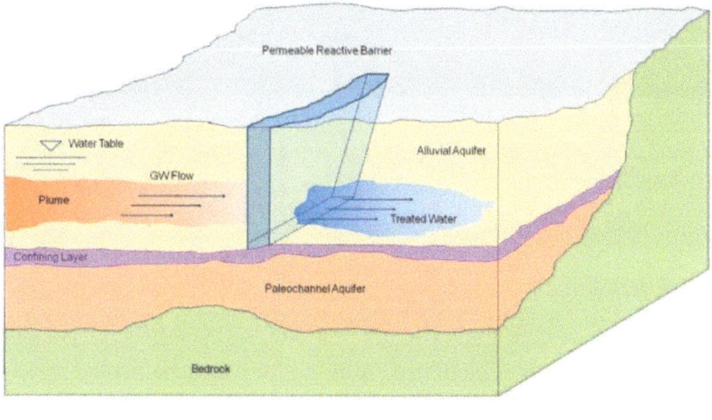

Fig.2. Schematic diagram of PRB, proposed for containment of mined metals of paleo-channel hosted uranium deposits in Namibia

Numerical Flow Modelling

Finite-elements or finite-differences based models are available for most existing and proposed mine sites, reproducing groundwater flow in the bedrock, paleo-channel and alluvial aquifers. On the one hand flow models are an essential tool to investigate hydraulic interactions between aquifers, i.e. groundwater flow from bedrock towards the alluvial aquifer, influence of confined conditions on down-stream water levels, and the effect of faults and fractures. Fig.3 shows the prismatic element mesh of the exemplarily aquifer structure with simulated groundwater heads and calculated streamlines. On the other hand, discharge volumes can be derived from the numerical models. An estimation of water budgets as input for a proper calculation of the design and dimensioning of PRB's is essential. From the authors point of view there is no preferential approach and both finite-element and finite-differences solutions are adequate to calculate flow budgets (Richardson et al. 2012). For this showcase there are two numerical flow models (finite-element and finite-differences) available. However, flow volume calculations were done with the software Feflow (finite-element approach, Fig.3).

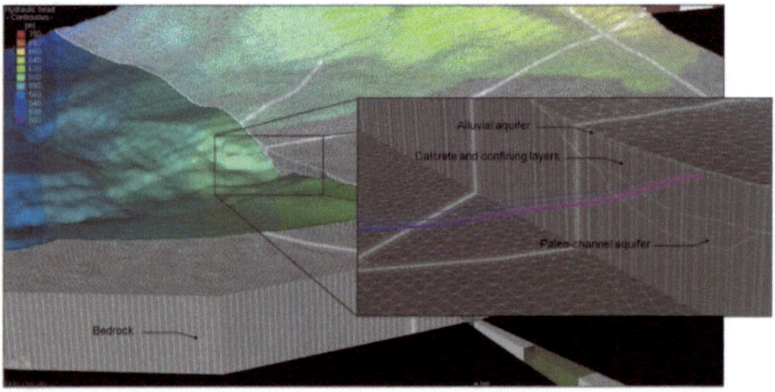

Fig.3. Prismatic element mesh of the aquifer structure, simulated hydraulic heads, and streamlines of the alluvial aquifer system

PhreeqC Modelling

Mineral saturation indices were calculated from groundwater chemistry in 46 representative sampling sites in an area containing high concentrations in uranium and vanadium. The showcase sampling site is characterized by neutral pH (6.9), high concentrations in sulphate (Ø 1,350 ppb), uranium (Ø 26,000 ppb), strontium (Ø 8,800 ppb) and vanadium (Ø 200 ppb).

Speciation and minerals saturation were calculated using the geochemical speciation program PHREEQC (Parkhurst and Appelo 1999). For modelling the llnl.dat database (Lawrence Livermore National Laboratory, 2010) was used. An iteratively approach was used to estimate the amount of ZVI to reduce uranium concentrations. After the reduction of U(VI) to U(IV) by Fe, Uranium precipitation as (probably amorphous) UO_2 was enforced.

Results

Since the requirement was to reduce uranium at least to the overall background value of 200 ppb. At least 6.67 mmol/l of Fe has to be available. Assuming a daily though flow of 0.45 l/m^2 which is characteristic for low permeable calcrete layers in the area, which is equal to a daily average of 251 mg/m^2. Furthermore, an effective operation time for about 25 years is assumed, hence 2.3 kg of Fe per m^2 must be used. Assuming a PRB horizontal extension of 120 m and a depth of 9 m.

Besides, the reactive material has to be combined with gravel to maintain adequate permeability. Laboratory investigations can be used to verify modelling results and identify the best composition for the PRB.

Conclusions

It has been shown that results from numerical flow modelling can be taken as a basis for a first estimation of the dimensioning of PRB's. For the showcase ZVI has been taken as reactive material. Performing PhreeqC calculations led to a first estimation of the amount of ZVI needed to achieve treatment. Further, more detailed results will be gained by numerical flow, particle tracking, and multispecies contaminant transport simulations implemented using Feflow. As mentioned above, on-site materials (quartz sand and iron hydroxide coated clays) have shown significant retention potential in laboratory scale, yet haven't been considered in geochemical modeling until now. Further effort has to be made to clarify on-site retention potential and dimensioning of PRB's with on-site materials.

References

Merkel, B and Schipek, M (2011). The New Uranium Mining Boom – Challenge and Lessons Learned. Preface.

Külls, C and Bittner, A (2013). Passive Barriers for Long-Term Containment of Uranium and Vanadium. Proceedings of the International Mine Water Association Annual Conference 2013: Reliable Mine Water Technology: 1017-1022

Bittner, A and Zingelmann, M (2013). Groundwater Use and Protection as Part of the Namibian Mining Application. Proceedings of the International Mine Water Association Annual Conference 2013: Reliable Mine Water Technology: 9-14

MAWF (Ministry for Agriculture, Water and Forestry, Namibia) (1956). The Water Act and its Requirements in the Terms of Water Supplies for Drinking Water and for Waste Water Treatment and Discharge into the Environment.

Langford, FF (1974). A Supergene Origin for Vein-Type Uranium Ores in the Light of the Western Australian Calcrete-Carnotite Deposits. Economic Geology 69(4), 516-526

Mann, AW and Deutscher, RL (1978). Genesis Principles for the Precipitation of Carnotite in Calcrete Drainages in Western Australia. Economic Geology 73(8) 1724-1237

Naftz, DL, Davis, CC, Fuller, SJ, Morrison, GW, Feltcorn, EM, Wilhelm, RG, Piana, MJ, Joye, J, and Rowland, RC (1999). Field Demonstration of Permeable Reactive Barriers to Control Radionuclide and Trace-Element Contamination in Groundwater from Abandoned Mine Lands. Proceedings of the Technical Meeting Charleston South Carolina 1999. Vol. 1 of 3. U.S. Geological Survey Toxic Substances Hydrology Program.

Richardson, S, Barnett, R and Evans, R (2012) Australian Groundwater Modelling Guidelines. Waterlines Report Series 82: 36-56. Australian Government, National Water Commission.

Parkhurst DL, Appelo CAJ (1999) User's guide to PHREEQC (Version 2) -- a computer program for speciation, batch-reaction, one-dimensional transport, and inverse geochemical calculations.- U S Geological Survey Water-Resources Investigations Report 99-4259: pp 312

Uranium in 31 Swedish ashes – differences between boiler type and fuels

Naeem Saqib[1], **Mattias Bäckström**[1]

[1]Man-Technology-Environment Research Centre, Örebro University, 701 82 Örebro, Sweden

Abstract. From 14 Swedish boilers (grate fired as well as fluidized bed) 31 different ashes were collected and analysed for uranium. Uranium concentrations ranged from 0.32 to 22 mg/kg dw. Average uranium concentration in the bottom ash and fly ash was 1.3 and 2.7 mg/kg dw, respectively, indicating that uranium in the fuel is quite volatile during combustion. Highest concentration of uranium was found in a fly ash from a boiler burning peat indicating that peat is a natural source of uranium.

Introduction

Radioactivity is a known problem in coal ashes (Menon et al. 2011) since many coals contain radioactive isotopes naturally. In Sweden the use of coal for heat or electricity production is very uncommon. Instead most facilities for power production use either different wastes or bio fuels. However, due to the incident in Chernobyl in 1986 several Swedish ashes have had significantly elevated levels of radioactivity. This is mainly a problem for facilities using bio fuel in mid Sweden and the reason is almost entirely due to cesium.

This paper looks into the concentrations of uranium in several Swedish ashes as a function of both boiler types and fuels used.

B.J. Merkel and A. Arab (eds.), *Uranium – Past and Future Challenges*,
DOI 10.1007/978-3-319-11059-2_86

Methods

Sampling

Study includes 13 incineration facilities with 4 grate fired, 6 circulating and 3 bubbling fluidized bed boilers. Table 1 report the information summary of the plants. 1 kg for each ash was sampled by staff on facilities at four occasions during two days to collect representative samples. Sample jars were sealed with tight lids to avoid any contact with air to prevent oxidation and carbonation. During the sampling period operation was stable and was functioning with normal fuel mixture as reported. Bottom ashes were collected from the falling stream at the bottom of the furnace while fly ash sampling locations are described in Table 1. Dry samples were collected, however some wet samples due to ash quenching were dried at 40°C and course ashes were crushed down to <4 mm to get homogenous and manageable materials.

Chemical analysis

Uranium concentrations were determined through digestion in concentrated nitric acid using a CEM MARS 5 microwave. Analytical determination was performed using an ICP-MS (Agilent 4500).

Digestion using concentrated nitric acid will not provide the total uranium concentration, but rather the acid leachable fraction. This is a more relevant parameter when it comes to leaching and potential risk for human health and the environment.

Results and discussion

Facility data

Summary of the technical data for the different facilities are presented below in Table 1. In a grate fired boiler 18 % of the total ash production is fly ash and in a bubbling fluidized bed (BFB) and a circulating fluidized bed (CFB) the average fly ash fractions are 42 % and 59 %, respectively.

Table 1. Sampled facilities with information about boiler type and fuel use

Facility	Boiler type	Fuel	Temp. (°C)	Amount ash (bottom/fly) (%)	Fly ash sampling location
Umeå	Grate	Municipal	1 087	83/17	Bag filter
Kiruna	Grate	Municipal	1 091	92/8	ESP
Sundsvall	CFB	Municipal	860	50/50	ESP + Bag filter
Lidköping	BFB	Municipal + Industrial	874	73/27	Bag filter
Linköping	Grate	Cont wood chips, bark, plastic reject	NA	85/15	Bag filter
Händelö P13	CFB	Wood chips	899	34/66	ESP
Händelö P14	CFB	Municipal + industrial	860	52/48	Bag filter
Söderenergi	Grate	Industrial + fuel pellets	1 000	81/19	ESP
Nynäshamn	BFB	Cont wood chips	850	34/66	Bag filter
Mälarenergi	CFB	Cont wood chips + peat	782	22/78	Bag filter
Munksund	CFB	Bark + wood chips	809	38/62	ESP
Braviken	Grate	Bark + cont wood chips + DIP sludge	1 023	67/33	ESP
Eskilstuna	BFB	Wood chips	870	68/32	ESP
Högdalen	CFB	Industrial waste	NA	50/50	Bag filter

[a] BFB: bubbling fluidized bed, CFB: circulating fluidized bed, NA: not available, cont: contaminated, ESP: electrostatic precipitator

Uranium concentrations

Uranium concentrations in all ash samples can be found in Table 2 below. Uranium concentrations ranged from 0.32 to 22.2 mg/kg dw. Average uranium concentration in bottom ashes was 1.23 mg/kg dw and average concentration in fly ashes was 2.65 mg/kg dw. When excluding the incinerator with the highest uranium concentrations average uranium concentrations for bottom and fly ash, respectively, was 0.93 and 1.35 mg/kg dw. This indicates that uranium concentrations in general are higher in the fly ash than in the bottom ash and that uranium possibly is volatile during combustion.

Table 2. Uranium concentrations (mg/kg dw) in the different ash samples

Facility	Ash sample	Uranium conc (mg/kg dw)	Ratio fly/bottom
Umeå	Bottom	1.18	1.06
	Fly	1.25	
Kiruna	Bottom	1.11	
Sundsvall	Bottom	0.44	4.57
	Fly	2.01	
Lidköping	Bottom	0.33	4.45
	Fly	1.47	
	Cyclone	1.69	
	Turn over shaft	0.92	
Linköping	Bottom	0.74	
Händelö P13	Bottom	0.40	2.38
	Fly	0.95	
Händelö P14	Bottom	0.32	4.94
	Fly	1.58	
	Turn over shaft	1.52	
Söderenergi	Bottom	2.09	0.71
	Fly	1.48	
Nynäshamn	Bottom	0.46	2.80
	Turn over shaft	1.29	
Mälarenergi	Bottom	4.81	4.62
	Fly	22.2	
Munksund	Bottom	0.72	1.24
	Fly	0.89	
Braviken	Bottom	2.67	0.77
	Fly	2.06	
Eskilstuna	Fly	0.89	
Högdalen	Bottom	0.72	2.17
	Fly	1.56	

Impact of boiler types

When looking at the uranium concentrations in ashes from grate fired boilers and from fluidized bed boilers (both circulating and bubbling beds) there are some differences as shown in Table 3. There is almost no difference between uranium concentrations in bottom and fly ashes from the grate fired boilers while the concentration of uranium is around three times higher in the fly ash compared to the bottom ash from fluidized boilers. It should be noted that the average temperature in the grate fired boilers are 1 050 °C compared to the average temperature of 850 °C for the fluidized boilers. Usually a higher temperature results in a higher transfer of elements to the fly ash (Morf et al. 2000; Yoo et al. 2002). This indicates that there is another process aside from temperature that increase the uranium concentrations in fly ash compared to the bottom ash from fluidized bed boilers. A fluidized bed boiler generates a lot more fly ash than a grate fired boiler due to the movement in the bed. It is possible that uranium is associated to the smaller particles moved into the fly ash by the movement in the bed (high turbulence).

Average concentrations in ash from grate fired boilers compared to ash from fluidized bed boilers are not significantly different from each other when the facility with the highest uranium concentration is excluded.

From a mass balance perspective 81 % of the total uranium can be found in the bottom ash from a grate fired boiler while the corresponding fraction for a fluidized bed is only around 24 %. Earlier studies have reported 30 % in the bottom ash from a rotary kiln (Schoch 2000).

Table 3. Average uranium concentrations (mg/kg dw) in the ashes dependent on boiler type

Ash type	Uranium conc (mg/kg dw)
Grate bottom	1.56
Grate fly	1.60
CFB/BFB bottom	1.03
CFB/BFB fly	3.08
CFB/BFB bottom*	0.48
CFB/BFB fly*	1.34

* the facility with the highest uranium concentrations excluded

Impact of different fuels

Uranium concentrations as a function of different fuel types are presented in Table 4 below. All ashes were somewhat arbitrarily divided into four different fuel types; wood (both contaminated and clean), municipal waste, plastic waste and wood/peat. Most facilities used a combination of fuels and the category should be considered a best estimate.

Most fuels generate low uranium concentrations. It is, however, quite obvious that the ashes with the highest uranium concentrations have been found at the facility using peat as one of the fuels. Peat is known to accumulate trace metals (Boyle 1982) and during incineration the concentration is increased in the ash. Uranium concentrations in ashes from firing of peat are comparable to concentrations in coal ashes (10-30 mg/kg dw; USGS 1997).

Table 4. Average uranium concentrations (mg/kg dw) in the ashes dependent on fuel type

Fuel type	N	Uranium conc (mg/kg dw)
Wood bottom	4	1.06
Wood fly	5	1.22
Municipal bottom	5	0.68
Municipal fly	7	1.49
Plastic bottom	3	1.18
Plastic fly	2	1.52
Wood/peat bottom	1	4.81
Wood/peat fly	1	22.2

Conclusion

Uranium concentrations in Swedish ashes are fairly low with an average of 1.23 and 2.65 mg/kg dw in bottom and fly ashes, respectively. In grate fired boilers 81 % of the uranium is found in the bottom ash while the corresponding fraction for fluidized boilers is 24 %. The only fuel significantly impacting the uranium concentration in ash was peat.

References

Boyle R (1982) Geochemical prospecting for thorium and uranium deposits. Elsevier

Menon R, Raja P, Malpe D, Subramaniyam K, Balaram V (2011) Radioelemental characterization of fly ash from Chandrapur Super Thermal Power Station, Maharashtra, India. Curr Sci 100: 1880-1883

Morf L, Brunner P, Spaun S (2000) Effect of operating conditions and input variations on the partitioning of metals in a municipal solid waste incinerator. Waste Manag & Res 18: 4-15

Schoch B (2000) Metal partitioning in the department of energy-Oak Ridge TSCA Incinerator. WM′00 Conference, February 27-March 2, 2000, Tuscon, AZ, USA (22 pp)

Yoo J, Kim K, Jang H, Seo Y, Seok K, Hong J, Jang M (2002) Emission characteristics of particulate matter and heavy metals from small incinerators and boilers. Atmos Enviro 36: 5057-5066

USGS (1997) Radioactive elements in coal and fly ash: abundance, forms and environmental significance. US Geological Survey Fact Sheet FS-163-97

Characterization of natural phosphates and phosphogypsum

Fatima Zahra Boujrhal[1,2]

[1]Moulay Slimane University, Faculty of Sciences and Techniques, Laboratory of Management and Valorization of Natural Resources, Beni Mellal, Morocco
[2]Mohamed V University, Faculty of Sciences, Laboratory of Nuclear Physic, Rabat, Morocco

Abstract. Sedimentary phosphate from Moroccan deposit and its corresponding phosphogypsum where characterized by several physic-chemical techniques as well as by nuclear method.

Uranium and radium contents were determined by gamma spectrometry. Specific characterizations (density and specific surface area) are determined by standard liquid immersion and BET techniques. Phase transformations were studied by differential thermal analysis. Structure and microstructure where investigated by XRD, IR, SEM/TEM. A comparison studies were done between the two different materials; phosphate and phosphogypsum.

Introduction

Moroccan sedimentary phosphate contains 50 to 120 ppm of uranium. Recent studies in sedimentary phosphates (Berrada et al. 1995; Boujrhal et al. 1999; 2004; 2005; 2007; Boujrhal and Cherkaoui 2009; 2011), in the fossil bones and dentine (Boujrhal and Cherkaoui 2009; 2011; Reyes-Casga et al, 1999; Allard and al. 1997; Berrada et al. 1995; Naudet 1991) and in other natural phosphates (Tucotte et al. 1982; Weber 1982) showing that the phosphate rock is mainly composed of apatite ($Ca10(PO_4, CO_3)_6(F, Cl, OH)_2$.

Phosphogypsum is a waste by-product of phosphate fertilizer industry. It derived from dissolution of phosphate ore in H_2SO_4 in order to produce phosphoric acid and It contains various toxic elements (radioactive isotopes and heavy metals).

$Ca_{10}(PO_4)_6\,F_2 + 10\,H_2SO_4 + 20\,H_2O$ ------> $6\,H_3PO_4 + 10\,CaSO_4.2H_2O + 2FH$
Phosphate ores   Phosphogypsum

B.J. Merkel and A. Arab (eds.), *Uranium – Past and Future Challenges*,
DOI 10.1007/978-3-319-11059-2_87

Three tons of natural phosphates generate five tons of phosphogypsum which requires better management and valorization of this by-product gypsum ($CaSO_4.2H_2O$), since in most countries; this by-product is spilled in the oceans, seas and rivers. Currently, the growing awareness in the world of this water pollution leads to file it on the land and to treat it for applications in civil engineering; construction, and in particular road.

All phosphate and phosphogypsum contain various toxic/radioactive elements which can be transferred to environment and induce artificial pollution (Gaudry et al. 2007).

Materials and Methods

Materials

Moroccan sedimentary phosphate and its corresponding phosphogypsum were characterized by nuclear analysis and conventional techniques.

Methods

Gamma spectrometry

Gamma spectrometry was used to determine the uranium and radium content as well as radon emanation power using the method developed in (Boujrhal and Cherkaoui, 2011).

Liquid immersion and BET techniques

Standard liquid immersion technique was used to measure the density of studied specimens. BET Technique based on the adsorption of gas molecules (He) on the solid surface, was served to measure the specific surface area these materials.

X-ray diffraction

X-ray diffraction was collected on an automated Philips PW 1877 diffract-meter, operating in step scan mode using Co-Kα radiation in order to exam the structure of the two specimens.

SEM

Scanning Electron Microscopy (SEM) observations on the microstructure of the two specimens were done on a Philips instrument (25 KV, 10nA) and Transmission Electron Microscopy observations was used on Joel 2000.

IR spectrometry

The powdered samples were ground under acetone by mixing 1mg of sample and 300 mg of KBr. A transparent pellet was prepared after been pressed at 10 ton.cm^2. IR spectra were recorded in absorbency mode with Nicolet Magna 560 spectrometer.

Thermogravimetric and differential thermal Analysis (TG-DTG)

TG-DTG measurement was used to study the thermal behavior of the specimens using Sataram (TGA92) TD-DTG apparatus in air with a heating rate of 10°C/min.

Results and discussions

Radiometric analysis determined by gamma spectrometry and specific characterizations; density and specific surface area measured respectively by liquid immersion technique and BET method of the two specimens studied are regrouped in Table 1.

The contents of U-238 and Ra-226 of the phosphate studied here is similar, whereas, the content of U-238 in the corresponding phosphogypsum is lower to detection limit. On the other hand, the radon emanation power is largely important in phosphogupsum than in phosphate.

Table 1. U-238, Ra-226, Rn-222 analysis and physical characterizations of phosphate and the corresponding phosphogypsum

	U-238 ppm	Ra-226 ppm eU	Rn-222 Emanation %	Specific density g/cm^3	Specific surface area m^2/g
Phosphate	124+9	137+5	8.4+0.7	2.9	12.27
Phosphogypsum	< DL	70+4	29+1.6	2.7	2.5

X-ray diffraction, infrared spectrum and SEM are used to characterize the structures and the microstructures of these tow specimens whereas thermogravimetric analyses were served to study their thermal properties and phase transformations (Table 2).

XRD analysis show the principal phase in each specimen; Apatite mineral for phosphate and hydrated calcium sulfate ($CaSO_4.nH_2O$, n=0.5, 2) for phosphogypsium, associated with secondary phases as SiO_2, $CaCO_3$, $CaSO_4$ …

Infrared spectrum of phosphate shows the bands corresponding to apatite mineral (PO_4^{2-}, HPO_4^{2-} and OH^- bands) and IR spectrum of phosphogypsum shows the bans corresponding to the hydrated calcium sulfate (SO_4^{2-}, H_2O bands). These results are in good agreement with X-ray diffraction results.

On the other hand, X-ray analysis using SEM coupled with EDS also confirm this conclusion, the obtained spectra show the main elements the principal compound phase of each specimen.

In addition, SEM observations show the form of crystallites; hexagonal in the phosphate and tubular in phosphogypsum (Fig. 1, Fig. 2).

In the TG-DTG curves of phosphate, a large endothermic process accompanying a rapid and large weight loss occurs at 780°C, due to the decomposition of the carbonate and to reorganization of apatite structure (Fig. 3). The two first endothermic processes correspond to the loss of water (at 150°C) and organic matter decomposition (at 350°C).

In the TG-DTG curves of phosphogypsum, an endothermic process was observed at 150°C corresponding to the loss of structure's water or to the conversion of calcium sulphate dihydrate to hemihydrate as shown as follows (Greenwood and Earnshaw, 1989):

$$\underset{\text{Dihydrate}}{CaSO_4.2H_2O} \xrightarrow{150°C} \underset{\text{Hemihydrate}}{CaSO_4.1/2H_2O} \xrightarrow{600°C} \underset{\text{Anhydrite}}{CaSO_4}$$

At 600°C, no pick was observed but a second endothermic process was occurred at 870°C certainly corresponding to the transformation of other phases (impurities).

Table 2. Structure, microstructure and thermal analysis

	DRX principal phase	SEM/TEM Crystallite form	IR bands	TG-DTG picks (°C)
Phosphate	Apatite	Hexagonal	H_20, OH^-	150°C
			CO_3^{2-}	350°C
			PO_4^{2-}, HPO_4^{2-}	790°C
Phosphogypsum	$CaSO_4.nH_2O$	Tubular	H_20, SO_4^{2-}	150°C
	n = 0.5, 2			870°C

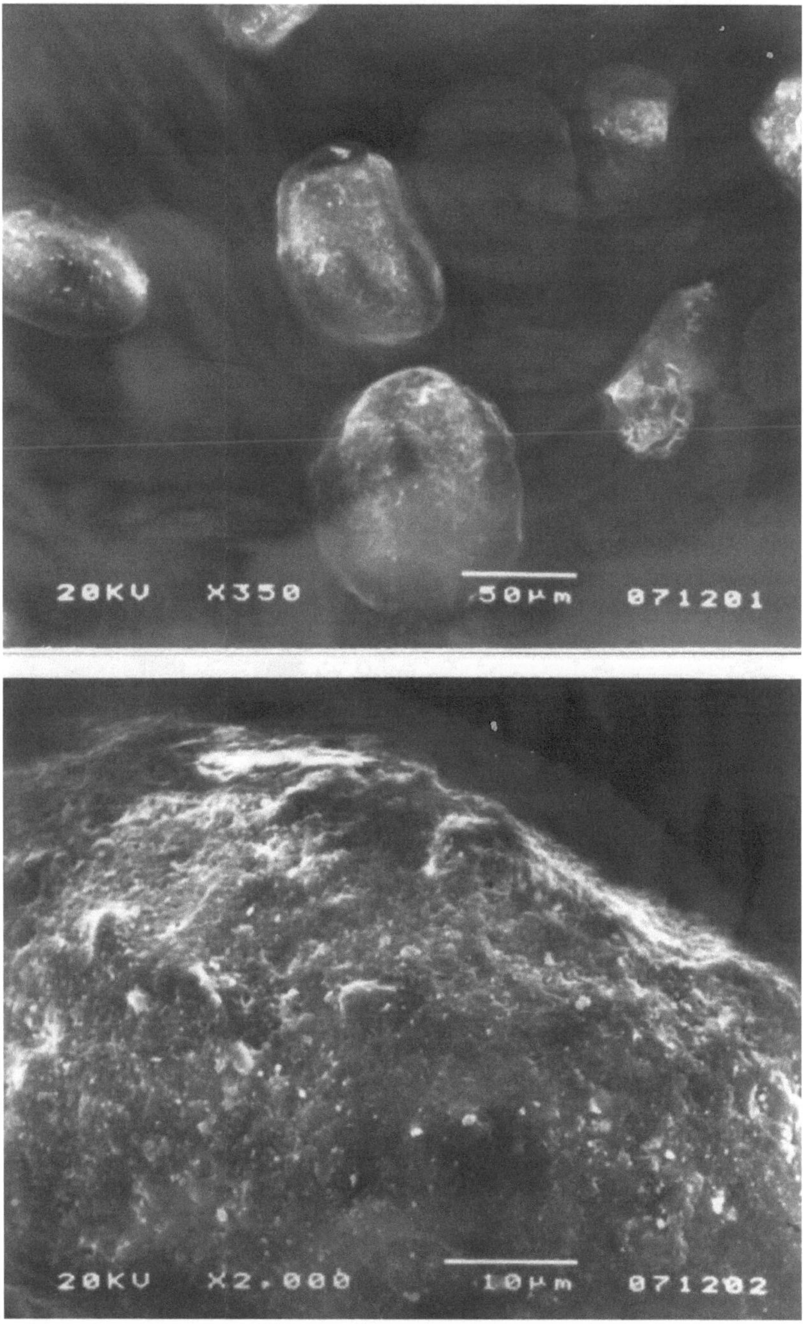

Fig.1. SEM photograph s of phosphate. In the 1st photograph: Granular form of phosphate (grains about 200µm). In the 2nd photograph: External morphology/microstructure of one grain

Fig.2. SEM photograph of phosphogypsum. In the 1st photograph: Tubular shapes (platelets) of phosphogypsum. In the 2nd photograph: Microstructure of the strand

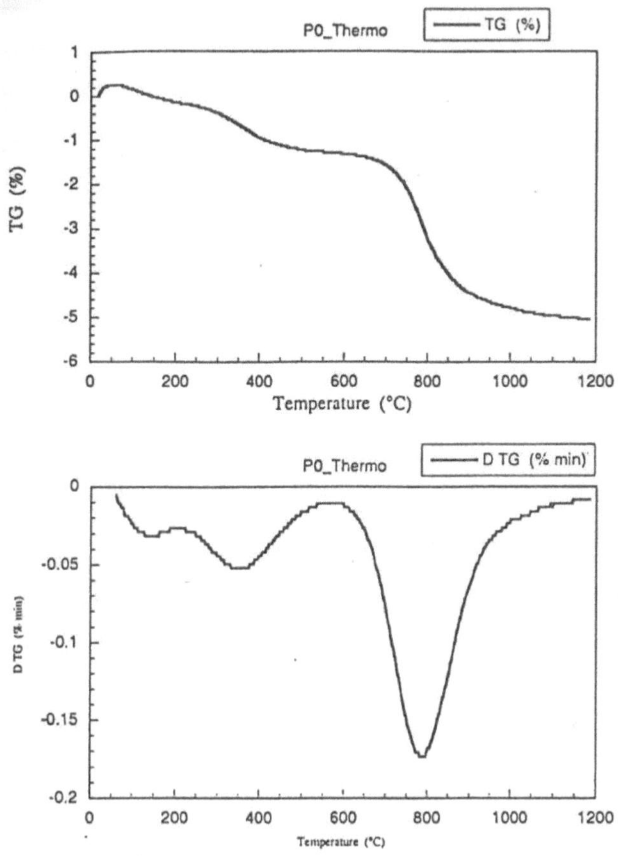

Fig.3. TG-TDG analysis of phosphate

Conclusion

Phosphate and corresponding phosphogypsum contain radioactive isotopes Ra-226 and U-238. The last one is lower than the limit detection in phosphogypsum. Radon emanation power is more important in phosphogypun than in phosphate. Apatite is the principal compound of the phosphate, whereas the calcium sulphate hydrates ($CaSO_4.nH_2O$, n= 2, 0.5) are the mixture phases of phosphogypsum, results confirm by various physic-chemical techniques used.

On the other hand, an endothermic process occurs at 790°C in phosphate corresponding to the carbonate decomposition and apatite structure reorganization, whereas, the endothermic process occurs at 150°Cin phosphogypsium corresponds to the dehydration of the phosphogypum (calcium sulphate hydrate).

References

Allard T, Calas G, Ildefone P, Morin G, Muller J.P (1997). C.R. Acad. Sci. Paris, Earth and Planetary Sciences, 325: 1-10

Berrada M, Boujrhal F.Z, Couchot P, Chambaudet A, Mercier R (1995) Effet de la température de cuisson sur le potentiel d'émanation et le taux de dégazage en radon de phosphates sédimentaires, Dans : Gaz Geochimestry (C. Dubois, D. Clein, A. Chambaudet, M. Rebetez). Science Reviews: 335- 358

Boujrhal F.Z, Hlil E.K, Cherkaoui El Moursli R (2004) Study of apatite behaviour in the presence of radionuclides U and Rn and local modification of their crystalline and electronic structure. Radiation Physics and Chemistry, 69: 1-6

Boujrhal F.Z, Hlil E.K, Cherkaoui El Moursli R, El Khoukhi T, Sghir B (2005) A comparative study of radon retention ability of crystalline apatite and amorphous oxide materials. Materials Science Forum, 480-481:169-174

Boujrhal F.Z, Sghir B, Ossama S, Cherkaoui El Moursli R (2007) Investigation of the micro/nanostructure and the structure defect of sedimentary phosphates by electron microscopy. Acta Cryst. A63: s148

Boujrhal F.Z, Cherkaoui El Moursli R (2009) Morphologie externe et interne des phosphates sédimentaires marocains: effet de recuit. International Journal of Environmental Studies, 66 Issue 2: 229-249

Boujrhal F.Z., Cherkaoui El Moursli, R (2011) Direct and Indirect Effects of Uranium on Microstructure of Sedimentary Phosphate: Fission Tracks and Radon Diffusion. Merkel B. Schipek M. Eds. The New Uranium Mining Boom, DOI 10.1007/978-3-642-221224 Springer: 743-748

Boujrhal F.Z., Cherkaoui El Moursli, R (2011) The Routine Determination of Uranium Activity in Natural Rocks and Minerals by Gamma Spectrometry. Merkel B. Schipek M. Eds. The New Uranium Mining Boom, DOI 10.1007/978-3-642-221224. Springer: 477-481

Gaudry A, Zeroual S, Gaie-Level F, Moskura M, Boujrhal F.Z, Cherkaoui El Moursli R, Guessous A, Mouradi A, Givernaud T, Delmas R (2007) Heavy metals pollution of the atlantic marine environment by the Moroccan phosphate industry, as observed through their bioaccumulation in Ulva Lactuca. Water Air & Soil Pollution, Vol.178, n°1-4: 267-285.

Greenwood N.N and Earnshaw A (1989) Chemistry of the Elements. Pergamon Press, Elmsford, New york: 133

Naudet R (1991). Eyrolles, Paris: 695p

Reyes-Casga J, Garcia G.R, Alvarez-Fregoso O, Chavez-Carvayar J.A, Vargas-Ulloa L.E (1999) Journal of Materials Science, 34: 2183-2188

Turcotte R.P, Wald J.W, Roberts F.P, Rusin J.M. Lutze W. (1982) Journal of American Ceramic Society, 65/12: 589-593

Wheber W.J. (1982) Journal of the American Ceramic Society, 65/11: 544-593

Radioactivity assessment at the site of historical radium salts factory

Fernando P. Carvalho[1], **João M. Oliveira**[1], **Margarida Malta**[1]

[1]Instituto Superior Técnico/Laboratório de Protecção e Segurança Radiológica, Universidade de Lisboa, Estrada Nacional 10, km 139, 2695-066 Bobadela LRS, Portugal (E-mail: carvalho@itn.pt)

Abstract. A radium production factory was operated in the Barracão village, near Guarda, Portugal, and from 1908 to 1941 produced radium-barium sulphate salts for use in radiotherapy. The factory site has been closed for decades and surveillance was kept by the uranium mining company. In the early 80s of past century, the site was monitored and partly cleaned up with transfer of ore remainders and waste heaps to a waste disposal mine site elsewhere. New clean-up operation took place in 2007 and the facilities were fenced. The radioactivity levels in and around the Factory were recently re assessed and results are reported herein. Concentrations of uranium series radionuclides were determined in agriculture soils nearby, in water from irrigation wells, in the surface stream that received waste water from the factory, and in water from a public fountain in the village. Vegetables from agriculture plots near the factory were analyzed also and the effective radiation dose to inhabitants through consumption of these products was assessed. Radiation exposure of members of the public and results of clean-up operation are discussed.

Introduction

The Radium Salts Factory, located in the village of Barracão, near Guarda in Portugal, was owned by a private bank and later by the Companhia Portuguesa do Rádio (Portuguese Company of Radium, CPR). This Radium Factory was in operation from 1908 to 1941 to process radioactive ores from several mines in the centre-North of the country to producing radium, as radium-barium sulphate, for export. This Factory produced several curie of radium, before displacement of the chemical processing to the Urgeiriça mine facilities, and abandon of facilities from 1944 onwards. Following discontinuation of activities, the Radium Salts Factory was maintained as a property of CPR, later transferred to the Empresa Nacional de

B.J. Merkel and A. Arab (eds.), *Uranium – Past and Future Challenges*,
DOI 10.1007/978-3-319-11059-2_88

Uranio (Uranium Mining Company, ENU) and, although not completely closed, the site was of restricted access.

The Radium Salts Factory is located inside the perimeter of the village and is surrounded by houses and agriculture plots. In the 80s of past century the site was monitored by the Laboratory of Radiological Protection and Safety and several samples were analyzed by gamma spectrometry (Teixeira and Elias, 1982). Reports made at that time identified abandoned waste heaps containing elevated radioactivity reported up to about 1000 Bq/kg in 1982 (Teixeira and Elias, 1982). Following this first assessment, the ENU company transferred about 3000 tons of waste materials to the milling tailings of Bica mine in order to reduce the environmental contamination and radiation risks for the population of Barracão village. A survey of the same area reported in 1988 up to 83.9 Bq/g of ^{226}Ra in waste materials still left in the area, and ambient radiation doses up to 20 µGy/h inside the Radium Factory property (Bettencourt et al. 1985, 1988). In 2007, additional cleanup was undertaken by EDM, mine holding that inherited the assets of the former uranium mining company ENU, and more contaminated materials were removed. Warning signs and a wire fence delimiting the site were redone, aiming at improving radiological safety and avoiding use of contaminated materials.

As part of the national monitoring programme of radioactivity in former uranium mine regions, this site was surveyed in September 2008 and visited again in July 2014 to assess the ambient radiation dose and the radionuclide concentrations in environmental samples (Carvalho 2014). Results and conclusions of this site assessment are presented herein.

Materials and Methods

The property of the old Radium Salts Factory occupies an area of about 1.5 hectares limited by a road and railways on East side and a road on the West side (Fig. 1). Inside the property there are ruins of administration, laboratory, and warehouse buildings, and ore leaching and radium precipitation tanks. Most waste heaps were removed in 2007, the top soil layer of the yard was scrapped and a layer of gravel was put as a cover. In the Factory yard, several stone carved leaching tanks still contain several tons of solid waste materials.

The area of Radium Salts Factory and the Barracão village were monitored for the ambient radiation dose using hand held radiation dose rate meters, namely an ionization chamber Smart-Ion and a Geiger-Muller Thermo FH-40, calibrated by a certified Metrology Laboratory. Ambient radiation measurements were performed at about 1.2 meter above the ground in triplicate at each point and readings averaged.

Samples of waste heaps were collected inside the perimeter of the old Radium Salts Factory. Agriculture soil, vegetables from the available production, and water from wells and boreholes were collected. Surface water and sediments from a stream, Ribeira de Noeime, used in the past by the Factory to discharge process water were collected also. Water from a public fountain built in 1921 near the Factory to supply water to the population, was collected for analyses.

Samples were prepared for separation of radioelements by radiochemical procedures and radionuclide activity concentrations were determined by alpha spectrometry according to procedures described elsewhere and validated (Carvalho et al. 2007a, 2007b; Carvalho and Oliveira 2009; Oliveira and Carvalho 2006).

Fig.1. Radium Salts Factory site (delimited by a dashed line). Circles with numbers indicate sampling sites. The stream Ribeira de Noeime is located about 500 m to the North

Results and Discussion

Ambient radiation dose in the agriculture fields around the Barracão village averaged 0.5 μSv/h and raised to 1.0 μSv/h in the agriculture plots and trails at the perimeter of the Radium Salts Factory property. At the gates of the Factory over the access road the radiation dose was of 1.2 to 2.2 μSv/h, while on the inside yard dose rates ranged from 1.5 to 6.3 μSv/h, and were even higher inside the old buildings that served as chemical laboratories and warehouse, ranging from 3.0 to

7.0 μSv/h. The highest dose rates from 7.0 to 12.0 μSv/h were measured above the solid waste material left in the leaching tanks and nearby at sampling point #16.

In the streets of the Barracão village, by the inhabited houses and railway station the ambient dose rate was generally at 0.3-0.4 μSv/h.

The average radiation ambient dose rate for the region, reported as the outdoor terrestrial gamma radiation absorbed dose rate in the air for this region, Guarda (mine areas not included) was 122.9 nGy/h (60.4-193.4 nGy/h), and is the natural ambient radiation background for this region. The annual effective dose for this region (outdoor and indoor) from ambient radiation was calculated at 0.96 mSv/y, which should be considered as a regional baseline value (Amaral et al., 1992)

Activity concentrations of radionuclides determined in solid waste materials, surface soils and stream sediments are shown in Table 1. Waste materials left in the tanks used for dissolution of the ore showed high concentrations of ^{230}Th, 26847±1537 Bq/kg, about 5 times higher than those of ^{226}Ra and ^{238}U isotopes, indicating that these were wastes from radium extraction. Soils from the neighbor agriculture plots (#10 and #17) showed a clear contamination with particularly elevated radionuclide concentrations in soil #10, while soil #17 displayed concentrations only slightly above other soils of the region analyzed for comparison (#23). Soil #10, taking into account the radionuclide ratios, was probably contaminated by the process water released by the Factory into a streamlet flowing into the Ribeira de Noeime stream. Sediments of the stream Ribeira de Noeime (#12 and #13) were contaminated with discharges of process water and after all these decades still displayed 226Ra concentrations higher than uranium and both radionuclides higher than sediments from other streams of this region (#24) (Table 1).

Water from the public fountain at the railway station square (#11), not far from Radium Factory, and supplied from a borehole nearby, displayed activity concentrations of radionuclides very low which indicated that process water from the Radium Factory did not contaminate this local groundwater source (Table 2). Radioactivity in water samples from the surface stream Ribeira de Noeime (#12 and #13) was slightly higher than in water from the public fountain. Water from wells of the agriculture plots were particularly high for dissolved uranium, especially well #10, in the same agriculture plot with contaminated soil. Radionuclide concentrations in the suspended particulate matter of the same water samples showed highly contaminated radioactive particles particularly in well #10 for uranium isotopes, ^{230}Th and ^{210}Pb. Water from the public fountain was consistently low in concentrations in suspended particulate matter (Table 2).

Activity concentrations of radionuclides in several agriculture products from the area are shown in Table 3. Vegetable and fruit products were available in four areas in and around the Radium Factory. Wild berries, spontaneously growing inside the perimeter of the Factory, displayed the highest ^{226}Ra concentrations, 15672±815 mBq/kg wet weight, much above any other radionuclide accumulated in berries. Outside the Factory, in the agriculture plot #10, lettuce, tomato and carrot displayed enhanced concentrations in comparison with agriculture products from the areas #15 and #17, and in line with contaminated soil and irrigation water from the contaminated well both containing enhanced radionuclide concentrations. These radionuclide concentrations in products from the area #10 were higher also than in similar products from the region, and were comparable or higher than concentrations determined in agriculture products grown in the neighborhood of uranium mines such as Quinta do Bispo and Cunha do Bispo with significantly contaminated mine drainage (Carvalho et al., 2009a, 2009b).

In Barracão village, water consumption by the population using either the public water distribution network or the public fountain does not give a significant contribution to the internal radiation dose due to low radioactivity in the water. Nowadays, water from wells generally is used for irrigation only. However, if water from wells is used for human consumption the dose rate could be significantly elevated.

Although not all agriculture plots produce vegetables with enhanced radionuclide concentrations, the consumption of products from area #10 conduces to higher internal radiation dose than from other areas around the Barracão village. Consumption of such vegetables, for example through preparation of potato and vegetable soup, a common item in the local diet, could deliver an ingestion dose rate of about 4 mSv/y. Internal radiation doses of this order were calculated for similar scenarios in the neighborhood of Cunha Baixa uranium mine, where agriculture plots were irrigated with contaminated water (Carvalho et al., 2014a). Collection of wild berries in the Radium factory site could aggravate the radionuclide intake but it is assumed that in the perimeter of the factory the production of spontaneous berries would not be significant. Taking the external and internal radiation doses into account, the total radiation effective dose for a member of the public, in this critical group, could attain 4-6 mSv/year.

Comparing the current situation with assessments made in the past, the removal of waste heaps from the factory site contributed to decrease the ambient radiation dose and environmental health risks. The removal of radioactive waste that could be a source of environmental contamination by dust dispersal and surface runoff prevented further contamination of the surface stream and agriculture soils nearby. Nevertheless, the remaining facilities, including old buildings, tanks, and milling tailings that are contaminated materials, still need final disposal. Contaminated areas, including agriculture plots nearby and contaminated water in surface wells need full characterization and delimitation.

The emplacement of this old Radium Factory in the perimeter of the village may require further intervention to conclude site remediation up to a satisfactory level of radiation protection. In the terrestrial environment it is frequently found in contaminated areas that radium is the main contributor to the radiation dose in humans (Carvalho et al. 2014b). Similar contaminated radium factories were dismantled and remediated, such as in Ollen, Belgium, and Nogent, France, contributing to solve radiation exposure in legacy sites (IAEA, 2014).

Table 1. Activity concentration of radionuclides in soils and sediments (Bq/kg dry weight)

Sample		Id.	^{238}U	^{235}U	^{234}U	^{230}Th	^{226}Ra	^{210}Pb=^{210}Po	^{232}Th
Radium Salts Factory	Tailings	#16	5891±256	267±18	6022±262	26847±1537	4624±316	19109±895	89±6
Agriculture plot	Soil	#10	2832±76	129±5	2930±78	1034±64	1478±91	1082±41	100±7
Agriculture plot	Soil	#17	526±14	24±2	543±14	550±29	738±54	489±16	163±9
Stream R. Noeime	Sed.	#12	522±16	23±2	537±16	439±28	969±104	645±22	99±7
Stream R. Noeime	Sed.	#13	778±22	33±2	779±22	875±48	1146±89	897±31	165±10
Freixinho village	Soil	#23	475±15	18±2	446±14	470±32	620±52	382±15	340±24
Stream R. da Pêga	Sed.	#24	453±15	22±2	470±15	503±26	573±45	413±15	125±7

Table 2. Activity concentration of radionuclides in the soluble phase (< 0.45 μm) and suspended particulate matter (> 0.45 μm) of water samples

Sample	Nr	^{238}U	^{235}U	^{234}U	^{230}Th	^{226}Ra	^{210}Pb	^{210}Po	^{232}Th
Dissolved (mBq/L)									
Barracão, irrigation well	#10	549±14	26.6±0.9	579±14	0.13±0.03	59.7±4.3	105±4	37.8±0.9	0.04±0.02
Barracão, public fountain	#11	21.1±0.6	0.93±0.08	19.4±0.5	0.10±0.02	40.7±1.8	142±5	<0.08±0.3	0.09±0.04
Stream R. de Noeime	#12	120±3	5.5±0.2	151±3	1.0±0	33.3±1.3	20.8±1.0	98.7±2.6	0.07±0.03
Stream R de Noeime	#13	36.1±0.9	1.7±0.1	37.6±0.9	0.6±0.1	14.8±1.2	42.1±1.6	9.8±0.3	0.03±0.02
Barracão, borehole	#15	437±12	19.7±0.8	663±17	0.44±0.08	16.2±0.8	81.8±3.1	5.0±0.2	0.07±0.03
Suspended particulate matter (Bq/kg dry weight)									
Barracão, irrigation well	#10	46023±1216	2206±106	47853±1261	14669±783	6252±614	154830±8364	24401±1269	342±38
Barracão, public fountain	#11	2933±86	143±10	3223±94	1859±89	763±43	8357±285	4962±257	55.1±5.6
Stream R. de Noeime	#12	7859±253	339±32	7364±240	7844±402	3376±280	18182±671	5054±129	217±8
Stream R. de Noeime	#13	1299±41	63.6±5.8	1286±41	1882±104	2925±227	3297±125	2909±70	45.1±5.0
Barracão, borehole	#15	4202±120	196±17	4572±128	11756±22	7620±827	33715±1231	10953±278	291±22

Table 3. Activity concentrations of radionuclides in vegetable samples from local agriculture (mBq/g wet weight)

Sample	Station Nr	Dry/wet weight	^{238}U	^{235}U	^{234}U	^{230}Th	^{226}Ra	^{210}Pb	^{210}Po	^{232}Th
Lettuce	#10	0.05	2034±52	87.5±5.2	2063±52.4	732±43	2183±110	429±15	431±15	58.9±5.7
Tomato	#10	0.06	67.2±4.8	2.3±1.1	64.3±4.9	24.7±2.0	179±11	24.7±2.1	8.9±0.6	1.3±0.3
Corn cob	#10	0.4	24.7±1.4	0.95±0.27	24.2±1.4	14.9±2.1	499±34	120±16	71±9	<2.5
Corn leaf	#10	0.19	367±9	14.8±1.1	354±9	229±13	838±52	880±37	550±18	23.0±2.1
Carrot	#10	0.12	383±11	18.7±1.5	398±11	64.6±5.0	3096±138	623±15	204±5	4.7±1.0
Apple	#10	0.17	5.4±1.2	1.3±0.8	8.7±1.4	4.7±0.9	464±35	53.9±6.2	103±4	0.35±0.16
Onion	#15	0.17	0.09±0.01	0.006±0.004	0.10±0.01	0.09±0.03	3.3±0.3	49.0±4.7	47.4±3.9	0.08±0.02
Potato	#15	0.24	0.035±0.003	0.002±0.001	0.022±0.002	0.16±0.02	1.02±0.08	15.0±1.8	3.2±0.2	0.026±0.007
Wildberry	#16	0.25	686±20	32.3±2.9	662±20	324±23	15672±815	1520±40	3202±66	5.6±1.7
Tomato	#17	0.06	3.5±0.8	<3.6	3.1±1.0	6.5±1.1	131±9	44.3±3.2	17.3±1.0	<2.2
Onion	#17	0.1	2.9±0.3	0.25±0.09	3.2±0.3	4.2±0.7	216±18	20.1±1.8	4.1±0.5	<1.0
Carrot	#17	0.11	1.30±0.07	0.05±0.01	1.38±0.07	0.59±0.06	8.6±0.5	423±16	157±6	0.07±0.02

Conclusions

It is worth noting that radioactive contamination in this Radium Salts Factory operated in the first half of the 20th century, before the development and introduction of radiation protection regulations and radioactive waste management guidelines, has been restricted to the Radium Factory area and caused limited radioactive contamination outside.

Near 70 years after the closure of the Radium Salts Factory and two interventions made by the inheriting company to clean up the Factory site and relocate solid waste, radioactivity levels in a few spots still are higher than natural background in the surrounding environment. Further cleaning and removal of solid waste (milling tailings) abandoned in the ore leaching tanks and dismantling the ruins of the facilities, would reduce further the ambient radiation dose at the area and would remove a potential source of contaminants. In the close vicinity of the Factory, measures to prevent the use of contaminated water from wells and contaminated soil are needed to reduce transfer of radionuclides through the terrestrial food chain to humans and curtail radioactivity exposure of local population members to the natural levels.

References

Amaral EM, Alves JGA, Carreiro JV, 1992. Doses to the Portuguese population due to natural gamma radiation. Radiation Protection Dosimetry 45(1/4) 541-543.

Bettencourt AO, Teixeira MMGR, Madruga MJ, Faísca MC, 1988. Dispersion of 226Ra in a contaminated environment. Radiation Protection Dosimetry 24(1/4): 101-108.

Bettencourt, AO, Elias MD, Teixeira, MMGR, 1985. Radiological Impact Assessment of an Abandoned Radium Salts Factory, Science Total Environment 45:173-180.

Carvalho F. P., D. Chambers, S. Fesenko, W.S. Moore, D. Porcelli,H. VandenhovE,T. Yan Kovich (2014b). Radium Environmental Pathways and Corresponding Models. In: The Environmental Behaviour of Radium: Revised Edition. Technical Reports Series No. 476, pp. 106-172. International Atomic Energy Agency. Vienna.

Carvalho F. P., Oliveira J. M., Faria I. (2009a). Alpha Emitting Radionuclides in Drainage from Quinta do Bispo and Cunha Baixa Uranium Mines (Portugal) and Associated Radiotoxicological Risk. Bulletin Environmental Contamination and Toxicology, 83, 668-673.

Carvalho F.P., Oliveira J.M. (2009). Performance of alpha spectrometry in the analysis of uranium isotopes in environmental and nuclear materials. Journal Radioanalytical and Nuclear Chemistry, 281, 591-596.

Carvalho F.P., Oliveira J.M., Lopes I., Batista A. (2007a). Radio nuclides from past uranium mining in rivers of Portugal. Journal of Environmental Radioactivity, 98, 298-314.

Carvalho F.P., Oliveira J.M., Malta M. (2009b). Analyses of radionuclides in soil, water and agriculture products near the Urgeiriça uranium mine in Portugal. Journal Radioanalytical and Nuclear Chemistry, 281, 479-484.

Carvalho FP, Oliveira JM (2007b). Alpha emitters from uranium mining in the environment. Journal Radioanalytical and Nuclear Chemistry, 274, 167-174.

Carvalho FP, Oliveira JM, Malta M (2014a). Intake of Radionuclides with the Diet in Uranium Mining Areas. Procedia Earth and Planetary Science 8: 43–47.

Carvalho, FP (2014). The National Radioactivity Monitoring Program for the Regions of Uranium Mines and Uranium Legacy Sites in Portugal. Procedia Earth and Planetary Science: 33–37.

Teixeira MMGR and Elias MD (1982). Efeitos ambientais da indústria da extracção do rádio. (In Portuguese).LNETI/DPSR, B, 70 (SRA).

IAEA (2014). The Environmental Behaviour of Radium: Revised Edition. Technical Reports Series No. 476, pp. 106-172. International Atomic Energy Agency. Vienna.

Oliveira, JM, Carvalho, FP (2006). A Sequential Extraction Procedure for Determination of Uranium, Thorium, Radium, Lead and Polonium Radionuclides by Alpha Spectrometry in Environmental Samples. (Proceedings of the 15th Radiochemical Conference). Czechoslovak Journal of Physics 56 (Suppl. D): 545-555.

Hydrogeochemistry of Uranium in the Groundwaters of Serbia

Marina Ćuk[1], Maja Todorović[1], Petar Papić[1], Jovan Kovačević[2], Zoran Nikić[3]

[1]University of Belgrade, Faculty of Mining and Geology, Đušina 7, 11000 Belgrade, Serbia, +381(0)113219235
[2]Geological Survey of Serbia, Rovinjska 12, 11000 Belgrade, Serbia
[3]University of Belgrade, Faculty of Forestry, Kneza Višeslava 1, 11000 Belgrade

Abstract. For the present study, 172 groundwater samples (collected from wells and natural springs in Serbia) were analyzed for different chemical parameters including uranium. For each groundwater sample, information was compiled about the basic parameters of its chemical composition. Most of the samples were found to be of the Na – HCO_3 and Ca, Mg – HCO_3 water types, while only a few were classified as the Ca, Mg – SO_4 or Na – Cl type. HR–ICP–MS is an analytical technique used for uranium determinations. The measured concentrations of uranium were in the range from 0.0005 μg/L to 25.34 μg/L.

Introduction

Uranium (U) is a naturally-occurring radioactive element that is supplied to the Earth's crust along with magmatic melts (Omel'yanenko 2007). Typical abundances of U are in silica–rich igneous rocks (Gascoyne 1989). Granites and pegmatites have high U and Th concentrations. Significant departures may occur in sedimentary environments due to water–rock interactions (Langmuir 1997).

The most important U mineral is a simple oxide, uraninite, which is a common accessory mineral in pegmatites and peraluminous granites, and is probably the most important source of dissolved U in groundwaters emanating from weathered granite terrains. The U silicate, coffinite, is of secondary economic importance, with most major coffinite–bearing deposits restricted to low–temperature deposits (Finch and Murakami 1999). The transport of uranium in geologic fluids is a function of its speciation (Shock et al. 1997). The most important oxidation states of uranium in nature are the uranous [U(IV)] and uranyl [U(VI)] oxidation states. The most oxidized state of U in nature is U^{6+}, which forms the uranyl ion, UO_2^{2+} that can migrate many kilometres from its source in altered rocks. At pH values higher than 5, the uranyl ion forms a number of aqueous hydroxide complexes. In

B.J. Merkel and A. Arab (eds.), *Uranium – Past and Future Challenges*,
DOI 10.1007/978-3-319-11059-2_89

most groundwaters, dissolved carbonate combines with UO_2^{2+} to form uranyl carbonate solution complexes (Langmuir 1997, Finch and Murakami 1999). Complexation with carbonate anions greatly increases the solubility of uranium minerals. U^{4+} is dominant ion in reducing groundwater (Weiner 2007) and dissolved U concentrations are kept low (Sheppard et al. 2005). Uranium has a low specific activity and, as such, its chemical toxicity is of greater concern than its radiotoxicity (Stalder et al. 2012). The recommended U.S. Environmental Protection Agency drinking water standard for U is 30 μg /L (US EPA 2011), while according to Serbian standards the permissible U concentration in drinking water is 50 μg/L (Official Gazette of SFRY 98, 99). U concentrations in groundwater are usually low (Gómez et al. 2006) and spatially and genetically associated with granitic rocks and can be deposited in clastic sediments (Nikić 2008).

Methods

One hundred and seventy-two groundwater samples were collected in 2012 across Serbia, from water wells and natural springs (Fig. 3). The oxidation/reduction potential (Eh), pH values, electrical conductivity and water temperatures of the samples were determined in–situ using a portable Field Case (WTW pH 3110 SET 2 and Electrode Sen Tix ORP 100 °C). Major cations (Ca, K, Mg and Na) in the groundwater samples were measured by inductively-coupled plasma optical emission spectrometry (ICP–OES). Major anions were determined by ion chromatography (Dionex ICS–3000 DC). U concentrations in the groundwater samples were measured using high resolution inductively-coupled plasma mass spectrometry (HR ICP–MS). Geochemist's Workbench (GWB) software was used for water type determinations and for establishing the uranium ionic forms. The results were statistically processed using IBM software SPSS 17.0, and graphically interpreted by ESRI ArcGIS 9.3.

Results and Discussion

The statistical values of the major hydrochemical parameters and uranium concentrations in the 172 groundwater samples are shown in Table 1. Broad range of solute concentrations suggests that multiple sources and/or complex hydrochemical processes affected the spatial variability of the hydrochemical properties (Masoud 2014).

Table 1. Statistical summary of major hydrochemical parameters and uranium concentrations in groundwater samples

	Min	Median	Max	Mean
T (°C)	8	21.55	77.9	26.99
EC (μS/cm)	95	1408	28925	2189.29
pH	5.75	7.28	9.72	7.32
EMS (V)	–0.226	0.1865	0.4265	0.15
Ca (mg/L)	0.1	47.2	731.46	66.87
Mg (mg/L)	0.006	20.255	199.82	31.11
Na (mg/L)	0.55	149.77	6647	435.18
K (mg/L)	0.005	4.62	113	16.73
HCO_3 (mg/L)	38.2	582.14	6588	1159.08
SO_4 (mg/L)	0.35	20.2	1298	69.93
Cl (mg/L)	0.9	30.49	10209	173.85
U (μg/L)	0.0005	0.043	25.34	0.76

The hydrocarbonate ion was dominant in almost all the samples, which is closely related to natural dissolution of the soil and rock. A Piper diagram of the groundwater samples (Fig. 1) indicates that the dominant hydrochemical types of groundwater were Na–HCO_3 and Ca–Mg–HCO_3, with only a few groundwater samples classified as the Ca, Mg – SO_4 or Na – Cl type.

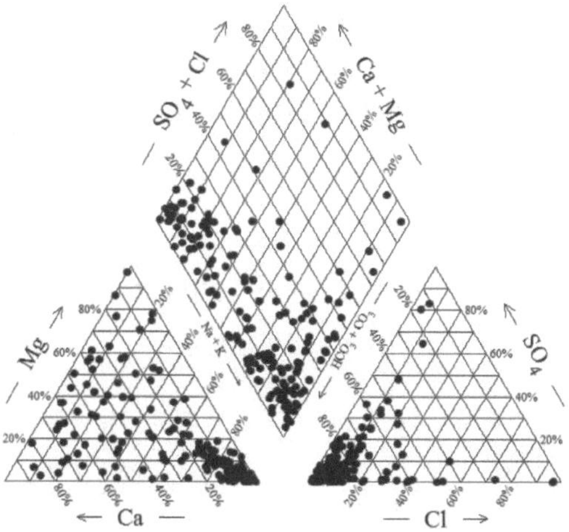

Fig.1. Piper diagram of groundwater samples

The collected groundwater samples were slightly acidic to highly alkaline, with pH levels in the range 5.75 – 9.72, and the redox potential (Eh) from −226 mV to 426 mV. Most of the Na–HCO_3 type waters originate from aquifers in granitoid rocks, with elevated concentrations of dissolved solids and CO_2. Past studies of carbonated groundwaters have shown that groundwaters with elevated CO_2 concentrations tends to occur in areas of large tectonic faults, in zones of Tertiary and Quaternary volcanism and within regional metamorphism (Stevens 2005). In some water types, the SO_4 anion is dominant in reduction conditions and the Cl anion in marine sedimentary environments.

In the 172 groundwater samples, U concentrations were in the range 0.0005 – 25.34 μg/L (median 0.0429 μg/L, mean 0.756 μg/L). The percentiles of the full uranium dataset are shown in Table 2.

Table 2. Percentiles of the full dataset of uranium concentrations in groundwaters

	10%ile	25%ile	50%ile	75%ile	80%ile	90%ile	95%ile	99%ile
U (μg/L)	0.0011	0.00653	0.0429	0.344	0.57	1.18	3.838	12.57

Eh–pH diagrams are useful for understanding the geochemical behavior of elements; they depict the dominant aqueous species and stable solid phases and are thus essential for understanding solute and radionuclide transport in groundwater (Geological Survey of Japan 2005). A diagram was produced using GWB software for a total concentration of uranium of $10^{-7.97}$ mol/L (25.34 μg/L U) in the absence of dissolved complexing ligands (Fig. 2). Under oxidizing conditions, the uranyl ion (UO_2^{2+}) was dominant at low pH, while at higher pH levels, $UO_2OH^+$and $(UO_2)_3(OH)^-_7$ became dominant (Fig. 2). Under reducing conditions, uraninite (UO_2) tends to be predominant over a wide pH range, while uranous (U^{4+}) and $U(OH)^-_5$ may be present under extremely acidic and alkaline conditions, respectively. Pitchblende (U_3O_8) and U_4O_9, oxidized or partly oxidized species of U, could form at intermediate Eh values and high pH levels. Under reducing conditions, dissolved U concentrations are low because of the low solubility of uraninite (Langmuir 1979, Wu et al. 2014).

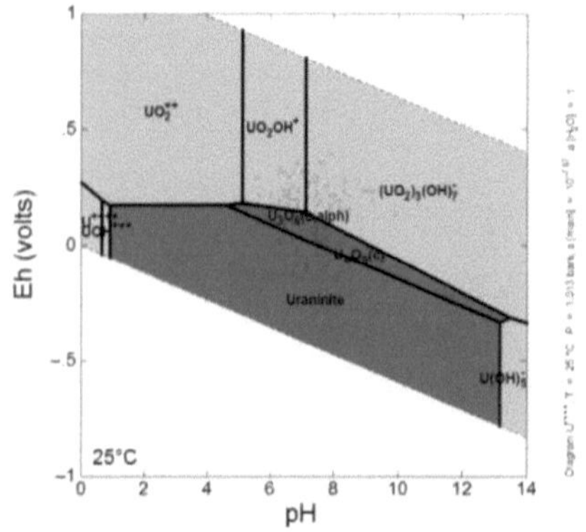

Fig.2. Eh-pH diagram for U-H_2O system ($10^{-7.97}$ mol U/L, 25 °C, 1 bar)

An overview of the geographic distribution of uranium and areas with the highest uranium concentrations are shown in Fig. 3. The featured high values are associated with the underlying geology and uranium concentrations attributable to Tertiary volcanic activity within the territory of Serbia. According to elevated concentrations of uranium in groundwater samples three areas have been selected in the Vardar Zone. These three areas are identified as Area 1: Miocene granitoid rocks of Bukulja Mt.; Area 2: Tertiary volcanic rocks of Rudnik - Borač – Kotlenik complexes; and Area 3: Paleogene granitoid rocks of Jastrebac Mt. (Fig. 3). Table 3 shows the physico–chemical properties of the groundwater samples collected from selected areas.

Area 1: Miocene granitoid rocks of Bukulja Mt. Miocene S–type plutons with a typical peraluminous affinity are mostly situated along the southern Pannonian margin (Koroneos et al. 2011). There is a hypothesis that the Bukulja Pluton is related to tectonomagmatic events controlled by the early extensional phases in the opening of the Pannonian Basin (Cvetković et al. 2007). Endogenic uranium minerals are spatially and genetically associated with granitic rocks of Bukulja Mt. Uranium minerals can also be found deposited in clastic sediments (sandstones and conglomerates with infrequent coaly interbeds) (Nikić et al. 2008). At the contact of crystalline schists and intrusives, there are a large number of mineral springs, generally of different water types, with elevated concentrations of dissolved solids and CO_2, and with U concentration in the range from 0.98 to 5.8 μg/L (Table 3).

Table 3. Hydrochemical characteristics of spring water samples from selected areas

	No	Water type	Eh (mV)	pH	U (μg/L)	U ionic form	Carbonate complexes
Area 1	1	$Ca-Mg-HCO_3$	383	7.58	5.8	UO_2OH^+	$UO_2(CO_3)_2^{2-}$, $UO_2(CO_3)_3^{4-}$
	2	$Ca-Na-HCO_3$	363	6.4	1.5	UO_2OH^+	$UO_2(CO_3)_2^{2-}$
	3	$Mg-Ca-HCO_3$	336.5	7.07	4.3	UO_2OH^+	$UO_2(CO_3)_2^{2-}$, $UO_2(CO_3)_3^{4-}$
	4	$Ca–Mg–Na–SO_4$	210	6.7	0.98	UO_2OH^+	$UO_2(CO_3)_2^{2-}$, $UO_2(CO_3)_3^{4-}$
Area 2	5	$Na-HCO_3$	260	8.84	14	$(UO_2)_3(OH)^-_7$	$UO_2(CO_3)_3^{4-}$
	6	$Na-HCO_3$	305.3	8.66	11.76	$(UO_2)_3(OH)^-_7$	$UO_2(CO_3)_3^{4-}$
	7	$Mg-HCO_3$	359	7.46	7.74	$(UO_2)_3(OH)^-_7$	$UO_2(CO_3)_3^{4-}$
Area 3	8	$Mg–Na–Ca–HCO_3$	216	6.4	5.97	UO_2OH^+	$UO_2(CO_3)_2^{2-}$
	9	$Mg-Na-HCO_3$	350	6.5	5.21	UO_2OH^+	$UO_2(CO_3)_2^{2-}$
	10	$Mg–Ca–HCO_3$	316	5.75	25.34	UO_2OH^+	$UO_2(CO_3)_2^{2-}$

Area 2: Tertiary volcanic rocks of Rudnik - Borač - Kotlenik. Volcanic eruptions of the Rudnik and Borač – Kotlenik complexes have produced mainly dacitic, andesitic and basaltic extrusions of high – K calc – alkaline affinity in Oligocene–Miocene times (Cvetković et al. 2004). Groundwaters in these areas are of the $Na-HCO_3$ and $Mg-HCO_3$ types, neutral to alkaline, with uranium concentrations ranging from 7.74 to 14 μg/L (Table 3).

Area 3: Paleogene granitoid rocks of Jastrebac Mt. was formed by melting of crustal material buried into the deeper reaches of the convergent Serbian–

Macedonian–Vardar (SMCT – VZCT) suture and subsequently intruded into the Cretaceous–Paleogene complex (Marović et al. 2007). The south–west intrusive border forms a young regional fault. Along this rupture, the granitoid contacts Lower Paleozoic schists that have led to the formation of various types of aquifers (Pavlović et al. 2003). The Mg (Ca, Na) – HCO_3 water type is dominant in this area. Groundwaters are slightly acidic, with dissolved CO_2, and uranium concentrations ranging from 5.21 to 25.34 µg/L.

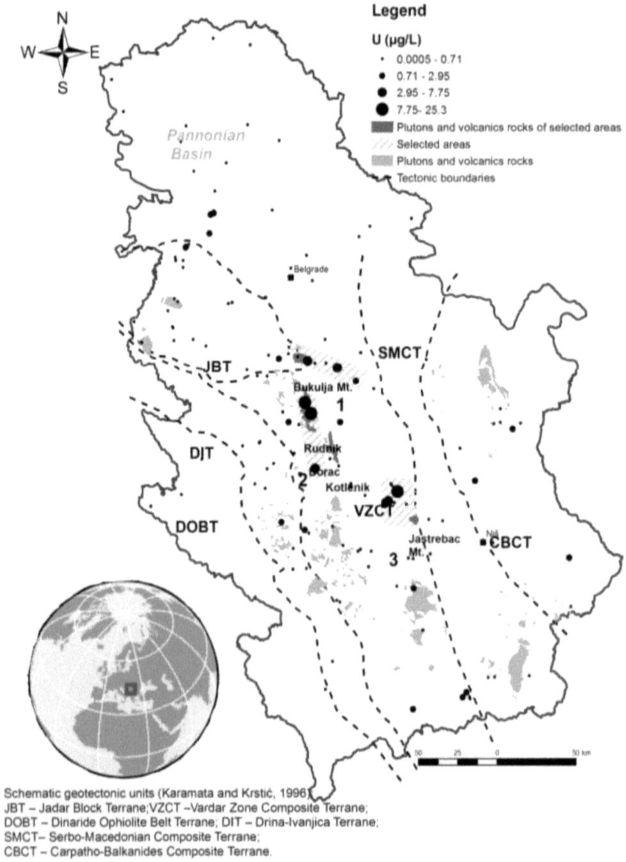

Fig.3. Map of Serbia showing the distribution of plutons and volcanic rocks, and U concentrations in groundwater samples

Ten groundwater samples from the selected areas reflected oxic aquifer conditions (Eh: 210 – 383 mV), with a pH range from 5.75 to 8.84. The groundwater samples were found to contain UO_2OH^+ as the major ionic form in acidic groundwater and $(UO_2)_3(OH)^-_7$ in neutral to alkaline groundwaters (Table 3). In oxidized waters (Eh > 200 mV), uranyl ions tend to form carbonate complexes as a function of the pH level and partial pressure of $CO_{2(g)}$ (Gomez et al. 2006). Table 3 shows the main uranyl carbonate complexes. These complexes are important because they increase the solubility of uranium minerals and improve uranium mobility in groundwater systems (Langmuir 1997). In the groundwater samples, uranium was found to exist predominantly in its hexavalent state and the uranium-carbonate species, $UO_2(CO_3)_2^{2-}$ and $UO_2(CO_3)_3^{4-}$, were expected to be the dominant in solution. $UO_2(CO_3)_2^{2-}$ was dominant in acidic and neutral groundwater samples, while $UO_2(CO_3)_3^{4-}$ was dominant in alkaline samples.

Conclusion

Uranium concentrations in groundwater samples acquired from 172 wells and natural springs in Serbia were in the range from 0.0005 to 25.34 μg/L. The samples were collected from all geotectonic units in Serbia. Groundwater occurrences with the highest uranium concentrations were found in the Vardar Zone. These elevated values (0.98 – 25.34 μg/L U) are attributable to the geological complexes of Tertiary magmatism: granitoid rocks of Bukulja Mt., volcanic rocks of Rudnik - Borač- Kotlenik, and granitoid rocks of Jastrebac Mt. There are certain rules that seem to apply to the occurrence of the ionic forms of uranium in the selected areas: all groundwater samples came from oxidizing environments (Eh > 200 mV); in acidic waters dissolved uranium occurred as uranyl ion (UO_2OH^+), while in neutral to alkaline waters dissolved uranium occurred as $(UO_2)_3(OH)^-_7$. These uranium forms are able to form uranyl carbonate solution complexes $UO_2(CO_3)_2^{2-}$ and $UO_2(CO_3)_3^{4-}$, which increase uranium mobility in groundwater systems. The low solubility of U^{4+} aqueous complexes leads to low concentrations in reduced waters, corroborating that an oxidizing environment is suitable for uranium migration.

Acknowledgement This work is supported by Ministry of Education, Science and Technological Development of Serbia within the Project III 43004 and OI 176018.

References

Cvetković V, Giampiero P, Christofides G, Koroneos A, Pécskay Z, Resimić–Šarić K, Erić V (2007) The Miocene granitoid rocks of Mt Bukulja central Serbia; evidence for Pannonian extension–related granitoid magmatism in the northern Dinarides. European Journal of Mineralogy 19(4): 513–532

Cvetković V, Prelević D, Downes H, Jovanović M, Vaselli O, Pécskay Z (2004) Origin and geodynamic significance of Tertiary post–collisional basaltic magmatism in Serbia (Central Balkan Peninsula). Lithos 73: 161–186

Finch R, Murakami T (1999) Systematics and paragenesis of uranium minerals. In: Burns PC, Ewing RC (eds) Uranium: Mineralogy, Geochemistry and the Environment. Mineralogical Society of America and Geochemical Society Reviews in Mineralogy and Geochemistry 38: 91–179

Gascoyne M (1989) High levels of uranium and radium in groundwaters at Canada's Underground Research Laboratory, Lac du Bonnet, Manitoba, Canada, Appl. Geochemistry 4: 577–591
Geological Survey of Japan Open File Report No. 419 (2005) Atlas of Eh – pH diagrams. Intercomparison of thermodynamic databases. Naoto Takeno: National Institute of Advanced Industrial Science and Technology
Gómez P, Garralón A, Buil B, Turrero J, Sánchez L, Cruz B (2006) Modeling of geochemical processes related to uranium mobilization in the groundwater of a uranium mine. Science of the Total Environment 366: 295–309
Karamata S, Krstić B. (1996): Terranes of Serbia and neighboring areas. In: Knežević – Đorđević V, Krstić B (ed) Terranes of Serbia. Faculty of Mining and Geology, University of Belgrade, Belgrade: 25 -44
Koroneos A, Poli G, Cvetković V, Christofides G, Krstić D, Pécskay Z (2011) Petrogenetic and tectonic inferences from the study of the Mt Cer pluton (West Serbia). Geological Magazine 147:1–23
Langmuir D (1979) Uranium solution–mineral equilibria at low temperatures with applications to sedimentary ore deposits. Geochimica et Cosmochimica Acta 43: 1989–1990
Langmuir D (1997) Aqueous Environmental Geochemistry. Prentice Hall, Upper Saddle River, New Jersey, USA
Marović M, Đoković I, Toljić M, Spahić D, Milivojević J (2007) Extensional Unroofing of the Veliki Jastrebac Dome (Serbia). Annales geologiques de la Peninsule Balkanique: 21–27
Masoud A (2014) Groundwater quality assessment of the shallow aquifers west of the Nile Delta (Egypt) using multivariate statistical and geostatistical techniques. Journal of African Earth Sciences 95: 123–137
Nikić Z, Kovačević J, Papić P (2008) Uranium minerals of Bukulja mountain controls on storage reservoir water. Uranium, Mining and Hydrogeology, Springer: 645–652
Official Gazette of SFRY No 42/98 and 44/99.Regulation of Hygienic Correctness of Drinking Water
Omel'yanenko B, Petrov V, Poluektov V (2007) Behavior of uranium under conditions of interaction of rocks and ores with subsurface water. Geology of Ore Deposits 49: 378–391
Pavlović R, Nikić Z, Trivić B (2003) The role of ground and underground waters in evolution of the ring structure of Jastrebac M. Materials and Geoenvironment 50: 293–29
Sheppard S, Sheppard M, Gallerand M, Sanipelli B (2005) Derivation of ecotoxicity thresholds for uranium. J Environ Radioact 79: 55–83
Shock E, Sassani D, Betz H (1997) Uranium in geologic fluids: Estimates of standard partial molal properties, oxidation potentials, and hydrolysis constants at high temperatures and pressures. Geochimica et Cosmochimica Acta: 4245–4266
Stalder E, Blanc A, Haldimann M, Dudler V (2012) Occurrence of uranium in Swiss drinking water. Chemosphere: 672 – 679
Stevens S (2005) Natural CO_2 Fields as Analogs for Geologic CO2 Storage, Chapter 3. In Carbon Dioxide Capture for Storage in Deep Geologic Formations, by David Thomas, Elsevier: 687–697
US EPA (2011) edition of the drinking water standards and health advisories.Washington DC
Weiner E (2007) Applications of Environmental Aquatic Chemistry: A Practical Guide. Taylor & Francis Group; USA
Wu Y, Wang Y, Xie X (2014) Occurrence, behavior and distribution of high levels of uranium. Science of the Total Environment 472: 809–817

Seed crops: Alternative for non-remediable uranium mine soils

Gerhard Gramss[1], **Klaus-Dieter Voigt**[2]

[1]Friedrich-Schiller-University Jena, Burgweg 11, D-07749 Jena, Germany
[2]Food GmbH Jena, Orlaweg 2, D-07743 Jena, Germany

Abstract. Excluder seed crops of wheat and pea down-regulate the uptake of heavy metals from excessively metalliferous soil to equip their seeds with essential trace metals in quantities that correspond with "normal plant concentrations" and ensure thus their vigorous and stress-free germination. Due to this uptake blockage, non-remediable soils can be used to grow food- and forage-quality seeds. Breeding and providing certified heavy-metal excluder (seed) crops should thus be given precedence to flopping soil phytoextraction measures.

Introduction

The world-wide loss of cropland to expanding urban communities, to sealing, mining, and areas of industrial heavy metal (HM) and radionuclide emissions is completed by agricultural mismanagement, land erosion (Nair 2014; UNEP 2014), and the use of AsCd contaminated mineral fertilizers (Severin 2007). Contemporarily, food and forage production compete with rising demands for industrial and energy crops. Lasting renunciation of the use of insufficiently remediable metalliferous soils in face of a phytoextraction technology claiming hundreds to thousands of years for a 10- to 50-% reduction of the soils` *aqua regia* heavy metal (HM) stock (Duquène et al. 2009; Gramss and Voigt 2014) is thus intolerable.

Potential food and forage crops chosen for contaminated cropland should thus be able to repress the uptake of those minerals which are the soils` dominating contaminants. Relative to HM excluder plants (been, cereals, lupin, maize), HM sequestering crops (buckwheat, beet root species) accumulate up to the 18-fold of trace metals in their shoot tissues (Bergmann et al. 2006). Herbage from U mine soils varied in the concentrations of As (32x); Cd (60x); Cu (5x); Ni (7x), Pb (7x); U (27x); and Zn (27x) and modified the cultivar-specific uptake of Cd by another 2.5 times. Uptake of CdCoCuMnNiZn was further repressed by a restricted nitrogen supply in favour of organic carbon (Bergmann et al. 2006; Gramss et al. 2004).

B.J. Merkel and A. Arab (eds.), *Uranium – Past and Future Challenges*,
DOI 10.1007/978-3-319-11059-2_90

The passage of elevated shoot concentrations in Zn (Wang et al. 2011) as well as in most of the trace minerals to the seed is strictly regulated and optimized to enable the vigorous and stress-free development of the potential germling (Gramss and Voigt 2013). In this way, food-quality seed crops such as cereals are widely devoid of a detrimental HM load and can thus even be grown on seriously HM contaminated soils (Il'in 2007).

The inclusion of an insufficiently remediated open-cast U mine into non-contaminated cropland in the Ronneburg district (East Thuringia, Germany) created a gradient of diminishing HM concentrations in geologically related soils. From locations A to C within the gradient, soil samples and seeds of winter wheat and field pea were collected and analyzed for 20 minerals. It was the goal to determine the seed:soil transfer factors (TFs) of the (potentially toxic) elements, the resulting crop quality in the light of legislative contamination limits, and the mode of TF regulation from soil to seed. Possible concessions to the restrictive regulations of the soil hygiene standards for seed crops are proposed.

Materials and Methods

Mature seeds were obtained from field cultures of the winter wheat cv., Bussard (E quality rating, http://www.aglw.de/sorten_weizen.htm) in 2009, and from field pea cv. Rocket in 2013. The crops were taken from identical locations (A to C) along a gradient of diminishing HM concentrations (refer to Table 2). The geologically related clay-loam soils A/B/C of C_{org} 3.53/3.84/4 % and pH_{aqu} 7.28/7.30/6.18 comprised the HM hot spot formed by mine overburden soil at location A and soil C which was not impaired by mining. In early May, the sum of NH_4/NO_3-N at the locations A to C amounted 234/87/60 mg kg^{-1} soil DW for wheat, and 53/61/116 for pea. Mineral concentrations were determined by ICP-MS for two composite soil samples from each location after *aqua-regia* extraction, and from duplicate 300-mg samples of microwave-digested whole-seed homogenates prepared from at least 20 g of two independent samplings per location. Statistics based on the use of SPSS 8.0 software.

Results

In spite of the wide range of soil HM concentrations with elevated values in AsCdCuMnPbUZn at the locations A and B (Table 1, bold face font), their associated seeds of wheat and pea met the EU legislative food hygiene limits for AsPb but could only serve as forage grains in regard of their content in Cd (Table 1). Several sources of wheat > pea did not satisfy the Chinese food hygiene standards in (CuNi)Zn. As recently stated, variations in soil mineral concentrations were

Table 1. Mineral content (mg kg^{-1} dry weight) of the soils A (HM hot spot), B (moderately contaminated), and C (non-contaminated) in 2009/2013, and of the respective wheat and pea seeds (italics). In parentheses, mean of mineral transfer factors seed:soil x10^{-3}

	Winter wheat cv. Bussard from 2009			Field pea cv. Rocket from 2013			Limits[a]
	Soil A	B	C	Soil A	B	C	
Essential elements							
Ca	7,235	5,040	2,296	6,648	4,503	2,400	
	495 (68.4)	*281* (55.8)	*332* (144.6)	*932* (140)	*854* (190)	*847* (353)	
Co	27.3	23.2	11.4	25.8	21.4	18	
	< *0.031*	< *0.031*	< *0.031*	*0.063* (2.44)	*0.056* (2.62)	*0.083* (4.61)	
Cu	**283**	**205**	37.7	**261**	**216**	50.2	2-20
	11.4 (40.3)	*11.5* (56.1)	*12.3* (326)	*8.14* (31.1)	*7.80* (36.2)	*7.63* (152)	LLC 10
Fe	20,790	23,390	17,200	18,510	22,670	31,750	
	39.6 (1.90)	*54* (2.31)	*55.6* (3.23)	*49.8* (2.69)	*51.1* (2.25)	*50.3* (1.58)	
Mg	3,240	2,268	2,250	3,655	2,746	5,110	
	1,520 (469)	*1,397* (616)	*1,240* (551)	*1,120* (306)	*1,103* (402)	*1,153* (226)	
Mn	**2,130**	**1,400**	**1,490**	**1,776**	**1,145**	**1,538**	14-30
	26.4 (12.4)	*27.7* (19.8)	*28.4* (19.1)	*11.8* (6.64)	*8.55* (7.47)	*10.5* (6.83)	
Ni	40.8	35.6	17.1	43	36.4	37.9	0.1-3
	1.41 (34.6)	*0.694* (19.5)	*0.180* (10.5)	*0.950* (22.1)	*0.626* (17.2)	*0.858* (22.6)	LLC 1
Zn	**3,005**	**1,525**	119	**2,889**	**1,388**	208	10-100
	132 (44)	*111* (72.8)	*35.8* (301)	*70.2* (24.3)	*53.5* (38.5)	*40.1* (193)	LLC 50
Non-essential elements as toxicants							
As	**156**	**153**	23.5	**139**	**152**	28.8	0.01-1
	< *0.080*	< *0.080*	< *0.080*	*0.049* (0.35)	*0.044* (0.29)	*0.051* (1.77)	LL 0.5 LLC 0.7
Cd	**41.3**	**23.2**	1.31	**40.4**	**23.8**	1.38	0.05-0.4
	1.37 (33.2)	*0.990* (42.7)	*0.057* (43.5)	*0.302* (7.48)	***0.333*** (14)	*0.030* (21.7)	LL 0.1 LLC 0.1 1[b]
Cr	20.3	16.1	25.7	16.4	13.6	30.7	0.1-1
	0.029 (1.43)	*0.038* (2.36)	*0.046* (1.79)	*0.181* (11)	*0.073* (5.37)	*0.196* (6.38)	LLC 1
Pb	**150**	**131**	44.3	**148**	**143**	53.9	0.1-6
	0.057 (0.38)	*0.078* (0.60)	*0.092* (2.08)	*0.040* (0.27)	*0.030* (0.21)	*0.036* (0.67)	LL 0.2 LLC 0.4 10[b]
Stable isotopes of long-lived radionuclides							
Ba	29.1	46.5	107	35.9	36.1	108	
	0.500 (17.2)	*0.702* (15.1)	*3.05* (28.5)	*0.379* (10.6)	*0.438* (12.1)	*2.12* (19.7)	
Cs	61.8	49.5	6.35	26.3	22.2	8.30	
	0.043 (0.70)	*0.035* (0.71)	*0.019* (3.0)	*0.099* (3.76)	*0.105* (4.73)	*0.013* (1.57)	
Sr	64.8	57.9	74.7	53.8	42.1	43.6	
	1.80 (27.8)	*1.10* (19.0)	*2.46* (32.9)	*4.42* (82.2)	*4.36* (104)	*7.11* (163)	
Th	14.9	15	6.68	6.80	6.80	2.90	
	< *0.170*	< *0.170*	< *0.170*	*0*	*0.001* (0.15)	*0*	
U	**40.5**	**34.4**	4.75	**29.9**	**39**	6.70	0.002-
	< *0.016*	< *0.016*	< *0.016*	*0.004* (0.13)	*0.004* (0.10)	*0.001* (0.15)	0.015

[a] Normal plant heavy-metal concentrations (Auermann et al. 1980; Schachtschabel et al. 1998); LL, legislative limits for food grains (Decree [EU] No. 466/2001 from March 8, 2001); LLC, tolerance limits of Chinese standards for wheat grains (Huang et al. 2008).
[b] Expanded contamination limits for forage crops (Severin 2007).
Bold face font/shaded digits denote non-permissible and unusually elevated HM concentrations in soils/seeds. Standard deviations of mineral concentrations determined by duplicate soil/seed extractions range within 0.1-3/0.6-4 % of value.

compensated by up-regulation of the seed:soil transfer factor (TF) on mineral-poor, and its down-regulation on mineral-rich soils (Table 1; Gramss and Voigt 2013). Consequently, ranges in seed mineral concentrations (Table 2, columns 3 and 6) to those in soil (Table 2, columns 2 and 5) were lower in the case of CaCdCu(Fe)MgMnZn whereas the deposition of BaNiSr increased with the concentrations in soil for both plants. The TF regulatory potential was best expressed by quotients of the respective soil range to seed range factors (Table 2, columns 4 and 7) and was pronounced in regard of CuZn, insufficient in Cd, but ineffective in BaNiSr. Seed concentrations in the predominantly non-essential As(Co)CrCsPbThU persisted below 0.1 mg kg^{-1} DW and were apparently impaired by TF regulatory modes, too (Table 1).

Table 2. Quotients of the ranges of mineral concentrations in soils A to C and the associated seeds as the optimum indicator of the plant's TF regulatory potential

	Winter wheat cv. Bussard			Field pea cv. Rocket		
	Soil range	Seed range	Quotient	Soil range	Seed range	Quotient
Essential elements						
Ca	3.15	1.76	1.79	2.77	1.10	2.52
Cu	7.51	1.08	6.95	5.20	1.06	4.91
Fe	1.36	1.40	0.97	1.72	1.03	1.67
Mg	1.44	1.23	1.17	1.86	1.05	1.77
Mn	1.52	1.08	1.41	1.55	1.38	1.12
Ni	2.39	7.83	0.30	1.18	1.52	0.78
Zn	25.3	3.69	6.86	13.9	1.75	7.94
Non-essential elements as toxicants						
Cd	31.5	24.0	1.31	29.3	11.1	2.64
Stable isotopes of long-lived radionuclides						
Ba	3.68	6.10	0.60	3.03	5.59	0.54
Sr	1.29	2.24	0.58	1.28	1.63	0.79

Data of AsCoCrCsPbThU denoted by seed mineral concentrations ≤ 0.1 mg kg^{-1} are omitted.

Seed mineral concentrations were thus stabilized in a narrow (and optimum?) range of variation. Figure 1 demonstrates once more how the wide variability of mineral concentrations in soils shrinks in seeds by TF adaptations. Seed concentrations in CaCu were widely congruent in face of moderate fluctuations in soil concentrations. In regard of Zn/Cd, the TF regulatory potential did not suffice/was not activated to compensate the extreme variations in soil concentrations.

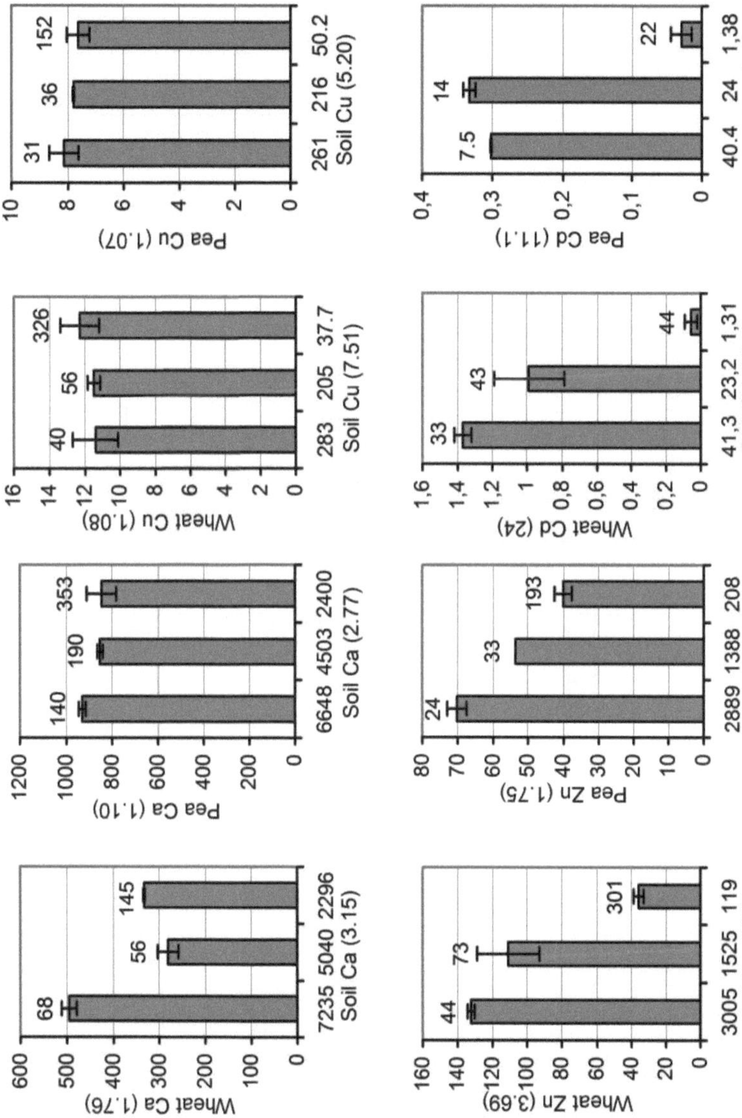

Fig.1. The drastic changes (by factors…x) in total HM concentrations of the soils A-C diminish significantly but with variable efficacy in seeds of wheat and pea due to the down-regulation of the seed:soil transfer factors (figures heading the columns) on higher-concentrated soils. Error bars indicate confidence intervals of 95 %

Discussion

Heavy-metal excluder properties of the crops and regulatory adaptations of the seed:soil TFs (Tables 1 and 2; Fig. 1) effected that seed HM concentrations in wheat and pea did predominantly not surpass the EU legislative food/forage hygiene standards (Table 1), that AsCoCrCsPbThU remained at a negligible level and persisted with all elements in the limits of "normal plant concentrations" (Table 1; Auermann et al. 1980; Schachtschabel et al. 1998) which are typical of herbage devoid of HM stress. "Normal concentrations" especially in the essential traces CoCuFeMnNiZn prevent malnutrition in man and livestock (McDowell 2003; Yang et al. 2007) as well as damages to health and productivity of germling and crop themselves (Kirchmann et al. 2009; Marschner 1995). Pea>wheat surprised with uptake blockage of the excessive Cd from the aged soil.

Table 3. Usual and permissible ranges of HM concentrations (mg kg^{-1} DW) of cropland in Central Europe and the upper HM limits of the partially ingested pasture soils compared with (expanded) HM values of uranium mine soil planted to food seed crops

Mineral	Usual HM range of cropland [a]	Permissible HM range [b]	Pasture soil: intervention threshold [c]	Recommended extension: wheat/pea [d]
Micronutrients				
Co	1-40 (30)	50		
Cu	2-40 (30)	100 (50-100)	200	300/300
Mn	40-1,000 (550)			2,000/> 2,000
Ni	3-50 (30)	100 (40-50)		35/45
Zn	10-80 (90)	300 (200-500)		500/1,500
Nonessential elements				
As	1-20 (6)	20 (30-40)	50	>200/>200
Ba	(500)	2,000		
Cd	0.1-0.6 (0.35)	3 (0.3)	10	2/7 (food) 25/>80 (forage)
Cr	5-100 (30)	100 (150-200)		
Pb	2-80 (35)	100 (250-300)		
Th	(9)			
U	(2)		23	>40/>40

[a] Concentration ranges after Schachtschabel et al. (1998); ( ), after Bowen (1979).
[b] Concentration ranges after Kloke (1979; ( ), after Huang et al. (2008).
[c] Intervention thresholds for pasture soils after Severin (2007) and Bermudez et al. 2010 (for U).
[d] Recommended extension of the permitted HM range specific for the soil type and the excluder seed crops employed, considering the restrictive Chinese LLC standards for CuNiZn in wheat grains (Huang et al. 2008; refer to Table 2)

The observation of the permissible arable-soil concentrations in potentially toxic trace metals (Table 3) combined with CdPb solubility tests in ammonium nitrate extractant (Severin 2007) stand for quality wheat grains whose mean AsCdPb concentrations were as low as 0.021-0.024/0.037-0.047/0.018-0.054 mg kg^{-1} FW within 1997 to 2006 (BVL Weizenkörner.mht). These values should not be notably surpassed if one is forced to grow excluder seed crops on higher contaminated and non-remediable soils. The wheat cv. used in the present case study was lower in AsU (> 6x); Cd (2.5x); and Pb (2x) but also somewhat in FeZn than two comparable high-performance wheat cvs. (Gramss and Voigt 2013). Observing EU and Chinese legislative limits for food crops (Table 1), production of the cvs. Bussard and Rocket can be expanded to a Ronneburg-type soil which is considerably higher in AsCdCuMnUZn but not Ni (Table 3). The recommended HM limits do thereby not exceed the actual concentrations of test soil A (compare Table 1), may thus be in practice higher than proposed but potentially confined by upcoming soil phytotoxicity effects (Sauerbeck 1983).

No concessions can be made to the permissible HM load of pasture soil (Table 3) which is ingested with root and shoot material of the herbage. Grazing a soil type A by a 65-kg wildlife or domestic herbivore around the year would surpass the permissible daily intake rates of 0.65/11.6 mg in CdPb by the 11-27/0.7-4.7-fold (Gramss and Voigt 2014).

Conclusions

Seed crops with their out-regulated HM balance recommend themselves for, and reduce the pressure on the quality of detoxification to, non-remediable metalliferous soils. This includes at least the availability of actively selected and bred HM excluder (food and forage) crops with certified uptake restrictions for the main HM contaminants, determined under standard conditions. Attempts to provide the respective crops should be given precedence to fruitless soil phytoextraction procedures. Electing a plant cv. which is restrictive in the uptake of the main HM contaminants of a given soil comprises then an inevitable case study to substantiate the quality of the resulting crop.

References

Auermann E, Dässler H-G, Jacobi J, Cumbrowski J, Meckel U (1980) Untersuchungen zum Schwermetallgehalt von Getreide und Kartoffeln. Die Nahrung 24: 925-937

Bergmann H, Voigt K-D, Machelett B, Gramss G (2006) Variation in heavy metal uptake by crop plants. In: Merkel BJ, Hasche-Berger A (eds.) Uranium in the environment. Springer, Berlin, pp 459–468

Bermudez GMA, Moreno M, Invernizzi R, Plá R, Pignata ML (2010) Heavy metal pollution in topsoils near a cement plant: The role of organic matter and distance to the source to predict total and HCl-extracted heavy metal concentrations. Chemosphere 78: 375-381

Bowen HJM (1979) Environmental chemistry of the elements. Academic Press, London

Duquène L, Vandenhove H, Tack F, Meers E, Baeten J, Wannijn J (2009) Enhanced phyto-extraction of uranium and selected heavy metals by Indian mustard and ryegrass using biodegradable soil amendments. Sci Total Environ 407: 1496-1505

Gramss G, Voigt K.-D (2013) Regulation of heavy metal concentrations in cereal grains from uranium mine soils. Plant Soil 364: 105-118

Gramss G, Voigt K.-D (2014) Forage and rangeland plants from uranium mine soils: long-term hazard to herbivores and livestock? Environ Geochem Health 36: 441-452

Gramss G, Voigt K-D, Bergmann H (2004) Plant availability and leaching of (heavy) metals from ammonium-, calcium-, carbohydrate-, and citric-acid-treated uranium-mine-dump soil. J Plant Nutr Soil Sci 167:417–427

Huang M, Zhou S, Sun B, Zhao Q (2008) Heavy metals in wheat grain: Assessment of potential health risk for inhabitants in Kunshan, China. Sci Total Environ 405: 54-61

Il'in VB (2007) Heavy metals in the soil-crop system. Eurasian Soil Sci 40: 993-999

Kirchmann H, Mattsson L, Eriksson J (2009) Trace element concentration in wheat grain: Results from the Swedish long-term soil fertility experiments and national monitoring program. Environ Geochem Health 31: 561-571

Kloke A (1979) Contents of arsenic, cadmium, chromium, fluorine, lead, mercury and nickel in plants grown on contaminated soil. Paper presented at United Nations-ECE Symposium on Effects of air-borne pollution on vegetation, Warsaw

Marschner H (1995) Mineral nutrition of higher plants (2nd ed.). Academic Press, London

McDowell LR (2003) Minerals in animal and human nutrition (2nd ed.). Elsevier Science, Amsterdam

Nair PKR (2014) Grand challenges in agroecology and land use systems. Front Environ Sci 2: 1

Sauerbeck D (1983) Landwirtsch Forsch Special Issue 39: 108-129

Schachtschabel P, Blume HP, Brümmer G, Hartge KH, Schwertmann U (1998) Lehrbuch der Bodenkunde (14th ed.) Enke, Stuttgart

Severin K (2007) merkblatt_anbauempfehlungen_sm-belastete_boeden_20070615[1].pdf. Landwirtschaftskammer Niedersachsen

UNEP (2014) Bringezu S et al.: Assessing global land use: Balancing consumption with sustainable supply. A report of the Working Group on Land and Soils of the International Resource Panel. ISBN: 978-92-807-3330-3

Wang YX, Specht A, Horst WJ (2011) Stable isotope labelling and zinc distribution in grains studied by laser ablation ICP-MS in an ear culture system reveals zinc transport barriers during grain filling in wheat. New Phytol 189: 428–437

Yang XE, Chen WR, Feng Y (2007) Improving human micronutrient nutrition through biofortification in the soil-plant system: China as a case study. Environ Geochem Health 29: 413-428

Using high temperature reactors for energy neutral phosphate fertilizer and phosphogypsum processing

Nils Haneklaus[1], Harikrishnan Tulsidas[2], Frederik Reitsma[1], Ewald Schnug[3]

[1]International Atomic Energy Agency, Division of Nuclear Power, Section of Nuclear Power Technology Development, Vienna International Centre, PO Box 100, 1400 Vienna, Austria
[2]International Atomic Energy Agency, Division of Nuclear Fuel Cycle and Waste Technology, Section of Nuclear Fuel Cycle and Materials, Vienna International Centre, PO Box 100, 1400 Vienna, Austria
[3]Technical University Braunschweig – Faculty 2 Life Sciences, Pockelsstraße 14, 38106 Braunschweig, Germany

Abstract.. This paper presents the conception to employ High Temperature Reactors (HTRs) to power a combined phosphate fertilizer/phosphogypsum (PG) processing plant that produces phosphate fertilizer by the wet acid process while treating the main associated waste product PG to sulfuric acid/Portland cement. The idea is based on a past plant design proposed by Consolidated Minerals Inc. (CMI) and promotes the usage of a lean greenhouse gas emission energy source instead of the previously suggested coal-fired plant.

Introduction

The inherent advantages of using High Temperature Gas-cooled Reactors (HTGRs) to produce the process heat and electricity needed for elemental phosphorous production that is suitable as fertilizers or food grade phosphoric acid whilst recovering the accompanying uranium in phosphate rocks to power the greenhouse gas lean energy source employed or other nuclear power plants has been shown previously (Haneklaus et al. 2014). First estimates show that already a relatively small HTGR (less than 100 MW_{th}) delivers sufficient energy to power most of the present phosphate fertilizer plants. Since the profitability of HTRs is higher if nuclear units are produced in larger quantities and allocate higher amounts of energy, there will be a considerable energy surplus if HTRs are employed to power phosphate rock processing plants. A representative HTGR which is currently under construction in China (HTR-PM) provides 500 MW_{th} per unit

B.J. Merkel and A. Arab (eds.), *Uranium – Past and Future Challenges*,
DOI 10.1007/978-3-319-11059-2_91

(Sun 2013). Given that the HTR-plant is employed at or in the vicinity of an existing phosphate fertilizer plant employing the wet acid process the energy surplus could be efficiently used to process the currently accumulating waste product PG. Worldwide approximately 160 million tons PG are produced annually as a side-product of phosphate fertilizer production in the sulfuric acid wet process (IAEA 2013). The International Atomic Energy Agency (IAEA) estimates that with increasing phosphate fertilizer production 200-250 million tons PG could accrue annually within the next 10 to 20 years (IAEA 2013). It is estimated that for every ton of phosphoric acid produced using wet acid processes, 4-6 tons of PG (dry mass) accrue (IAEA 2013). In practice the material mass is much higher since original PG contains at least 20 % of free water by weight (UNIDO/IFDC 1998).

Current use of byproduct phosphogypsum

PG is currently stacked near fertilizer plants and labelled as "radioactive waste" in many countries. Concentrations of uranium and radium-226 in PG samples taken in Florida were about 10 times higher than the background levels in soils for uranium and 60 times higher than the background levels for radium-226 (EPA 2012). The radionuclides appearing in PG are restricting its use as additive in agriculture, road construction material or other operational areas. At present only a relatively small amount (less than 5 %) of the produced PG is actively used (IAEA 2013). However, PG can for instance be used as a source for sulfuric acid production that could be re-used for wet acid phosphate fertilizer manufacturing. Maintaining PG reservoirs (both wet and dry) is expensive and was estimated to be as high as US$ 25 per ton in 2010 (Hilton 2010). Due to increased environmental regulations this value can be assumed to be even higher today. In addition, considerable costs to set up new piles and close old ones may once occur. The total costs for closing the PG repositories of Mulberry Phosphate Inc. (Florida) that was operating two phosphate mining facilities until its bankruptcy in 2001 was estimated to be as high as US$ 164 million (Oppaga 2005).

Production of sulfuric acid from phosphogypsum

Producing sulfuric acid from PG is a well-known chemical process that has been studied intensively (Austin 1980; Kouloheris 1981; Wheelock 1988). Portland cement is a byproduct of the process and is produced roughly in the same amount as sulfuric acid. Plants that used PG besides natural anhydrite and natural gypsum have been operated in Austria, England, Germany, Poland and the Republic of South Africa (UNIDO/IFDC 1998). None of the plants ran exclusively on PG for an extended period of time, although some did for short periods (UNIDO/IFDC 1998). Considerable amounts of energy are required to produce sulfuric acid from calcium sulfate making the process presently uneconomical since sulfur is available in most parts of the world as an inexpensive byproduct (mainly) from oil and natural gas refining (IAEA 2013). Compared to other naturally occurring calcium sulfates (e.g. natural anhydrite) that are relatively dry after mining the use of PG that exits the fertilizer plant with considerable amounts of moisture (more than 20 % by weight) is associated with even larger energy requirements (Ashburner 1968). Nevertheless, CMI proposed plans for a cement/sulfuric acid plant as part of an integrated complex consisting of a phosphate fertilizer plant producing 1.0 million tons P_2O_5 per year, a Portland cement plant producing 2.35 million tons cement per year and a 650 MW (presumably MW_{th}) coal-fired power station. Figure 1 shows a brief overview of the Pine Level Project proposed by CMI in Florida (CMI 1991).

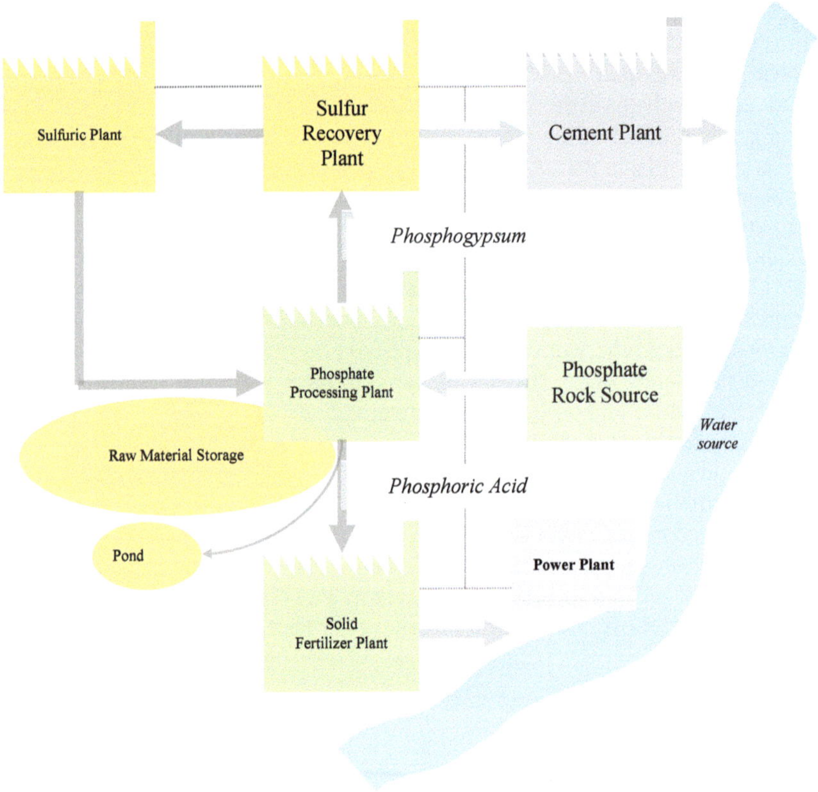

Fig.1. Brief overview of integrated plant proposed by Consolidated Minerals, Inc. (CMI)

CMI estimated that due to the integrated approach capital costs as high as 25 % of the estimated total capital costs could be saved. Table 1 provides an overview of these estimates (CMI 1991)

Table 1. Estimated savings of capital costs due to the integrated approach of the Pine Level Project (CMI 1991)

Area	Savings
Transport	10 %
Energy	2 %
Other Infrastructure	13 %
Total	**25 %**

Production of sulfuric acid from phosphogypsum using high temperature reactors

Past plants, experiments and attempts show that the recycling of PG to value-added products is possible, but presently not practiced as it is highly energy consuming. High temperatures of 1050-1450 °C (1050-1100 °C Gruncharov et al. 1988; 1150 °C Wheelock et al. 1988; 1450 °C Clur 1988) are required to produce desired commodities such as sulfur, sulfuric acid, fluorine, ammonium sulfate, sodium sulfate, calcium carbonate, strontium carbonate, Portland cement, non-shrinking cement, hydraulic binder, mineralizer and fiber boards or other construction materials from PG. Undoubtedly, such an approach is sustainable as precious metals are recovered and resources are saved. In the present framework with increasing energy costs and inexpensive sulfuric acid available conventional technologies are not feasible because of economic restraints. Using HTRs as an energy source that could generate electricity and process heat (relatively) independent from fluctuating fossil fuel prices may be a viable option in the near future. Several HTGR designs (e.g. INET 2014, JAEA 2014, STL 2014) are available today and could provide the energy for integrated phosphate fertilizer/PG treatment plants such as the Pine Level Project (650 MW_{th}) proposed by IMC or even larger endeavors. Other HTR concepts that are presently under development such as the Fluoride-salt-cooled High-temperature Reactor (FHR 2014) or the Compact High Temperature Reactor (Dulera and Sinha 2010) that are presently under development may be added to this list in the near future.

Present and upcoming efforts regarding "greener" production methods from regulation authorities may incentivize combined plants. New phosphate fertilizer plants in China for instance are asked to use 30 % of their produced PG by 2025 while in Guizhou Province (China) a goal for zero PG accumulation is set for the same year (Zhang 2013). In addition, reduced infrastructure- and energy needs (see Table 1) as well as the chance to employ a greenhouse gas lean technology that reliably provides large amounts of energy results in a desirable reduced environmental footprint. HTGRs reach outlet temperatures of up to 1000 °C (GEN IV 2014) and may also be used for cogeneration (combined heat and electricity generation) at such a facility since parts of the combined process to produce sulfuric acid/cement require a temperature of less than 1000 °C, for example when converting calcium sulfate from PG to calcium sulfide (Margaret et al. 1988). More efficient direct heat applications would further reduce environmental loads and may favor HTRs over other reactor technologies that reach lower outlet temperatures. Finally, a combined plant may promote the recovery of Rare Earth Elements (REEs) from PG. Processes that require additional amounts of sulfuric acid are available (Abramov et al. 2012; Rusin et al. 1980) and may be used to recover and sell precious REEs at various phosphate rock processing facilities throughout the world, thus diversifying the global REE market.

Conclusions

Various environmental benefits such as a reduced environmental footprint, the chance to recover valuable resources from PG while solving a major disposal issue can be associated in a combined phosphate fertilizer/PG processing plant as it is described here. However, further investigations with view to the technical implementation and the economic feasibility are needed to evaluate the overall viability of the proposed conception.

References

Abramov YK, Veselov VM, Zalevsky VM, Argunov ND, Bogdanova LP, Gukasov NA, Evdokimov D, Tamurka VG, Motovilova LV (2012), Method for extracting rare earth elements from phosphogypsum, Pub.No.: US 2012/0114538 A1

Ashburner DS (1968), The Production of cement and sulfuric acid by the Marchon Process, Proceedings of the 18th Annual Meeting of the Fertilizer Round Table, Washington D.C., pp 72-77

Austin RD (1980) Sulfur from Gypsum – Laboratory, Bench-Scale and Pilot-Plant Studies, Proceedings of the International Symposium on Phosphogypsum, Florida

Clur DA (1988) FEDMIS Sulphuric acid/cement from phosphogypsum process, Proceedings of the second international Symposium on phosphogypsum – Vol. II, Miami, Publication No. 01-037-055

CMI (1991) Clean, White Phosphogypsum and a way to use it, Phosphorous and Potassium 172:28-32

Dulera I V, Sinha R K, Indian High Temperature Reactor Programme: An Overview, BARC Newsletter, ISSUE NO. 315 SEP.-OCT. 2010

EPA (2012) United States Environmental Protection Agency – About Phosphogypsum, www.epa.gov/rpdweb00/neshaps/subpartr/about.html, accessed 21.06.2014

FHR (2014) Fluoride-salt-cooled High-temperature Reactor research at the University of California, Berkeley, http://fhr.nuc.berkeley.edu/, accessed 20.06.2014

GEN IV (2014) Generation IV International Forum, VHTR, https://www.gen-4.org/gif/jcms/c_9362/vhtr, accessed 20.06.2014

Gruncharov I, Pelovski Y, Dombalov I, Kirilov P, Videnov N (1988) Processing of phosphogypsum to sulphuric acid, Proceedings of the second international Symposium on phosphogypsum – Vol. II, Miami, Florida, Publication No. 01-037-055

Haneklaus N, Schnug E, Tulsidas H, Reitsma F (2014) Using high temperature gas-cooled reactors for greenhouse gas reduction and energy neutral production of phosphate fertilizers, Annals of Nuclear Energy *(submitted)*

Hilton J (2010) Phosphogypsum (PG) Uses and current handling practices worldwide, 25 th Annual Lakeland Regional Phosphate Conference, October 13-14, 2010

IAEA (2013) Safety Report Series No. 78, Radiation Protection and Management of NORM Residues in the Phosphate Industry, IAEA, ISBN 978-92-0-135810-3

INET (2014) Institute of Nuclear and New Energy Technology – Tsinghua University, http://www.inet.tsinghua.edu.cn/publish/inet/index.html, accessed 20.06.2014

JAEA (2014) Japanese Atomic Energy Agency, High Temperature Engineering Test Reactor, http://httr.jaea.go.jp/index.html, accessed 20.06.2014

Kouloheris A P (1981) Evaluation of potential commercial processes for the production of sulfuric acid from phosphogypsum – final report, Florida Institute of Phosphate Research, Publication No. 01-002-001

Margaret M, Ragin M, Brooks D R (1988) Recovery of sulfur from phosphogypsum Part 1. – conversion of calcium sulfate to calcium sulfide, Proceedings of the second international Symposium on phosphogypsum – Vol. II, Miami, Publication No. 01-037-055

Oppaga (2005) Department Strengthens Financial Responsibility Requirements for Phosphate Mining Companies, Report No. 05-27, http://www.oppaga.state.fl.us/MonitorDocs/Reports/pdf/0527rpt.pdf, accessed 20.06.2014

Rusin N F, Deyneka G F, Andrianov A M (1980), Extraction of sum rare earth metals from phosphogypsum, Proceedings of the International Symposium on Phosphogypsum, Florida

STL (2014) Steenkampskraal Thorium Limited, Th-100, http://www.thorium100.com/, accessed 20.06.2014

Sun Y (2013) HTR Development Status in China, IAEA TWG-GCR Meeting, VIC, Vienna, 5-7 March 2013, http://www.iaea.org/NuclearPower/Downloadable/Meetings/2013/2013-03-05-03-07-TWG-NPTD/Day_1/3.Sun.pdf, accessed 20.06.2014

UNIDO/IFDC (1998) Fertilizer Manual (3rd edition), Kluwer Academic Publishers, ISBN 0-7923-5032-4

Wheelock TD, Fan C-W, Floy KR (1988) Utilization of phosphogypsum for the production of sulfuric acid, Proceedings of the second international Symposium on phosphogypsum – Vol. II, Miami, Publication No. 01-037-055

Zhang P (2013) Comprehensive Recovery & Sustainable Development of Phosphate Resources, Presentation International Symposium on Innovation and Technology in the Phosphate Industry (SYMPHOS 2013), Agadir, Morocco

In case of emergency
SUPER FAST
RUN
Break the glass

On-Line X-Ray Fluorescence Analysis of Uranium and Thorium Materials in Mining and Processing Industry

E. Hasikova[1]**, A. Sokolov**[1]**, V. Titov**[1]

[1]Baltic Scientific Instruments, Ganibu dambis 26, Riga LV 1005, Latvia; office@bsi.lv

Abstract. The introduction of on-line X-ray fluorescence (XRF) analysis in mining and processing industry of uranium and thorium-bearing materials is discussed. Potential applicability of industrial XRF analyzer CON-X series is shown for continuous measurement of uranium (U) and thorium (Th) in various materials (rutile and zircon sands, phosphate rock and fertilizers, carbonatite and granite as unconventional U and Th resources, monazite concentrate as Th and U associated resources and etc.) at the wide range of U and Th concentrations (from less than 100 ppm to percents). U and Th detection limits (DL) are estimated as less than 50 ppm at concentrations below then 0.1 %.

Introduction

Currently economically viable U ores contain 0.05-0.07 % uranium oxide (U3O8). One thousand metric tons of U ore that contains < 0.1 % U3O8 has to be withdrawn from the depth of the earth (or dissolved with in-situ-leaching process) to produce each 1 ton of U3O8, without considering enormous amount of dead rock from the open pits or shuffles (Reilly et al.1991).

Thorium (Th) as nuclear fuel is clean and safe and offers significant advantages over uranium. The technology for several types of thorium reactors is proven but still must be developed on a commercial scale. In the case of commercialization of thorium nuclear reactor Th raw material will be on demand (2).

With all this, mining and processing companies producing uranium, thorium and rear earth elements will require prompt and reliable methods and instrumentation for U and Th quantitative on-line analysis.

Baltic Scientific Instruments develops and fabricates conveyor X-ray fluorescence analyzer (XRF) CON-X series for the on-line analysis of bulk solids transported by the conveyor in mining and process operations (3).

B.J. Merkel and A. Arab (eds.), *Uranium – Past and Future Challenges*,
DOI 10.1007/978-3-319-11059-2_92

These on-line analyzers are based on the non-destructive XRF technology. They intended to be mounted directly above conveyor belt to measure and analyze crushed U- and Th-bearing materials (uranium ore and monazite concentrate as U- and Th-associated resources; rutile and zircon heavy mineral sands, carbonate minerals and phosphate rock and fertilizers as unconventional resources, U ore residues after leaching as waste product etc.).

Accurate and timely information of the uranium content on the conveyor belts in the production chain enables optimizing the yield of uranium production process and gives great monetary savings by reducing the consumption of energy, raw materials, chemicals and water in mines and concentration plants.

Results of online analysis are not to be affected by variation of external parameters: temperature and humidity of material and industrial environment, variation of the material lump size, variation in height flow, etc. XRF analyzer CON-X series has been designed and manufactured accounting for these features. Changes of distance between moving product flow and analyzer and humidity variation are compensated with special mathematical algorithm.

Below the results of the laboratory studies and small-scale industrial tests of CON-X analyzer applicability for potential on-line measurements in U and Th industry are presented and discussed.

Mining and Processing Applications

CON-X analyzer is applicable starting with the primary enrichment (sorting and rejection of uranium ore) to the final control of the content of U ore residue after heap-leaching process.

CON-X on-line analyzer is able to measure U and Th content in real time even if these elements are only moderately concentrated compared to the average crust composition. Phosphate rock and fertilizers (50-200 ppm U), rutile and zircon sands (50-350 ppm U and 100-600 ppm Th), carbonate minerals, U ore residues ($\leq$ 100 ppm U), monazite concentrate (0.2% U_3O_8 and 5-7% ThO_2) and others were studied.

Fig.1. XRF analyzer CON-X02 installed on the industrial conveyor (traces of iron in quartz sand)

Phosphate rock and fertilizers

Uranium can be found as one of the components in phosphate ores in the amount of 50-200 ppm. The world U resources in phosphate rock are estimated as 9x106 metric ton. In production U is left in produced fertilizer as a radioactive contaminant.

Fig.2 below shows spectra of phosphate Moroccan rock and phosphate fertilizers (mono-calcium=MCP and mono-ammonium=MAP phosphates) produced form the rock. U lines are identified in all these materials even in 5-10 minute measurements.

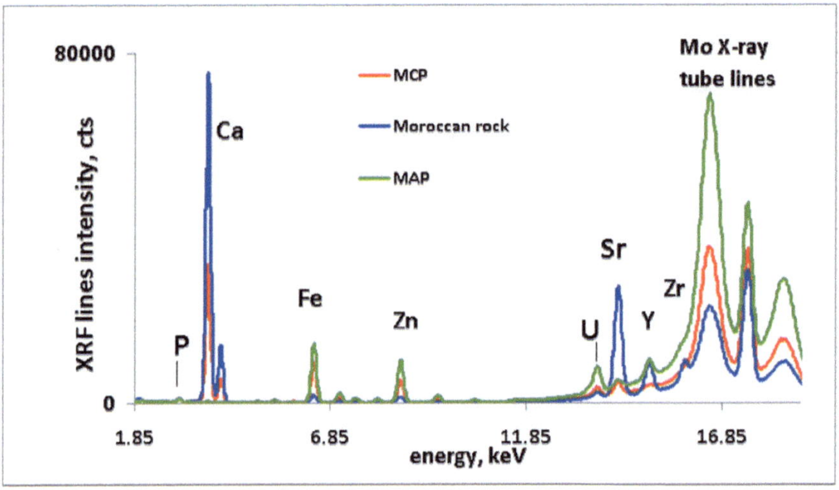

Fig.2. Spectra of phosphate rock and MCP and MAP fertilizers. Parameters of measurement: X-ray tube voltage - 39 kV, current – 100 μA, 100 μm thick Mo filter of primary radiation, measurement time 600 seconds, distance between sample and detector 50 mm

DL of U in phosphate rock was quantitatively estimated in line with the procedure 40CFR part 136 as 30 ppm U in the range of concentrations 50-300 ppm.

Estimated accuracy of U determination was ≤ 10 ppm (2σ). This level of quantification allows analyzer to make sorting of raw phosphate rock or monitoring of uranium level in commercial fertilizers.

Uranium ore residues after heap-leaching

On-site test of the CON-X analyzer for on-line measurement of uranium in ore residues after heap-leaching was implemented on the pilot conveyor at Areva Mines. The study showed that U may be determined in the matrix containing a number of interfering elements (rubidium, molybdenum and zircon) with accuracy ≤ 10% (2σ) at measurement time 600s. Range of U concentration studied was 30-200 ppm.

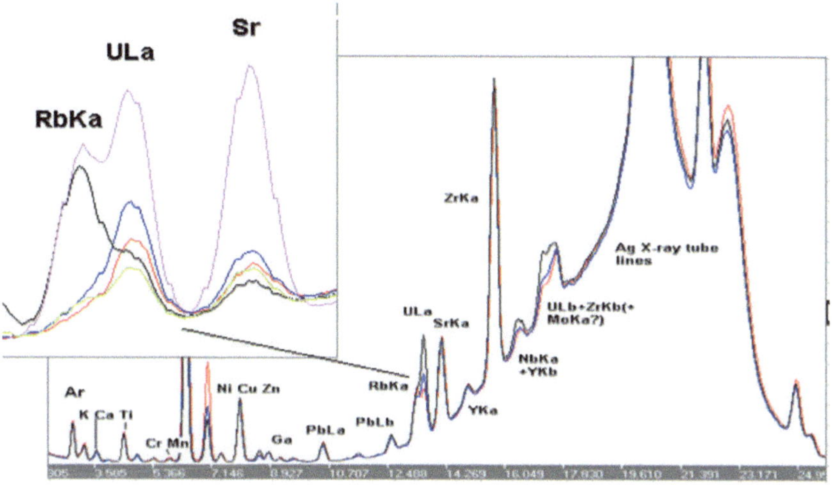

Fig.3. Spectra of U ore residues after heap-leaching process. Parameters of measurement: Ag X-ray tube voltage - 36 kV, current – 500 μA, 100 μm thick Ag filter of primary radiation, measurement time 600 seconds, distance between sample and detector 50 mm

Overview of XRF spectra of U ore residues after heap-leaching is shown on the Fig.3: U La and Lb lines are proportional to U concentration.

Detection limit (DL) is estimated at 30-50 ppm U in 5-10 minute measurements depending on interfering element.

Dedicated software BSIKit based on the Method of Fundamental Parameters was successfully used to solve the problem of the overlapping lines Rb Ka and U La and calculate U concentrations in the solid residues after U extraction with accuracy $\leq 10\%$.

U and Th in rutile and zircon heavy mineral sands

In 2009 industrial test of CON-X02 analyzer applicability was carried out at Richard Bay Minerals heavy mineral sand separation plant.

The target elements of this application were Zr and Ti, but capability of analyzer to monitor the level of minor elements (such as U and Th) was also proven. Spectrum of the sand after separation from silica containing rutile as major compound and zircon as a minor one is shown below on Fig.4.

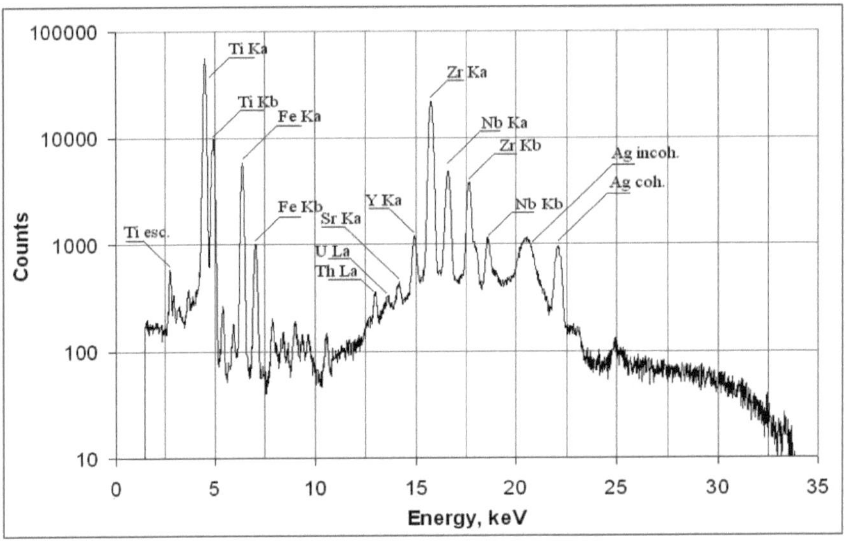

Fig.4. Spectrum of rutile sand after separation from silica. Parameters of measurement: Ag X-ray tube voltage – 39 kV, current – 100 μA, 50 μm thick Rh filter of primary radiation, measurement time 300 seconds, distance between sample and detector 50 mm

After 5 minute measurement U and Th XRF lines are clearly visible in the rutile sample mostly consisting of the TiO2 sand. Concentration range of U in heavy sands is 50-350 ppm.

DL proven for this study is 50 ppm U; accuracy was ≤ 15% relative.

Th concentration range in zircon sand is 100- 600 ppm. XRF spectra of zircon sand containing traces of ThO2 are shown on the Fig.5 below. After 10 minute measurements Th and U lines are clearly visible.

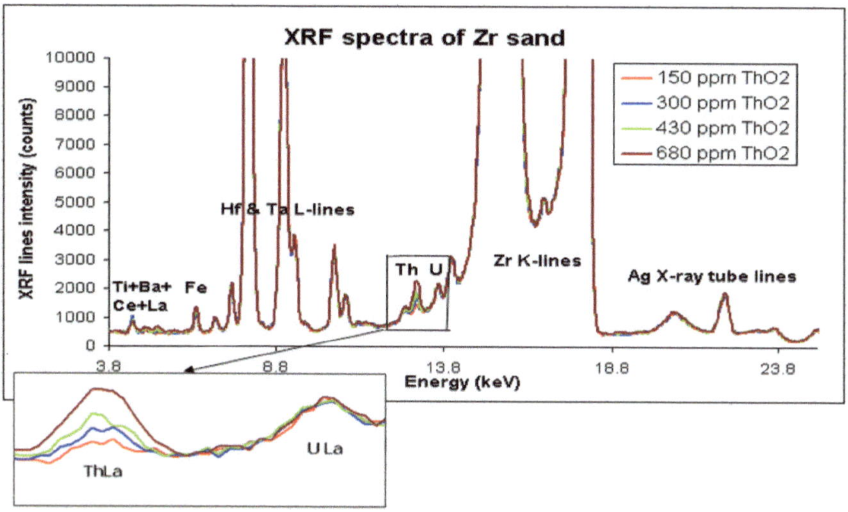

Fig.5. Spectra of zircon sand spiked with increasing Th amounts. Parameters of measurement: Ag X-ray tube voltage – 36 kV, current – 100 μA, 60 μm thick Ag filter of primary radiation, measurement time 600 seconds, distance between sample and detector 50 mm

Analyzer was calibrated at BSI XRF application laboratory with zircon sand samples provided from mining site and spiked with know amounts of Th-containing material. Linear calibration curve and screen shot of Th concentration trend measured for several zircon sand samples are shown on Fig.6 and 7. Linear correlation of calibration curve is high enough (R2 = 0.98).

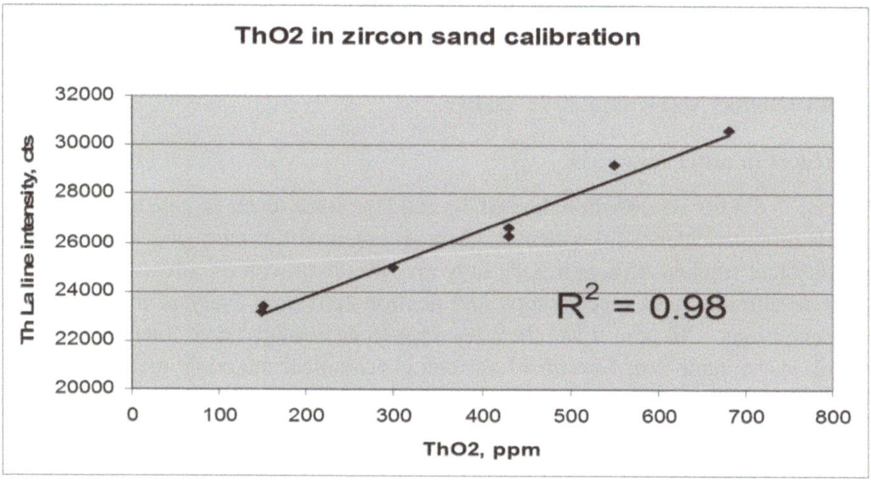

Fig.6. Calibration curve for Th traces in zircon sand. Intensity of Th La line is plotted vs. ThO2 concentration in calibration samples

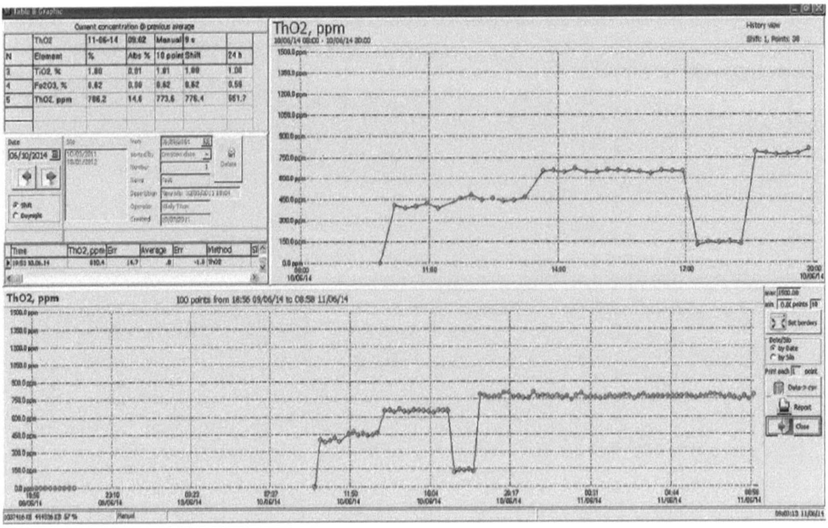

Fig.7. Trend of measured Th concentrations in laboratory on-line imitation test

Th DL is estimated at 25-50 ppm in 10 minute measurements depending on the type of material and interfering elements. Statistical precision is estimated as 25 ppm.

Industrial tests of the CON-X analyzer for continuous analysis of zircon and rutile sands were conducted at heavy mineral sand separation plant. The analyzer was installed above the conveyor after separation of silica. Along with positive results for Ti and Zr (target elements), test has shown reliable correlation of Th and U on-line measurements with laboratory results. Difference between on-line and plant laboratory was within 20 % relative.

Th traces in other minerals

Precise and accurate determination of Th and U at trace levels in geochemical materials (granite, calcite, ankerite and others) is essential for many geochemical and geophysical studies. Although XRF only provides data with the precision comparable to dilution mass spectroscopy and neutron activation analysis at Th and U concentrations > 10 ppm it has the advantage to determine many other useful elements in the same run. Spectra of various geochemical materials measured with CON-X analyzer are shown below on the Fig.8 and 9.

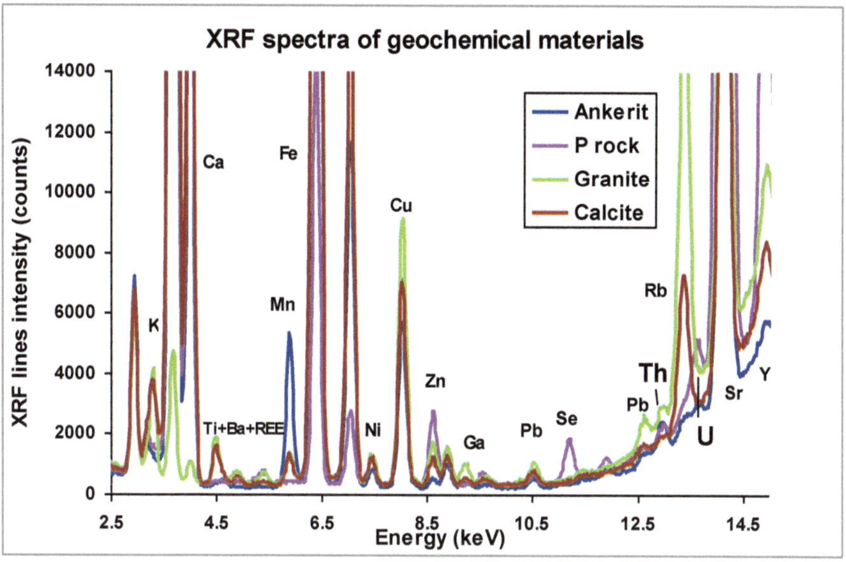

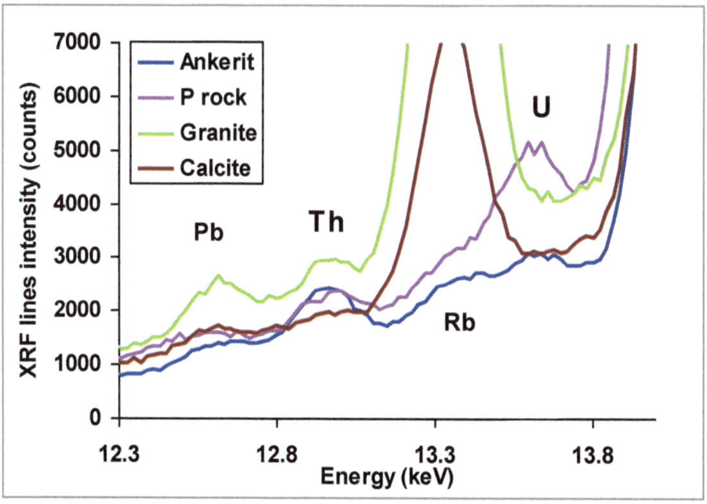

Fig.8 and 9. Overview (8) and zoom into weak lines (9) of XRF spectra of geological samples indicated in the legend. Parameters of measurement: Ag X-ray tube voltage – 36 kV, current – 100 μA, 60 μm thick Ag filter of primary radiation, measurement time 600 seconds, distance between sample and detector 50 mm

Th XRF line is clearly identified in all mineral geological products indicated on the pictures in 10-15 min measurements. It is know that generally granite contains 10-40 ppm of Th. Provided that Th concentration in the measured granite sample is on the level of 40 ppm Th detection limit in 10 min measurements is estimated as 20 ppm.

This measurements show that CON-X analyzer may be applied as a core drill analyzer. Geochemical studies may benefit from on-site core analysis in field mapping with a transportable XRF unit. For example, it can help in rapid decision-making on where to drill next.

Analysis of Th and U from associated materials

The amount of U and Th in monazite ore may be as high as 0.2 and 4-7%wt respectively. XRF spectrum of such monazite ore concentrate is shown on the Fig.10.

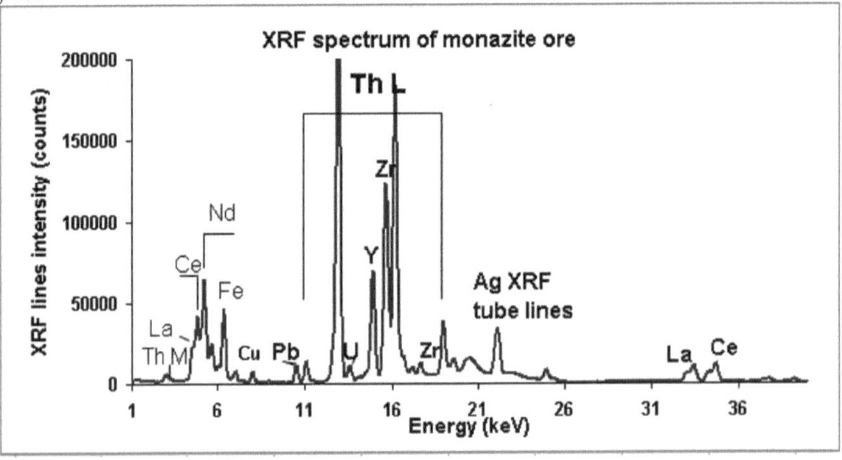

Fig.10. Spectrum of Monazite. Parameters of measurement: Ag X-ray tube voltage – 45 kV, current – 100 μA, 100 μm thick Ag filter of primary radiation, measurement time 600 seconds, distance between sample and detector 50 mm

In this study higher voltage of X-ray tube excitation was used to enhance the high energy detection efficiency to the level needed for the analysis of the rear-earth elements (La, Ce, Pr, Nd etc.) using their K-series lines. High energy resolution of the silicon drift detector allows detecting uranium XRF lines (at ~ 0.2%) alongside with the significantly more intense Th lines. Their partial overlap does not significantly affect the precision due to spectral deconvolution algorithm used in the software.

Estimated DL's are on the level of 20 ppm both for Th and for U.

Conclusion

Industrial XRF analyzer CON-X series fabricated by Baltic Scientific Instruments is capable to measure U and Th content in various mining and processing materials containing U ad Th both as trace, minor and major components.

In materials where U and Th present as trace or minor components (below 0.1%wt.) detection limit is estimated on the level of 30-50 ppm at 10 minute measurement depending on matrix and interfering elements. Accuracy is estimated as 10 % relative whereas at high U and Th concentrations accuracy is estimated as 1 % relative.

Small scale industrial test of CON-X analyzer applicability for on-line U measurements was carried out at uranium ore processing plant. Test was conducted with uranium ore residues after heap-leaching on the pilot conveyor (U content in residues studied 50-200). Difference between on-line and plant laboratory results is shown to be within 10 % relative.

Reference

Reilly D Ensslin N, Smith H, Kreiner S (1991) Passive nondestructive assay of nuclear materials. Los Alamos National Laboratory, US NRC

Investing in the future of thorium. Special report, Energy and Capital 2013: 11

http://www.bsi.lv/online_xrf_conveyor_analyzer_con_x_eng.html

What are you looking at?
It's the endangered common German nuclear power plant.

Nanofiltration of uranium-contaminated water – focus on separation mechanisms

Michael Hoyer[1], Roland Haseneder[1], Robin Steudtner[2], Vinzenz Brendler[2], Jens-Uwe Repke[1],

[1]Institute of Thermal, Environmental and Natural Products Process Engineering, TU Bergakademie Freiberg, 09599 Freiberg, Germany
[2] Institute of Resource Ecology, Helmholtz-Zentrum Dresden-Rossendorf e.V., 01314 Dresden, Germany

Abstract. Nanofiltration offers new perspectives for the treatment of mine drainage from former uranium mines, e.g. in East Germany. In this study the performance of various commercial nanofiltration membranes for a real mine water sample was determined experimentally and modelled. Experimental data is key to membrane selection and is necessary to validate modelling results. Mathematical modelling offers deeper insights into the interaction of uranium species with solid surfaces which also has the potential to extrapolate to other research fields.

Introduction

Nanofiltration offers new perspectives for the treatment of mine drainage, as has been shown by many authors, e.g. (Al-Zoubi et al. 2010). This paper explores the application of membrane filtration for the treatment of mine drainage from former uranium mines. The experiments are performed with commercial membranes and real mine waters, consequently, the results have implications on a theoretical and a practical level.

Membrane processes are a separation process, which means that a separation of the mine water into a purified and a concentrated stream is possible. The purified stream can then be discharged while the concentrated stream will require further treatment, such as precipitation.

The two main objectives of this work is to A) compare different membranes for their performance in order to select one membrane with the best performance based on the selection criteria, and to B) apply a mathematical model for a sensitivity analysis in order to gain deeper understanding of the properties of membranes which are advantageous for the treatment of uranium-contaminated water. The performance criteria that are chosen for membrane evaluation are their uranium retention, also in comparison to other feed constituents, in order to reach a

B.J. Merkel and A. Arab (eds.), *Uranium – Past and Future Challenges*,
DOI 10.1007/978-3-319-11059-2_93

highly selective process, as well as their permeability, which is a measure of the water quantity that is purified and thus an indicator for process economics.

The model is based on the Donnan Steric Pore Partitioning and Dielectric Exclusion model (DSPM&DE) which has been developed, improved, and investigated by many researchers, e.g. (Bowen and Mukhtar 1996), (Bandini and Vezzani 2003), (Yaroshchuk 1998).

Materials and Methods

The membrane experiments were performed in a cross-flow test unit in continuous mode with feed and permeate recirculation to ensure stable feed concentrations. Details on the membrane filtration unit can be found elsewhere (Hoyer et al. 2013).

Experimental procedure

To find the best-performing membrane from the initial 18 membranes, the two steps in Table 1 were performed, which include a pre-selection of membranes based on their KCl-retention and a subsequent performance evaluation of the pre-selected 7 membranes with the real mine water.

Table 1. Experimental protocol for the membrane experiments

	Membrane screening	Performance evaluation
Feed solution	1mM KCl solution	Real mine water
Membranes	18 membranes	7 membranes
Trans-membrane pressure	15 bar	15 bar
Temperature	25°C	25°C
Cross-flow velocity	1.1m/s (Reynolds number approx. 5000)	1.1m/s (Reynolds number approx. 5000)
Objective	Pre-select membranes	Compare the selected membranes for their performance

Fresh membrane samples were prepared for each experiment and immersed in deionized water for 24 h. Samples of the mine water feed were taken before the particle filtration, before and after each experiment. Flux was measured with a balance with automatic data recording. Prior to each experiment the facility was flushed with deionized water until the conductivity of the flush water did not increase anymore. After cleaning the facility the fresh membrane was inserted so it would not be impaired by the cleaning procedure.

Experimental results: Membrane screening

The selection of membranes needs to be based on characteristics such as retention and flux. Membrane screening for all membranes has been performed to get consistent data, as literature and manufacturer data on membrane performance can be inconsistent due to differences such as, e.g. the membrane test facility used. The screening results for the initial 18 membranes are shown in Table 2.

Table 2. Membrane screening results

Membranes	Screening result
NF1 – NF7	Selected for the performance evaluation (for results see next section)
NF11 – NF14	Redundant, as they exhibited similar retention and fluxes as the selected membranes
NF8 – NF10, UF1	Retentions were relatively low
RO1 – RO3	Fluxes were relatively low

Experimental results: Performance evaluation

The selected membranes from the screening experiments were then used for experiments with the real mine water. The obtained retentions for the main water constituents are shown in Figure 1.

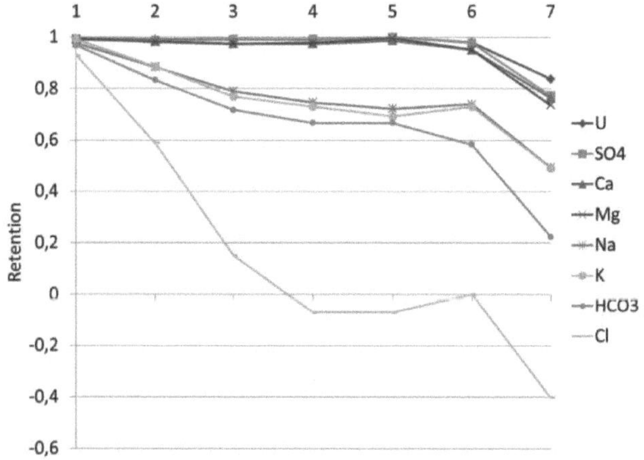

Fig.1. Retention for all main water constituents for each membrane. The x-axis shows the ID numbers of the 7 selected membranes. The lines that connect the individual data points are for illustrator purposes and are not intended to indicate a trend

As can be seen in Figure 1, all membranes except membrane NF7 exhibit high retentions, and can thus produce purified water of high quality, i.e. of low uranium concentration. Additionally, Membranes NF3 to NF6 show a clear separation between multi- and monovalent ions. Such selectivity has the potential to decrease the amount of chemicals that are needed in the subsequent precipitation treatment of the mine water, depending on the kind of precipitation that is being used.

The final selection of the best performing membrane needs to also consider another criterion: the flow of water produced per membrane area and trans-membrane pressure, i.e. membrane permeability. This data is shown on the x-axis of the diagram in Figure 2, and related to the corresponding uranium retention, shown on the y-axis, for each of the 7 membranes.

Figure 2 can be used to select the best performing membrane as both criteria, the quality as well as the quantity of the purified water can be inferred. The best-performing membrane is membrane NF3.

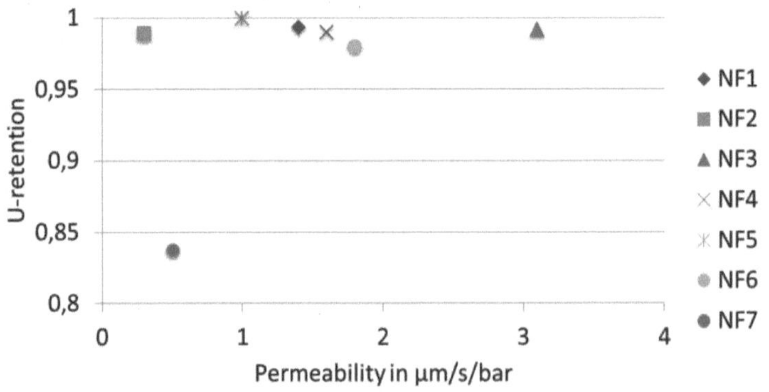

Fig.2. Uranium retention for the 7 selected membranes. The x-axis shows the membranes' permeability which is a measure for the amount of purified water

Experimental evaluation of the membrane properties which make the membrane perform well is difficult to achieve. The applicability of independent methods, such as the molecular weight cut-off (MWCO) is limited as many membrane properties together (membrane pore radius, membrane thickness, membrane charge, membrane relative permittivity) determine a membrane's performance, and these properties influence on each other. So, the MWCO which is a measure for the "openness" of the membrane cannot predict the membrane performance. As shown in Figure 3, the order of the membranes is completely different from the order of the membranes in Figure 2. Thus, direct conclusions cannot be drawn for performance prediction of a membrane based on its MWCO.

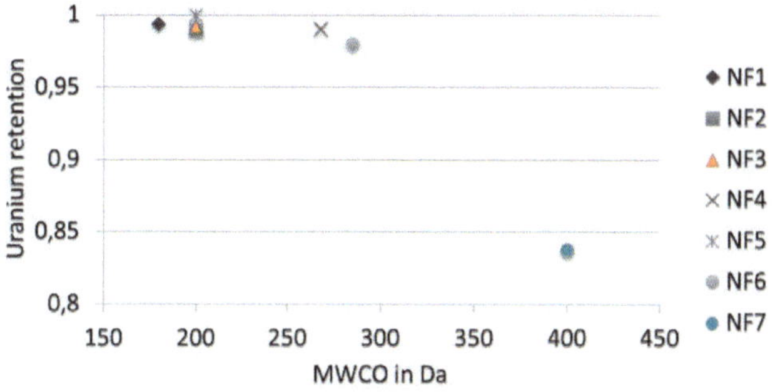

Fig.3. Uranium retention for the 7 selected membranes. The x-axis shows the membranes' molecular weight cut-off which is the molar mass of an theoretical solute that is retained by the membrane by 90%

A solution to this problem of complex interactions is mathematical modelling which takes the interaction between the individual parameters into account.

Modelling results

The advantage of modelling data is shown here using a preliminary model. This means the model is still under development but the fact that it already produces powerful output at this stage shall be documented. In Figure 4 the effect of pore radius variation is shown for sulfate and uranium. All other membrane parameters are kept constant and the isolated influence of the pore radius is shown.

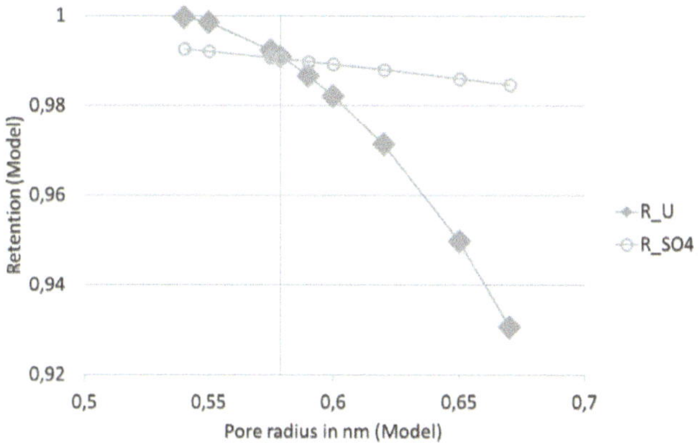

Fig.4. Modelled uranium and sulfate retention of membrane NF3 for varying pore radii

As can be seen from Figure 4, uranium retention depends strongly on pore radius while the retention of sulfate does not. This means, the main separation mechanism for uranium is size exclusion while for sulfate, e.g. electrical effects will play a role.

Conclusions

The performance of 18 different membranes was compared for a potential applicability in the treatment of uranium-contaminated water. A best-performing membrane was selected based on water quality and quantity. Modelling was introduced as a method to investigate the influence of individual membrane properties in an isolated form. Long term studies to evaluate stability of membranes as well as model development and validation are currently running.

References

Al-Zoubi H, Steinberger P; Rieger A, et al. (2010) Nanofiltration of Acid Mine Drainage. Journal of Desalination and Water Treatment 21: 148–161.

Bandini S, Vezzani D (2003) Nanofiltration modeling: the role of dielectric exclusion in membrane characterization. Chemical engineering science 58: 3303 – 3326

Bowen W R, and Mukhtar H (1996) Characterisation and prediction of separation performance of nanofiltration membranes. Journal of Membrane Science 112: 263-274

Hoyer M, Haseneder R, Repke J-U (2013) Membrane filtration of uranium contaminated water - focus on speciation. Wissenschaftliche Mitteilungen des Instituts für Geologie der TU Bergakademie Freiberg 44: 37 – 41.

Yaroshchuk A E (1998) Rejection mechanisms of NF membranes. Membrane Technology 100: 9 – 12

Investigations of uranium and trace elements in groundwater of the Tanjero Area, Kurdistan Region, Iraq

Aras Kareem[1,*], Broder Merkel[1], Omed Mustafa[1]

[1]Institute of Geology, TU Bergakademie Freiberg, Gustav-Zeuner-Str. 12, 09599 Freiberg, Germany, *Corresponding Author/ email:arasomer@yahoo.com.

Abstract. Geochemical investigations and modeling of groundwater chemistry was performed in Tanjero area, Sulaimani city, Kurdistan region, Iraq. Uranium in waters of the area shows low concentrations, however, uranium is positively correlated with other trace elements which are typical indicators for man-made contaminations. Two main point sources of contamination show higher level of uranium and V, Cr, Zn and Fe: the municipality dumping site of Sulaimani city and industrial dumping sites. Comparing ^{235}U with ^{238}U reveals that the uranium source is neither depleted uranium (DU) nor enriched uranium. Thus it is very likely that scrap iron from hydrocarbon production and storage facilities dumped in the land fill is the source of the slightly elevated uranium concentrations. Depending on the uranium and other trace elements three different classes of water can be distinguished: contaminated water, non-contaminated waters and mixed water. Dominance of U (VI) and As (V) with the positive correlation between U and nitrate indicate oxidizing environment.

Introduction

The growing population in Sulaimani city (Kurdistan region, Northern Iraq) has led to high appearance of solid waste, reaching up to 1,000 tons pair day (Rashid 2010) figure (1). In the last decade, the industry zone in Tanjero area has been characterized by an increasing number of factories for industrial products. The area is located to the southwest of Sulaimani city (12 km from the city center). Waste, warehouses and fuel tanks are spread randomly without any controlling or a concept for environment protection: the industrial remains are dumped uncontrolled in a fenced but non-supervised and a landfill site has not been constructed. The area of the landfill is about 50 hectares, allocated for the municipal waste dumping. The wells around the waste dump area have been contaminated by certain chemical elements, causing changes in the quality of the pumped water and

B.J. Merkel and A. Arab (eds.), *Uranium – Past and Future Challenges*,
DOI 10.1007/978-3-319-11059-2_94

the suitability for human consumption. The seasonal weather changes are very distinguished in Kurdistan region. The climate is characterized by cold and snowy winters and long, warm dry summers, while autumn and spring are short. During the summer, daily temperatures can reach 45 °C. The main characteristic of seasonal rainfall distribution is the absence of precipitation during the summer (June-September). The rainy season usually starts in mid-October and ends at the beginning of May. The month with the greatest amount of precipitation in the whole region is January (Stevanovic and Markovic 2003).The area of interest is composed by clastic rocks (Tanjero formation) with a thickness of approximately 140 m and clastic Alluvium containing gravel, sand and clay (Budy et al. 1980)(Ali 2003). Tanjero is a permanent river flowing along the southern margin of Sulaimani city and feeding in east-southeast direction the Darbadikhan lake. This small river is polluted from the sewage of the city and some villages around (NI 2008).

Fig.1. Map of the study area

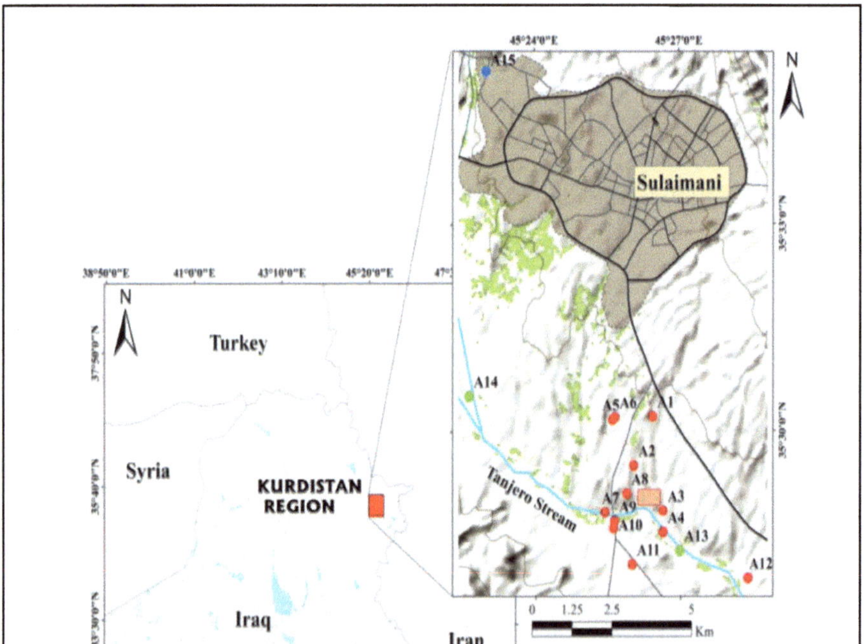

Materials and Methods

Fifteen water samples were collected from several wells around the landfill site. (Figure 1). Samples 13 (southeast from the waste dump site and factories) and 14 (northwest from the waste dump site and refinery factories) were collected from Tanjero River. A15 was collected from Sarchnar spring, 10 km from the dump site. The remaining samples were collected from water wells in the industry area at varying distances from the waste dump site. pH, redox potential, electrical conductivity, dissolved oxygen (DO) and temperature were measured in during the field work by using portable advices (WTW: pH320, LF320, Multiline P4) and (HACH:HQ 40D). 100 ml water sample was collected in pre-cleaned polythene for major and cations and anions analysis. 30 ml water sample was additionally collected and filtered with 0.2 µm membrane filters. Subsequently, ultra-pure HNO_3 (30%) acid was added to these samples for ICP-MS analysis. Trace elements were analyzed by ICP-MS (Thermo Scientific xSeries 2). Cations and anions were measured by ion chromatography (Metrohm: 881 CompactICapro AnionMCS-2.881.0030 and 85Professional IC 2.850.1010). Total inorganic carbon TIC was determined by LiquiTOC (Elementar Analysensysteme GmbH). Thermodynamic modeling and hydrochemical calculations were computed by PHREEQC (Parkhurst and Appelo 2013) and Wateq4F database. However, results may be very different when using other databases (Merkel 2011).

Results and discussion

Uranium and other trace elements

Uranium (U) in the water samples taken is correlated with the other trace elements (Fig.2). U is positively correlated with Sr, V and the sum of all trace elements (Al, Mn, Fe, Ni, Zn, As, Se, Sr, Ba, Cr, V and Pb). This relation provides an indication about the similarity in sources and refers to contamination of the water resources of the area by these elements. Mg, Sr and V also shows a correlation with U (Fig.1d); on contrary Ca does not correlate with U. Therefore, U seems to be bound in Mg-containing complexes and minerals. Two main point sources of contamination show higher levels of U, V, Cr, Zn and Fe (Table 1), namely the municipality (A3 and A4) and industrial dumping site (A5) of Sulaimani city.

Table 1. Physiochemical characteristics of water samples in the studied area

	A-1	A-2	A-3	A-4	A-5	A-6	A-7	A-8	A-9	A-10	A-11	A-12	A-13	A-14	A-15
Tem	24	20.7	20.9	21.3	21.2	21.6	25.5	21.4	20.7	24.6	20.1	22.6	24.1	21.7	19.8
pH	7.8	8.4	7.1	7.2	7.3	7.8	7.2	8.8	6.8	7.3	7.4	7.4	7.9	7.7	7.8
pe	5.22	5.83	6.00	5.31	6.11	5.70	5.12	5.36	5.01	5.72	5.88	6.21	4.54	4.74	6.05
Ca	27.6	6.9	117	117	102	21.9	121	10.4	138	56.5	27.4	73.9	86.2	80.7	74.4
Mg	6.4	1.0	21.3	21.2	29.8	4.7	21.4	5.2	26.0	19.2	11.6	11.8	14.1	13.2	10.5
K	0.5	0.5	1.9	2.2	0.8	1.0	3.9	0.4	3.7	1.0	0.6	0.6	7.8	10.1	1.0
Na	161	135	20.7	22.7	26.3	222	58.4	140	41.7	80.6	145	148	41.6	63.9	4.6
HCO_3	292	257	336	385	365	337	450	222	443	303	336	279	358	433	236
SO_4	28.2	45.1	50.7	51.0	105	118	80.3	96.8	97.9	67.2	111	144	74.9	84.5	38.6
Cl	110	33.5	54.4	54.8	57.2	87.9	69.9	36.3	55.6	27.4	22.2	13.8	44.4	66.6	5.7
NO_3	3.1	7.8	11.5	5.6	25.9	2.5	4.5	6.7	6.5	4.2	2.8	2.4	7.3	11.7	10.4
NO_2	0.003	0.01	0.03	0.06	0.02	0.01	0.12	0.03	0.05	0.04	0.02	0.03	0.35	0.40	0.04
PO_4	1.00	1.14	0.68	0.59	33.5	0.45	0.51	0.25	1.14	0.61	0.85	0.41	1.56	2.80	0.15
Si	9.77	20.4	14.2	14.0	17.8	18.5	11.7	13.5	11.1	14.8	14	13	7.56	7.06	4.24
F	1.65	0.48	0.13	0.13	0.10	0.21	0.18	0.31	0.18	0.24	0.26	0.19	0.11	0.13	0.10
Li*	24.8	4.4	3.01	3.23	2.18	8.98	5.99	4.33	5.82	6.25	6.68	3.12	4.49	4.29	2.59
B*	3144	926	58.3	53.1	82.5	624	340	435	63	1242	633	47.8	48.8	48.8	19.6
Al*	13.4	3.59	4.68	1.79	1.53	3.78	1.08	42	1.19	0.52	1.56	1.79	19.7	16.7	7.18
Mn*	6.26	1.98	0.41	206	0.74	0.44	911	0.74	492	153	0.86	1.85	286	276	2.76
Fe*	15.1	2.8	7.74	9.6	9.92	3.09	68.4	8.3	4.43	2.66	13.3	1.11	52.7	36.8	6.52
Ni*	0.48	0.11	0.79	4.09	0.78	0.39	2.43	0.22	5.21	8.32	3.18	0.14	5.64	5.52	0.53
Zn*	51.1	49.3	28.7	40.6	35.3	77.1	42.2	58.6	50.2	145	41.2	58.7	58	62.3	39.6
As*	5.78	0.76	0.94	1.18	0.67	1.54	1.61	7.8	1.1	0.19	1.49	0.47	2.84	2.41	0.29
Se*	0.91	0.03	0.09	0.1	0.33	0.57	0.14	0.11	0.21	0.15	0.62	0.18	0.23	0.31	0.32
Sr*	770	319	1257	1231	1757	1033	1028	294	1032	677	564	823	622	566	503
Ba*	196	2.2	133	146	77.4	16.6	215	20.2	171	76	28	63	113	138	114
Pb*	0.23	0.05	0.03	0.17	0.11	0.09	0.05	0.05	0	0.02	0	0.02	0.13	0.12	0.07
V*	2.76	0.89	14.4	13.7	15.6	4.27	0.67	4.44	6.17	2.73	7.13	6.65	8.44	6.42	1.99
Cr*	0.12	0.02	0.38	0.32	1.46	0.28	0.03	0.5	0.07	0.02	0.04	0.17	0.43	0.58	0.42
U*	0.39	0.28	1.8	1.82	3	1.5	0.4	0.36	1.57	1.49	1.48	1.52	0.99	0.63	0.65
ΣTE	1.06	0.38	1.45	1.65	1.9	1.14	2.27	0.44	1.76	1.06	0.66	0.96	1.17	1.11	0.68

Tem = water temperature; concentration of elements and ions are in mg/L

ΣTE = sum of trace elements Al, Mn, Fe, Ni, Zn, As, Se, Sr, Ba, Cr, V and Pb .

*: µg/L

Uranium and saturation indices of minerals

U shows a significant correlation with the saturation index (SI) of celestite, gypsum (Fig.3) and apatite minerals group which is not surprising because U is contained in apatite (phosphate minerals) and gypsum (Fred and Wesley 1982). Thus, correlation of U with SI of gypsum and apatite minerals indicates contamination with phosphate bearing materials. Decomposition of apatite leads to release of U

(VI), although this is accompanied by the presence of pyrophosphates (Fred and Wesley, 1982). U (VI) and $UO_2(HPO_4)_2$ were the dominant species and complex of uranium in studied samples (Fig.4), and thus the presence of pyrophosphates in the contamination sources is very likely. A positive correlation was noticed between U and P, which also refers to the presence of phosphate in the contamination sources (Fig.5b).

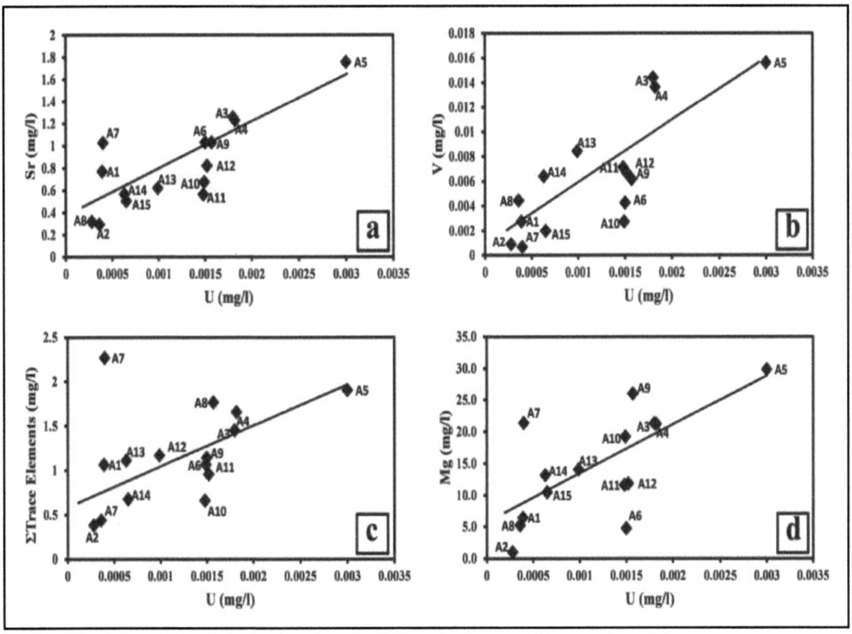

Fig.2. Correlation of uranium and other elements

Considering the U and $SI_{Calcite}$ relationship, sample A11 was under-saturated and samples A4, A5, A7, A8, A13, A14 and A15 were super-saturated with respect to calcite (Fig.3c). Considering the U and $SI_{Dolomite}$ relationship, sample A2, A9 and A11 were under-saturated while the others were saturated and super-saturated with dolomite (Fig.3d).

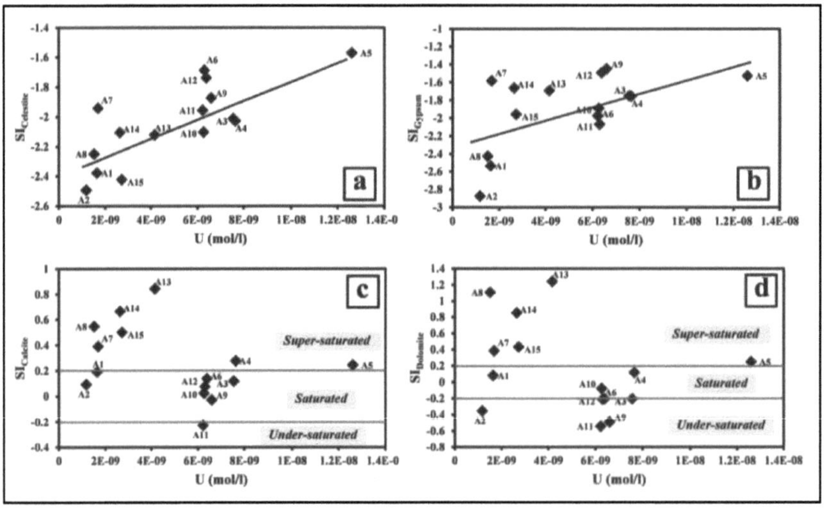

Fig.3. Correlation of U and saturation index of minerals

Therefore, it can be concluded that contamination sources leads to increased calcite and dolomite dissolution. Generally, contaminated samples (A3, A4 and A5) show a lower degree of saturation compared to the non-contaminated samples.

Table 2. Saturation indices of selected minerals in waters of the area

	Al(OH)3(a)	Aragonite	Barite	Calcite	Celestite	Dolomite	Gibbsite	Gypsum	Hematite	Uraninite(c)	Fluorapatite
A-1	-2.24	0.05	0.19	0.19	-2.38	0.08	0.47	-2.54	15.9	-13.7	4.0
A-2	-3.28	-0.05	-1.46	0.09	-2.49	-0.35	-0.55	-2.88	14.5	-15.9	2.7
A-3	-1.91	-0.03	0.21	0.12	-2.01	-0.21	0.82	-1.75	15.0	-13.6	2.3
A-4	-2.43	0.13	0.24	0.28	-2.03	0.12	0.29	-1.76	15.2	-12.2	2.5
A-5	-2.61	0.10	0.24	0.24	-1.57	0.25	0.12	-1.53	15.4	-17.1	7.6
A-6	-2.70	-0.07	-0.32	0.08	-1.69	-0.21	0.02	-2.07	14.6	-13.8	1.3
A-7	-2.81	0.25	0.48	0.39	-1.94	0.39	-0.13	-1.58	16.7	-12.5	2.6
A-8	-2.64	0.40	-0.26	0.55	-2.25	1.11	0.08	-2.43	15.2	-15.7	2.3
A-9	-2.27	-0.18	0.53	-0.03	-1.87	-0.49	0.46	-1.45	13.02	-11.78	1.99
A-10	-3.21	-0.12	0.07	0.02	-2.10	-0.08	-0.52	-1.89	14.18	-13.32	2.07
A-11	-2.62	-0.37	-0.08	-0.23	-1.96	-0.54	0.11	-1.97	15.78	-13.97	1.23
A-12	-2.68	-0.01	0.27	0.14	-1.74	-0.21	0.04	-1.49	13.53	-14.06	2.14
A-13	-2.19	0.70	0.27	0.84	-2.12	1.24	0.51	-1.69	16.97	-12.27	5.74

A-14	-1.96	0.52	0.45	0.66	-2.11	0.85	0.76	-1.66	16.72	-13.13	5.72
A-15	-2.34	0.35	0.15	0.50	-2.42	0.43	0.39	-1.96	15.33	-14.11	2.19

Classification of water resources

Three different classes of water can be distinguished depending on U, V, Zn, Sr and Mg: contaminated water representing groundwater in the vicinity of the land fill and the industrial dumping area, non-contaminated waters referring to the Sarchinar spring and some wells water and a mixed zone contain samples from Tanjero River and nearby well waters.

Uranium and oxidation environment

Al uranium occurs in the oxidation state (VI) and As is mainly arsenate, which both indicate oxidizing environment (Fig.4a). The positive correlation between U and nitrate (NO_3) also refers to the oxidizing conditions; therefore, oxidation processes are assumed to be dominant in waters of the area. A negative anomaly (lower value compared to the other samples) of UO2 $(HPO_4)_2$, Mg^{2+} and Sr^{2+} was observed in samples A2, A6, A8 and A15 (Fig.4b). These anomalies indicate that non-contaminated water show lower UO_2 $(HPO_4)_2$, Mg^{2+} and Sr^{2+} concentrations.

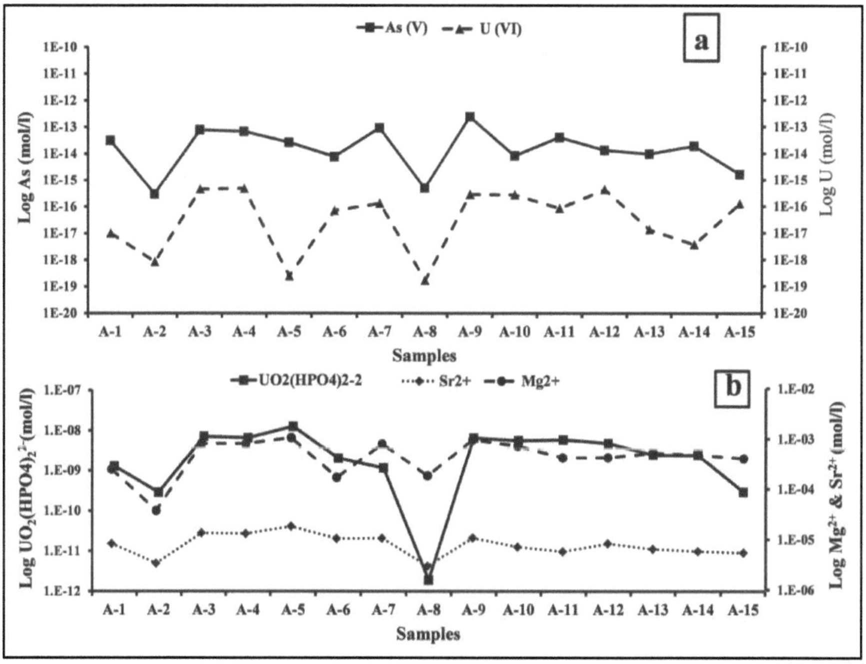

Fig.4. Dominant species of U, As, Mg and Sr in the studied samples

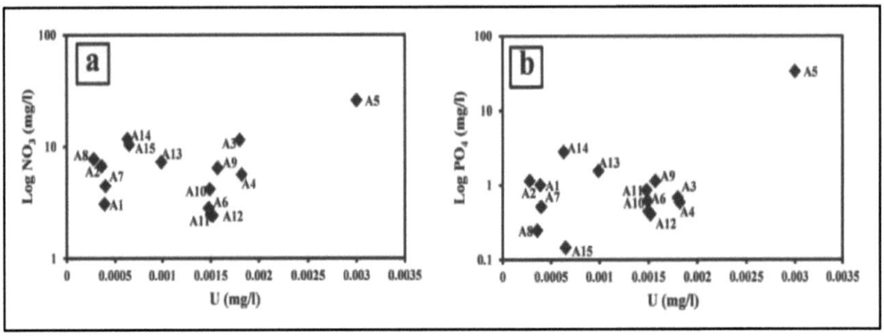

Fig.5. Relation between U and NO_3 and PO_4. Y-axis in both (a) and (b) are in logarithmic scale

Conclusions

U is positively correlated with Sr, V and the sum of trace elements which indicate contamination of the water and similarity in two main point sources of contamination (municipality and industrial dumping site of Sulaimani city) recognized by higher levels of uranium and other trace elements. Correlation of U with SI of gypsum, celestite and apatite minerals group is an indication for contamination with phosphate bearing materials. Three different classes of water can be distinguished depending on uranium V, Zn, Sr and Mg: contaminated water representing groundwater in the vicinity of the land fill and the industrial dumping area; non-contaminated waters are referring to the Sarchnar spring and some wells water and a mixed zone contain samples from Tanjero River and well waters nearby. The contaminated waters are supersaturated with hydroxyapatite, barite, gibbsite, and hematite, while not contaminated waters show super-saturation with respect to calcite.

References

Ali S.S, (2007) Geology and hydrogeoloy of Sharzoor-Piramagroon basin-Sulaimani, Belgrade Univeristy, unpublished Ph.D thesis: 330.

Budy T., Jassim, S.Z., Kassab, I.I (1980) Regional geology of Iraq, Vol. 1, Stratigraphy. Min. investigation report: 445.

Hurst, Fred J., Wesley D. (1982) a discussion of uranium control in phosphogypsum. In Hydrometallurgy 9 1: 69–82.

Merkel, B.J., 2011. Thermodynamic data dilemma, in: Merkel, B., Schipek, M. (Eds.), The New Uranium Mining Boom. Springer Berlin Heidelberg, 627–633.

NI (2008) survey of water quality, sediment, benithic macro invertebrate and fisheries for Qara Ali dam irrigation project (QDIP) enviromental impact assessment. Nature Iraq field and lab. Report: 15-16.

Rashid Kh. (2010) enviromental Implication of Tanjero waste disposal site of Sulaimani, Sulaimani University, unpublished PhD. Thesis: 2-3.

Stevanovic Z., Markovic M (2003) Hydrogeology of northern Iraq, Vol.1, FAO, Erbil, Climate, Hydrology, Geomorphology, Geology, Ed. "Field documents":130.

People don't want nuclear power?
Problem solved!

Radon measurement along faults in the Upper Rhine Graben with standardized methods

Georg Kuhn[1], **Rouwen Lehné**[1], **Andreas Hoppe**[1]

[1]TU Darmstadt, Institut für Angewandte Geowissenschaften, Schnittspahnstr. 9, 64287 Darmstadt

Abstract. Radon measurements are currently under way along newly discovered faults in the northern Upper Rhine Graben near Groß-Gerau. Measurements of radon activity in soil air should show if the disruptions are recently active. In the process, it is difficult to compare the measurements of soil types, with different permeability, without logging the actual air flow. It is intended therefore to develop a standardised method which combines the measurement of radon with the detection of CO_2 in soil air. Outcomes will be used for correlation with radon concentrations in the indoor air of buildings in the region of Darmstadt (Kuhn 2013), which is situated in the tectonically active northern Upper Rhine Graben, in order to evaluate geogenic radon in the area.

Introduction

Radon is a radioactive noble gas which originates in decay series. Its natural occurrence in soil as well as ambient air is strongly dependent on the local mineralogical composition of rocks and soil. High radon concentrations in and around Darmstadt are due to the crystalline rocks of the Odenwald Mountains and migration paths (natural such as tectonic faults and anthropogenic, e.g. openings in foundations) which enable radon to overcome larger distances and accumulate into radon anomalies.

They can be an indicator for recent tectonic activities. High concentrations of radon, e.g. in poorly ventilated cellars, may also pose a health hazard as related alpha emissions are supposed to be the second most frequent cause of lung cancer (Darby et al., 2005).

In Darmstadt, radon concentrations in soil air have been measured on the university´s campus at Lichtwiese and Botanischer Garten as well as along the eastern master fault of the Upper Rhine Graben in the city center, accompanied by measurements of indoor room air in order to find possible connections with migration paths from the underground.

B.J. Merkel and A. Arab (eds.), *Uranium – Past and Future Challenges*,
DOI 10.1007/978-3-319-11059-2_95

In addition, radon measurements are currently underway along newly discovered faults in the northern Upper Rhine Graben near Groß-Gerau in order to verify recent activities. Measurements are being conducted in line with a project for setting up a geothermal power plant.

The measuring method described by Kemski et al. (1998) is widely used. However, it measures the activity of radon only, without determining the quantity of soil air flowing through the measurement chamber. Consequently, it is difficult to compare soils with different permeability. It is intended to develop a standardized method which combines the measurement of radon with the detection of CO_2 in soil air.

Furthermore, a flow-meter will be installed in the measurement setup to log the actual air flow. This is necessary to compare the different results of measurements of several soil types. Field measurements of soil samples are complemented with gamma ray spectrometry analysis in order to determine the concentration of radium, the parent nuclide of radon. This is necessary to differentiate between radon which is formed in situ and the radon which migrates along the fault zones.

Results

Darmstadt

Radon concentrations in the city are high due to the underground geological conditions. While the area under the Lichtwiese and Botanischer Garten campuses is related to Permian volcanic rocks, the center lies on gabbro in the east and is dissected by the eastern master fault of the Upper Rhine Graben.

Campus Botanischer Garten

Several measurements of radon in the soil air show concentrations of > 50.000 Bq/m³; one measurement exceeds 100.000 Bq/m³ (Fig. 1), which are very likely related to Permian basalts. Therefore, radon concentration in indoor air has been measured in several campus buildings. Results show noncritical concentrations.
However, varying concentrations in different parts, i.e. the Institute of Applied Geosciences, indicate a pattern between renovated and non-renovated areas (Fig. 1). Outcomes are used for correlation with radon concentrations in the indoor air of buildings in the region of Darmstadt (Kuhn 2013), which is situated in the tectonically active northern Upper Rhine Graben.

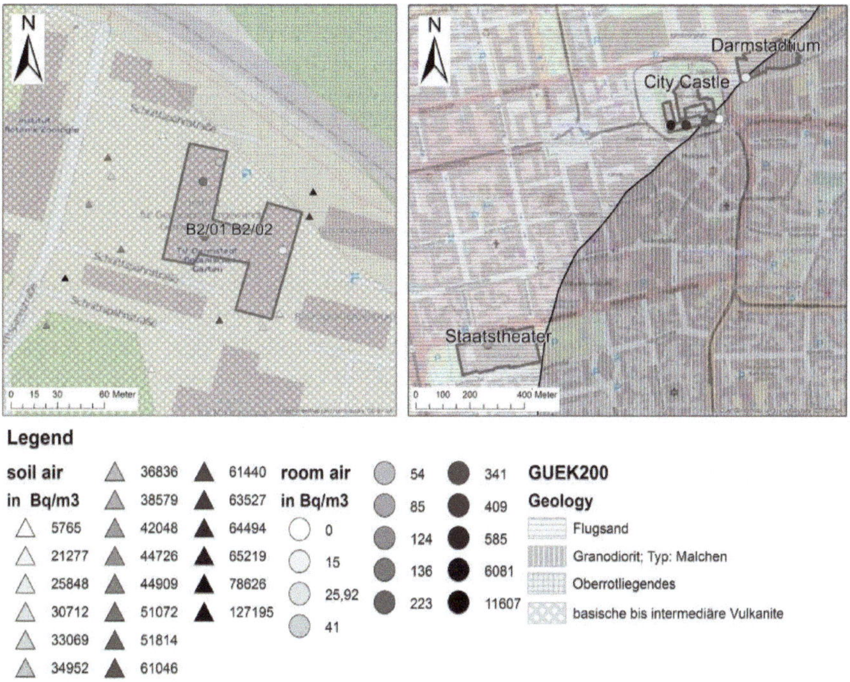

Fig.1. Concentrations of radon in both soil and ambient air in the area of the campus Botanischer Garten of Technische Universität Darmstadt (left) in the vicinity of the eastern master fault of the Upper Rhine Graben (right)

While part B2/01 of the building erected in the 1960s shows higher radon concentrations in the indoor air, the concentrations in the recently modernized part B2/02 are very low, indicating that either old migration paths have been closed or ventilation is effective in the reduction of radon concentrations. In building B2/01 both average radon concentrations of 136 Bq/m³ and maximum radon concentrations of more than 300 Bq/m³ significantly exceed the average concentration of radon in the air of closed rooms in Germany (49 Bq/m³, Menzler et al. 2006). Therefore it is recommended (i) to ensure regular ventilation and (ii) to repeat measurements because staff should not be exposed to concentrations of more than 150 Bq/m³ over a long period.

Darmstadtium

The darmstadtium is the city´s congress centre. It is located immediately on the eastern master fault of the Upper Rhine Graben which separates fluvial Tertiary in the west from gabbros in the east (Hoppe & Lang 2007). Here measurements of radon in soil air have not been conducted. Ambient air in the building though shows very high concentrations of radon in two of three rooms. This is due to favorable migration conditions along the master fault (Fig. 1).

Groß-Gerau

Planning for a geothermal power plant near Trebur raised the question of whether tectonic faults in the vicinity might have been active recently, thus enabling migration paths. In the frame of a running co-operation with the local energy supplier (Überlandwerke Groß-Gerau) radon in soil air is currently being measured. The co-operation partner provided the results of the seismic survey, which was conducted during preliminary reconnaissance.

As pictured in Fig. 3 unknown faults 380 m below ground level were located. Based on this information the faults outcrop on the surface is projected.
The dashed line in Fig. 2/3 shows the position of the fault 380 m below surface. In the study area altogether 12 faults have been mapped. All faults are normal faults and show offsets of 0 to 2400 meters and most of all dip towards west with 80° (Table 1). Since both continuation and angle of dip towards the surface is not clear, assumptions regarding a possible outcrop and thus recent activity of faults have to be made. Therefore the faults have been projected to the surface considering different angles of dip, ranging between 70° and 85°.

Table 1. Known and discovered faults in the area

Name	Dip direction	offset
Westliche Grabenrandstörung	ESE	2400
Nauheim-Wallerstätten-Störung	W	930
Niederwaldstörung a	ENE	Unproven
Niederwaldstörung b	ENE	Unproven
GRS-Antitheter 1 Nord	W	Unproven
GRS-Antitheter 2 Nord	W	Unproven
GRS-Antitheter 3	WSW	Unproven
GRS-Antitheter 4	W	Unproven
GRS-Antitheter	E	Unproven
Riedelbruch Neuwiese	W-WSW	Unproven
Riedelbruch Neuwiese Antithete	W-WSW	Unproven
Treburer Störung	W-NW	Unproven

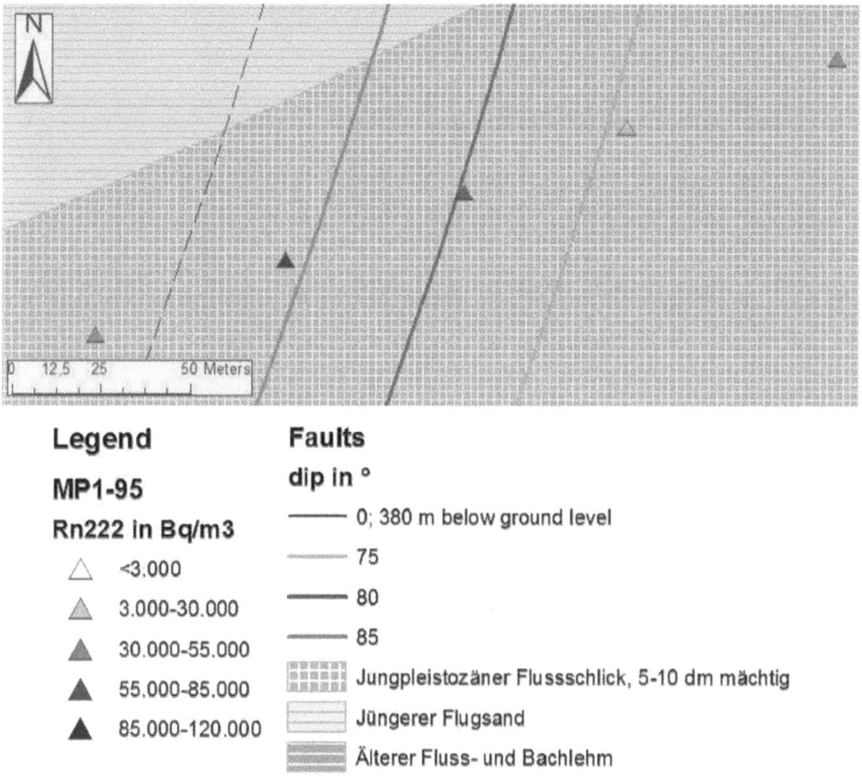

Fig.2. Zoom of the framed area in Fig.3.Measuring points and fault. 0° Line shows the fault trend 380 m below surface, the straight lines show the projected outcropping of the faults by adoption of different dips. Higher values of the radon concentration are located in areas of soil with high clay percentages or as shown here in the fault zones. Base map: HLUG, GK 25, Steuer et al., 1974

As pictured in Fig. 3, 96 soil gas measurements were taken. Soil samples for every single measuring point will be analyzed for the presence of radium, the parent nuclide of radon, in order to differentiate between radon that developed in situ and radon that migrated to the measuring environment. To increase the reliability of detected concentrations, CO_2 is also measured. CO_2 could be a carrier gas for radon, accelerating the rise of radon from deeper underground. Furthermore CO_2 concentrations above 0,6 Vol. % are a proof that soil gas is measured. A flow meter is therefore integrated in the measurement setting to enable it to log the permeability of the researched soil in order to compare measurements in different types of soil.

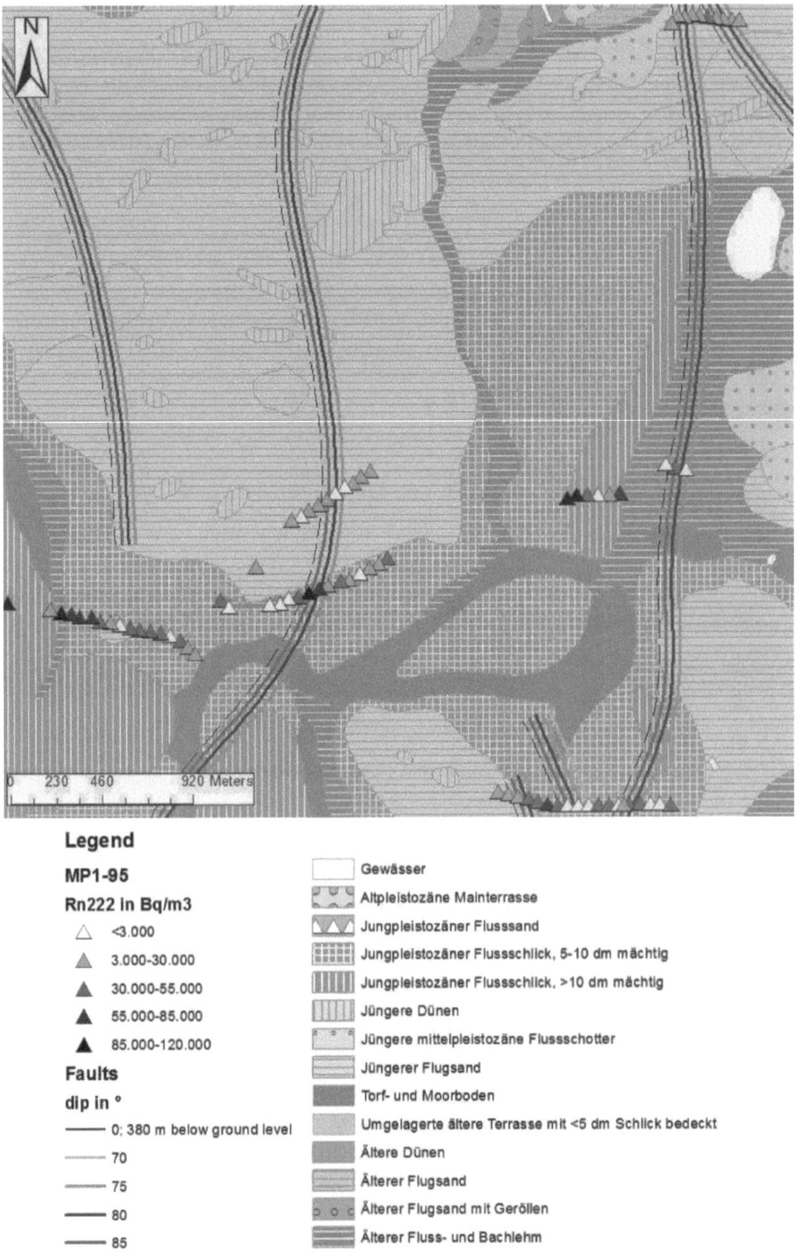

Fig.3. Measuring points and fault trends in the area of interest near Groß-Gerau. 0° Line shows the fault trend 380 m below surface, the other lines show the projected outcropping of the faults by adoption of different dips. The high values of the radon concentration are located in areas of soil with high clay percentages or the rift zones. Base map: HLUG, GK 25, Steuer et al., 1974

The project thus progressively leads to both a deeper understanding of the geological inventory and an improved methodology for analyzing and interpreting soil gases. The measurements show, that depending on the soil type in which they were executed, the radon concentration is directly linked to the soil. The CO_2 concentration does not show such a link.

Conclusion

The results of the radon measurements show a direct link between the type of soil and the concentration of radon. If the percentage of clay rises in the soil, the radon concentration rises and vice versa. On the one hand the percentage of minerals which can release radon is higher in clay soils and on the other hand stiff soils act like traps for the rising radon. The CO_2 concentration does not show such a link to the soil types.

In general it can be concluded that the concentration of radon und CO_2 show higher values in the fault zones than in areas apart. Based on the high cover of quaternary sediments in the area of interest the possible peak in the radon concentration is disposed over the area. A clear peak above faults as shown in the paper of Ioannides, 2003 could only be measured in areas with little cover of sediments over the crystalline basement.

References

Darby, S., Hill, D., Auvinen, A., Barros-Dios, J-M., Baysson, H., Bochicchio, F., Deo, H., Falk, R., Orastiere, F., Hakama, M., Heid, I., Kreienbrock, L., Kreuzer, M., Lagrange, F., Mäkeläinen, Muirhead, C., Oberaigner, W., Pershagen, G., Ruano-Ravina, A., Ruosteenoja, E., Schaffrath Rosario, A., Tirmarche, M., Tomascaron, L., Whitley, E., Wichmann, H-E., Doll, R. (2005): Radon in homes and risk of lung cancer: collaborative analysis of individual data from 13 European case-control studies, BMJ 2005;330:223

Hoppe, A. & Lang, S. (2007): The eastern master fault of the Upper Rhine Graben below the Science and Conference Centre in Darmstadt (Germany). Z. dt. Ges. Geowiss. 158/1: 113-117, Stuttgart.

Ioannides, K. et al. (2003): Soil gas radon: a tool for exploring active fault zones. - Applied Radiation and Isotopes, 59: 205-213, Amsterdam

Kemski, J., Siehl, A, Stegemann, R., Valdivia-Manchego, M. (1998): Geogene Faktoren der Strahlenexposition unter besonderer Berücksichtigung des Radonpotentials. Abschlussbericht zum Forschungsvorhaben ST. Sch. 4106. Geologisches Institut der Universität Bonn.

Kuhn, G., Lehnè, R., Hoppe, A. (2014): Radon measurements in soil and ambient air in and around Darmstadt – an investigation in a geological context. Annual Report 2013 Faculty of Materials and Geo Sciences, Technische Universität Darmstadt, 209-211, Darmstadt

Menzler, S., Schaffrath-Rosario, A., Wichman, H.E., Kreienbrock, L. (2006): Abschätzung des attributablen Lungenkrebsrisikos in Deutschland durch Radon in Wohnungen. Ecomed-Verlag, Landsberg.

Steuer, A., Schmitt, O., Sonne, V. (1974): Groß-Gerau 6016: Geologische Karte von Hessen 1:25000, Hessisches Landesamt für Bodenforschung

Mulde River - A Uranium Mining Archive

Kay Nestler[1], Broder Merkel[2]

[1] nestler-kai@web.de
[2]TU Bergakademie Freiberg, Hydrogeology Department, 09596 Freiberg

Abstract. River sediments are archives for past water contaminations. This is shown for the catchment of the Mulde River draining major parts of the Erzgebirge, Saxony, Germany. Mining activities have been impacting this part of Germany for over 850 years. It started with silver, then changing to lead, cadmium, zinc, tin, tungsten and copper. Finally, after 1945 a big uranium boom took place in this area which was led by the Soviet Union and the company SAG Wismut / SDAG Wismut, respectively. Until 1990 this area and parts of Thuringia produced 231,400 t of uranium and for this time period made the GDR the third biggest uranium producer of the world. Field work in 2011 revealed that both water and sediments still contain significant concentrations of uranium, cadmium, zinc, arsenic and other mining-related elements. Factor and cluster analysis has been used to visualize the data. Certain areas such as the former Freiberg mining area are high in lead, arsenic and cadmium but low in uranium. On Contrary, those areas where uranium ore was mined still show high uranium concentration in both water and sediments.

Introduction

The Erzgebirge is a tectonic structure stretching NE-SW exposing Variscian crystalline basement and is part of the Bohemian Massif. It contains various ore deposits in granitic rocks and phyllits. For understanding the formation and evolution of the manifold ore bodies Tischendorf (1986) gives a good overview. The catchment has a size of 7600 km^2 and the elevation ranges from 1244 to 56 (m.a.s.l.) at the discharge point to the Elbe River.

Mining for silver and later for zinc, lead, cadmium and other elements took place in the Erzgebirge since the 11th century. Besides mining activities, smelters and small- and medium-sized metal industry over centuries produced waste which was often discharged to the Mulde River and its tributaries. This had already been recognized by Agricola (1556). Uranium was mined around the 1900s in rather small

B.J. Merkel and A. Arab (eds.), *Uranium – Past and Future Challenges*,
DOI 10.1007/978-3-319-11059-2_96

quantities for the production of uranium glass in particular in the region of Jáchymov (Sankt Joachimsthal, Czech Republic). This rapidly changed in 1945 when after WW II the Soviet Union started mining uranium in the Erzgebirge. The SAG Wismut was founded in 1947 and renamed to SDAG Wismut in 1954. Until the early 1960s uranium mining, milling and processing was done in many places. Environmental issues had not been considered at all leading to severe environmental contamination by means of spills and dam failures. After 1960 the Wismut SDAG focused on a few big mines (deep and open pit mines) and did the treatment in two centralized milling and treatment plants (Seelingstädt and Crossen). Step by step at least certain environmental protection measures were implemented. With the unification of West and East Germany uranium mining was terminated officially at the end of 1990. SDAG Wismut was reconstructed as Wismut GmbH (owned by the Federal Republic of Germany) and is responsible for the restoration and environmental cleanup of the former mining and milling sites.

Beuge et al. (1999), Kluge 1997 and Hoppe (1995) investigated river sediments of the Mulde River in 1992 and found that the Mulde at that time was by far the biggest contributor to arsenic and metals loads to the Elbe River and the North Sea. Further Investigations with respect to transport and accumulation of metals were performed in the Mulde catchment by Geller et al. (2004).

Methods

Sediment and water samples were taken during November 2011 (Fig. 1). Sediments were taken by a liner or a grabber from the riverbed. In most cases one representative sediment sample was created by mixing over the entire depths; only in some cases sampling and analysis by RFA was performed depth depending. For RFA analysis only the fraction <63 µm was used. The determination was performed with a SPECTRO XEPOS (Bruker) using a standard multi-element calibration procedure. Water samples were taken from the surface of the rivers. Samples for ion chromatography (IC) and ICP-MS were filtered with poly-carbonate filters (200 nm, Sartorius). Samples for ICP-MS were acidified with 1 mL supra pure HNO_3 prior to determination with an X Series 2 (Thermo Scientific) using PlasmaLab version 2.5.9. Calibration was done by external multi-element standards and internal standards (germanium, rhodium, and rhenium). Some elements were read in the direct mode, others with the KED mode (Kinetic Energy Discrimination). An 881 Compact IC pro (Metrohm) was used for determining anions. Cations were determined with an 850 Professional IC (Metrohm).

Evaluation of the data was performed by means of cluster and factor analysis using the public domain Excel add-on XLSTAT and the open source program PSPP. In cases of more than 30% missing values, the parameter was excluded from cluster and factor analysis. In all other cases missing values were replaced by appropriate values (median, detection limit multiplied by 0.5 etc.).

For further statistical evaluation all parameters and element concentrations that were rated to be relevant were tested for normal distribution. The check was carried out both by subjective evaluation of the graphical representation of the histogram, as well as by checking with the Kolmogorov-Smirnov distribution test. Only pH, electrical conductivity and nitrate in water were normal distributed. The sediment data was not normal distributed. Therefore the logarithm of the readings was taken and re-checked with Kolmogorov- Smirnov test. Transformed or untransformed data was used depending on which data showed the greater significance. For the trace elements the logarithmically transformed data was always given preference.

To avoid an impact by different scales data was standardized by means of the following equation.

$$z_{ij} = \frac{x_{ij} - x_i}{s_i} \text{xij}$$

x_{ij} value i of sample j
x_i mean of variable i from all samples
si standard deviation of variable i from all samples
zij standardized value

After performing the main component methods VARIMAX Rotation was applied and factor scores were calculated.

With respect to cluster analysis Ward´s-method was found to be most suitable for the data given.

Fig.1. Sampling points in the Mulde River catchment. SW = Schwarzwasser, LB = Lungwitzbach, ZM = Zwickauer Mulde, FM = Freiberger Mulde, ZSCH = Zschopau, VM = Vereinigte Mulde (Nestler 2012)

Results

Fig. 2 displays the uranium concentrations in river sediments of the Mulde catchment area. Significantly elevated concentrations of up to 130 mg/kg were found in the sediments of the Zwickauer Mulde and Schwarzwasser which drain the area of the former uranium mining and milling area. On the contrary the Freiberger Mulde which drains the huge Freiberg mining area has sediments with very low uranium concentrations in the range of natural geogenic sediment concentrations (<1 to 10 mg/kg).

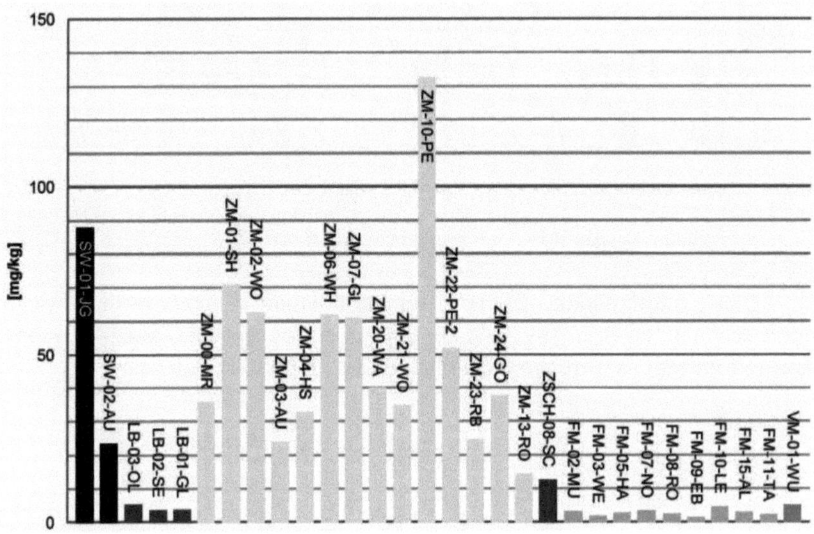

Fig.2. Uranium content in sediments of the Mulde River and tributaries. SW = Schwarzwasser, LB = Lungwitzbach, ZM = Zwickauer Mulde, FM = Freiberger Mulde, ZSCH = Zschopau, VM = Vereinigte Mulde. (Nestler 2012)

However, if one looks at cadmium (Fig. 3), lead and arsenic (the latter two elements not shown here) these elements are significantly elevated in the Freiberger Mulde especially in the vicinity of the city of Freiberg. Here all mines have been closed in 1969 after 850 years of mining activities.

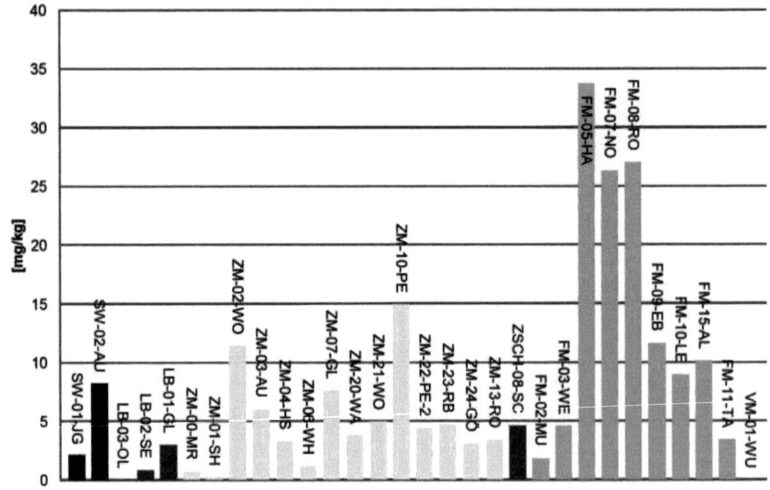

Fig.3. Uranium content in sediments of the Mulde River and tributaries. SW = Schwarzwasser, LB = Lungwitzbach, ZM = Zwickauer Mulde, FM = Freiberger Mulde, ZSCH = Zschopau, VM = Vereinigte Mulde (Nestler 2012)

Fig 4 shows a significant difference for the uranium contents in the Zwickauer Mulde sediment taken in 1992 and 2012. With one exception (ZM-02-WO) the uranium concentrations were more than halved over a period of 20 years.

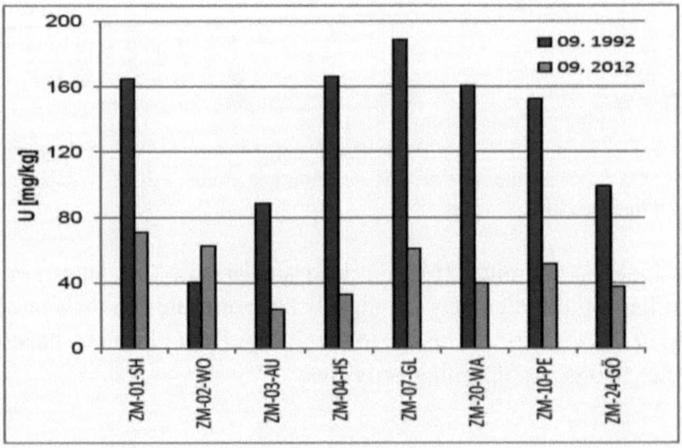

Fig.4. Uranium content in Zwickauer Mulde sediments in 1992 (data from Beuge et al. (1999)) and 2012 (Nestler 2012)

Fig. 5 shows that if one looks at greater depth still significantly elevated uranium concentrations can be found. This is an evidence for the high contamination load during the uranium mining activities in the 1950s, 1960s and 1970s. It is rather likely that the high concentrations in the sediments will be eluted over time and therefore the water in the Zwickauer Mulde will have significant uranium concentrations for decades.

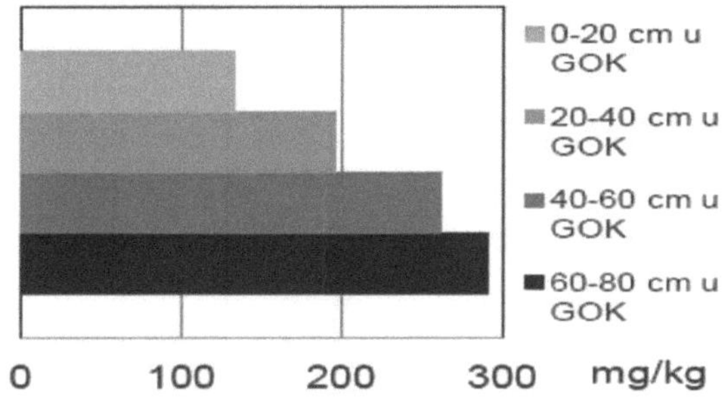

Fig.5. Depth distribution of uranium concentration below surface at the sampling site in Penig (ZM-10-PE) showing an increase with depth. (Nestler 2012)

Table 1 and Table 2 show the factor loadings 1 to 4 and 5 to 7 respectively of water chemistry parameters after VARIMAX rotation.

Table 1: Factor loading of water chemistry data after Varimax rotation; V= variance

Variable	Factor 1 V=46.86%	Factor 2 V=18.65%	Factor 3 V=10.29%	Factor 4 V=6,76% %	Communa lity
ln (Cr)	0.236	0.324	0.594	-0.002	0.513
ln (Mn)	0.087	0.671	0.446	-0.428	0.840
ln (Fe)	-0.795	0.136	0.419	-0.124	0.842
ln (Co)	0.212	0.709	0.469	-0.041	0.769
ln (Ni)	0.175	0.123	0.780	0.309	0.749
ln (Cu)	-0.011	0.591	0.222	0.600	0.759
ln (Zn)	0.169	0.919	0.019	0.053	0.877
ln (As)	0.192	0.093	0.254	0.813	0.771
ln (Cd)	0.103	0.921	-0.115	0.070	0.876
ln (Pb)	0.054	0.840	-0.050	0.230	0.765
ln (U)	0.155	-0.392	**0.782**	0.174	**0.820**
ln(Na)	0.775	0.305	0.287	0.154	0.800
ln(K)	0.863	0.209	0.311	0.016	0.886
ln(Ca)	0.958	0.120	0.080	0.054	0.942
ln(Mg)	0.890	-0.027	0.199	0.017	0.833
ln(Cl)	0.884	0.269	0.195	0.115	0.905
NO_3	0.932	0.119	-0.166	-0.066	0.915
ln (SO_4)	0.887	0.206	0.269	0.031	0.903
ln (HCO_3)	0.946	0.028	0.179	0.022	0.929
ln(PO_4)	0.666	-0.307	0.284	-0.233	0.673
pH	0.822	0.019	-0.062	0.409	0.847

LF	0.920	0.190	0.244	0.083	0.949

Table 2: Factor loading of sediment data after Varimax rotation; V= variance

Variable	Factor 5 V=33.64%	Factor 6 V=28.98%	Factor 7 V=15.55%	communality
ln (Cr_s)	0.083	-0.137	-0.950	0.928
ln (Mn_s)	-0.027	0.779	0.244	0.667
ln (Fe_s)	0.217	0.866	-0.152	0.821
ln (Co_s)	-0.211	0.727	0.024	0.574
ln (Ni_s)	0.140	0.507	-0.795	0.909
ln (Cu_s)	0.853	0.162	-0.298	0.844
ln (Zn_s)	0.866	-0.074	-0.167	0.783
ln (As_s)	0.857	0.167	0.280	0.841
ln (Cd_s)	0.733	-0.088	-0.220	0.594
ln (Pb_s)	0.858	-0.417	0.197	0.949
ln (U_s)	-0.072	**0.774**	-0.293	**0.690**

Fig.6 (left part) displays a result of the factor analysis after VARIMAX rotation. The following parameters were used as variables: pH, electrical conductivity, concentrations of HCO_3^-, Na^+, Ca^{2+}, K^+, Mg^{2+}, Cl^-, NO_3^- SO_4^{2-}, PO_4^{3-}, Fe, Mn, Co, Cu, Zn, As, Cd, Pb, Cr, Ni and U in water. Cr, Mn, Fe, Co, Ni, Cu, Zn, As, Cd, Pb and U in the sediment was used as well but is not shown here.

The yellow marked parts of the Zwickauer Mulde are impacted by uranium, nickel, and chromium in water. The occurrence of these elements is due to different sources source, although, this sources are spatially close toeach other. Occurrence of uranium is mainly the result of uranium mining of the SDAG Wismut. Significant positive factor loadings are recorded for the Zwickauer Mulde from Aue, the Schwarzwasser and the Schlemabach (Fig.7). Uranium mining in the 50´s and 60´s under city of Johanngeorgenstadt felt is the reason for elevated uranium content in the Schwarzwasser creek. The former uranium mining provides as well elevated arsenic concentrations of the river. However, a correlation of arsenic and uranium could not be identified. It is assumed that the elevated uranium and arsenic concentrations of the Zwickauer Mulde are not necessarily due to a common source. Factor 2 represents large parts of the Freiberger Mulde where other ore type prevails. Factor 1 interacts with a variance of 46.9% with the exception of the river headwaters in almost all samples (positive factor values). Factor 1 represents those parts of the rivers in the catchment where the major cations and anions in the river water are dominant. Fig 6 (right) shows the result of cluster analysis with the same set of data (water chemistry). The pattern of clusters 1 to 4 is rather similar to that of factor analysis. As well the variables that dominate the 4 factors and the 4 clusters respectively are very similar. This proves that both statistical techniques may be used to evaluate the water chemistry in river systems. Factor and cluster analysis performed for sediment samples showed very similar results but are not displayed here.

Investigating water and sediments in the catchment of the Mulde River showed that river sediments are meaningful archives for past industrial use and in particular mining activities. The water quality is even after decades of mine closure still significant affected. This is on the one side caused by high concentrations of certain elements in the river sediments and drainage of contaminated mine water from abandoned mines.

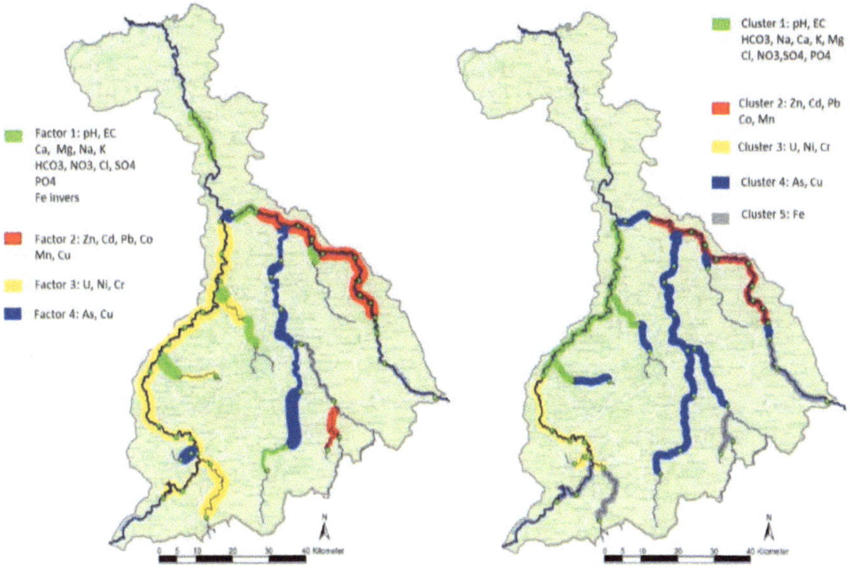

Fig.6. Comparison of factor (left) and cluster (right) analysis for water samples. Both multiple parameter techniques give rather similar results. Factor 2 and cluster 2(red) is dominated by Zn, Cd, Pb, and Mn and show the impact on water quality in the Freiberger Mulde while factor 3 and cluster 3 (yellow) dominate the Zwicker Mulde with the elements U, Ni and Cr. (Nestler 2012).

The highest contaminations of sediments and thus as well factor scores were found in the mining and processing region of Freiberg where the elements As, Cd, Pb, Zn, and Cu prevail. On contrary miainly U and As werc found to be elevated in sediments in the area of former uranium mining and processing (Aue, Niederschlema, Hartenstein, Crossen) within the Zwickauer Mulde catchment.

Investigating water and sediments in the catchment of the Mulde River showed that river sediments are meaningful archives for past industrial use and in particular mining activities. The water quality is significant affected even after decades of mine closure. This is on the one side caused by high concentrations of certain elements in the river sediments and drainage of contaminated mine water from abandoned mines.

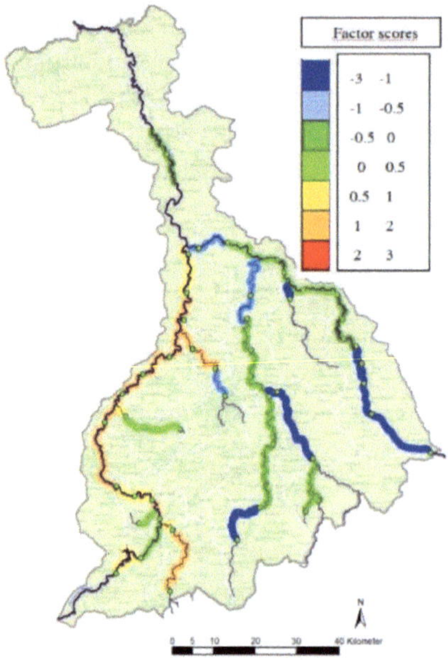

Fig.7. Factor scores for factor 3 (U, Ni, Cr) showing high concentration for water of the Zwickauer Mulde (Nestler 2012)

References

Agricola G (1556): De re metallica libri XII. Translated by H.C. Hoover and L.H. Hoover. Dover Publications, New York, 1950

Beuge P, Greif A, Hoppe T, Kluge A, Klemm W, Martin M, Mosler U, Starke R, Alfaro J, Anders B, Behrens K, Grunwald N, Haurand M, Knöchel A, Meyer A, Potgeter H, Staub S, Stocker M (1999): Die Schwermetallsituation im Muldesystem, Bd. 1 - 3 Freiberg, Hamburg, ISBN 3-924330-21-2

Geller W, Ockenfeld K, Böhme M Knöchel A (2004): Schadstoffbelastung nach dem Elbe-Hochwasser 2002. http://www2.ufz.de/data/HWEnd1333.pdf

Hoppe, T. (1995): Geochemische Untersuchungen an Gewässern und ihren Sedimenten im Einzugsgebiet der Mulde, Dissertation, Freiberg; p. 97

Kluge, A. (1997): Anwendung multivariater statistischer Verfahren auf geochemische Fließgewässerdaten des Muldensystems. Wissenschaftliche Mitteilungen 3; Freiberg; p. 140

Nestler K (2012): Erstellung eines geochemischen Profils der Zwickauer- und Freiberger Mulde. Unpublished Master thesis, TU Bergakademie Freiberg

Tischendorf G (1986): Variscan ensialic magmatism and metallogenesis in the Ore Mountains - Modelling of the process. Chemie der Erde 45, 75–104

Mitigation of radon exposures caused by uranium mining legacies at WISMUT sites

J. Regner[1], **P. Schmidt**[1]

[1]Wismut GmbH, Chemnitz, Germany

Abstract. Uranium mining activities of the former Soviet-German stock company SDAG Wismut in East Germany caused significant radiological environmental impact. At the Wismut sites in the Ore Mountains the radioactive noble gas radon is of significant relevance for the environmental impact. The paper quantifies the different contributions of the uranium mining legacies to elevated outdoor and indoor radon concentrations. This is exemplified by the local radon situation at the Wismut site of Schlema-Alberoda. The mitigation measures taken by Wismut GmbH to reduce the radon exposure of the general public are presented.

Introduction

A considerable number of uranium mining legacy sites are located in the German Federal States of Saxony and Thuringia, resulting from the mining activities of the former Soviet-German stock company SDAG Wismut in the period from 1946 to 1990. Since 1990 the remediation of these legacies is conducted by the state-owned company Wismut GmbH. In this effort, an essential objective of the remediation is to reduce the radiological environmental impact in line with the ALARA principle (as low as reasonably achievable). A significant part of the environmental impact by uranium mining legacies is caused by the radioactive noble gas radon as a part of the uranium series. In the case of the uranium mining legacies in Saxony, relevant radon exposures may generally result from mine ventilation systems, dumped waste rock material and diffuse radon emission from the natural ground above near-surface mine workings.

Situation at the Wismut site of Schlema-Alberoda

By the end of operations in 1990, a total of some 80,000 t of uranium had been mined at the Wismut site of Schlema-Alberoda. Mining operations created a vast system of workings reaching down to a depth of 1,800 m. In their present state,

B.J. Merkel and A. Arab (eds.), *Uranium – Past and Future Challenges*,
DOI 10.1007/978-3-319-11059-2_97

some 97 % of the workings are flooded. Air-filled mine voids are for the most part encountered near the ground surface. In the context of ongoing underground remedial works, mine ventilation is maintained.

Waste rock generated by uranium mining in the order of some 45 million m³ was dumped on a total footprint of approximately 313 ha. Waste rock dumps were piled up within the town of Bad Schlema in such a way that dwellings were only a couple of meters away from the rim of the mine dump. With a few exceptions, mine dumps were included in the remedial program starting in 1990. While some mine dumps were completely removed and relocated, most dumps were regraded and subsequently capped.

An assessment of the exposure situation of the general public at the Schlema-Alberoda site revealed that the uranium mining induced effective dose is dominated by the exposure pathway „inhalation of radon/radon decay products". Analysis of this exposure pathway has to focus prominently on occupancy hours (88 % of exposure time according to BMU (2010)). Relevant processes which have an effect on indoor radon levels are the change of air with radon-laden outdoor air and the entry of radon from the underlying soil.

The legacies of uranium mining operation at the Schlema-Alberoda site cause the following relevant radon emissions:

1. Radon exhalation from uranium mine waste rock dumps;
2. Radon emission to the atmosphere from die mine ventilation shafts;
3. Radon releases from near-surface mine voids.

Figure 1 schematically illustrates radon emissions at the Schlema-Alberoda site.

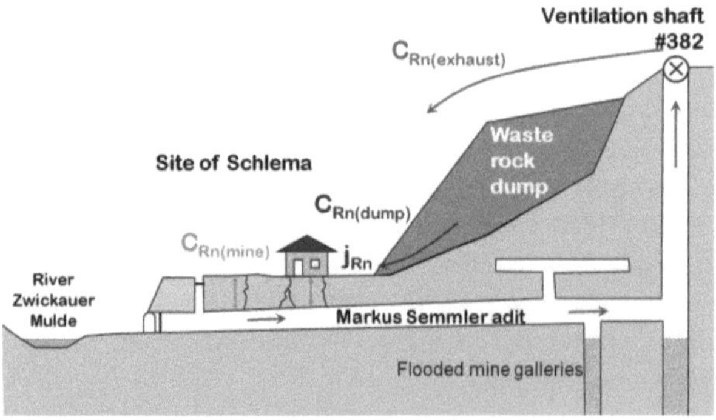

Fig.1. Radon emission and imission situation at the Schlema-Alberoda site

Radon emission from waste rock dumps

Thermally driven convective soil air flows similar to natural mine ventilation occur within the waste rock dumps at the Schlema-Alberoda site. In summer conditions, when atmospheric temperatures are above the relatively stable air temperature within the body of the mine dump, the air in the dump flows from the upper parts of the dump down to the toe of the dump where it exits. When the air passes through the waste rock dump it gets enriched in radon because it moves across a significantly greater volume of waste rock as compared with purely diffuse radon migration. During winter months when atmospheric temperature falls below the dump's internal temperature the summer flow direction is reversed (see Fig. 2).

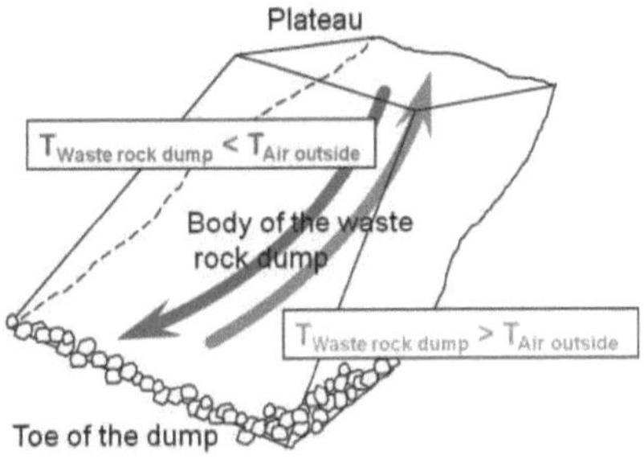

Fig.2. Schematic of thermally driven convective air flows in waste rock dumps located in the Ore Mountains

The specific relevance of thermally driven convective air flow at the Schlema-Alberoda site is due to the fact that in summer conditions radon release from a waste rock dump is funneled to a small surface area which in the majority of cases is in close vicinity to residential areas. This gives rise to significantly elevated levels of atmospheric radon as long as the waste rock dumps were not remediated.

Elevated rates of radon exhalation due to convective processes, for example, have caused annual average radon concentration in near surface air of up to 1,000 Bq/m³ for a home located near waste rock dump #66/207. Basically, similar conditions were encountered at other waste rock sites. Average radon concentration levels in the vicinity of other waste rock dumps were in the order of 200 Bq/m³ to 400 Bq/m³. Recorded radon concentration levels in the vicinity of waste rock dumps show a distinct seasonal cycle which is related to typical patterns of radon release from waste rock dumps.

Radiation exposure to residents in close vicinity to waste rock dumps due to the inhalation of radon and its short-lived decay products was significantly in excess of 1 mSv/a and justified appropriate remedial measures being taken to reduce radon release from waste rock dumps. Such reduction was pursued by completely covering the regraded dump surface with cohesive mineral soil. Standard covers at the Schlema-Alberoda site have a thickness of 1 m, with the lower lift of 0.8 m minimum thickness acting as barrier element to diminish radon release. Cover placement and vegetating of waste rock dumps in the Schlema-Alberoda township has to a large extent been completed and resulted in a fundamental improvement of the radon situation in that locality.

Figure 3 shows the evolution of radon concentration levels over time exemplified by waste rock dump #66/207 both for the unremediated initial condition and the state during the remediation.

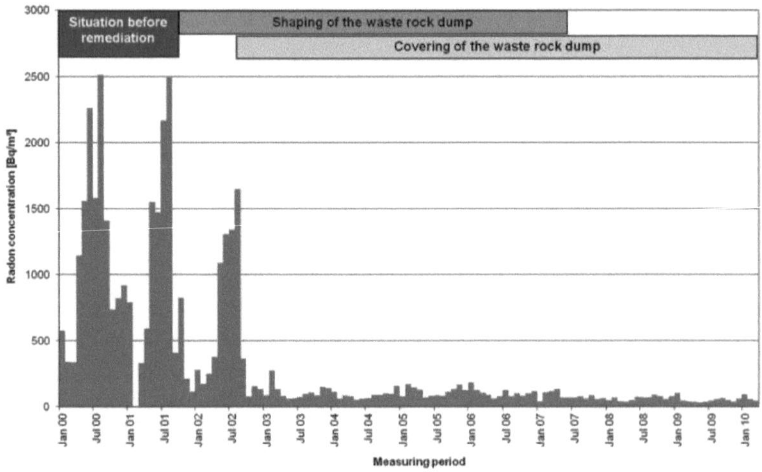

Fig.3. Monthly average values of radon concentration at a home close to waste rock dump #66/207

Radon emission by mine ventilation

During active uranium mining, the required mine ventilation emitted substantial radon activities to the atmosphere. For the last full year of mining operations (1989), radon emissions from the Schlema-Alberoda mine totaled approximately 44 MBq/s.

After the termination of uranium ore mining at the Schlema-Alberoda mine site, ventilation of the open, non-flooded mine workings had to be continued. On the one hand, ventilation had to ensure radiation protection for the continuation of underground remedial work and to control the local radon situation, on the other. With that said, radon emitted to the atmosphere from ventilation shaft exhausts will continue to be unavoidable in the foreseeable future.

Following several mine ventilation rearrangements and adjustments, mine exhausts are emitted to the atmosphere since 1997 from ventilation shaft #382. That shaft is located in a hilltop location providing favorable dispersal conditions outside town limits.

Development of ventilation shafts' radon emission over time was to a great extent influenced by the progressive flooding of the mine. As a consequence of the decrease of air-filled mine voids, the radon emission rate was on the decline from 1995 onwards, diminishing from approximately 32 MBq/s to below 3 MBq/s in 2009. During the same period, radon concentration levels in exhaust air decreased from approximately 180 kBq/m³ to circa 70 kBq/m³.

Figure 4 depicts the chronological evolution of emissions. From this it follows that while flooding of the greater part of mine voids was successful in significantly diminishing radon emissions. Otherwise radon emission will have to continue in the long term due to the ventilation of near-surface mine voids that cannot be flooded. For this reason, selection of an optimum site for radon emission from mine ventilation exhausts was of crucial importance.

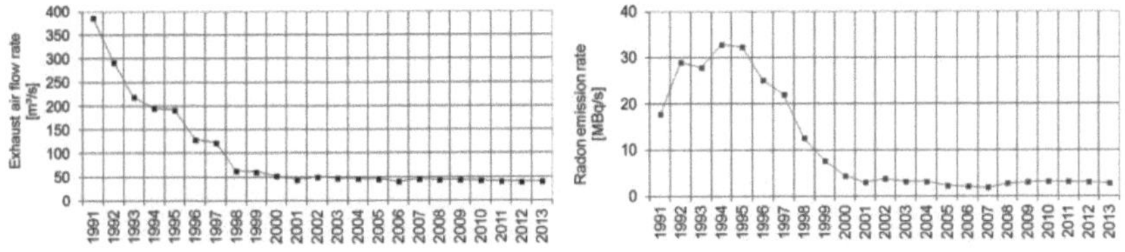

Fig.4. Chronological evolution of mine ventilation exhausts and radon emission rate of upcast ventilation shafts at the Schlema-Alberoda site

While previous configurations of mine ventilation patterns had still produced significantly elevated atmospheric radon concentrations in the order of circa 400 Bq/m³ across large parts of the township of Niederschlema, immission loads to the public could be diminished to values of < 3 Bq/m³ (as of 2013). Neither present nor future radon emissions via ventilation shaft #382 will yield relevant radiation exposure to the residents of homes in the surrounding localities.

Radon emission by diffuse radon migration from near-surface mine galleries

At the end of the flooding process of the Schlema-Alberoda mine to be accomplished in a few years, the water level will have reached the elevation of the Markus Semmler adit. As a consequence, mine voids located at higher elevations will be permanently non-flooded and constitute a potential radon source in the long term, in particular for homes sitting above. Figure 5 shows the locations of near-surface mine workings in the Bad Schlema area. The historical Markus Semmler adit (see also Fig. 1) extends along the valley location and is running immediately below the homes sitting above. The adit and its pathway are a distinctive feature of the Schlema site.

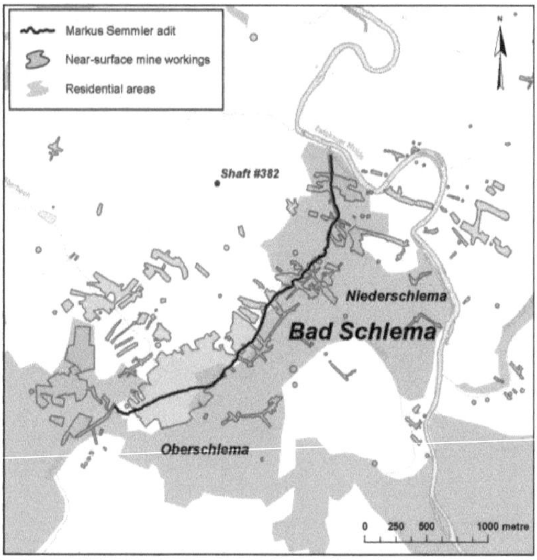

Fig.5. Locations of near-surface mine workings within the town of Bad Schlema

Many years of complex investigations into the impact of near-surface mine workings on the radon situation in Schlema have provided evidence that in the presence of natural buoyancy conditions, convective gas transport processes may carry radon from the mine into houses through bedrock disturbed by mining even over extended distances (representing a multiple of the radon diffusion length). Under natural ventilation conditions, radon concentrations within the mine workings may rise to very high levels of several hundred kBq/m³, in extreme cases of up to 1 MBq/m³. In this way, convective gas transport processes would in particular cases provoke indoor radon concentration levels significantly exceeding 10,000 Bq/m³ which would result in an inacceptable radiation exposure to the residents. In order to prevent such a situation, the Schlema-Alberoda site is equipped with a sucking mine ventilation system which uses a fan installation to produce a negative-pressure condition whereby radon transport is being directed toward the mine. It appears from investigations that the indoor radon situation in Schlema can in principle be efficiently controlled by maintaining the negative-pressure condition. Modell calculations have shown that the generation of a negative-pressure condition in the order of circa 100 Pa in the mine will require a relatively low fan power of <10 kW in the long run. Centralized generation of negative-pressure condition has the key advantage of large-scale action which, as a rule, dispenses with the need for specific mitigation to be performed at individual homes. The concrete technological design holds considerable potential for optimization. With this, establishment of sustainable safeguards for an acceptable radon situation is among the priority long-term tasks.

In preparation for designing an optimum ventilation approach to mitigate the indoor radon situation, a number of time-limited experiments were conducted to investigate the mechanisms and impact of convective gas transport processes through bedrock disturbed by mining. Use of tracer gas (SF6) in combination with real-time radon concentration measurements was successful in providing experimental evidence of radon transport from the mine into homes and quantifying that phenomenon in individual cases. The investigations also revealed that the impact of the mine is focused mainly on a corridor running along the Markus Semmler adit and on adjacent areas as well as on the Oberschlema subsidence area and more or less distinct in each of the buildings (Löbner et al. 2012).

Fig. 6 illustrates by way of example the impact of mine ventilation on radon concentration levels in a basement.

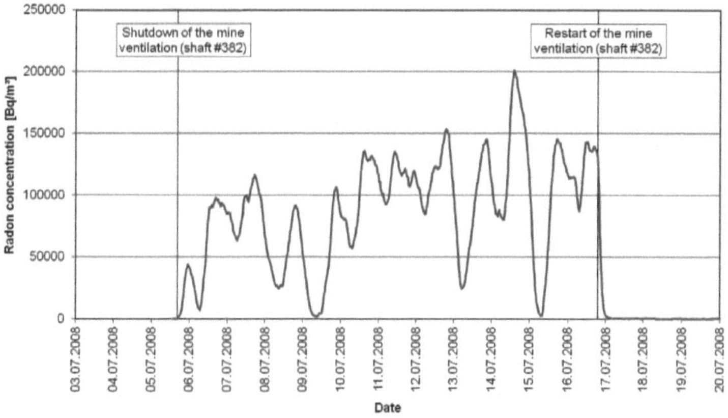

Fig.6. Impact of mine ventilation on radon concentration levels in a basement of a dwelling house in Niederschlema

Summary

The specific radon situation in the Schlema-Alberoda region is predominantly characterized by mining legacies. Remediation of uranium mine dumps at the site has made a substantial contribution to improving the situation with regard to the radon-induced radiation exposure of the public. Future action here will focus on post-remedial care and long-term cover performance with respect to the reuse of rehabilitated objects. Some waste rock dumps will require additional measures to be taken locally in order to curtail convective flows within the dump and thereby diminishing radon emission rates.

As flooding of the Schlema-Alberoda mine advances, radon emission from mine ventilation will decrease and in this context the improvement of the radon situation in the township of Bad Schlema is to be expected to continue till the end of mine flooding.

However, control of the indoor radon situation due to convective radon transport from mine voids through the disturbed bedrock will require the preparation of a technical solution, both optimized and sustainable, involving lowest possible spending, energy consumption, and monitoring and care efforts. In addition, it became apparent that the indoor radon situation in Schlema cannot in each and every case be mitigated by mine ventilation. There are a few scattered places which would require disproportionate efforts in order to be mitigated by the centralized mine ventilation. Therefore, individual approaches might be required to mitigate the mining-induced radon situation. There will be need for further studies on that issue.

References

BfS (2011) Calculation Guide Mining: Calculation Guide for the Determination of Radiation Exposure due to Environmental Radioactivity Resulting from Mining Department. Radiation Protection and Environment. BfS-SW-09/11, Radiation Protection and Environment. Salzgitter. urn:nbn:de:0221-201109056212

Löbner W et al. (2012). Application of the tracer gas technology to investigate convective radon transport processes between mine workings and houses. Kerntechnik: Vol. 77, No. 3, pp. 187-190.

Determination of uranium in mineral phosphate fertilizers using a low cost gamma spectroscope

Frank Jacobs[1,2], Sascha Riedl[1], Steven Sesselmann[3], Ewald Schnug[1]

[1]Institute for Crop and Soil Science, Federal Research Institute for Cultivated Plants, Bundesallee 50, D-38116 Braunschweig, Germany; frank.jacobs@jki.bund.de
[2]Geoscience Advice & Support in Geoscience, Glockengiesserstrasse 1, D-38640 Goslar, Germany; f-jacobs@t-online.de
[3]Bee Research, Suite 314, 247 Coward Street, Mascot NSW 2020, Australia; steven@beejewel.com.au

Abstract. Uranium (U) is a natural but undesired heavy metal in rock phosphates and processed phosphorus fertilizers. Analytical methods employing the direct determination of U through α-spectroscopy or ICP mass spectrometry are tedious and expensive, require sophisticated laboratory equipment and skilled staff. This contribution describes an easy to perform test based on a low-cost gamma-ray spectroscopy system. The determination of U was carried out through its only gaseous decay product Radon (^{222}Rn). Seven straight and compound fertilizer samples with different U and K content were measured and the spectra compared to that of the U-containing mineral schrockingerite.

Introduction

U in rock phosphates and processed mineral phosphate fertilizer products can be found in a wide range of concentrations. U is an undesirable impurity in fertilizers (Schnug and Haneklaus 2014). 87% of phosphate-containing fertilizers are made of sedimentary rock phosphates. With sedimentation and diagenesis of phosphate rock, massive enrichment of U may have occurred by geochemical processes (Kratz and Schnug 2006). The U content of phosphate rocks varies greatly between individual deposits and vacillates in sedimentary deposits between 2.8 to 660 mg/kg U (Al-Shawi and Dahl 1995). Phosphates, which are the result of chemical terrestrial weathering, can reach up to 4000 mg/kg U (Ohly 2005).

Phosphate fertilizers are produced commonly by wet digestion of rock phosphates. Different types of fertilizer are produced depending on the concentrated acid used. A highly concentrated straight fertilizer (up to 48 % P_2O_5) is TSP (triple superphosphate). By adding various additives to TSP compound phosphate fertilizers such as NPK and PK fertilizer are manufactured. The U content in the ferti-

B.J. Merkel and A. Arab (eds.), *Uranium – Past and Future Challenges*,
DOI 10.1007/978-3-319-11059-2_98

lizer product is closely related to the U concentration in the used phosphate rock. Usually, a higher U content can be found with increasing phosphate concentration in the fertilizer (Kratz and Schnug 2006). Mineral phosphate fertilizer applications cause U contamination of agricultural soils where U accumulates. This U is prone to leaching into groundwater (Schnug and Haneklaus 2014).

Detection methods for uranium in fertilizers and theoretical background

Usually, α-spectroscopy and ICP-mass spectrometry are used to determine the total content of U in mineral fertilizers. However, such analyses require suitable laboratory equipment and highly qualified personnel. Besides this the analyses are time-consuming and expensive.

An alternative method for the detection of U in mineral fertilizers is gamma-ray spectroscopy. This is a rapid, easy and non-destructive analytical technique that can be used to identify various radioactive isotopes in different sample matrices. In gamma-ray spectroscopy, the energy of incident gamma-rays is measured by a scintillation detector and interpreted. By comparing the measured energy to the known energy of gamma-rays produced by individual radioisotopes, the identity of the element can be determined. Gamma-ray spectroscopy is commonly carried out in a laboratory with liquid nitrogen cooled Germanium detectors, which are expensive and not easily portable. In the present study a portable GSB-2812 Gamma Spectroscopy kit was employed to detect indirectly trace amounts of U in mineral fertilizers. The advantage of the kit is that it is available at low costs and the mobile unit can operate location-independently (Gammaspectacular 2014).

The system used consists of a photomultiplier tube (PMT) with a 2.8 "x 1.2" NaI (Tl) gamma scintillation detector and a GS-1102-PRO sound card spectrometry driver. The signal output is connected via an audio cable to the line input of a regular PC or laptop. There are at least four recently developed, free open source software programs, which run on a standard PC platform, and emulate a multichannel analyzer (MCA). These programs make use of the digitally sampled audio signal from the sound card to generate a high quality gamma spectrum. A typical 2" NaI(Ti) scintillation detector detects gamma radiation in the range of 25 keV to 3000 keV. Detector and sound card driver were developed and built in Australia by Bee Research Pty Ltd.

With this gamma spectrometry system, a standard Thallium doped Sodium Iodide scintillation crystal, NaI(Tl), is used to detect gamma rays. This detector consists of a scintillation crystal mounted to a photomultiplier tube (PMT), which amplifies the energy of the tiny flashes of light, by a factor of at least 10^6, and outputs an electronic pulse in the milli-ampere range. The pulse is proportional to the original energy of the gamma ray. The electronic pulses are then amplified and impedance matched in the sound card driver, and output as a line level audio signal in real time. The audio signal is then digitized by the PC sound card and analyzed by the MCA software (Gammaspectacular 2014), which calculates the pulse height and sorts each pulse into a pulse height histogram. Within minutes, the histogram will reveal the energy peaks of any persistent gamma rays coming from the sample. Once the apparatus has been correctly calibrated, the element can easily be identified.

^{238}U is an alpha particle emitter, and cannot easily be detected by gamma spectroscopy, but ^{238}U decays in a specific decay chain called the U-238 series, exhibiting a whole range of gamma emissions, of which the most notable ones are from ^{226}Ra, ^{214}Pb and ^{214}Bi which are some of the shorter lived isotopes. These easily detectable gamma rays will show up at 186, 242, 295, 351 and 609 keV, and may be immediately recognized by the more experienced operator.

^{238}U decays into ^{226}Ra, which decays with a relatively long half-life to ^{222}Rn, a short-lived noble gas (half-life 3.8 days), which further breaks down to Po and so on. Rn may escape as a gaseous radionuclide from contaminated fertilizers. Thus, the indirect detection of U, by the gamma emissions of its gaseous and non gaseous decay products, is easily achievable, and by applying a strict method, it is also quantifiable.

Because of the typical arrangement of the three small photo peaks from ^{214}Pb and the significant ^{214}Bi photo peak, this sequence of the spectrum is referred to as "Goldilocks and the three bears" from the English fairy tale of the same name. To detect ^{238}U in a fertilizer sample, one must therefore identify the daughter nuclides.

Description of the experimental setup and selection of fertilizer samples

The measurement of the fertilizer samples was performed in a lead castle to shield and thus minimize the natural gamma background radiation, because the natural background radiation would otherwise mask the weak gamma radiation coming from a very low U content of the fertilizer sample. The sample material was placed in a *Marinelli* beaker, which prevents the emanation of gaseous elements such as Rn. The *Marinelli* beaker provides a gas-tight sample container with a special annular configuration, which fits the crystal part of the detector to provide optimum detection efficiency. The *Marinelli* beaker combined with the lead castle enables the detection of trace amounts of U contamination and provides repeatable results. Figure 1 shows the test setup with laptop, sound card spectrometry driver, probe and *Marinelli* beaker in the open forward lead castle.

To ensure comparable measurement conditions, all investigated fertilizer samples were measured sequentially in a continuous measurement campaign over a period of 30 minutes.

Fig.1. The test setup, probe and *Marinelli* beaker in the open forward lead castle

Seven fertilizer samples each consisting of 300 g were examined. The test samples of different fertilizer types were selected from the worldwide largest collection of phosphate fertilizer samples at the Institute for Crop and Soil Science. The total U and potassium (K) content has been determined previously by standard laboratory methods (Table 1). For the validation of data measured by the GSB-2812 Gamma Spectroscopy kit fertilizer samples with U concentrations between 0.03 and 206 mg/kg U and K contents of up to 19.89% K were analyzed.

Table1. U and K content in selected straight and compound fertilizer samples

Sample ID	Fertilizer type	U (mg/kg)	K (%)
BMELV 42	PK-Fertilizer with S	206.0	9.132
BMELV 66	Triple-Superphosphate	153.0	0.226
BMELV 62	PK-Fertilizer with partly digested rock phosphate with Mg, S	102.0	13.93
MD 1	Triple-Superphosphate	52.25	not analyzed
MD 38	PK 18/18 (Hyperkali)	52.23	18.79
MD 39	NPK-Fertilizer 15/5/5	13.11	6.93
MD 24	NPK-Fertilizer 12/12/17	0.560	19.99

For the low-cost gamma spectroscopy are currently four free software programs ("free sound card spectrometry software" (Gammaspectacular 2014)) to choose from:

PRA: an Australian program with multi peak calibration enabling interpolation, long preset measurement times, and the graphical user interphase is not customizable, but data can be exported for processing in other programs.

BecqMoni: a Japanese program which allows very precise measurements, automatic peak detection and simultaneous presentation of comparative measurements, easy polynomial calibration (for non- linear detectors), however without an English manual.

Theremino: an Italian Program which provides very quick results enables peak recognition and simultaneous display of comparative measurements, but with less functions than PRA and Becqmoni.

Geigerbot: an American designed app for iPhone which requires a special iPhone adaptor cable produces a good gamma spectrum and can output the dose rate with GPS coordinates.

The measurements for the present study were performed with the program Theremino because it proved to provide the best results even after a short measurement period and is easy to use. Before starting the measurement, the system was calibrated with a test source emitting a known gamma spectrum. As a test source, the radionuclide ^{137}Cs was chosen with the typical 32 keV X-ray peak, the 662 keV gamma peak and ^{133}Ba with typical peaks at 81 keV and 356 keV. These isotopes were chosen because the calibration peaks are spread across our emission region of interest. Subsequently, a blank measurement was performed in order to examine the shielding for ambient radiation by the lead castle.

Measurement results

The selected seven fertilizer samples were measured in ascending concentration from the lowest U content to the highest. The graphical representation of the measurement results after a measuring time of 1800 seconds and a comparison spectrum of the U-containing mineral schrockingerite is shown in Figure 2 and 3.

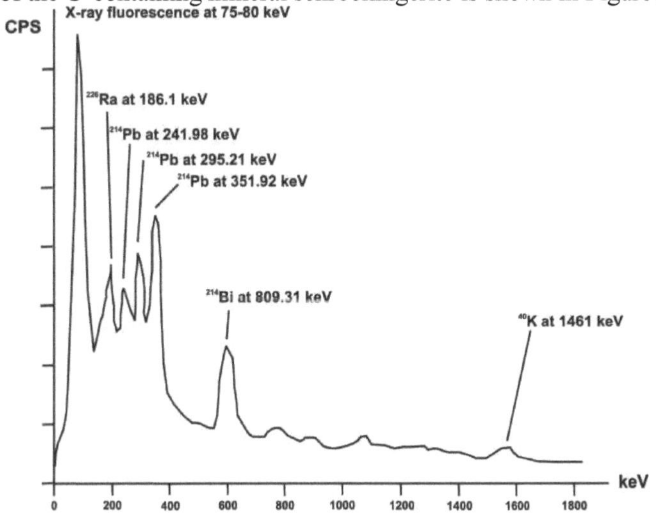

Fig.2. Spectrum of a schrockingerite U sample analyzed by radiation sensors SPD 38 NaI(Tl) GS1100A MCA Gamma Spectacular (modified, redrawn from Anti-Proton.com 2014)

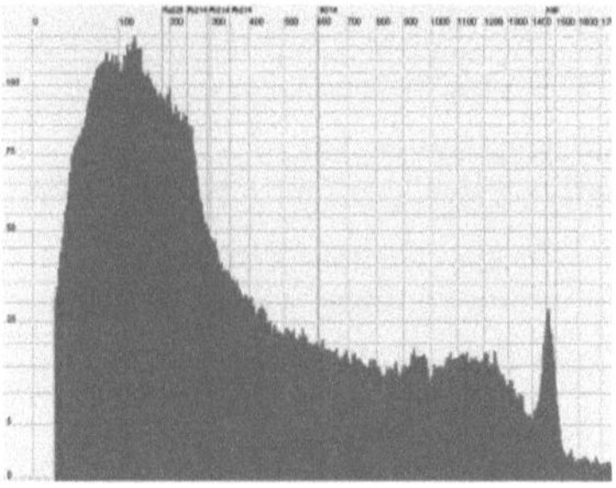

MD 24: no U; 19.99% K

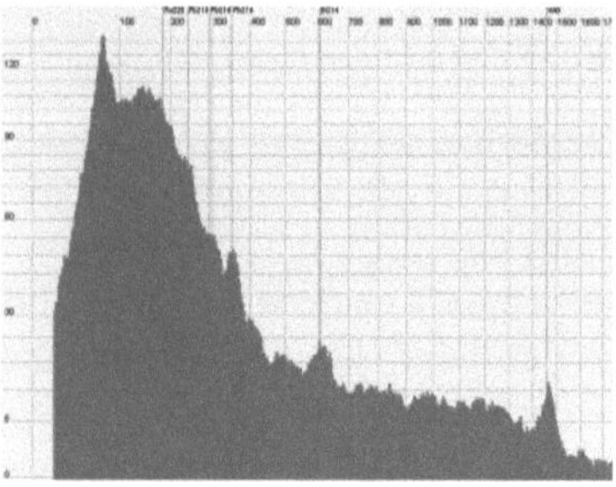

MD 39: 13.11 mg/kg U; 6.93 % K

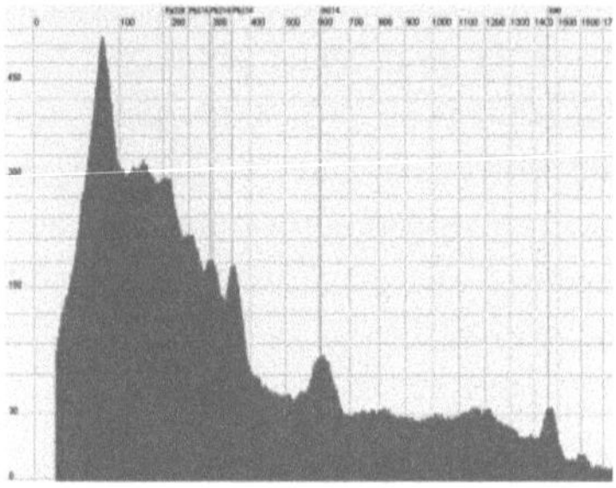

MD 38: 52.23 mg/kg U; 18.79 % K

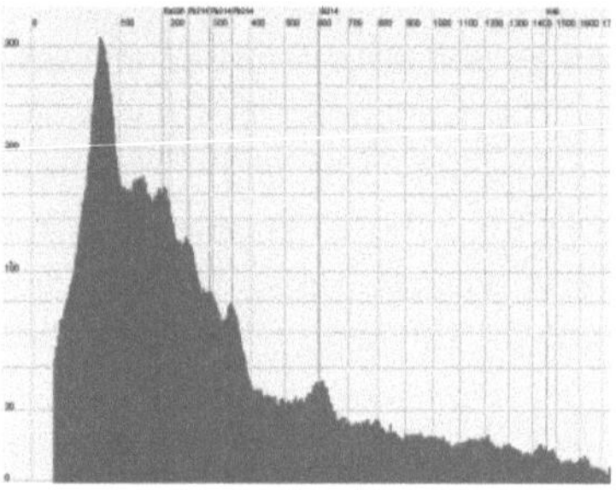

MD 1: 52.25 mg/kg U; no K

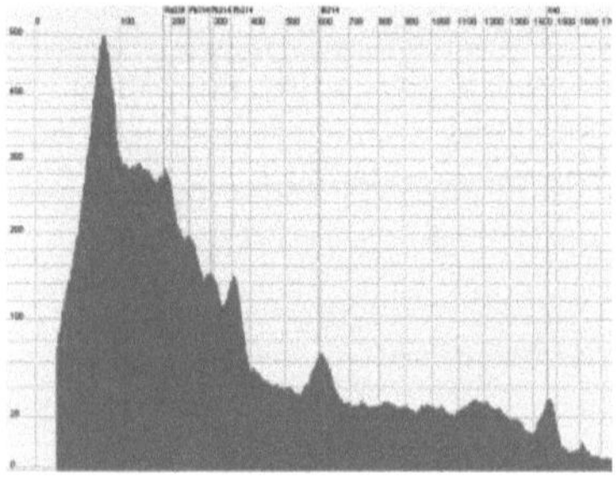

BMELV 62: 102.0 mg/kg U; 13.93 % K

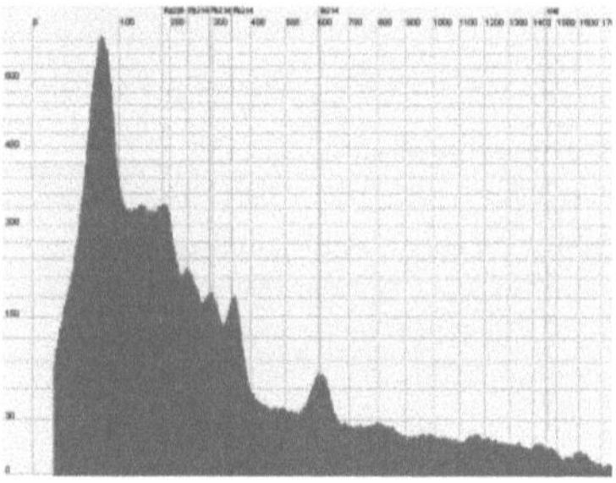

BMELV 66: 153.0 mg/kg U; 0.226 % K

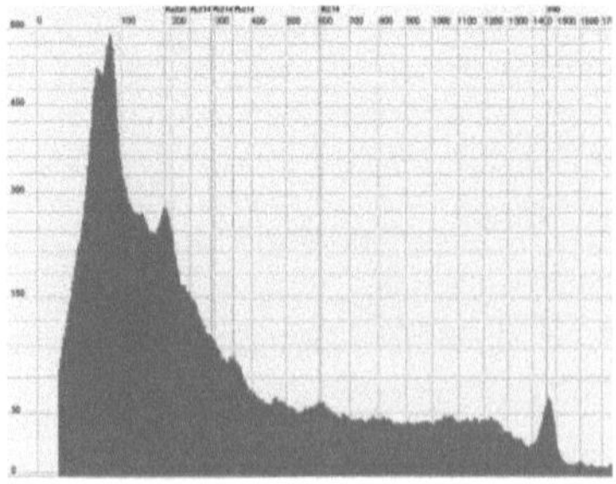

BMELV 42: 206.0 mg/kg U; 9.132 % K

Fig.3. Spectra of measured fertilizer samples (vertical solid lines from left to right indicate peaks of the isotopes ^{226}Ra, ^{214}Pb, ^{214}Pb, ^{214}Pb, ^{214}Bi and ^{40}K; for details of peak distribution see Figure 2)

The abscissa in the illustrated spectra in Figure 2 and 3 shows the energy in kilo electron volts (keV) and the ordinate shows the accumulated number of disintegrations per unit of time expressed as CPS (counts per second).

The comparison spectrum of the U-containing mineral schrockingerite (Anti-Proton.com 2014) shown in Figure 2 which is a much stronger radiation source compared to the investigated fertilizer samples, the characteristic peaks for the X-ray fluorescence from the Pb castle, can be seen at 75-80 keV, ^{226}Ra at 186 keV and ^{214}Pb triplet at 242, 295, 352 keV followed by the peak of ^{214}Bi at 609 keV. On the right side of the abscissa the K peak is visible at 1461 keV.

For U-bearing samples this is a quite typical sequence of peaks for the X-ray fluorescence and following peaks from ^{226}Ra, ^{214}Pb and ^{214}Bi. The isotopes ^{226}Ra and ^{214}Pb are very clearly visible in the measured fertilizer samples with U concentrations of 52, 102 and 153 mg/kg U. In the fertilizer sample with the highest U content of 206 mg/kg U this sequence was not so clearly visible at a measurement period of 30 minutes.

The peak of ^{214}Bi at 609 keV was very clear and proved to be suitable as a guiding parameter in the presence of U in all fertilizer samples. The ^{214}Bi peak was even more pronounced in the fertilizer sample with the lowest U content of 13.11 mg/kg U. These results show that low-cost Gamma spectroscopy provides a reliable qualitative detection of U with the existing calibration.

If the fertilizer sample contains larger amounts of K, the typical peak for ^{40}K 1460 keV is also clearly visible on the right edge of the spectrum. No ^{40}K peak was detected in the sample BMELV 66 with a very low K content and MD1 without K.

To summarize, short measurement times, good detection capabilities coupled with low cost and ease of use, make this gamma spectroscopy system a valuable tool in the identification of U contaminated fertilizers. The equipment and method can easily be used as laboratory setup, or as a portable RIID (Radio Isotope Identifier Detection) solution (Gammaspectacular 2014).

Quantitative analysis of the spectrum

The accurate quantitative analysis of the U content is possible, providing a calibration sample with certified activity is at hand. The method involves comparative analysis of the net counts in a region of interest, which may include one or more photo peaks. Such quantitative analysis in a region of interest is simple to carry out in PRA software or Becqmoni software, but not currently a standard feature of Theremino, as used in our experiment. Alternatively the histogram data from all of the available software can be exported and analysed in a spreadsheet such as Excel.

The net counts in the selected region of interest, normally a known photo peak for a particular isotope, is therefore proportional to the volume and concentration of the isotope in the sample.

References

Al-Shawi AW, Dahl R (1995) Determination of Thorium and Uranium in Nitrophosphate Fertiliser Solution by Ion Chromatography. Journal of Chromatography A 706(1-2): 175-181

Anti-Proton.com (2014) Spectrum of the uranium-bearing mineral schrockingerite. extracted from http://www.gammaspectacular.com/gamma_spectra/ra226-spectrum on 06/24 2014

Gammaspectacular (2014) Bee Research, Suite 314, 247 Coward Street, Mascot NSW 2020, Australia, http://www.gammaspectacular.com/

Kratz S, Schnug E (2006) Rock phosphates and P fertilizers as sources of U contamination in agricultural soils. In: Merkel B, Hasche-Berger A (eds.) Uranium in the environment. Springer, Berlin Heidelberg 2006, pp. 57-68

Ohly M (2005) Diplomkartierung: Einfluss der Phosphatdüngung auf den Urangehalt im Boden, MSc thesis TU Bergakademie Freiberg, 96 pp

Schnug E, Haneklaus N (2014) Uranium in phosphate fertilizers – review and outlook. This volume

The study of remediation activity of system plant-microorganisms in the model experiments of oil polluted soils

Yerlan Doszhanov[1,2], Aygerym Gabdualiyeva[1], Galym Umbetkaliev[1], Yerdos Ongarbaev[1,2], Azhar Zhubanova[1], Zhulkhair Mansurov[1,2]

[1]Al-Farabi Kazakh National University
[2]Institute of Combustion Problems

Abstract. In this study, results have been presented on remediation processes of oxidation of various petroleum hydrocarbons by *Pseudomonas* bacteria cells and by specific types of mixture between seed of legumes and cereals. It has been found that the maximum remediation activity of complex plant- microorganisms was characterized for saturated hydrocarbons.

Introduction

At present it has been observed widespread contamination of the environment (water, coastal zones, sediment and soil) by the petroleum hydrocarbons in the areas of oil production and refining due to increasing of production transportation and refining (Halimov et al. 1996).

The great interest of scientists and practitioners are attracted by bioremediation methods (reduction) of contaminated components in the environment by biological products on the basis of microorganisms, which are capable of using organic pollutants as carbon source. Due to a variety of metabolic capacity, these objects can undergo through a series of reactions catalyzed by enzymes, to carry out the splitting nature of the organic pollutants and includes products produced with the main microbial metabolism.

Recent studies have shown that strains-destructors, specially selected from the contaminated biological sites have to be used in order to make effective treatment of contaminated soils. Application of the consortium, which includes such strains allows to create an effective biological decomposition of complex organic compounds and recycling the major pollutants, thus providing the flow of remediation processes.

In this study, results have been presented on remediation processes of oxidation of various petroleum hydrocarbons by *Pseudomonas* bacteria cells and

B.J. Merkel and A. Arab (eds.), *Uranium – Past and Future Challenges*,
DOI 10.1007/978-3-319-11059-2_99

specific types of mixture between seed of legumes and cereals. It has been found that the maximum remediation activity of complex plant- microorganisms was characterized for saturated hydrocarbons.

Experimental

Studies in recent decades have shown that enhancing the effectiveness of biologics can be achieved by using cells of destructor microorganisms, which are immobilized on a variety of media. This can be explained by the fact that attachment of cells to solid surfaces provides a high concentration of microbial cells in their area of action, it prevents their leaching and protects from the toxic effect of high concentrations of oil components and makes it possible to increase the specific activity of specific destructive microorganisms. It is also important to know, that the applied mineral media can be used by microorganisms as a source of mineral nutrition. A various methods and media are used to construct immobilized biodestructors, the simplest and relatively cheap is – an adsorption of microbial cells at different chemically and biologically inert medium which are insoluble in water.

In connection with the foregoing, research dedicated to solve the problem of recovering oil polluted ecosystems through biological products on the basis of free and immobilized cells of microorganisms - oil destructors present a great scientific and practical interest. Therefore, search for effective microorganisms oil destructors, as way of enhancing the microorganisms, which are present in the soil to be cleaned and a full-fledged microbial community in technogenic substrates is an important issue. One of the methods to accelerate degradation processes of hydrocarbons can be the introduction of cleaning object together with microorganisms - oil destructors bacteria, contributing stimulation of all microbial life (Zhang et al. 2010).

Initially, it has been carried out a seeding of phytoremediant plants a mixture of seeds of leguminous and cereal plants into experimental soil, and then in the same primer was added a cell suspension *Pseudomonas*, where in the soil treatment is performed until the suspension concentration equal to 1-3 $\times$ 10^7 cells/g layer to a depth of 10 cm. Such procedures were designed to improve the efficiency of phytoremediation process (Mansurov et al. 2013).

It should be said that from the modern methods of cleaning the environment from petroleum hydrocarbons technology phyto-and bioremediation, ie use for the remediation of polluted plants and associated microorganisms with it, it is the most attractive because it provides a steady cleaning process due to high adaptive potential and mutualistic relationships of plants and microorganisms. It does not require high energy consumption, because operated by solar energy used in the process of photosynthesis, and as a result, a cost-effective and aesthetically appealing method to clean up contaminated sites.

Studies on the phytoremediation of petroleum hydrocarbons contaminated soils, conducted in laboratory conditions at the Institute of Combustion Problems al-Farabi Kazakh National University (Doszhanov et al. 2014). We have laid the model experiments with 9 samples which are most exposed to oil pollution of soils collected from West Kazakhstan region (Table 1), repeated experience - three times. As carbon sources for microorganism and raw material the oil was added, toluene and decane in the amount of 2, 5 and 10 ml per Petri dish 1. Quantitative analysis of the hydrocarbons in the environment, with growth of its biomass was determined by using IR spectroscopy data. Samples were determined using FT IR spectrometer Spectrum 65.

Plants vegetated within 15 days after sprouting at a humidity of 55-60% of field water-absorbing ability. Later on, a fresh growth of green mass have been recorded by means of cutoff and subsequent weighing.

At the next stage of the study, evaluation of the effectiveness of multi-crop for phytoremediation of soils that contaminated with oil, in the conditions of growing experience was carried out.

Table 1. Biocultural samples studied under laboratory experiments

№	Samples
1	Uncontaminated soil (control)
2	Soil + microorganisms
3	Soil + crop
4	Soil + hydrocarbons
5	Soil + microorganisms + crop
6	Soil + microorganisms + hydrocarbons
7	Soil + crop + hydrocarbons
8	Soil + microorganisms + hydrocarbons + crop
9	Soil + microorganisms + hydrocarbons + crop (crop in 10 days)

Results and discussion

Studied types of oil in the structure have a large quantity of hydrocarbons of different chain lengths. Oil composition comprises low molecular weight aromatic components with oxygen bonds, oil - polycyclic aromatic hydrocarbons and resin-asphalting compounds (Fig. 1).

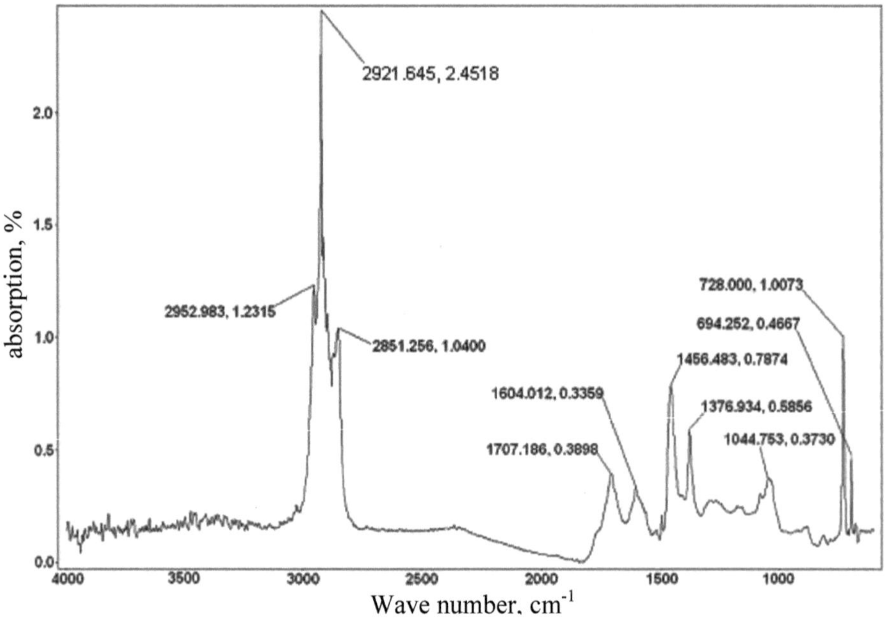

(a)

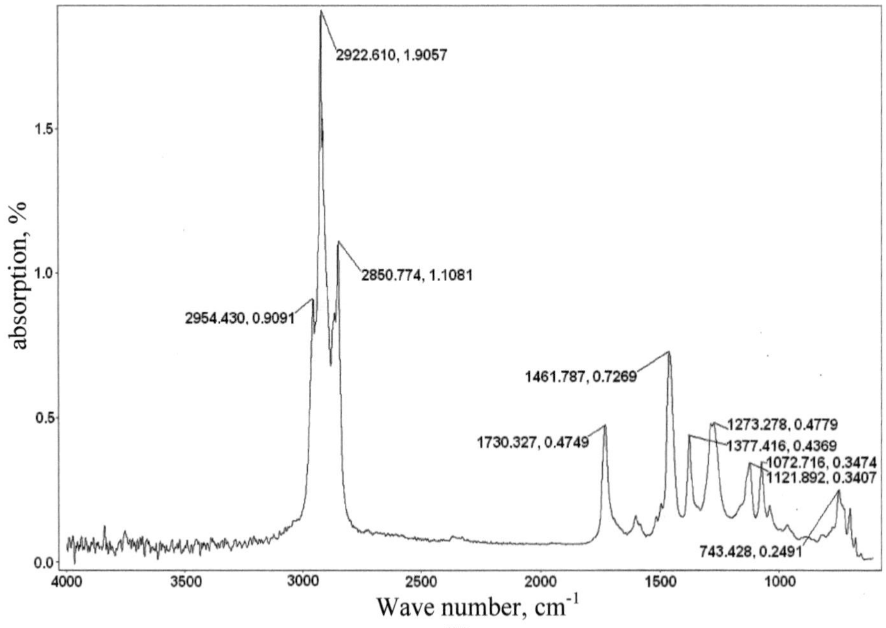

(b)

Fig.1. IR Spectrum before (a) and after (b) remidiation during 15 days

IR spectroscopic studies of samples containing different concentrations of hydrocarbons after growth of oil destructors-microorganisms and vegetable crops have yielded interesting results. It has been shown that normal paraffin and long parafin chains are dominated in the samples №2-9 (2952.983; 2921.645; 2851.256; 1456.483; 1376.934; 728.000 cm^{-1}).In addition, the presence of aromatic structures (1604.012 cm^{-1}) at minor level were identified. Uncontaminated control samples have an absorption band at 1707.186 and 1044.753 cm^{-1}, indicating a minor oxidation of the sample.

Analysis of bio-phytoremediation processes showed that recycling options petroleum and petroleum products during 15 days resulted some changes in IR spectrum data. IR data showed absorption bands (valence and deformation vibrations at 1377.416; 2850.774; 2922.610; 2954.430 cm^{-1} related to groups $-CH_2$ and $-CH_3$) related to compound of aliphatic structure after adding microorganisms and plants into polluted soil samples. The intensity of the characteristic absorption bands decreased in comparison to control samples. A weak band in the interval (1461.787 cm^{-1} -benzene ring group) can be attributed to absorption. The spectra also have a broad band in the range of 1072.716; 1121.892; 1273.278 cm^{-1} due to the absorption of various oxygen-containing compounds (C-O, C-O-C) alcohols, ethers and esters, which are intermediates in the metabolism of microbial oxidation n-paraffin. The absorption band at 1730.327 cm^{-1} may also be considered as an indication of oxygen-containing compounds presence.

Conclusion

Methods of phyto- and bioremediation can be used for decontamination of soils polluted by hydrocarbons. Such techniques can be of interest and used in all industries associated with the extraction, transportation, processing or storage of oil and oil products, as well as in the aftermath of accidental spills. The method can be used for cleaning soil and a soil concentration of hydrocarbon to 30 g/kg (Doszhanov et al. 2010).

References

Halimov JeM, Levin SV, Guzev VS (1996) Environmental and microbiological aspects of oil influence to soil properties . Vestn. Mosk. un-ta. Ser. Pochvovedenie 2: 59-64

Zhang BY, Zheng JS, Sharp RG (2010) Phytoremediation in Engineered Wetlands: Mechanisms and Applications. Procedia Environmental Sciences 2: 1315–1325.

Mansurov ZA, Doszhanov YeO, Ongarbaev YeK, Akimbekov NSh, Zhubanova AA (2013) The Evaluation of Process of Bioremediation of Oil-Polluted Soils by Different Strains of *Pseudomonas.* Advanced Materials Research, Vol. 647, 363-367.

Doszhanov YeO, Mansurov ZA, Ongarbaev YeK, Tileuberdi Ye, Zhubanova AA (2014) The study of biodegradation of diesel fuels by different strains of *Pseudomonas.* Applied Mechanics and Materials, Vol. 467, 12-15.
Doszhanov YeO, Ongarbaev YeK, Mansurov ZA, Zhubanova AA, Hofrichter M (2010) Bioremediation of Oil and Oil Products Bacterial Species of the Genus *Pseudomonas.* Eurasian Chemico-Technological Journal. Vol. 12, №2. 157-164.

Laboratory and field test Study on Sandstone Permeability Characteristics

Liu Zaibin[1], Jin Dewu[1], Dong Shuning[1], Liu Qisheng[1]

[1]Xi'an Research Institute Company of China Coal Technology and Engineering Group, Xi'an, Shaanxi Province, China, 710054

Abstract. Sandstone permeability is a key factor to study Jurassic coal field hydrogeological conditions and it's also an important factor to study sandstone uranium in-situ leaching hydrogeological conditions. Sandstone whole stress-strain process permeability characteristics under different confining and osmotic pressure were studied through laboratory tests. Sandstone permeability characteristics along vertical direction were studied through field tests. Relationships between permeability with failure process, confining pressure, osmotic pressure, bulk strain and buried depth were obtained respectively.

Introduction

Rock permeability is not only related to rock material and rock structure but also related to stress environment, deformation and failure process. Laboratory and field test can be conducted to study sandstone permeability characteristics. Permeability values of rock failure whole stress-strain process under different confining pressure and osmotic pressure can be obtained through laboratory tests to reveal rock permeability evolution mechanism. Rock permeability space characteristics can be analyzed through field test data to provide basis for mining designing and water quantity prediction.

Permeability test methods

Rock permeability test methods include in-situ field test and laboratory test (Miu 2006). Main in-situ test methods are Sectional pumping test and water injection test. In this study, transient test method was used in laboratory to measure sandstone permeability. In-situ pumping test method was used in a coal mine field to analyze permeability space characteristics.

Transient test method was first put forward by W. F. Brace at 1968 to measure westerly granite permeability (Brace 1968). The MTS electro-hydraulic servo

B.J. Merkel and A. Arab (eds.), *Uranium – Past and Future Challenges*,
DOI 10.1007/978-3-319-11059-2_100

test system has a complete set of techniques to measure rock permeability. Permeability calculation formula can be deduced by Darcy's law.

Permeability test system is consisted by three parts which are upstream pressure vessel, rock specimen and downstream pressure vessel, as shown in figure 1. Main measurement variables are pressures of upstream and downstream vessels.

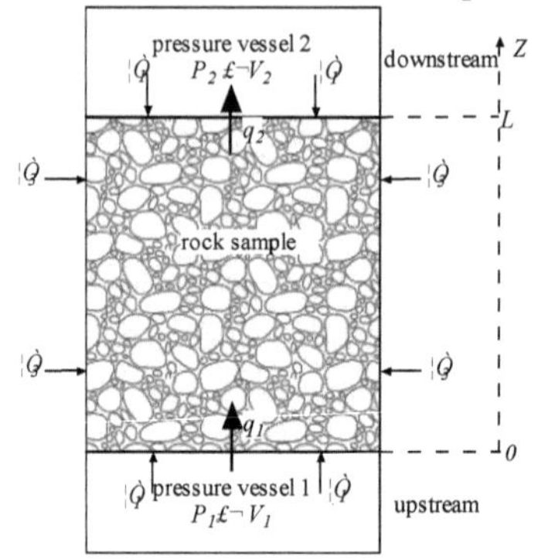

Fig.1. Permeability transient test method

Pressures of upstream vessel and downstream vessel are P_1 and P_2. Volumes of two vessels are V_1 and V_2. The seepage direction is from upstream to downstream. Mass flow from upstream vessel to the rock specimen is q_1. Mass flow from the rock specimen to downstream vessel is q_2. The final differential pressure is dP_f, permeability calculation formula can be simplified to:

$$k = \frac{u\beta VL}{2Adt} \ln \frac{dP_0}{dP_f}$$

In the formula, k stands for permeability (cm^2); μ stands for fluid viscosity coefficient (10^{-3}Pa.s); β stands for fluid compressibility ($4.53*10^{-10}Pa^{-1}$); V stands for pressure vessel volume (cm^3); L stands for rock specimen height (cm); A stands for rock specimen cross-sectional area (cm^2); dt stands for seepage duration (s); dP_0 / dP_f stands for the ratio of initial and final differential pressure.

Sandstone rock specimens were taken from Changhua coal mine in Shanxi Ningwu coal field in China. In order to load pore pressure inside the rock specimen, two holes were drilled in the specimen. In order to simulate mining rock mass characteristics, permeability of damaged rock specimens which were broken under confining pressure was measured. In the laboratory test, there are four confining pressures which are 4MPa, 8MPa, 12MPa and 16MPa and three osmotic pressures which are 3MPa, 6MPa and 9MPa. Through permeability tests under different confining pressure and osmotic pressure, permeability confining pressure effect and osmotic pressure effect can be analyzed.

Permeability laboratory test characteristics

Whole stress-strain permeability characteristics

The whole stress strain process can be divided to 5 phases which are marked by O/A/B/C/D/E in figure 2. In normal trixial compression test, the second principle stress is same as the third principle stress, $\sigma_2=\sigma_3$. At the beginning step, confining load is 4MPa, osmotic pressure is 3MPa. Axial load is same as confining load, $\sigma_1=\sigma_3$. Then the axial load is increased gradually to break the rock sample. The deviatoric stress, σ_1-σ_3, is the difference between axial load and the confining load.

OA phase is the initial compression phase. AB is the linear elastic phase. Deviatoric stress and strain fit to linear relationship. BC is nonlinear phase, the increase rate gradually decreases. Deviatoric stress achieves the peak value 71.05MPa at the C point. CD is Strain softening phase, the deviatoric stress decreases to 18.88MPa. DE is residual strength phase, at the end of the test the deviatoric stress is 18.75MPa.

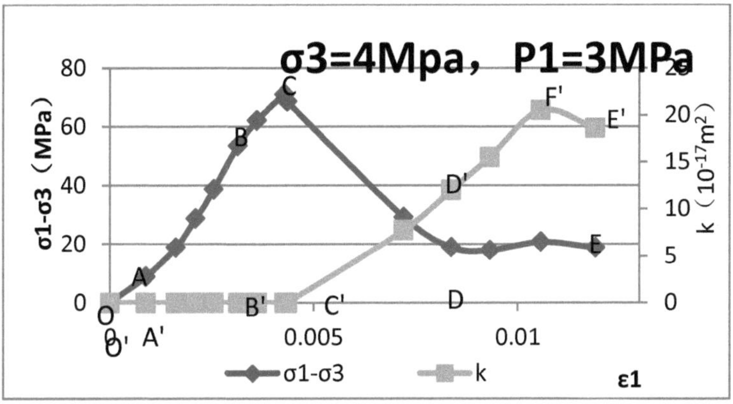

Fig.2. Sandstone permeability whole stress-strain curve

As similar to the whole stress-strain curve, the permeability curve also can be divided to 5 phases marked by O'/A'/B'/C'/D'/E'. At OA, AB and BC phases, sandstone permeability values are too tiny to be measured. After the stress peak point, permeability increases gradually. At the F' point in the residual strength phase, the permeability value increases to the maximum value $20.5\times10^{-17}m^2$. Then the permeability decreases to $18.6\times10^{-17}m^2$. The permeability increases when the rock strength achieves the peak point. After the strength peak point internal micro cracks evolve to macroscopic fractures, where the rock specimen is damaged.

Permeability confining pressure effect

Sandstone permeability values under different confining pressures are shown in figure 3. Rock permeability decreases with confining pressure increases. Confining pressure effect decreases with confining pressure increases as curves become flat with high confining pressures.

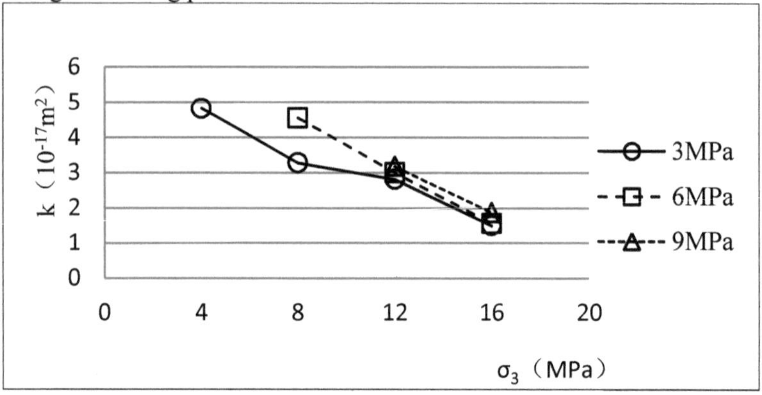

Fig.3. Permeability confining pressure effect

Confining pressure effect is caused by the mechanical effect of the confining pressure to rock skeleton. The rock permeability is influenced by rock skeleton deformation caused by the confining pressure. As confining pressure increases, lateral deformation is formed to compress rock pore and fracture space which leads effective permeable channels decreased. Permeability decreases while effective permeable channels decrease.

Influence of osmotic pressure to permeability

Hydraulic gradient inside the specimen is different under different osmotic pressure. Different osmotic pressure can influence permeability test values. Sandstone permeability test values under different osmotic pressure are shown in figure 4. Permeability increases with osmotic pressure increases. The curvature of permeability-osmotic pressure decreases with confining pressure increases which means the influence degree of osmotic pressure difference to permeability decreases.

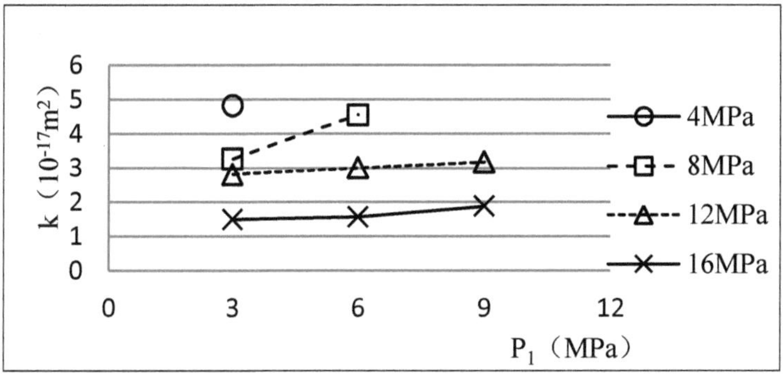

Fig.4. Permeability under different osmotic pressure

With the increase of osmotic pressure difference, effective permeable channels increase which means the increase of connected pore and fracture space. Osmotic pressure influences rock permeability from two kinds of effect which are hydrodynamic effect and hydrostatic pressure effect. Because the seepage velocity is very slow, the hydrodynamic effect can be neglected during the test process. Hydrostatic pressure forms the pore pressure inside the rock specimen to reduce the rock strength. Hydrostatic pressure also has hydraulic fracturing effect to accelerate the crack propagation then break the specimen. In the test process, the initial hydraulic pressure at the specimen bottom is P_1 while the hydraulic pressure at the top of the specimen is 0. The osmotic pressure gradient fits nonlinear distribution inside the specimen. Osmotic pressure concentrates at the bottom, while the failure zone also concentrates at the bottom as shown in figure 5.

Fig.5. Failure zone under osmotic pressure

The osmotic pressure effect is not very obvious under high confining pressure conditions. This phenomenon reflects the pore pressure effect to effective channels is not as strong as confining pressure effect. The comprehensive influence of confining pressure and pore pressure is the effective stress. Biot-Willis coefficient in effective stress equation also reflects pore pressure effect to rock deformation is less than confining pressure effect.

Permeability-bulk strain relationship

The relationship between permeability and bulk strain can be obtained through the permeability test data. Sandstone permeability at the strain softening phase has a very good correspondence relationship with bulk strain. Permeability increases with the bulk strain increases. The relationship of permeability and bulk strain under confining pressure 8MPa and osmotic pressure 6MPa fits the equation:

$$k = 1.519 \times 10^9 \times \varepsilon_V^{4.424}$$

In this equation, ε_V is the bulk strain.

Permeability field test characteristics

Pumping test includes steady flow pumping test and unsteady flow pumping test. Calculation formulas for single-hole steady flow pumping test of fully penetrated well and partially penetrated well for homogeneous and isotropic confined aquifers are (Du 2010):

$$K = \frac{Q\lg \frac{R}{r_W}}{2.73MS_W} \quad \text{Fully penetrated well}$$

$$K = \frac{Q\ln \frac{1.32l}{r_W}}{2plS_W} \quad \text{Partially penetrated well}$$

In these formulas, K stands for permeability coefficient of homogeneous rock mass (m/s); S_W stands for borehole water level drawdown (m); M stands for confined aquifer thickness (m); Q stands for steady flow pumping rate (m^3/s); R stands for radius of pumping influence; r_W stands for borehole radius; l stands for test section length.

There were 46 pumping tests of Jurassic sandstone in the study area, in which 9 pumping tests were implemented in the area affected by the river, and 37 pumping tests in normal area. In order to eliminate the influence of the river and alluvial aquifer, normal area's pumping data were selected to study vertical permeability pattern, the K-D (Permeability Coefficient-Buried Depth) curve was shown in the figure 6. Sandstone permeability coefficients decrease with buried depths increase which fits the relationship:

$$K = 0.8364e^{-0.011D}$$

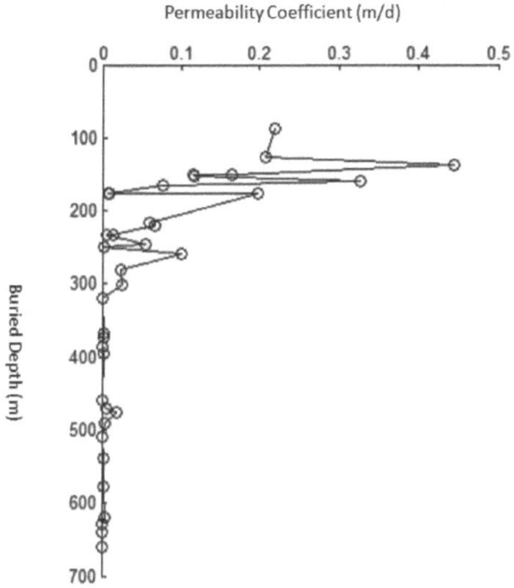

Fig.6. K-D curve

The K-D curve is not a smooth curve especially in the shallow rock mass. The curve shows a regular wave which is very obvious in shallow rock mass. The K-D curve also has a self-similarity change rule which fits the fractal rule.

Conclusions

Relationships among permeability and failure process, confining pressure, osmotic pressure, bulk strain and buried depth were obtained through laboratory tests and in-situ tests.

During rock failure process, permeability increases obviously after the rock strength achieves the stress peak point. Permeability decreases with confining pressure increases and osmotic pressure decreases while confining pressure effect is stronger than osmotic pressure effect. The permeability of damaged rock specimen has a exponential relationship with bulk strain.

According to in-situ test results, in vertical direction sandstone permeability coefficients decrease while buried depths increase.

References

Miu XX, Liu WQ, Chen ZQ (2006), Mining rock mass seepage theory. Science Press of China : 12-22

Brace WF, Walsh JB, Frangos WT (1968), Permeability of granite under high pressure. Journal of Geophysical Research 45(7): 82-87

Du X, Zeng YW, Tang DY (2010), Research on permeability coefficient of rock mass based on underwater pumping test and its applicaiont. Chinese Journal of Rock Mechanics and Engineering 29(Supp.2): 3542-3548.

FEFLOW
Groundwater modelling for mine sites
• Open pit mines and tailings
• Pit-lake simulation
• Underground mining
• Optimisation of dewatering schemes
• Seepage analysis
• Contaminant transport in mining environments
www.feflow.com
Software for WATER ENVIRONMENTS
MIKE by DHI

MIX
Papier aus verantwortungsvollen Quellen
Paper from responsible sources
FSC® C105338

If you have any concerns about our products,
you can contact us on
ProductSafety@springernature.com

In case Publisher is established outside the EU,
the EU authorized representative is:
Springer Nature Customer Service Center GmbH
Europaplatz 3, 69115 Heidelberg, Germany

Printed by Libri Plureos GmbH
in Hamburg, Germany